General Zoology Laboratory Guide, 16th Edition

NYU: Tandon School of Engineering

BMS 2001

Lytle, Meyer

ISBN-13: 9781307882759

ISBN-10: 1307882757

Contents

Preface

Many exciting advances have occurred since publication of the previous edition of this book. New data from molecular research and other new approaches have increased our knowledge and understanding of the biology of animals and their phylogenetic relationships. Increased appreciation for the importance of biodiversity and the challenges of environmental changes has stimulated new studies of animal ecology and behavior; discoveries of medically important compounds in marine animals have led to further investigations of biologically active compounds in a variety of invertebrates; and new biochemical and genetic techniques have provided new insights into the evolutionary relationships of animals. Major changes have been made in the classification of many groups of animals as a result of systematic studies. It is an exciting time to study zoology, and students should find zoology to be an interesting and rewarding subject.

It is important to remember that our knowledge of zoology, as of all science, is cumulative and builds on the work of previous scientists. Our understanding of animals and the processes by which they live is continuously changing as we gain new information from research. As additional knowledge is obtained, prior findings are confirmed, refuted, or revised as needed to reflect our current understanding. Since knowledge continues to be added, science is an ever-changing status report. We need always to keep our minds open for further change as new evidence is obtained and evaluated.

Zoology is the study of the evolution, classification, structure, and function of animals and how they live and interact with their environment. A good working knowledge of basic zoology is essential to understand the biology of animals. A solid foundation in the structure, function, and diversity of animals is critical for understanding the central issues in biology, such as the molecular mechanisms of cellular structure and function, mathematical analyses and modeling of animal processes, and the functioning of the nervous system.

Meaningful laboratory experiences are a vital part of learning zoology, and this book is designed to facilitate laboratory study of selected animals. In the laboratory, students learn the importance of careful observation, of following specific instructions, and of seeing relationships of structure and function. By carrying out well-designed scientific observations and experiments in the laboratory, students learn to **do** science instead of merely listening to someone **talk** about science.

In writing this book, we have tried to remember our own days as students and the kinds of questions we had while studying various kinds of animals for the first time. We have attempted to provide descriptions, illustrations, and appropriate guidance for meaningful laboratory study of animals. Although we agree with the spirit of Louis Agassiz' admonition, "Study nature, not books," we believe that students can benefit most in their study of nature when guided by a good instructor and aided by appropriate instructions and illustrations.

We can remember some of our own frustrations in zoology labs when we attempted to follow some vague verbal description of anatomical structures with no illustrations or other visual aids to help us locate important structures or to give us some appropriate orientation. We have written the exercises in this book with the intention of reducing such frustration and with the intention of making student experiences in the zoology laboratory interesting, rewarding, and meaningful.

We continue to emphasize the study of living and anesthetized animals whenever feasible because students should learn that zoology is the study of animal life rather than the study of dead animals. Live animals give students the opportunity to observe and experiment with behavior and to do simple physiological experiments as well as to see the natural color and texture of body parts. Preserved specimens serve well for many anatomical studies, but students should always have the opportunity to observe and to work with living animals whenever possible. Few of us would choose a stuffed or embalmed dog or cat for a pet if given the option.

This book is written to aid students and teachers in many colleges and universities operating with different schedules, resources, and preferences, so we have intentionally included more material than can reasonably be covered in the time available in a two-semester general zoology course. We expect instructors to select those parts of the guide and those animals they deem most appropriate for their own classes. With judicious selection of chapters and of animal types, this book can also be used successfully for one-semester and one-quarter zoology classes.

Changes in This Edition

Major changes in this sixteenth edition include extensive revisions of six chapters: Chapter 6 Protozoa; Chapter 9 Platyzoa: Platyhelminthes, Gastrotricha, and Rotifera; Chapter 10 Mollusca; Chapter 11 Annelida; Chapter 12 Nematoda and Nematomorpha; and Chapter 13 Arthropoda. We have also changed the sequence of Chapters 10, 11, and 12 to better reflect the current understanding of phylogenetic relationships among these phyla. Our revisions are based on the results of recent research, changes in classification, and new insights into evolutionary relationships.

All chapters have been updated to reflect studies of cellular function, molecular biology, ecology, physiology, and systematics. New illustrations have been added in nearly every chapter, including more than 50 new photographs and drawings. There are several new color photographs of animals as well as new diagrams for the life cycles of some invertebrates. We have improved some Questions for Critical Thinking to stimulate student interest and encourage practice in the analysis of relevant information. We hope you find these changes helpful in your studies.

Basic Features of This Manual

In this edition we have continued the basic organization and pedagogical features of the previous edition. Important pedagogical features of the book include **boldface headings** within each chapter to indicate the major divisions of each exercise.

We also use **boldface** in the text to identify important terms (ideas, structures, processes) that students should remember and understand. The most important of these boldface terms are included in the list of **key terms** at the end of the chapter in which they are first introduced.

Several chapters provide space for students to add their own drawings of particular animals or structures to aid them in learning and remembering things observed during their laboratory study. The book also provides several blank tables and pages of graph paper for students to record and plot data from their laboratory observations and experiments.

Each chapter begins with a list of specific **objectives** that identifies important principles, concepts, and facts that students should learn as a result of their laboratory study. We have found that a specific list of laboratory objectives helps students focus their attention on the important material in each lab. We also suggest that instructors modify and add to these lists of objectives as appropriate for their own classes. Such lists of objectives can be most helpful in ensuring that students understand what they will be tested on and that the tests actually focus on students' understanding of the important principles, concepts, and processes.

Most chapters in this book start with a brief **introduction** with pertinent background material to help orient students for the exercises to follow. A **materials list** is provided showing the specimens and other materials needed for each exercise. Most chapters have one or more lists of suggested **demonstrations,** which are suggested to supplement the main studies of each exercise.

Other important features in many chapters include cladograms to illustrate possible evolutionary relationships of animals, sections on the field collection and preservation of many kinds of invertebrates, and Fact Files providing interesting information about various species and groups of animals. For instance, did you know that fossil shells of some molluscs can provide important clues to the climatic conditions of ancient times?

Within each chapter, student-directed questions have been placed in boldface type to trigger students to stop and think about the animal structure or procedure being discussed.

At the end of each chapter is a list of **key terms** introduced in that chapter, as well as Internet Resources and Suggested Readings for further study. Each chapter also ends with a list of **Questions for Critical Thinking** to help students review the important concepts and processes of the chapter. Most chapters also have blank space provided for students to add their own notes and sketches. If students use these pages to record their observations, they will have a consolidated record of their laboratory work bound in a single place instead of a scattered bunch of papers and drawings likely to be lost.

We have tried to make this laboratory manual a convenient, user-friendly companion for laboratory study. We hope every student has as much fun and satisfaction in zoology lab as we have had.

Anatomy Videos

From our many years of teaching zoology laboratories, we have learned that it is very helpful to give students an overview of the anatomy of an animal to be studied and/or dissected before they undertake the anatomical study on an actual specimen themselves. It's a lot like football players viewing game films before facing a major rival football team. They might do all right without knowing what kinds of plays the opposition typically runs and who their key players are, but they are not likely to win the championship without some good scouting information.

An excellent way to prepare for a serious anatomical study of an animal is to view a good video of the anatomy of that animal prior to beginning work with an actual specimen. Such preparation for the lab study gives students a better perspective and orientation and greatly increases their confidence. It also aids in their identification of anatomical structures, facilitates their recognition of relationships among various organs, and assists them in relating structure and function. Good videos also help students review their laboratory work in preparation for a test and in comparing the anatomy of different animals.

We have collaborated with the staff of Carolina Biological Supply Company in the development of a series of videos specifically designed to aid in the study of nine of the more complex animals included in this manual. These videos illustrate the anatomy and dissection of these nine animals and parallel the descriptions of those animals in this book.

Each video illustrates the anatomy of the animal in detail, discusses the functions of various organs and systems, and demonstrates good dissection techniques. Each video is divided into sections according to organ system so that each system can be located easily and viewed separately if desired. Several of the longer videos are too long to be productively viewed in a single session.

The videos are now available on DVDs only from Carolina Biological Supply Company, 2700 York Road, Burlington, North Carolina, 27215. The videos and the corresponding chapters in this manual containing the exercises for the study of these animals are listed in the following table.

Chapter	Title	DVD
10. Mollusca	The Anatomy of the Freshwater Mussel	492365D
11. Annelida	The Anatomy of the Earthworm	492372D
13. Arthropoda	The Anatomy of the Crayfish	492403D
	The Anatomy of the Grasshopper	492404D
14. Echinodermata	The Anatomy of the Starfish	492369D
16. Shark Anatomy	The Anatomy of the Shark	492655D
17. Perch Anatomy	The Anatomy of the Perch	492662D
18. Frog Anatomy	The Anatomy of the Frog	492704A
19. Fetal Pig Anatomy	The Anatomy of the Fetal Pig	493075A

Acknowledgments

We are grateful to many people who provided information, suggestions, photographs, and other material for this edition, including Roger Phillips and his colleagues at Carolina Biological Supply Company; Carol Majors of Publications Unlimited; Dr. John Clamp of North Carolina Central University; Patty Aune, Linda Rudd, and Dr. Betty Black of North Carolina State University; Dr. Wayne Starnes of the North Carolina Museum of Natural Sciences; Dr. Burton Vaughn of Washington State University Tri-Cities Natural History Museum; Dr. Ben Hanelt of the University of New Mexico; Dr. Ash Bullard of Auburn University; Dr. Stanley King of Dalhousie University; Jody Jordan of Niles Biological Inc.; Marilyn Pendley of Caldwell Career Center, Middle College; Dr. Janet Taylor of Franklin County Schools; and Dr. James J. English, Gardner Webb University.

The following reviewers of the fifteenth edition provided excellent suggestions and feedback for this new edition. We carefully considered their suggestions and incorporated many of them into this new edition. Any errors that remain are our own.

Dr. Barbara Abraham, *Hampton University*
Dr. Amir Assadi-Rad, *Delta College*
Dr. Ann M. Barse, *Salisbury University*
Dr. Mark S. Blackmore, *Valdosta State University*
Barbara Kuehner, *Hawaii Community College*
Dr. Michael L. McMahan, *Union University*

Laboratory Safety

A zoology laboratory is a place for serious scientific work and study. Students and teachers must recognize that a number of potential safety hazards are present in all science laboratories. In a zoology laboratory, the principal safety hazards are electrical circuits, potentially dangerous chemicals, hot liquids and heat sources, broken glass, live animals, and sometimes infectious agents (pathogenic bacteria, viruses, and parasites). Achieving safety in the laboratory, as in other places, requires paying attention to potential hazards and observing appropriate safety measures. Nothing we do is without some degree of risk. For example, people sometimes fall out of bed and injure themselves; many people also drown in swimming pools each year. But science laboratories can provide a safe environment when both students and teachers are aware of potential hazards and follow appropriate safety procedures.

The following list of safety rules is offered as a good start toward safe practices in the lab. This is not a complete list of safety rules, and it is certainly not a substitute for proper safety awareness for the particular lab in which the student is engaged. It is essential that students understand the need and importance of safety. All actions in a laboratory have consequences. Be protected by paying attention, listening to your instructors, and knowing about the materials and procedures necessary to perform each laboratory investigation. You should also be mindful of the activities of other students around you. Frequently it is an accident caused by another person that endangers someone in a laboratory.

Some Basic Rules of Safety for the Laboratory

1. Use common sense.
2. Avoid horseplay in the laboratory.
3. Never eat, drink, or smoke in the laboratory.
4. Always wash your hands for at least 15 seconds and rinse them well after handling chemicals or live or preserved animals.
5. Always wear close-toed shoes in the laboratory. Sandals or open-toed shoes are not appropriate.
6. Be familiar with the location, operation, and proper use of fire extinguishers, eyewash fountains, safety showers, and other safety equipment in the laboratory.
7. Know the location of emergency exits and the evacuation routes to be used in case of an emergency.
8. Always be cautious when using electric hot plates and gas burners. You can get a serious burn by touching a hot surface or by spilling a hot liquid.
9. Use protective mittens or tongs to handle hot objects.
10. Be cautious when transferring liquids because aerosols may be formed, which can be dangerous to your eyes and lungs.
11. Wear safety goggles or other appropriate protective eye gear when performing or observing experiments or demonstrations.
12. Be familiar with the properties of, and hazards associated with, all chemicals used in the laboratory exercises. When you are in doubt about the hazards associated with any chemical, consult the Material Safety Data Sheets (MSDS) provided by the manufacturers. The appropriate MSDS for all potentially hazardous chemicals should be kept readily available in the laboratory. Your laboratory instructor should provide you with appropriate warnings for the materials to be used, but you may also ask to see the MSDS for any chemical to be used if you feel you need further information. Laboratory instructors should encourage all students to read the MSDS.
13. Beware of electrical equipment with frayed or bare wires or with faulty switches or plugs. Report such damaged items to your instructor.
14. Always work in a well-ventilated area when studying preserved specimens.
15. Make sure that all specimens you dissect are properly secured in a dissecting pan or appropriate surface. A specimen not properly secured might slip and lead to an injury from a scalpel or other sharp dissecting instrument.
16. Keep scalpel blades sharp to avoid slipping and possible injury.
17. All broken glass should be placed in a sharps container or one designated as a glass receptacle.
18. Any contact with human blood should be reported promptly to your instructor to limit your exposure to possible infection.
19. Clean all laboratory tables and other work surfaces after each use.

20. PRACTICE SAFETY AWARENESS, and remember that you are responsible for the safety of yourself and your coworkers.

Safety Precautions When Using Preserved Animals

Some of the chemicals used to preserve animals and parts of animals can be toxic, flammable, and/or dangerous if used improperly or under improper conditions. Ethanol, isopropanol, formaldehyde, and phenol are commonly used preservatives. Some specimens are held in propylene glycol–based preservatives such as Carosafe™ after tissue fixation has been accomplished using another material, for example, a formalin solution. Propylene glycol is a common cosmetic and food additive and is generally regarded as safe, so its use represents an improvement in safety by reducing exposure to other more hazardous materials. There are some new safer fixatives and preservatives in use today. Carolina's Perfect Solution™ is one such material. This nontoxic formula is used both as a tissue fixative and as an external preservative. It is very important to understand what chemicals are present in preserved specimens so safety precautions may be undertaken relative to the risk of those chemicals.

It is also very important for students and instructors working with preserved specimens to understand the proper precautions and conditions for safe usage of such materials. All instructors are responsible for implementing proper safety procedures when students will be using potentially hazardous chemicals and for communicating appropriate information about these materials to their students in accordance with applicable federal, state, and local regulations. In recent years these regulations have greatly increased in complexity as a result of increased public concern about environmental health and safety.

The following information, supplied through the courtesy of the Carolina Biological Supply Company, provides some excellent safety guidelines to follow when handling and dissecting preserved animal specimens. Other suppliers use similar chemicals for their preserved animal specimens. You should carefully study the safety information supplied with any preserved specimens before you begin to handle or dissect them.

Carolina provides specimens preserved in Carosafe™ (contains propylene glycol), Carolina's Perfect Solution™ (a safer proprietary fixative and preservative), and, when specially ordered, formalin solutions.

Information is provided in the Carolina Biological Supply catalog regarding which particular preservative is used in a certain type of specimen. Note that specimens are never provided in a formalin preservative unless this is specifically requested by the customer. Note also that specimens are provided with a specific Material Safety Data Sheet (MSDS) prepared according to the type of preservative used.

Regardless of the preservative that is used, we recommend you follow these safety tips whenever working with preserved specimens:

1. Wear appropriate protective eyewear at all times.
2. Wear appropriate protective equipment such as gloves and lab coats.
3. Work only in a well-ventilated area.
4. Prohibit eating, drinking, and smoking in the work area.
5. In the event of contact, wash skin with soap and water; flush eyes with water.
6. If overexposure to any chemical occurs, seek medical attention immediately.

Formalin-preserved or embalmed specimens should always be used in a well-ventilated area to prevent irritation to the eyes, skin, or respiratory tract. The use of goggles lessens eye irritation from formaldehyde vapors.

The components of Carosafe™ and those of Carolina's Perfect Solution™ may irritate eyes and skin in some people. In addition, the vapor of some components would be irritating if inhaled.

For all of these preservatives, wash eyes or skin with water if exposure occurs.

When working with preserved materials, be careful with sharp objects such as pins, scalpels, and the spines and teeth of specimens. When using a scalpel, we recommend cutting away from oneself and ensuring that fingers are kept out of the cutting path at all times.

Carolina-preserved specimens are available in Carolina's Perfect Solution™, a proprietary formula that has been tested and found to be nontoxic. It is both a fixative and a preservative. It keeps exposure to formaldehyde to an extremely low level while producing specimens of very high quality.

Carolina-preserved specimens are also available in Carosafe™, a propylene glycol–based shipping and holding fluid. Carosafe™ is not a fixative; it is a preservative designed to prevent mold and tissue deterioration after the tissue has been properly fixed with formalin or, for some small specimens such as insects, fixed with alcohol. Carosafe™ is an effective substitute for the standard formalin preservative and acts to hold the unpleasant odor of formaldehyde to an absolute minimum.

The following table contains further safety and health information regarding the three most common chemicals used by Carolina in the preservation process. This information is given in the form of a columnar table. Additional information may be obtained by calling Carolina during regular business hours at 336-584-0381.

Comparative Safety of Preservatives

	Formaldehyde	Carosafe™ (Propylene Glycol)	Carolina's Perfect Solution™
Physical Data			
Hazardous components (OSHA PEL)	Methanol (TWA 200 ppm) Formaldehyde (TWA 0.75 ppm)	Propylene glycol	Components proprietary. Not hazardous under normal conditions of use.
Flash point	184° F (combustible)	225° F	200° F
Lower explosion limits	7%	2.6%	
Upper explosion limits	73%	12.5%	
Fire extinguishing media	Alcohol foam, water fog, carbon dioxide, dry chemical	Water fog, carbon dioxide, dry chemical	Foam, carbon dioxide, dry chemical
Unusual fire or explosion	Vapor heavier than air, may travel along ground to distant ignition source and flash back	No unusual fire hazards noted.	No unusual fire hazards noted.
Threshold limit value (TLV) ACGIH	200 ppm (TWA) methanol 0.3 ppm ceiling formaldehyde	None known.	1000 ppm (TWA) one component
Effects of Overexposure			
Eyes	Vapor causes severe irritation, redness, tearing, blurred vision. Liquid may cause severe or permanent damage.	Direct contact may cause irritation.	Direct contact may cause irritation.
Skin (contact)	Irritation, dermatitis, strong sensitizer.	Direct contact may cause irritation.	Direct contact may cause irritation.
Inhalation	Irritation of respiratory tract, dyspnea, headache, bronchitis, pulmonary edema, gastroenteritis.	Vapor may cause irritation to respiratory tract.	Vapor may cause irritation to respiratory tract.
Ingestion	May be fatal or cause blindness if ingested. LD_{50} (oral-rat) = 500 mg/kg	Expected to be relatively nontoxic. Individuals with kidney problems may see more severe effects. LD_{50} (oral-rat) = 20,000 mg/kg	Nontoxic. May cause gastrointestinal discomfort.
Chronic effects	Listed by the National Toxicology Program (NTP) as reasonably anticipated to cause cancer in humans. Also listed by IARC and OSHA as possible human carcinogen.	Not listed as causing cancer by NTP, IARC, or OSHA. Gastrointestinal discomfort, nausea, vomiting, lethargy, and diarrhea have been cited for chronic exposure.	Primary components not listed as causing cancer by NTP, IARC, or OSHA. Contains a very small amount of a potential human carcinogen.
Target organs	If inhaled, eyes, nasal passages threat.	None.	None.
First aid measures	If inhaled, remove to fresh air. Call 911 or emergency medical service. If not breathing, give artificial respiration using proper respiratory medical device. If eye or skin contact, immediately flush with running water for at least 20 minutes. Seek medical attention for all instances of overexposure to this chemical.	If inhaled, remove to fresh air. If not breathing, give artificial respiration. If eye or skin contact, immediately flush with running water. Seek medical attention for all instances of overexposure to this chemical.	If inhaled, remove to fresh air. If not breathing, give artificial respiration. If eye or skin contact, immediately flush with running water. Seek medical attention for all instances of overexposure to this chemical.
Spill control measures	If a spill occurs, cleanup personnel should wear full protective clothing. Eliminate sources of ignition. Keep nonessential personnel away. Absorb spilled material on vermiculite or other suitable absorbent. Containerize for disposal.	Eliminate sources of ignition. Equipment used in handling must be grounded. Do not touch or walk through spilled material. Absorb material with dry earth, sand, or other suitable noncombustible absorbent. Containerize for disposal.	Cleanup personnel should wear proper protective clothing to avoid contact with liquid. Absorb material on vermiculite or other suitable absorbent material. Containerize for disposal. Flush area of spill with water.
Disposal	Dispose in accordance with all applicable local, state, and federal regulations. Contact local or state waste agencies if disposal questions arise.	Dispose in accordance with all applicable local, state, and federal regulations. Contact local or state waste agencies if disposal questions arise.	Dispose in accordance with all applicable local, state, and federal regulations. Contact local or state waste agencies if disposal questions arise.
Personal protection	Wear gloves, lab coat, splash goggles, and any other appropriate equipment suggested by the laboratory supervisor.	Wear gloves, lab coat, splash goggles, and any other appropriate equipment suggested by the laboratory supervisor.	Wear gloves, lab coat, splash goggles, and any other appropriate equipment suggested by the laboratory supervisor.
Storage information	Store tightly closed in a location suitable for general chemical storage.	Store in a location suitable for flammable liquid storage.	Store in a cool, dry, well-ventilated area. Store below 120°F.

TWA–time weighted average; ACOIH–American Conference of Governmental Industrial Hygienists; IARC–International Agency for Research or Cancer; OSHA–Occupational Safety and Health Administration; PEL–Permissible Exposure Limit; NIOSH–National Institute for Occupational Safety and Health; RTECS–Registry of Toxic Effects of Chemical Substances; LD_{50}–lethal dose for 50% of a population.

Handling and Care of Animals in the Laboratory

The study of anatomy and physiology of animals is fundamental to the training of zoology students. Many students find that working with living and preserved animals is one of the most interesting and beneficial aspects of their education. Prospective employers in business and industry, and admissions committees of graduate programs, as well as medical, dental, and veterinary schools, have frequently emphasized the importance of such practical experience.

Research with laboratory animals has led to important scientific advances in physiology, genetics, behavior, nutrition, ecology, and other fields. Advances in human medicine that are direct results of experimentation involving animals include immunization against polio, diphtheria, measles, and other diseases; insulin production and therapy; blood transfusions; chemotherapy; electrocardiography, open-heart surgery, and artificial heart valves; organ transplantation; and kidney dialysis.

Major advances in veterinary medicine resulting from experimentation with animals include the development of vaccines for rabies, distemper, swine cholera, and brucellosis; medication for dog heartworms; artificial insemination, in vitro fertilization, and embryo transfer technology; methods for preserving endangered species; and surgical techniques for hip replacement. These veterinary advances have saved thousands of animal and human lives and have contributed greatly to the human food supply and to the quality of life of farm and companion animals.

Studies of animals from textbooks, photographs, charts, models, and computer simulations are good supplements, but they are not adequate substitutes for actual laboratory experience with living and preserved animals. Zoology students need to learn and practice proper methods to observe, handle, care for, experiment with, and dissect laboratory animals. Consider the dilemma of a neurosurgeon who has never observed, handled, or dissected an actual brain, but who is about to do his or her first operation on you, or on a member of your family with a brain tumor.

The handling and treatment of vertebrate animals is regulated by federal law under the Animal Welfare Act enacted in 1966 and amended numerous times, most recently in 2007. The National Academy of Sciences recently published the eighth edition of its *Guide for the Care and Use of Laboratory Animals*, a comprehensive discussion of regulations and proper procedures to use with laboratory animals. It is listed at the end of this section along with several other helpful reports on this subject. The Animal Welfare Information Center (AWIC) established in the U.S. Department of Agriculture provides extensive information for improving animal care and use in research, testing, teaching, and exhibition. The Center can be found on the web at the following url: http://awic.nal.usda.gov/. Additional regulations governing the use and care of laboratory animals have been developed by the National Institutes of Health. Many individual states also have laws governing animal use. Invertebrate animals are generally not covered under these laws, but such animals should also be treated with care and respect as living creatures. Rare and endangered species are protected by special laws and may not be collected or used in laboratory studies except under special permits. All teachers and researchers must be familiar with these federal and state regulations and be responsible for using good judgment and for following appropriate procedures for handling and experimenting with all animals.

As a responsible citizen and a student of zoology, you should also handle living and preserved animals with care and respect. When working with both vertebrate and invertebrate animals, you should always take adequate precautions to avoid causing unnecessary stress or discomfort to the animals due to your handling or experimenting. Any animals kept in the laboratory must have a clean and appropriate environment, including adequate ventilation, food, water, and regular care. Be sure to follow the specific federal guidelines established for the care of animals kept in the laboratory for the duration of an experiment. At the end of the experiment, the animals must either be disposed of in an approved humane manner or returned to a permanent animal care facility as directed by your instructor.

Some people oppose the use of animals in the laboratory either for training or research because they believe it is unethical for humans to use animals in any way that might be harmful or detrimental to the animals for the benefit of humans or other animals. Appropriate usage of animals has been one of the most active controversies in the United States and elsewhere during the past several years.

Opponents of animal use seek to reduce or eliminate the use of animals in teaching and research based on their convictions. They often cite alternatives to the use of animals in research and testing, such as computer simulations, models, films or videos, tissue culture, and in vitro chemical tests, as effective substitutes. While many scientists agree that alternatives to the use of animals are effective in some cases, no adequate alternatives are available in many other cases. Most scientists agree that the rational use of animals for teaching and research continues to be essential for the progress of human health and welfare. This position has been endorsed by several prestigious scientific bodies, including the Society of Integrative Biology, the American Association for the Advancement of Science, the Society of Sigma Xi, the National Science Teachers Association, the National Association of Biology Teachers, and several state academies of science.

The continuing controversy over the use of animals for teaching and research, as well as the escalating costs of obtaining and caring for laboratory animals, has already resulted in substantial reductions in the number of animals used for study and in research and improvements in the care and handling of animals in the laboratory. Concerns over the use of animals have also led to numerous governmental regulations on the use and handling of animals in the laboratory. Therefore, in addition to learning about the animals themselves, zoology students must also learn the rules and methods for the proper care and handling of the animals.

Suggested References

Akins, C. K., S. Panicker, and C. L. Cunningham, eds., 2004. Laboratory Animals in Research and Teaching: Ethics, Care, and Methods. American Psychological Assn. Washington, DC, 274 pp

Lewbart, G.A. ed. 2012. Invertebrate Medicine, 2nd edition, Wiley-Blackwell, Chichester, UK. 488 pp

National Academies of Science, *Guide for the Care and Use of Laboratory Animals*. 2011. The Guide incorporates new scientific information on common laboratory animals, including aquatic species, and includes extensive references. National Academies Press, Eighth edition, Washington, DC, 248 pp

US Public Health Service 2002. Public Health Service Policy on Humane Care and Use of Laboratory Animals, Office of Laboratory Animal Welfare. http://grants.nih.gov/grants/olaw/references/phspol.htm retrieved 6/14/2011.

Hints for Dissection

Many of the animals you will learn about in this course will require dissection for the study of internal structures. In order to benefit from these experiences it is important that you follow the directions carefully and develop good dissection skills. Take special care to avoid damaging important structures before you have an opportunity to study them carefully. A good rule to remember is "investigate first, cut last" because once a body part is cut apart or separated from adjacent parts it is difficult to determine its real shape and relationship to other body parts.

Zoologists use a precise set of terms when they describe the anatomy, orientation, and location of body parts. We have provided a brief glossary of anatomical terms at the end of this section, some of which are illustrated in figure 1 below. We suggest that you study these terms and refer to them as you do the dissection animals in the exercises in the later chapters of this book.

The chief objective of the dissection of a specimen is to expose body parts for the study of their structure and their relationships to other parts. Therefore, it is important to proceed carefully with each dissection and to avoid cutting or removing anything, unless so instructed in the printed directions or by your laboratory instructor. Read through the instructions completely before you begin each dissection. This will save you time later and will often prevent disappointing results.

Body parts should be parted carefully along natural lines of separation and attachment wherever possible. This can often be done best (without cutting) with forceps, probes, the handle of a scalpel, or with a finger.

Good quality dissection instruments are essential to good laboratory work. Your dissection kit should include the following items:

- Scissors, fine points (or one fine point and one blunt point), about 4–6 inches long
- Forceps, fine points, about 4–5 inches long
- Scalpel with replaceable blade
- Extra scalpel blades
- Two teasing needles
- Probe, about 6 inches long
- Plastic centimeter ruler
- Dropping pipette

Scissors should be constructed of first-quality chrome or stainless steel and have the blades joined by a screw rather than a rivet. Some common dissecting instruments are illustrated in figure 2.

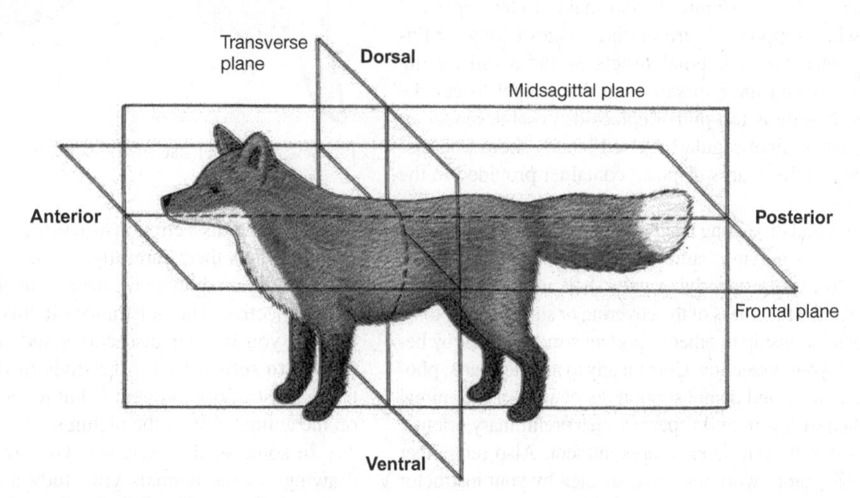

FIGURE 1 Basic anatomical directions and planes.

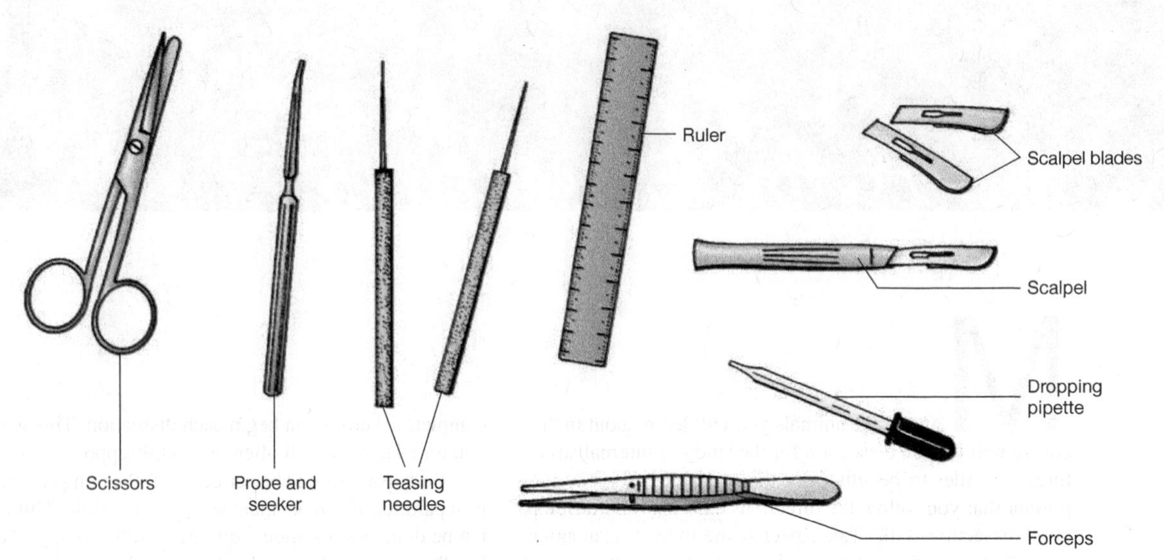

FIGURE 2 Dissecting instruments.

Dissecting instruments should be kept clean and sharp at all times. A small oilstone and a piece of emery cloth or fine sandpaper should be used to keep cutting edges and needle points sharp. Two types of scalpel are in common use. One type has a permanent blade that can be resharpened, and the second type has a separate handle and replaceable blades that are not usually sharpened. Blades on scalpels with replaceable blades should be replaced often to ensure a good edge. Sharp scalpels and scissors allow clean cuts and minimize damage to adjacent tissues. Always wash and fully dry your dissecting instruments with a paper towel after each use. With proper care, a good set of instruments will last for many years.

The scalpel is used for making incisions in the body wall and occasionally for sectioning interior structures. Hold the scalpel upright with the handle nearly perpendicular to the surface to be cut (figure 3) and make a clean forward incision while supporting surrounding tissues with your fingers. Be conscious of internal structures and avoid cutting too deeply. If you have to saw with the scalpel to cut the tissue, your blade is too dull. Replace disposable blades or sharpen your scalpel regularly. Used blades should be discarded only in the sharps disposal container provided in the laboratory.

Forceps and dissecting needles should be used for loosening, lifting, and moving various structures to facilitate their study and to expose underlying parts. It is important to gain some idea of the thickness of the covering or surface layer to be cut and its relationship to other important structures nearby before starting your dissection. Consult any available charts, photographs, models, and demonstration dissections of the animal to be studied. A few minutes spent in such preliminary orientation will often prevent later disappointment. Also remember that your laboratory work may be evaluated by your instructor partly on the basis of your care and skill in dissecting.

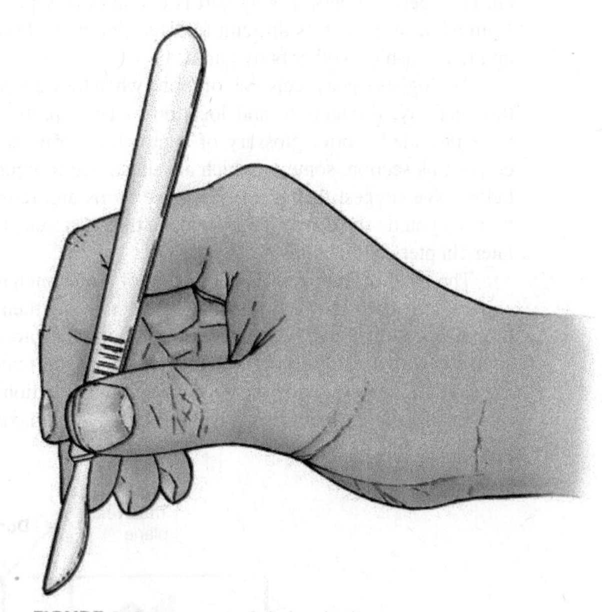

FIGURE 3 Proper use of the scalpel.

Read the dissection instructions in the laboratory manual and follow them carefully. Always remember that **direct study** of internal anatomy from your dissections is the primary objective. The illustrations in this manual are intended to **help** you in your dissections and study but are **not intended to substitute** for the study of real specimens. Refer to the illustrations frequently, but focus your study primarily on the animal, not on the pictures.

In some of the exercises, you are also asked to make drawings of the animals you study and dissect. Take care to be accurate and neat in your drawings so that they will be helpful to you for later study and review. Remember that your laboratory notes and drawings serve a purpose similar to your lecture notes. They provide a record to help you remember actual structures during your later study and reviews. They help you to retain and to recall mental pictures of the things actually seen through the microscope and during your dissections. Include a magnification scale in your drawings to help you remember size and dimensions.

You should also use the Notes and Sketches section at the end of each chapter for any additional things that you observe or need to remember.

A Glossary of Directional Terms Used in Zoology

Aboral away from the mouth; the opposite of oral.

Anal toward the anus or away from the mouth.

Anterior toward the front of the body; the opposite of posterior.

Axial toward the midline of the body; the opposite of radial.

Caudal toward the tail or tail end; the opposite of cephalic.

Cephalic of or pertaining to the head; the opposite of caudal.

Cranial relating to the skull or cranium.

Cross-section a cut made through the body along a transverse plane.

Deep away from the surface of the body; the opposite of superficial.

Distal away from the point of attachment; the opposite of proximal.

Dorsal relating to the back or upper surface; the opposite of ventral.

Frontal plane plane parallel to the dorsal and ventral surfaces of the body, which bisects a bilaterally symmetrical animal into upper and lower halves.

Lateral toward the side; the opposite of medial.

Longitudinal lengthwise; parallel to the long axis of the body.

Medial toward the center of the body; the opposite of lateral.

Oral toward the mouth; the opposite of aboral.

Peripheral toward the outer surface.

Posterior the hind part (rear) of the body; the opposite of anterior.

Proximal toward the point of attachment; the opposite of distal.

Radial away from the midline of the body; the opposite of axial.

Sagittal pertaining to the long axis of the body.

Sagittal plane any longitudinal plane passing from the head to the tail. The midsagittal plane bisects a bilateral animal into two symmetrical halves (mirror images).

Superficial located near the surface of the body; the opposite of deep.

Transverse perpendicular to the long axis of the body.

Transverse plane any plane perpendicular to the sagittal and frontal planes. Sections of the body cut on a transverse plane are called cross-sections.

Ventral relating to the belly or underside; the opposite of dorsal.

Chapter 1

Microscopy

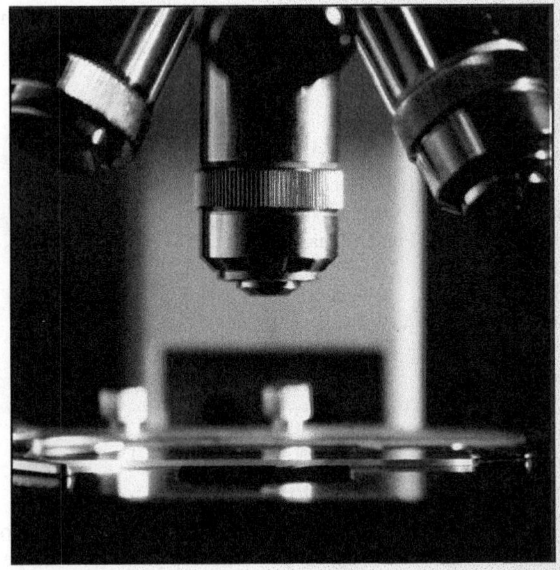
Sean Justice/Corbis.

Objectives

After completing the laboratory work in this chapter, you should be able to perform the following tasks:

1. Identify the main parts of a compound microscope and explain their function.
2. Define and explain focus, working distance, resolving power, and magnification.
3. Describe the proper use and care of both compound and stereoscopic microscopes.
4. Explain the difference between compound and stereoscopic microscopes and give examples of appropriate uses of each type.
5. Use both compound and stereoscopic light microscopes in the correct way.
6. Explain the operating principles of a phase contrast microscope and an interference microscope and give examples of their use.
7. Estimate magnification of a compound microscope and use the microscope to measure a microscopic object.
8. Describe the two main types of electron microscopes and give examples of their use.
9. Explain the importance of microscopes in biological studies.

Introduction

The light microscope is one of the most useful tools biologists have for viewing very small objects and minute details of larger ones. The first microscope was probably constructed in the late 16th century and is widely credited to the ingenuity of Hans and Zaccharias Janssen, two Dutch craftsmen who made and sold eyeglasses. But it was pioneers like Robert Hooke (1635–1703) and Antony van Leeuwenhoek (1632–1723) who first stimulated broad interest in exploring the smallest features of the natural world with their published descriptions of plant cells, spermatozoa, blood corpuscles, and "animalcules" swimming in drops of pond water. In their day, the best microscopes provided magnification up to about 270×.

A significant improvement in the optics of light microscopes occurred near the end of the 18th century with the development of specially coated lenses that refract white light without breaking it up into component colors of the spectrum. These achromatic lenses produce a sharper image that is free of prismatic aberrations. In recent years, inventors have devised many additional methods for improving illumination and increasing image contrast. Today, the best light microscopes are capable of nearly 2,000× magnification.

At first you might think that viewing a tiny object simply requires getting enough enlargement or magnification. Actually, as is usually the case, the problem is much more complicated. To achieve good observation of tiny living things and their parts, four main problems must be overcome: (1) getting sufficient contrast, (2) finding the right focal plane, (3) getting good resolution, and (4) recognizing the subject or part when you see it magnified. We will address the first three challenges in this chapter and we will help you with the fourth challenge as you work through the rest of the book.

Modern light microscopes generally fall into one of two categories: compound microscopes or stereoscopic microscopes. Each has its unique advantages and applications, as described in the following pages of this chapter. You will have opportunities to use both types of microscope during your study of the animal kingdom.

Because microscopes are used to view small objects, you need to be familiar with the units of measurement commonly used for the dimensions of microscopic structures. This book follows the convention of most science laboratories in using the metric system of linear measurement. These units are listed in table 1.1. Objects viewed with a stereoscopic microscope usually range from a few millimeters to a few centimeters in size. Objects viewed with a compound light microscope generally range in size from a few micrometers to a few hundred micrometers. Electron microscopes, first developed in the late 1930s, provide even higher resolution than the best light microscopes by illuminating the subject with a focused beam of electrons instead of visible light. Objects viewed with a typical transmission electron microscope usually range from several angstroms to several nanometers in size.

Table 1.1 Metric Units Used in Microscopy

1 angstrom (Å)	=	0.1 nanometer (nm)
10 angstroms	=	1.0 nanometer (formerly called millimicron [mμ])
1,000 nanometers	=	1.0 micrometer (μm) (formerly called micron[μ])
1,000 micrometers	=	1.0 millimeter (mm)
10 millimeters	=	1.0 centimeter (cm)

The Compound Microscope

The **compound microscope** is especially useful for the study of cells and cell parts, the organization of tissues, the structure of bone, the features of developing embryos, and many other important applications. Because many of the exercises in this course will require the use of the compound microscope, it is important to review some aspects of its construction, use, and care.

A modern compound microscope is illustrated in figure 1.1. Because there are numerous makes and models of compound microscopes in use, the microscope assigned for your use may differ slightly from the one illustrated. The operating principles and procedures, however, will be similar to those outlined later. Your laboratory instructor will point out any important differences between your microscope and the one illustrated.

A microscope is an expensive precision instrument and must be handled with care. Always carry your microscope **with both hands.** Grasp the arm of the microscope firmly with one hand and support the base with the other hand. Place the microscope carefully on your table with the arm facing you. **Do not "clunk" it on the table.**

Materials List

Living specimens
- *Artemia* (brine shrimp) larvae

Prepared microscope slides
- Letter "e"
- Frog blood

Audiovisual material
- Wall charts showing parts of compound and stereoscopic microscopes

Parts of the Microscope

Identify the principal parts and controls of the microscope with the aid of figure 1.1. At the top of the microscope is the **eyepiece,** or ocular lens, which is inserted in an inclined **body tube.** Most recent models of microscopes are **binocular,** having two eyepieces and two body tubes to allow viewing specimens with both eyes. On binocular microscopes the eyepieces are **movable** to accommodate differences in the distance between the viewer's eyes. Try moving the eyepieces together and apart until they match your eyes for good viewing.

Microscopes with a single eyepiece and body tube are **monocular.** Below the body tube find the **arm** that is attached to the base of the microscope. Also attached to the arm is the movable **stage,** which holds a microscope slide

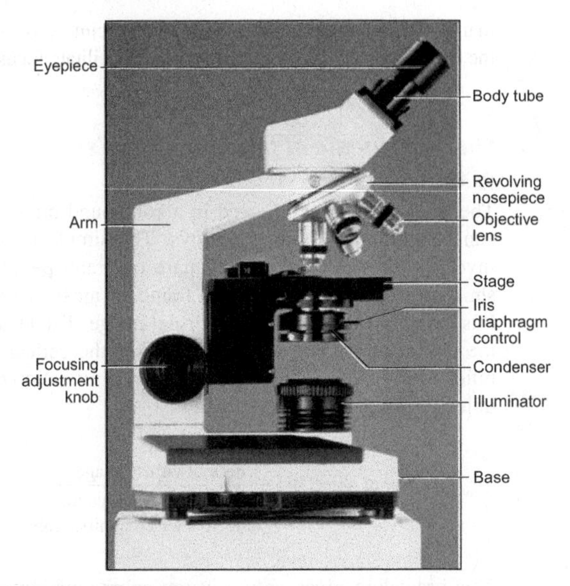

FIGURE 1.1 Compound microscope.
Courtesy of the Olympus Corporation.

or other object for viewing. **Stage (slide) clips** aid in holding the slide in position on the stage. Above the stage is the **revolving nosepiece** with two or more **objective lenses.** Within the stage is another lens system, the in-stage **condenser,** which serves to concentrate light rays from the built-in **illuminator.** On the base near the illuminator is the **light switch.** Instead of a built-in illuminator, some microscopes have a **substage mirror** to reflect light from an auxiliary light source.

Beneath the condenser, locate the **disc aperture diaphragm.** Rotating the disc diaphragm increases or decreases the amount of light on the specimen. More expensive microscopes often have an **iris diaphragm** with movable elements instead of a disc diaphragm. Raising and lowering the condenser also regulates the illumination of the specimen, although in most of your work in this course you will obtain satisfactory results by adjusting the condenser to the position that gives maximum illumination (usually near its uppermost position) and then making any further needed reductions in illumination with the disc diaphragm. This simplified method of controlling light does not produce the precise illumination required for advanced microscopy, but it produces results satisfactory for most routine purposes.

Accurate observation of a specimen requires positioning the objective lens at a specific distance from the specimen; this distance is determined by the specific construction of each objective lens and is called the **working distance** of the lens. When the objective lens is located at the proper working distance, the specimen will be in **focus.** The working distance of the lens varies **inversely** with the magnification of the objective; low-power objectives have longer working distances, and high-power objectives have shorter working distances.

Focusing the lens system is accomplished by mechanically changing the distance between the specimen and the objective lens. Coarse and fine **adjustment knobs** for this purpose are provided on the side of the arm near the base. On some modern microscopes both coarse and fine adjustments are controlled by a single knob with dual function, but most models have separate controls. ***How many focusing control knobs are there on your microscope?***

When you observe a specimen through a compound microscope, several images are formed by the optical system of the microscope. Three of these images are of special importance—the **real image,** the **virtual image,** and the **retinal image** (figure 1.2).

The real image is formed at a specific distance above the objective lens. What you actually see when you look through the objective lens is the virtual image, which appears both larger and farther away than the specimen on the stage. The retinal image is formed by the rays of light striking the retina of your eye. The virtual image can be helpful to you in estimating the size of an object viewed through a microscope as explained later.

Magnification

The principal purpose of a microscope is to magnify the image of an object. The **magnification** of an object is determined by the construction of the ocular and objective lenses of the microscope, and the total magnification is the product of the separate magnification of these two lenses.

Example:

10× ocular × 10× objective = 100× total magnification

Student microscopes used in introductory biology and zoology courses are commonly equipped with 10× ocular lenses, and both 10× and 43× objective lenses (some objective lenses are 40× or 44×). These two lenses are mounted on a revolving nosepiece and are called the **low-power** and **high-power** objectives, respectively. Often, a 3.5× objective lens is also used on student microscopes; this is called a **scanning lens.** It is useful for viewing relatively large objects or for preliminary location of a specimen on the microscope slide. Occasionally a 90× or 100× **oil immersion lens** may be present for viewing very small objects like bacteria. Special instructions and precautions are needed for the use of oil immersion lenses.

Magnification is changed in microscopes with multiple objectives by rotating the revolving nosepiece until the desired objective is in position below the body tube of the microscope. A newer type of compound microscope (zoom lens type) uses a system of movable lenses to produce a variable magnification rather than a series of fixed magnifications as in a microscope equipped with a rotating nosepiece.

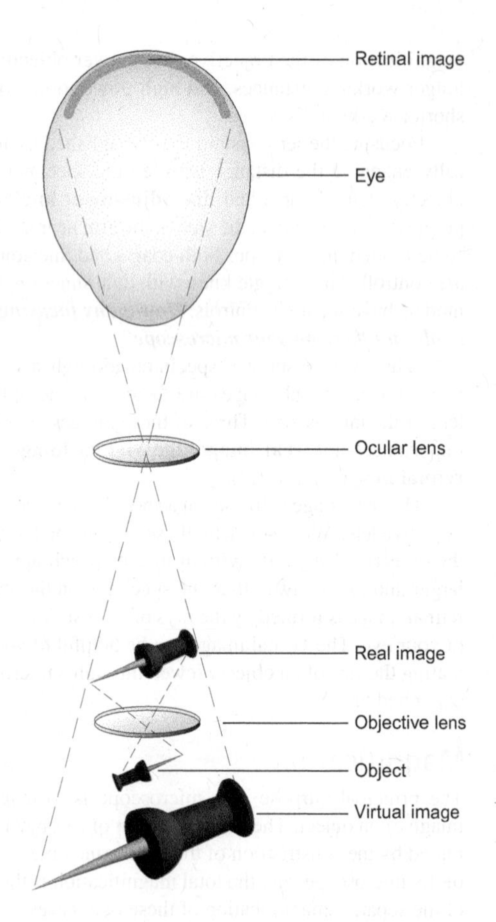

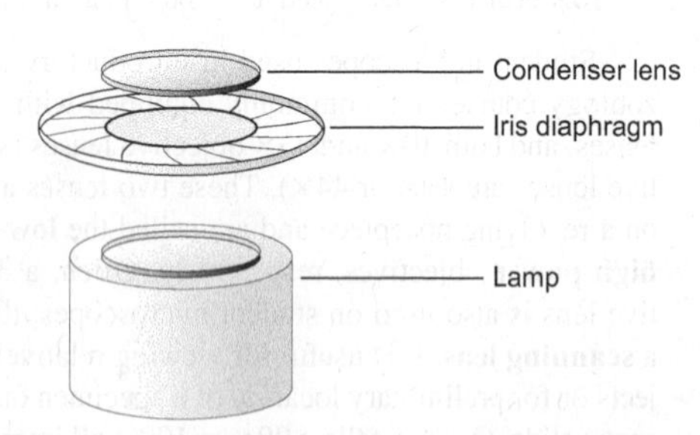

FIGURE 1.2 Comparison of virtual image, real image, and retinal image.

Observe the 10× objective on your microscope and note that it is also marked 16 mm (or some similar value). This is the working distance of that lens; thus, this lens will be in focus at approximately 16 mm above the surface of your specimen. ***What is the working distance of your high-power objective?***

All the other parts of the microscope are accessory to the main purpose of magnification by the lenses. They consist mainly of mechanical devices to hold the specimen, to regulate the light necessary for clear vision, and to facilitate focusing.

Quick Estimate of the Magnification of an Object

The size of an object viewed in a compound microscope can be estimated in several ways. The simplest method involves the use of a small square of graph paper or a small plastic ruler of the type found in most dissecting kits to measure the size of the virtual image. The length of the virtual image is then divided by the theoretical magnification of the ocular lens times the magnification of the objective lens.

$$\text{size of object} = \frac{\text{length of virtual image}}{\text{ocular lens magnification} \times \text{objective lens magnification}}$$

This procedure gives only a rough approximation of the magnification because the actual magnification of the ocular and objective lenses is slightly different from the stated theoretical magnification, and your estimate of the size of the virtual image is not precise. Detailed instructions for estimating the magnification of an object under your microscope are given on page 8.

You can use a similar procedure to estimate the size of an object you see under your compound microscope. If you observe a cell under the microscope that appears to be 25 mm in diameter, simply divide by the calculated magnification to determine the actual size of the cell.

Example:

$$\frac{25 \text{ mm (observed size)}}{103 \text{ (magnification)}} = 0.24 \text{ mm}$$

Estimating the size of a larger object viewed under high power is a bit more complicated, but follows the same principle. First, place the ruler on the stage and measure the apparent diameter of the microscope field at high power. Next, place the slide with millimeter square graph paper on it on the stage and orient the slide again so the sides of the square are parallel to the ruler. Switch to high power and move the millimeter square so that one corner is at one edge of the field. Locate some distinctive landmark on the baseline of the square, such as a bulge in the line or a large fiber, at the opposite edge of the field. Carefully move the slide to place your landmark directly across the field and find a new landmark on the opposite side of the field again. Continue this procedure until you determine how many microscope fields at high power it takes to traverse the one millimeter square.

Multiply the apparent field diameter as measured with the ruler by the number of diameters required to traverse a one millimeter square to obtain the magnification of your microscope at high power.

Example:

$$\text{apparent diameter of field} \times \frac{\text{number of fields}}{\text{to traverse 1 mm}} = \text{magnification}$$

or

$$175 \text{ mm} \times 2.4 \text{ field diameters} = 420\times$$

In a like fashion, you can estimate the actual size of an object by dividing its apparent size (measure the virtual image with a ruler) by the magnification under high power.

Example:

$$\frac{\text{apparent size}}{\text{magnification}} = \frac{160 \text{ mm}}{420\times} = 0.38 \text{ mm}$$

The values obtained by these procedures are only approximate but provide size and magnifications useful for most purposes.

Measurements Using an Ocular Micrometer

More precise measurement of microscopic objects and estimation of magnification require additional equipment. One common method requires an **ocular micrometer,** a graduated micrometer disc that is placed in one eyepiece of the microscope. An ocular micrometer contains a millimeter scale marked off into several divisions and must be calibrated with a stage micrometer. A **stage micrometer** is a glass slide with a precision millimeter scale usually divided into 100 parts. Superimposing the micrometer disc in the eyepiece over the stage micrometer allows precise calibration of the divisions in the eyepiece and permits accurate measurements of microscopic objects.

Resolving Power

While magnification is the increase in size of the object's image, the ability of a lens or a microscope to reveal the fine detail of a specimen is called its **resolving power.** In microscopy, an increase in the apparent size of a specimen is not always accompanied by an increase in the clarity of detail within the specimen. Beyond certain limits, further magnification simply makes the apparent image of a specimen become progressively fuzzy or indistinct as it becomes larger.

The resolving power of a light microscope depends largely upon the design and quality of its objective lenses. It is actually the resolving power rather than magnification that determines the useful magnification of a compound microscope. Furthermore, it is the wavelength of visible light that limits resolving power of a microscope. This is the reason that transmission electron microscopes are capable of much greater resolution than light microscopes. Electrons have much shorter wavelengths than the light rays of visible light.

Resolving power is defined as the shortest distance between two points that can be visually distinguished as two separate points. Additional magnification without increased resolution produces a larger but less distinct image. Imagine a photograph in a newspaper when the image is enlarged (magnified) to appear larger but in which the number of dots making up the photographic image remains the same. Each time the photo is enlarged, the dots are spaced farther apart, making the face in the photo appear increasingly fuzzy and indistinct (figure 1.3). This is an example of magnification without increased resolution.

Illumination

Proper illumination of the specimen is an extremely important matter, since improper lighting of the specimen can produce poor images, inaccurate observations, and/or unnecessary eyestrain.

Most modern student compound microscopes are equipped with in-base illuminators. If you have such a microscope, you can adjust the amount of light reaching the objective lens by rotating the disc (or iris) diaphragm. Experiment with the diaphragm control, and observe the changes in amount of light as the diameter of the diaphragm opening changes. You will find that the best illumination for viewing specimens is obtained with the smallest diaphragm opening that lights the entire microscopic field.

Sometimes illumination for a compound microscope is provided by a separate lamp and a substage mirror. If you are assigned a microscope of this type, examine the two surfaces of the mirror—one surface is flat, and one surface is concave. The flat surface of the mirror is the one used most often in normal laboratory situations. When the mirror is properly adjusted, light is reflected by the mirror and passes through the condenser, the specimen, and the lenses to your eye. The concave (curved) surface of the mirror is used less frequently and only in certain situations where there is a need to concentrate the light rays. Adjust your mirror to reflect the maximum light on the specimen, and reduce the illumination as necessary by adjusting the iris diaphragm.

Illumination is also controlled by the substage condenser. For routine work in an introductory laboratory, the condenser should not be used to adjust illumination, but should be set at or near its uppermost position and the light intensity adjusted by means of the diaphragm.

Careful adjustment of the light beam enables you to obtain the best images with your microscope. To obtain maximum resolution and the best results, scientists use a special type of light adjustment with parallel light rays illuminating the specimen. This technique, called **Koehler illumination,** requires a lamp equipped with a field diaphragm that allows you to center and control the diameter of the light beam.

Procedure for Koehler Illumination

1. Place a microscope slide on the stage and bring a specimen into focus using the low-power objective lens.
2. Close the field diaphragm on the light source until you can see its edges through the microscope.

(a) Magnification 1×

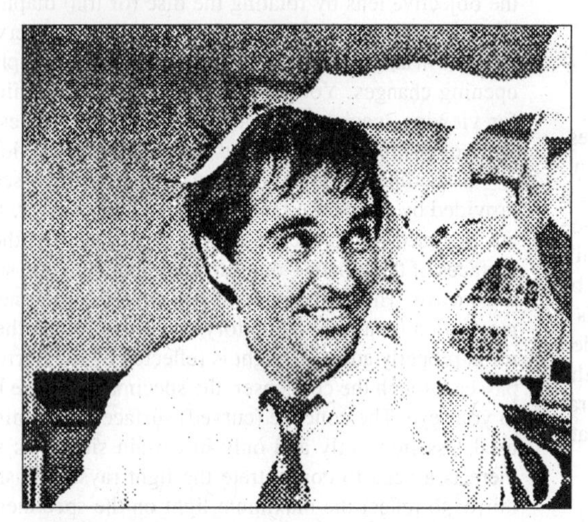
(b) Magnification 2×

(c) Magnification 4×

FIGURE 1.3 Magnification without increased resolution.
Photographs courtesy of the Raleigh News and Observer, Raleigh, NC.

3. Focus the condenser by moving it up and down until the outline of the field diaphragm is in focus.
4. Center the field diaphragm until it covers the full field of view.
5. Remove the eyepiece and observe the image coming from the objective lens. (Better results can be obtained with a centering telescope or phase telescope to replace the objective lens if one of these accessories is available.) Observe the illuminated circular field coming from the objective lens.
6. Slowly close the condenser diaphragm until you see its outline appear at the edge of the circular field. Adjust the condenser diaphragm to cover approximately the outer two-thirds of the image coming from the objective lens.
7. This setting of the field and condenser apertures produces parallel light rays to illuminate the specimen and usually gives the best balance between maximum resolution and contrast of an image.

Focusing

To obtain a clear image of a specimen, you must carefully adjust the distance between the lenses and the specimen. This adjustment of the distance between the lenses and the specimen is called **focusing.** When a clear image of the specimen can be seen through the ocular lens, the specimen is said to be **in focus.** At this adjustment of the lens system, the specimen is located precisely at the working distance of the objective lens.

- Locate once again the coarse adjustment knob on your microscope. Rotate the low-power lens in place and turn the control slightly in one direction. Observe that movement of the knob increases or decreases the distance between the stage and the objective lens.

  If you have a microscope with separate coarse and fine adjustment knobs, carefully try each control to see the difference in their actions. You will find that it takes several turns of the fine adjustment control to equal the effect of a single turn of the coarse adjustment control.

  While testing the action of the focusing controls, watch the objective lens and the specimen stage. ***Which part moves, and which part remains stationary?***

Sometimes you will find that the fine adjustment will not turn any farther and you cannot reach focus. When this occurs, you have reached the end of the range of the fine adjustment in that direction. To rectify the situation, simply turn the fine adjustment five full turns away from its stop, and refocus the image as best you can with the coarse adjustment. Now you can use the fine adjustment to improve the focus.

CAUTION!

Never allow the objective lens to touch a slide or the condenser lens, which may protrude through the hole in the specimen stage. Anything coming in contact with the objective lens presents a danger of scratching, cracking, or otherwise damaging this critical and expensive part of the microscope. Always remember that this expensive scientific instrument is assigned to you for your use and safekeeping. Be certain that you use it properly and carefully, and that you keep it in good condition. Promptly notify your laboratory instructor of any malfunctions or if you have any difficulty using your microscope.

Procedure for Use of the Compound Microscope

Good microscopy requires the adoption of work habits that facilitate observation, minimize fatigue and eyestrain, and protect the equipment from damage. Follow these steps each time you use the compound microscope.

1. Place the microscope directly in front of you on the laboratory table. Remember to carry the microscope by using both hands after removing it from the storage cabinet. Clean the objective and ocular lenses by wiping gently with a clean sheet of lens paper; never use anything but lens paper for cleaning the lenses.
2. Rotate the revolving nosepiece until the low-power (10×) objective lens clicks in place directly over the center of the condenser. While viewing the objective lens from one side of the microscope tube, **carefully** adjust the distance between the lens and the specimen stage, using the coarse adjustment control, to approximately 12 mm.
3. Select a microscope slide provided by your laboratory instructor and examine it to locate the position of the specimen on the slide. Then place the slide on the stage, with the specimen centered over the condenser lens. Make certain that the coverslip and specimen are on top of the slide.
4. Open the disc (or iris) diaphragm fully and turn on your microscope lamp. If your microscope is equipped with a substage mirror and an auxiliary lamp, adjust the positions of the lamp and the mirror until you obtain an evenly lighted, circular microscope field. Reduce the light on the specimen as necessary for clear observation by adjusting the diaphragm.
5. While observing the objective lens and the microscope slide from the side again, carefully readjust the distance between them to about 3 mm. **Do not look through the ocular while performing this step.**
6. Now look through the ocular and slowly **increase the distance between the objective lens and the specimen** until the specimen comes into focus. This may involve either raising the objective lens or lowering the specimen stage, depending upon the design of your particular microscope. **Never** focus in the opposite direction—by decreasing the distance between the objective lens and the specimen—**while looking through the ocular.** Further, center the specimen in your field of view as necessary, and bring the specimen into sharper focus with the fine focusing control.
7. Rotate the revolving nosepiece until the high-power objective lens (43×) clicks into position directly over the condenser lens. The objective lenses on most modern microscopes are factory installed and adjusted so only minor changes in focusing and centering are necessary when magnification is changed. Such lenses are **parfocal**—that is, their planes of focus and the center of their field of view are identical, or nearly so, although their working distances are different. You should find it necessary to make only a slight adjustment with the fine focusing control in order to achieve good focus after changing from the 10× objective to the 43× objective. **Use only the fine focus control while looking through the ocular with the high-power objective in position.** You will also find it necessary to open the iris diaphragm slightly when you switch from low to high power. ***Why?***

Returning the Microscope after Use

1. Rotate the revolving nosepiece to place the **low-power objective** (or scanning lens if your microscope has one) into position over the condenser lens.
2. Remove the microscope slide from the specimen stage and return it to its proper box or tray. If you have been using wet mounts, clean the specimen stage with a **clean cloth** or **cleaning tissues** as provided by your laboratory instructor.
3. Clean the objective and ocular lenses of the microscope with **lens paper.**
4. If your microscope has an in-base illuminator, turn off the power, disconnect the power cord, and carefully wind the cord as directed by your instructor.
5. Return the microscope to its storage cabinet. Remember to use both hands in carrying the microscope. Check once again to make certain that the low-power objective lens (or scanning lens) is in position **and that you have not left a slide on the stage.**

Special Precautions

1. **Never** focus down (raise the stage or lower the objective, depending on the type of your microscope) while looking through the microscope.
2. **Always** locate the specimen under low power before switching to high power.

3. **Never** turn your microscope upside down or lay it on its side. The ocular might fall out and could be damaged.
4. **Always** keep the microscope clean and dry. **Use only lens paper to clean the lenses.**
5. **Never** turn the microscope lamp on or off before turning the power down (if you have an adjustable lamp).
6. **Always** wrap the power cord carefully around the base of the microscope or the cord hanger before putting away your microscope.
7. **Never** use the coarse focusing control when the high-power lens is in position. Focus only with your fine focusing control when using high power.
8. **Always** try to relax and keep both eyes open when using the microscope. This helps to prevent undue eyestrain. With a little practice, you can learn to concentrate on the specimen and to disregard the image received by the other eye if you are using a monocular microscope.

Exercises Using the Compound Microscope

1. Obtain a microscope slide with the letter "e" on it. Place the slide on the stage of your compound microscope and observe under low power. ***What is the orientation of the letter "e" as you view it through the compound microscope? Why does it look different from a letter "e" printed on this page?***
   Move the microscope slide slightly to the left while you look through the eyepiece. ***Which way does the letter "e" seem to move?***
   Move the slide slightly away from you. ***Which direction does the letter "e" move this time?*** Change the objective lenses to high power by rotating the nosepiece or by changing the zoom adjustment if you have a zoom-type microscope. Observe the letter "e" under high power. ***How much of the letter "e" can you see under high power?*** Note that the letter seems larger and that your field of view is smaller at high power than at low power.
2. Obtain a slide of frog blood and study it first under low power and later under high power. ***What is the shape of the cells? What structures can you observe in the cell?*** Draw a frog blood cell on the first page of the Notes and Sketches section at the end of this chapter. Label the cell parts you recognize. You will learn more about cells and their parts in the next chapter.

Estimating Magnification

You can estimate the magnification of your microscope by using a small square of millimeter graph paper mounted on a clean microscope slide under a coverslip.

1. Carefully move the slide to center a 1 millimeter square in the microscope field with two sides of the square parallel to the edge of the microscope stage.
2. Place a small plastic metric ruler on the right edge of your microscope stage so that you can see the ruler with one eye while you view the slide with the other eye.
3. Keep both eyes open and focus your eyes on both images with the square of graph paper in one eye and the scale of the ruler in the other eye. It will seem strange at first to look at two different images simultaneously, but if you relax and continue to observe both objects, they will soon become superimposed in your vision.
4. When you see the two superimposed images, measure the apparent size of the superimposed millimeter square as it appears on the ruler. Since the lines surrounding the square appear relatively thick, be sure that you measure from the upper edge of one line to the upper edge of the next line to include one thickness of the line in your measurement of the square.
5. Record your measurement of the millimeter square in the Notes and Sketches section at the end of this chapter. Also in this space sketch the superimposed images that you see using both eyes.
6. Calculate the magnification under low power by dividing the apparent size of the millimeter square by its actual known size (1 mm).

Example:

$$\frac{103 \text{ mm (apparent size)}}{1 \text{ mm (known size)}} = 103\times$$

Hair Thickness Microscope Exercise

Human hair shows many variations; hair might be straight or curly, it can be course or thin, it might be round in cross section or flattened. Just look around at your classmates and teachers; think of your family and friends. Much of the variation in appearance of hair is the result of cosmetics and styling, but the basis for every hairdo is the hair itself that grows on your body. Scientists have identified several genes that affect various characteristics of human hair. This variability of hair is what makes it an important material as evidence in many crime investigations.

Thickness of hair is one of these characteristics found to be genetically controlled. Thickness of hair varies from individual to individual. Scientists have also found that hair thickness varies among hair taken from various parts of the body and among individuals of different ethnic backgrounds. Now that you are learning to use microscopes and to make measurements of small objects, you can apply some of your new skills to investigate the thickness of human hair in this exercise.

1. Obtain several samples of hair from your instructor or by collecting them from members of your family, your classmates, or your friends. Look around the classroom

and observe the different kinds of hair among your classmates. Arrange to trade hair samples with several other people. Collecting samples from a diverse group of people will provide more information about the variability of hair thickness.
2. Label each sample to identify its source. Take a piece of a hair sample about 1–2 cm in length and place it in the center of a clean glass microscope slide in a small drop of water to hold it in place. Carefully place a coverslip over the hair to keep it flat.
3. Observe the hair sample under your microscope. Go to high power and use an optical micrometer to measure the diameter of the hair sample as described previously. Record the result and its source. Make a table on the page at end of the chapter designated for Notes and Sketches for your data.
4. Repeat with two additional hair samples from the same source.
5. Continue this procedure with other hair samples that you have collected from other individuals until you have sufficient data to make some comparisons. How much variation in hair thickness did you find? Thickness of human hair usually varies from about 50–120 micrometers. Make notes to keep a record of your observations so you can share them with other members of the class.
6. What other microscopic investigations of variability in human hair can you think of? Make some notes on other possible studies of hair and its variation.

The Stereoscopic Microscope

The **stereoscopic microscope,** also frequently called the dissecting microscope or stereomicroscope, is another common and extremely useful laboratory instrument (figure 1.4). Stereoscopic microscopes are useful for viewing objects such as small insects, frog eggs, or large protozoa at low magnifications.

The image seen through a stereoscopic microscope is not inverted, as is the image seen through a compound microscope. Also, the internal configuration of lenses and prisms provides dual light paths and thus produces a stereoscopic or three-dimensional image. The effective magnification of such a lens system is more limited than is that of compound microscopes. Most stereomicroscopes provide magnifications in the range of 5–50×, although useful magnification of up to 100–200× can be obtained in the best quality research stereomicroscopes.

Two designs of stereomicroscopes commonly used in teaching, as well as in research laboratories, employ different mechanical and optical systems to achieve changes of magnification. One of these types of stereomicroscopes

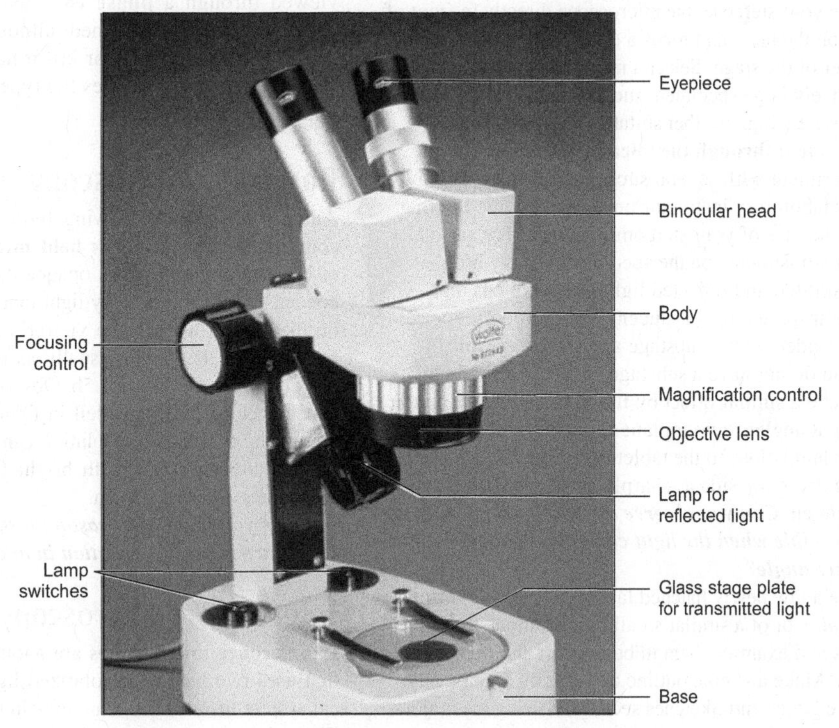

FIGURE 1.4 Stereoscopic microscope.
Microscope photograph by H. F. Holloway. Courtesy of Wolfe Sales Corporation and Carolina Biological Supply Company, Burlington, NC.

employs a fixed-position ocular, pairs of objective lenses, and an internal rotating drum on which several paired prisms are mounted. Two large knobs located on opposite sides of the microscope head serve to rotate the internal drum and thus to change magnification. Focusing is accomplished by raising or lowering the objective, as in a compound microscope. The **focusing controls** are located on the sides of the microscope arm.

The other type of stereomicroscope uses a zoom-type lens system and provides a continuously variable magnification by the rotation of a cylinder located above the objective lens (figure 1.4). Focusing is accomplished in this type of stereomicroscope by means of two lateral focusing control knobs as in the previously discussed type of stereomicroscope.

Stereomicroscopes may be equipped with either an opaque or a transparent glass stage plate or disc. When equipped with a transparent stage plate and a substage beneath the regular stage, a stereomicroscope can be used to view objects in transmitted as well as in reflected light. This is often a very useful feature for biological studies.

Exercises Using the Stereomicroscope

1. Remove your stereomicroscope from its storage cabinet and determine which type has been provided for your use in this course. ***Are there other types of stereomicroscopes present in the laboratory?*** Compare them with the type you have.
2. Place your stereoscopic microscope directly in front of you on the table and focus a concentrated light on the center of the stage. Select a microscope slide with a relatively large specimen, such as a fluke, a tapeworm, an insect wing, or other suitable specimen, and **examine it through the microscope.**
3. Experiment with several such slides as provided by your laboratory instructor until you become familiar with the use of your stereomicroscope. Try various light adjustments on the specimen, and try both transmitted and reflected light on specimens that are transparent or translucent if your microscope is provided with a substage and a substage mirror. If you do not have a substage, you may be able to achieve a similar effect by illuminating the specimen at right angles to your plane of viewing. Place your lamp close to the tabletop and focus the light beam from the side as sharply as possible on your specimen. ***Can you observe any structural details not discernible when the light comes from above or at a higher angle?***
4. Place a few newly hatched larvae of the brine shrimp *Artemia,* or of a similar small living animal, in a watch glass and examine them in both direct and transmitted light. Make a simple outline drawing on the first page of the Notes and Sketches section showing the major structures that you see.
5. When you have completed your study with the stereoscopic microscope, wipe the stage clean of any spills using a clean cloth or cleaning tissue, and return the microscope to its storage cabinet.

Other Types of Microscopes

Compound and stereoscopic microscopes are only two of the several different types of microscopes that have been developed by scientists in their continuing efforts to observe small objects more closely and in greater detail. Three other kinds of microscopes of particular importance in zoology are the phase contrast microscope, the interference microscope, and the electron microscope. In their design, these microscopes employ physical principles that differ significantly from those employed in ordinary light microscopes.

Phase Contrast Microscopy

Living cells and many other biological structures have low contrast and are therefore difficult to see in ordinary light microscopes. Phase contrast microscopes take advantage of the difference in thickness and in refractive index of light by various parts of a specimen. The light passing through the specimen is manipulated in the lens system in such a way that these minor physical differences within the specimen are transformed into varying degrees of brightness and darkness. Thus, a living cell viewed through a **phase contrast microscope** has the appearance of being stained, although no chemicals that might alter its structure or kill it have been added to the cell. Figure 1.5a illustrates the type of image obtained in this kind of microscopy.

Dark Field Microscopy

Another method for viewing living cells and other low-contrast materials is **dark field microscopy.** In this type of microscopic system, an opaque disk is placed below the condenser lens so that only light that is scattered by objects on the slide can reach the viewer's eye. Everything in the field of view is visible, usually as bright white on a dark background, as in figure 1.5b. Observe the difference in the appearance of the cheek cell in figures 1.5a and 1.5b. It is interesting that higher resolution can be obtained with dark field microscopy than with bright field microscopy, even using the same lens system.

Can you think of a reason for this, based on what you have learned about resolution in microscopy?

Interference Microscopy

Interference microscopes are another type of instrument that uses two beams of polarized light in a complex optical system to visualize structures in living, unstained cells. They also permit quantitative measurements of dry mass

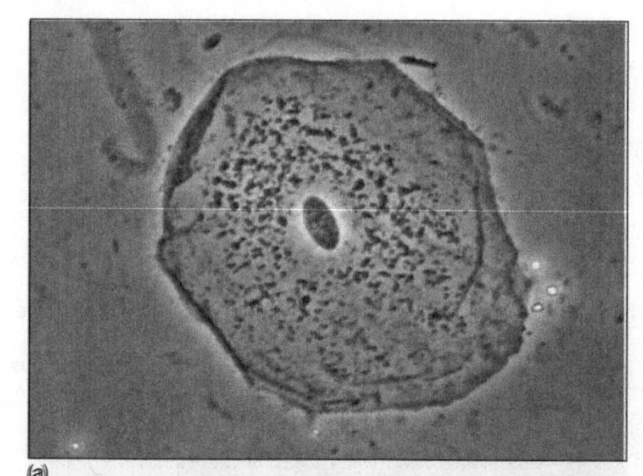

(a)

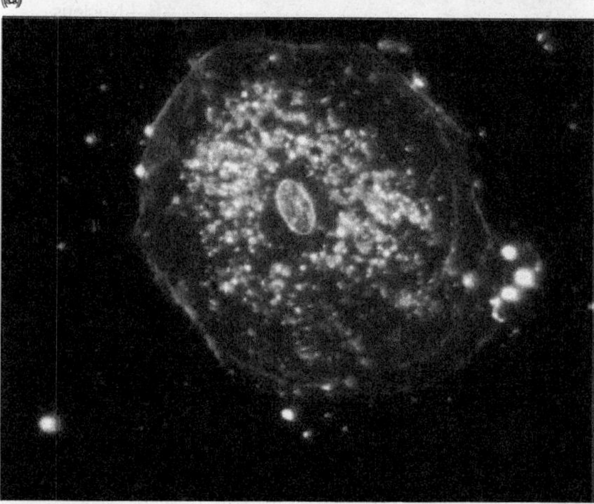

(b)

FIGURE 1.5 Human cheek cell. (*a*) Phase contrast microscope. (*b*) Dark field microscope.
© Stephen Durr.

per unit area of microscopic objects. A special kind of interference microscope, the Nomarski interference microscope, has been especially useful in the study of cells and tissues.

The **Nomarski interference microscope** can provide excellent images of unstained materials with a three-dimensional effect (figure 1.6). This type of microscope uses polarized light, which is split into two light beams of slightly different wavelength. Interactions of these two light beams emphasize the hills and pits within the specimen to yield the three-dimensional effect. Interference microscopes are found mainly in research laboratories since they are expensive due to the complex system of prisms and lenses necessary to produce these optical effects.

Electron Microscopy

Electron microscopy has become a very valuable research technique in zoology in the last few decades. In electron microscopy, a specimen is irradiated with a concentrated **beam of electrons** rather than with rays of visible light, as in a light microscope. The two main types of electron microscopes commonly used by zoological researchers are transmission electron microscopes and scanning electron microscopes.

Transmission electron microscopes (figure 1.7) are capable of much greater resolving power and, therefore, much higher magnifications than are ordinary light microscopes. Direct magnifications up to 200,000 diameters are commonly achieved with electron microscopes

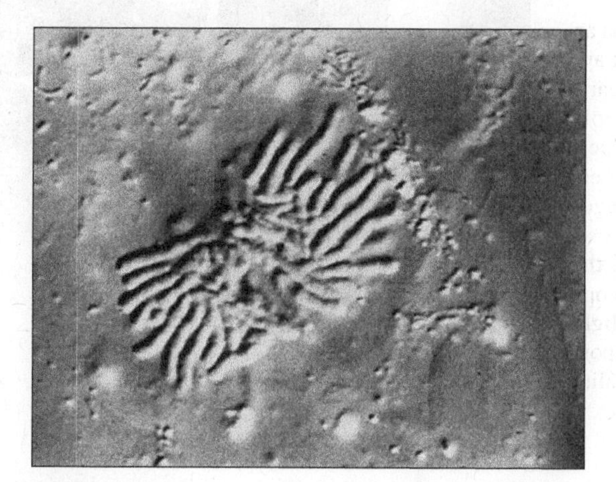

FIGURE 1.6 Chromosomes in a newt lung cell.
Nomarski differential interference micrograph courtesy of Southern Micro Instruments.

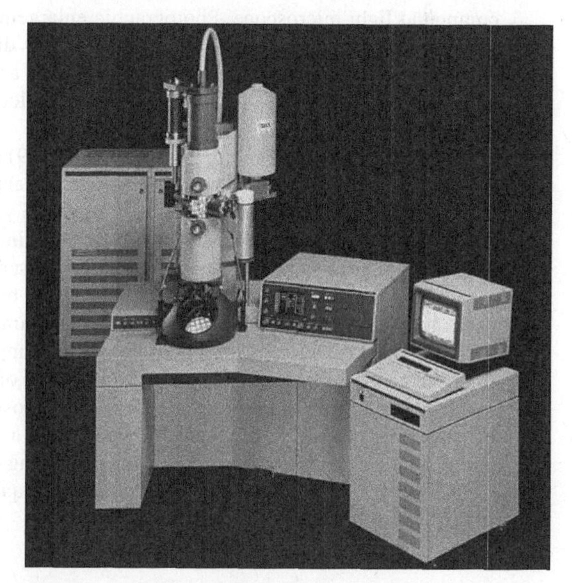

FIGURE 1.7 Transmission electron microscope equipped with a special X-ray analyzer on the right.
Photograph courtesy of Philips Electronic Instruments, Inc.

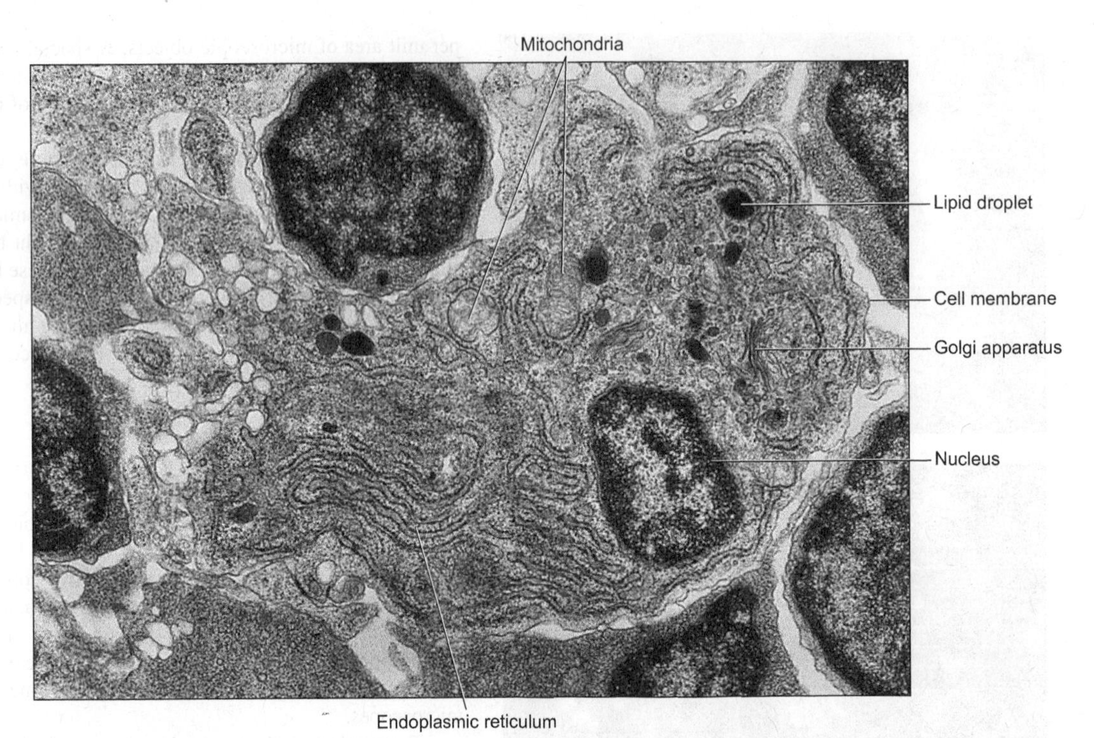

FIGURE 1.8 Transmission electron micrograph of a plasma cell.
Photograph by Kenneth E. Muse.

compared to a maximum of about 2,000 diameters with a compound light microscope. Photographic enlargements yield final magnifications greater than 1,000,000 diameters in many electron micrographs. Figure 1.8 is an example of a photograph taken with a transmission electron microscope.

The **scanning electron microscope** (Figure 1.9) is an even newer research tool and has had many important applications in biological research during the past 35 years. The scanning electron microscope differs in principle from both light and transmission electron microscopes. This kind of electron microscope has proved to be especially useful in providing three-dimensional images of small objects, information about chemical composition, electrical properties, and structural details of the surface of specimens. The scanning electron microscope has a great depth of field (7 to 10 times that of a light microscope at comparable magnifications), making possible photographs of excellent three-dimensional quality (figures 1.10a and 1.10b).

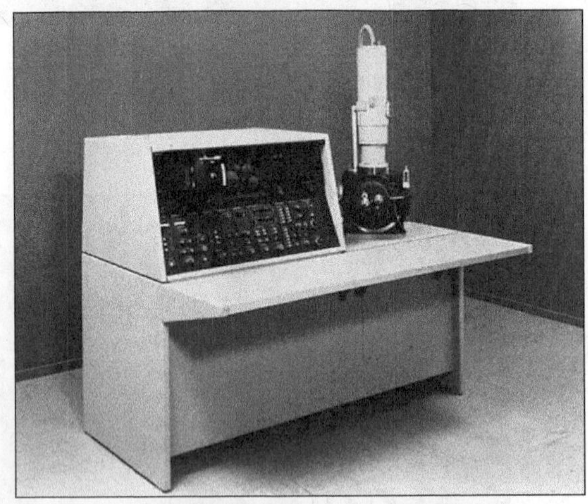

FIGURE 1.9 Scanning electron microscope.
Photograph courtesy of Philips Electronic Instruments, Inc.

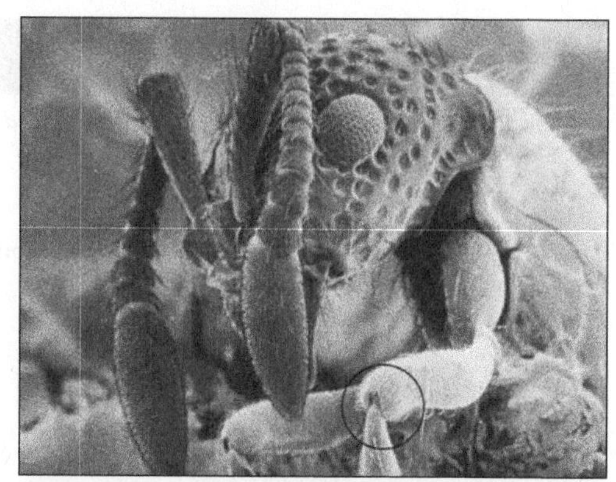

(a)

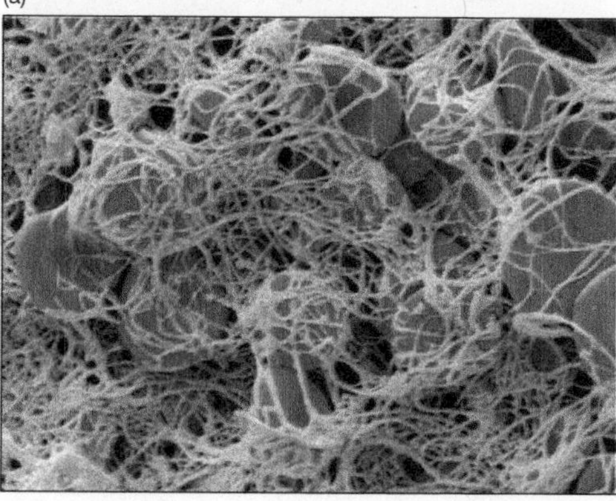

(b)

FIGURE 1.10 Scanning electron micrographs. (*a*) Ant. Magnification 300×. (*b*) Human blood clot with red blood cells and fibrin strands. Magnification 3,750×. Electron micrographs produce black and white images; however, such images can subsequently be colored using computer programs to distinguish different objects.

(a) Photograph by Kenneth E. Muse. (b) Science Photo Library RF/Getty Images.

Key Terms

Compound microscope a type of light microscope with two separate lens systems, an eyepiece and an objective lens, which together serve to magnify the image of an object. Provides magnification to about 1,000–2,000 diameters.

Dark field microscopy a type of light microscopy in which an opaque disk is placed in the light path to block light transmission through the specimen. This results in a light-colored image on a black background. Used to visualize living cells and other unstained objects.

Interference microscope a type of light microscope that uses beams of polarized light to allow visualization of low-contrast specimens such as living cells and also the determination of mass or dry weight of specimens. Nomarski differential interference microscopy employs slight differences in the wavelengths of two beams of polarized light to enhance the contrast of unstained specimens and to give an image that looks three-dimensional.

Magnification the ratio of the apparent size to the actual size of an object when viewed through a microscope.

Ocular micrometer a graduated disc placed in the eyepiece of a microscope that can be used to measure the size of microscopic objects.

Parfocal lenses objective lenses constructed and mounted so that their focal planes are approximately the same; a specimen remains in focus and centered in the field of view when you change from one parfocal lens to another.

Phase contrast microscope a special type of light microscope that permits the observation of thin, unstained materials. Special lenses and illumination techniques increase the contrast in an unstained specimen due to variations in the refractive indices of its parts.

Resolving power the ability of a microscope to reveal fine detail in a specimen. More precisely, resolving power is defined as the shortest distance between two points that allows them to be distinguished as separate points.

Scanning electron microscope a type of electron microscope that provides three-dimensional images of very small objects. A concentrated beam of electrons is focused and moved along the surface of a specimen and induces the emission of secondary electrons from the specimen. These secondary electrons produce a magnified image of the specimen on a cathode-ray tube.

Stereoscopic microscope a type of light microscope with two separate optical paths that provide a magnified three-dimensional image. Provides useful magnifications to about 100–200 diameters.

Transmission electron microscope a type of electron microscope used to view specially prepared thin specimens at magnifications to about 200,000 diameters. A concentrated beam of electrons passes through the specimen and produces a pattern of light and dark areas on a phosphorescent screen because of the differential passage of electrons through portions of the specimen. Darker areas represent areas of greater electron density within the specimen.

Working distance the distance from the front of the objective lens of a compound microscope to the top of the specimen.

Internet Resources

There are many valuable Internet sites with information about zoology. Several sites containing pertinent zoological information for this chapter can be found on the McGraw-Hill Zoology web site at http://www.mhhe.com/zoology. Just click on this text's title.

Questions for Critical Thinking

1. List three examples of biological materials or specimens that would best be studied with each of the following types of microscope: (a) compound microscope, (b) stereoscopic microscope, (c) scanning electron microscope, and (d) transmission electron microscope. Explain why you chose the specific type of microscope for each example.
2. Define focus, resolving power, and magnification. Explain how they are different and why each is important in microscopy.
3. Describe two methods for measuring objects viewed with a compound microscope and give examples for the use of each method in a biological study.

Suggested Readings

Murphy, D.B. 2001. *Fundamentals of Light Microscopy and Electronic Imaging*. New York: Wiley-Liss. 360 pp

Spector, D.L., and R.D. Goldman. 2005. *Basic Methods in Microscopy: Protocols and Concepts from Cells, a Laboratory Manual.* Cold Spring Harbor, ME: Cold Spring Harbor Laboratory Press. 375 pp

Notes and Sketches

Chapter 2

Animal Cells and Tissues

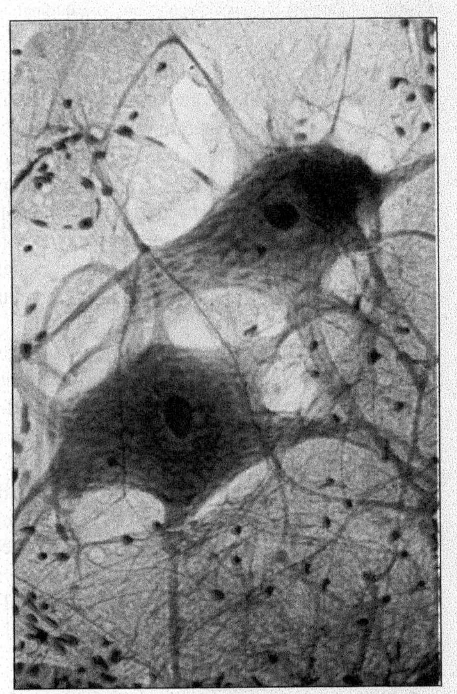

Courtesy of Carolina Biological Supply Company, Burlington, NC.

Objectives

After completing the laboratory work in this chapter, you should be able to perform the following tasks:

1. Describe the cell theory and explain its importance in zoology.
2. Describe the principal organelles of a typical animal cell, as represented by a sea star egg, that are visible in a light microscope.
3. List six organelles typically seen in a transmission electron micrograph of a generalized (typical) animal cell that are not generally visible with a compound microscope.
4. Distinguish among a cell, a tissue, an organ, and an organ system.
5. List four main types of animal tissues and give examples of each. Identify typical examples of each in microscope slides.
6. Describe the histological structure of a compact bone and explain the role of a Haversian canal, lacuna, canaliculi, and lamellae.
7. Distinguish among smooth muscle, cardiac muscle, and striated muscle from microscopic slides or photographs.
8. Describe the structure of a vertebrate neuron and give the function of each main part.
9. Describe the composition of human blood and give the functions of erythrocytes, leucocytes, and platelets.
10. Identify erythrocytes, leucocytes, and blood platelets in microscopic preparations.

The Cell Theory

Cells are the fundamental structural and functional units of virtually all living organisms. The body of an animal typically is made up of many different kinds of cells that are organized into tissues, organs, and organ systems that carry out certain essential functions. Thus, a knowledge of cell structure and function is essential for the proper understanding of reproduction, growth, heredity, and all other normal and abnormal animal functions.

The importance of cells is summarized in a statement called the **cell theory,** which is one of the most important unifying concepts in biology. This theory was developed in the 19th century, when naturalists were experimenting with their latest technology—the compound microscope. The cell theory is generally attributed to two German scientists, botanist Matthias Schleiden and zoologist Theodor Schwann, who published their ideas in 1838 and 1839; however, other scientists also have contributed to the modern version of the theory.

The main points of the cell theory can be summarized as follows:

1. All organisms are composed of cells.
2. All cells come from other cells.
3. All vital functions of an organism occur within cells.
4. Cells contain the hereditary information necessary for regulating cell functions and for transmitting information to the next generation of cells.

Materials List

Prepared microscope slides
- Sea star eggs or sea urchin eggs
- Cuboidal epithelium (rabbit kidney)
- Columnar epithelium (rabbit kidney or amphibian intestine)
- Stratified epithelium (human skin)
- Loose connective tissue
- Mammalian hyaline cartilage, cross section
- Human bone, ground, cross section
- Smooth muscle, teased
- Striated muscle
- Neurons (smear from spinal cord of cow)
- Blood film (human), Wright stain
- Ciliated epithelium (demonstration)
- Adipose tissue (demonstration)
- Cardiac muscle (demonstration)
- Amphibian blood (demonstration)

Fresh cartilage (from frog sternum, ends of long bone, etc.)

Cross sections of long bones from pig, cow, or other mammal

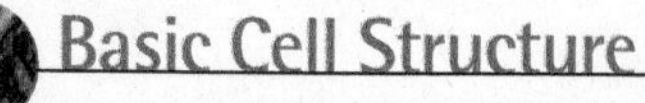

Basic Cell Structure

Our knowledge of cell structure has increased dramatically in the last few decades because of the availability of electron microscopes with much higher magnification and resolution than light microscopes, as described in Chapter 1. Figure 2.1 illustrates the appearance of a relatively simple cell seen in an ordinary compound light microscope of the type usually available in general zoology laboratories. Figure 2.2 is a diagram illustrating the parts of an animal cell seen in a transmission electron micrograph. Note the much greater structural detail in the electron micrograph. We know that animal cells are composed of many kinds of **organelles,** the distinctive parts of cells that carry out specific functions. Several types of cellular organelles can be visualized only in electron micrographs because they are too small to be seen with a compound microscope.

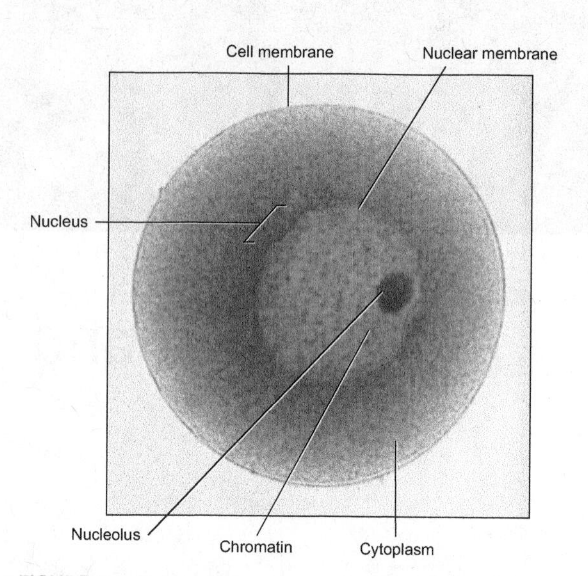

FIGURE 2.1 Sea star egg.
Courtesy of Carolina Biological Supply Company, Burlington, NC.

Since electron microscopy is a very complex process, in this exercise we will concentrate on those aspects of cell and tissue structure observable with a compound microscope and compare our findings with information available from more sophisticated techniques and instruments.

Good illustrations of generalized (unspecialized) animal cells are provided by the unfertilized eggs of many animals. For our introductory study of cells, we will use the unfertilized eggs of the sea star. Sea urchin eggs are very similar to sea star eggs and serve equally well. Prepared microscope slides containing many stained sea star eggs will be provided for your study.

- Obtain a prepared microscope slide and, under low power on your compound microscope, observe the numerous sea star eggs. Select a well-stained cell similar to that illustrated in figure 2.1 and center the cell in your field of view. Rotate the high-power objective into position, regulate the light as needed, and readjust the focus. Observe the two well-differentiated parts of the **cell:** the central **nucleus** and the surrounding **cytoplasm.** Externally, the cytoplasm is bounded by a thin **cell membrane.** Although the cell membrane of animal cells is seen only as a thin outer boundary of the cell, it plays a very important role in the functions of the

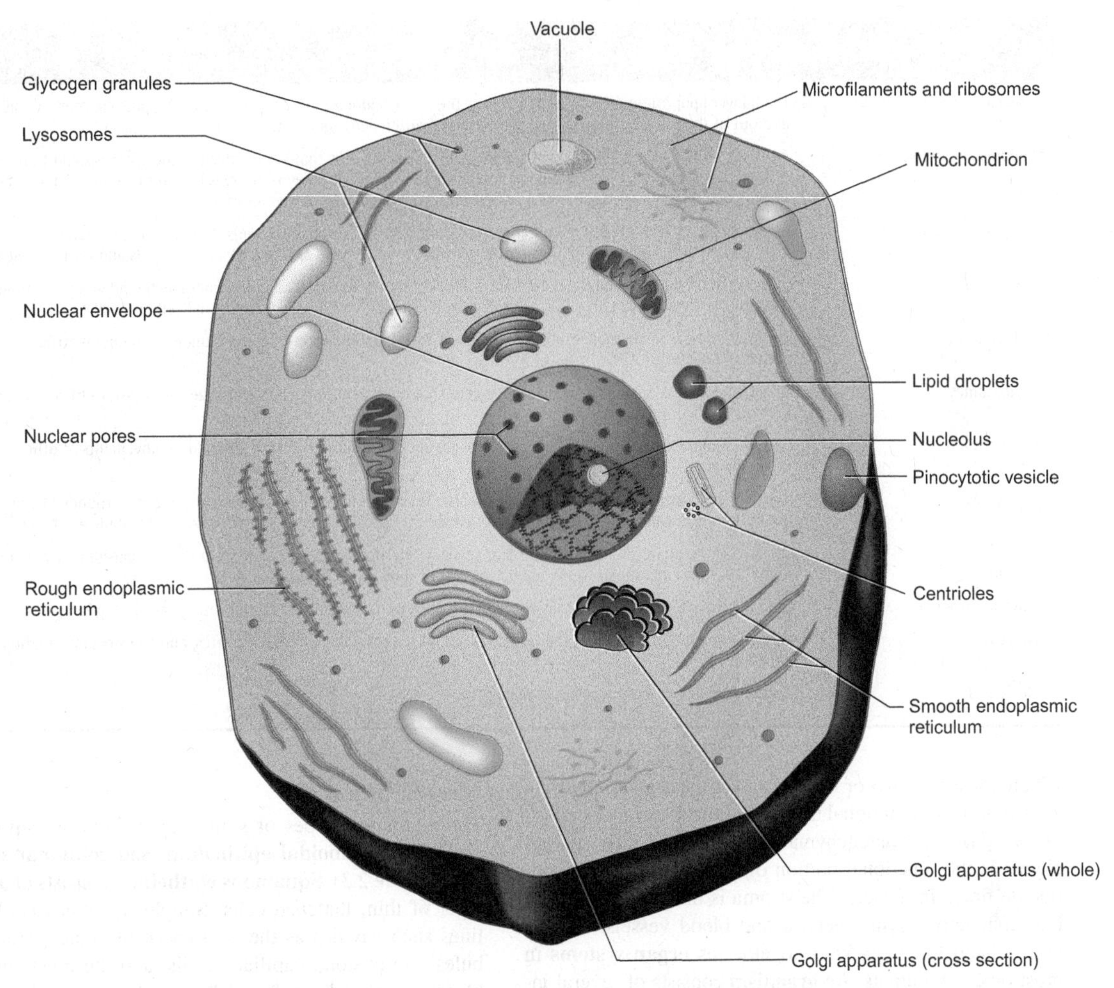

FIGURE 2.2 Generalized animal cell, showing organelles identifiable in a transmission electron micrograph.

cell. The special properties of the cell membrane control the passage of materials into and out of the cell. Animal cells lack the thickened cellulose cell walls outside the cell membrane that are usually found in plant cells.

- Also on your slide, identify the spherical nucleus, bounded by the nuclear membrane and containing numerous darkly staining masses of chromatin material. Within the nucleus of the sea star egg, find the darkly staining nucleolus. In table 2.1, list the cell organelles that you are able to identify with your compound microscope.

Several additional cell organelles can be seen in micrographs of cells obtained with a **transmission electron microscope** as well as many more structural details of those organelles visible with a light microscope. The drawing in figure 2.2 illustrates an animal cell as seen in a transmission electron micrograph, and table 2.2 lists several of these organelles and their functions.

Table 2.1 Cellular Organelles Observed with Light Microscope
1.
2.
3.
4.
5.
6.

Animal Tissues

Tissues are groups of cells with a common embryonic origin that work together to perform a certain function or functions. Tissues, in turn, are organized into **organs,**

Table 2.2 Principal Cell Organelles Seen in Transmission Electron Micrographs

Organelle	Description
Plasma membrane	The bilayer lipoprotein membrane that forms the outer boundary of a cell, regulates the passage of materials into and out of the cell, and allows the cell to interact with its environment.
Endoplasmic reticulum	A system of membrane-bound compartments in eukaryotic cells that are involved in the synthesis and transport of materials through the cell. Rough endoplasmic reticulum has many ribosomes bound to its membranes, and smooth endoplasmic reticulum has no ribosomes associated with its membranes.
Golgi apparatus	A complex stack of flattened membranous sacs and vesicles in eukaryotic cells that modifies, sorts, and packages proteins and other cellular products synthesized in the cell. It also aids in transport of lipids and forms lysosomes.
Lysosomes	Specialized products of the Golgi apparatus in eukaryotic cells consisting of membrane-bound vesicles containing hydrolytic enzymes that can digest foreign materials or aid in the breakdown of old cell organelles.
Mitochondria	Ovoid or cylindrical organelles of eukaryotic cells with a double membrane surrounding an inner matrix. Serve as the principal site for ATP synthesis.
Peroxisomes	Membrane-bound organelles in eukaryotic cells that contain enzymes that catalyze the transfer of hydrogen and break down hydrogen peroxide.
Cytoskeleton	System of minute tubules and fibrils that provide structural support for the cell, aid in movements of other organelles, and function in cell movement.
Centrioles	A pair of cylindrical structures found in animal cells composed of nine triplet microtubules surrounded by an amorphous area called the centrosome with which the centrioles appear to organize microtubule assembly.
Ribosomes	Particles made up of RNA and protein that serve as the site of protein synthesis in the cytoplasm; may either be bound to the membrane of rough endoplasmic reticulum or dispersed the cytoplasm.
Nuclear envelope	Double lipoprotein membrane that encloses the nucleus during interphase but disappears during mitosis.
Nucleus	Large organelle surrounded by the two-layered nuclear envelope. Stores, replicates, and transfers information stored in DNA.
Nucleolus	Cluster of ribosomes in the interphase nucleus.

which consist of two or more kinds of tissues grouped into a structural and functional unit. Functionally, organs consist of two parts, the parenchyma and the stroma. The parenchyma is the functional portion of the organ, like the contractile fibers in muscle. The stroma is the supporting part, like connective tissue, nerves, and blood vessels. Organs with related functions work together as **organ systems** in most types of animals. An **organism** consists of several integrated organ systems.

Animal tissues are usually divided into four main types: (1) **epithelial tissue,** (2) **connective tissue,** (3) **muscular tissue,** and (4) **nervous tissue.** Each of these tissue types shares certain common functions, and each type is represented by two or more subtypes found in various tissues and in various kinds of animals. The study of tissues, including their structure and function, is called **histology.**

Epithelial Tissue

The principal function of epithelial tissues is to cover and protect surfaces. In addition to covering the outside of the body like the outer layers of your skin, other kinds of epithelial tissues line internal cavities and ducts, form glands, and aid in the transport of materials through, from, and into ducts and canals. Various kinds of epithelia (singular: epithelium) are characterized mainly by the shape and arrangement of their cells. **Simple epithelia** consist of single layers of cells, and **stratified epithelia** contain several layers of cells.

Simple Epithelia

Three common types of simple epithelium are **squamous epithelium, cuboidal epithelium,** and **columnar epithelium** (figure 2.3). **Squamous epithelium** consists of a single layer of thin, flattened cells. Simple squamous epithelium lines such cavities as the air sacs of the lungs, kidney tubules, the coelom, capillary walls, and the inner lining of blood vessels where the exchange of materials by the absorption and the diffusion of gases is especially important to the underlying cells. For example, amphibians have a thin, moist skin covered with squamous epithelium, which permits easy passage of oxygen and carbon dioxide (figure 2.4).

- Obtain a prepared slide of squamous epithelium cells and draw a few adjacent amphibian epithelial cells in the space provided in figure 2.5 showing the structures you can observe under the light microscope.

Cuboidal epithelium is another type of epithelial tissue found lining several kinds of ducts such as kidney tubules, salivary glands, and the secretory follicles of thyroid glands. These cells are like little hexagonal boxes and are sturdier than simple squamous epithelial cells. They can withstand the abrasions from the passage of materials through the ducts.

- Examine a slide of cuboidal epithelium (figure 2.6) from the kidney tubules of a rabbit or other appropriate tissue. The height and width of the single layer of cuboidal cells are about equal. Observe the prominent nuclei and the darkly staining **basement membrane** found at the base of the cells.

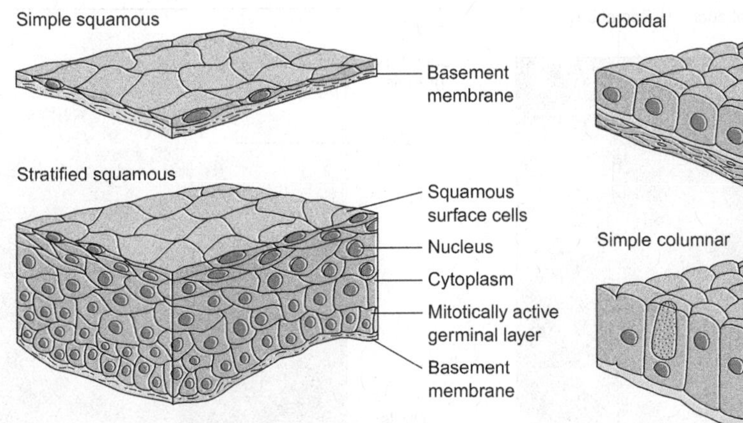

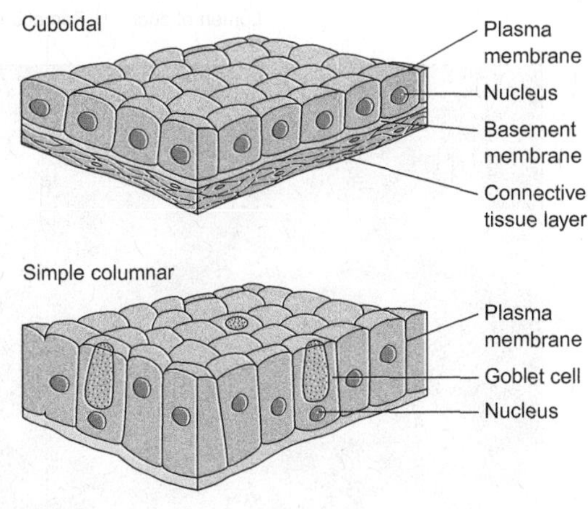

FIGURE 2.3 Types of epithelium.
Courtesy of Carolina Biological Supply Company, Burlington, NC.

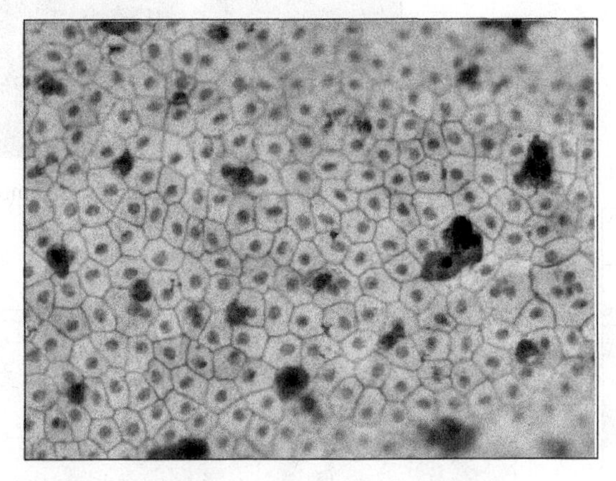

FIGURE 2.4 Squamous epithelium, frog skin.
© McGraw-Hill Companies, Inc./Al Tesler, photographer.

FIGURE 2.5 Student drawing of amphibian epithelial cells.

Electron microscopic studies of the basement membrane have revealed it to be a complex structure consisting of a network of very fine collagen filaments and other complex protein molecules with carbohydrate side chains. The basement membrane attaches to and stabilizes the overlying epithelial cells. It also serves as a barrier for epithelial cells that become malignant and hinders them from penetrating to deeper lying cell layers. The basement membrane of certain specialized cells has other important functions.

Columnar epithelium (figure 2.7) consists of a layer of tall, closely packed cells that line most of the digestive tract of many vertebrate animals and are also found lining the excretory ducts in many glands.

- Obtain a slide with columnar epithelium from the intestine of a frog or other animal and identify the nuclei, the darkly staining outer brush border, and the basement membrane. The brush border consists of many tiny microvilli, tiny fingerlike extensions of the cells that extend into the lumen or opening of the intestine. ***What do you think the function of the microvilli might be?***

Columnar epithelia from the intestine and from the trachea often contain goblet cells, which release their secretions directly into the lumen (figure 2.8). Some columnar epithelia, such as those lining the oral cavity of the frog and

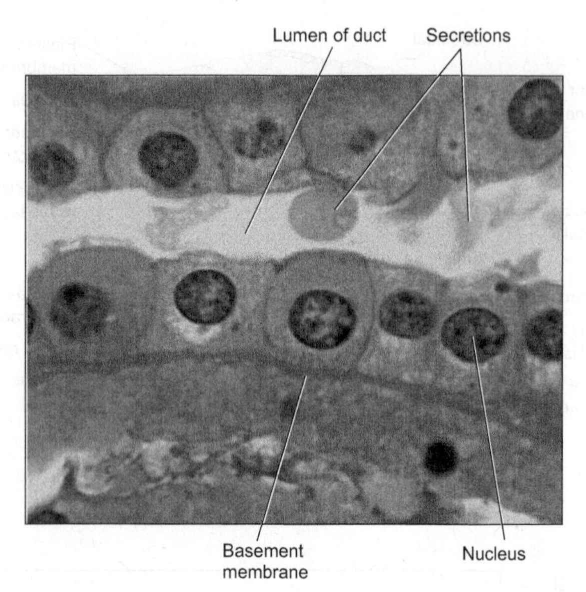

FIGURE 2.6 Cuboidal epithelium.
Courtesy of Carolina Biological Supply Company, Burlington, NC.

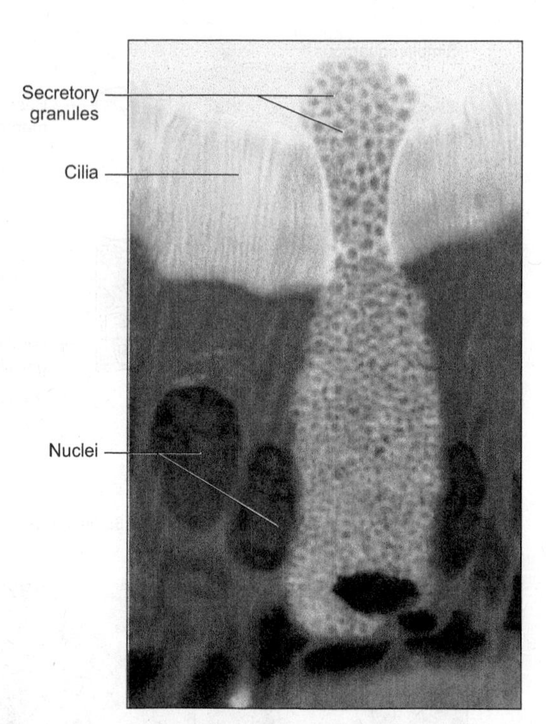

FIGURE 2.8 Goblet cell in columnar epithelium of monkey trachea.
Courtesy of Carolina Biological Supply Company, Burlington, NC.

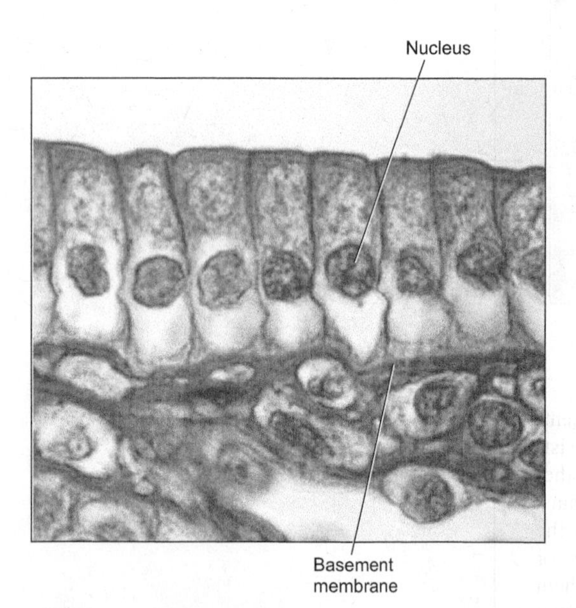

FIGURE 2.7 Columnar epithelium.
Courtesy of Carolina Biological Supply Company, Burlington, NC.

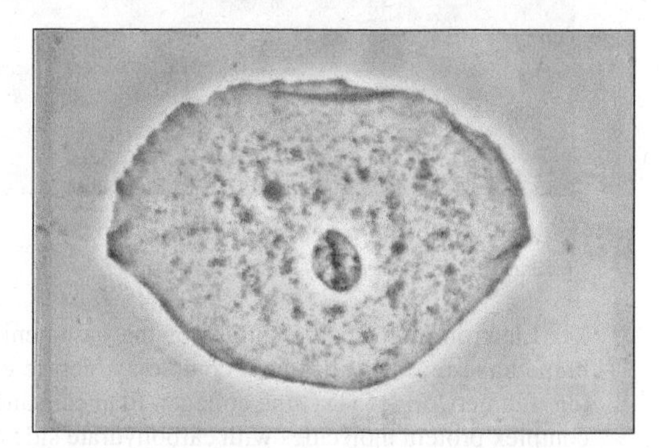

FIGURE 2.9 Phase contrast photograph of a living cell from human cheek epithelium.
Courtesy of Carolina Biological Supply Company, Burlington, NC.

the trachea of mammals, also bear numerous cilia. ***What function might these cilia have?***

Stratified Epithelia

Stratified epithelia are made up of several layers of cells stacked on top of each other. They are found on surfaces in which wear and abrasion occur. The most common type is stratified squamous epithelium like that found lining the mouth, esophagus, and vagina. The cells lining your mouth cavity are a good example of this tissue type. Figure 2.9 is a phase contrast micrograph of a single living human squamous epithelium cheek cell. This cell was removed from the outer layer of the stratified epithelium lining the mouth.

Human Skin

Some types of stratified epithelia are hardened by the secretion of **keratin,** a tough protein. Human skin provides a good example of stratified epithelium in which the outer layers of cells have been keratinized (figures 2.10 and 2.11). This type of epithelium consists of several layers of squamous epithelium that become progressively flattened and hardened as they rise to the surface and are constantly worn and sloughed off. This outer layer of skin is the **epidermis.** Underlying the epidermis is the **dermis** containing nerves, muscle, glands, hair follicles, and connective tissue. Beneath the dermis is the **hypodermis,** consisting mostly of fatty (adipose) tissue and other connective tissue. The dermis actually is not part of the skin.

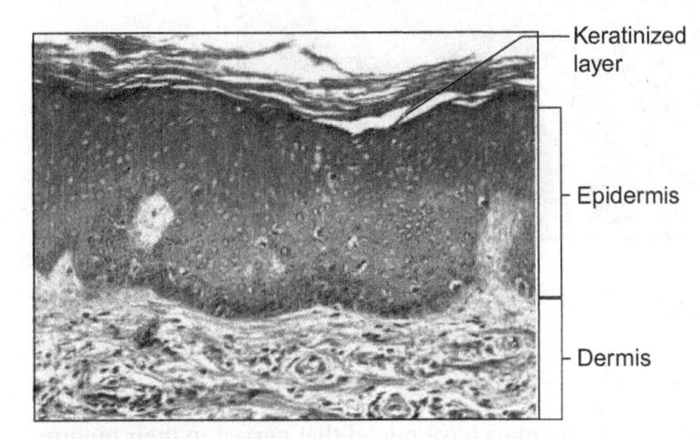

FIGURE 2.10 Stratified squamous epithelium, human skin, cross section.
The McGraw-Hill Companies, Inc./Al Telser, photographer.

Skin is a good example showing how a variety of cell types together form **tissues,** groups of cells of common origin working together with a particular function. Tissues also combine to form organs; in this case, the skin is a component of the integumentary system, which protects the body.

Other types of stratified epithelia found on other surfaces are made up of columnar or cuboidal cells.

- Obtain a slide of human skin and study the complex structure of the multilayered integument, or outer covering, of the human body. ***Use your observations from the skin cross section and figure 2.10a to identify and to label the structures in figure 2.10b.***

Connective Tissue

There are several types of connective tissue and all exhibit large amounts of **intercellular substance** secreted by the living cells. The intercellular substance may be liquid, solid, or semisolid. You will study six types of connective tissue in this exercise, including **blood, loose connective tissue, dense connective tissue, adipose tissue, cartilage,** and **bone.**

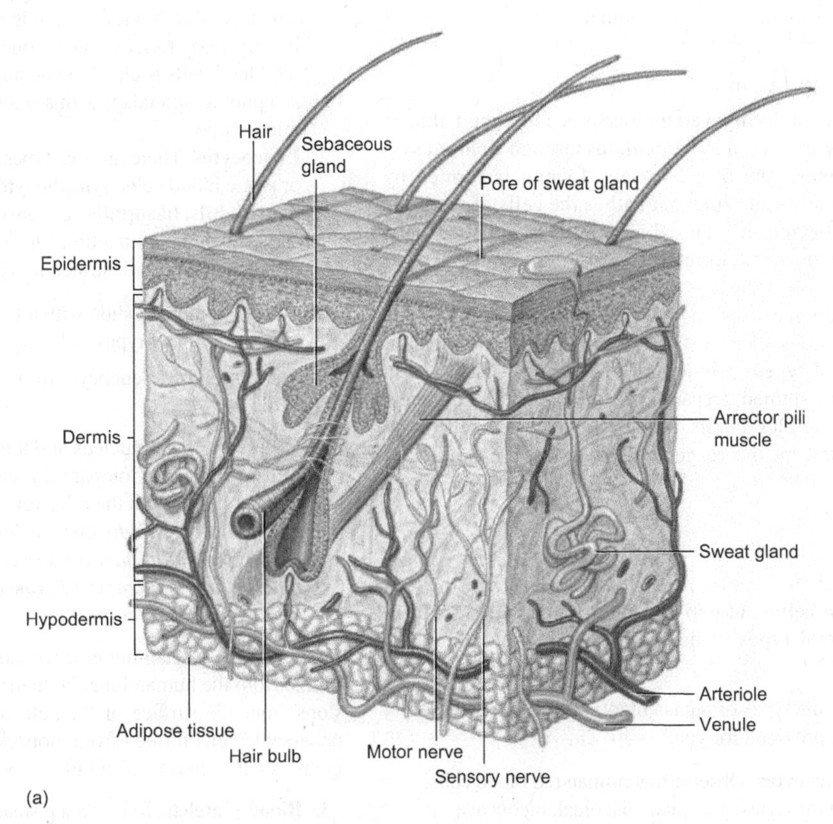

FIGURE 2.11a Human skin, cross section, drawing. Observe the several types of tissues that make up the skin; therefore, the skin is an organ rather than a single tissue.

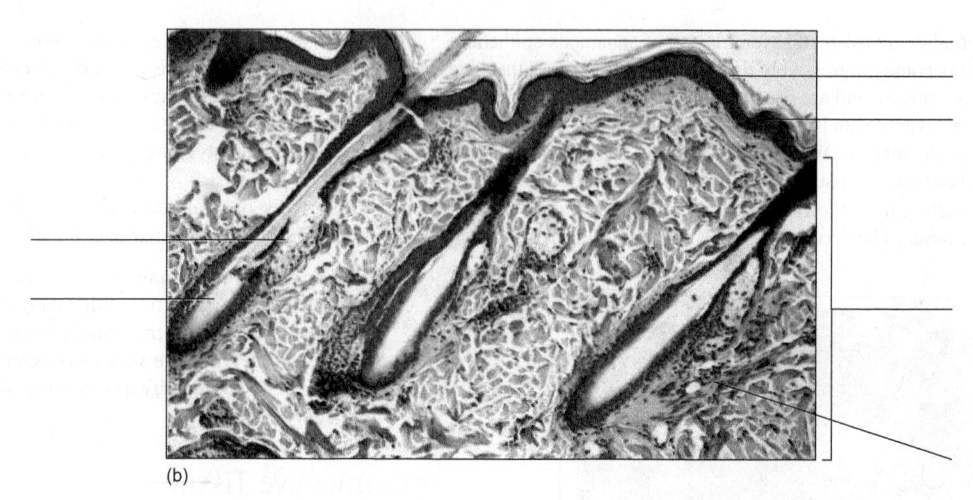

(b)

FIGURE 2.11b Human skin, cross section for students to label.
(b): © Victor P. Eroschenko.

Blood and lymph are rather different connective tissues because of their important role in transporting materials to and from cells and tissues, but they do connect parts of the body in a very real physiological sense. The other types of connective tissues mainly provide physical support and protection for various parts of the animal body.

Blood and Lymph

The blood of living vertebrates is a red liquid that is constantly in motion as it circulates through a closed system of tubes—the blood vessels. Lymph is formed by intercellular tissue fluid that bathes the cells of the body and is collected in a closed system of lymphatic ducts. The lymphatic system returns the fluid to the veins near the heart.

Blood comprises about 7 percent of the human body weight. It consists of a straw-colored liquid, the **plasma,** and several types of **blood cells** suspended within it. In permanent, stained preparations, the plasma is not seen. Many of the blood cells, in particular the red cells, may be distorted by the reagents used in the preparation of the slides.

Human Blood

Suspended in the plasma of human blood are the **red** and **white cells** and cell fragments called **blood platelets.** Several types of human blood cells are shown in figure 2.12.

- Study the types of human blood cells in the prepared slides provided for you.

1. **Erythrocytes.** Observe that human red blood cells, or erythrocytes, are small, discoidal, biconcave (concave on both sides), and **lack nuclei** like those of all mammals (figure 2.12). Most other vertebrate animals have nuclei that persist in their mature erythrocytes as in the blood of amphibians and snakes (figure 2.13). The chief function of the erythrocytes is to carry oxygen to cells in the body. They contain large amounts of the iron-containing protein **hemoglobin,** which combines with oxygen in the lungs and releases it in the body tissues. The biconcave shape of human red blood cells is clearly shown in figure 2.14, which is a photograph taken with a scanning electron microscope.
2. **Leucocytes.** There are five types of leukocytes or white blood cells: **lymphocytes, monocytes, neutrophils, basophils,** and **eosinophils.** Each type serves to combat infections in its own way. These types of leucocytes are illustrated in figure 2.15.

- Obtain a prepared slide with a blood smear and identify as many types of leucocytes as you can find.

A cross section of a leucocyte of a mouse is shown in figure 2.16.

- Observe its large nucleus and the complex structure of its cytoplasm. Compare this complex structure of the cytoplasm with the relatively simple appearance of the erythrocytes. ***Why do you think that mammalian leucocytes would have a more complex structure than mammalian erythrocytes? How is their structure related to their function?***

Figure 2.17 is a scanning electron micrograph of a **macrophage** from the human lung. Note the many irregular extensions from the surface of the cell. Macrophages are large phagocytic cells formed from monocytes, which engulf and break down bacteria and cellular debris.

3. **Blood platelets.** Platelets are small, nonnucleated bits of cytoplasm that bud off from large cells (megakaryocytes) in the bone marrow. They are about

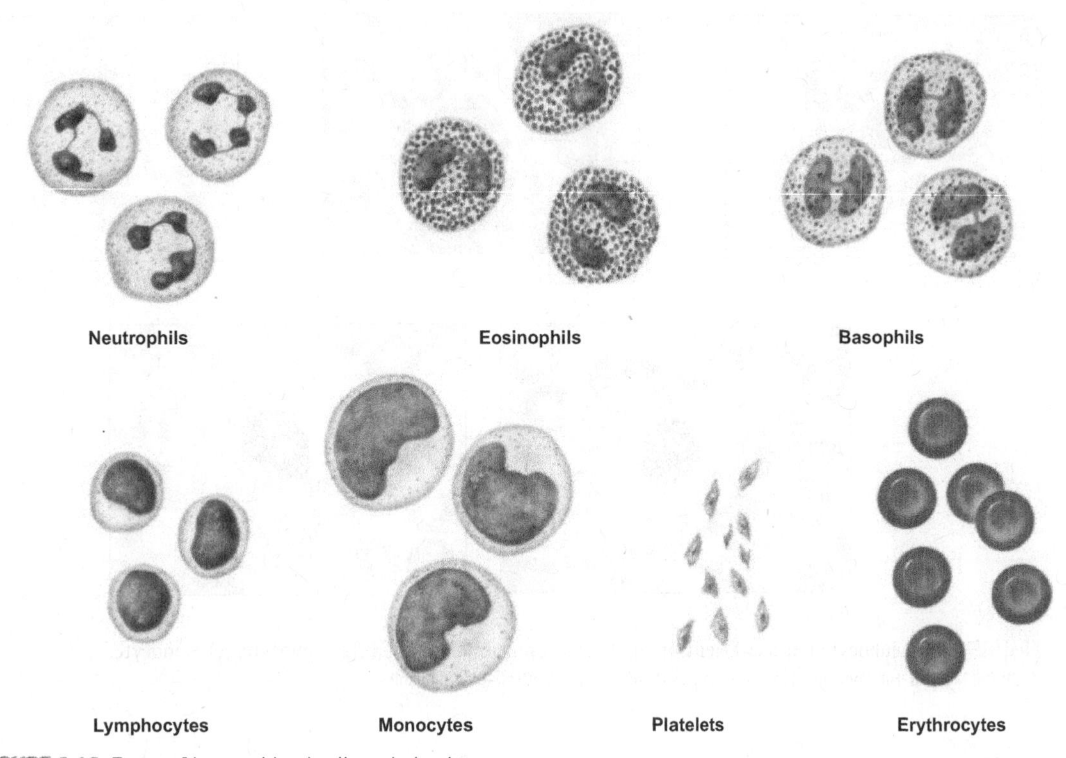

FIGURE 2.12 Types of human blood cells and platelets.

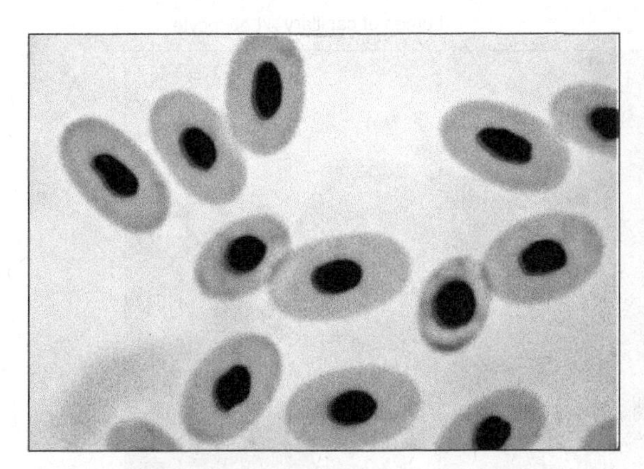

FIGURE 2.13 Red blood cells of a snake. Observe the large central nucleus.
Courtesy of Carolina Biological Supply Company, Burlington, NC.

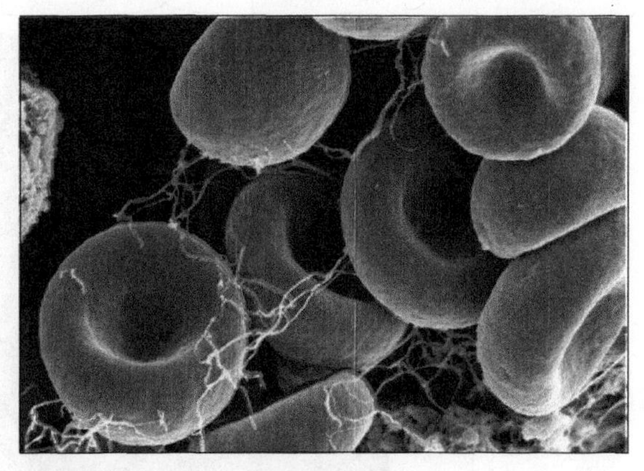

FIGURE 2.14 Human erythrocytes in a blood clot with fibrin fibrils, scanning electron micrograph, 4000×.
Science Photo Library RF/Getty Images.

one-third the size of erythrocytes and are not usually preserved in prepared microscope slides. Platelets function in **blood clotting** by temporarily plugging punctures that may occur in blood vessels and also by releasing substances that trigger later chemical reactions necessary for clotting.

Loose Connective Tissue

Loose connective tissue (figure 2.18a) consists of scattered cells surrounded by a clear, jellylike **ground substance** and two types of fibers: thin **elastic fibers** and thicker bundles of **nonelastic** (collagenous) **fibers.** Loose connective tissue is found in the tip of the nose, the outer ear, and the epiglottis.

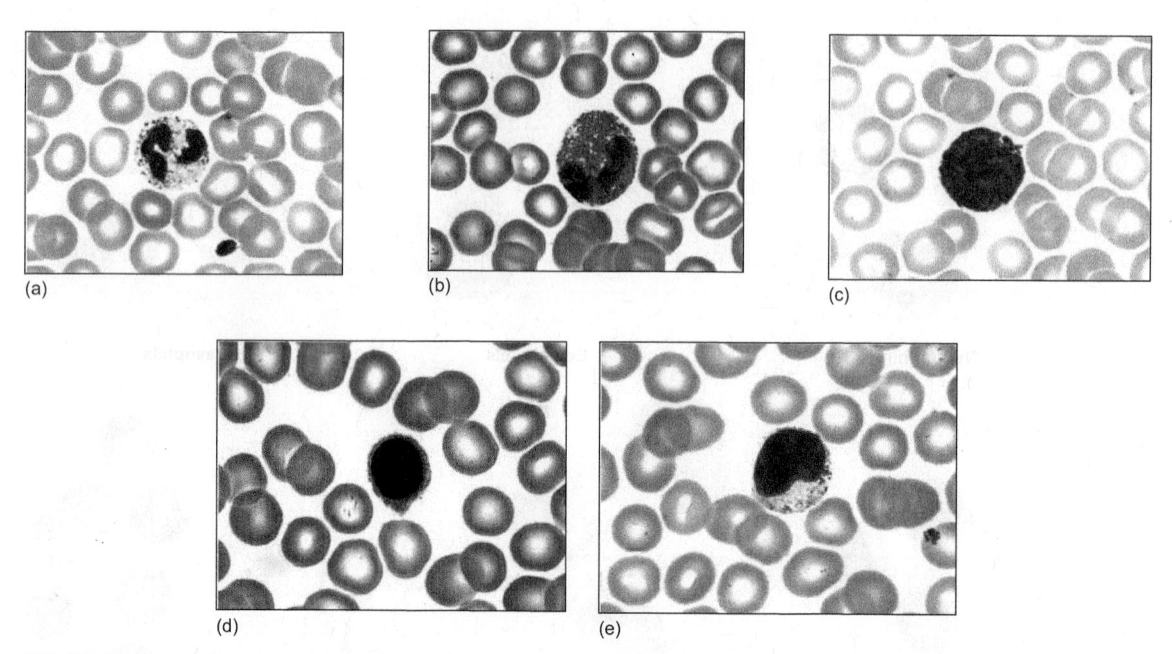

FIGURE 2.15 Leukocyte types. (*a*) Neutrophil, (*b*) eosinophil, (*c*) basophil, (*d*) lymphocyte, (*e*) monocyte.
(a-d): The McGraw-Hill Companies, Inc./Al Telser, photographer. (e): McGraw-Hill Companies.

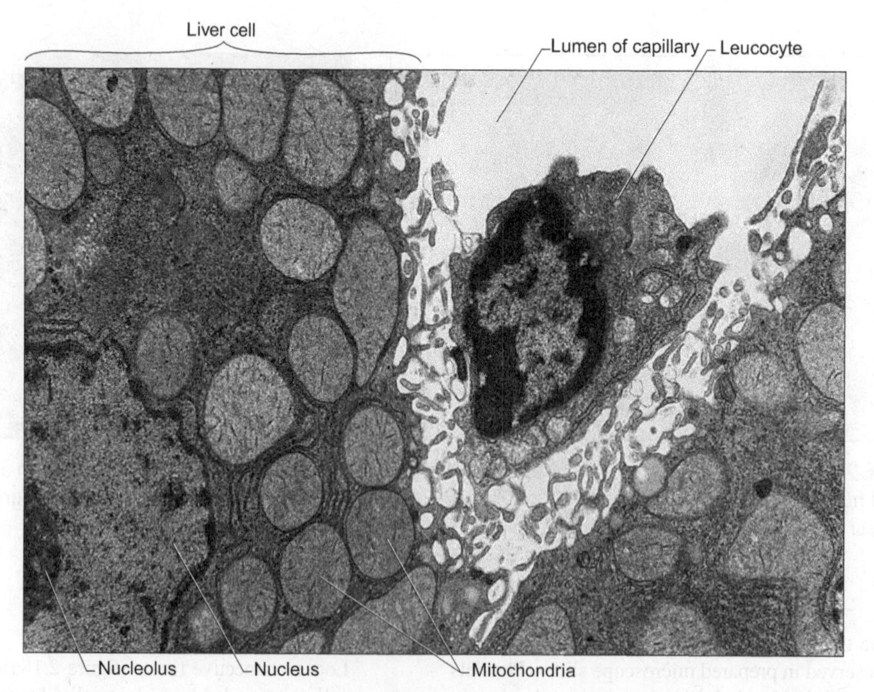

FIGURE 2.16 Leucocyte in capillary of a mouse. Magnification 21,000×.
Electron micrograph by Kenneth E. Muse.

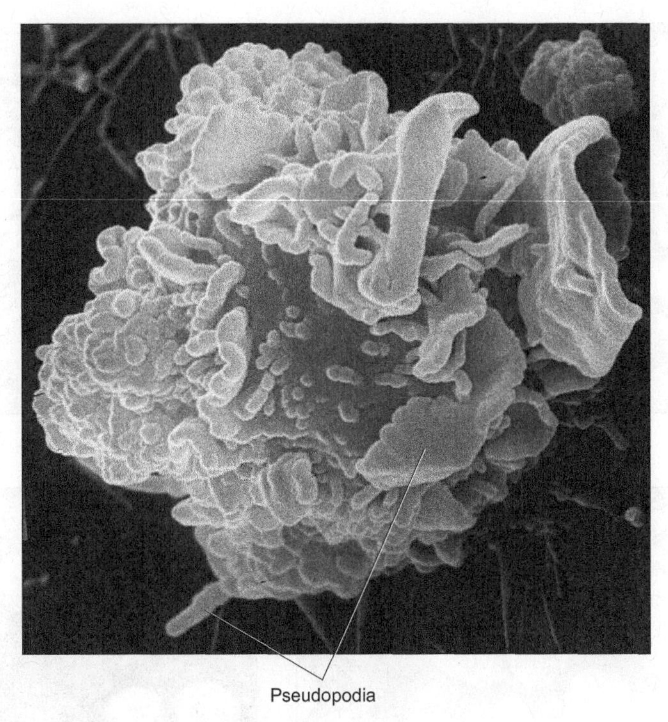

FIGURE 2.17 Macrophage from human lung. Magnification 20,000×.
Scanning electron micrograph by Kenneth E. Muse.

- Obtain a slide of loose connective tissue and observe the cells, the elastic and nonelastic fibers, and the apparently open spaces where the ground substance has been dissolved during the preparation of the slide. Identify the thin, wavy, elastic fibers and the thicker nonelastic fibers.
- Draw examples of elastic and nonelastic fibers in figure 2.19 and label each type.

Dense Connective Tissue

Dense connective tissue (figure 2.18b) consists of many nonelastic collagen fibers with few scattered cells as in a tough, fibrous tendon that connects a muscle to bone. Its function is support.

Adipose Tissue

Adipose, or fat, tissue is a loose connective tissue comprised of cells called adipocytes (figure 2.18c). The chief function of adipose tissue is to store energy in the form of fat, a highly reduced compound that releases much energy when oxidized. Fat also cushions and insulates the body. Recently, scientists have discovered that adipose tissue also secretes several hormones. Adipose tissue occurs in two forms, white adipose tissue and brown adipose tissue. Brown adipose tissue provides energy for temperature regulation and is found in young children rather than adult humans and also in animals preparing for hibernation.

Cartilage

Cartilage (figure 2.18d) consists of a firm but elastic **matrix** secreted by numerous cartilage cells embedded within the matrix. **Hyaline cartilage** is found at the ends of long bones. Cartilage cells are found in open spaces or **lacunae** (singular: lacuna) scattered within the matrix (chondrin), which they secrete. Most cartilage cells are isolated in a lacuna separated by a relatively large amount of intercellular material. Occasionally, however, you may find two recently divided cells within a single lacuna. Following their division, the two new cartilage cells begin to secrete more chondrin, and each new cell creates its own new lacuna. This is how cartilage tissue grows and how it may be repaired after injury. Athletes often suffer torn cartilages in their knees. If the injury is severe, it may require surgery to remove the damaged cartilage. Small tears in the cartilage, however, may be partially repaired by limited regrowth and the formation of scar tissue by the cartilage.

- Study a microscope slide of hyaline cartilage and observe the scattered cells, lacunae, and chondrin matrix. ***Can you find two recently divided cartilage***

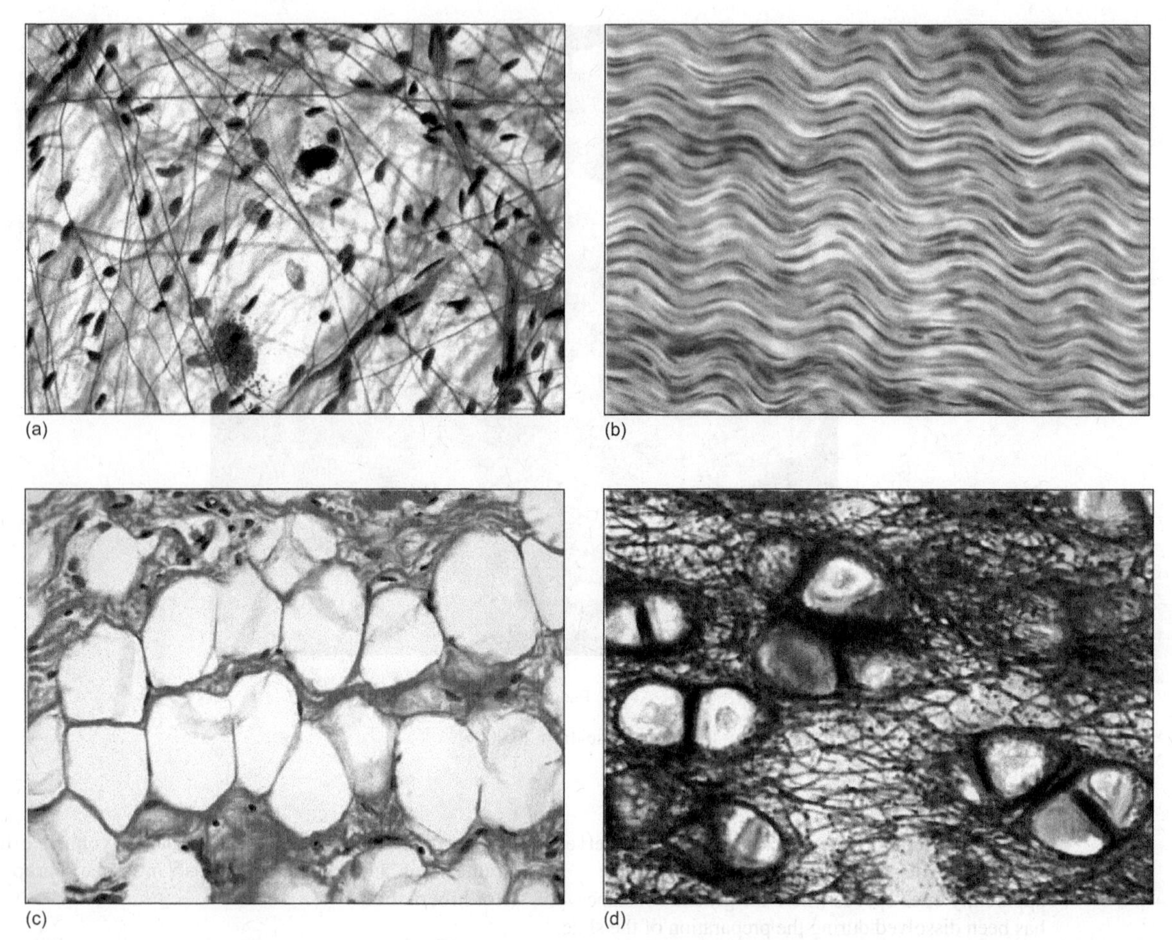

FIGURE 2.18 Connective tissue types: (*a*) Loose connective tissue, (*b*) dense connective tissue, (*c*) adipose tissue, (*d*) cartilage.
(a, c, d): The McGraw-Hill Companies, Inc./Al Telser, photographer. (b): The McGraw-Hill Companies, Inc./Dennis Strete, photographer.

cells in your microscope slide? How would you identify them as recently divided cells?

- Also study the demonstration of fresh cartilage from the end of a bone or the sternum. Feel its tough rubbery nature. ***How is this physical property related to the function of cartilage?***

Bone

Bone plays an important role in the mechanical support of the body and in protecting vital parts from injury. The skull and the vertebral column of humans, for example, play dual roles in the support and protection of the brain and the spinal cord as well as support for the main part of the body. The heart and the lungs are also protected by the bony framework of the rib cage embedded in the thoracic wall.

The intercellular matrix of bone also plays an important physiological role in the storage of calcium, which can be withdrawn from the bone and returned to circulation in soluble form when the calcium level in the blood is lowered. This is one reason why it is important for growing children and older persons to maintain an adequate intake of dietary calcium to avoid osteoporosis, the weakening of bone from excessive loss of calcium.

The bone marrow found in long bones contains tissue that plays an important role in the formation of blood elements and in immunity.

- Examine a thin section of **compact bone** (figure 2.20) prepared from a long bone, such as the humerus or femur of a human or other mammal, and identify the following structures: (1) the central **Haversian canal** through which passes many small blood vessels and nerves; (2) the **concentric layers** of bone (lamellae); (3) the **lacunae,** or spaces that house the bone cells or osteocytes; and (4) the numerous fine **canaliculi,** which serve to interconnect the lacunae and the Haversian canals.

FIGURE 2.19 Student drawing of elastic and non-elastic fibers from loose connective tissue.

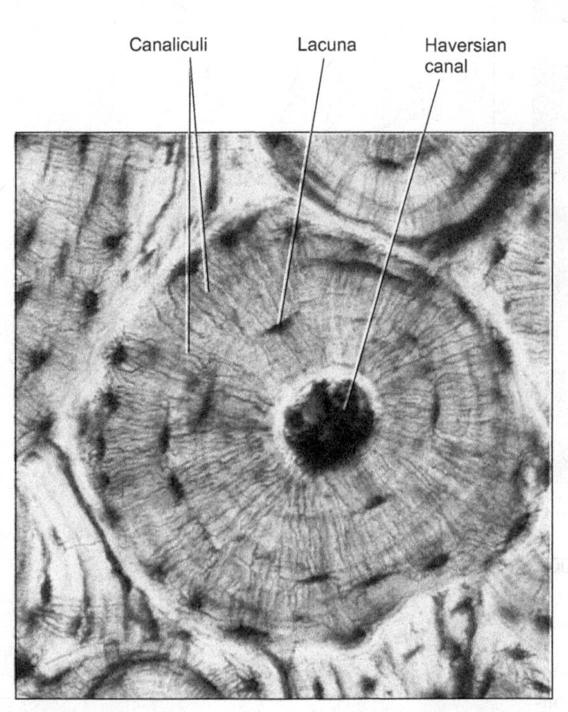

FIGURE 2.20 Compact bone, ground, cross section.
Courtesy of Carolina Biological Supply Company, Burlington, NC.

Muscular Tissue

Muscular tissue is specialized for **contraction** and therefore has the capacity to perform mechanical work. Three different types of muscular tissue are distinguished: **smooth muscle, skeletal (striated) muscle,** and **cardiac** (heart) **muscle.**

Smooth Muscle

Smooth muscle consists of elongated, spindle-shaped cells with a single, central nucleus (figure 2.21a). This type of muscle is sometimes referred to as nonstriated muscle because it lacks the cross striations seen in both striated and cardiac muscle.

Smooth muscle forms the simplest type of muscle tissue and is generally found in parts of the body not under voluntary control where rapid movement or contraction is not essential, such as in the walls of the digestive tract, in the walls of blood vessels, and in the walls of the urinary bladder and the uterus.

- Examine a slide of smooth muscle that has been teased apart to show the individual spindle-shaped cells or a section through some smooth muscle that shows the individual cells cut longitudinally. Locate the tapered smooth muscle cells and note the location of the nucleus. Identify the cell membrane and the fine longitudinal threads, the **myofibrils,** in the cytoplasm. Observe also the demonstration slides of smooth muscle from other types of material.

  Draw several smooth muscle cells in figure 2.22 and label the nucleus, cell membrane, and myofibrils.

Striated Muscle

The large muscles attached to various parts of the skeleton are composed of skeletal, or voluntary, muscle (figure 2.21). This type of muscular tissue is made up of long cylindrical fibers containing many nuclei. ***What is unusual about the location of the nuclei in skeletal muscle as compared to most other kinds of cells?*** The multinucleate (syncytial) condition in skeletal muscle arises during embryonic development of the muscle as a result of the fusion of many mononucleate cells.

- Note the conspicuous cross striations in these fibers and the outer limiting membrane, called the **sarcolemma.** Observe also the fine longitudinal **myofibrils** running lengthwise through the skeletal muscle fibers. ***Draw some myofibrils of striated muscle in figure 2.23 and label the important parts.***

The individual myofibrils of striated muscle also have a very distinctive banded or striated structure. Because of their small size, individual myofibrils can be studied only with an electron microscope. Figure 2.24 is an electron micrograph showing parts of four adjacent myofibrils.

The functional unit of the myofibril is the **sarcomere,** which extends between two adjacent Z-lines. During muscle contraction, the sarcomeres shorten (the distance between

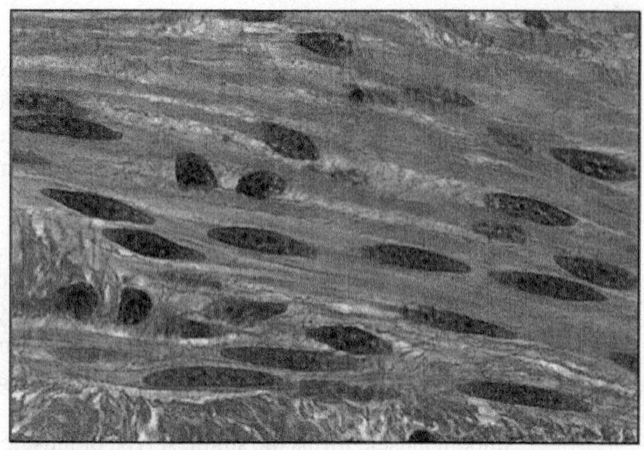

(a)

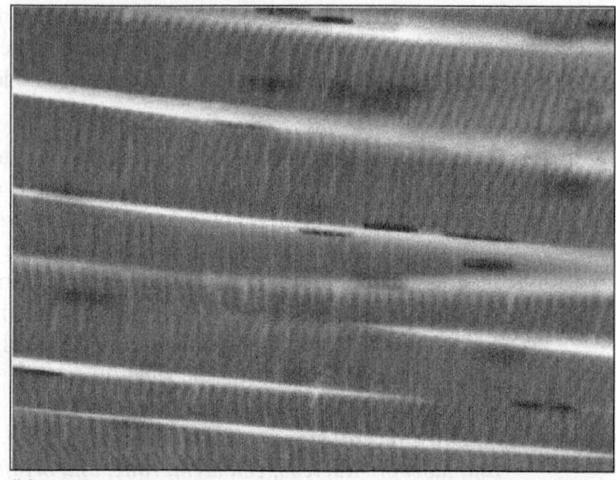

(b)

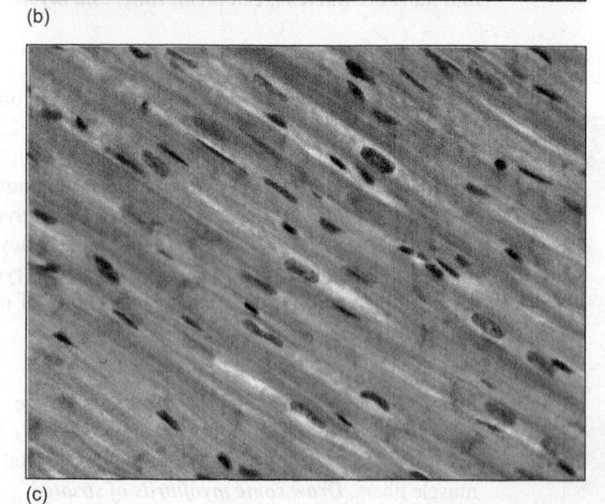

(c)

FIGURE 2.21 Muscle tissue types: (*a*) Smooth muscle, (*b*) striated muscle, (*c*) cardiac muscle Magnification (*a*-*c*) 500X.
(a): The McGraw-Hill Companies, Inc./Dennis Strete, photographer.
(b-c): The McGraw-Hill Companies, Inc./Al Telser, photographer.

Z-lines decreases). Investigations have shown that the contraction of muscle is due to interactions between two contractile proteins, **actin** and **myosin.** These two proteins make up a substantial portion of each myofibril.

FIGURE 2.22 Student drawing of smooth muscle.

FIGURE 2.23 Student drawing of myofibrils of striated muscle.

The large **mitochondria** provide energy in the form of ATP (adenosine triphosphate) to power the contraction.

Cardiac Muscle

A third type of muscle tissue, **cardiac muscle,** is found in the walls of the heart of vertebrate animals. Cardiac muscle (figures 2.21c and 2.25) consists of striated muscle fibers, which branch and reunite (anastomose) with other fibers to form a continuous network of muscle fibers. Cardiac muscle contains cross striations like striated muscle, but the fibers of cardiac muscle are divided into cell-like units by many **intercalated discs,** which partially divide the fibers. This type of muscle, like smooth muscle, is involuntary, and throughout the life of the organism, cardiac muscle contracts and relaxes rhythmically and automatically. In fact, the heart of some of the lower vertebrates, such as a frog or a turtle, can be removed from the body and placed in a physiological salt solution where it will continue to beat for many hours or even days.

- Study the demonstration slides of human and other vertebrate cardiac muscle and observe: (1) the branching and anastomosing network of muscle fibers, (2) the cross striations, (3) the scattered nuclei, (4) the outer sarcolemma, and (5) the intercalated discs.

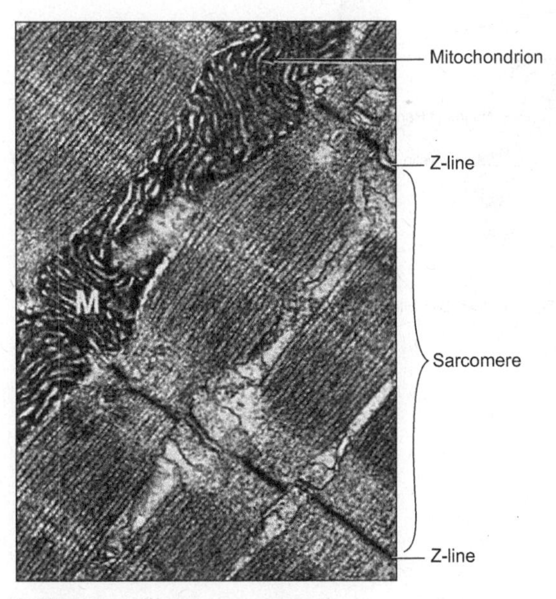

FIGURE 2.24 Striated muscle fibrils.
Electron micrograph by Kenneth E. Muse.

Nervous Tissue

The nervous system of vertebrate animals consists of the brain, spinal cord, and nerves. Nervous tissue consists of highly specialized cells that carry impulses from one part of the body to another. A nerve cell, together with its branches or processes, which in some cases are many centimeters long, is called a **neuron.** A neuron consists of a cell body containing the nucleus and two or more elongated nerve processes (figure 2.26). These processes are **axons** and **dendrites.** Impulses in vertebrate neurons can travel only in one direction. Dendrites carry impulses **toward** the cell body and axons carry impulses **away** from the cell body.

Nerve fibers are covered by one or more sheaths. All peripheral nerves are covered by a thin neurolemma, and most peripheral nerves are also covered by a myelin sheath, which may be several layers thick. Myelinated nerves are capable of rapid transfer of impulses because of their special electric properties.

Sensory neurons carry impulses from sensory receptors to the spinal cord or other parts of the central nervous

Demonstrations

1. Ciliated epithelium (microscope slide)
2. Adipose (fat) tissue (microscope slide)
3. Fresh cartilage (from the sternum of a frog, ends of a long bone, etc.) to illustrate gross structure, toughness, and flexibility
4. Cross sections (1–2 inches thick) of long bone to illustrate gross structure of bone
5. Cardiac muscle (microscope slide)
6. Amphibian blood (microscope slide)

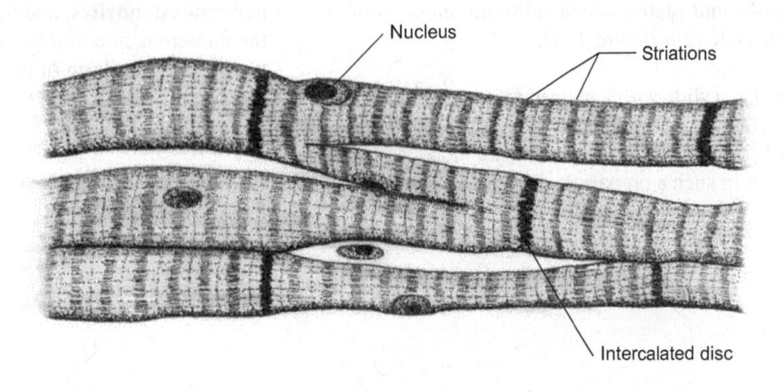

FIGURE 2.25 Cardiac muscle.

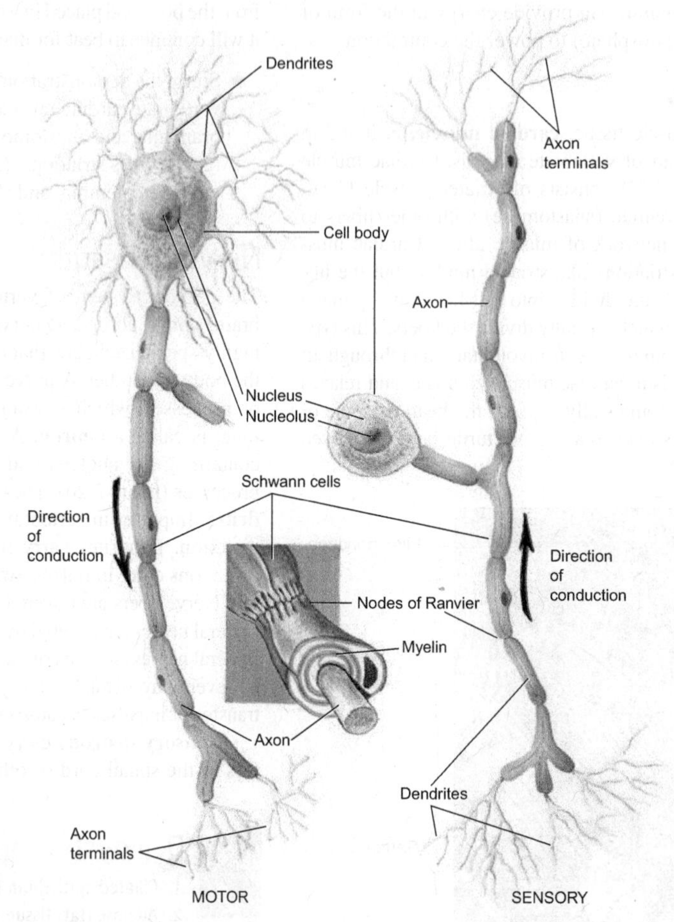

FIGURE 2.26 Motor and sensory neurons.

system. **Motor neurons** carry impulses away from the spinal cord or brain and end at muscle cells or other effectors. Motor nerves that end at muscle cells have specialized endings called **motor end plates,** which aid in transmission of stimuli to the muscle cells (figure 2.27).

- Examine also a slide with a stained smear preparation of the gray matter from the spinal cord of a cow. The details of the individual neurons are more readily seen in such a preparation than in sections.

Observe the large neurons (easily seen under low power). Compare with figure 2.28. Select a typical cell and note the **cell body, nucleus, nucleolus,** numerous **dendrites,** and the longer **axon.** Observe the numerous fine fibrils, the **neurofibrils,** found within the cytoplasm of the cell body and extending into the processes.

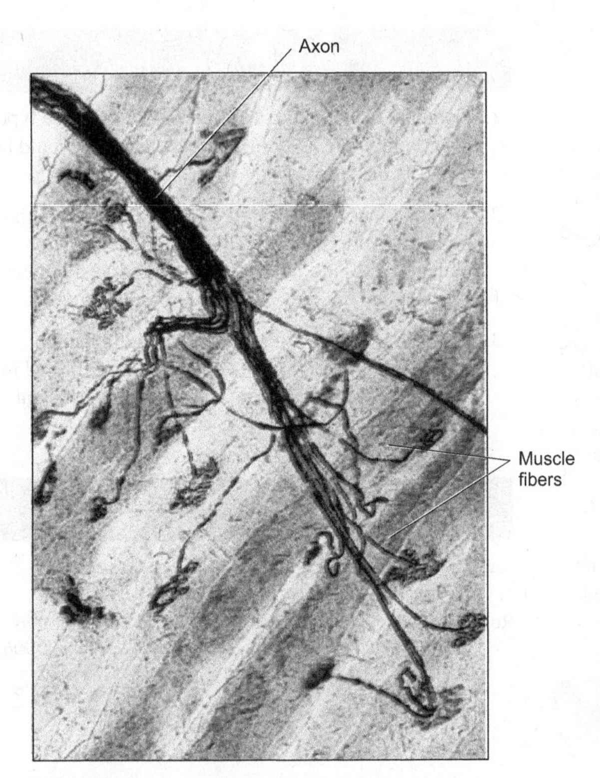

FIGURE 2.27 Motor end plate, teased, snake.
Courtesy of Carolina Biological Supply Company, Burlington, NC.

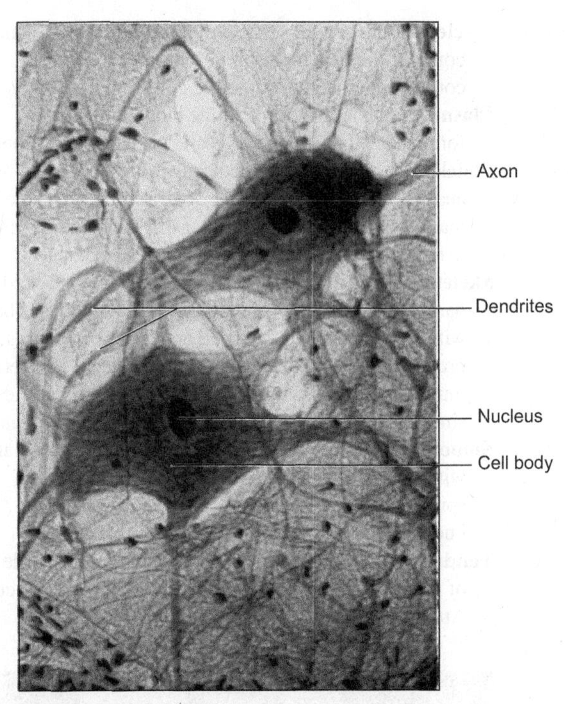

FIGURE 2.28 Neurons from spinal cord smear from a cow.
Courtesy of Carolina Biological Supply Company, Burlington, NC.

Key Terms

Adipose tissue loose connective tissue comprised of adipose or fat cells. Consists of two types, white adipose tissue and brown adipose tissue. Both store fats and other lipids that serve as energy reserves; brown fat is found mainly in infants and also in animals preparing for hibernation and aids in temperature regulation.

Bone hardened, mineralized type of connective tissue that provides support and protection in vertebrate animals.

Cardiac muscle type of muscle found in the heart wall of vertebrate animals. Consists of a branching and anastomosing network of striated, multinucleate muscle fibers.

Cartilage a type of connective tissue with a gelatinous matrix containing numerous fibers. Provides support in vertebrate animals.

Cell membrane outer limiting membrane of an animal cell; composed of phospholipids, proteins, cholesterol, and carbohydrates.

Connective tissue type of animal tissue that binds, supports, and protects other body parts. Includes several kinds of tissue, such as loose connective (areolar) tissue, adipose (fat) tissue, cartilage, and bone.

Epithelium type of animal tissue that covers the outer surface, and lines inner cavities and ducts of the body. May be simple or stratified (multilayered) and may be hardened with keratin.

Erythrocytes red blood cells whose chief function is the transport of oxygen; contain large amounts of hemoglobin. May be nucleated (as in frogs and salamanders) or without a nucleus in the mature stage (as in humans).

Leucocytes white blood cells that are usually colorless in life and that perform many essential functions, including engulfment of foreign particles, production of antibodies, and wound healing. Several distinct types of leucocytes are usually present in the blood of an animal.

Motor end plate specialized ending of a motor neuron on a muscle cell; serves to transmit stimulus from nerve to muscle.

Muscular tissue type of tissue specialized for contraction; contains fibrils constructed of contractile proteins. Three types are distinguished—smooth muscle, skeletal muscle, and cardiac muscle.

Nervous tissue type of tissue specialized for the conduction of electrical impulses; important in the coordination of body activities. The basic functional unit is the neuron, or nerve cell.

Neuron the basic, functional unit of the nervous system of animals. The neuron is a single nerve cell, which consists of a cell body with a nucleus and two or more long extensions, or processes (axons and dendrites).

Nucleus the central organelle of an animal cell, which contains the genetic material (chromosomes) and controls the metabolism of the cell.

Plasma the liquid portion of the blood in which the formed elements (blood cells and platelets) are suspended.

Platelet a small, nonnucleated body in the blood of humans and other mammals. Formed by megakaryocytes in the bone marrow and serves mainly to plug leaks in blood vessels and to release chemical substances that initiate clotting.

Skeletal muscle type of muscle tissue found attached to parts of the skeleton. Consists of long, cylindrical fibers with a cross-banded or striated appearance in microscopic preparations; each fiber contains many nuclei. Skeletal muscle is also known as striated or voluntary muscle since many skeletal muscles are under voluntary control.

Smooth muscle type of muscle tissue found associated with internal organs in higher animals. Consists of spindle-shaped cells with a single central nucleus. Also known as involuntary muscle.

Tendon type of tough dense connective tissue made up of many parallel collagen fibers and few scattered cells. Attaches muscles to bone.

Internet Resources

There are many valuable Internet sites with information about zoology. Several sites containing pertinent zoological information for this chapter can be found on the McGraw-Hill Zoology web site at http://www.mhhe.com/zoology. Just click on this text's title.

Questions for Critical Thinking

1. Compare the structure and function of the three types of muscle tissue. Describe where each type would be found in the human body.
2. Several kinds of tissues play a protective role in the body of a vertebrate animal such as a dog, cat, or human. List four kinds of tissue that serve a protective role and explain how each kind serves to protect.
3. List four different kinds of cells found in human blood, briefly describe how they can be identified in a microscopic preparation, and give their principal function.

Suggested Readings

Karp, G. 2007. *Cell and Molecular Biology: Concepts and Experiments*, 5th ed. New York: John Wiley. 864 pp

Ross, M., and W. Pawlina. 2006. *Histology: A Text and Atlas*, 5th ed. Philadelphia: Lippincott, Williams & Wilkins. 906 pp

Notes and Sketches

Chapter 3

Mitosis and Meiosis

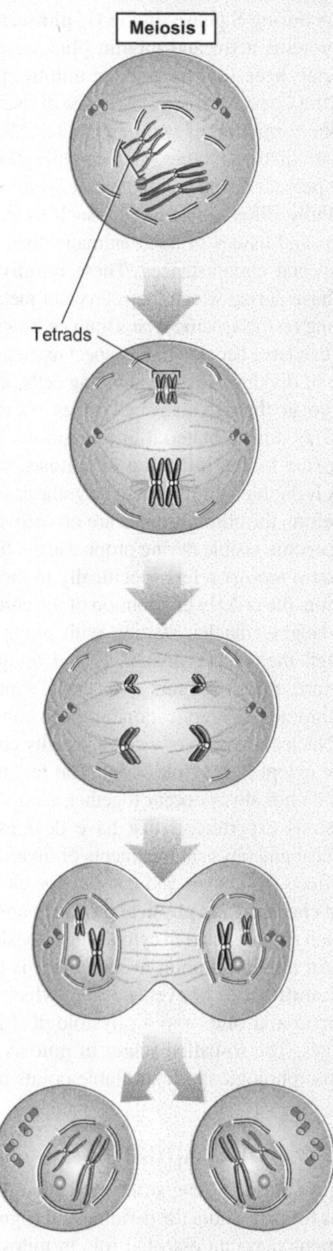

Objectives

After completing the laboratory work in this chapter, you should be able to perform the following tasks:

1. Explain the basic differences between mitosis and meiosis.
2. Describe the cell cycle and explain the principal events of its four main stages.
3. Briefly describe the structure and function of chromosomes, centromeres, spindle fibers, aster rays, centrioles, and centrosomes.
4. Identify the principal mitotic stages of an animal cell in microscope slides or photographs.
5. Explain the chief events that occur in prophase, metaphase, anaphase, and telophase of mitosis.
6. Define random sample and explain its significance in biological research.
7. Describe a method for estimating the relative duration of the various stages of mitosis in a population of animal cells.
8. Define the terms haploid and diploid and explain how they relate to the process of sexual reproduction in animals.
9. Distinguish between a chromosome and a chromatid.
10. List the principal stages of meiosis and identify each stage in microscopic preparations or illustrations of animal cells.
11. Explain how the process of meiosis contributes to the genetic variability of offspring in the next generation.

Introduction

All animals depend on cell division for their growth and repair processes. Each cell has a precise set of genetic information built into its chromosomes. This information is essential for the proper functioning of each cell and for prescribing the characteristics of the next generation of cells. The division of the **nucleus** of the cell and the precise distribution of the duplicated chromosomes between the two new cells in this type of cell division is called **mitosis.**

Meiosis is a specialized type of cell division that usually occurs during the formation of the gametes, or sex cells, of multicellular animals. During meiosis, the normal diploid (2n) chromosome number of the somatic (body) cells is reduced by half to the typical (n) chromosome number of the gametes. Meiosis is extremely important for the survival and evolution of animals because it provides for recombinations of genes during each generation. Thus, variations occur among the offspring of each generation, and natural selection can operate to select the better adapted individuals.

Materials List

Prepared microscope slides
- Whitefish blastula (mitosis) (or *Ascaris* embryo)
- *Ascaris* ova (meiosis)

Audiovisual materials
- Color transparencies of mitosis in whitefish blastula
- Color transparencies of meiosis in *Ascaris* ova
- Wall charts illustrating mitosis and meiosis
- Film or video of meiosis

Mitosis

The division of nuclei by mitosis is exhibited by the somatic or body cells in most plants and animals. During its life span, a cell passes through a regular sequence of physiological events called the **cell cycle** (figure 3.1). This sequence includes several distinct stages, each characterized by certain metabolic activities of the cell. In actively dividing cells, the cycle may last only a few hours; in other cells, the cycle may last for days or weeks. At the completion of the cell cycle, a new generation of cells is produced by division of the parent cell.

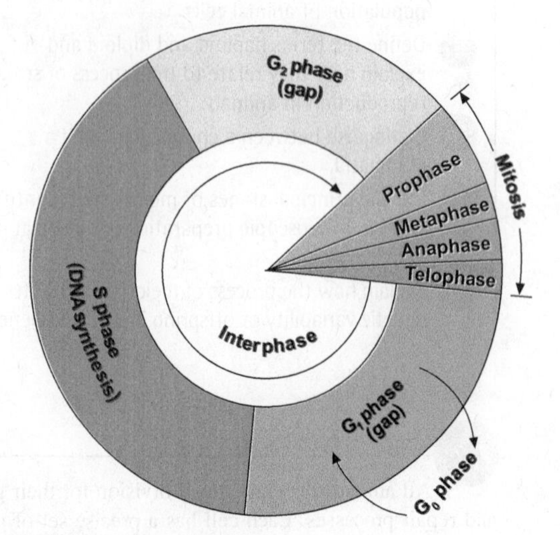

FIGURE 3.1 The cell cycle.

The cell cycle in actively dividing cells, like early embryos, consists of four phases: Gap_1, Synthesis, Gap_2, and Mitosis, abbreviated as G_1, S, G_2, and M. The **G_1 phase** is a period of active protein synthesis and the formation of new cell organelles like mitochondria, Golgi, ribosomes, and endoplasmic reticulum. It is a period of rapid cell growth. The **S phase** is the time when DNA and other molecules making up the chromosomes, such as histones, are synthesized. Replication of the DNA strands and of the chromosomes also occurs during S phase. In the **G_2 phase,** the cell synthesizes the proteins actin and tubulin plus the enzymes and other materials necessary to form the mitotic spindle. **Mitosis** follows the G_2 phase and involves the division of chromosomes and the formation of two new nuclei. Mitosis itself consists of four distinct stages, but actually takes up only about 5–10 percent of the complete cell cycle in most cells.

Fully differentiated cells, such as those found in most organs and tissues of adult animals, divide rarely or only under special circumstances. These **nondividing cells** enter a **G_0 phase** during which they carry out metabolic functions depending on the specific type. Brain cells are a good example of cells that have become highly specialized and are no longer capable of dividing. Rapidly dividing cells, like those in an early embryo, go through many cell cycles in a short time.

The genetic material in the nucleus is duplicated in the cell **prior to the initiation of mitosis.** Thus the amount of DNA is doubled through active synthesis of DNA at this time. Therefore, the chromosomes are **already doubled** when they first become visible during prophase, the first stage of mitosis. The term *mitosis* refers specifically to the process of nuclear division, the orderly distribution of the chromosomes between two daughter nuclei, starting with prophase and continuing through metaphase, anaphase, and telophase. Technically, therefore, mitosis occurs *after* DNA synthesis is completed and chromosome replication has been completed.

Nuclear division of a cell is usually coupled with division of the cytoplasm **(cytokinesis).** The fact that these two processes do not always occur together, along with evidence from numerous experiments that have demonstrated that various chemical and physical treatments of dividing cells have different effects on mitosis and cytokinesis, clearly shows that different **chemical** and **physical** processes are involved in nuclear division (mitosis) and in cytoplasmic division (cytokinesis).

An important point to remember is that cell division is a **dynamic** series of events during which the cell undergoes dramatic and often rapid physiological and morphological changes. The so-called stages of mitosis merely represent a few morphologically identifiable points in this continuum.

The Mitotic Apparatus

During mitosis, a new structure, called the **mitotic apparatus,** is formed within the dividing cell (figure 3.2). The mitotic apparatus plays an essential role in mitosis and provides the

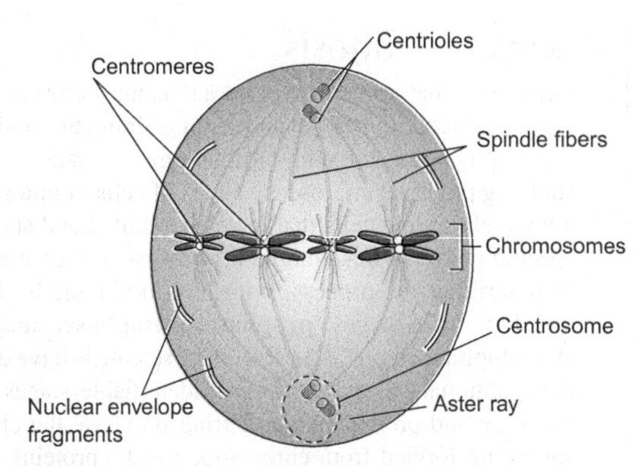

FIGURE 3.2 Mitotic apparatus.

mechanism by which the replicated chromosomes of the dividing cell are distributed between the daughter cells. A knowledge of the structure of the mitotic apparatus and its principal parts will help you to understand the process of mitosis and to learn how genetic continuity is maintained throughout many generations of cells. The mitotic apparatus is of vital importance to you; without it, a cut finger would not heal, a broken leg would not mend, and you would be unable to have children.

There are six main components of the mitotic apparatus: the asters, the chromosomes, the centromeres, the centrioles, the centrosomes, and the spindle fibers. Most dividing animal cells exhibit all six of these components. Most plant cells and certain invertebrate animal cells, however, lack **centrioles** and **aster rays.** Five of these six parts of the mitotic apparatus appear to play important roles in the process of mitosis, but the function of one of these structures, the asters, is still uncertain. The functions of these various components will be considered later in this exercise.

Several important events take place in a living cell during the early stages of mitosis. These events include:

1. The breakdown of the **nuclear membrane** and the mingling of the nuclear contents with the cytoplasm of the cell.
2. The condensation of the chromosomal material within the nucleus to form discrete, visible chromosomes.
3. The separation of the centrioles and their migration to opposite sides of the cell.
4. The formation of spindle fibers and aster rays.

The spindle fibers and aster rays are formed by the coalescence or condensation of relatively small protein molecules already existing within the cytoplasm of the cell. Thus, the formation of the components of the mitotic apparatus represents a recombination of molecules preexisting in the parent cell rather than the synthesis of new protein molecules. There is relatively little synthesis of new molecules during the process of mitosis; instead, most of the new molecules formed within the cell are synthesized during the interphase.

Chromosomes

Chromosomes are more or less elongated structures present throughout the life cycle of a cell. Their structure, however, appears quite different when observed at various times during the life cycle of a cell. During mitosis, the chromosomes can be seen as short, rodlike structures that are formed by condensation or contraction of very fine, threadlike structures present in the nucleus prior to mitosis (figure 3.3). Thus, the appearance of chromosomes, like the appearance of the other components of the mitotic apparatus during the early portions of mitosis, results mainly from the reorganization of preexisting materials rather than from the actual synthesis of new materials.

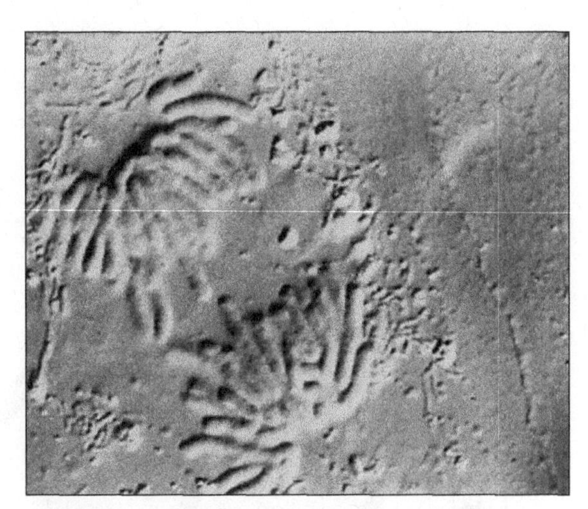

FIGURE 3.3 Anaphase chromosomes in a newt lung cell. Nomarski differential interference photomicrograph.
Courtesy of Southern Micro Instruments.

Figure 3.4 is a photograph of the metaphase chromosomes from a male human cell with the typical 46 chromosomes. These chromosomes have been spread out so they are more clearly visible. Observe the different sizes and shapes of the chromosomes. Each species of animal (and plant) has a characteristic number of chromosomes. In humans, this number is 46; in the fruit fly *Drosophila,* it is 8; in the dog, it is 78; and in the cat, it is 60.

The unique number, size, and shape of an individual's chromosomes is called a **karyotype.** In human genetics the use of karyotypes is very helpful in identifying hereditary genetic conditions.

The chromosomes contain the hereditary units, or genes, which specify each of the characteristics of a cell and determine all of its capabilities. Thus, the chromosomes are responsible for the control of cell metabolism and enable cells to differentiate and to organize into specialized tissues, organs, and organ systems.

Centromeres

A **centromere** is a specific segment of a chromosome that serves to attach the chromosome to a spindle fiber by the formation of a protein complex called the kinetochore. During the latter portion of mitosis, the separated chromatids move toward opposite **poles** of the **mitotic spindle.** Research studies indicate that the chromosomes appear to be both pulled and pushed toward the poles by separate actions of the microtubular proteins of the spindle. Other studies have demonstrated that damaged

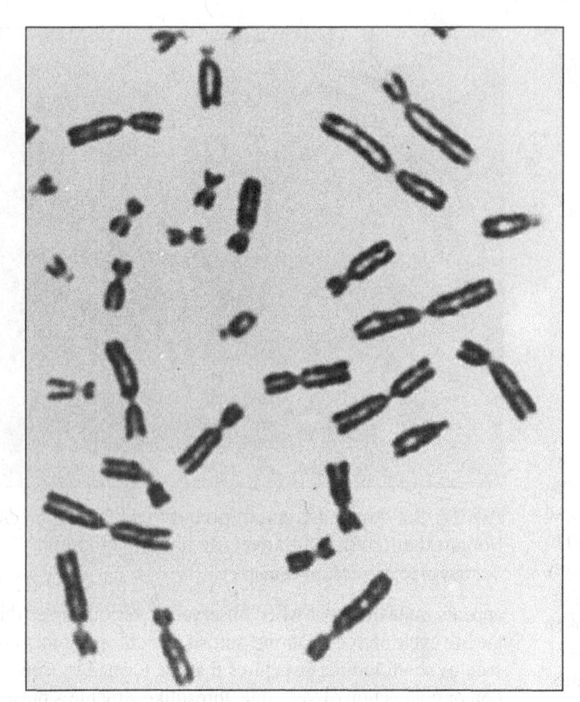

FIGURE 3.4 Spread of chromosomes from a male human lymphocyte in metaphase.
Photograph by Wendell Mackenzie.

chromosomes in cells injured by X rays or by chemical agents sometimes separate into two or more fragments. Those chromosomal fragments that lack centromeres do not become attached to the chromosome and do not move toward the poles. Such observations clearly demonstrate the important functional role played by the centromeres in mitosis.

Spindle Fibers and Asters

Special research techniques have been devised that allow scientists to isolate and to study the mitotic apparatus from living cells. Biochemical studies on such preparations of isolated mitotic apparatuses have demonstrated that the **spindle fibers** and **asters** are microtubules constructed mainly of two common contractile proteins: actin and tubulin. These proteins are involved in the movements of all eukaryotic cells. Some research evidence indicates that the asters may influence the location of the cleavage plane in cytokinesis.

Centrioles and Centrosomes

Centrioles are tiny structures usually found in pairs near the nucleus in nondividing animal cells with a well-formed nucleus. The centrioles migrate toward opposite sides of the cell during the early stages of mitosis and seem to form centers, or foci, for the spindle fibers and aster rays at each end of the mitotic apparatus. Surrounding each centriole is an amorphous area with no visible structures called the **centrosome,** which serves as the organizer for the formation of microtubules into the mitotic spindle. Most plant cells have centrosomes that serve the same purpose, but plant cells typically lack centrioles.

Stages of Mitosis

Remember that mitosis involves a dynamic series of events during which the cell is undergoing dramatic, and often rapid, physiological and morphological changes. When studying prepared microscope slides of cells in mitosis, you are merely seeing cells that have been killed and stained at specific points in this continuous process. For convenience in describing the process, mitosis is traditionally divided into four main stages: **prophase, metaphase, anaphase,** and **telophase** (figure 3.5). Recently, scientists have divided the traditional prophase into two identifiable stages called prophase and prometaphase. During prophase the chromosomes are formed from chromatin, spindle proteins are assembled, and the centrioles separate. Prometaphase begins with the breakdown of the nuclear envelope and continues with the assembly of the mitotic spindle and the attachment of the chromosomes to the spindle.

The stage between successive mitotic divisions is called **interphase.** Knowing the names of the stages is not as important as understanding what happens within the nucleus of the cell as it divides.

- Many rapidly growing tissues provide good material for the study of mitotic cell division. Obtain a microscope slide prepared from some appropriate tissue such as the whitefish blastula (figure 3.6), early embryos in the uterus of the roundworm *Ascaris,* or the skin of an amphibian tadpole. Study first one or more cells in interphase. Observe the structure of the nucleus and the arrangement of the chromatin material. ***Can you identify definite chromosomes? How many nucleoli do you find? Is the number of nucleoli the same in all interphase cells?***

  Observe other cells on the slide and select cells in each of the main stages of mitosis for further study. During your study, try to follow the actual sequence of stages as described and try to visualize the changes that occur during the transition from one "stage" to the next.

1. **Early prophase stage** during which the **chromatin** material shortens to form long, coiled, threadlike chromosomes.
2. **Middle prophase stage** with relatively thick chromosomes.
3. **Late prophase stage** in which the chromosomes are further shortened and thickened. Under high magnification, late prophase chromosomes can be seen to consist of two separate strands, the chromatids, joined by a single centromere.
4. **Metaphase stage** showing the chromosomes arranged in a disclike pattern on the **equatorial plane** and attached to fibers of the mitotic spindle.
5. **Anaphase stage** during which the centromeres divide and the two chromatids of each chromosome move apart to opposite poles. The anaphase stage is relatively brief.
6. **Early telophase stage** showing the full set of chromosomes at each end of the elongated cell and the beginning of a **cleavage furrow** around the middle of the cell.

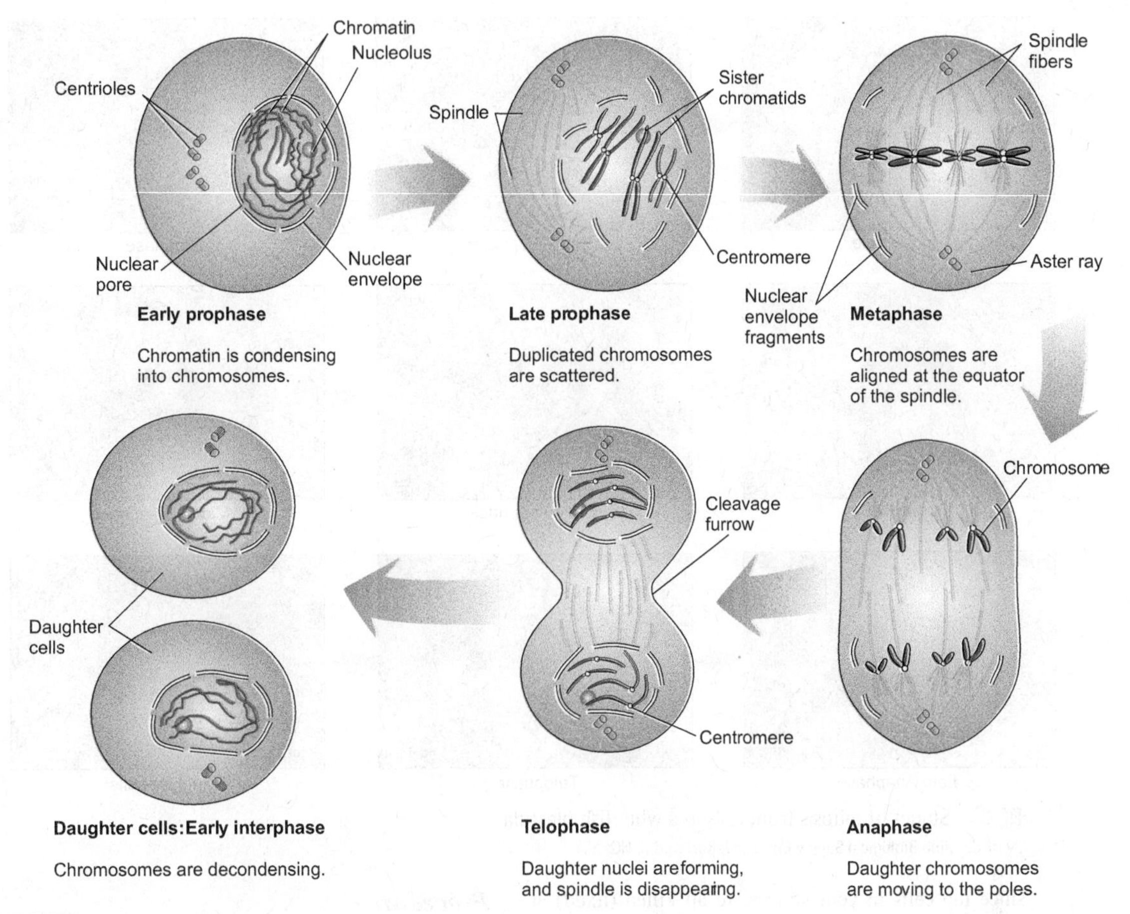

FIGURE 3.5 Stages of mitosis.

7. **Middle telophase stage** in which the individual chromosomes start to uncoil and lengthen, and begin to appear less distinct.
8. **Late telophase stage** during which the nuclei in the two daughter cells reorganize. The chromosomes disappear, nuclear membranes and the nucleoli reappear, and the separation of the daughter cells is completed.

Timing in the Cell Cycle

The different stages in the cell cycle are not of equal duration; some stages are relatively long, while others are very brief. The actual duration of mitosis and the relative duration of individual stages of mitosis vary widely from one cell type to another. Cells of a particular type also vary in their rates of division depending upon numerous physiological and environmental conditions. Nevertheless, you can obtain an estimate of the relative duration of the various stages in a population of cells and learn something about the kinetics of cell division by employing some relatively simple experimental techniques. You will need a whitefish blastula slide for this experiment.

Consider the whitefish embryo from which your mitosis slide was prepared as a **population of dividing cells** and see what can be determined about the kinetics of cell division, the relative duration of various stages of mitosis, and the relationship between mitosis and interphase in the life cycle of a cell. The whitefish embryo represents a population of relatively homogeneous cells; during the earliest stages of development, the embryo consists of few cells, which divide more or less synchronously. As development continues and the number of cells in the embryo increases as a result of cell division, the degree of synchrony decreases, and division becomes progressively more randomized.

Your whitefish embryo slide represents a slice through a fish embryo containing several thousand cells. It therefore represents a sample taken from a larger population of cells. Still, the number of cells on your slide is too great for you to count readily in the limited time available in the lab. Select a still smaller sample of cells from the embryo to yield some numbers that you can use to estimate some characteristics of the population of cells that made up the original whitefish embryo.

- Take your whitefish embryo slide and select a random sample of 50 cells. ***How can you be sure that you obtain a random sample of cells?*** Record the number of cells in each of the four mitotic stages and those in interphase and carefully record the results of your count in the Notes and Sketches section at the end of this chapter.

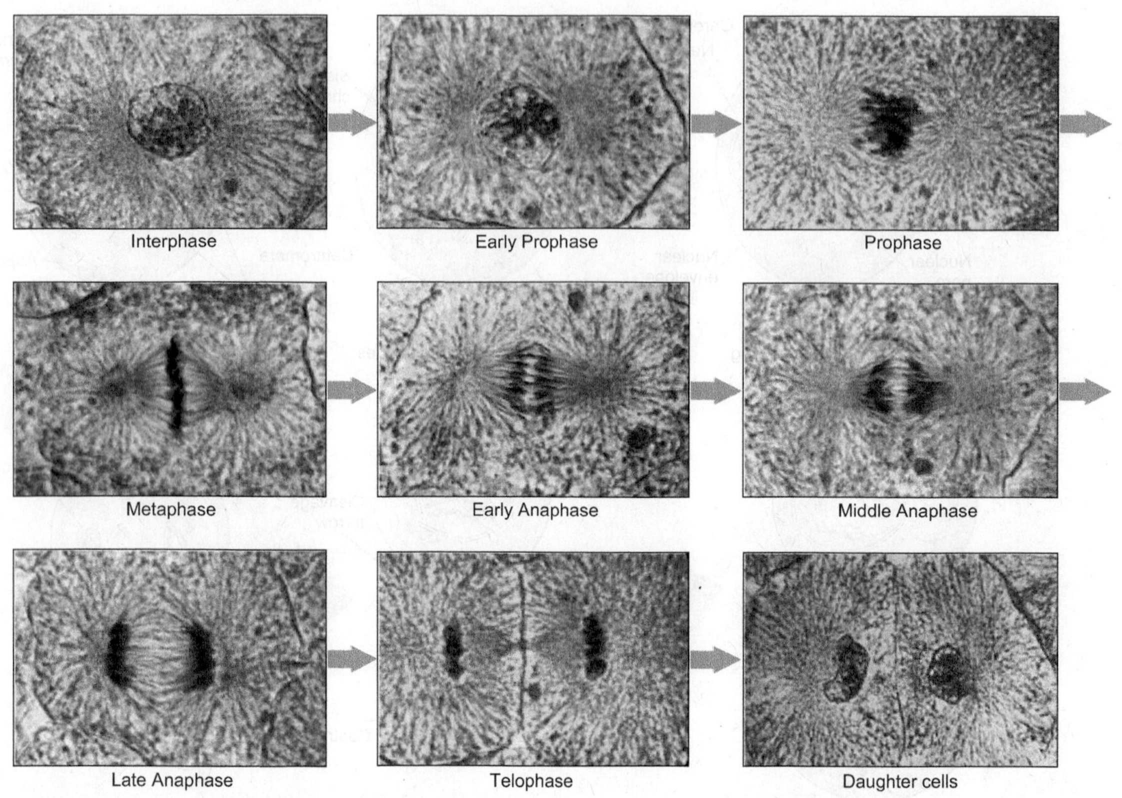

FIGURE 3.6 Stages of mitosis from cells in a whitefish blastula.
Courtesy of Carolina Biological Supply Company, Burlington, NC.

Since the cells in your slide were all killed (fixed) at approximately the same time, they represent a sample of the fish embryo cells stopped in action at a particular point in time. **Thus, the frequency of cells in the various stages is proportional to the relative duration of the stages.**

Random Sample

A ***random sample*** *can be defined most simply as a sample from a population in which* ***every member*** *of the population has an* ***equal chance*** *to be included. Thus, a sample in which the individuals selected are determined by use of a table of random numbers (or the random number generator in a computer) would be a random sample. A sample in which the individuals selected are determined by taking every tenth individual would not necessarily be a random sample.* ***Why?***

NOTE!:

Count only cells in which the nucleus appears in the section and in which enough nuclear material (chromosomes, mitotic spindle, etc.) can be seen to allow accurate identification of the stage.

Procedure

1. Select a random sample of 50 cells from your whitefish embryo slide.
2. Identify the stage of each cell in your random sample (prophase, metaphase, anaphase, telophase, or interphase).
3. Record the stage of each cell in your sample in the Notes and Sketches section at the end of this chapter.
4. Construct a bar graph on figure 3.7 to illustrate the results of your count showing the number of cells in each stage.

In which stage did you find the most cells? The fewest cells? Which stage, therefore, would you conclude is the longest in duration? The shortest? What are the relative lengths of the other stages as estimated from your sample?

- Compare the data from your slide with those obtained by other students from their slides. ***How do their counts compare with yours? How much variation does there appear to be in the counts on different slides? What are some of the sources of variation that may account for the differences in your counts? Are your conclusions regarding the relative lengths of the various stages the same as those of other students in the class, or do they differ? Why?***

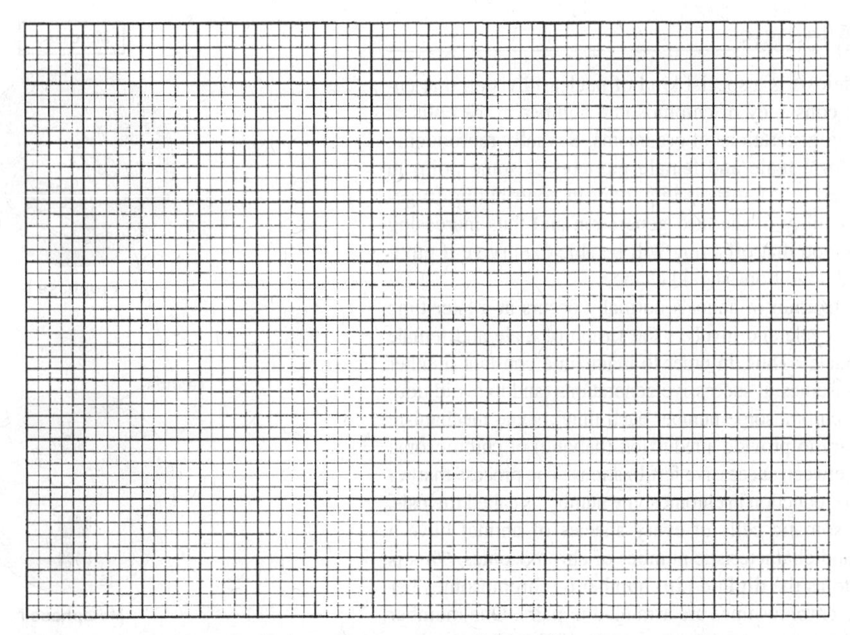

FIGURE 3.7 Graph of the distribution of cells in various stages of division. Data from your slide.

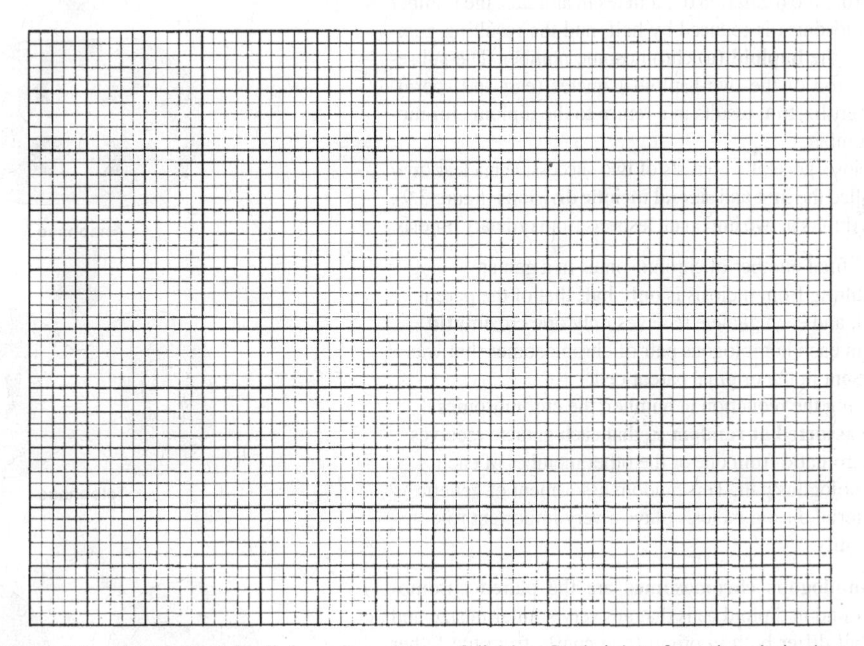

FIGURE 3.8 Graph of the distribution of cells in various stages of division. Pooled data from the whole class.

Now, combine the data from your count with those of the other students in your lab section to obtain estimates of the characteristics of another population of cells. The population represented by these pooled data consists of the cells in several (20–24) different whitefish embryos, and the 50 cells counted by each student represent a series of subsamples taken from that larger population of cells.

From these pooled data, construct another bar graph on figure 3.8 to show the number of cells in each stage. Estimate the relative lengths of the various stages in the new cell population from these pooled data and compare the results with those from your count of a single slide. ***How do your results compare with and how do your conclusions differ from those based on data from a single whitefish blastula slide? Explain.***

Meiosis

Meiosis is a special kind of nuclear division that ensures the constancy of chromosome number in the cells of succeeding generations of organisms. Sexually reproducing animals form male and female **gametes** at some point in their life cycle. Meiosis is important in contributing to genetic variability of the next generation and thus increasing the likelihood that offspring will be able to survive if environmental conditions should change.

Fertilization in animals normally occurs at a later time in the life cycle and involves the fusion of male and female gamete nuclei. Thus, in order to maintain a constant number of chromosomes in successive generations (and to avoid doubling the chromosome number each time), some mechanism is necessary to provide a reduction (halving) of chromosome number between successive fertilizations. The process that results in the reduction in chromosome number is called **meiosis.**

The somatic cells of every species of animal have a definite and characteristic number of chromosomes. This is referred to as the **diploid** (2n) number of chromosomes because the chromosomes are arranged in pairs. One member of each chromosome pair came from the father and one chromosome came from the mother.

During the formation of gametes in animals, the number of chromosomes is reduced by half, and the resulting gametes have the **haploid** (n) chromosome number. The subsequent fusion of the two haploid gametes (egg and sperm) during fertilization results in a return to the diploid chromosome number.

Meiosis generally consists of two successive nuclear divisions called the **first** and **second meiotic divisions** (figure 3.9). Meiosis differs in two important respects from ordinary mitosis.

1. The final number of chromosomes in a gamete resulting from meiosis is **only half** that of the parent cell, and each gamete receives only **one chromatid** from each homologous pair of chromosomes that was present in the original parent cell.
2. During the reduction in number, the chromosomes are **assorted at random** so that each gamete receives a chromatid from one or the other member of each homologous pair. This random assortment of genetic material during meiosis plays a very important role in heredity.

Homologous chromosomes are the paired chromosomes found in diploid cells that are very similar in size and shape, but differ both in origin (one comes from the father and one from the mother) and in genetic composition. (The father and mother usually contribute different sets of alleles to the offspring.)

Meiosis, like mitosis, is a dynamic process during which the cells are undergoing continuous change. Nonetheless, a good understanding of the process can be achieved by describing it as a sequence of two nuclear divisions, each with four distinct stages and with an intervening interkinesis stage between the first and second meiotic divisions.

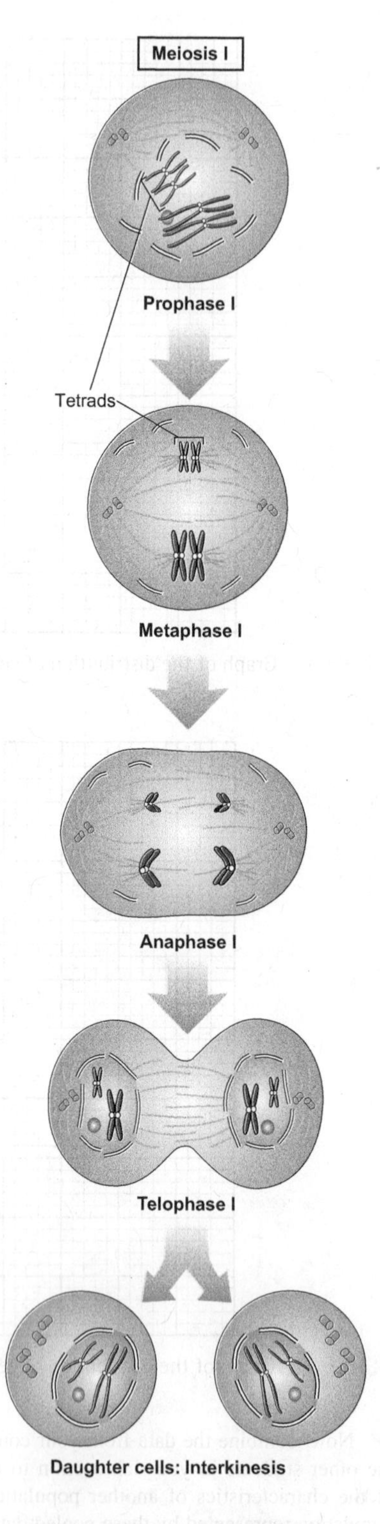

FIGURE 3.9 Meiosis. (*a*) Reduction division. Red chromosomes are from one parent; blue chromosomes are from the other parent. (*b*) Nonreduction division.

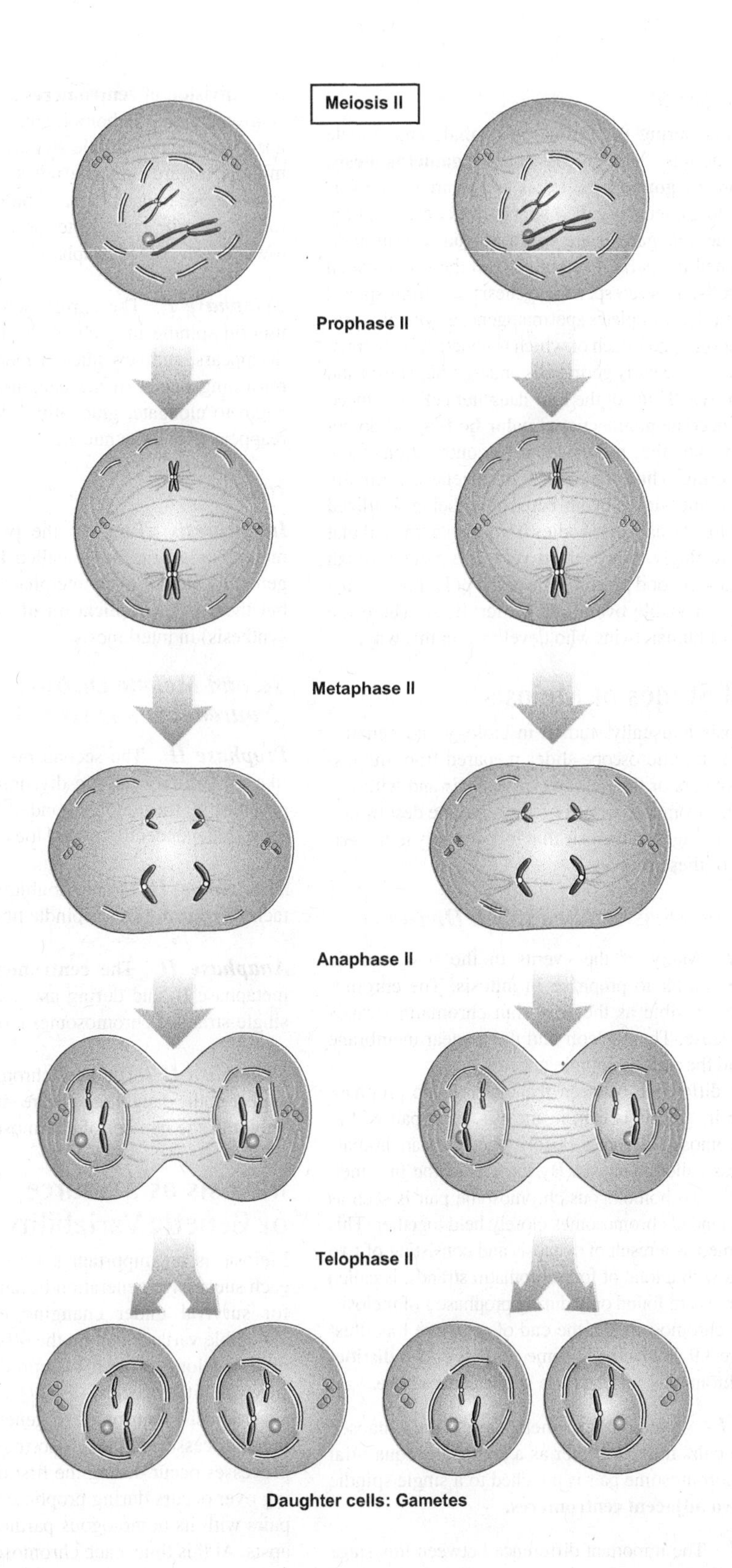

Meiosis II
Prophase II
Metaphase II
Anaphase II
Telophase II
Daughter cells: Gametes

(b)

Gametogenesis

Meiosis occurs during the formation of male and female gametes in animals. This process, called **gametogenesis,** takes place in the **gonads,** the testes and ovaries. The formation of male gametes is called **spermatogenesis,** and the products of spermatogenesis are **spermatozoa,** a term commonly shortened to *sperm.* Certain cells in the testis, called early **germ cells,** undergo spermatogenesis and form sperm. Each germ cell that completes spermatogenesis typically produces four spermatozoa, each of which is genetically distinct.

In the ovary, the early germ cells undergo **oogenesis** and form eggs, or **ova.** Three of the four daughter cells produced in oogenesis become nonfunctional **polar bodies** and do not form ova. Typically they make no genetic contribution to any resulting offspring. Thus, the process of oogenesis normally produces only one viable ovum capable of being fertilized and three nonfunctional polar bodies from each germ cell that matures. Interestingly, however, in very rare cases through some poorly understood process, one of the polar bodies may also form into a viable ovum and be fertilized. There are known cases of human twins who developed in this way.

Principal Stages of Meiosis

Animal meiosis is usually studied in biology and genetics laboratories using microscope slides prepared from the testes of grasshoppers or the roundworm *Ascaris* and with microscope slides from the ovaries of *Ascaris*. The description of the principal stages of meiosis that follow apply to materials from any of these sources.

First Meiotic Division (Reduction Division)

Prophase I Many of the events in the first meiotic prophase are similar to prophase in mitosis. The chromosomes become visible as the very thin chromatin strands coil and condense. The nucleoli and the nuclear membrane disappear, and the mitotic spindle appears.

The key difference between the first meiotic prophase and prophase in mitosis is that, in meiosis, each pair of homologous chromosomes comes closely together in an intimate pairing process called **synapsis.** By the end of the first meiotic prophase, each homologous chromosome pair is seen as two double-stranded chromosomes closely held together. This structure, formed as a result of synapsis and consisting of two chromosomes with a total of four chromatin strands, is called a **tetrad.** Tetrads are found only during prophase I of meiosis. Note the two chromosomes at the end of prophase I as illustrated in figure 3.9. Each chromosome consists of two **distinct strands,** or chromatids, attached to a **single centromere.**

Metaphase I At the onset of metaphase I, the synapsed chromosome pairs move together as a unit to the equatorial plane. Each chromosome pair is attached to a single spindle fiber by its **two adjacent centromeres.**

Anaphase I The important difference between this stage and the corresponding stage in ordinary mitosis is that there is **no division of centromeres** in anaphase I of meiosis. The centromeres of the homologous chromosomes simply move apart, and the two double-stranded chromosomes of each pair migrate toward opposite poles. Thus, **half** of the chromosomes move to one pole, and **half** of the chromosomes move to the opposite pole. Note the two chromosomes migrating toward each pole in anaphase I, as shown in figure 3.9.

Telophase I The chromosomes reach the poles of the mitotic spindle in each of the daughter cells, the spindle disappears, and new nuclear membranes appear around the reforming nuclei in the daughter cells. The chromosomes begin to elongate, gradually fade from view, and nucleoli reappear within the nuclei.

Interkinesis

Interkinesis Between the two successive divisions in meiosis is a brief stage called **interkinesis.** This stage is generally similar to an interphase between mitotic divisions, but there is **no replication of genetic material** (no DNA synthesis) in interkinesis.

Second Meiotic Division (Nonreduction Division)

Prophase II The second meiotic division is essentially like an ordinary mitotic division. There is no **synapsis** in prophase II; the double-stranded chromosomes reappear and move independently toward the equatorial plane.

Metaphase II Each double-stranded chromosome attaches **separately** to a spindle fiber.

Anaphase II The **centromeres divide** at the end of metaphase II, and during anaphase II the newly separated, single-stranded chromosomes move toward opposite poles.

Telophase II The new chromosomes in the nuclei that appear in the daughter cells are **single-stranded** and contain only **half** the number of chromosomes as in prophase I.

Meiosis as a Source of Genetic Variability

Meiosis is an important source of genetic variability for each successive generation because it increases the chances for survival under changing environmental conditions. Favorable variants among the offspring have the best chance of living long enough to reproduce and contribute to the survival of the species.

Meiosis contributes to genetic variability through two main processes: random assortment and crossing over. Both processes occur during the first division of meiosis. Crossing over occurs during prophase I, when each chromosome pairs with its homologous partner, an event known as synapsis. At this time, each chromosome has already replicated and consists of two **chromatid** strands attached at their centromere. When two homologous chromosomes pair off, they form a group of four chromatids known as a tetrad. During this close pairing in prophase I the chromatids within a tetrad often become entangled. This entanglement is known as a **chiasma** (pl. chiasmata). When the chromatids separate at the end of metaphase I, they often will have exchanged segments because of these entanglements. This exchange of segments between chromatids is called **crossing over** (figure 3.10) and it results in daughter cells with different combinations of genes (alleles).

Random assortment occurs when the already doubled chromosome pairs line up on the metaphase plate of meiosis I. Half of these chromosomes carry genetic information from the mother and the other half come from the father. They are assorted randomly at the metaphase plate so that some of the maternal chromosomes go to one pole of the mitotic spindle and thus into one daughter cell while others are pulled to the opposite pole and into the other daughter cell. The paternal chromosomes are randomly assorted in the same way with some going into one daughter cell and some going into a different daughter cell when the doubled chromosomes split in anaphase. Thus, each daughter receives some chromosomes from the mother and some chromosomes from the father producing a new set of genetic instructions. During meiosis II, the chromosomes separate again, so that each of the four daughter cells produced has a different set of genetic instructions.

Ways to Study Meiosis

The actual observation of meiosis is often difficult for an inexperienced observer because of the small size of the chromosomes in most kinds of cells. Also, it is difficult to make good microscopic preparations showing cells in clearly recognizable stages of meiosis. The chromosomes in most cells are small, they are often numerous, and it may require study of several slides to find good examples of meiotic stages. For these reasons, meiosis is often studied in introductory biology and zoology courses using a series of carefully selected demonstration microscope slides.

Microscope Slides

- Study the demonstration materials on meiosis provided in the laboratory and **draw** selected stages of meiosis in the space provided in figure 3.11 as directed by your laboratory instructor.

Chromosome Simulation Kits

- Chromosome simulation kits are another good way to study meiosis. Several types of kits are available, but one common type consists of a string of beads with a magnet in the middle that represents the centromere of a chromosome. Different colors of beads can be used to help students distinguish different chromatids, to simulate the movements of chromatids and chromosomes during the process of meiosis, and to see the effects of crossing over. Ask

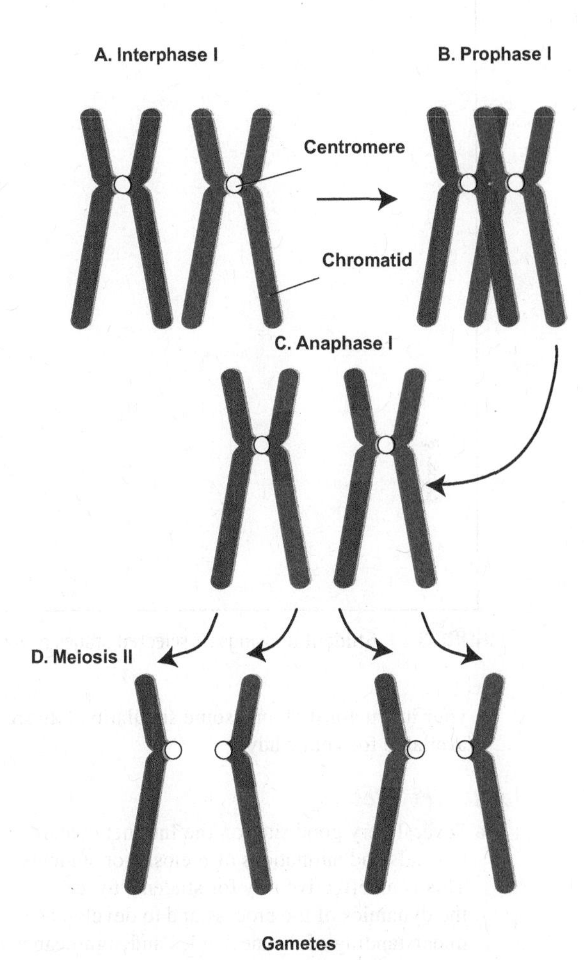

FIGURE 3.10 Crossing over between chromatids during meiosis may result in genetic diversity of offspring. Major steps include the following:

A. **Interphase I.** Chromosomes replicate to form two genetically identical sister chromatids. B. **Prophase I.** Homologous chromosomes come together as pairs, forming a tetrad consisting of four chromatids, each chromosome with two chromatids. Portions of chromatids of homologous chromosomes may become entangled or crossed at some points. These crossings are called **chiasmata** (singular: chiasma). C. **Anaphase I.** Portions of the crossed chromatids may be exchanged as the chromatids separate to become single-stranded chromosomes. D. **Meiosis II.** Separation of sister chromatids in Meiosis II, the second meiotic division, then forms new combinations of alleles in some of the gametes formed. This results in genetic diversity among the offspring.

FIGURE 3.11 Student drawings of selected stages of meiosis.

your instructor if chromosome simulation kits are available for your study.

Internet Sites

- Several very good sites on the Internet contain tutorials and animations of meiosis for students. This is an effective way for students to see the dynamics of the process and to develop an understanding of the mechanics and significance of the process. A few good Internet sites for meiosis are listed here.
  http://www.biology.arizona.edu/cell_bio/tutorials/meiosis/page3.html
  http://www.pbs.org/wgbh/nova/baby/divi_flash.html
  http://www.biology.iupui.edu/biocourses/n100/2k4ch9meiosisnotes.html
  http://highered.mcgraw-hill.com/sites/0072495855/student_view0/chapter28/animation__how_meiosis_works.html

Films and Videos

- One of the best ways to gain an understanding of the process of meiosis is to study a film or video showing time-lapse sequences of meiosis. Several good films and videos are available that show the dynamic nature of meiosis, and they can be studied both in the laboratory and in libraries or learning centers where available. Ask your instructor what films and videos may be available.

Comparsion of Mitosis and Meiosis

An excellent way to review and to check your knowledge and understanding of mitosis and meiosis is to consider the similarities and differences between these two processes. A common question appearing on biology examinations asks students to compare and contrast mitosis and meiosis. When asked to compare and contrast, you should always discuss **both** the similarities and differences between the two processes or structures cited in the question. Table 3.1 provides a summary of similarities and differences between mitosis and meiosis.

Table 3.1 Comparison of Mitosis and Meiosis

Description	Mitosis	Meiosis
Type of cells involved	somatic	reproductive
Number of nuclear divisions	1	2
Number of cells produced	2	4
Chromosome number		
Before	diploid (2n)	diploid (2n)
After	diploid (2n)	haploid (n)
Synapsis occurs (close pairing of homologs)	no	yes
Crossing over occurs	no	yes
Genetically identical cells produced	yes	no

Key Terms

Aster includes all of the aster rays surrounding one pole of the mitotic apparatus in an animal cell. May serve to influence the location of the cleavage plane during cytokinesis. Absent in plant cells.

Aster ray one of the fibrils, or rays, making up an aster.

Centriole self-replicating tubular organelles usually found in pairs adjacent to the nucleus of an interphase cell surrounded by an amorphous centrosome. Also found centered in the asters of the mitotic apparatus of most animal cells.

Centromere a short region of a chromosome that binds the two chromatids together prior to separation in meiosis and attaches the spindle fibers to the chromosome by the formation of a kinetochore during mitosis.

Centrosome the amorphous area surrounding a centriole in animal cells. Serves as the organizing center for the mitotic spindle.

Chiasma (plural: chiasmata) the crossings of chromatids observed in microscopic preparations during synapsis in prophase I of meiosis.

Chromatid one strand of a replicated chromosome.

Chromatin genetic material in the interphase nucleus; represents the chromosomes in a long, thin, threadlike form.

Chromosome filamentous structure that carries the genetic material of the cell (DNA). Chromosomes can be very long and uncoiled in interphase, shorter with double strands (chromatids) in prophase of mitosis, and still shorter with a single strand in anaphase.

Cleavage furrow indentation of the cell membrane around the equator of an animal cell at the beginning of cytokinesis.

Crossing over The exchange of segments of chromatids during the separation of homologous chromosomes following metaphase I of meiosis. Such exchanges result from the entanglements of chromatids in prophase I. Crossing over is an important source of genetic diversity in sexually produced offspring.

Cytokinesis division of the cytoplasm of a cell.

Diploid (or **2n**) cells containing both members of each homologous pair of chromosomes.

Equatorial plane by analogy with the earth, a plane that passes through the middle of the cell, equidistant from the poles and perpendicular to the line connecting the poles.

Gametes the specialized haploid reproductive cells produced by sexually reproducing animals; the ova and spermatozoa (eggs and sperm).

Gametogenesis the production of male and female gametes (eggs and sperm) in sexually reproducing animals.

Germ cells unspecialized cells found in the gonads of sexually reproducing animals that develop into the gametes (eggs and sperm).

Gonads the reproductive organs of sexually reproducing animals, the testes and ovaries, that produce gametes.

Haploid (or **n**) cells containing only one member of each homologous pair of chromosomes.

Homologous chromosomes chromosomes with the same size and shape, and carrying genetic material for the same characteristics. One member of each homologous pair comes from each parent.

Interkinesis the period intervening between the first and second meiotic divisions. No chromosomes are replicated and no DNA is synthesized during interkinesis.

Interphase the stage between successive nuclear (mitotic) divisions, consisting of G_1, S, and G_2 phases. It is the stage during which the cells are metabolically active.

Karyotype the unique chromosome complement of an individual organism characterized by the number, size, and configuration of the chromosomes.

Meiosis a special type of nuclear division in which the chromosome number is reduced from 2n to n by separating the members of the homologous pairs of chromosomes.

Mitosis nuclear division resulting in two new nuclei with the same genetic complement as the original nucleus.

Mitotic apparatus a special structure formed during mitosis consisting of the spindle fibers, asters, and centrioles.

Mitotic spindle all of the spindle fibers collectively.

Nonreduction division the second meiotic division. This division follows interkinesis and resembles a mitotic division. The chromatids separate, but there is no reduction in chromosome number.

Nuclear membrane (nuclear envelope) the double membrane surrounding the nucleus in an interphase cell.

Nucleus a membrane-bound organelle containing the genetic material of a cell and which controls cell metabolism.

Oogenesis the formation of ova or eggs. Includes meiosis and the further development into functional ova.

Ovum (plural: ova) the haploid female gamete or egg.

Polar bodies nonfunctional nuclei produced in the process of oogenesis that migrate to the periphery of the functional ovum. Three polar bodies and one functional ovum are usually formed during oogenesis.

Poles opposite ends of a cell where spindle fibers converge during mitosis and meiosis.

Reduction division the first meiotic division, during which the number of chromosomes in a cell is reduced by half.

Spermatogenesis the formation of the spermatozoa in sexually reproducing animals.

Spermatozoon (plural: spermatozoa) the haploid male gamete produced by spermatogenesis. Often shortened to "sperm."

Spindle fiber one of the microtubular filaments extending between the poles of the cell in mitosis and meiosis; made of contractile protein.

Synapsis the pairing of homologous chromosomes in prophase I of meiosis; the centromeres adhere to each other at this time.

Tetrad a group of four chromatids of a pair of homologous chromosomes formed by synapsis during prophase I of meiosis.

Internet Resources

There are many valuable Internet sites with information about zoology. Several sites containing pertinent zoological information for this chapter can be found on the McGraw-Hill Zoology web site at **http://www.mhhe.com/zoology**. Just click on this text's title.

Questions for Critical Thinking

1. Compare the four phases of the cell cycle. Explain how each phase prepares the cell for the next phase.
2. What would happen if you treated a cell actively undergoing mitosis with a chemical that inhibited cytokinesis but did not affect mitosis? What would be produced?
3. Why is it important to use a random sample in an experiment if we wish to compare two populations? Describe two ways you could select a random sample from a population of dogs.
4. Where in your body would you most likely find cells in active mitosis? Where would you find cells undergoing meiosis?
5. Compare and contrast the first and second divisions of meiosis. Why is it important that animals have two meiotic divisions rather than just one division as in mitosis?
6. What is the adaptive value of the process of crossing over as a source of variation in a species?
7. The process of spermatogenesis usually produces four viable sperm from each parent cell, while the process of oogenesis usually produces only one viable egg. Considering that the meiotic processes in both types of gamete formation are essentially the same, do you think that spermatogenesis is a more **important source of variation** than oogenesis? Why or why not?

Suggested Readings

Harris, H. 1999. *The Birth of the Cell*. New Haven: Yale University Press. 212 pp

Karp, G. 2009. *Cell and Molecular Biology: Concepts and Experiments*. 6th edition, 832 pp

Morgan, D.O. 2007. *The Cell Cycle: Principles of Control*. Sunderland, MA: Sinauer Associates. 297 pp

Snustad, D.P., and M.J. Simmons. 2005. Principles of Genetics, 4th ed. New York: Wiley, Chap. 2.

Notes and Sketches

Chapter 4

Development

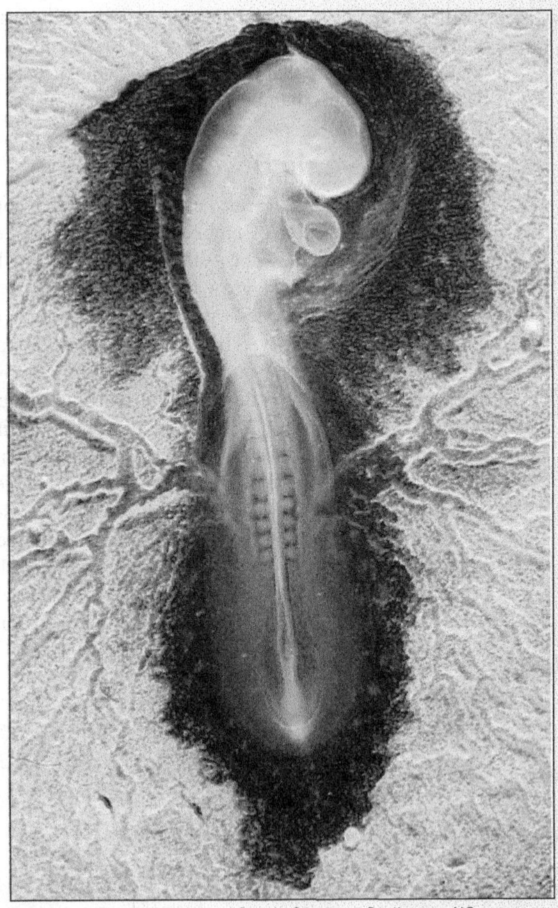

Courtesy of Carolina Biological Supply Company, Burlington, NC.

Objectives

After completing the laboratory work in this chapter, you should be able to perform the following tasks:

1. List and explain the component processes of development.
2. List and describe the main types of cleavage observed in animal embryos.
3. Distinguish between determinate and indeterminate development.
4. Describe the structure of a spermatozoan of a frog or other representative animal and identify its principal structures visible in a light microscope.
5. List and describe the major events in a sea star or sea urchin development from fertilization to the gastrula stage and identify representative stages in microscopic preparations.
6. Identify the principal structures in the blastula and gastrula stages of a sea star or sea urchin.
7. Discuss the organization of a frog egg and tell how it differs from that of a sea star egg.
8. Describe the major events in frog development from fertilization to the tadpole stage and identify representative stages from living or preserved specimens.
9. Discuss the organization of a chick egg and its adaptations for development on land.
10. Identify the four extraembryonic membranes surrounding a chick embryo and explain the function of each.
11. Identify the principal structures seen in whole mounts of 24- and 48-hour chick embryos.

Introduction

Animal development usually begins with the fertilization of an egg by a sperm. The nuclei of the egg and sperm fuse, and the male and female parents' genes share in determining the characteristics of the offspring. Compared to other biological processes, embryonic development is relatively slow. New cells, tissues, and organs make their appearance in the embryo over a period of hours, days, or weeks.

Embryonic development can be divided into five major phases: (1) **gametogenesis,** the formation of the haploid male and female gametes (sperm and eggs); (2) **fertilization,** activation of the egg and fusion of the sperm and egg nuclei to form the diploid zygote; (3) **cleavage,** the subdivision of the zygote into many cells by mitosis; (4) **gastrulation,** the formation of germ layers; and (5) **organogenesis,** the initiation and differentiation of specific organs.

In this exercise, we shall study examples of development from three different animals to illustrate different aspects of embryonic development and some variations in the development of different kinds of animals. The **sea star** illustrates the earliest stages of development, the **chick** serves to illustrate the later stages of development and adaptation for terrestrial life, and the **frog** is well suited for the study of development from fertilization to hatching.

Component Processes of Development

Embryonic development consists of a complex series of processes by which a new organism arises from an egg. Following the fertilization of an egg with a sperm, the fertilized egg or **zygote** grows and increases in the complexity of its structure, function, and behavior. Different kinds of animals exhibit many differences in the details of their development, but there are some important basic similarities in the development of all animals.

We can identify four major component processes of development in organisms: growth, determination, differentiation, and morphogenesis.

Growth

Growth is the increase in mass of the organism through the addition of new cells and/or an increase in size of existing cells.

Determination

An unfertilized egg has the potential to form a complete embryo with many kinds of cells, tissues, and organs. As development proceeds, and the egg becomes fertilized and divides to form new cells, individual cells and tissues become progressively restricted in the structures that they are able to form. This progressive limitation of the prospective fate of a cell or tissue is called **determination.** Determination precedes the process of differentiation.

Differentiation

The progressive increase in the complexity of organization and specialization of individual cells and tissues is called **differentiation.** A fertilized egg of a frog, for example, forms many new cells and tissues during its development. This progressive increase in the number of biochemically and morphologically specialized cells and tissues is differentiation.

Morphogenesis

Another important process of development is **morphogenesis.** A living organism is not merely a bag of assorted parts heaped together in a random fashion. The parts of every living organism are arranged in a specific pattern and bear definite relationships to one another. Morphogenesis includes those **movements** of cells and tissues through which the characteristic structures (both external and internal) of an organism are formed. For example, the movements of cells and masses of cells in an embryo to form a wing or a limb are important processes of morphogenesis. Similarly, the heart of a chick and of a human embryo first forms as a simple tubular structure that twists into the shape of an S before becoming the four-chambered organ we know as the heart.

Postembryonic Development

Most animals begin life as fertilized eggs and continue through a progression of embryonic stages. Their development, however, does not usually end at birth or hatching. Most young animals grow rapidly, and in many cases, they will pass through one or more larval or juvenile stages before reaching maturity. These changes involve many of the same developmental processes seen during embryonic development: growth, differentiation, and morphogenesis. Many insects and other animals have a complex life cycle with a series of developmental stages that differ from one another in form, and each stage may be adapted for specialized modes of life. Some animals undergo several conspicuous and abrupt changes of form. Mosquitoes, for example, live under water as larvae and feed by filtering organic matter from their environment. Later, they will develop into the familiar flying adult stage that feeds on your

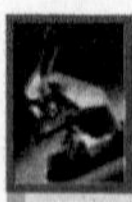

Materials List

Living specimens
- Frog embryos at various stages
- Frog sperm

Prepared microscope slides
- Sea star embryos
- Chick embryo, 24-hour, whole mount
- Chick embryo, 48-hour, whole mount
- Frog ovary with ova, cross section (demonstration)
- Frog testis, cross section (demonstration)
- Selected slides of early frog development (demonstration)

Chemicals
- Amphibian Ringer's solution

blood. Transitions between subsequent life stages are largely regulated by changes in gene expression. This process is called metamorphosis (change of form) and is discussed more extensively in Chapter 13, Arthropoda.

Gametes

Gametes are the mature germ cells, **eggs** and **spermatozoa.** Both living sperm and stained microscope slides should be available for your study of male gametes. Observe the rapid movements of the living sperm in a wet mount.

- On a stained microscope slide, identify the anterior **head,** the narrower **midpiece,** and the long posterior **tail.** The sperm of different animal species vary considerably in size and shape, particularly in the shape of the head. Draw a sperm cell in figure 4.1 and label each of these parts.

Mature spermatozoa have little cytoplasm and are highly specialized for transport and penetration of the egg. They contain numerous mitochondria that provide energy for their locomotion and have a specialized cap that facilitates penetration of the egg membrane. Sperm cells have two main functions in the process of fertilization: (1) they activate development of the egg as they penetrate the egg membrane, and (2) they contribute their nuclear material to the fertilized egg. Eggs, or ova, are much larger than sperm and contain varying amounts of stored food materials for the nourishment of the developing embryo. Since the eggs of the sea star, frog, and chick differ substantially in this regard, they will be described separately as we study the development of each animal.

FIGURE 4.1 Student drawing of sperm.

Embryonic Cleavage

Shortly after fertilization, the zygote undergoes a series of rapid mitotic cell divisions. This process of dividing the zygote into many cells is called **embryonic cleavage.** The pattern of cleavage differs among various groups of animals, and these patterns have been conserved during the course of evolution. For this reason, scientists have long been interested in cleavage patterns as possible evidence of evolutionary relationships among animals. One of the factors that influences the pattern of cleavage is the amount and distribution of nutrients or **yolk** in the egg.

Influence of Yolk

Eggs can be classified into four main types based on their yolk contents: (1) **isolecithal eggs,** as found in sea stars and humans, which have relatively little yolk and in which the yolk is uniformly distributed throughout the egg; (2) **mesolecithal eggs,** as found in the frogs, toads, and salamanders, which have a moderate amount of yolk, and have a concentration of yolk in the vegetal (lower) hemisphere; (3) **telolecithal eggs,** as found in birds and reptiles, which have a large amount of yolk, and the cleaving portion of the embryo is restricted to a small disc at one end of the egg; and (4) **centrolecithal eggs,** as found in insects, which have much yolk and in which the actively developing portion of the embryo forms a thin layer of cells around the outside of the large central yolk mass.

Among the animals chosen for this exercise, therefore, we have examples of three egg types:

Sea star egg—isolecithal
Frog egg—mesolecithal (although some textbooks classify frog eggs as "moderately telolecithal")
Chick egg—telolecithal

Patterns of Cleavage

Embryonic cleavage is influenced both by the amount of yolk in the egg and by the orientation of the mitotic spindle in dividing embryonic cells. Rotation of the mitotic spindle between successive cell divisions leads to spiral and radial cleavage in certain types of invertebrate animals (figure 4.2).

Eggs with little or moderate amounts of yolk, such as a sea star or a frog, divide completely. This is **holoblastic cleavage** because the whole embryo is divided into cells. Eggs with large amounts of yolk, like insects, birds, and many mammals exhibit **incomplete cleavage.** They do not divide completely into cells because of the thick, viscous nature of the yolk mass. In such eggs, cleavage is limited to some portion of the surface of the embryo with a large underlying mass of yolk that provides nutrients and energy for the developing embryo.

Spiral Cleavage

Animals with holoblastic cleavage also demonstrate differences in the directions of subsequent cleavage planes. Many flatworms, nematodes, most molluscs, and many annelids

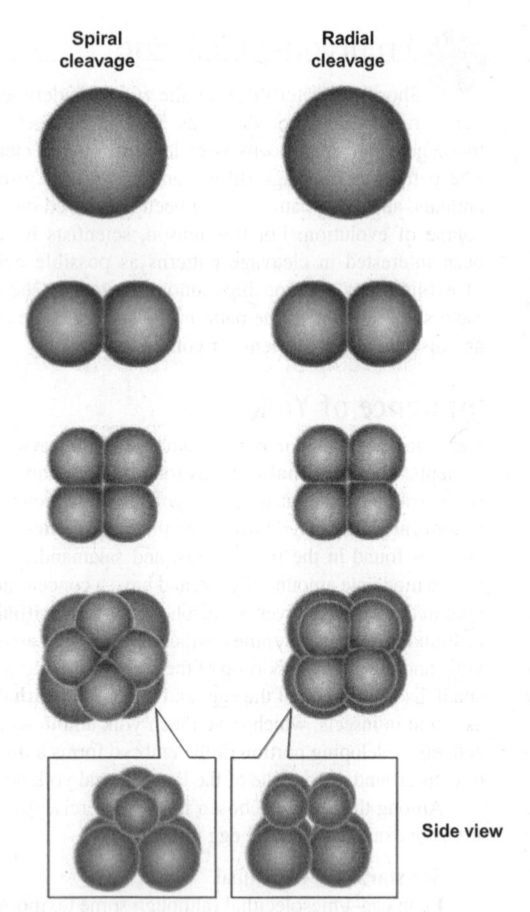

FIGURE 4.2 Spiral and radial cleavage.

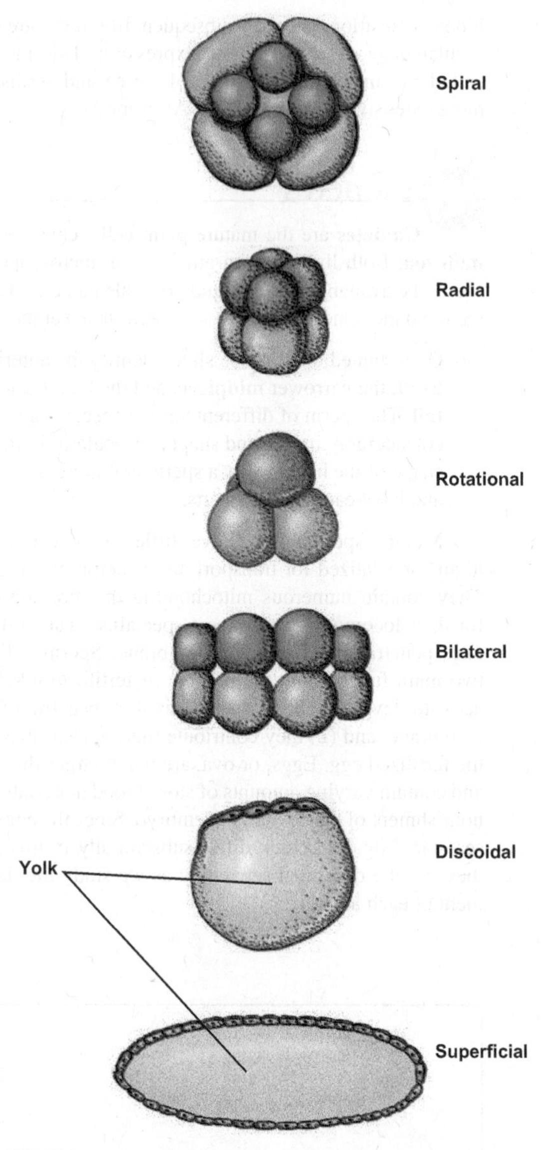

FIGURE 4.3 Cleavage patterns in different types of eggs.

exhibit **spiral cleavage.** The first two cleavage planes are vertical and produce blastomeres of equal size. The third cleavage plane is horizontal and produces four blastomeres of unequal size, four smaller micromeres nearest the **animal pole** on the upper surface, and four larger macromeres nearest the **vegetal pole** on the lower surface of the egg. Prior to the third cleavage the mitotic spindle of the blastomeres rotates 45 degrees, causing the micromeres to lie over the furrows between the macromeres. Following the third cleavage and continuing after several subsequent cleavage divisions, the mitotic spindles rotate 90 degrees in the opposite direction. This produces several layers of blastomeres with each layer lying over the furrows between the blastomeres just below.

Radial Cleavage

Sea stars, sea urchins, and frogs (also sponges, cnidarians, and cephalochordates) exhibit **radial cleavage** (figure 4.2). In this type of cleavage, the first two cleavage planes are horizontal and produce blastomeres of equal size. The third plane is horizontal and produces four smaller micromeres and four larger macromeres. The four micromeres contain less yolk and surround the animal pole, which remains uppermost; the four macromeres contain more yolk and surround the vegetal pole opposite the animal pole. The blastomeres remain radially arranged around the animal-vegetal axis, and there is no rotation of the blastomeres as in spiral cleavage.

Other Types of Cleavage

Mammals (including humans) also have holoblastic cleavage, but their cleavage pattern is **rotational** (figure 4.3). After the second cleavage, one pair of blastomeres comes to lie at right angles to the other. Squid, octopi, and other cephalopod molluscs have a **bilateral cleavage** (figure 4.3) pattern in which the first cleavage plane divides the embryo into right and left halves and establishes a plane of bilateral symmetry.

In other animals with abundant yolk in the eggs, only part of the embryo is divided, and much of the yolky part of the embryo remains undivided. One example is the telolecithal eggs of birds, which have **discoidal cleavage** (figure 4.3) with cleavage restricted to a small disc of cells at one end of the embryo. Another example is found among the arthropods in which yolk is concentrated in the center of the egg and cleavage is restricted to the outer surface. This is known as **superficial cleavage** (figure 4.3).

Sea Star Embryology

Sea star embryos are often used for introductory studies of development because they illustrate clearly the basic pattern of early development of multicellular animals (figure 4.4). Sea urchin embryos are also frequently used, and their development is very similar to that of sea stars in these early stages. Sea urchins also have the added advantage that living eggs and embryos are relatively easy to obtain for laboratory

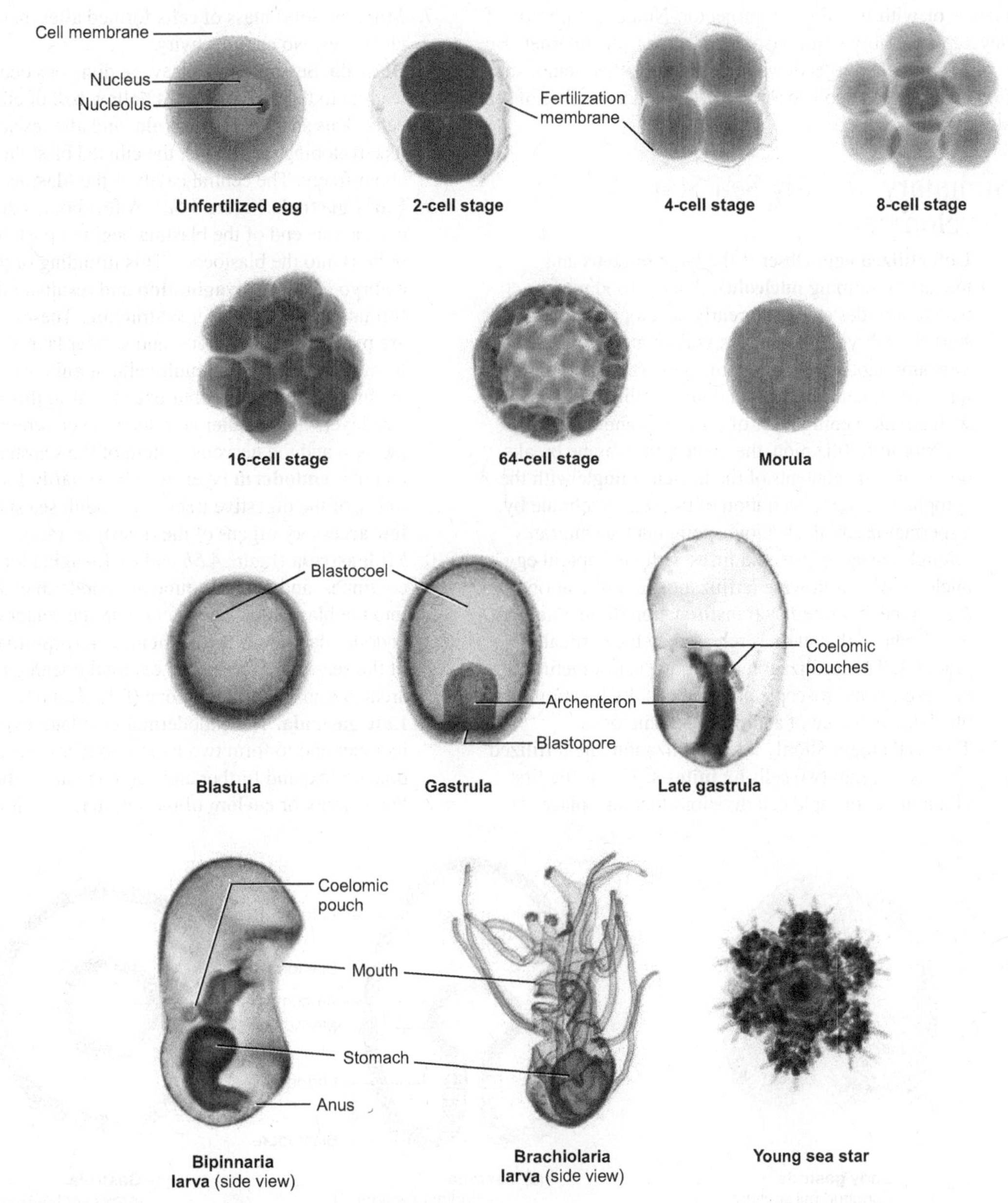

FIGURE 4.4 Sea star development.
Courtesy of Carolina Biological Company, Burlington, NC.

study. With living sea urchin embryos, it is possible to observe most of the early developmental events described later with an ordinary compound microscope.

- Obtain a whole-mount microscope slide with stained sea star embryos at various stages of development. Examine the slide first under the low power of your microscope and select good representatives of the main stages described in the following section.

Color transparencies are excellent aids for the study of early embryology. They can be viewed easily in a small slide viewer or with the aid of a projector. Numerous photos of developing embryos are also available on the Internet. Be sure to study the projection slides and any other demonstration materials available to assist in your understanding of sea star development.

Summary of Early Sea Star Development

1. **Unfertilized egg.** Observe the large **nucleus** and the darkly staining **nucleolus.** A large food reserve, which provides energy for early development of the starfish embryo, is present as yolk in the cytoplasm. Surrounding the egg, or ovum, are two membranes, an inner cell membrane and an outer vitelline membrane with an intervening layer of cortical granules.

   Prior to fertilization, the nuclear membrane breaks down, and the contents of the nucleus mingle with the cytoplasm. After penetration of the cell membrane by a spermatozoan, the haploid sperm nucleus migrates through the cytoplasm and fuses with the haploid egg nucleus. Also following fertilization, a fertilization membrane is formed by transformation of the vitelline membrane with materials released by the cortical granules. The fertilization membrane can sometimes be observed in microscopic preparations. Its function is to block the entrance of additional spermatozoa.
2. **Two-cell stage.** Shortly after fertilization, the fertilized egg divides into two cells by **mitosis.** This is the first of a number of rapid cell divisions that take place during the next several hours. This series of cell divisions is called **embryonic cleavage.**
3. **Four-cell stage.** Observe the fertilization membrane enclosing the four **blastomeres** (embryonic cells).
4. **Eight-cell stage.** ***What is the orientation of the cleavage plane that produced this stage relative to the last cleavage plane?***
5. **Sixteen-cell stage.**
6. **Thirty-two-cell stage.** Note the rapidly diminishing size of the blastomeres. ***What has happened to the overall size of the embryo?***
7. **Morula.** Solid mass of cells formed after many cleavages. No central cavity.
8. **Blastula.** Several more cleavage divisions occur, leading to the formation of a hollow ball of ciliated cells. This stage is the **blastula,** and after escaping from its enveloping membrane, the ciliated blastula swims about freely. The central cavity is the **blastocoel.**
9. **Early gastrula** (figure 4.5*a*). A few hours later, the cells at one end of the blastula begin to push (or be pulled) into the blastocoel. This infolding of the embryo is called **invagination** and results in the formation of a two-layered structure. These two layers are **primary germ layers,** and similar layers are formed by virtually all multicellular animals. (Sponges are one of the few peculiar exceptions to this rule.) The layers are an outer **ectoderm** layer, which forms the skin and the nervous system of the sea star, and an inner **endoderm** layer, which primarily forms the lining of the digestive tract of the adult sea star (and a few accessory organs of the digestive tract).
10. **Midgastrula** (figure 4.5*b* and *c*). Invagination continues, and a hollow tube of endoderm extends into the blastocoel. The opening in the center of the endodermal tube is the **archenteron** (primitive gut) of the sea star embryo. The external opening of the archenteron is the **blastopore** (future anus).
11. **Late gastrula.** The endodermal tube later expands at its inner end to form two lateral pouches. These lateral pouches expand further and become part of the internal body cavity or **coelom** of the sea star. The tissue

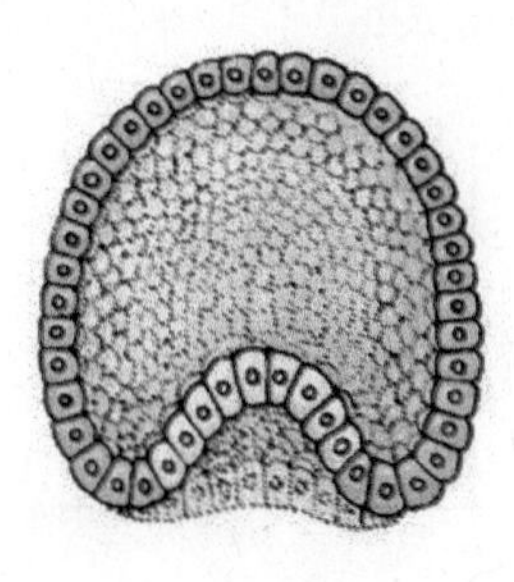

(a) **Early gastrula** (longitudinal section)

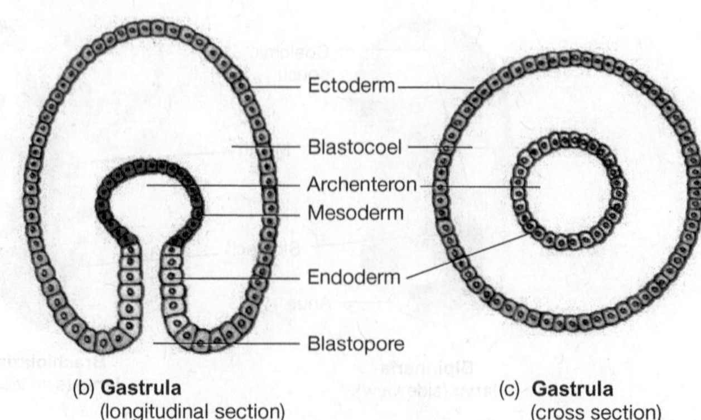

(b) **Gastrula** (longitudinal section)

(c) **Gastrula** (cross section)

FIGURE 4.5 Sea star gastrula.

lining these lateral pouches differentiates into the third primary germ layer, the **mesoderm.**

12. **Larval stages.** Following gastrulation, the digestive tract is completed with the formation of an anus and the embryo is transformed into a **bilateral bipinnaria larva.** Later, the bipinnaria develops several long armlike structures. This larval stage is called the **brachiolaria larva** which settles onto a suitable substrate and metamorphoses into a **radially symmetrical young sea star.**

Frog Development

Ripe, unfertilized eggs of the grass frog, *Rana pipiens,* are relatively large (averaging about 1.75 mm in diameter) and have a moderate amount of yolk (mesolecithal type of egg). Remember that the frog is a semiaquatic animal; its eggs must be deposited in water, and development of the embryos and tadpole larvae can take place only in water.

- Obtain a few unfertilized frog eggs in water and study them under your stereomicroscope (figure 4.6). Observe that the eggs have some dark pigmentation. ***How is the pigment distributed on the surface of the eggs? Is the distribution of pigment uniform or nonuniform?***

The center of the dark portion of the surface of the egg is called the **animal pole.** The center of the lighter surface (opposite the animal pole) is called the **vegetal pole.** The line connecting these two poles is the **animal-vegetal axis.** Shortly after fertilization the egg rotates so that the animal pole is uppermost. ***How does this differ from the orientation of the animal-vegetal axis of unfertilized eggs?***

Closely applied to the surface of the unfertilized egg is a transparent **vitelline membrane,** which lifts from the surface of the egg following fertilization and is transformed into the **fertilization membrane.** Surrounding the vitelline membrane are three **jelly coats** made up largely of albumen, a protein secreted by the oviduct during the passage of the egg from the ovary (figure 4.6). The jelly coats serve to protect the egg from injury and infection.

Fertilization is external in the frog. During mating, the male frog grasps the female with his forelimbs, and as the female discharges her ripe eggs into the water, the male releases sperm over them as they are released. This mating posture is called **amplexus** (figure 4.7).

Shortly after fertilization, the vitelline membrane rises to form the fertilization membrane, and later a **gray crescent** area appears on one side of the egg along the margin of the heavily pigmented zone. The gray crescent is a more lightly pigmented zone of the egg cortex and appears on the side opposite the point of sperm entrance. The center of the gray crescent area marks the future **posterior end** of the embryo; thus, the future **anterior-posterior axis** of the frog embryo is determined at the time of sperm entry into the egg.

Development is fairly rapid following fertilization (figures 4.8 and 4.9), and within a few hours at normal temperatures, the fertilized egg becomes divided into many cells, a hollow blastula is formed, and gastrulation is completed. Cleavage is **holoblastic** (all of the egg cytoplasm becomes divided) and **unequal** (the cells nearest the animal pole are smallest in size and those nearest the vegetal pole are largest in size). ***How is this related to the distribution of yolk in the frog egg?***

A few hours after gastrulation, the first visible elements of the nervous system appear, first with the **neural plate** followed by the elevation of the **neural folds** on each side of the neural groove to form the **neural tube** within which the dorsal nerve cord begins to differentiate. These steps begin the process of organogenesis.

Following the neural tube stage, the posterior portion of the embryo elongates to begin formation of the embryonic tail. This is the **tail bud stage,** which occurs at about 84 hours postfertilization at 18°C (figure 4.10).

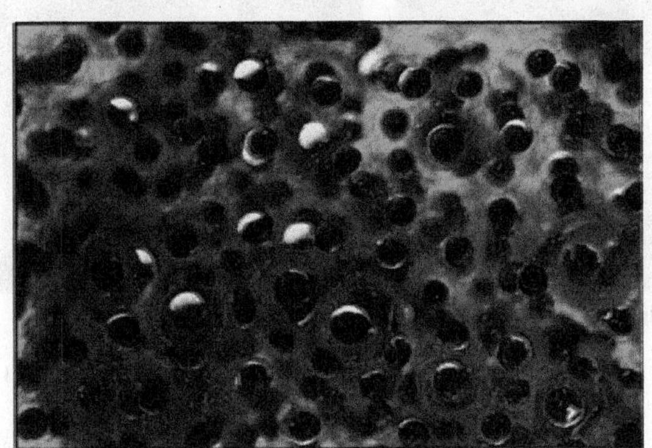

FIGURE 4.6 Unfertilized frog eggs surrounded by jelly coats. Note the random orientation of the pigmented area of the eggs. After fertilization, the dark area becomes the animal pole and rotates to the top.
Courtesy of Carolina Biological Supply Company, Burlington, NC.

FIGURE 4.7 Male and female frogs in mating position.
Courtesy of Carolina Biological Supply Company, Burlington, NC.

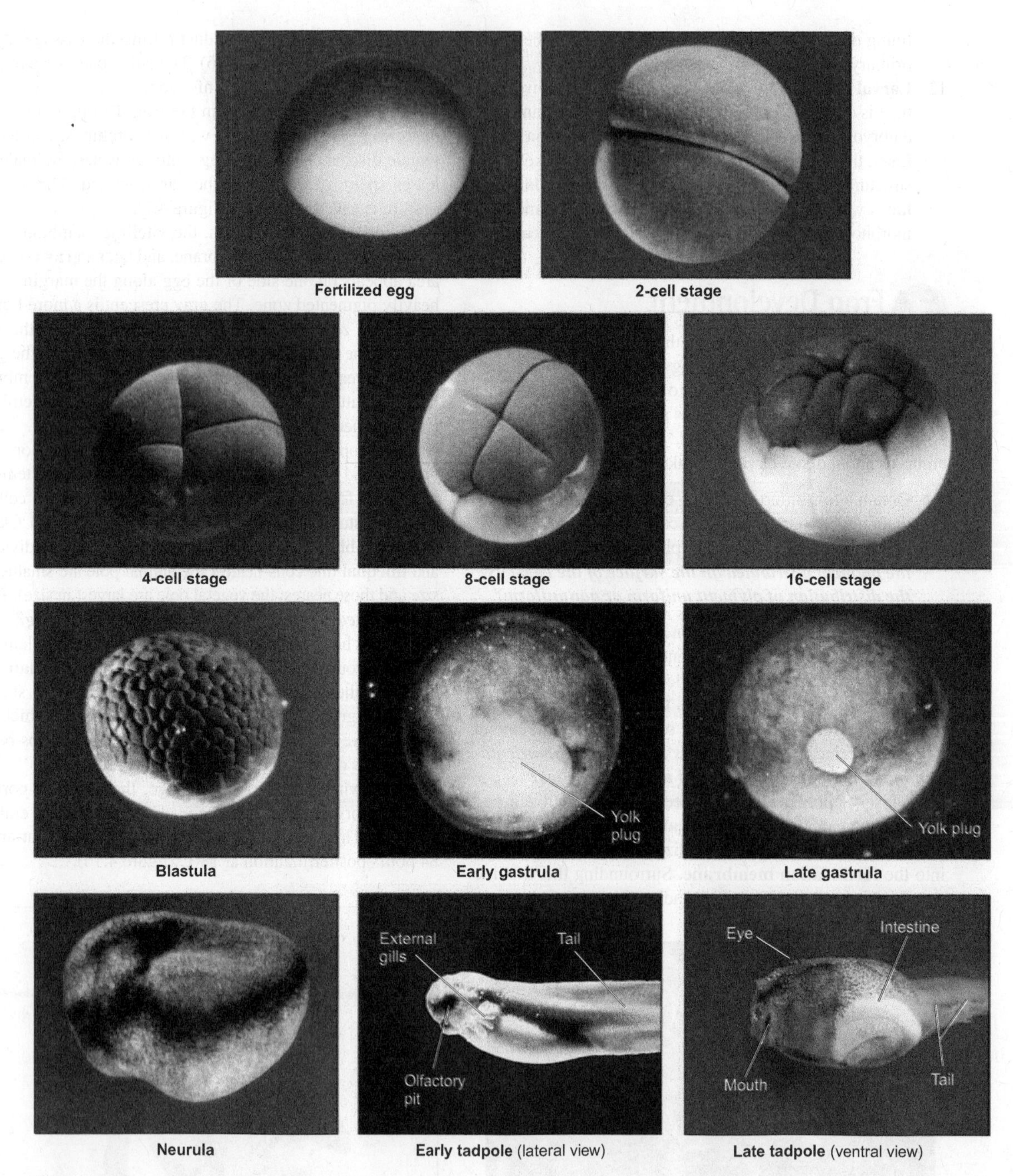

FIGURE 4.8 Development of the grass frog *Rana pipiens*.
Courtesy of Carolina Biological Supply Company, Burlington, NC.

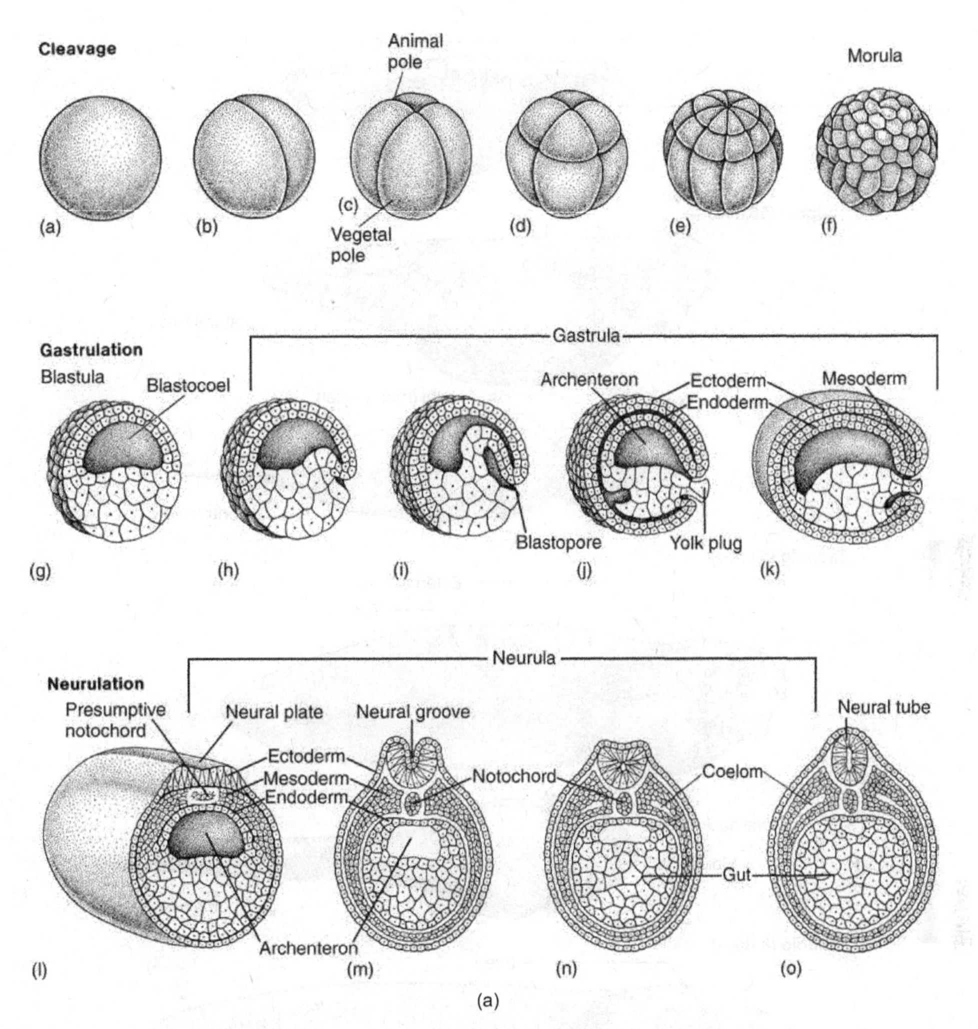

FIGURE 4.9 Frog development.

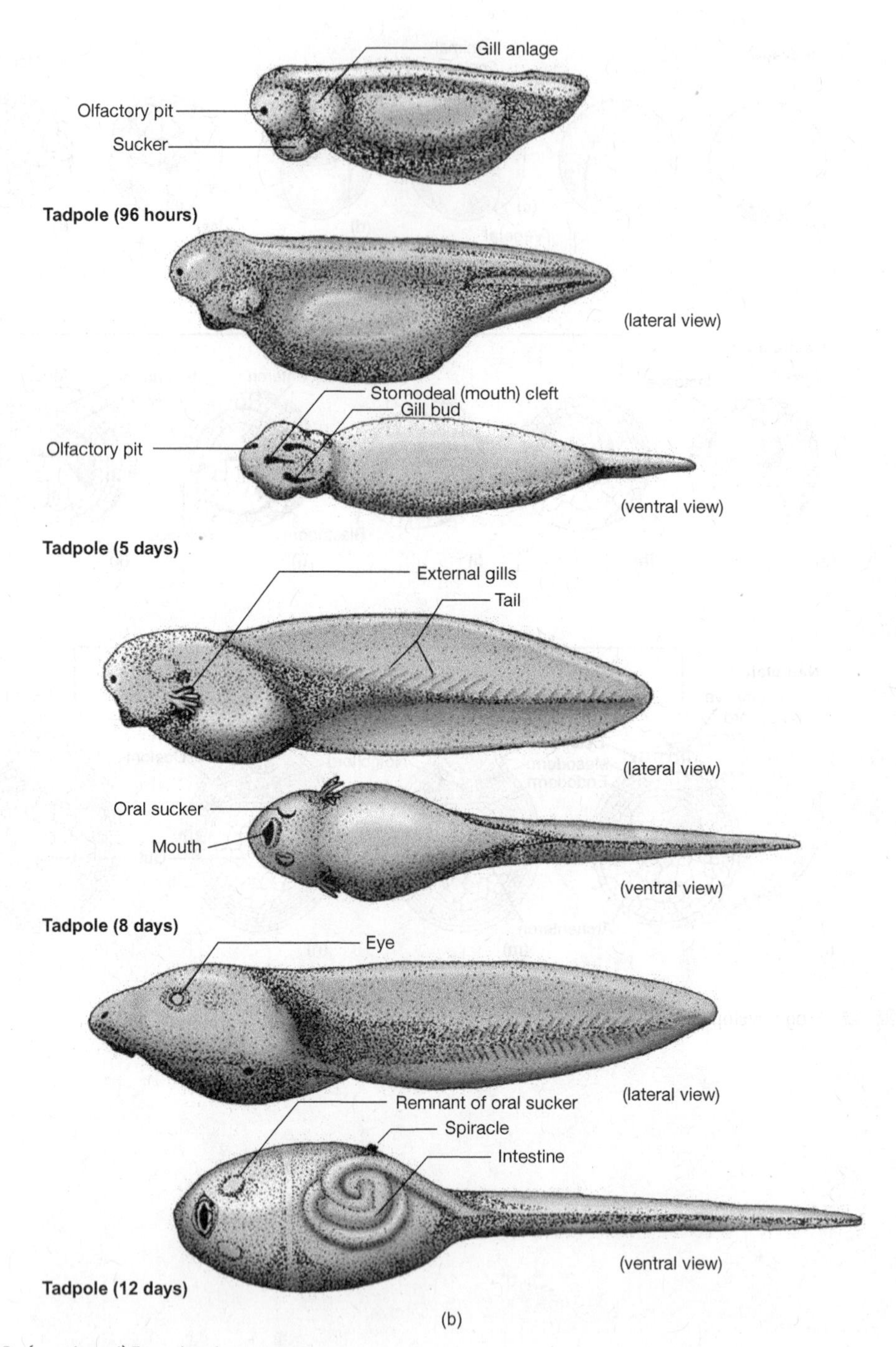

FIGURE 4.9 *(continued)* Frog development.

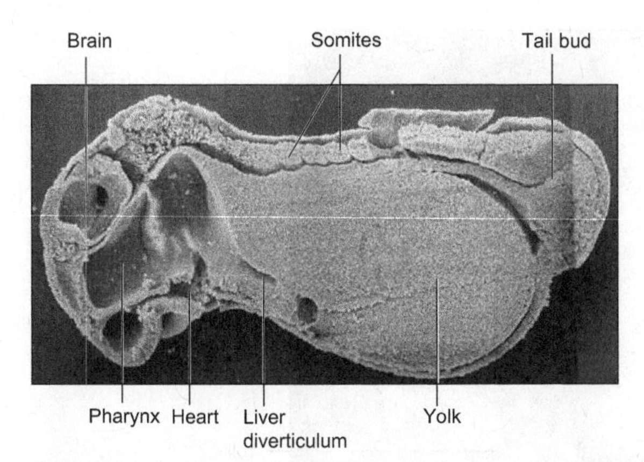

FIGURE 4.10 Frog development. Sagittal section of the tail bud stage of a developing frog embryo.
Scanning electron micrograph of a freeze fracture preparation by Louis de Vos.

Table 4.1 Schedule of Development in *Rana pipiens* at 18°C

Stage	Hours
Fertilization	0
Formation of gray crescent	1
Rotation (animal pole now uppermost)	1.5
First cleavage (2 cells)	3.5
Second cleavage (4 cells)	4.5
Third cleavage (8 cells)	5.5
Blastula	18
Early gastrula (dorsal lip stage)	26
Midgastrula	34
Late gastrula (yolk plug stage)	42
Neural plate	50
Neural folds	62
Ciliary movement	67
Neural tube formation	72
Tail bud stage	84
Muscular contractions	96
Heartbeat	5 days
Gill circulation, hatching (ruptures fertilization membrane)	6 days
Circulation in tail fin	8 days
Internal gills formed, opercular fold present	9 days
Operculum closed, covers internal gills	12 days
Metamorphosis into frog	3 months

After Shumway and Rugh.

Table 4.1 illustrates the schedule of development in *Rana pipiens* at 18°C. Other common species of frogs, toads, and salamanders follow a generally similar pattern of development but differ in the times required to reach the various stages. Several representative stages of development in *Rana pipiens* are shown in figure 4.8.

- Study the frog embryos provided in your laboratory and compare the various stages with figures 4.8 and 4.9. Living embryos are best for study if they are available, although plastic-embedded and preserved embryos can also be used. Supplement your study of the living or preserved whole embryos by observing the microscopic demonstrations provided to illustrate the internal structure of the embryos at different stages. When you have completed your study, you should be able to describe and explain the major events in the development of a frog from an unfertilized egg to an adult.

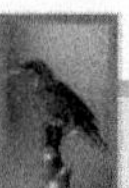

Demonstrations

1. Microscope slide showing developing eggs in the uterus
2. Microscope slide showing frog testis with developing sperm
3. Wet mount of living frog sperm in 10% amphibian Ringer's solution
4. Microscopic slides with cross sections and/or sagittal sections of selected early developmental stages of the frog through hatching
5. Procedures for inducing ovulation in the frog by injection of pituitary extract and artificial insemination of frog eggs in vitro

Chick Development

The eggs and embryos of birds exhibit several major differences from those of amphibians. Birds have eggs adapted for development on land. This type of egg is called an **amniotic** (or cleidoic) **egg** because the embryo is enclosed in special extraembryonic membranes that protect and support the development of the embryo in a liquid environment. Other important adaptations of the eggs of birds (and reptiles) that allow them to develop on land are the inclusion of a large food supply and a tough covering that protects the egg and reduces water loss. The eggs of chickens and most other birds are enclosed in a tough outer membrane with hardened calcium layers on the outside. These protective coverings are added to the egg as it passes down the oviduct during the process of laying.

The chick embryo develops from a small disc of cells, called the blastoderm, which lies on top of the large mass of yolk. Developing embryos can be observed in the laboratory by carefully removing the yolk from an incubated fertilized egg and placing the whole interior contents of the egg in saline solution in a culture dish. This simulates the aquatic environment within the egg. The brain, beating heart, nerve cord, and other large organs can be observed with the naked eye or with a magnifying lens (figure 4.11).

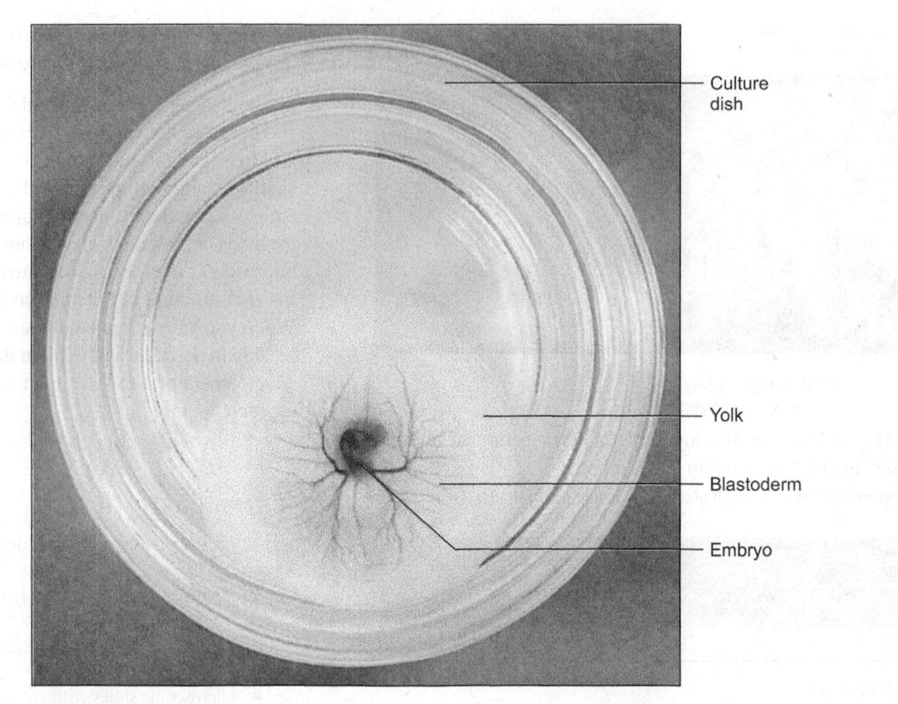

FIGURE 4.11 Chick development. Living embryo removed from shell and placed in a culture dish with saline solution.
Courtesy of Carolina Biological Company, Burlington, NC.

Extraembryonic Membranes

Four **extraembryonic membranes** are formed by a developing chick embryo: yolk sac, chorion, amnion, and allantois. Each of these membranes forms in a specific way, and each performs a distinctive role in protecting the embryo (figure 4.12).

The **yolk sac** forms as a pouchlike outgrowth from the developing gut. It grows around the yolk, releases enzymes to digest the yolk, and transports the digested yolk products through its blood vessels to the developing embryo.

The **chorion** and **amnion** are sheets of living tissue that grow out of and around the embryo. These sheets ultimately join above the embryo and enclose the embryo in a double sac. Both the chorion and the amnion consist of two tissue layers, ectoderm and mesoderm. After closure of the amnionic folds, the amnion becomes filled with a watery fluid. Thus, the amnion maintains an aqueous environment to protect the growing embryo from desiccation and serves as a physical protection for it.

The **allantois,** like the yolk sac, arises as a saclike outgrowth from the ventral surface of the gut. It has quite a different function, however. In birds, the allantois collects and stores metabolic waste products (largely crystals of uric acid). The allantois also grows and fuses with the chorion during later development to form the **chorioallantoic membrane.** The chorioallantoic membrane is highly vascularized and facilitates the exchange of gases between the embryo and the external environment.

Whole Mount of 24-Hour Chick Embryo

- Study first a whole mount of a 24-hour chick embryo. Consult figure 4.13, and study the whole-mount slide under your stereomicroscope.
- Note the approximate size and shape of the embryo. Observe the transparent area immediately surrounding the embryo and an opaque outer area. The inner portion of this opaque area contains many blood vessels that bring nutrients to the developing embryo.

> **CAUTION!**
> **Whole mounts of embryonic stages are relatively thick and should never be viewed under high power on your compound microscope.**

- Locate the **headfold** of the chick at the anterior end of the embryo, lying free and slightly elevated above the underlying membrane. Within the head, observe that the brain at this stage consists of a **neural tube,** which is continued posteriorly with the open **neural plate** consisting of a median **neural groove** and two lateral **neural folds.** Note also that the tissue of the neural folds is continuous laterally with the surface

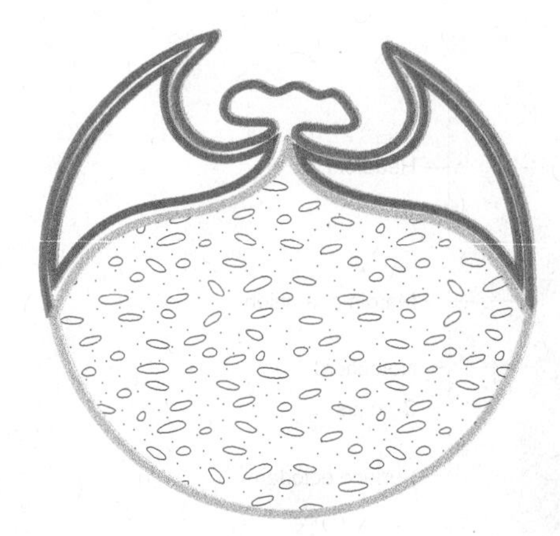
Two-day embryo

Five-day embryo

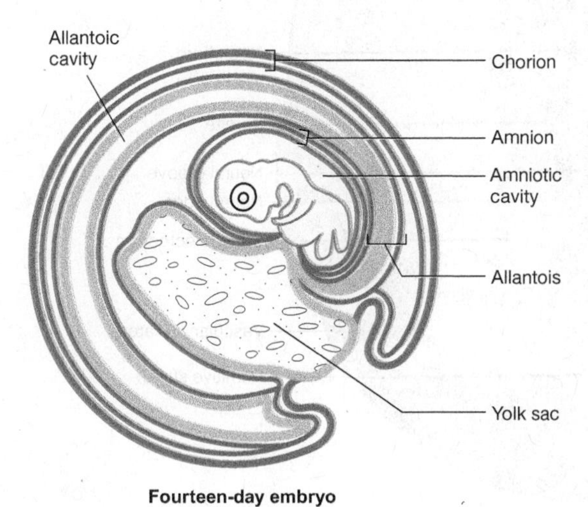

Fourteen-day embryo

FIGURE 4.12 Chick development. Formation of extraembryonic membranes. Ectoderm is blue, endoderm is yellow, and mesoderm is red.

ectoderm, thus demonstrating the ectodermal origin of the nervous tissue.

Posterior to the neural plate is the **primitive streak,** the site of invagination of the surface ectoderm and the formation of the mesoderm layer. The primitive streak, thus, is functionally similar to, or analogous to, the blastopore of the frog and other amphibian embryos.

- Find the **notochord,** which lies beneath the neural tube and the neural groove. Note that anteriorly, the notochord appears as a well-defined rod, but more posteriorly it appears as a wide band of less dense tissue.

Differentiation of the notochord, as of the neural structures, proceeds from anterior to posterior. Likewise, the **somites,** blocks of developing muscle tissue, differentiate from anterior to posterior. New somites are formed by the coalescence and differentiation of mesoderm cells behind previously formed somites.

Whole Mount of 48-Hour Chick Embryo

- Study a whole mount of a 48-hour chick embryo and observe some of its major features (figure 4.14). Note that the **head** and the **heart** are much more well-developed than at 24 hours. Observe that the **eyes** are clearly differentiated and that they exhibit a developing lens.

The brain is enlarged and is now divided into five divisions: the anterior **telencephalon** followed by the **diencephalon,** the **mesencephalon,** the **metencephalon,** and the **myelencephalon.** On the sides of the myelencephalon are a pair of **otic vesicles** that later form the ears.

The heart in a 48-hour embryo is a twisted tube with two well-established chambers: a **ventricle** and an **atrium.** The **sinus venosus** leading to the atrium is also beginning to develop. By 48 hours of development, the heart is actively pumping, and circulation of the blood within the embryo and the surrounding vitelline area is well established. The heart connects with three pairs of large arteries, which represent the first three **aortic arches.** A large **dorsal aorta** extends posteriorly and connects laterally with two large **vitelline arteries.** ***Why is the circulation of blood in the vitelline area important to the development of the chick?***

The head and the anterior part of the embryo are now covered by the **amnion,** which continues to grow posteriorly. The number of **somites** has increased to 24 pairs, which lie along each side of the neural tube. At the posterior end of the embryo, the caudal fold of the amnion has begun to grow forward to enclose the posterior portion of the embryo. Much later, the anterior and posterior portions of the amnion grow together and completely envelop the embryo.

The head of a chick embryo at 48 hours is turned to its right side. Beginning at about 38 hours of incubation, a

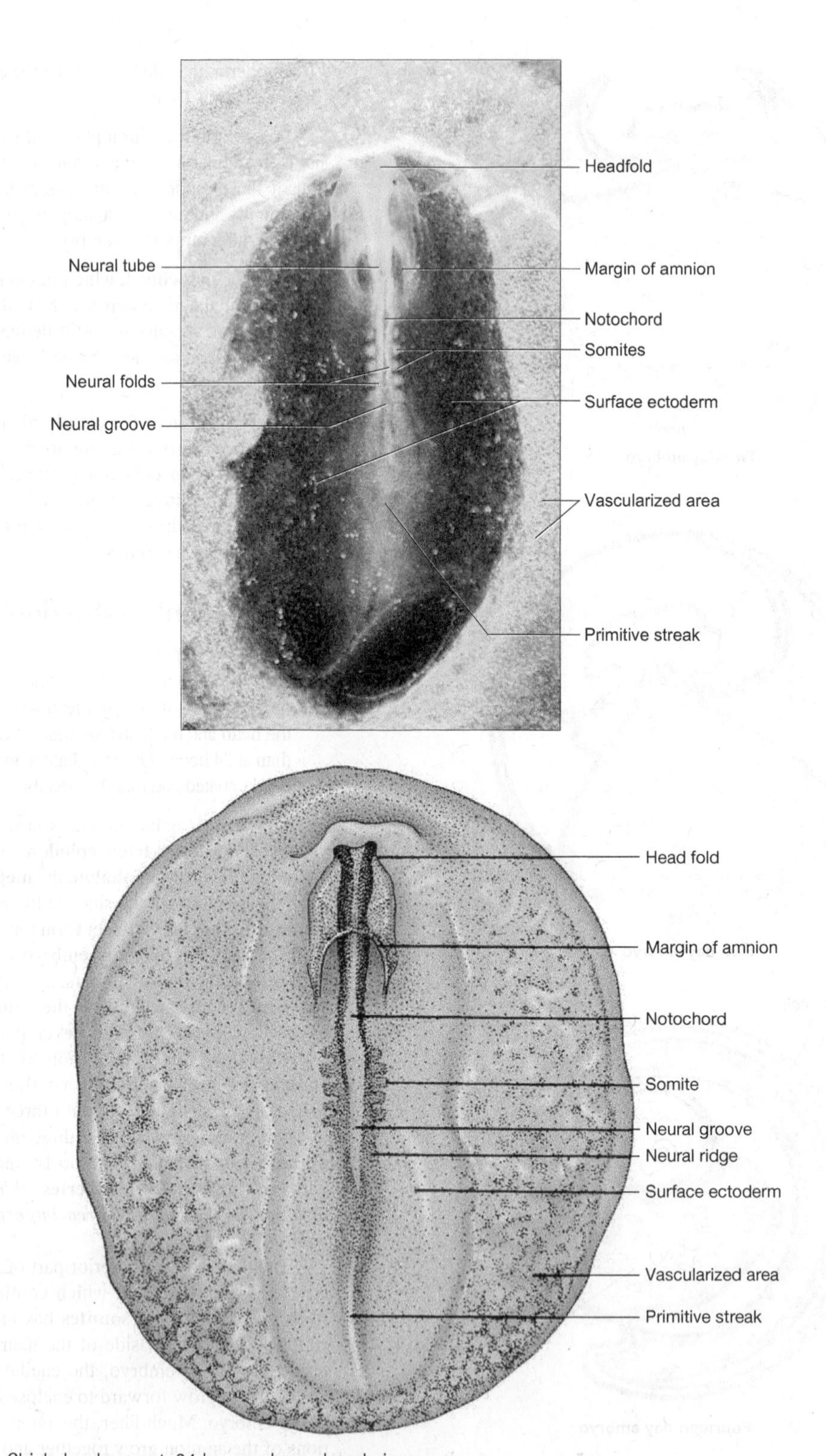

FIGURE 4.13 Chick development. 24 hour embryo, dorsal view.
Photograph courtesy of Carolina Biological Supply Company, Burlington, NC.

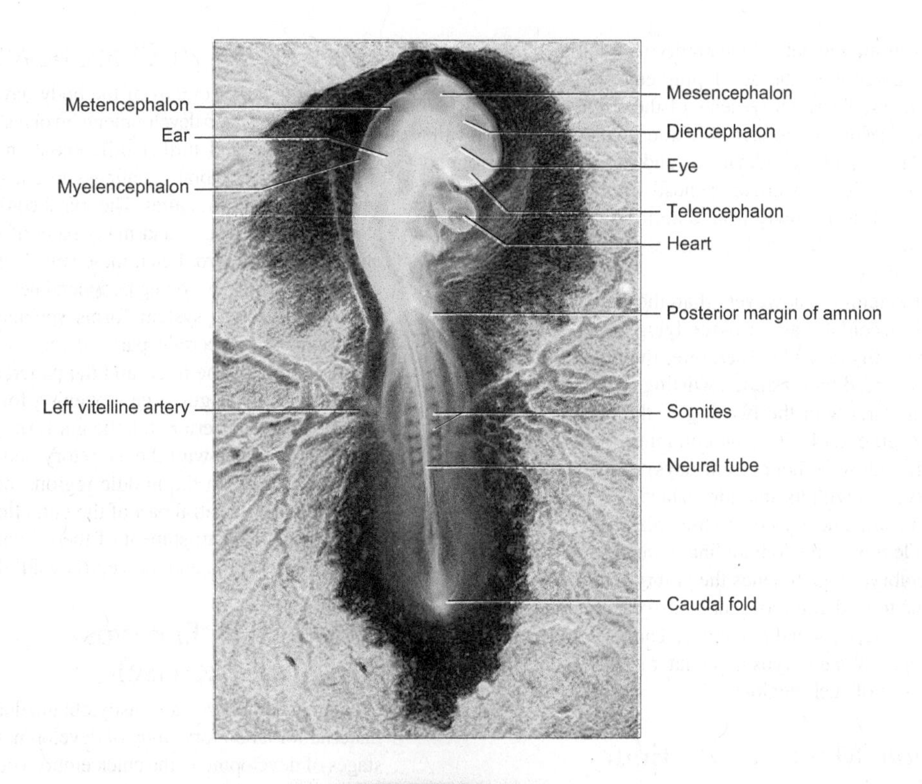

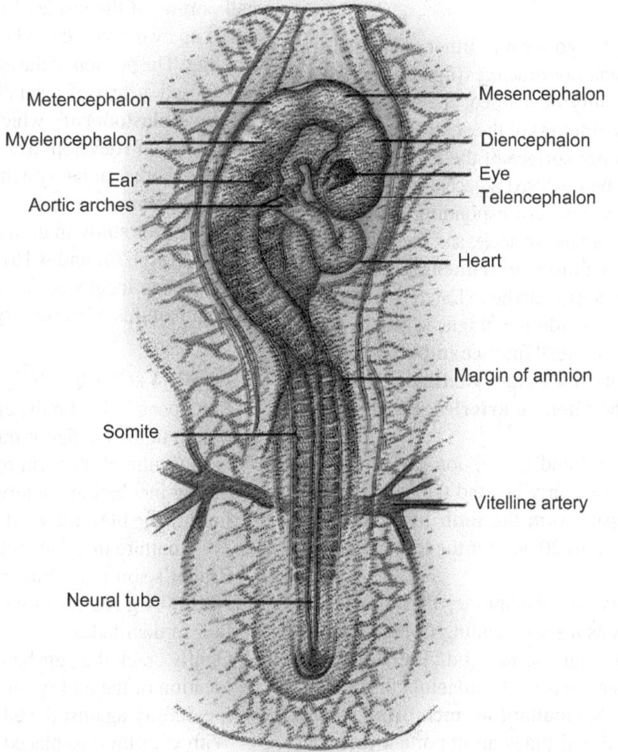

FIGURE 4.14 Chick development. 48 hour embryo, lateral view.
Photograph courtesy of Carolina Biological Supply Company, Burlington, NC.

series of movements of the embryo are initiated that change the orientation of the developing embryo to the underlying yolk mass. These movements of the embryo consist of two types, flexion and torsion. Flexion involves bending of the body along its longitudinal axis and starts at the anterior end with a bending forward of the head. **Flexion** (bending) of the body continues slowly from anterior to posterior so that in later stages the longitudinal axis of the body is bent into the shape of a C.

Remember, however, that the embryo consists of a small discoidal mass of tissue lying on top of a large mass of yolk (figure 4.11). Therefore, the flexion of the body is accompanied by **torsion,** a twisting of the longitudinal axis, which starts with the twisting of the head toward its right side (figure 4.14). Torsion continues from anterior to posterior, and by 96 hours (four days) of incubation, the entire embryo lies with its anterior surface directed to the right and with its left side adjacent to the yolk.

Flexion of the longitudinal (spinal) axis of the developing embryo, which bends the embryo into a C shape, is not unique to bird embryos, but is also characteristic of the embryos of reptiles and mammals. The well-known fetal position of human embryos in the later stages of development is the result of such flexions.

Whole Mount of 72-Hour Chick Embryo

Study of a 72-hour chick embryo shows further enlargement of the head, heart, and other anterior structures (figure 4.15). The eyes are larger and more fully developed, the **otic vesicle** (ear) is enlarged, and the beginnings of the nostrils, the **nasal pits,** are formed on the ventral surface of the telencephalon. Several **nerve ganglia** can be observed adjacent to the spinal cord, and four **gill arches** with corresponding **gill clefts** are present. In addition to the atrium, ventricle, and sinus venosus, the heart has developed a muscular **truncus arteriosus** that carries blood to the aortic arches. Later, partitions develop inside the heart to divide the single ventricle and the single atrium into the typical **four-chambered heart** of adult birds and mammals. Two large **vitelline veins** can be observed adjacent to the **vitelline arteries.** ***What is the function of these veins?***

Somites now number 36, and the torsion (twisting) of the body has progressed posteriorly beyond the level of the heart. Small masses of tissue form the **limb buds** for the wings adjacent to somites 15 to 20 and for the legs adjacent to somites 27 to 32.

The fourth extraembryonic membrane, the **allantois,** appears in a 72-hour chick as a sac extending from the ventral surface of the hindgut near the tail bud. The allantois continues to grow, and later, parts of the allantois fuse with the chorion to form the **chorioallantoic membrane** that lines the inside of the shell and plays an important role in gas exchange for the embryo. The allantois also serves as a reservoir for the deposit of waste products formed by the metabolism of the developing embryo.

Later Stages of Chick Development

Most of the major organs of the body have been established by 72 hours. Later development involves continued growth of the embryo and further differentiation of various organs (figure 4.16). The brain continues to enlarge and to develop several accessory structures. The spinal cord becomes enclosed along its entire length, and many pairs of spinal nerves form along the spinal cord. Later, these spinal nerves grow out and connect with the developing peripheral nervous system.

The digestive system forms specialized regions, and outgrowths from certain parts of the gut form the thyroid gland, the lungs, the liver, and the pancreas. At the anterior end of the tubular gut, a new opening forms to become the mouth. At the posterior end, the gut forms an opening into a common chamber with the excretory and reproductive system, the **cloaca.** In the middle region, the yolk sac is connected with the ventral part of the gut, allowing yolk to pass into the gut for nourishment of the growing embryo. Later, the yolk is used up, and the ventral wall of the gut closes.

Living Chick Embryos (Optional Exercise)

Living chick embryos are easily obtained and make excellent material for laboratory study of development. During its early stages of development, the chick embryo occupies a relatively small portion of the inside of the eggshell. Most of the space is taken up by the stored food materials, the yolk and the white of the egg. The portion of the egg that will become the chicken is composed at first of a small disc of rapidly dividing cells (called the **blastoderm**), which later undergoes a process of gastrulation, and develops a head, nervous system, circulatory system, and other organ systems.

Stages from two to four days (48–96 hours) are most satisfactory for study in an introductory zoology laboratory (figures 4.14, 4.15, and 4.16)). Obtain some fertilized eggs that have been incubated for two to four days and study the principal features of the embryo at each of these stages.

Directions for Opening Eggs

1. Put a penciled X on the egg to mark the side that was on top in the incubator tray. This step is important because the blastoderm rotates to the top of the egg during incubation. Otherwise, you may have difficulty finding the blastoderm after you crack the egg.
2. Fill a culture dish half full of warm (37°C) chick Ringer's solution. Ringer's solution contains 9.0 g NaCl, 0.4 g KCl, 0.24 g CaCl, and 0.2 g $NaHCO_3$ in a liter of distilled water.
3. Gently crack the eggshell on the side opposite the location of the embryo (i.e., opposite the X) by striking it carefully against the edge of the finger bowl.
4. With your thumbs placed over the X, lower the egg and your fingers into the solution and carefully pry open the cracked surface with your fingers, using your thumbs as a pivot.

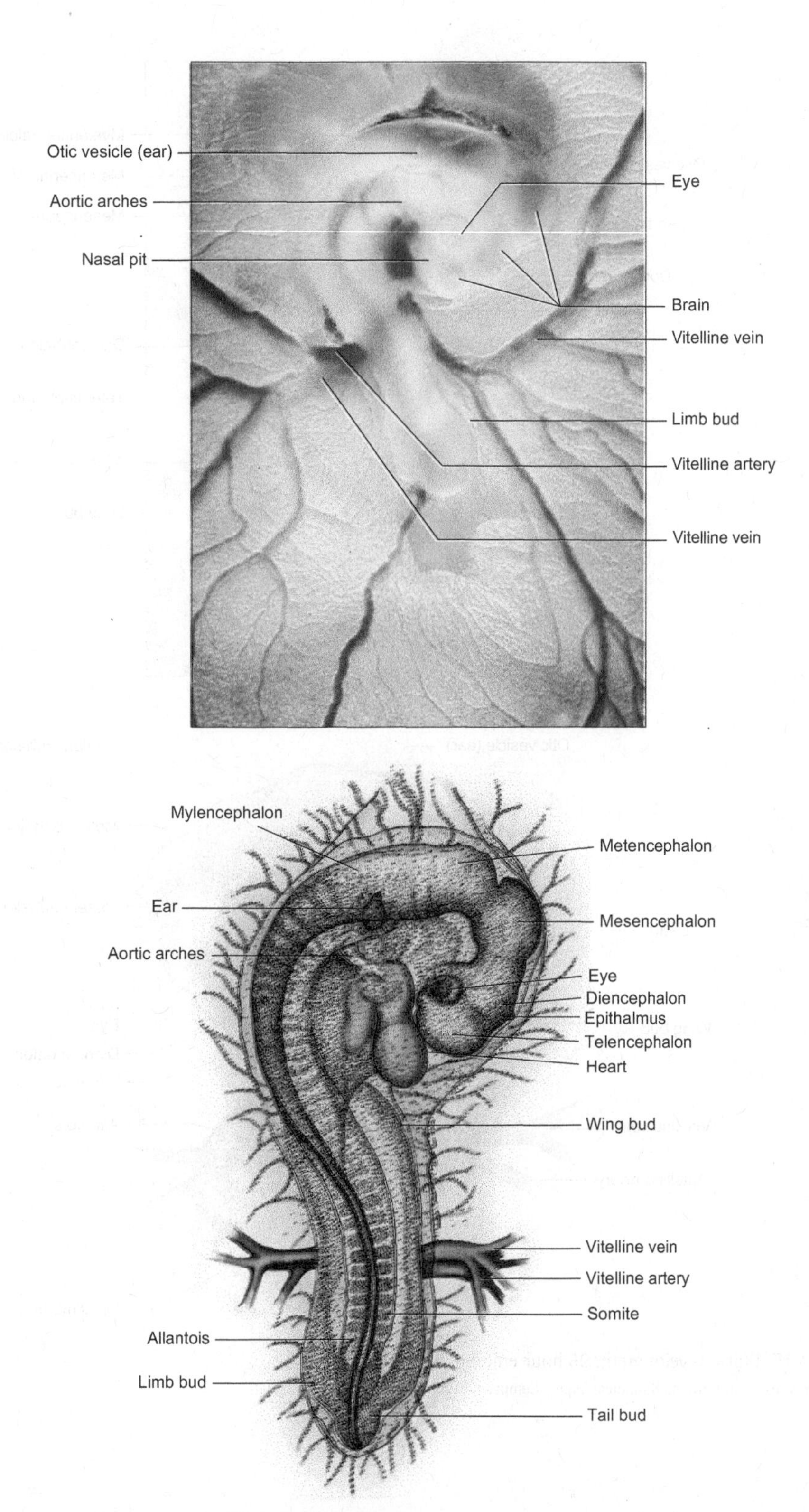

FIGURE 4.15 Chick development. 72 hour embryo, lateral view.
Photograph courtesy of Carolina Biological Supply Company, Burlington, NC.

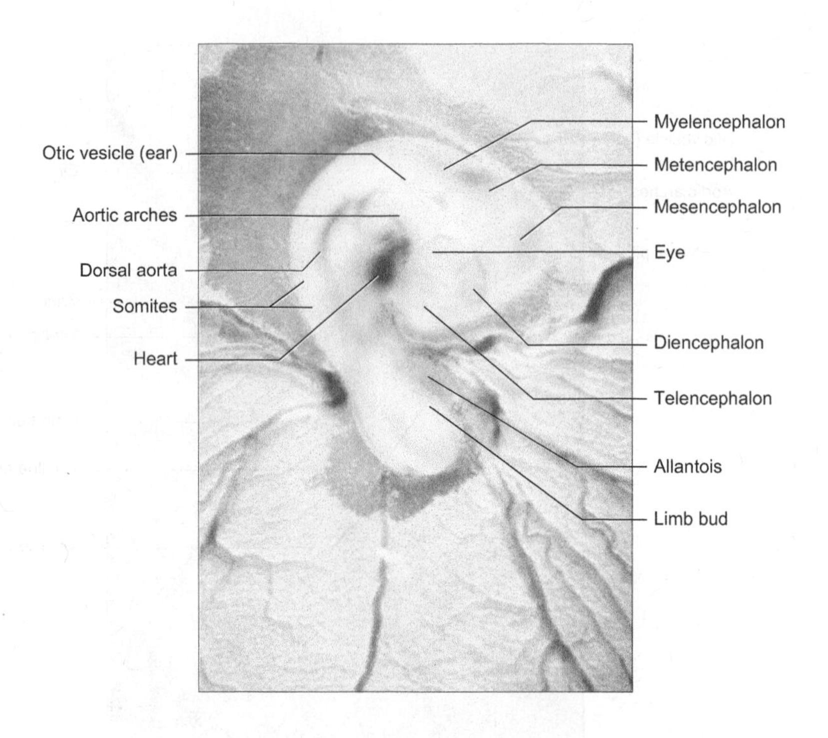

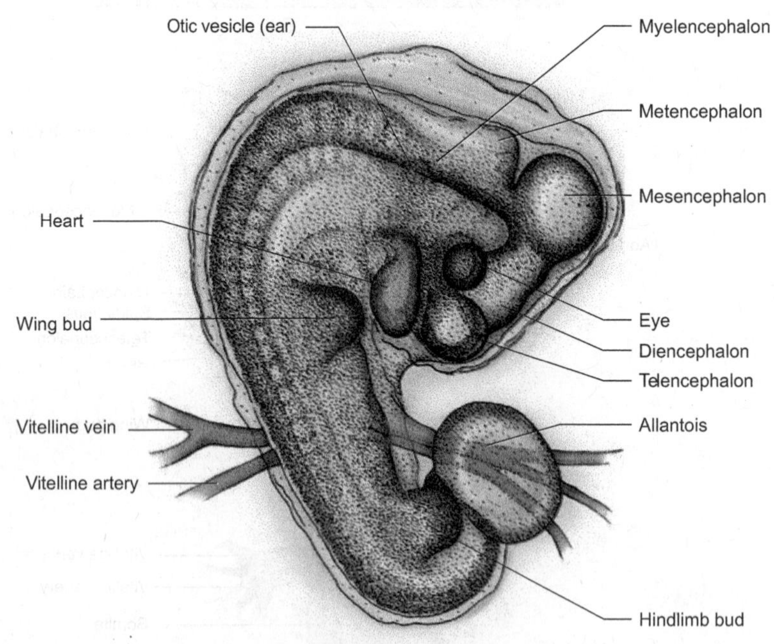

FIGURE 4.16 Chick development. 96 hour embryo, lateral view.
Photograph courtesy of Carolina Biological Supply Company, Burlington, NC.

5. **Caution:** If you open the shell too slowly, the sharp edges of the shell may sever the delicate membranes of the embryo as it slides out of the shell. Likewise, if you open the shell too quickly, you can also damage these membranes.
6. If there is no embryo on the surface of the yolk, try another egg. Be sure to look on the sides and bottom of the egg also, in case the egg rotated as you opened the shell. A plastic spoon is a good instrument to aid in rotating the egg.

- Study the embryo and identify as many structures of the embryo as possible using figures 4.14, 4.15, and 4.16 as a guide.

  Repeat your study with other stages of development as available. Make notes on your observations in the Notes and Sketches section at the end of the chapter. Make a list of the most obvious features of each stage that you study.

CAUTION!

Be sure to dispose of all eggs, eggshells, and other waste at the end of your study as directed by your instructor. Also, remember to wash your glassware when you have finished with it.

Demonstrations

1. Living chick embryos after 33, 56, 72, and 96 hours of incubation (in chick Ringer's solution)
2. Microscope slides with selected cross sections of 33-hour chick embryos
3. Microscope slides with whole mounts of 72- and 96-hour chick embryos

Key Terms

Allantois one of the extraembryonic membranes found in the chick; forms as an outgrowth of the gut. Functions in gas exchange and for the storage of metabolic wastes.

Amnion an extraembryonic membrane found in the chick; consists of layers of ectoderm and mesoderm. Encloses the developing embryo and provides a fluid environment to protect the embryo.

Archenteron the primitive gut of an embryo; lined by endoderm tissue.

Blastopore the opening in the gastrula through which the ectodermal cells invaginate. Often becomes the mouth or the anus of the adult, depending on the type of animal.

Blastula hollow-ball stage in early embryonic development following the morula stage and preceding the gastrula stage.

Centrolecithal egg type of egg with abundant, centrally located yolk, such as an insect egg.

Chorion an extraembryonic membrane composed of ectoderm and mesoderm layers; protects the developing embryo.

Cleavage the period of rapid cell divisions in an embryo following fertilization; leads to the formation of the morula and blastula stages.

Determination the progressive limitation of the developmental potential of an embryo, tissue, or cell.

Differentiation formation of a specialized tissue or cell type from a simpler tissue or cell type.

Ectoderm primary germ layer found on the exterior of an early embryo; forms the nervous system, integument, and certain other tissues in the adult.

Endoderm primary germ layer lining the gut of an early embryo; forms the lining of the digestive tract, liver, pancreas, thyroid, and certain other organs in the adult.

Extraembryonic membranes membranes that form external to the embryo of the chick and most other terrestrial vertebrate animals; serve to enclose and protect the developing embryo. Includes the chorion, amnion, allantois, and yolk sac in the chick.

Fertilization the fusion of the male and female nuclei, which initiates embryonic development. The sperm cell must first penetrate the cell membrane of the egg and migrate to the egg nucleus.

Gametes specialized sex cells, eggs and sperm, necessary for sexual reproduction.

Gametogenesis the formation of eggs and sperm; involves meiosis and cellular differentiation of the male and female sex cells to form gametes.

Gastrula developmental stage following the blastula. Invagination or inward movement of cells at this stage leads to differentiation of the three primary germ layers: endoderm, ectoderm, and mesoderm.

Gastrulation formation of the gastrula.

Growth increase in size or mass of an embryo or individual.

Invagination infolding of cells through the blastopore during gastrula formation; leads to the differentiation of endoderm and mesoderm tissues.

Isolecithal egg type of egg with a small amount of yolk, as in a starfish egg.

Mesoderm one of the primary germ layers; formed from invaginated ectoderm cells or as outpockets of the gut. Develops into the muscles, bone, connective tissues, circulatory system, and many other structures in the adult.

Mesolecithal egg type of egg with a moderate amount of yolk, as in the frog egg.

Morphogenesis the molding of a structure during embryonic development by cell and tissue movements.

Morula stage formed during the latter part of embryonic cleavage; consists of a solid ball of cells.

Notochord cartilaginous supporting rod parallel and interior to the dorsal nerve cord of all chordates. May be replaced by a vertebral column in later stages of development.

Organogenesis the formation of a specific organ, such as the heart, during embryonic development.

Primary germ layer one of the three tissue layers differentiated early in development—ectoderm, mesoderm, and endoderm.

Radial cleavage pattern of cleavage in which the blastomeres are arranged radially around the central animal-vegetal axis; characteristic of deuterostomes.

Somite a block of mesodermal tissue along the nerve cord in an embryo; differentiates into segmental muscles and other mesodermal tissues during development.

Spiral cleavage type of cleavage in which the blastomeres rotate in a spiral fashion after the third and several subsequent cleavage divisions because of shifts in the axis of the mitotic spindle; characteristic of protostomes.

Telolecithal egg type of egg with a large amount of yolk and an embryo restricted to one side (for example, a chick egg). The embryo develops on the surface of the yolk.

Yolk stored food reserves in an egg and embryo; rich in lipids and proteins.

Yolk sac one of the extraembryonic membranes of the chick and many other vertebrates; forms as an outgrowth of the gut and encloses the yolk.

Internet Resources

There are many valuable Internet sites with information about zoology. Several sites containing pertinent zoological information for this chapter can be found on the McGraw-Hill Zoology web site at http://www.mhhe.com/zoology. Just click on this text's title.

Questions for Critical Thinking

1. Compare the contributions of the egg and sperm to the developing embryo. Are they both essential? Why? What evidence can you cite for your answer?
2. Discuss the roles of growth, determination, differentiation, and morphogenesis in development. Why is each process important for the normal development of an embryo?
3. Discuss the contributions of each of the three primary germ layers to the body of an adult animal.
4. Discuss the principal similarities and differences in the embryonic development of a frog and of a chick. Why do you think the development of these animals is so different?

Suggested Readings

Gilbert, S.F. 2010. *Developmental Biology*, 9th ed. Sunderland, MA: Sinauer. 711 pp

Wolpert, L. and C. Tickle. 2010. *Principles of Developmental Biology*, 4th ed. New York: Oxford University Press. 720 pp

Notes and Sketches

Chapter 5

Systematics and Morphology

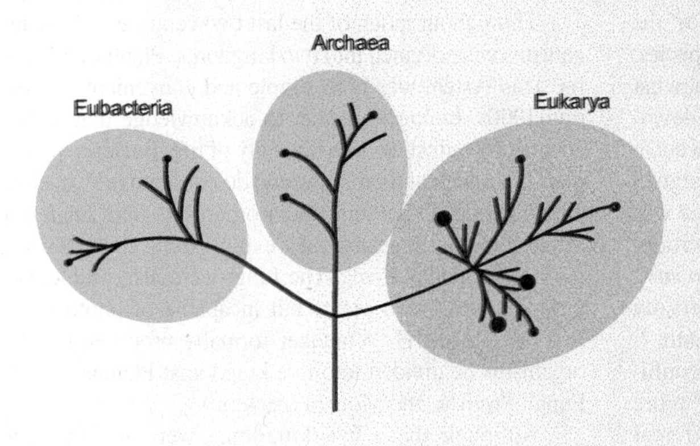

Objectives

After completing the laboratory work in this chapter, you should be able to perform the following tasks:

1. Describe the binomial system of nomenclature. Identify the components and correct grammatical style of a scientific name.
2. Identify the seven major categories in the traditional Linnaean system of classification.
3. Describe recent changes in the five-kingdom system of classification and explain what has prompted those changes.
4. Define a clade and explain how a typical cladogram illustrates relationships among animals.
5. Describe five types of evidence that biologists use to infer phylogenetic relationships among animals.
6. Name and describe three types of symmetry exhibited by animals.
7. List and describe three grades of tissue construction exhibited by animals.
8. Define the terms *acoelomate*, *pseudocoelomate*, and *coelomate*.
9. Briefly discuss cephalization and segmentation in animals and explain their significance.
10. Explain the importance of embryological features for understanding phylogenetic relationships in animals.
11. Distinguish between the Protostomia and the Deuterostomia and list the characteristics of each.
12. Identify six important evolutionary trends in the animal kingdom.

Introduction

Interest in the diversity of animals and in their evolutionary relationships has increased greatly in recent years in response to the degradation of our environment and the extinction or threat of extinction of species in many parts of the world. Although changes in the species composition of various habitats is a part of the natural process of evolution, the rate of extinction of species has increased at an alarming rate. Fortunately, the general public and the scientific community have shown concern about this threat, prompting a new interest in the conservation of our natural heritage.

Interest in **systematics** and the evolutionary relationships of animals has also increased due to the emergence of new research methods and investigations. Each year we learn more about human ancestry as well as new insights into the phylogenetic relationships of animals and other organisms. We are now able to use a number of chemical and physical techniques to unravel relationships at the molecular level with sophisticated genetic and biochemical methods. Understanding these relationships of animals, however, requires some knowledge of the basic structure of animals, and the kind of features that have been found to be useful in deducing evolutionary relationships. These are the topics of this chapter.

Table 5.1 Taxonomic Categories

Kingdom Animalia
Phylum Chordata
Subphylum Vertebrata
Class Mammalia
Order Carnivora
Family Canidae
Genus *Canis*
Species *lupus*
Subspecies *familiaris*
Common name Domestic dog

Systematics

Planet Earth is teeming with living organisms. They live in nearly every imaginable habitat from polar ice caps to geothermal hot springs, from open oceans to freshwater ponds, from desert sands to tropical rain forests, and from the highest snow-capped mountains to the deepest ocean trenches. Although more than 1.6 million species have already been described and named, hundreds of new species are added each year. Biologists speculate that a complete catalog of the Earth's biota may ultimately total more than 5 million species.

Keeping track of all these species is made somewhat easier by a hierarchical classification system that was developed in the 18th century by Carl von Linné. This Swedish botanist (his name is often anglicized to Linnaeus) assigned "scientific" names to many different species. Each name was a unique combination of two Latin words: a **genus** (group) name, followed by a **species** identifier (e.g., *Homo sapiens*). This system of **binomial nomenclature** is now universally accepted by biologists. By international convention, Latin is always used for scientific names, thereby avoiding the confusion and controversy that arises when an organism has more than one common name or has different names in different languages. The Latin name is always underlined or italicized with only the first letter of the genus name capitalized.

In the Linnaean system of classification, physical (morphological) similarity is the principal basis for a hierarchy of taxonomic groupings. The most similar species are grouped together and given the same genus name (e.g., *Rana pipiens* and *Rana catesbiana*); similar genera are grouped into a **family,** similar families are grouped into an **order,** similar orders are grouped into a **class,** similar classes are grouped into a **phylum** (or division), and similar phyla are grouped into a **kingdom** (Table 5.1). These higher taxonomic categories are established and defined on the basis of evidence considered and evaluated by scientists who specialize in each of the groups. As new similarities and differences are discovered and the evidence is reevaluated, higher taxonomic categories may be added, changed, or eliminated according to the best professional judgment of those specialists.

Species are generally defined as discrete, natural categories made up of all individuals that share a common gene pool (i.e., are capable of interbreeding). By its very nature, the species is a relatively stable taxon, less subject to revision than higher taxonomic categories due to a new discovery or a change in professional opinion. Indeed, taxonomists spend much of their time and effort pondering the merits of "splitting" or "lumping" species within genera, genera within families, families within orders, and so forth.

Throughout much of the last two centuries, all living organisms were divided into two kingdoms: Plantae and Animalia. This system was both simple and convenient, but by the mid-1900s, biologists began to acknowledge that it did not adequately reflect the true diversity of life. Bacteria, for example, lack a nuclear membrane and do not seem to fit in either of the kingdoms. Other unicellular organisms (like *Euglena*) are difficult to place because they exhibit some characteristics of *both* plants and animals. The fungi were also problematic—grouped with green plants but incapable of photosynthesis. In 1969, Robert H. Whittaker formally proposed that living organisms be divided into five kingdoms: Plantae, Animalia, Fungi, Protista, and Monera (bacteria).

Although these five kingdoms were widely adopted by biologists during the late 1900s, there has been controversy over kingdom Protista from the very start. Diversity of form and function is much greater within this kingdom than within any other, leading some biologists to favor splitting it into two or more kingdoms. Several alternative classification schemes have been proposed, but no consensus has emerged over the past two decades. Protozoa, one of the largest groups within Protista, exhibit many animal-like characteristics and, by tradition, are still covered in many zoology courses.

In the late 1970s, research by Carl Woese and his colleagues at the University of Illinois revealed a surprising dichotomy within the world of bacteria-like organisms: many

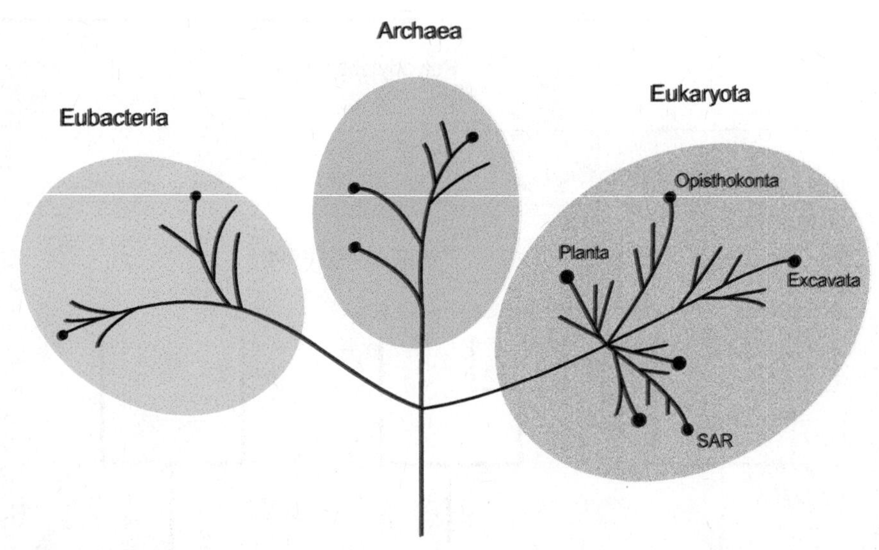

FIGURE 5.1 Possible evolutionary relationships of living organisms showing origin of three Domains: Eubacteria, Archaea, and Eukaryota. Note that Eukaryota has recently been divided into four genetically distinct groups or clades: (1) **Plants** (including green and red algae); (2) **Opisthokonts** (including amoebae, fungi, and animals); (3) **Excavates** (free-living organisms and parasites); and (4) **SAR** (Stramenophiles, Alvaeolates, and Rhizaria: diatoms, brown algae, water molds, foraminifera)

species that manage to survive in the absence of oxygen, in high salt concentrations, or at extreme high or low temperatures were strikingly different from other "normal" bacteria. When the ribosomal RNA of these microbes was sequenced, it became apparent that they represent a distinctive life form, more closely related in some respects to eukaryotes than to other prokaryotes. In fact, many scientists believe that these creatures (now collectively known as the **Archaea**) may be very similar (biochemically and genetically) to the first life forms that evolved on earth 4 to 6 billion years ago.

The concept of Archaea as a group of genetically distinct organisms so different from other known organisms quickly changed many biologists' perspective of the living world and necessitated yet another revision of the system of classification. The **Domain** was established as a new level in the Linnaean classification hierarchy above the level of kingdom. Proponents of this new system of classification regard the Archaea as one of three recognized domains along with the **Eubacteria** and the **Eukaryota.** The Eubacteria includes most of the previously known "normal" bacteria and the Eukaryota includes all the organisms with a compact nucleus and cellular organelles composed of lipoprotein membranes such as plants, animals, protists, and fungi.

Figure 5.1 provides an overview of the evolutionary relationships of living organisms according to the most recent evidence. Our views of animal evolution continue to change as we obtain new evidence. One of the most recent changes concerns relationships among the major groups of eukaryotes that are now believed to represent four distinct groups. Table 5.2 lists some representative members of each of these four large eukaryote groups.

Table 5.2 Major Clades of the Eukaryotes

Plants
Examples:
Plants, green algae, red algae
Opisthokonts
Members with single trailing flagellum
Examples:
Animals, fungi, slime molds
Excavates
A large group of protists with many free-living and symbiotic forms including some important human parasites. Most members lack mitochondria or have modified forms of mitochondria.
Examples:
Euglenids, trypanosomes, slime molds, trichomonads
SAR (for Stramenophiles, Alvaeolates, and Rhizaria)
Examples:
Diatoms, golden algae, brown algae, water molds, downy mildews, foraminifera, radiolaria

All of these changes have spawned widespread debate and controversy. Many biologists view the Linnaean hierarchy as obsolete and argue that it should be completely abandoned. Their long-standing dissatisfaction is rooted in a paradigm shift that has swept through the worldwide community of biologists over the past 50 years. Thanks to the work of Charles Darwin and other proponents of evolutionary theory, a species is no longer viewed as a fixed, static entity. Over time, it may change through the process of natural selection, losing old traits or acquiring new ones,

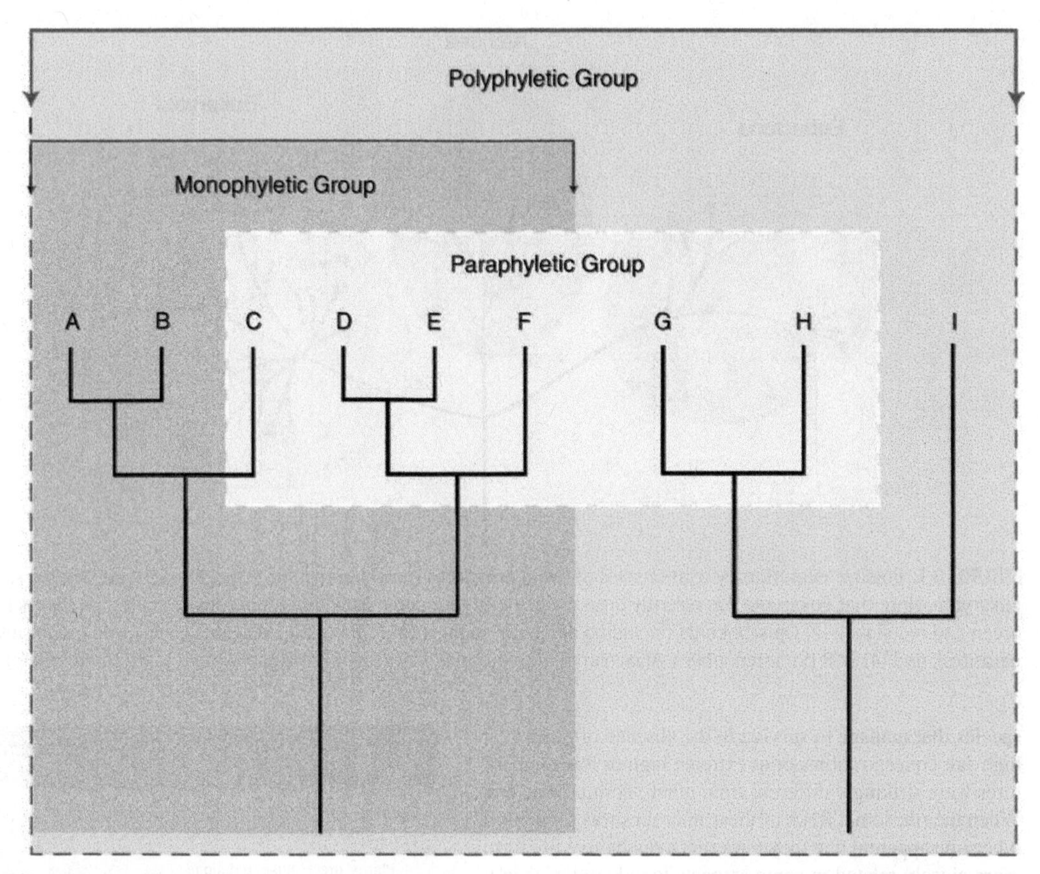

FIGURE 5.2 Relationships between taxonomy and phylogenetic groups. Monophyletic group: all members have a single common ancestor. Paraphyletic group: members lack a single common ancestor. Polyphyletic group: all members are descendants of two or more common ancestors.

even splitting into two or more independently evolving species. Traditional emphasis on the physical (morphological) characteristics of species is being replaced by a desire to chart the evolutionary history of a species and to construct a classification system that more accurately reflects its ancestry (phylogeny). One approach is known as the **cladistic** method of classification (figure 5.2). The name derives from the term **"clade,"** which is defined as the entire group of species that share a single common ancestor.

A typical **cladogram** is a branching diagram that represents the "family tree" of related species (or higher taxa). It is constructed by comparing the distribution of primitive and derived character states in an effort to illustrate evolutionary relationships. It requires that distinctions be made between **homologous** traits (those inherited from a common ancestor) and **analogous** traits (similar characters with different evolutionary origins). The cladogram is a hypothesis. It can be tested against additional traits and reevaluated in the light of new data. In some cases, cladistic analysis may confirm the traditional Linnaean pattern of organization or suggest only minor changes. But in other cases (birds, reptiles, and dinosaurs, for example), the Linnaean system fails to recognize the overwhelming evidence of common ancestry.

It seems incredible that we can even think of reconstructing events that happened so long ago. But biologists and paleontologists now have a wide array of tools and techniques that allow them to probe the world of today for clues that illuminate events of the distant past. Such clues form an ever-growing patchwork of data that has begun to merge into a rough outline for the tree of life, including the following:

- **The fossil record.** A database preserved in stone or in amber (the petrified resin [sap] of prehistoric trees). Teeth, bones, and exoskeletons are most often preserved, but even soft-bodied organisms can be found in sedimentary rock strata and in deposits of coal, shale, or volcanic ash.
- **Radiometric dating.** Physical or chemical techniques that reveal the age of fossils or their rock strata by measuring the concentration of a radioactive isotope (such as carbon 14) or by determining the ratio between a radioactive element (uranium 235, thorium 232, or potassium 40) and its spontaneous decay product.
- **Numerical taxonomy.** Statistical techniques, such as principal components analysis, cluster analysis, factor analysis, multidimensional scaling, and discriminant

analysis can digest large data sets and help taxonomists quantify the degree of overall similarity among groups of organisms.

- **Biochemistry.** Enzymes and metabolic pathways can reveal inherent patterns in nutrient processing, chemical defense, locomotion, intercellular and intracellular communication, or homeostatic mechanisms. Similarities and/or differences in these basic life processes are often consistent within an evolutionary lineage.
- **Nucleotide sequencing.** The evolutionary "distance" between two organisms can be inferred from the number and frequency of changes in the sequence of base pairs in corresponding regions of DNA (or RNA). By assuming that mutation rates have been relatively constant throughout geological history, it is possible to estimate how long it has been since any two groups diverged from a common ancestor.

Although most of our hypotheses about evolutionary pathways are tentative and controversial, they provide a valuable framework in which to study similarities, differences, and relationships among all surviving taxa. Our understanding of each phylogenetic group is enhanced by learning how it has been shaped by the selective pressures of the past, and how it differs from its nearest relatives of the present.

In general, animals have gradually become more complex and more diverse over the course of evolution. New cell types, new tissues, new organs and organ systems have appeared, disappeared, or been remolded by the process of natural selection. Each new feature, appearing as a result of mutation or a novel arrangement of genetic material, has persisted or failed to persist as a consequence of its survival value for the species. Similar selective forces have shaped physiological processes, behavior, and all other facets of life.

Even before zoologists knew about DNA and the genetic code, they recognized that certain similarities or differences in form and function could be used to infer relationships among taxa. Among the basic features considered important by most biologists are (1) body symmetry, (2) grades of tissue construction, (3) type of body cavity, (4) segmentation, (5) cephalization, and (6) patterns of embryological development.

Morphology

Body Symmetry

The symmetry or arrangement of body parts in relation to other parts is a fundamental characteristic of all animals. Three different kinds of symmetry are found among animals and protists: (1) **asymmetry,** (2) **radial symmetry,** and (3) **bilateral symmetry** (figure 5.3).

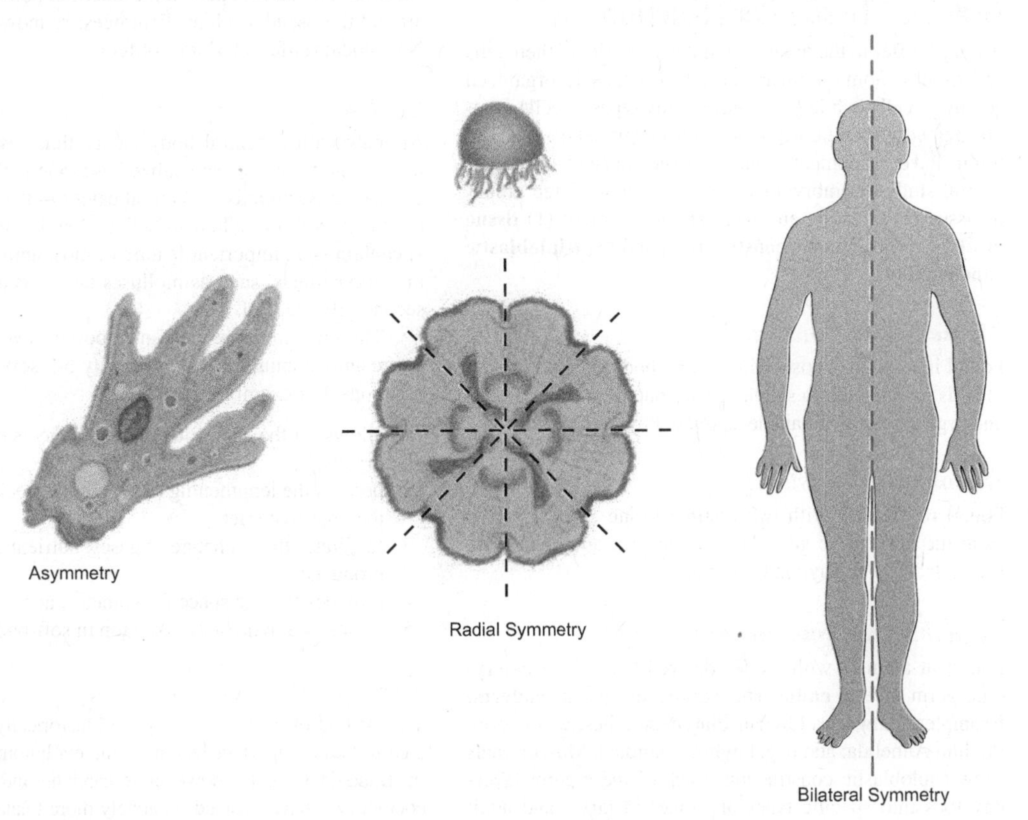

FIGURE 5.3 Types of symmetry.

Asymmetry

Irregular arrangements of body parts. Such animals have no plane of symmetry that divides the animal into similar halves. Example: many sponges.

Radial Symmetry

Body parts arranged around one central axis; any plane passing through the central axis divides the body into similar (mirror image) halves. Examples: most medusae (jellyfish), sea stars, sand dollars.

Bilateral Symmetry

Body parts divided into equal (mirror image) halves by a single plane of symmetry. Examples: flatworms, fish, humans.

Most animals are bilaterally symmetrical; radial symmetry is most common among the Cnidaria and the Echinodermata. In the cnidarians, radial symmetry is primary; both the larvae and adults usually show obvious radial symmetry. In the echinoderms, however, radial symmetry is secondary; the larvae are typically bilaterally symmetrical during their development, and only after metamorphosis do adults become radially symmetrical. This change in symmetry is commonly attributed to the sedentary habits of ancestral adult echinoderms.

Grade of Tissue Construction

Animals differ in the organizational complexity of their cells and tissues. Some animals consist of a loosely organized colony of cells, while others have many types of cells organized in very specific ways into well-integrated tissues. The origin and development of animal tissues can be observed by careful study of embryological development. Three grades of tissue organization are found among animals: (1) **tissue grade,** (2) **diploblastic construction,** and (3) **triploblastic construction.**

Tissue Grade Construction

Found in animals whose multicellular bodies are composed of cells organized into simple tissues but are lacking organs and organ systems. Example: sponge, Phylum Porifera.

Diploblastic Construction

Found in animals with two distinct tissue layers derived from the embryonic germ layers ectoderm and endoderm. Example: *Hydra,* Phylum Cnidaria.

Triploblastic Construction

Found in animals with tissues derived from three embryonic germ layers: **endoderm, ectoderm,** and **mesoderm.** Examples: flatworm, Phylum Platyhelminthes; earthworm, Phylum Annelida; and dog, Phylum Chordata. Most animals show triploblastic construction. Each of these germ layers develops into specific types of tissues in larval and adult organisms. Endoderm forms the lining of the gut and several internal glands derived from the gut. Ectoderm forms the outer layer of the body (integument) and all nervous tissue including the brain. Mesoderm gives rise to muscle, bone, connective tissue, and blood.

Body Cavity

Most large animals have a central body cavity, while some small animals lack a central cavity but have their central space filled with loosely packed cells. Further study of anatomy reveals three basic groups of animals based on their central body cavity: (1) **acoelomate,** (2) **pseudocoelomate,** and (3) **coelomate** (figure 5.4).

Acoelomate

Animals whose central space is filled with loosely packed cells **(mesenchyme)**. Body fluids percolate through irregular spaces between cells, carry nutrients to the cells, and assist in removing wastes from the cells. Example: flatworm, Phylum Platyhelminthes.

Pseudocoelomate

Animals with a central body cavity, the **pseudocoelom,** derived from the embryonic blastocoel. In the adult organism, this cavity lies between the endoderm and mesoderm tissues; there is no specialized mesodermal peritoneum lining around the pseudocoelom. Examples: roundworm, Phylum Nematoda; rotifer, Phylum Rotifera.

Coelomate

Animals with a central body cavity that develops within mesoderm tissue. A specialized mesodermal peritoneum completely surrounds the central cavity in the adult. Examples: earthworm, Phylum Annelida; dog, Phylum Chordata. A **coelom** is an important feature of most animals, although in some animals, such as molluscs and insects, it becomes secondarily reduced.

The coelom represents an important evolutionary advance among animals, and this cavity has several important functions. For example, the coelom:

1. allows for the expansion and movements of internal organs;
2. permits the lengthening and regional specialization of the digestive tract;
3. facilitates the exchange of gases, nutrients, and waste products;
4. provides storage space for gametes; and
5. serves as a hydrostatic skeleton in soft-bodied animals.

Phylogenetic Significance

The method of coelom development historically was considered to be an important feature in the evolutionary history of the Bilateria. Recently, however, research has indicated that the coelom may have evolved separately more than once and that its value in indicating evolutionary relationships is in question.

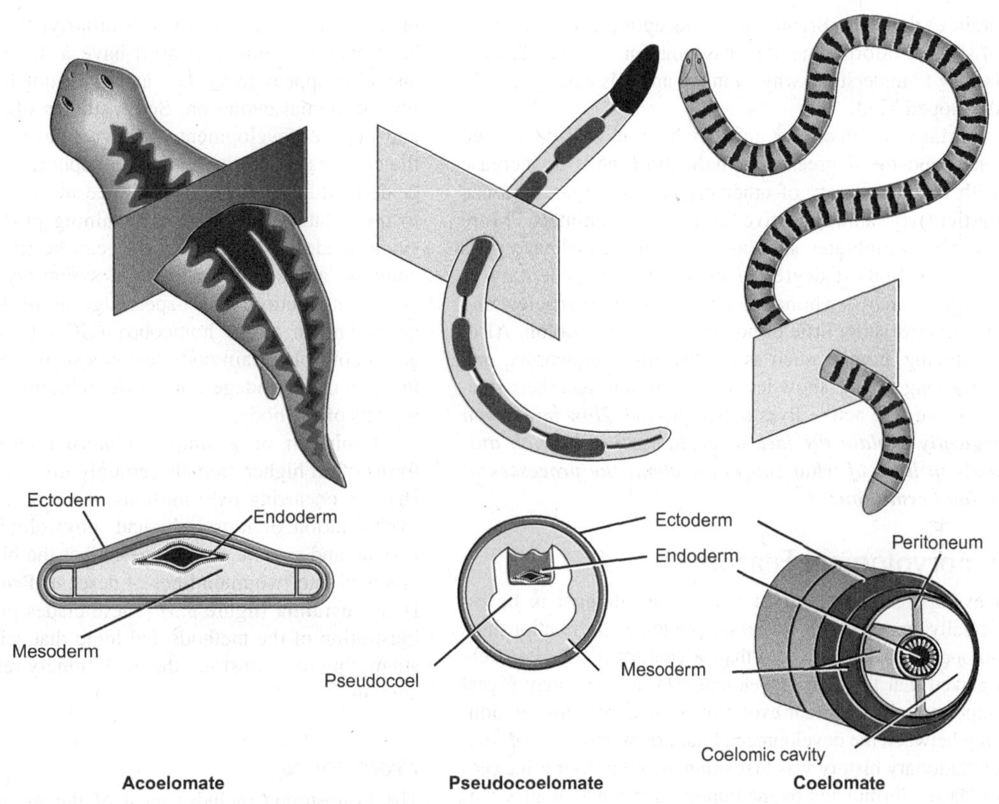

FIGURE 5.4 Types of body cavity.
Courtesy of Carolina Biological Supply Company, Burlington, NC.

Metamerism or Segmentation

Several animal groups exhibit repetition of body parts along the length of the body similar to a line of boxcars in a freight train. Such serial repetition of parts is called **metamerism** or **segmentation.** This condition is most obvious in three large phyla, including Annelida, Arthropoda, and Chordata, although it occurs in several other groups including cnidarians, turbellarians, cestodes, and rotifers.

Historically, many scientists have considered metamerism as an important character indicating **common ancestry** among certain animal groups, most commonly the annelids, arthropods, and chordates. Its occurrence in several diverse groups of animals with different grades of body design—acoelomate, pseudocoelomate, and coelomate—as well as in both protostomes and deuterostomes clearly suggests that **segmentation arose more than once** during the evolution of animals. Numerous writers have discussed segmentation among the various groups and have attempted to differentiate types of serial repetition seen in simpler invertebrates, cnidarians (siphonophores, strobila of scyphozoans), platyhelminths (cestodes), and some rotifers, from that observed among the annelids, arthropods, and chordates. Further complicating the matter, different workers have used different definitions of metamerism and segmentation. Unfortunately, no real consensus has been reached and the matter is yet unresolved.

At present there does seem to be general agreement that the **segmentation in annelids and arthropods probably was of common origin.** It seems doubtful that the segmentation in chordates had its origin from ancestors in common with the annelids and arthropods. More caution also seems warranted in viewing evidence of segmentation in other groups, such as molluscs, as indicating common ancestry with the annelid-arthropod line.

Cephalization

Most animals have a definite front end **(anterior)** and a rear end **(posterior).** This differentiation of the anterior/posterior body axis and the related concentration of nervous and sensory structures to form a head at the anterior end is called **cephalization.** Externally, we recognize a head by the presence of sensory structures like eyes, ears, nose, antennae, and other features. These sense organs provide an animal with information about the environment immediately ahead of it. Usually, but not always, the mouth is located on the head since much of an animal's activity is directed toward locating and obtaining an adequate supply of appropriate food. Internally, the head also contains a concentration of nervous tissue, organized into a brain in higher animals, that receives and processes the sensory information received by the sense organs. Proximity of the brain to the sense organs speeds reception and processing of sensory information often essential for survival. Thus, it is easy to understand why so many animals exhibit a well-developed head!

Many multicellular animals have a well-developed head, and the degree of cephalization tends to increase with the complexity of other organs and organ systems, particularly among active and motile animals. More complex vertebrates such as frogs, birds, and mammals show the highest degree of cephalization. Sedentary or sessile (attached) animals such as oysters, barnacles, and sea urchins show little or no signs of cephalization. Also, burrowing animals such as earthworms, shipworms, and burrowing snakes show less cephalization than their relatives that live active lives above ground. ***How might you logically explain the lack of cephalization in such animals in light of what you know about the processes of animal evolution?***

Embryological Features

Developmental features of animals are thought to be especially important in revealing phylogenetic relationships among animals. For more than a century it has been recognized that the early development of animals may reveal important clues to their evolutionary history. This relationship between the developmental pattern of animals and their evolutionary history was first summarized in four principles or "laws" formulated by the famous German zoologist Karl Ernst von Baer in 1828:

1. The general features of animals appear before the special features that distinguish them from other related species.
2. The special features of an animal develop from its general features. That is, the general features provide the basic form from which the special features of each species develops.
3. Different kinds of animals are more similar in earlier stages than in later stages of development.
4. The developing stages of an animal closely resemble the young stages of more primitive animals but do not necessarily resemble the adult stages of those animals.

Although these four general principles of von Baer have been somewhat modified and extended (and sometimes misused!) by more recent workers, they still represent a good summary of the importance of embryological characters in revealing the evolutionary relationships of animals.

Evolution of the Metazoa

Multicellular animals are thought to have arisen from some ancestral protozoan, perhaps a colonial flagellate. The sponges (Phylum Porifera) exhibit a substantially different body form from other animals. They appear to represent an early evolutionary experiment that led to no other more advanced phyla. Similarly, the jellyfish and anemones (Phylum Cnidaria) have a unique body form and also appear to be an early offshoot from the main line of animal evolution. Some studies of gene function and muscle development, however, have suggested that the cnidarians and the related ctenophores might actually be derived from a triploblastic ancestor that also gave rise to the Bilateria. All of the remaining phyla represent a single clade (the Bilateria) that can be traced back to a common ancestor with bilateral symmetry, triploblastic body construction, and a special group of developmental genes known as the homeobox (HOX complex). These genes control certain body features such as segmentation, location of appendages, and differentiation within various regions of the body.

Evolution of a simple bilaterian into the diverse forms of all higher animals certainly involved many small changes occurring over millions of years. The prevailing interpretation of morphological, physiological, embryological, and genetic data suggests that the bilaterian clade diverged into two main lines of descent: **Protostomia** and **Deuterostomia** (figure 5.5). These clades provide a good illustration of the methods and logic that scientists use in attempting to reconstruct the evolutionary relationships of animals.

Protostomia

The Protostomia includes most of the invertebrate phyla such as annelid worms, arthropods, molluscs, flatworms, nematodes, rotifers, and their relatives. To some extent, all of these animals exhibit the following characteristics:

1. **Mouth.** Forms from the embryonic blastopore.
2. **Determinate development.** Determination occurs early in development so that each blastomere from the two-cell stage is capable of forming only one half of an embryo. Experimental separation of these first two blastomeres yields two half-embryos.
3. **Schizocoelous coelom formation.** Certain cells from the endodermal lining of the archenteron migrate to the blastocoel and form a mass of mesoderm cells. This mesodermal mass later splits apart, forming a central cavity, the coelom. The mesodermal cells surrounding the coelom later expand, and the coelom becomes the main internal cavity of the adult animal.
4. **Spiral cleavage.** The pattern of cleavage in the early embryo proceeds in a spiral pattern. After the four-cell stage, the mitotic axis, which determines the plane of cleavage in each cell, rotates 45 degrees before the next cleavage. Thereafter, the mitotic axis rotates 90 degrees between several successive divisions to produce a spiral pattern.
5. **Trochophore larva.** A common larval type exhibited by most marine species among the protostomes. Not present in freshwater or terrestrial species.

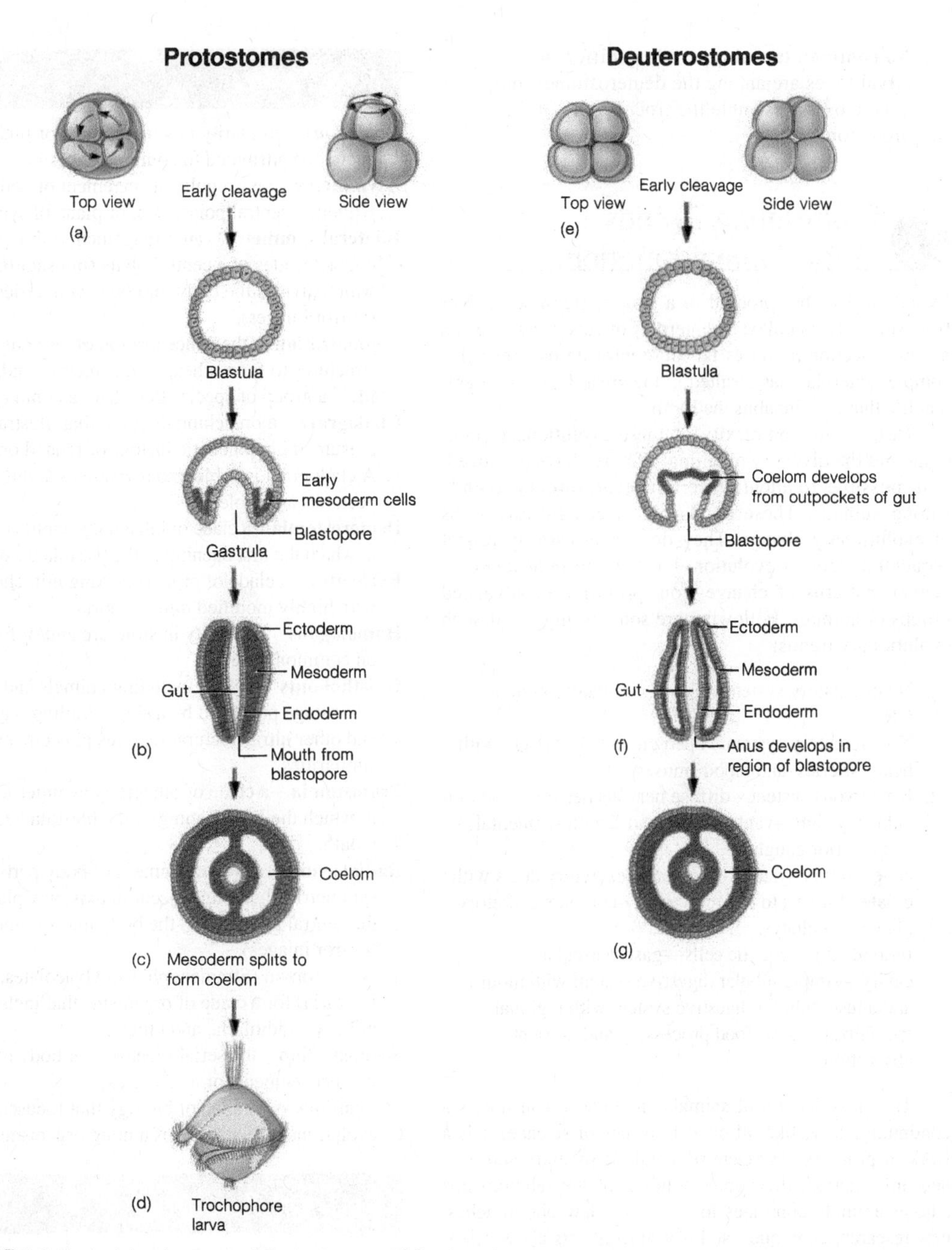

FIGURE 5.5 Typical development of Protostomes and Deuterostomes.

Deuterostomia

Deuterostomia is the other evolutionary branch of the Bilateria. It includes the phyla Echinodermata, Hemichordata, and Chordata. These animals share the following characteristics:

1. **Anus.** Forms from the blastopore.
2. **Indeterminate development.** The embryonic fate of blastomeres is determined relatively late in development. Thus, experimental separation of the first two blastomeres leads to the formation of two complete but half-size embryos.
3. **Enterocoelous coelom formation.** The primary mesoderm cells are formed as lateral outpockets or pouches from the archenteron. Thus, the coelom arises within these pouches as they separate from the archenteron.
4. **No common cleavage pattern.** Embryonic cleavage is radial or various in pattern, but never spiral as in protostomes.

5. **No common larval type.** Many different larval types are among the deuterostomes, but none closely resemble the trochophore of the protostomes.

Evolutionary Trends in the Animal Kingdom

Evolution does not proceed in a simple, stepwise manner from simple to complex; the interplay of genetic change and natural selection produces far more intricate patterns. This complex interplay has resulted in the great diversity of animal life that now inhabits the Earth.

Despite the complexity of these evolutionary processes and the diversity of animal life they have produced, it is possible to identify certain **evolutionary trends** among animals. These trends reflect general directions of evolutionary change. They do not necessarily reflect sequential steps in evolution, but they do indicate some general patterns of change from primitive to advanced groups of animals. Following are some examples of such evolutionary trends:

1. No circulatory system—closed circulatory system.
   OR
   No circulatory system—open circulatory system with heart, arteries, and blood sinuses.
2. No nervous system—diffuse nervous network—ventral ladder system—ventral nerve chord with segmental and anterior ganglia.
3. No excretory system—segmental excretory ducts with ciliated funnels to collect wastes—complex excretory glands or kidneys.
4. Individual phagocytic cells—gastrovascular cavity—simple tubular digestive system with mouth and anus—tubular digestive system with regional specializations for food processing and nutrient absorption.

The classification of animals and other organisms is a continuing story: like all other branches of science, it is a work in progress. As scientists continue to study and gain new information, their understanding of the relationships among animals continues to increase. New appproaches, new research techniques, and new instruments allow scientists to gain new insights. From this increased knowledge comes the need to modify the system of classification to reflect current understanding.

A phylum or a class or a domain is not a physical entity; rather, each is a concept shaped by scientists applying the most current accumulated knowledge to the challenge of classifying organisms in a way that reflects their heritage. So, to answer the question, How many kingdoms (or phyla, or divisions, etc.)? Tune in again tomorrow!

Key Terms

Analogous similarity in structure and/or function that cannot be attributed to common ancestry.

Asymmetry an irregular arrangement of body parts; without a central point, axis, or plane of symmetry.

Bilateral symmetry an arrangement of body parts on opposite sides of a central plane (midsagittal plane), which divides the body into two symmetrical halves (mirror images).

Cephalization the concentration of nervous and sensory structures to form a head at the anterior end.

Clade a group of species that share a common ancestor.

Cladogram a branching diagram that illustrates the presumed phylogenetic history of related organisms. A cladogram is subject to revision as additional data becomes available.

Deuterostomia a clade of bilaterally symmetrical animals in which the first opening of the blastula becomes the anus.

Excavates a clade of protists lacking mitochondria or with highly modified mitochondria.

Homologous similarity in structure and/or function based on common ancestry.

Opisthokonts a clade including animals and fungi, with a life stage propelled by a single trailing flagellum and other ultrastructural features plus other genetic similarities.

Protostomia a clade of bilaterally symmetrical animals in which the first opening of the blastula becomes the mouth.

Radial symmetry arrangement of body parts symmetrically around a central axis; any plane through the central axis divides the body into symmetrical halves (mirror images).

SAR acronym (**S**tramenophiles, **A**lvaeolates, and **R**hizaria) for a clade of organisms that includes diatoms, mildews, radiolaria, and others.

Segmentation the serial repetition of body parts into distinct segments or metameres.

Systematics a branch of biology that focuses on evolutionary relationships among organisms.

Internet Resources

There are many valuable Internet sites with information about zoology. Several sites containing pertinent zoological information for this chapter can be found on the McGraw-Hill zoology web site at **http://www.mhhe.com/zoology**. Just click on this text's title.

Questions for Critical Thinking

1. Explain why many biologists consider the Linnaean system of classification to be inherently flawed and wish to abandon it.

2. How does the existence of homologous and analogous traits complicate the study of animal systematics?
3. What assumptions do biologists make when they use changes in DNA (or RNA) sequences to infer relationships among different animal species?
4. Discuss the role cephalization may have played in the evolution of complex animals.
5. Describe the coelom and explain how the coelom may have contributed to the evolution of more complex animals.
6. Compare the following symmetry types: asymmetry, radial symmetry, and bilateral symmetry. Might there be some advantages of one over the other?
7. Discuss how higher grades of tissue organization. coelom formation, and segmentation have contributed to the great success of the more complex animals.

Suggested Readings

Baum, D., and S. Offner. 2008. Phylogenies and tree thinking. *American Biology Teacher* 70(4): 222–229.

Cracraft, J., and M.J. Donoghue. 2004. *Assembling the Tree of Life*. New York: Oxford University Press. 592 pp

Fedonkin, M.A., and J.G. Gehling, et al. 2008. The Rise of Animals: Evolution and Diversification of the Kingdom Animalia. Baltimore, MD: Johns Hopkins Press. 344 pp

Hall, B. 2007. *Phylogenetic Trees Made Easy*, 3rd ed. Sunderland, MA: Sinauer Associates. 232 pp

Lecointre, G., H.L. Guyader, et al. 2007. *The Tree of Life: A Phylogenetic Classification*. Cambridge, MA: Belknap Press. 560 pp

Neilson, C. 2001. *Animal Evolution: Interrelationships of Living Phyla*. Oxford: Oxford University Press. 563 pp

Offner, S. 2001. A universal phylogenetic tree. *American Biology Teacher* 63(3): 164–170.

Thanukos, A. 2008. A name by any other tree. *Evolution: Education and Outreach* 2(2):303–309, DOI: 10.1007/s12052-009-0122-7Open Access

University of California Museum of Paleontology. "From prokaryotes to eukaryotes." *Understanding Evolution*. 3 April 2011 http://evolution.berkeley.edu/evolibrary/article/_0/endosymbiosis_03

University of California Museum of Paleontology. "Understanding evolution: Phylogenetic systematics, a.k.a. evolutionary trees." *Understanding Evolution*, 3 April 2011 http://evolution.berkeley.edu/ogy.evolibrary/article/_0/endosymbiosis_03

Notes and Sketches

Notes and Sketches

Chapter 6

Animal-like Protista (Protozoa)

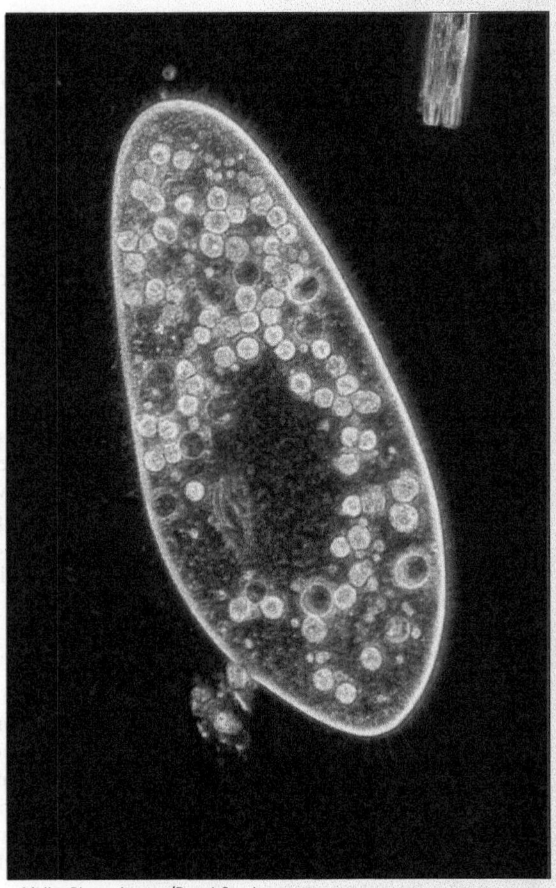

© Melba Photo Agency/PunchStock.

Objectives

After completing the laboratory work in this chapter, you should be able to perform the following tasks:

1. Discuss the possible role of protozoans in the evolution of animals.
2. Compare the structure of a protozoan with that of a typical animal cell.
3. List and describe three types of locomotion in protozoans.
4. Identify the major structures found in *Euglena.*
5. Identify the major structures in a specimen of *Amoeba proteus.*
6. Demonstrate the techniques for preparing a wet mount and a hanging drop for microscopy and explain appropriate uses of each.
7. Identify the main structures found in *Paramecium* and explain the function of each part.
8. Briefly describe the similarities and differences between cilia and flagella.
9. Compare the locomotion of *Euglena, Amoeba,* and *Paramecium.* If they were in a race, which would win?
10. Briefly describe the life cycle of *Plasmodium* and identify the main stages in microscope preparations.
11. List three important human parasites among the protozoans.
12. Identify three different protozoans, or kinds of protozoans, that are of economic importance (other than health related) and briefly explain why they are important.

Introduction

The protozoans are a large and diverse group of unicellular, eukaryotic organisms that exhibit certain animal-like features. For many years they were considered as members of a single, very diverse phylum called Protozoa. Later, as scientists began to accept the six kingdom system of classification, protozoa were included in the larger group of unicellular organisms called kingdom Protista. Members of this group were classified largely on the basis of types of locomotion (pseudopodia, cilia, flagella, etc.), and/or modes of nutrition (autotrophic, heterotrophic, or mixotrophic), and cellular morphology.

The term *protozoa* means "first animals," which is very appropriate since evidence now strongly suggests that they were among the first eukaryote organisms. Research has shown that eukaryotes evolved from prokaryotes in response to changing environmental conditions some 1.2–2 billion years ago during the Proterozoic era. One key event in the evolution of these new forms appears to have been the invagination of lipoprotein membranes from the outer cell membrane that allowed some partitioning of the interior of the cell. This process allowed partitioning of the cytoplasm for specialized functions. Other key events include incorporation of certain symbiotic prokaryotes with special properties beneficial to the host cells. The **serial endosymbiotic theory** is now widely accepted and postulates that such incorporations of other prokaryotes into the newly evolving eukaryotes led to the evolution of plastids, mitochondria, and possibly other organelles found in modern eukaryotic cells. Although our understanding of the details of early eukaryote evolution is far from complete, continuing research studies improve our understanding of the processes.

Materials List

Living specimens
- *Amoeba proteus*
- *Euglena*
- *Paramecium caudatum*

Prepared microscope slides
- *Arcella* (demonstration)
- *Difflugia* (demonstration)
- *Entamoeba histolytica* (demonstration)
- *Actinosphaerium* (demonstration)
- *Globigerina* (demonstration)
- *Peranema* (demonstration)
- Symbiotic flagellates from termite or wood cockroach (demonstration)
- Dinoflagellates (demonstration)
- *Paramecium*, pellicle (demonstration)
- *Paramecium*, trichocysts (demonstration)
- Representative ciliates (demonstration)
- *Plasmodium* (demonstration)
- *Eimeria* (demonstration)
- *Trypanosoma* (demonstration)

Chemicals and other supplies
- Lugol's solution
- Methylene blue, 0.1% solution
- Protoslo (or methyl cellulose solution)
- Modeling clay
- Congo red stained yeast cells

Classification and Phylogeny

The term protozoa is no longer considered a taxonomic category and is currently used in reference to animal-like unicellular protists; that is, heterotrophic nonfilamentous forms like *Amoeba* and *Paramecium.* It is important to recognize that recent research has demonstrated that protozoans are a much more complex and diverse group of organisms than anyone expected and that their classification is still a work in progress. Modern studies of morphology, biochemistry, and molecular phylogenetics are generally complementary but sometimes are also conflicting. The classification scheme currently proposed by current specialists recognizes some groups that are evolutionarily related, some groups moved to other taxa, and still other groups whose evolutionary relationships are still in doubt.

Since there is no final consensus on their classification, we can present here only a simplified example of a few groups of the most familiar forms such as *Amoeba, Paramecium, Euglena,* and a few other scientifically, economically, or medically important members. Several types of protozoans serve as important organisms for basic research, like *Paramecium* and *Tetrahymena.* Others are important pathogens of humans and animals, like *Plasmodium*, *Leishmania*, and *Entameba.* Still others are important for economic reasons, like foraminifera and radiolaria.

Flagellated Protozoa

The modern classification of unicellular animals is based on molecular evidence rather than morphological and physiological similarities that biologists relied on in the past. Therefore, we often find that members of a clade have few observable features in common. Members of this group, however, are all heterotrophic and have one or more flagella. We shall study two common flagellates as representatives of this clade: *Euglena* and *Peranema.*

Phylum Euglenozoa: *Euglena*

Euglena (figure 6.1) is a common green flagellate often found in the greenish surface scum of freshwater ponds or streams. *Euglena* is an unusual organism with a curious mixture of plant and animal characteristics. Its small size makes it difficult to observe its internal structures with the light microscopes usually available in student laboratories. You can, however, observe some of its features and behavior with a student microscope.

- Prepare a wet mount from a culture of living *Euglena* and observe the locomotion of an active specimen under your compound microscope.

The active swimming movements result from the beating of the long flagellum, which pulls the organism through the water. A second, shorter flagellum is present within the flagellar pocket, but does not aid in the swimming movements.

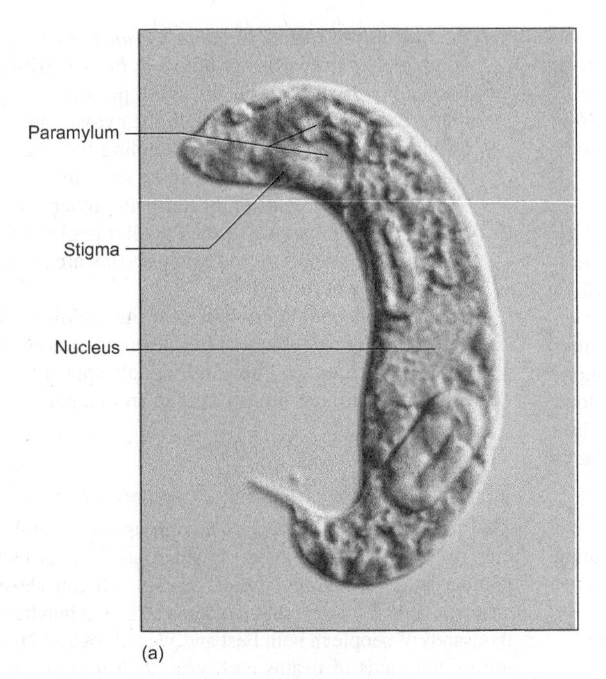

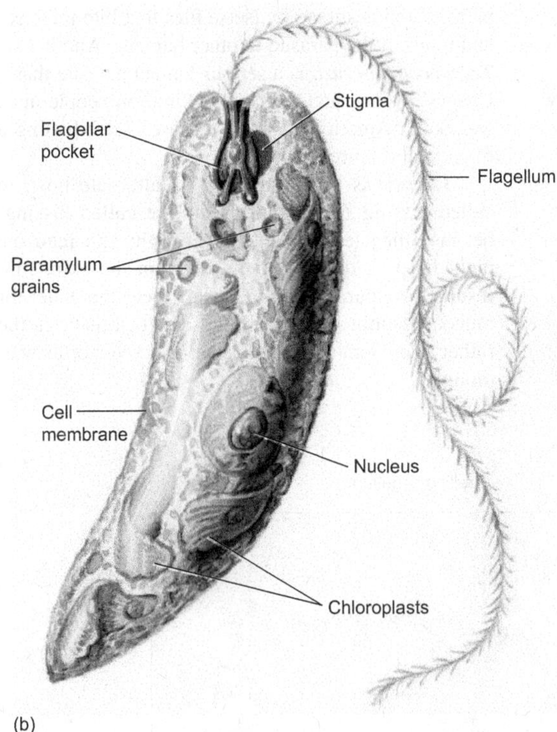

FIGURE 6.1 *Euglena.* (*a*) Photograph of a living *Euglena.* (*b*) Drawing of *Euglena.*
(a) © Stephen Durr.

At certain times *Euglena* also exhibits another type of wormlike locomotion during which waves of contraction pass along the body in a characteristic fashion. This type of locomotion is peculiar to *Euglena* and related organisms and is appropriately termed **euglenoid movement** or **metaboly.** It appears to result in part from the elasticity of the layer of microtubules making up the pellicle underlying the cell membrane.

- After the wet mount begins to dry out, temporarily immobilizing some of your specimens, study the anatomy of a stationary *Euglena.* You will also find it useful to supplement your observations with the study of a prepared microscope slide.
- Identify the following structures under high power on your compound microscope: (1) **cell membrane,** the outer covering of the body; (2) **chloroplasts** with green chlorophyll; (3) **nucleus,** exhibiting a large central **endosome** in stained preparations; (4) **flagellar pocket;** (5) **contractile vacuole;** (6) a light-sensitive red **stigma,** or eye spot; (7) the long anterior **flagellum;** and (8) **paramylum grains,** a type of starch that represents stored food materials.

The pellicle is a layer of parallel microtubules that lies just beneath the cell membrane. The microtubules spiral around the cell and support the characteristic shape of the body. The elasticity of the microtubules making up the euglenid pellicle enables the euglenloid movement. The pellicle is not actually visible in light microscope preparations without special staining.

The flagellar pocket is the one area on the surface of the body with no microtubules beneath the cell membrane. This allows the ingestion of bacteria and the formation of food vacuoles. ***What does this suggest about the nutritional capabilities of Euglena?***

Euglena is quite sensitive to light, and changing light intensity affects its behavior. In weak light, *Euglena* tends to be **positively phototactic** (attracted to the light); in strong light, *Euglena* tends to be **negatively phototactic** (repelled by the light); and sudden changes in light intensity often stun *Euglena* so that it remains stationary.

- After you have completed your observations of the living specimen, add a drop of Lugol's solution (iodine and potassium iodide). This solution will kill the specimen and stain the flagellum to make it more visible.

As suggested by the presence of chloroplasts, *Euglena* is normally autotrophic; organic molecules (sugars) are synthesized from inorganic nutrients absorbed from the medium. Light from the sun provides the energy necessary for this process.

Biochemical tests have shown the paramylum granules to be a form of starch similar to that found in plants. Thus, the presence of chloroplasts and the storage of a plantlike form of starch both indicate that *Euglena* and its relatives may be related to members of the plant kingdom. Some species of *Euglena* are also able to survive, grow, and reproduce in the dark with no visible evidence of chloroplasts, chlorophyll, or stored food materials. ***How might such organisms obtain their food under these conditions?***

Phylum Euglenozoa: *Peranema*

Peranema (figure 6.2) is a common flagellate often studied because of its prominent anterior flagellum, which is larger than the flagellum of most species of *Euglena*. *Peranema* (25–80 microns in length) is slightly smaller than most species of *Euglena* (50–100 microns in length) and is often found in stagnant pools of fresh water along with species of *Euglena*. Like *Euglena*, *Peranema* is a solitary flagellate, but unlike *Euglena*, *Peranema* lacks chloroplasts.

- Make a wet mount of *Peranema* on a clean microscope slide adding some Protoslo or methyl cellulose solution to minimize movement of your specimens, and observe a living specimen. Observe the prominent anterior **flagellum** that extends straight out from the anterior end. Note that most of the anterior flagellum is stiff except for a short portion near the anterior tip, which tends to be very flexible and active. *Peranema* actually has two flagella, but one of them adheres closely to the pellicle and is difficult to observe in most ordinary microscopic preparations.

  Observe how *Peranema* usually moves along the substrate in a gliding fashion, not clearly resulting from the action of its flagellum. The actual mechanism of its locomotion is not really understood.

  Find the flask-shaped **flagellar pocket** near the base of the two anterior flagella. Food particles pass through a tiny **cytostome** into this pocket before being enclosed in **food vacuoles** formed from membrane in its wall and ingested into the cytoplasm. Unlike *Euglena*, *Peranema* is completely carnivorous or holozoic in its nutrition. Food consists of bacteria and small protozoa that are engulfed whole through the cytostome. Supporting the reservoir and extending into the adjacent cytoplasm are two rodlike structures. When *Peranema* feeds on bacteria and other small organisms, the opening to the flagellar pocket can open widely, and these rods support this area while a food vacuole is formed.

  Locate the single **nucleus** near the middle of the body. Sexual reproduction has never been described in this flagellate, and new individuals appear to arise only through **binary fission** from a parent organism.

Phylum Kinetoplastida: *Trypanosoma*

The most familiar members of this group are several species of *Trypanosoma*, some of which are serious human pathogens. *Trypanosoma brucei* causes African sleeping sickness, a devastating human disease affecting hundreds of thousands of people in both East and West Africa and resulting in thousands of deaths each year. *T. brucei* is a blood parasite and is spread by tsetse flies that bite infected hosts and transmit the parasite to other humans. Another species, *Trypanosoma cruzi*, is a serious human parasite that causes Chagas' disease and affects 8–10 million people in Central and South America. Chagas' disease is debilitating and is often fatal if untreated.

T. cruzi is transmitted by its alternate host, insects called kissing bugs. The insects are called kissing bugs because they tend to emerge at night and feed on people's faces, biting and defecating on the skin. Infection results from parasites in the insect feces that enter through mucous membranes or through cuts or breaks in the skin rather than from the bites of the insect vector as was once thought.

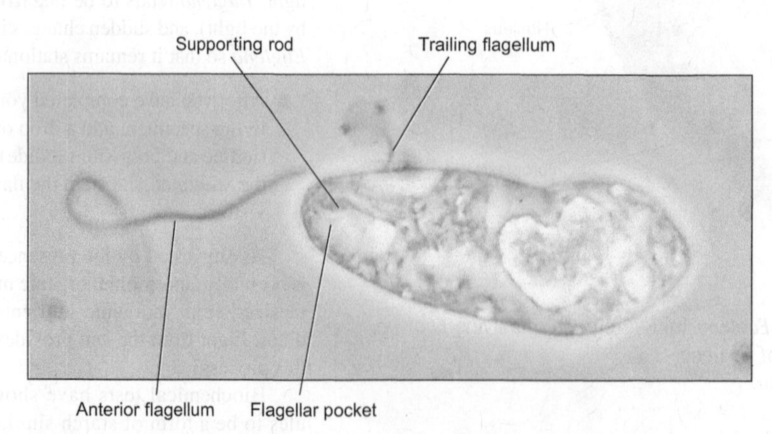

FIGURE 6.2 *Paranema*, phase contrast photograph.
Courtesy of Carolina Biological Supply Company, Burlington, NC.

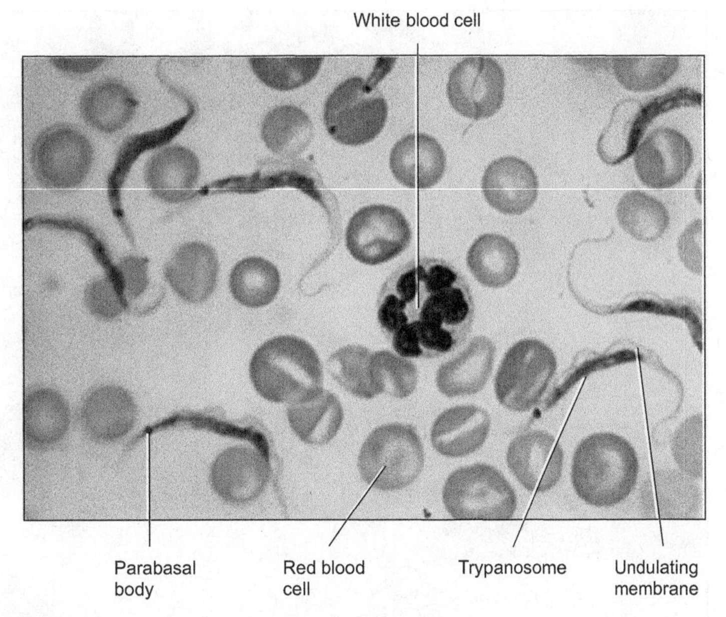

FIGURE 6.3 *Trypanosoma,* blood smear.
Courtesy of Carolina Biological Supply Company, Burlington, NC.

Trypanosoma has a thin, undulating membrane connecting its single, long, whiplike flagellum to the surface of the cell body (figures 6.3 and 6.4). The undulating membrane extends along the entire length of the body and consists of part of the cell membrane that wraps around the flagellum. As it ripples with the beating of the flagellum, the undulating membrane guides the trypanosome through blood plasma, acting like a fin.

- Observe a microscope slide of trypanosomes on demonstration in the laboratory. Since trypanosomes are very small, they must be viewed with high magnification and an oil immersion lens to see the characteristic undulating membrane with its flagellum.

Ancient Flagellates

This group of unicellular eukaryotes includes a variety of free-living predators and parasitic forms, some of which have both flagella and pseudopodia. Members of this group either lack mitochondria or have non-functional mitochondria that lack the usual biochemical systems. These flagellates were formerly thought to be primitive, but recent evidence indicates that these groups have probably undergone secondary loss of functional mitochondria resulting from long existence as internal parasites in anaerobic environments. As such, they would be considered specialized rather than primitive in nature. Two distinct groups, often considered phyla, have been identified within this group.

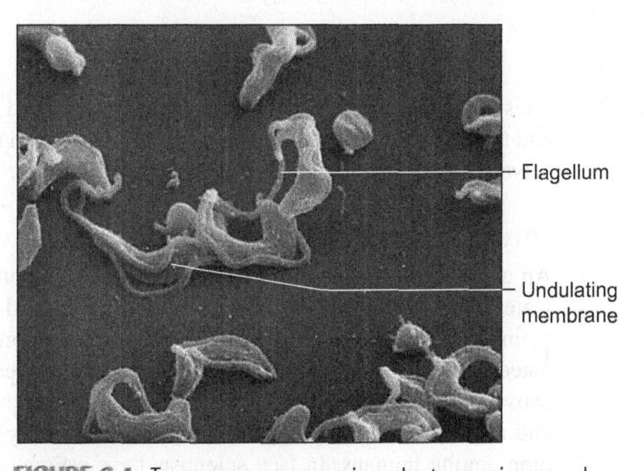

FIGURE 6.4 *Trypanosoma,* scanning electron micrograph, magnification 3525×.
Scanning electron micrograph by Louis de Vos.

Phylum Parabasalida

Trichomonas vaginalis is a sexually transmitted pathogen of humans, and infections cause vaginitis in women and can damage the urinary and reproductive tracts. Males are less susceptible to the parasite and rarely show clinical symptoms, although they can be involved in its transmission and infection of female partners.

Two symbiotic parabasalids are often found in human digestive tracts: *Trichomonas tenax* is found in the mouth and *Pentatrichomonas hominis* lives in the colon. Both appear to be harmless and cause no disease. *Trichonympha* and *Spirotrichonympha* (figure 6.5) are symbionts in the

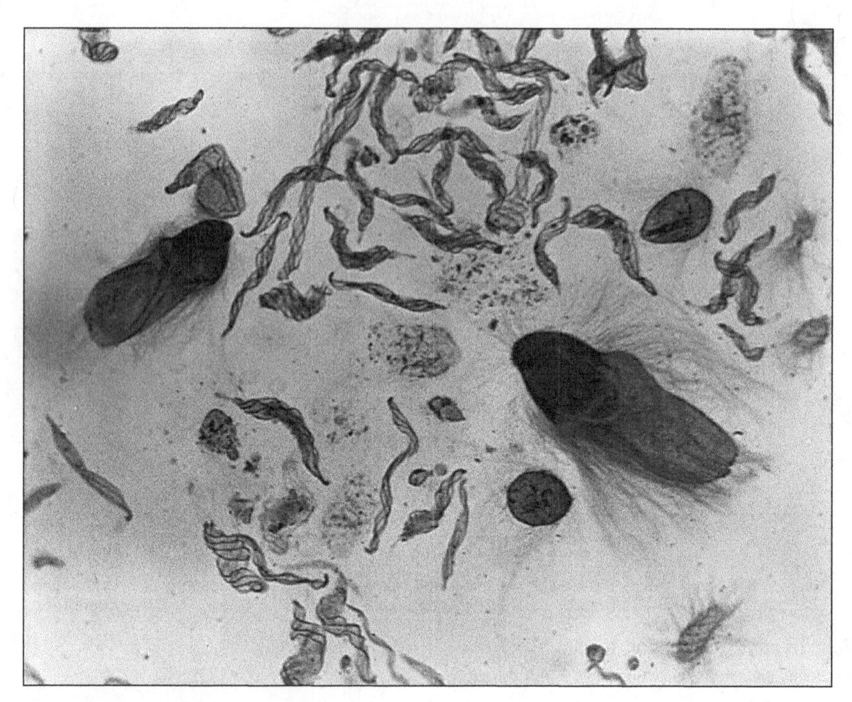

FIGURE 6.5 Symbiotic flagellates from termite gut.
Courtesy of Carolina Biological Supply Company, Burlington, NC.

intestine of termites. They digest cellulose from wood fibers and form sugars that are shared by the termites, which themselves lack the enzymes necessary to digest cellulose.

Phylum Diplomonada

An example of this phylum is *Giardia lamblia,* a common waterborne pathogen of humans, cattle, beavers, and other mammals. Infections result from drinking water contaminated by human or animal waste from infected sources. Because of the use of human and animal wastes for fertilizer and other poor sanitary conditions, giardiasis is very common among humans. In fact, scientists have estimated that as many as 20 percent of humans are infected with *Giardia* at any given time. Typically, the parasite causes only mild intestinal upsets, but serious infections can sometimes result in gastroenteritis.

Demonstrations

1. Microscope slide of *Paranema* stained to show flagella
2. Microscope slide with blood smear containing trypanosomes
3. Microscope slide of termite symbiotic flagellates
4. Microscope slide of *Trichomonas vaginalis*
5. Microscope slide of *Giardia lamblia*

Amoeboid Protozoa

The amoeboid protozoa are thought to have originated from flagellated forms that have lost their flagella. They move by forming pseudopods and by cytoplasmic streaming.

Classification of amoeboid protozoans has long been controversial and there is yet no real consensus on their classification. We can, however, distinguish several distinct groups of undetermined taxonomic rank. Some of these groups are considered to represent phyla and some are considered to represent classes. More study is needed before the classification can be resolved. Among these groups are the Rhizopoda (naked amoebas formerly called Sarcodina), the Foraminifera (amoebas with perforated calcareous shells), and the Actinopoda consisting of two subgroups: the Heliozoa and the Radiolaria.

Rhizopoda

Rhizopods (formerly called sarcodines) are widely distributed in damp soil and freshwater ponds, lakes, and streams as well as in saltwater habitats. They are called naked amoebas since they do not form tests or shells, but many secrete some form of test for protection of the cell body. The vast majority of amoebae in other groups are testate. Some rhizopods are facultative parasites, but most species are free-living and feed by engulfing bacteria and other small protozoa by phagocytosis. We shall first study *Amoeba proteus* to introduce some basic features of the amoebas.

An Amoeba: ***Amoeba proteus*** *Amoeba proteus* (figure 6.6) is a protistan found in ponds and streams. It often occurs on the undersides of plant leaves and among diatoms and desmids. The transparent amoeba constantly changes shape by extending pseudopodia, foot-like extensions of the cytoplasm, which serve for locomotion and in food capture. *Amoeba proteus* feeds on bacteria, small algae, and smaller protistans.

In feeding, an advancing pseudopodium flows over one or more food organisms to trap the food in a water-filled cup. The opening of the food cup then narrows until the food is completely enclosed in a food vacuole.

- Prepare a wet mount to study living amoebae under your compound microscope.

Preparing a Wet Mount and Hanging Drop Obtain a clean microscope slide and add a drop of amoeba culture solution to the center of the slide. Take care to withdraw the drop of culture solution from the bottom of the culture dish or jar with a clean eyedropper or pipette. The amoebae are slightly heavier than the culture solution and are usually concentrated on the **bottom** of the vessel. To avoid crushing the delicate organisms under the weight of the coverslip, you will need to add some support to the underside of the coverslip. Have a small container, like a small petri dish, with a layer of modeling clay at hand. Simply drag each corner of the coverslip across the clay to get a small amount on the corners. The clay will then support the coverslip when placed on the slide and you can press gently on the coverslip, taking care not to crush the organisms in the droplet. This preparation is called a **wet mount.**

Another method of studying living protozoa is to prepare a **hanging drop.** With this method you place a drop of the amoeba culture on a coverslip and invert the coverslip over the cavity of a depression slide. Both types of preparations can be observed for long periods of time if the outside edge of the coverslip is coated with petroleum jelly before it is placed in position on the microscope slide. Be sparing

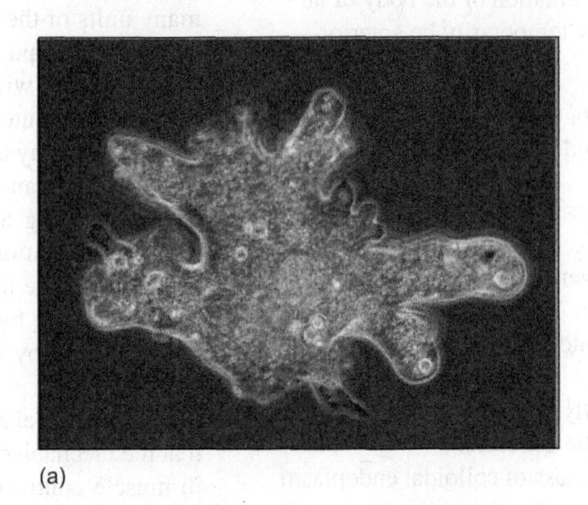
(a)

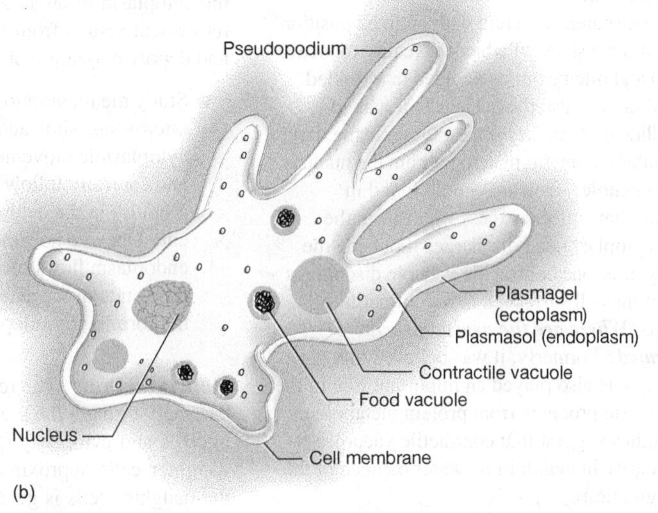

(b)

FIGURE 6.6 *Amoeba proteus.* (*a*) Photograph of a living amoeba. (*b*) Drawing of amoeba.
(a) © Melba Photo Agency/PunchStock.

in your use of the petroleum jelly; avoid getting it into the culture droplet and on your microscope lens.

Study your preparation under the low power of the compound microscope to observe the general appearance of the amoeba. For best observation of a living amoeba, reduce the illumination to a minimum with the iris diaphragm since living specimens are nearly transparent and almost invisible in bright light. Search your slide carefully to locate a specimen before discarding it or asking your instructor for a new preparation.

A phase contrast microscope is an excellent tool for viewing living amoebae and other protozoa since various parts of the cell appear lighter or darker depending on differences in their refractive indices. This type of a microscope reveals many structural details in thin preparations of living cells and/or tissues without staining.

- Locate an actively moving amoeba and note its constantly changing shape. The long, fingerlike projections are **pseudopodia** ("false feet"). Observe the lack of permanent orientation of the body of an amoeba; any portion may temporarily be anterior, posterior, right, or left.

- With the aid of figure 6.6, identify and study the following structures found in the amoeba.

1. **Endoplasm**—the inner granular region that forms the bulk of the cytoplasm.
2. **Ectoplasm**—the thin layer of clear cytoplasm that surrounds the endoplasm.
3. **Cell membrane**—the outer membrane surrounding the amoeba.
4. **Plasmagel**—the stiff, jellylike, granular outer layer of colloidal endoplasm in the **gel state.**
5. **Plasmasol**—the central mass of colloidal endoplasm in a fluid, or **sol state.** Note the streaming movements within the plasmasol.
6. **Nucleus**—a transparent structure with no fixed position in the cell. It has the shape of a biconcave disc (concave on two sides) and often exhibits a folded or wrinkled appearance. Examine also the nucleus in a stained microscope slide of *Amoeba proteus.* Observe the darkly staining granular chromatin material within the nucleus.
7. **Contractile vacuole**—a clear vacuole found in the endoplasm that collects excess water from the surrounding cytoplasm and discharges it outside the body. Shortly after one contractile vacuole discharges its contents at the cell surface, a new contractile vacuole forms. ***Where are the new contractile vacuoles formed?*** Formerly, it was believed that the contractile vacuole also played an important role in the excretion of waste products from protein metabolism, but recent studies suggest that contractile vacuoles function primarily in maintaining water balance in the cell (osmoregulation).
8. **Food vacuoles**—vacuoles containing bits of ingested food and the digestive enzymes that act to break down these food materials into soluble materials that can be utilized by the amoeba. ***How are the food vacuoles formed? How are the undigested contents of a food vacuole disposed of after digestion has taken place?***

Amoeboid Movement An amoeba moves about by extending pseudopodia into which some of the innermost cell contents flow. Various kinds of amoebae form pseudopodia of different size and form. Pseudopodia are important in feeding, support, and locomotion. The mechanism of amoeboid movement has been studied by many scientists because of its intriguing nature and because similar movements occur in many other kinds of cells, including human leucocytes. Also, scientists believe that amoeboid movement may be closely related to the phenomenon of cytoplasmic streaming, movement of cell contents that occur in virtually all kinds of living cells.

The movement of an amoeba is a result of the actions of components of the cytoskeleton, particularly microtubules and microfilaments. Microtubules are made up of many units of the structural protein tubulin, and microfilaments are composed of actin, the same contractile protein that combines with myosin in muscle cells to cause contraction. The outer portion of an amoeba is relatively stiff because of many actin/myosin complexes that are linked to the cell membrane. These actin/myosin complexes are continuously being assembled and disassembled as contraction and relaxation occur in the outer region of a growing pseudopod. The inner, more fluid portion of the cytoplasm is pulled along by the forward actin/myosin and it is also pushed along by contraction of the cell membrane at the rear of the cell.

Biochemical and biophysical studies have demonstrated that the mechanism of amoeboid movement is similar to that in muscle contraction. **Contractile proteins** similar to the actin and myosin found in vertebrate muscles are present in the cytoplasm of an amoeba. We now know that amoeboid movement results from folding, unfolding, polymerization, and depolymerization of these proteins.

- Study the locomotion of an *Amoeba* on your microscope slide and the pattern of its internal cytoplasmic movements. In an active specimen, locate and carefully follow the forward movement of some granules in the cytoplasm at the temporary posterior end. Observe how the more fluid plasmasol of the endoplasm flows forward and changes into the less fluid state of the ectoplasm again, just behind the tip of the forming pseudopodium.

Reproduction The reproduction of *Amoeba proteus* occurs only through the asexual process of binary fission. The nucleus and cytoplasm of a parent cell divide to form two daughter cells approximately equal in size. Thus, each of the daughter cells is genetically identical to the parent cell, excluding the rare occurrence of a mutation in one of the daughter cells.

A Parasitic Rhizopod: *Entamoeba histolytica* *Entamoeba histolytica* is an important human parasite in certain parts of the world (figure 6.7). Most infections come from contaminated drinking water in areas with poor sanitary conditions or from eating raw vegetables contaminated by human wastes.

Testate Rhizopods Many rhizopods, like *Arcella* (figure 6.9a), have tests or shells. In fact, there may be more testate than non-testate rhizopods. The tests may be secreted, or constructed with grains of sand (as by *Difflugia,* figure 6.9b) or other material, or a combination of the two. Sometimes past environmental history can be determined by analysis of the composition of fossil shells.

Foraminifera

Foraminifera are amoeboid protozoa that secrete a calcareous, multi-chambered test with many perforations through which many thin, stringlike pseudopodia extend (figure 6.8). These numerous pseudopodia often branch and anastomose to form a network that traps bacteria and other small organisms upon which they feed. Foraminifera are abundant in the fossil record and often serve as useful indicators of geological formations that may contain petroleum deposits.

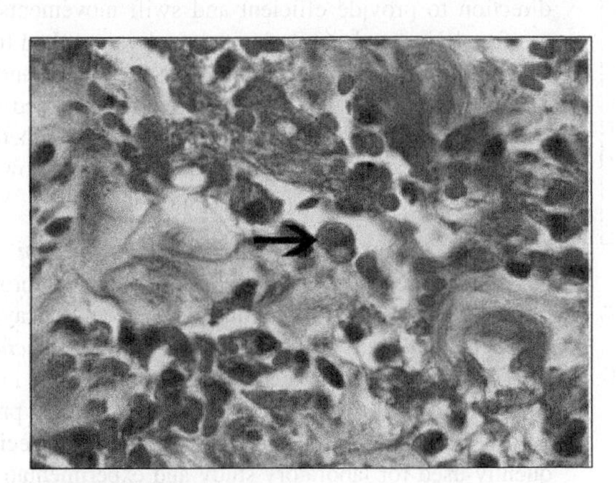

FIGURE 6.7 *Amoeba histolytica* in fecal smear.

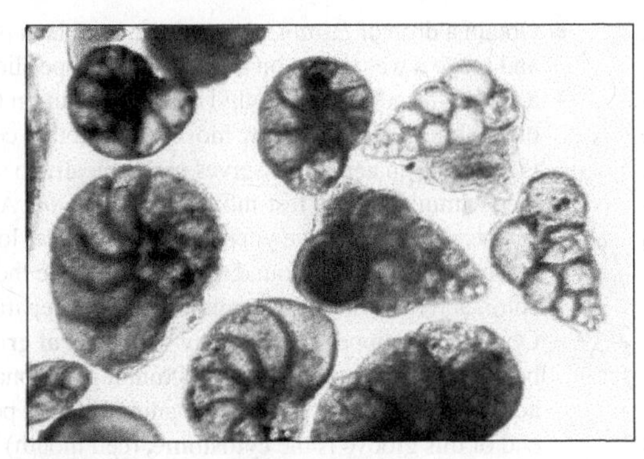

FIGURE 6.8 Foraminiferan tests.

Actinozoa

Heliozoa Heliozoans are amoeboid protozoa that form thin pseudopodia that radiate outward like the rays of the sun from a central body. They form a test of silicon and are also abundant in the fossil record.

Each of the thin pseudopodia is supported by a central bundle of microtubules. The microtubules also act as "tracks" for transport of food vacuoles to and from the surface of the pseudopodium using motor molecules like cytoplasmic dynein. The pseudopodia collect bacteria and other small food organisms and carry them to the central body where they are digested. Heliozoans are often found in freshwater plankton but some species are marine. *Actinosphaerium* is a common freshwater form (figure 6.10a).

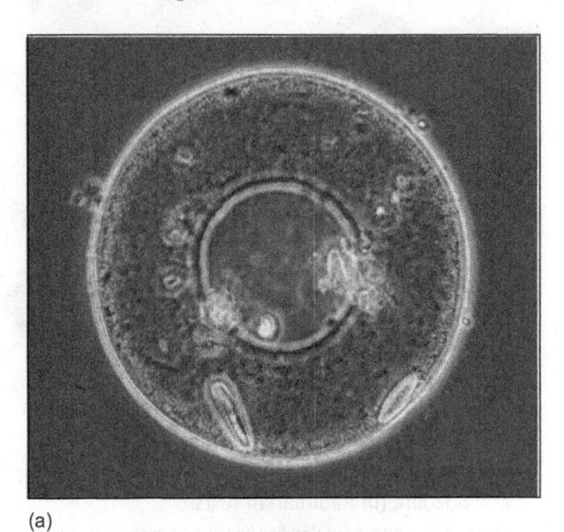

(a)

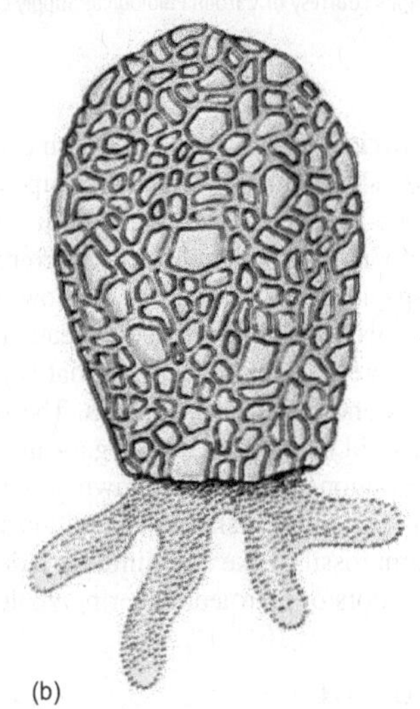

(b)

FIGURE 6.9 Testate rhizopods. (*a*) *Arcella*, living. (*b*) *Difflugia*, drawing.

(a)

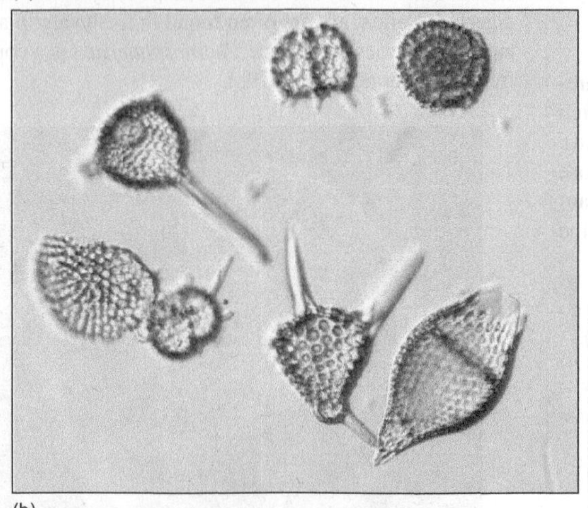

(b)

FIGURE 6.10 Heliozoa and Radiolaria. (*a*) *Actinosphaerium*, a heliozoan. (*b*) Radiolarian tests.
Photographs courtesy of Carolina Biological Supply Company, Burlington, NC.

Radiolaria Radiolarians are marine amoebas that secrete complex shells made of silicon compounds (figure 6.10b). They also accumulate strontium from seawater and thus are useful in monitoring radioactive strontium in the oceans. Most species are planktonic in shallow water but some species are also found in the deep ocean. The body is divided into two areas, a central capsule that is perforated and from which extend numerous axopodia. The sticky axopodia collect bacteria and other small organisms for food. Radiolarians are among the oldest known protozoa because of the composition of their skeletons, and most species are known only from fossils. Like foraminifera, radiolarians can be useful indicators of petroleum-bearing geological formations.

Alveolata

This group includes three important groups of unicellular eukaryotes, the ciliates, the apicomplexans or sporozoans, and the dinoflagellates, considered by some scientists as separate phyla. An important common feature is the presence of numerous, tiny, membrane-bound sacs (alveoli) located beneath the outer membrane. The alveoli are filled with water to form a fluid skeleton (hydrostatic) supporting the shape of the cell.

Demonstrations

1. Models and charts showing *Amoeba proteus*
2. Living *Amoeba* culture under a stereoscopic microscope
3. Microscope slide of *Amoeba histolytica*
4. Microscope slides of heliozoans
5. Microscope slides of foraminiferans
6. Microscope slides of radiolarians

Phylum Ciliata

These protozoans have numerous cilia extending from their surface that propel the organisms through their watery habitat. The cilia beat synchronously in a back and forth direction to provide efficient and swift movements. Most are free-living and motile but others live attached to some substrate. Some members form simple colonies and about one-third of the species are symbiotic. Specialized cilia in some species are used in feeding or for other functions in addition to locomotion. *Paramecium* is a well-known form that has been studied in many research studies.

Phylum Ciliata: *Paramecium caudatum* *Paramecium* (figure 6.11) is a large, common, ciliated protozoon often found in water containing bacteria and decaying organic matter. There are several species of *Paramecium* that differ in various details of structure and that range in length from about 120–300 microns. The description provided here is based upon *Paramecium caudatum,* a species frequently used for laboratory study and experimentation, but it will also apply, with minor variations (such as body size and number of micronuclei), to the study of other species of *Paramecium.*

- Obtain a drop of *Paramecium* culture in a clean pipette and make a wet mount on a clean microscope slide with a similar-sized drop of methyl cellulose solution (or other similar agent) to slow movement. Methyl cellulose is a viscous material and serves mechanically to slow the swimming of the fast-moving *Paramecium.* Add a coverslip and observe your preparation under low power with your compound microscope. Note the form, color, and behavior of the animals in your preparation. Observe the slipper-shaped body with an **oral groove** beginning at the anterior end and running diagonally across the anterior portion of the animal. At the posterior end of this groove is the **cytostome,** (cell mouth) to which food particles are passed as a result of the action of the specialized oral cilia lining the oral groove.

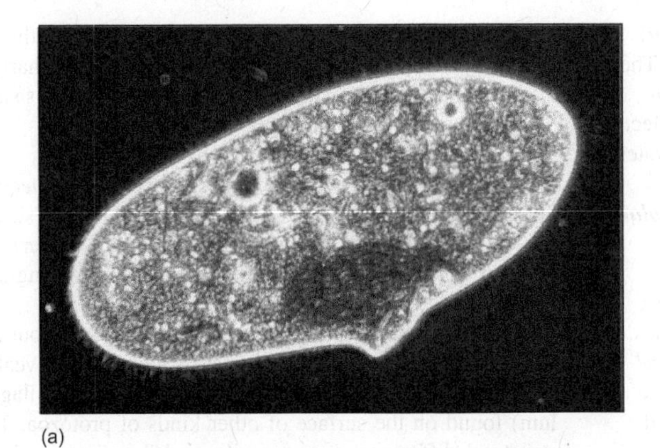

(a)

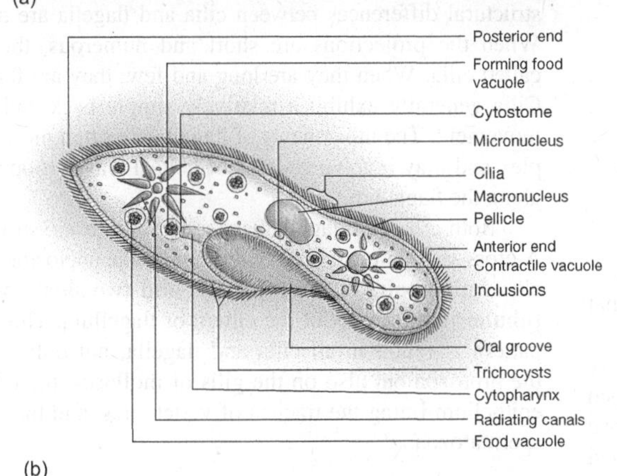

(b)

FIGURE 6.11 *Paramecium.* (*a*) Photograph of a living *Paramecium.* (*b*) Drawing of *Paramecium.*

(a) © Melba Photo Agency/PunchStock.

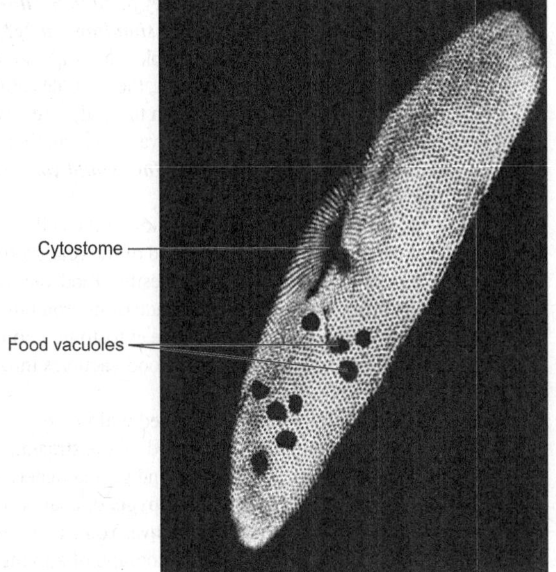

FIGURE 6.12 *Paramecium,* nigrosin stain to show structure of pellicle.

Photograph by Barbara Grimes.

Observe that *Paramecium* is much more complex in its structure than *Amoeba*. Select a large, immobile, or slowly moving specimen, and with the aid of figure 6.11, identify and study the following structures. The cytostome is not an actual opening but a section of the cell membrane with no alveoli beneath it, allowing food vacuoles to be formed and ingested.

1. **Cilia**—the numerous cylindrical cytoplasmic extensions that cover the surface of the *Paramecium* and that function in locomotion and in food gathering.
2. **Pellicle**—the layer of alveoli (membranous sacs filled with fluid or, sometimes, with more rigid protein fibers) that lie just beneath the cell membrane and function as a fluid-filled hydrostatic skeleton that maintains the shape of the cell. The pellicle and its system of associated fibers cannot be observed directly with a light microscope, but the pattern of alveoli can be seen with special stains. Figure 6.12 shows the pattern of surface depressions (dark areas) that lie between the small bulges created by the alveoli (light areas).
3. **Trichocysts**—tiny, rodlike structures embedded in the cortical (outer) cytoplasm beneath the pellicle. When properly stimulated, the trichocysts discharge their contents and form long threads. There is some evidence that the trichocysts may serve as a defense against predators, and they also serve to anchor the animal during feeding. In other types of ciliated protozoa, trichocysts have been found to have additional functions. Observe the microscopic demonstration of discharged trichocysts.
4. **Macronucleus**—the large nucleus located near the center of the cell. Since it is transparent in a living animal, the structure of the macronucleus is best studied in a stained preparation. Experiments have demonstrated that the macronucleus controls most metabolic functions of the cell.
5. **Micronucleus**—a smaller nucleus located close to and lying partly within a depression on the oral side of the macronucleus. The micronucleus is involved primarily in the reproductive and hereditary functions of the animal. This presence of two distinct types of nuclei is called **nuclear dimorphism** and is a condition found only in the Phylum Ciliata. *Paramecium caudatum* has only a single micronucleus, but some other species of *Paramecium* have two or more micronuclei. As with the macronucleus, the structure of the micronucleus is best studied in a prepared microscope slide.
6. **Contractile vacuoles**—two clear, slowly pulsating vesicles located near each end of the body. Each contractile vacuole is surrounded by several **radiating canals** (not often seen in ordinary student preparations), which collect water from the surrounding cytoplasm. Observe the behavior of

the contractile vacuoles. ***Are they fixed in position? Do they contract alternately or simultaneously?*** The function of the contractile vacuoles in *Paramecium* is similar to that in *Amoeba* (i.e., the vacuoles collect and discharge excess water from the cell). Freshwater protozoa often have contractile vacuoles; marine protozoa generally lack them. ***How would you explain this difference?***

7. **Cytostome** (cell mouth)—a region of the cell membrane near the posterior end of the oral groove that is used for forming and ingesting food vacuoles.
8. **Cytopharynx**—a short tube made of microtubules that extends posteriorly from the cytostome into the cytoplasm and serves to guide food vacuoles into the cell after ingestion.
9. **Food vacuoles**—vacuoles located within the cytoplasm where they are carried by the streaming movements of the cytoplasm. Undigested materials are discharged through the **cytopyge,** or anal pore, located posterior to the oral groove. You can observe discharged materials in the photograph of a living *Paramecium* on the first page of this chapter, page 79.

Feeding *Paramecium* is a filter-feeding organism that normally feeds on bacteria and yeast cells collected by a specialized food-collecting apparatus. An **oral groove** extends diagonally back along the body to a funnel-shaped **cytopharynx.** Food particles are swept along the oral groove by the action of specialized cilia lining the groove, collected into a mass at the circular cytostome, enveloped in a food vacuole that forms as a cup-shaped invagination from the membrane of the cytostome, and ingested into the cytoplasm within the food vacuole. Once inside the cytoplasm, the food vacuole is guided further into the cell by the microtubules of the **cytopharynx.**

- Prepare a wet mount with a drop of *Paramecium* culture to study the feeding process. Add a small amount of Congo red stained yeast with the tip of a toothpick or clean dissecting needle. Try to pick up the smallest amount of yeast possible on the toothpick; too much yeast will cloud your preparation and obscure the *Paramecium.* Remember to add a bit of modeling clay to the underside of the coverslip as recommended earlier in our discussion of a wet mount for amoeba (see page 85).

  With this preparation, you can study the movement of the food particles, the formation of food vacuoles, and the subsequent movement of the food vacuoles within the cytoplasm. After the food vacuoles are formed, digestive enzymes are released into them, and chemical digestion of the food particles begins. Note the color change in the vacuoles as the enzymes work. The color change is due to a change of pH in the vacuoles. ***Where do the digestive enzymes come from? Why don't they digest the other materials in the cell such as mitochondria and ribosomes?*** The diffusible products of digestion are released into the cytoplasm, and the undigestible remains are discharged at a specific site on the surface of the animal. This site is the **cytopyge,** or anal pore.

Cilia and Flagella Most of the surface of *Paramecium* is covered by thin, hairlike projections called **cilia** (singular: cilium). Cilia are extensions of the cortical (outer) cytoplasm of the cell and play important roles in feeding and locomotion.

A great deal has been learned in recent years about the structure and function of cilia. These studies have revealed that cilia are closely related to the **flagella** (singular: flagellum) found on the surface of other kinds of protozoa. The structural differences between cilia and flagella are minor. When the projections are short and numerous, they are called cilia. When they are long and few, they are flagella. Cilia generally exhibit a relatively simple back and forth movement. The movements of flagella are often more complex and may involve a series of helical waves propagated along the flagellum.

Both cilia and flagella have a common basic structure. A cross section reveals an outer membrane enclosing a circle of **nine pairs of microtubules and two single microtubules** in the center of the cilium or flagellum. This basic pattern is found in all cilia and flagella, not only among the protozoa but also on the gills of molluscs, the ciliated epithelium lining the trachea of vertebrates, and the tail of spermatozoa.

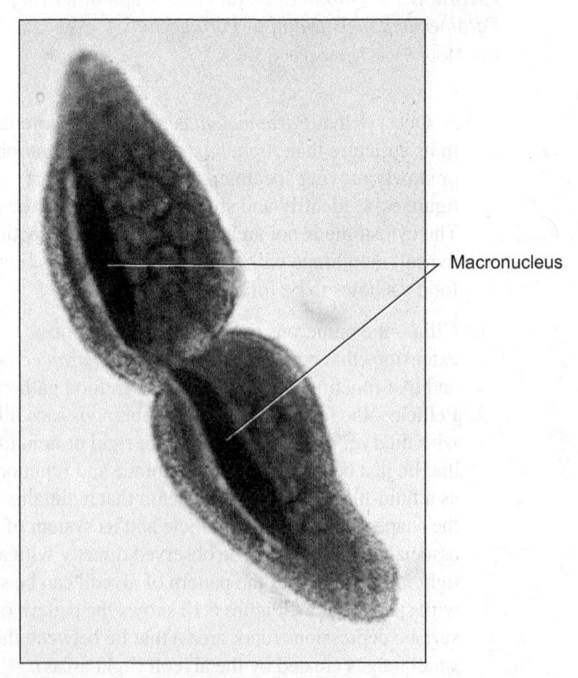

FIGURE 6.13 Binary fission in *Paramecium.*
Courtesy of Carolina Biological Supply Company, Burlington, NC.

Biochemical studies have also demonstrated that the movements of cilia and flagella involve **contractile proteins** similar to those found in striated muscle. This is another important illustration of the basic similarity of all living organisms.

Reproduction *Paramecium* reproduces by a simple type of asexual reproduction in which the parent divides into two equal daughter cells. This type of asexual reproduction is termed **binary fission** and is found in many kinds of protozoa. Living specimens are occasionally seen in the process of fission, but the details of fission are best studied in a stained microscope slide (figure 6.13).

- Obtain a prepared slide of *Paramecium* in fission and observe the nuclei. During fission, the micronucleus first divides by **mitosis,** and the macronucleus later divides by **amitosis.** The macronucleus is stretched apart by contractions of internal microtubules to form two halves that are subsequently split by the cleavage furrow that divides the cell into two daughters. The process of fission may be completed rapidly, and under optimal conditions, *Paramecium* can reproduce asexually two or more times per day.

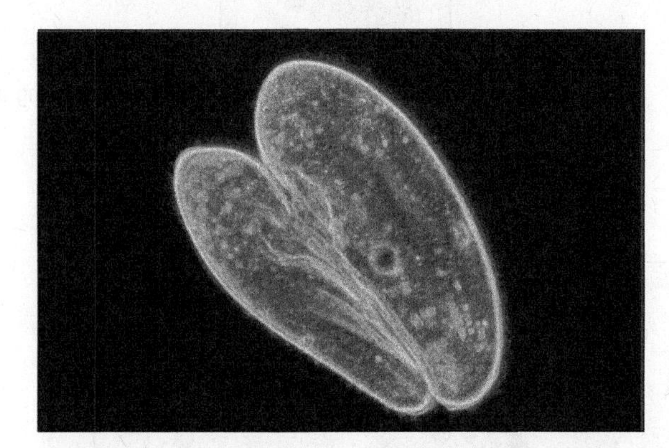

FIGURE 6.14 *Paramecium,* in conjugation.
© Melba Photo Agency/PunchStock.

Unlike *Amoeba, Paramecium* can also reproduce sexually. The specialized type of sexual process exhibited by *Paramecium* is called **conjugation** (figure 6.14). During this process, two individuals come together, adhere by their oral surfaces, undergo a complex series of changes in both the macronuclei and the micronuclei, exchange a single pair of micronuclei (one from each cell), separate, and resume asexual reproduction. Following the exchange of micronuclei in each *Paramecium,* the newly introduced micronucleus fuses with another (nonmigrating) micronucleus. Thus, there is an exchange of hereditary material and a subsequent fusion of hereditary material from the two parents, a situation analogous to that of ordinary sexual reproduction studied in Chapter 4.

- Examine a prepared slide or a demonstration of *Paramecium* in conjugation.

Phylum Ciliata: Other Ciliates The ciliates are a large and diverse group of protozoa with many important and well-known species and genera. One well-known form is *Stentor* (figure 6.15), a large, trumpet-shaped ciliate common in freshwater lakes, streams, and ponds where it temporarily attaches to submerged sticks, stones, and vegetation. Some common species of *Stentor* contain a blue or green pigment. *Stentor* has been the subject of many experimental studies. Its macronucleus looks like a string of beads. Another characteristic feature is a spiral array of complex ciliary organelles (membranelles) leading from its apical end to the cytostome.

Euplotes (figure 6.16) is another well-known ciliate that has been used in many experimental studies. Found in both fresh and salt water, *Euplotes patella,* the most common American species, has a flattened ovoid body averaging about 90 μm by 52 μm in size. Cilia in *Euplotes* are restricted

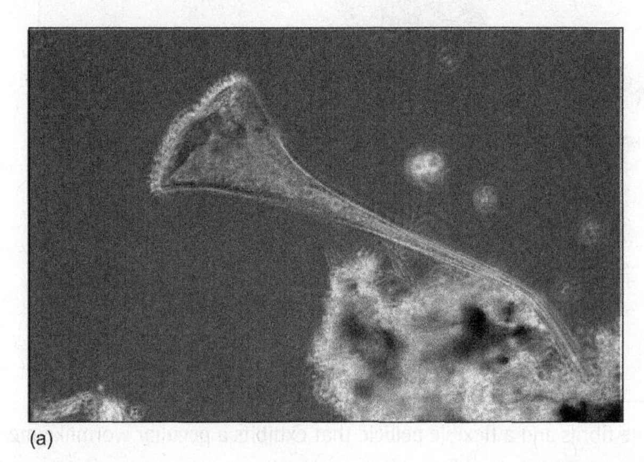

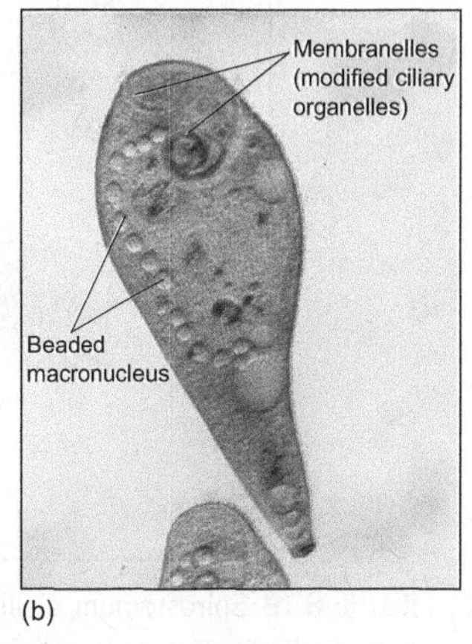

FIGURE 6.15 *Stentor.* (*a*) Photograph of a living *Stentor.* (*b*) Stained microscope slide of *Stentor.*
(a) © Melba Photo Agency/PunchStock. (b) Courtesy of Carolina Biological Supply Company, Burlington, NC.

to certain regions of the body, and groups of adjacent cilia fuse together to form a row of triangular **membranelles** along the oral groove leading to the cytostome. Each membranelle is composed of a short row of fused cilia and the membranelles are arranged in a long row that beats rhythmically to capture food particles. On the ventral surface, tufts of cilia are fused to form stiffened cirri that are used for walking on the substrate. Other rows of cilia on the dorsal surface are used for swimming.

These ciliary structures allow *Euplotes* to run along the substrate to capture particles like bacteria and small flagellates from a biofilm and then take off swimming to alight on another surface to continue feeding, much like a mobile vacuum cleaner!

Euplotes has a ribbonlike, C-shaped macronucleus and a small spherical or ovoid micronucleus. **Conjugation** occurs in fashion similar to that described for *Paramecium;* asexual reproduction occurs by binary fission.

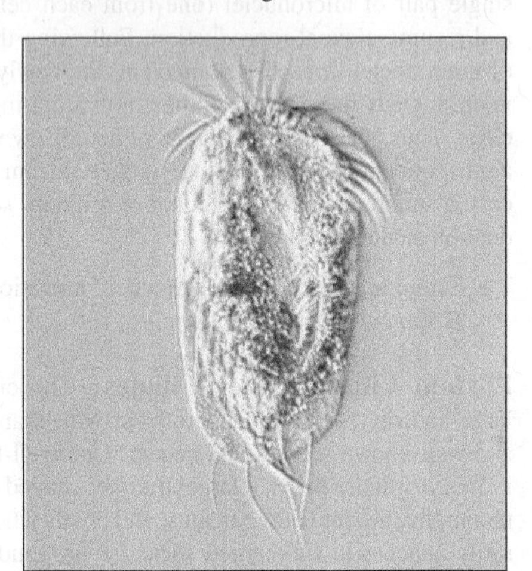

FIGURE 6.16 *Euplotes,* living, unstained.
Courtesy of Carolina Biological Supply Company, Burlington, NC.

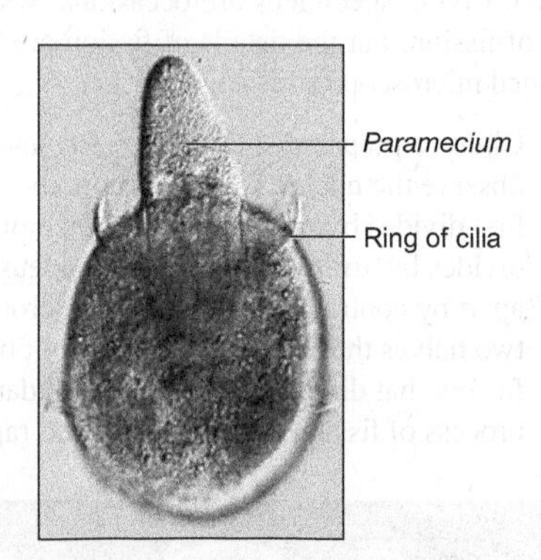

FIGURE 6.17 *Didinium,* a carnivorous ciliate ingesting a *Paramecium.*
Courtesy of Carolina Biological Supply Company, Burlington, NC.

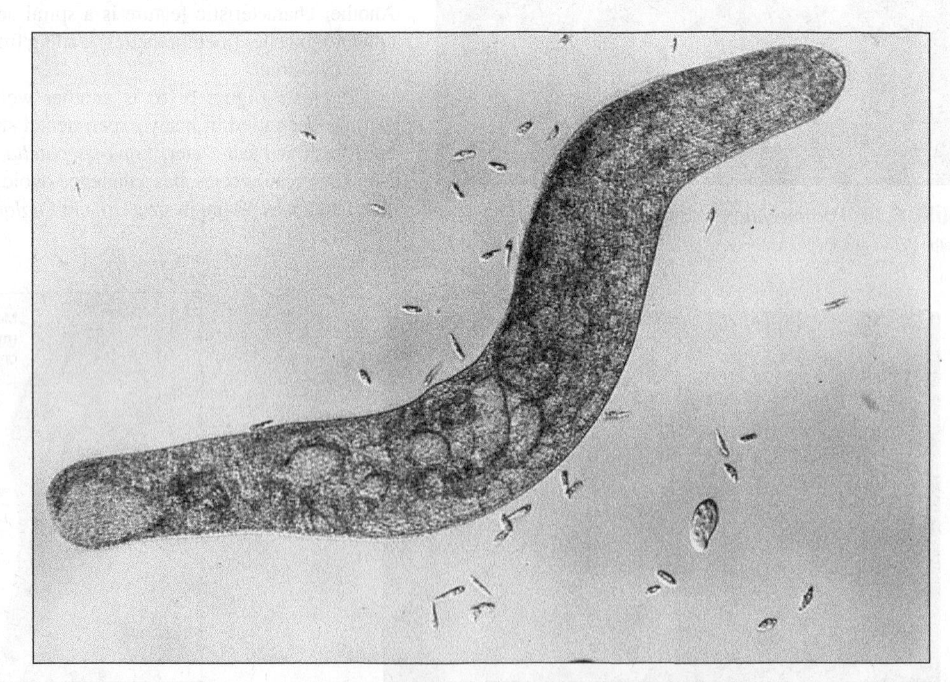

FIGURE 6.18 Spirostomum, a ciliate with contractile fibrils and a flexible pellicle that exhibits a peculiar wormlike movement.
Courtesy of Carolina Biological Supply Company, Burlington, NC.

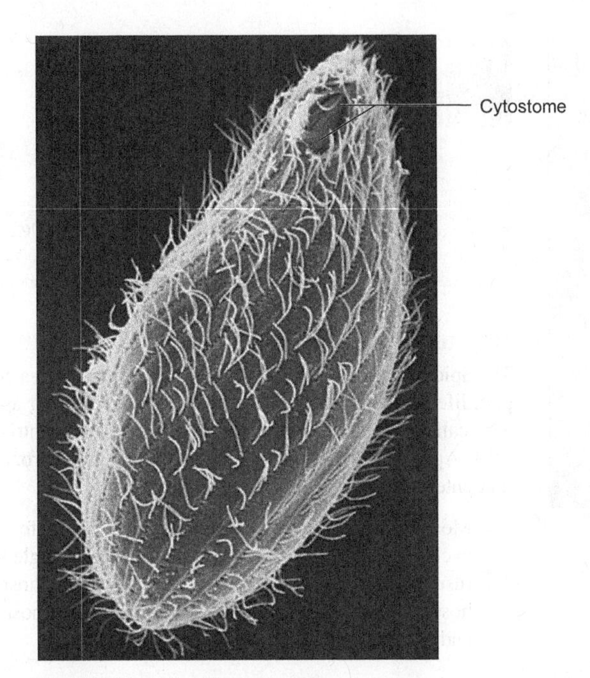

FIGURE 6.19 *Tetrahymena.*
Scanning electron micrograph by Jolanta Nunnley.

Didinium (figure 6.17) is a barrel-shaped predaceous ciliate with a voracious appetite. It feeds on other ciliates including *Paramecium.* A hungry *Didinium* can eat a *Paramecium* every two hours.

Spirostomum (figure 6.18) is a long, wormlike ciliate that has contractile fibrils that seem to function in a way similar to striated muscle fibrils. *Tetrahymena* (figure 6.19) is a small, ovoid ciliate that has been used in many experimental studies of biochemistry and genetics.

Vorticella is a solitary, sessile form with a long, contractile stalk that attaches to submerged stones, shells, plants, animals, and other objects. Other relatives of *Vorticella,* such as *Epistylis, Carchesium,* and *Zoothamnium* (figure 6.20) are similar but form colonies with branching stalks.

Podophyra (figure 6.21) is a suctorian, a specialized group of sessile ciliates that have **suctorial tentacles** in their mature stages and are predators of other ciliates. *Podophyra* has an interesting adaptation that aids its predatory habits. Each tentacle has a miniature cytostome on its tip that lets the suctorian ingest bits of the prey piecemeal into food vacuoles that pass into the tentacle and down into the cytoplasm. The cytostome region at the tip of the tentacle is surrounded by small, cytoplasmic organelles like trichocysts, except they are called **haptocysts.** When the haptocysts explode they shoot like little darts into the prey and hold it fast to the tentacle. Any ciliate that blunders into the tentacles will thus be immobilized and then eaten alive bit by bit.

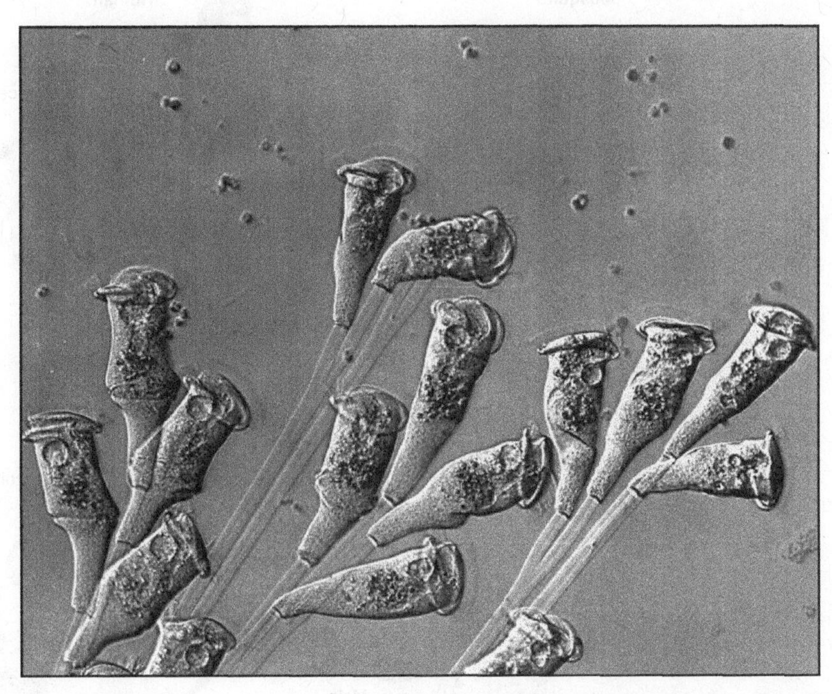

FIGURE 6.20 *Epistylis,* a sessile colonial ciliate with a long, branched stalk for attachment and a ring of cilia surrounding the mouth of each individual.
Courtesy of Carolina Biological Supply Company, Burlington, NC.

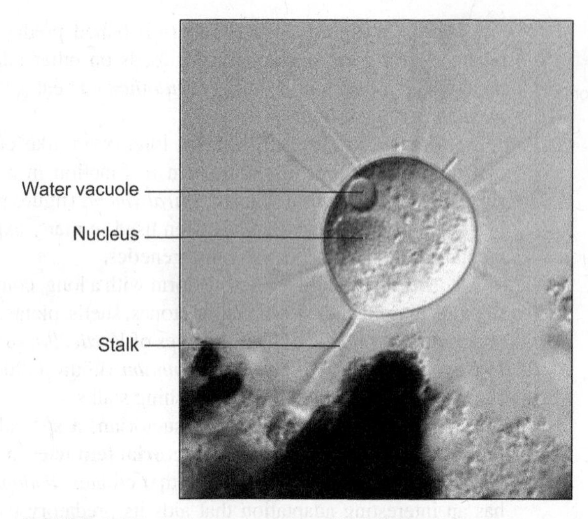

FIGURE 6.21 *Podophyra*, a suctorian, living; note the delicate tentacles extending out from the main cell body. Differential interference contrast photomicrograph.

Demonstrations

1. Models and charts of *Paramecium*
2. Stained slide showing pellicle of *Paramecium*
3. Stained slide to show discharged trichocysts
4. Stained slides with representative members of the Class Ciliata, such as *Stentor, Euplotes, Tetrahymena, Vorticella, Didinium, Blepharisma, Trichodina*, and *Podophyra*

Phylum Apicomplexa

The apicomplexans are parasitic alveolates that have a complex life cycle and possess a special structure at their apical end, called an apical complex, that facilitates entry into host cells. Apicomplexans form infective cells called sporozoites that enter host cells to continue the life cycle.

- Most apicomplexans have multiple hosts in the life cycle, but some of them are gut parasites of a single host (usually an insect or worm) and are passed from host to host by infective cysts that are defecated by one host and ingested by another to spread the infection.

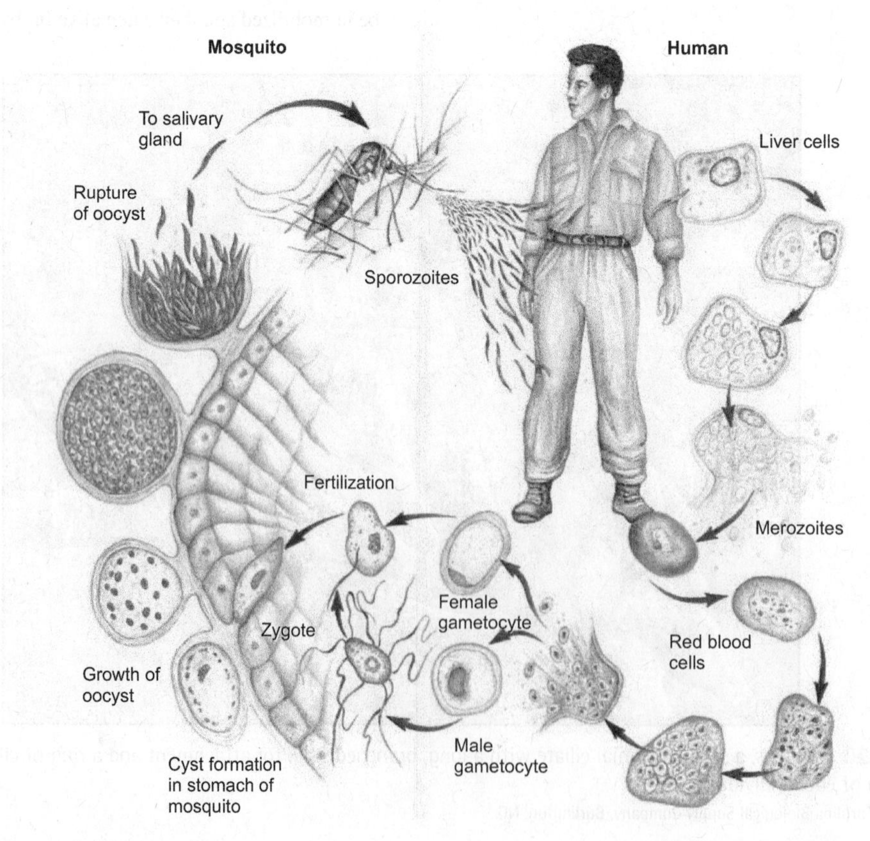

FIGURE 6.22 Life cycle of *Plasmodium*.

The phylum includes several species that cause serious diseases of humans and other animals including *Plasmodium* (malaria), *Toxoplasma* (toxoplasmosis), and *Cryptosporidium* (cryptosporidiosis) in humans and *Eimeria* (coccidiosis in rabbits and poultry).

- Study the demostrations illustrating the life cycles and importance of *Eimeria* and coccidiosis.

The Malaria Parasite: *Plasmodium* More than 50 species of *Plasmodium* have been described. All are parasites of vertebrate animals, including amphibians, reptiles, birds, and mammals. Human **malaria** is contracted during the feeding bout of an infected mosquito (*Anopheles* spp.; figure 6.23) and can be caused by any of five species of *Plasmodium*: *P. falciparum, P. vivax, P. ovale, P. malariae*, and *P. knowlesi*.

Malaria is one of the most serious and debilitating of human diseases. It has played an important role in human history from the fall of the Roman Empire to the war in Vietnam. Although modern medicine has made some progress in treatment, malaria remains prevalent in many parts of the world and still causes more than a million human deaths each year. *P. falciparum* is the most virulent parasite. It is most common in sub-Saharan Africa where it accounts for more than 90 percent of all human deaths attributed to malaria. Infections of *P. vivax, P. malariae*, and *P. ovale* are more common in West Africa, Asia, and South America, but cause lower rates of mortality. *P. knowlesi* is primarily a disease of monkeys in Southeast Asia, but since 2003 it has been recognized as an emerging cause of human malaria in the region. The life cycles of all five species are basically similar.

Life Cycle of Plasmodium Like other members of their phylum, *Plasmodium* parasites have a complex life cycle (figure 6.22) that includes multiple hosts and generations with both sexual and asexual reproduction. The life cycle can best be followed by starting inside an infected mosquito where a *Plasmodium* zygote (diploid) rests against the inner wall of the mosquito's digestive system.

The zygote becomes motile (it is now called an ookinete) and burrows through the intestinal epithelium. When the ookinete reaches the outside of the gut wall it stops moving and forms an oocyst. After several days of growth, the oocyst undergoes repeated meiotic divisions to form several hundred haploid sporozoites. The sporozoites escape by rupturing the external wall of the oocyst and migrate through the hemocoel to the salivary glands of the mosquito.

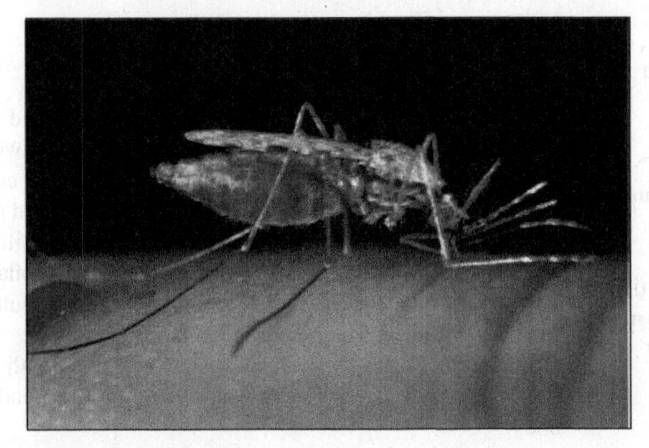

FIGURE 6.23 *Anopheles* mosquito feeding on arm.
CDC/Jim Gathany.

When a mosquito bites a human host to withdraw blood, sporozoites are injected under the skin along with salivary secretions. Sporozoites that find their way into the human bloodstream are carried to the liver where they quickly invade liver cells (hepatocytes) and begin the first phase (exoerythocytic stage) of human infection. It usually takes no more than 30 minutes for all of the sporozoites to settle into a liver cell and begin their transformation into amoeboid cells, called schizonts, that feed on the contents of the hepatocytes. The schizonts grow, differentiate, and (about 14 days later in the case of *P. falciparum*) reproduce asexually by multiple fission to form tens of thousands of haploid merozoites.

In *P. vivax* and *P. ovale,* some schizonts may remain in a dormant state (known as a hypnozoite). After many months (or even years) a hypnozoite may reactivate and produce a fresh crop of merozoites. These hypnozoites are thought to be responsible for cases of malaria that "reappear" weeks, months, or even years after treatment. (Since this type of outbreak results from parasites acquired in the initial infection, it is known as a "recrudescence" of the disease, rather than a "relapse".)

The merozoites produced by each schizont (or hypnozoite) migrate out of the liver and into the bloodstream where they invade red blood cells (erythrocytes) and begin the second phase (erythrocytic stage) of human infection. Once inside a red blood cell, the *Plasmodium* (now called a trophozoite) forms a vacuole and consumes most of the surrounding hemoglobin as a food source.

After a brief period of growth, each trophozoite undergoes another round of asexual reproduction, producing 12-16 more merozoites that burst free, enter the bloodstream, and invade other erythrocytes for yet another round of asexual reproduction. This multiplication process can be repeated several times so that an enormous number of parasites may be produced within the host. Release of merozoites tends to be synchronized, depending on species: every 24 hours for *P. knowlesi* (quotidian malaria), every 48 hours for *P. falciparum, P. vivax,* and *P. ovale* (tertian malaria), and every 72 hours for *P. malariae* (quartan malaria). This simultaneous rupture of erythrocytes also releases accumulated toxic wastes from the parasites, causing recurrent bouts of fever and chills that are a characteristic symptom of all human malaria infections.

Clinical diagnosis of malaria is accomplished by looking for the distinctive morphology of trophozoites in Giemsa-stained preparations of human blood. The blue cytoplasm and red nucleus of a *Plasmodium* parasite stands out in sharp contrast to the pale pink of normal erythrocytes. Young trophozoites form a circular ring inside the erythrocyte. This "ring stage" (figure 6.24a) gradually becomes a dark mass of chromatin as the cells divide to form new merozoites (figure 6.24b).

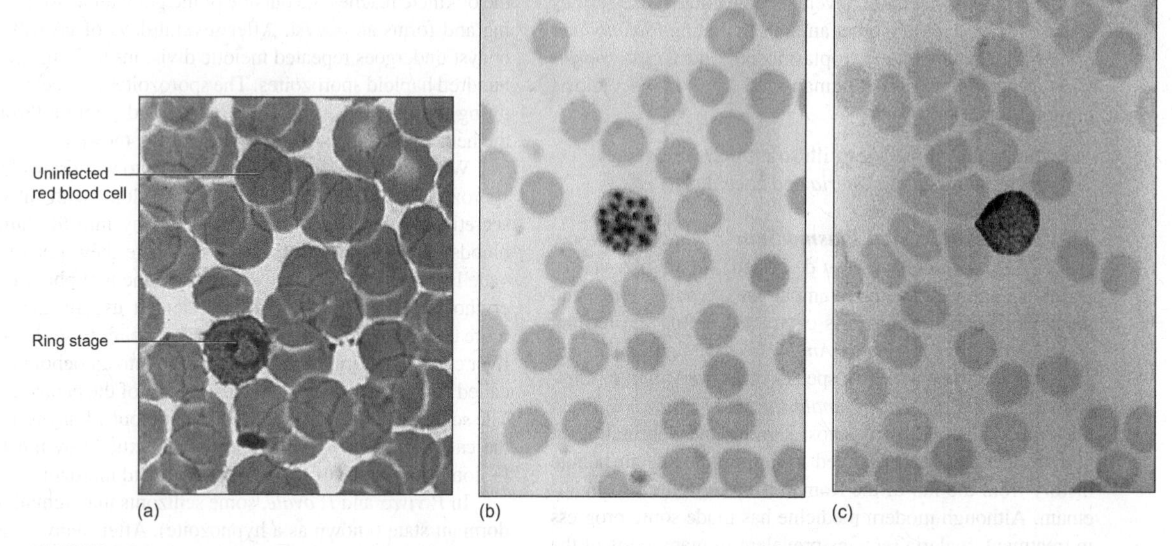

FIGURE 6.24 Some erythrocytic stages of *Plasmodium* life cycle. (*a*) Ring stage. (*b*) Mature schizont with 18 merozoites. (*c*) Megagametocyte.

(a) courtesy of Carolina Biological Supply Company, Burlington, NC. (b, c) CDC/Dr. Mae Melvin.

Some of the merozoites develop into sexual forms instead of repeating the asexual merogony. These sexual forms develop within the erythrocytes and become microgametocytes (male) or macrogametocytes (female) (figure 6.24c). These stages are the progenitors of the male and female gametes, and represent the start of the sexual portion of the life cycle.

If a mosquito bites an infected host and ingests infected erythrocytes, the male and female gametocytes pass into the mosquito's midgut. Male gametocytes divide three times to form eight microgametes which break free, become motile, and fertilize macrogametes. The fertilized macrogametes, or zygotes, then invade the gut wall of the mosquito to start the cycle again.

- Study the microscope slides with blood smears prepared with blood from humans infected with *Plasmodium vivax* or a similar species. Identify as many stages in the life cycle of *Plasmodium* as you can from your own slides and the demonstration slides. Observe the changes in morphology of the parasite during its development in the human erythrocytes. With the aid of figure 6.22, try to relate the portion of the *Plasmodium* life cycle in human erythrocytes to the other parts of the life cycle completed in the mosquito and in the exoerythrocytic stages in the human. ***Why do you think* Plasmodium *requires two hosts to complete its life cycle? What special adaptations for life as a parasite can you identify in* Plasmodium*?***
- List several of these adaptations for parasitism in table 6.1.

Table 6.1 Adaptations for Parasitism Exhibited by *Plasmodium*

Phylum Dinoflagellata

Dinoflagellates are a group of about 1,200 described species, many of which are abundant in the oceans of the world. Some species are important symbionts of reef-forming corals while population booms of other species can cause red tides that kill fish and other marine animals and can cause illness in humans eating shellfish contaminated with the dinoflagellates. Many others are harmless species commonly found in freshwater and saltwater environments.

Dinoflagellates typically have two flagella with one leading and one occupying a transverse groove around the body.

The body of most dinoflagellates is enclosed in a test composed of cellulose plates, each one of which is formed

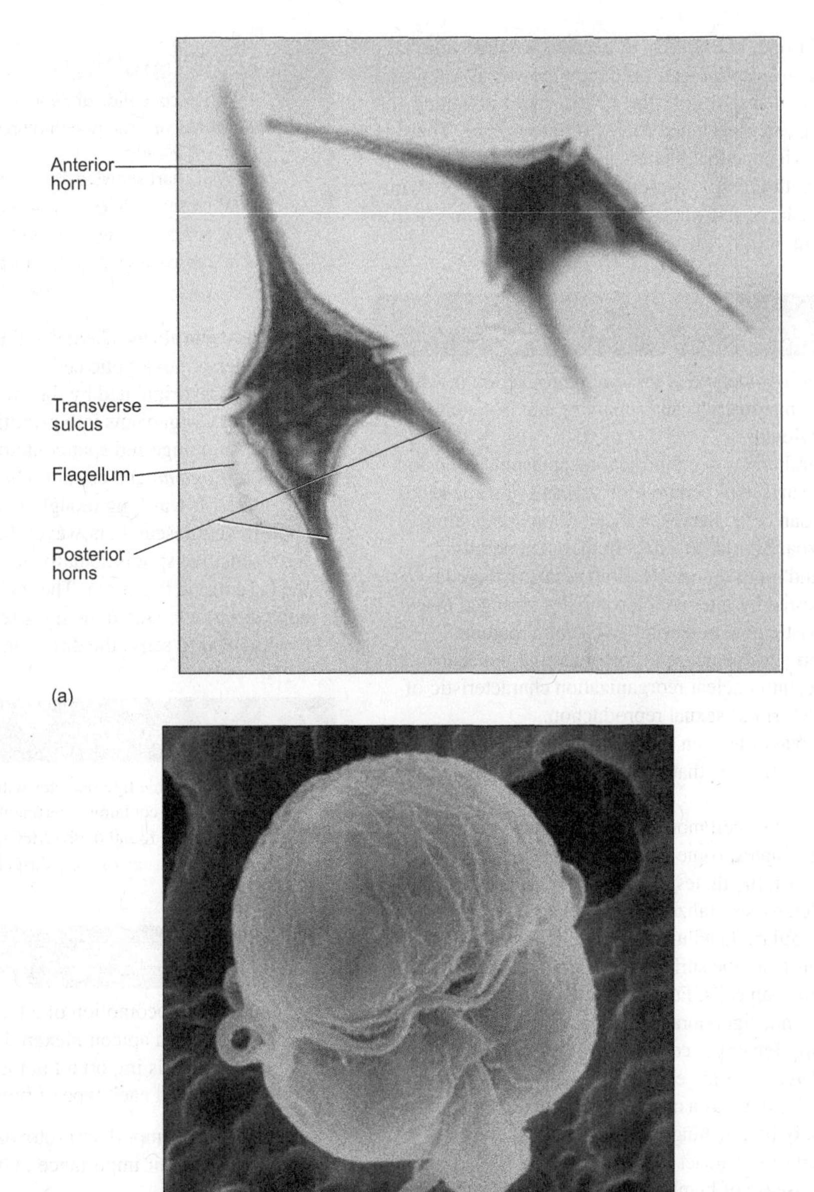

FIGURE 6.25 (*a*) *Ceratium*, a dinoflagellate. (*b*) *Pfisteria piscida*, a toxic dinoflagellate responsible for fish kills and a human health threat in Atlantic coastal waters.
(a) Courtesy of Carolina Biological Supply Company, Burlington, NC. (b) Scanning electron micrograph courtesy of Dr. JoAnn Burkholder.

within an alveolus lying beneath the cell membrane. *Ceratium* is a common dinoflagellate found in both freshwater and saltwater habitats (figure 6.25a). It has a characteristic shape with one long, pointed anterior horn and one to four posterior horns. A transverse groove around the body houses one of its two flagella and a longer flagellum tails behind the body as it swims. *Gonyaulax* and *Gymnodinium* are two marine dinoflagellates often associated with red tides in North

America, Europe, and Africa. A recently studied dinoflagellate, *Pfisteria piscida,* can produce a powerful toxin and has been found responsible for several fish kills along the Atlantic coast of the United States (figure 6.25b). There is substantial evidence that blooms of *Pfisteria* also constitute a serious health threat to persons fishing, boating, or swimming in or near coastal waters with a large population of this dinoflagellate.

Key Terms

Binary fission method of asexual reproduction in which an organism constricts and separates into two smaller new individuals.

Cilium (plural: cilia) cylindrical cytoplasmic extension from the surface of certain protozoa and of some kinds of metazoan cells. Serves in locomotion and feeding of protozoa. Similar to a flagellum, but generally shorter and more numerous. Both cilia and flagella are supported by internal microtubules arranged in a characteristic nine outer plus two central pattern.

Conjugation a specialized type of mating, nuclear exchange, and nuclear reorganization characteristic of ciliates; a form of sexual reproduction.

Contractile vacuole an organelle found in many freshwater protozoa that serves in osmoregulation (water balance).

Cytostome the "cell mouth" found in many protozoa, including ciliates, some flagellates, and some apicomplexa. In ciliates, the cytostome is often surrounded by specialized ciliary feeding organelles.

Flagella (singular: flagellum) cylindrical cytoplasmic extensions from the surface of certain protozoa and some metazoan cells. Function in locomotion and feeding of mastigophorans. Similar to cilia but longer and usually fewer per cell.

Macronucleus the large metabolic nucleus typical of ciliates. Often has a characteristic shape. Divides amitotically by pinching in two. Contains many duplicated sets of genes (polyploid).

Malaria disease of humans and other animals. Caused by members of the protozoan genus *Plasmodium,* which invade the blood and other tissues of the hosts.

Micronucleus the small reproductive nucleus in ciliates. Some ciliates have more than one micronucleus. Usually divides by ordinary mitosis.

Nuclear dimorphism having two distinct types of nuclei within the same cell. Characteristic of the ciliated protozoa, Phylum Ciliophora.

Pseudopodium (plural: pseudopodia) protoplasmic extension of an amoeboid cell; the "false foot" of the Sarcodina used for feeding and locomotion. Various Sarcodina have pseudopodia specialized for specific purposes. Also present in other kinds of amoeboid cells such as leucocytes or white blood cells in many kinds of animals.

Serial Endosymbiosis Theory theory that certain organelles of eukaryotic cells, particularly mitochondria and plastids, originated by the incorporation of prokaryotic organisms into primitive eukaryotic cells.

Stigma an orange-red spot containing carotenoids, found in *Euglena* and certain other photosensitive flagellates. It was long thought to be a light-sensitive spot. Recent research, however, has shown that the light-sensitive spot is actually another body located at the base of the flagellum. The real function of the stigma appears to be to shield the light-sensitive spot and allow the *Euglena* to sense the direction of the light source.

Demonstrations

1. Microscope slide of *Eimeria*
2. Microscope slides with representative stages of *Plasmodium*
3. Wall chart showing *Plasmodium* life cycle
4. Microscope slides of heliozoans
5. Microscope slides of radiolarians
6. Microscope slides of foraminiferans

Internet Resources

There are many valuable Internet sites with information about zoology. Several sites containing pertinent zoological information for this chapter can be found on the McGraw-Hill Zoology web site at http://www.mhhe.com/zoology. Just click on this text's title.

Questions for Critical Thinking

1. Compare the locomotion of a flagellate, a ciliate, an amoeba, and an apicomplexan. Explain how the type of locomotion is important in the nutrition, ecology, and behavior of each type of protozoan.
2. Identify three important protozoan parasites of humans and discuss their importance in human health.
3. Discuss the possible role of protozoa in the evolution of multicellular animals.

Suggested Readings

Anderson, R.O., and M. Druger. (eds.) 1997. *Explore the World Using Protozoa.* Arlington, VA: National Science Teachers Association. 240 pp

Jahn, T.J. et al. (eds.). 1978. *How to Know the Protozoa.* New York: McGraw-Hill. 304 pp

Lee, J.J., and P. Bradbury (eds.). 2000. *Illustrated Guide to the Protozoa,* 2nd ed. New York: Wiley-Blackman. 2 vols.

Patterson, D.J. 1996. *Free-living Freshwater Protozoa: A Color Guide.* Washington, DC: ASM Press. 223 pp

Chapter 7

Porifera

Objectives

After completing the laboratory work in this chapter, you should be able to perform the following tasks:

1. Briefly characterize the Phylum Porifera.
2. Describe the basic organization of the sponge body.
3. Explain the skeletal elements of sponges and their composition.
4. List and distinguish the three classes of sponges.
5. Differentiate among asconoid, syconoid, and leuconoid types of canal systems found in sponges.
6. Describe the structure of a choanocyte and explain the importance of choanocytes in sponges. Explain where choanocytes are found in asconoid, syconoid, and leuconoid sponges.
7. Describe the pattern of water flow in a syconoid sponge and explain the importance of water currents in sponges.
8. Explain the role of gemmules in sponges.
9. Explain the significance of experiments on the reaggregation of sponges from dissociated sponge cells.
10. Discuss the evolutionary origin of sponges and their possible phylogenetic relationships with other groups of organisms.

Introduction

The Phylum Porifera includes the sponges, a group of sedentary aquatic animals so different from other types of animals that they were long thought to be plants. Sponges are among the most primitive of multicellular animals. They have a simple type of body organization, with a porous body permeated by a system of water canals through which water is pumped by action of special flagellated cells, the choanocytes (figure 7.1).

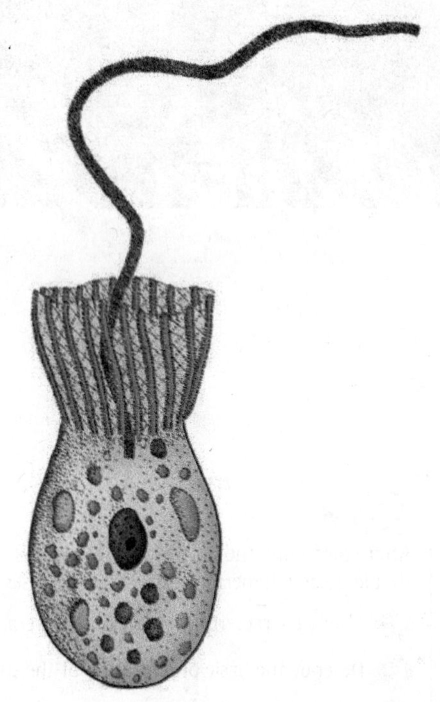

FIGURE 7.1 Choanocyte.

FIGURE 7.2 Bath sponge. One-fourth natural size. Class Demospongiae.
Courtesy of Carolina Biological Supply Company, Burlington, NC.

The body of a sponge consists of a jelly-like extracellular matrix bordered by two layers of cells.

Key characteristics of the sponges are (1) presence of choanocytes, (2) a water canal system with external pores, (3) skeleton of mineral spicules and/or spongin fibers, (4) highly mobile cells, and (5) the presence of some totipotent cells capable of differentiating into any other cell type.

Approximately 15,000 species are grouped within the Phylum Porifera. Most of these live in marine habitats but there are also about 150 species that occur in fresh waters. Natural sponges (cleaned and dried) have been used since ancient times for bathing and holding water for drinking. About 17 different species are still harvested and sold commercially. Sponge fishing was probably started by Greeks in the Aegean Islands. It later spread throughout the Mediterranean Sea, the Gulf of Mexico, and the Caribbean Sea where warm, shallow waters and rocky bottoms favor survival and growth of bath-type sponges (figure 7.2). During the early 1900s, Greek immigrants to the United States developed a large sponge fishing industry in Tarpon Springs, Florida. By 1935, sponges were one of Florida's most important fishery products.

Overfishing, sponge diseases, and pollution have taken their toll on sponge beds around the world. In the early 1940s, a large percentage of Florida's sponge population was killed by bacterial contamination of offshore waters and the sponge industry was nearly destroyed. Synthetic sponges, made from cellulose, were developed about the same time and contributed to the market's collapse. Fortunately, a hurricane in 1985 stirred up waters off the west coast of Florida and the sponge beds began to grow again. Although there is still a demand for natural sponges in Europe where people seem to appreciate their texture and durability, the old sponge market on the waterfront in Tarpon Springs is now mostly a tourist attraction.

Why do you think a hurricane might have an effect on the growth of sponges and other marine life?

Fact File

Sponges

- Sponges were considered to be plants by the ancient Greeks and Romans. This misconception persisted until after the microscope was invented and zoologists were able to observe the beating of flagella on the choanocytes.
- A living bath sponge filters water through millions of tiny chambers. There may be 7,000–18,000 chambers per cubic centimeter of sponge, and each chamber can filter up to 1,200 times its own volume of water daily.
- Sponges in the family Cladorhizidae use Velcro-like spicules to trap small crustaceans as prey. Amoeboid cells in the sponge then surround the prey and digestion occurs extracellularly.
- Sponges are excellent synthetic chemists. They manufacture a wide range of unusual substances for protection against bacterial infection, predation, ultraviolet radiation, and so forth. Some of these chemicals have proven useful in human medicine, including compounds that have anticancer, antiinflammatory, and antibiotic properties.

Phylogeny

Sponges are generally regarded as primitive animals because they lack organized organs and tissues, such as nerves and muscles. They also lack the "tight junctions" (desmosomes and hemidesmosomes) that are found between adjacent cells in other animal phyla. Individual sponge cells are largely independent of their neighbors: they retain the ability to move, and under some circumstances may change into another functional cell type. Because of these peculiarities, sponges have long been regarded as an early offshoot from the main line of metazoan evolution and, therefore, not closely related to more advanced animal taxa.

Many years ago, scientists noted an obvious morphological similarity between sponge choanocytes and a distinctive group of protozoa known as the **choanoflagellates.** Both the choanoflagellates and the sponge cells have a collarlike ring of microvilli surrounding a flagellum. This observation led to speculation that sponges may have evolved from choanoflagellates (or both arose from some common ancestor). The argument is further supported by the fact that some choanoflagellates form simple colony aggregations.

Scientists have also believed for many years that sponges arose from some ancestral protistan **before** the other metazoans although various researchers in the past have differed over what type of ancestral protist might have given rise to the sponges. Although sponges are indeed multicellular, their body organization is quite different from most other metazoans. Metabolic functions are basically carried out at the cellular level (e.g. intracellular digestion, respiration, and excretion by diffusion). Sponges lack body symmetry, have no true tissues or organs, and show little evidence of coordination within the sponge body.

The presence of choanocytes appears to be an ancestral feature, connecting sponges with choanoflagellates. Recent biochemical and genetic studies have supported the connection with ancestral choanoflagellates and suggest that sponges arose from this ancestral line prior to the origin of other more complex metazoans, the so-called **true metazoa.**

- Review the discussion of the evolution of multicellularity in Chapter 5, page 74.

Morphology

The body of a sponge is basically a loose assemblage of cells embedded in a noncellular matrix, called **mesohyl,** which is supported by an internal skeleton of organic fibers and/or mineralized elements called **spicules.** The fibers are tough, structural proteins: mostly **collagen** and/or **spongin** (chemically similar to collagen but unique to sponges). The spicules are hard slivers of calcium carbonate (lime) or silica (glass). They occur in a variety of shapes and sizes (figures 7.3 and 7.4), and are important in the classification of sponges.

The external surface of the sponge body is covered by an epithelial layer of flattened cells, the **pinacocytes.** Certain of the inner canals and chambers are lined by **choanocytes.**

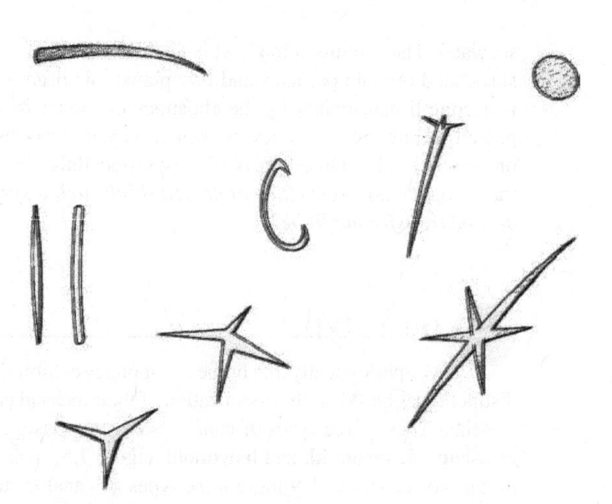

FIGURE 7.3 Examples of spicule shapes.

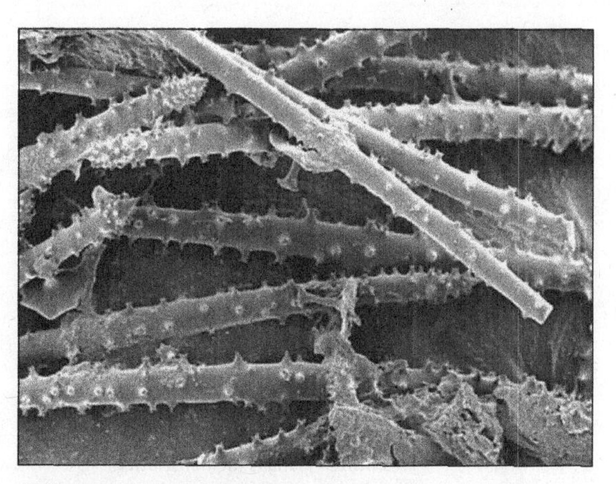

FIGURE 7.4 Spicules from a freshwater sponge. Magnification 7,000×.
Scanning electron micrograph by F.W. Harrison.

Between the inner and outer cell layers lies the mesohyl (formerly called mesenchyme), a gelatinous matrix containing several types of amoeboid cells, fibrils, and skeletal elements. The mesohyl resembles a type of connective tissue. Most textbooks state that sponges lack true tissues. Some specialists, however, have claimed that the **pinacoderm** and mesohyl do represent tissues, despite the fact that these layers are not homologous with the tissue layers found in higher animals. An inner layer of flagellated choanocytes generates water currents through the internal canal systems. These water currents are essential in the life of the sponge because they carry food particles and oxygenated water into the sponge and waste products as well as gametes and/or larvae out of the sponge.

Sponges are sedentary filter-feeders, able to capture tiny food particles measuring from about 0.5 to 50 μm from the seawater. The capture of food, which consists chiefly of fine, suspended organic particles and tiny planktonic organisms, is accomplished mainly by the choanocytes. Some food is passed to internal amoebocytes, which may in turn pass it on to other cells. Digestion is usually intracellular (within individual cells). ***How efficient do you think such a system of food transfer might be?***

Body Types

Morphologically, the bodies of sponges exhibit three distinct types based on the organization of their internal canal systems. These three types of canal systems are designated as **asconoid, syconoid**, and **leuconoid** (figure 7.5). It is important to recognize that these three types of canal systems represent morphological organization and are not the basis for differentiating the three classes of extant sponges. Actually, most extant sponges are leuconoid.

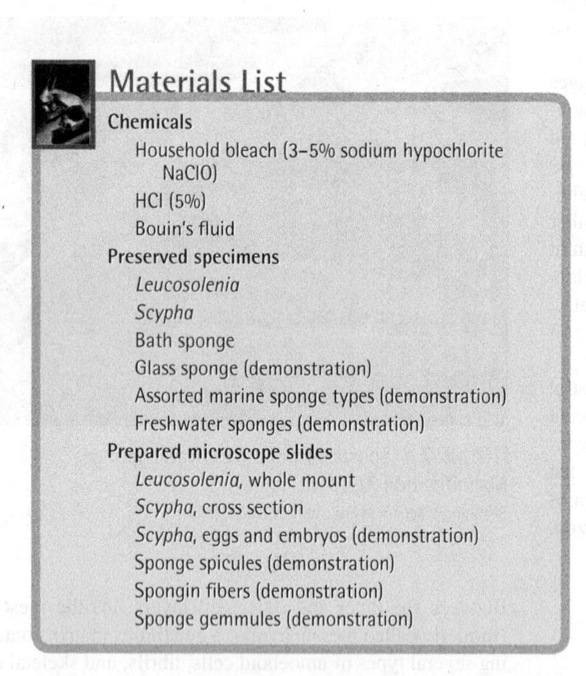

Materials List

Chemicals
- Household bleach (3–5% sodium hypochlorite NaClO)
- HCl (5%)
- Bouin's fluid

Preserved specimens
- *Leucosolenia*
- *Scypha*
- Bath sponge
- Glass sponge (demonstration)
- Assorted marine sponge types (demonstration)
- Freshwater sponges (demonstration)

Prepared microscope slides
- *Leucosolenia*, whole mount
- *Scypha*, cross section
- *Scypha*, eggs and embryos (demonstration)
- Sponge spicules (demonstration)
- Spongin fibers (demonstration)
- Sponge gemmules (demonstration)

Classification

The phylum is divided into three distinct classes traditionally based on the nature of the skeleton. Neither chemical studies of spicules nor modern molecular studies have yet been useful in determining sponge relationships.

One way that scientists illustrate the possible evolutionary relationships among animals is by constructing a **cladogram.** A cladogram is a branching family tree that shows how groups of organisms may be related to each other based on the available evidence. Refer to the discussion of cladograms in Chapter 5.

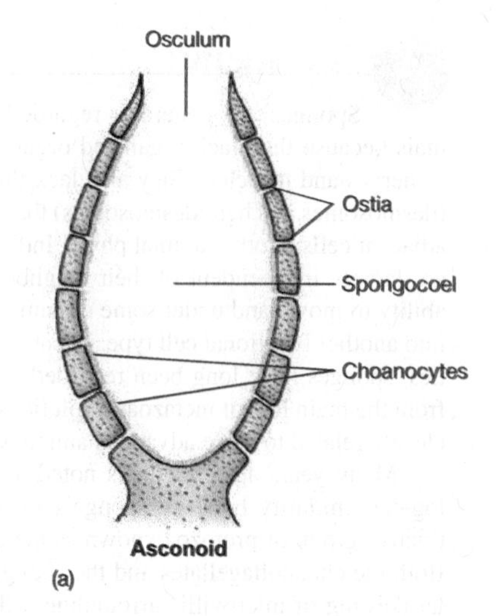

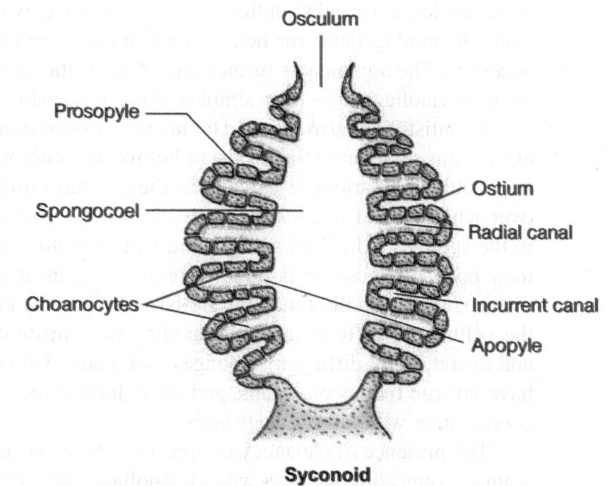

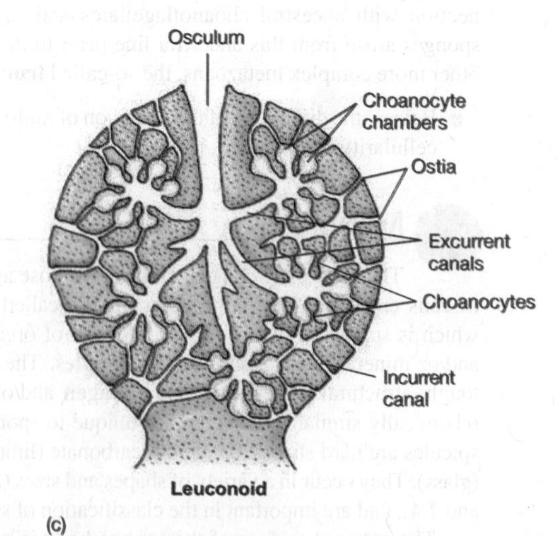

FIGURE 7.5 Canal systems: (*a*) asconoid body type, (*b*) syconoid body type, and (*c*) leuconoid body type.

Cladogram of Porifera

- Class **Hexactinellida** (siliceous sponges)
- Class **Calcarea** (calcareous sponges)
- Class **Demospongiae** (horny sponges)

Class Demospongiae

Horny Sponges

Members of this class possess a skeleton made up of a network of spongin fibers (a structural protein secreted by certain sponge cells), siliceous spicules, both, or neither. Most members of this class are marine, but two families are found in freshwater streams, ponds, and lakes. All commercial sponges belong to this class. This is the largest class of sponges and all members have a leuconoid canal system. Examples: *Spongia, Haliciona, Microciona*, and the deep-sea sclerosponges (all marine); *Spongilla* (freshwater).

Class Calcarea (Calcispongiae)

Calcareous Sponges

Sponges with a skeleton consisting of many small spicules made of calcium carbonate embedded in a loose, jellylike matrix. All species are marine. All three types of canal systems (asconoid, syconoid, and leuconoid) are found among members of this class. Examples: *Scypha, Leucosolenia.*

Class Hexactinellida (Hyalospongiae)

Glass Sponges

Sponges with a skeleton composed of siliceous spicules, usually with six rays, as the class name implies. The spicules are often fused together into a continuous network. The tissues of glass sponges are a syncytial network of fused amoeboid cells. All glass sponges are marine, and most species are found in deep areas of the world oceans. Because of their remote habitat, little is known about their biology. Members of this class have either syconoid or leuconoid canal systems. Examples: *Euplectella* (Venus's flower basket) and *Hyalonema.*

Leucosolenia: An Asconoid Sponge

Leucosolenia (figure 7.6) is a small colonial sponge of the asconoid type. Examine a preserved specimen or a microscopic whole mount with your stereoscopic microscope and observe the following features.

1. A system of horizontal tubes that bears numerous upright branches.
2. The upright branches that represent individual sponges of the colony.
3. Buds formed on the sides of the individual sponges.
4. The terminal opening, or **osculum,** at the upper end of each sponge. Water passes out of the sponge through this opening.

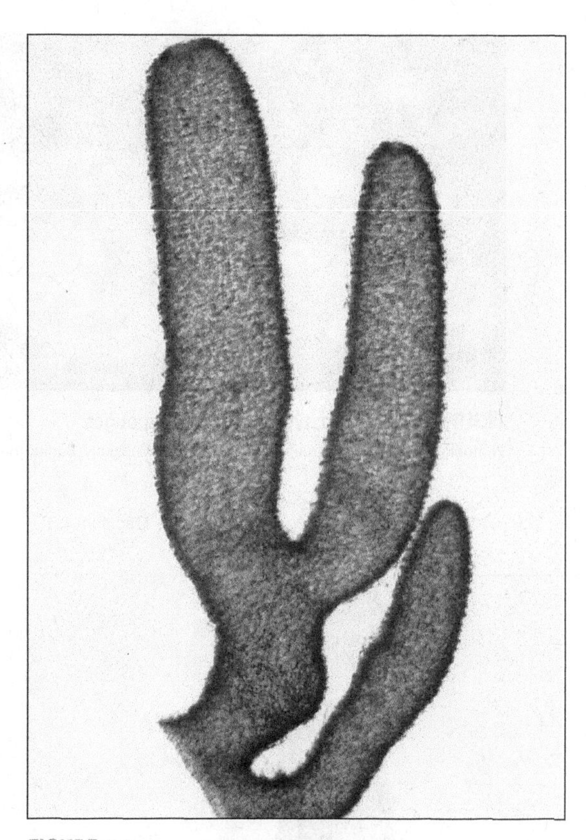

FIGURE 7.6 *Leucosolenia.*
Photograph courtesy of Carolina Biological Supply Company, Burlington, NC.

5. The **spongocoel,** a large central cavity within the sponge. This cavity is lined by the specialized, flagellated collar cells or **choanocytes,** which create water currents within the sponge.
6. Water enters the sponge through many **ostia,** tiny pores that penetrate the body wall.
7. Numerous triradiate (three-rayed) **spicules** may be seen embedded in the wall.

Scypha: A Syconoid Sponge

Scypha (figures 7.7 and 7.8), formerly also called *Sycon* and *Grantia*, is a small, slender sponge of the syconoid type.

- Obtain a preserved specimen and note the size and shape of the body. Examine specimens cut in half longitudinally and observe the spongocoel, the radial canals extending from the spongocoel into the body wall, and the short collar region, which leads to the exterior through a cylinder of giant spicules surrounding the osculum. On the external surface, observe the numerous cortical spicules that extend outward through the pinacocyte layer, which covers the external surface of the sponge. Also on the external surface of the sponge are many ostia that open into the

FIGURE 7.7 *Scypha*, typical cluster of sponges.
Photograph courtesy of Carolina Biological Supply Company, Burlington, NC.

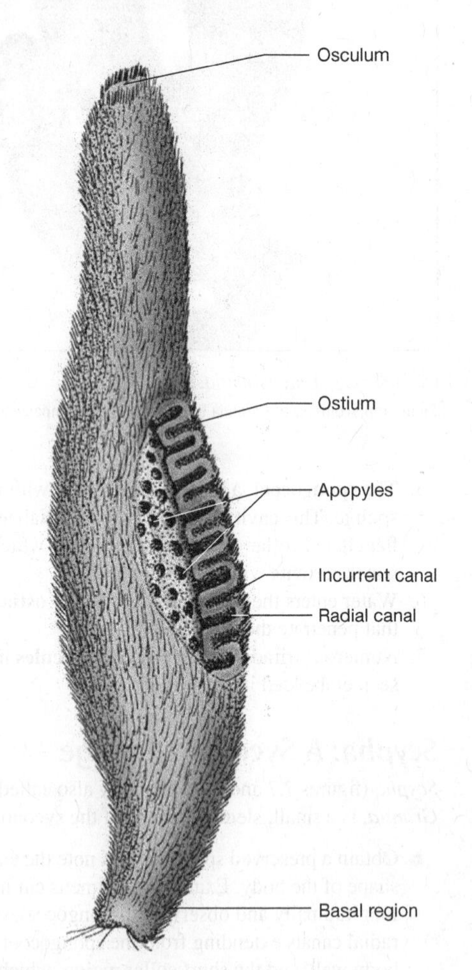

FIGURE 7.8 *Scypha*, typical cluster of sponges.

incurrent canals that carry water into the sponge. Water moves from the incurrent canals to the adjacent radial canals via numerous small openings, the **prosopyles**.

- Study a stained cross section of *Scypha* and identify the **spongocoel,** the **radial canals,** and the **incurrent canals.** Water drawn by currents created by the choanocytes enters through small incurrent pores into the incurrent canals. Note in the cross sections, that the radial and incurrent canals are often cut diagonally so that you see only a small portion of each canal. ***How can you tell which are incurrent canals and which are radial canals?***

The openings between the radial and incurrent canals are the **prosopyles;** the radial canals empty into the spongocoel through the **apopyles.** Much of the substance within the sponge body consists of a loose noncellular mesohyl. Details of the cross section of the body wall of *Scypha* are illustrated in figure 7.9.

The external surface of the sponge is covered by a layer of thin, flat cells, called **pinacocytes.** A similar layer of pinacocytes lines the spongocoel. The **choanocytes** (figure 7.9) are found lining the radial canals, which empty into the spongocoel through the apopyle (figures 7.9 and 7.12). These choanocytes are small and difficult to identify in most microscopic preparations.

You should be able to find some large, undifferentiated **amoeboid cells** within the mesohyl. Some of these amoeboid cells develop into eggs. Sponges lack differentiated gonads, and eggs and embryos are often found embedded in the body wall in microscopic cross sections of mature sponges (figure 7.11). **Sperm** develop from choanocytes and are carried by water currents into other sponges. *Scypha,* like many sponges, is **hermaphroditic,** but the eggs and sperm develop at different times.

- Examine your cross section and see if you can identify some eggs and embryos.

Sponges reproduce both sexually and asexually. Marine sponges may reproduce asexually by fragmentation, in which part of a sponge colony may break off and be washed away to grow into a new colony elsewhere. In sexual reproduction, the fertilized eggs of sponges develop into a ciliated larva that is released into the plankton. Later, if a larva settles onto a suitable substrate, it may develop into a small sponge.

Leuconoid Sponge

Leuconoid sponges are structurally the most complex and also the most common body type among living sponges. All freshwater sponges and most marine sponges, including *Hippospongia communis* and *Spongia officinalis*, are leucon-type.

- Study a portion of a preserved specimen of a bath sponge and note its rubbery texture and the complex system of branching canals.

Located between the incurrent canals and the excurrent canals are numerous small, spherical **flagellated chambers.** Collar cells are found lining only these tiny flagellated chambers in leuconoid sponges.

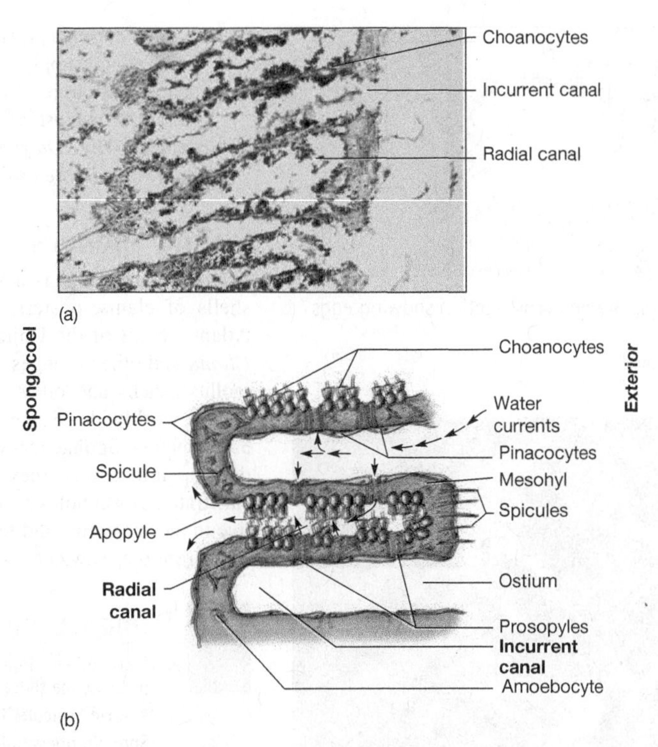

FIGURE 7.9 *Scypha.* (*a*) Stained cross section. (*b*) Drawing with detail of cross-section showing path of water flow.
(a) Photo by John Meyer.

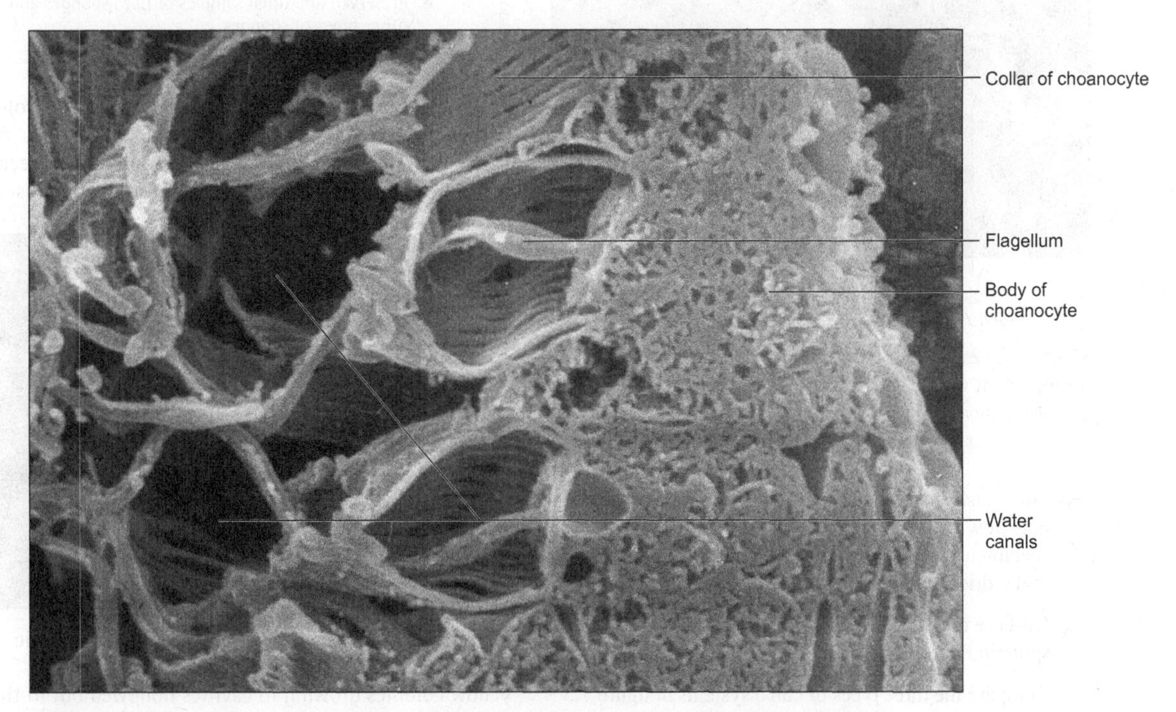

FIGURE 7.10 *Ephydatia*, a freshwater sponge. Cross-section through choanocytes. 21,000X.
Scanning electron micrograph of a freeze fracture section by Louis de Vos.

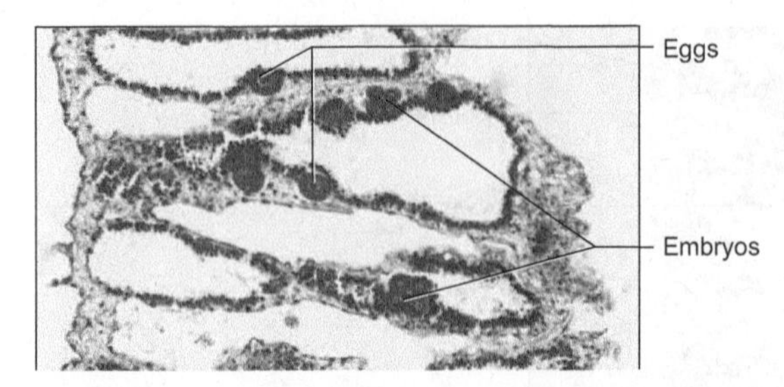

FIGURE 7.11 *Scypha.* Stained cross section showing eggs and embryos.
Photo courtesy of Betty Black.

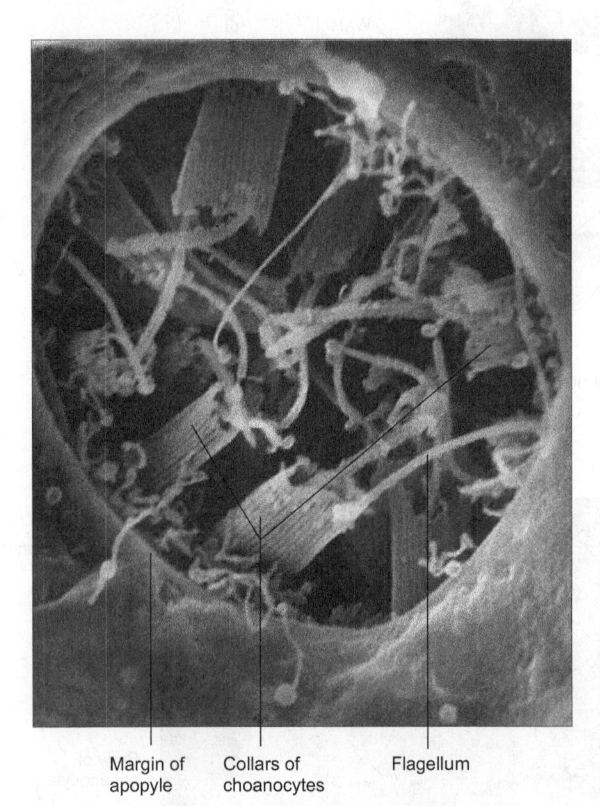

FIGURE 7.12 *Ephydatia,* opening of apopyle showing arrangement of choanocytes within 10,500×.
Scanning electron micrograph by Louis de Vos.

- Study also a dried specimen of a bath sponge and note how its texture differs from that of the preserved specimen. Only the network of **spongin fibers** remains in the dried sponge.
- Observe the microscopic demonstration slide of spongin fibers.
- Compare the three types of canal systems in figure 7.5 and in the sponge materials provided in the laboratory. Remember, sponges are sedentary and, thus, dependent on water currents passing through their canal systems for their food supply and for gas exchange. Even sponges have to respire. ***Which type of canal system is likely to be most efficient in gas exchange and food capture? Why? Do you think this might have some importance in the evolution of sponges?***

Boring Sponges

Cliona (figure 7.13) is a sponge that makes holes in the shells of clams, oysters, and other molluscs along the Atlantic coast of the United States. Numerous species of *Cliona* and other sponges live in cavities they make in the mollusc shells and other calcareous substrates in shallow waters worldwide. Some species occur at depths up to 3,000 meters. Sponge larvae settle on the shells and as they develop into adults, they secrete enzymes that etch the substrate to form holes as much as 1 cm deep. The sponges live in these cavities and feed on suspended materials they filter from the seawater.

Demonstrations

1. Eggs and developing embryos of *Scypha* (microscope slide)
2. Assorted spicules (microscope slide)
3. Spongin fibers (microscope slide)
4. Glass sponge and examples of other types of sponges
5. Skeletons of glass sponges
6. Preserved and dried samples of bath sponges and other types of sponges

Boring sponges play an important ecological role by aiding in **recycling** insoluble calcium compounds from the shells. They are also important economically because they also grow on living oysters and other molluscs as

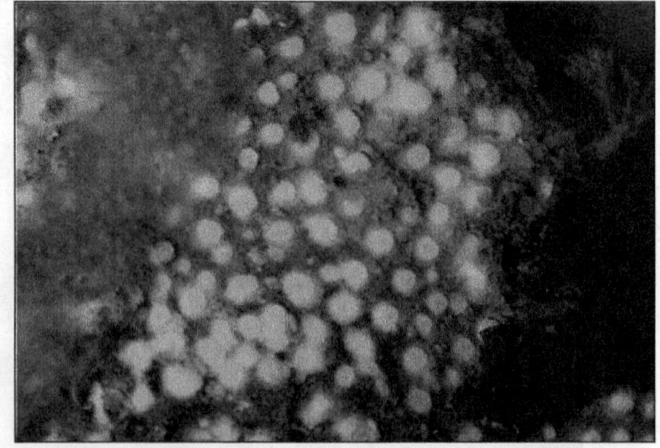

FIGURE 7.13 *Cliona sulphurea,* a common Atlantic coast boring sponge. Note the numerous individual yellow colonies growing in cavities hollowed out in the mollusc shell.
Courtesy of Carolina Biological Supply Company, Burlington, NC.

well as corals and may damage or kill the host organisms. Scientists have reported on destructive effects of boring sponges in commercial oyster beds and on coral reefs in certain areas.

Freshwater Sponges

Although most sponges are marine, a few species live in freshwater streams, ponds, and lakes. Freshwater sponges (figure 7.14) are much less prominent than their ubiquitous marine relatives.

All freshwater sponges (and a few marine sponges) form internal asexual buds called **gemmules** (figures 7.15 and 7.16). Gemmules are resistant stages, which serve to carry the sponge through the winter or aid in survival during a drought. Observe the microscope slide of gemmules on demonstration. ***What important features of the gemmule aid in the survival of the species?***

Demonstrations

1. Gemmules (microscope slide)
2. Living or preserved freshwater sponges

FIGURE 7.14 *Spongilla*, Freshwater sponge.
Courtesy of Carolina Biological Supply Company, Burlington, NC.

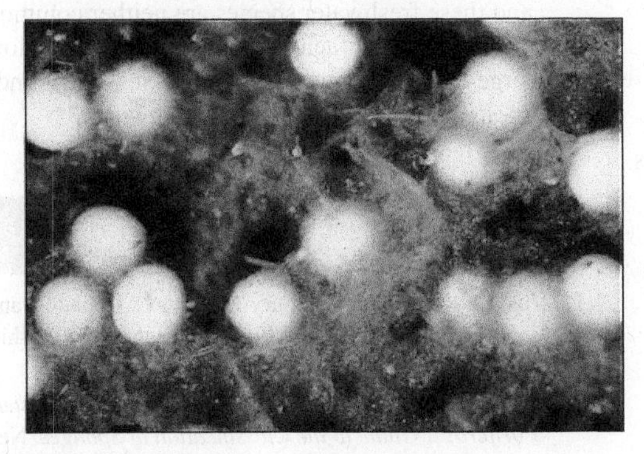

FIGURE 7.15 Gemmules in freshwater sponge.
Courtesy of Carolina Biological Supply Company, Burlington, NC.

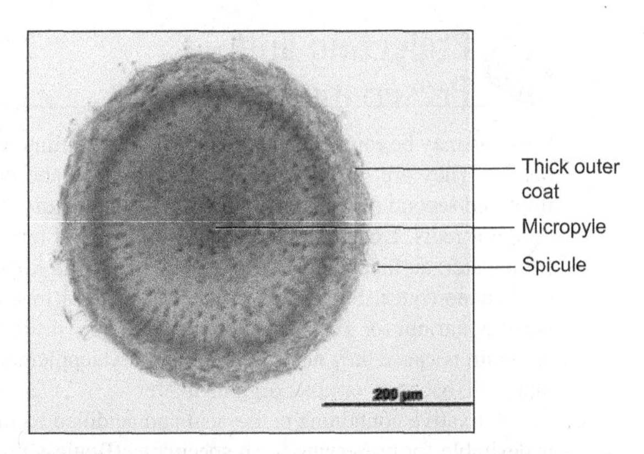

FIGURE 7.16 Gemmule of freshwater sponge showing detail.
Photo by John R. Meyer.

Regeneration and Reconstitution (Optional Exercise)

Sponges are especially noted for their powers of regeneration. A small part of a sponge can regenerate a complete sponge. Many years ago, scientist H.V. Wilson showed that pieces of a living sponge pressed through a fine cloth mesh to separate the sponge into individual cells and clusters of cells could reassemble and develop into a complete new sponge. More experiments with the separation and reassociation of sponge cells have provided important information about the nature and mechanism of cell-recognition processes and the organization of tissues during development.

If living sponges are available in your laboratory, you may be able to repeat this famous experiment. *Microciona*, a common red sponge on the Atlantic coast of the United States, was used in the original experiments. You will need a small finger bowl, a watch glass, a clean microscope slide, a pipette, some silk mesh cloth, and some seawater for this experiment.

Procedure

1. Fill the finger bowl about two-thirds full of seawater of the proper ionic strength (depending on the source of the sponges) and place the watch glass on the bottom of the dish. Place the microscope slide on top of the watch glass.
2. In a separate bowl of seawater, prepare a suspension of cells and fragments of *Microciona* by pressing small pieces of sponge through a fine silk bolting cloth or nylon mesh of about 150–200 micrometers in diameter.
3. Pipette a small amount of this cell suspension onto the slide and allow the cells to settle.
4. Carefully lift out the slide and observe the cells. Make similar observations at intervals during the next 24–48 hours. Watch for the initiation of cellular aggregation and the thin protoplasmic extensions (filopodia) put out by the small clumps of aggregating cells. Sketch the cells and clumps of cells as a record of your observations.

Collecting and Preserving Sponges

Sponges may be found in marine habitats at all latitudes and depths. They are the most common animals on the ocean floor and second only to corals in abundance (and diversity) on coral reefs. Living specimens should never be removed from water: air that becomes trapped internally will kill delicate choanocytes. Small individuals can be held in a saltwater aquarium for a few days, but they are very difficult to maintain because they need a regular dietary supplement of plankton, bacteria, or other organic matter.

A fixative containing picric acid and acidified formalin is desirable for preserving fresh specimens (Bouin's fluid is a 3:1 mixture). Fix specimens for 24 hours, then rinse and store in 95% ethanol.

Spicules can be separated from their organic matrix by immersing a small piece of fresh or preserved sponge in ordinary household bleach solution, which is about 3–5 percent sodium hypochlorite (NaClO). After the organic matter dissolves, rinse with water, transfer the residue to a microscope slide, and examine it under 40× magnification with a compound microscope. Determine the chemical composition of the spicules by adding a drop of 5% HCl. Calcareous spicules dissolve in a frenzy of bubbles as their calcium carbonate reacts with the acid to form carbon dioxide gas and soluble calcium chloride. Siliceous spicules are unaffected by the acid.

Freshwater sponges can often be found in unpolluted streams. They grow as tufts or small irregular masses encrusting sticks, stones, or submerged plants. Most species are yellow or brown, but a few species are green because of symbiotic algae living within the sponge. All freshwater species are very susceptible to pollution.

Key Terms

Asconoid the simplest type of sponge canal system with a central spongocoel lined with choanocytes and with many ostia opening directly into the spongocoel.

Choanocyte a special type of flagellated collar cell characteristic of sponges that lines the flagellated water chambers.

Gemmule a dormant stage in the life cycle of freshwater sponges and a few species of marine sponges. Formed as internal asexual buds, they aid the sponges in surviving adverse environmental conditions.

Leuconoid a complex type of sponge body form with an intricate system of internal water canals. Choanocytes line certain small chambers within the system.

Mesohyl the loose, gelatinous matrix in a sponge body containing several types of amoeboid cells, fibrils, and skeletal elements (spicules and/or spongin fibers) located between the outer pinacoderm and the inner choanocyte layer.

Osculum (plural: oscula) an excurrent opening from the spongocoel.

Ostium (plural: ostia) an incurrent pore or opening.

Pinacocyte a thin, flattened type of cell that lines the external surface of sponges and some inner surfaces not lined by choanocytes.

Pinacoderm external cell layer of sponge body made up of pinacocytes.

Spicules individual skeletal elements that make up the skeleton of most sponges. Consist mainly of calcium carbonate or silicon salts and exhibit many different shapes.

Spongin fibers a loose network of protein fibers that form part or all of the skeleton of most horny sponges (Class Demospongiae); also occur in the mesohyl of some other sponges and in the walls of gemmules.

Spongocoel The large central cavity in syconoid sponges that opens through the osculum.

Syconoid a type of sponge canal system that exhibits a central spongocoel into which numerous radial canals lined with choanocytes empty.

Internet Resources

There are many valuable Internet sites with information about zoology. Several sites containing pertinent zoological information for this chapter can be found on the McGraw-Hill Zoology web site at **http://www.mhhe.com/zoology**. Just click on this text's title.

Questions for Critical Thinking

1. Many years ago, sponges were often thought to be plants rather than animals. Discuss why early naturalists might have made such an error.
2. Discuss how natural selection may have influenced the development of complex canal systems in sponges.
3. Sponges are very common and abundant in marine waters but only a few species are found in fresh water, and these freshwater species are neither common nor abundant. What factors might be responsible for this large difference in number of species and abundance of sponges between the sea and fresh water?

Suggested Readings

De Vos, L., K. Rutzler, N. Boury-Esnault, C. Donadey, and J. Vacelet. 1991. *Atlas of Sponge Morphology*. Washington, DC: Smithsonian Institution Press. 258 pp

Hooper, J.N.A., and R.W.M. Van Soest (eds.). 2002. *Systema Porifera: A Guide to the Classification of Sponges*. New York: Kluwer Academic/Plenum. 2 vols

Mueller, W.E.G. 2003. *Sponges*. New York: Springer-Verlag. 258 pp

Chapter 8

Cnidaria

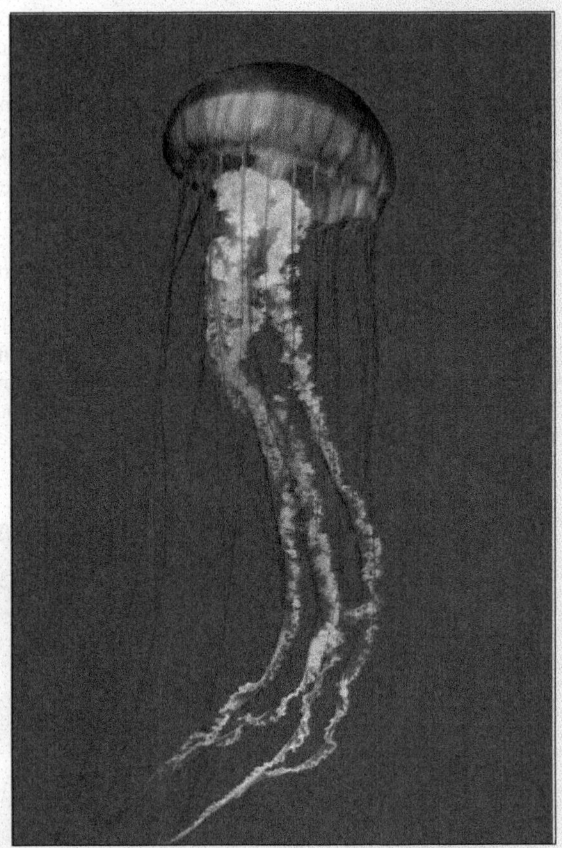

Kevin Schafer/Alamy.

Objectives

After completing the laboratory work in this chapter, you should be able to perform the following tasks:

1. Briefly give five distinguishing characteristics of the phylum Cnidaria.
2. List the four classes of cnidarians and briefly characterize each class.
3. Discuss polymorphism in cnidarians and its significance.
4. Describe the general morphology of *Hydra* as a model cnidarian polyp.
5. Discuss the structure and function of nematocysts and explain their importance in *Hydra* and other cnidarians.
6. Describe the reproductive processes of *Hydra* and be able to identify mature male and female individuals.
7. Identify the chief morphological features of the medusa of *Gonionemus* and explain the function of each part.
8. Describe the life cycle of *Gonionemus*.
9. Identify the main morphological features of a mature *Obelia* colony and explain the function of each.
10. Describe the life cycle of *Obelia* and explain how it illustrates the alternation of generations commonly found in cnidarians.
11. Describe the main morphological features of the scyphozoan medusa *Aurelia* and explain the function of each.
12. Describe the life cycle of *Aurelia*.
13. Describe the structure of the sea anemone *Metridium* and explain the function of its principal parts.
14. Explain three major differences between hard corals (Hexacorallia) and the soft corals (Octocorallia).

Introduction

Cnidarians (formerly called coelenterates) are the simplest animals with definite tissues. The cnidarian body exhibits **diploblastic construction;** that is, it consists of two well-defined tissue layers and a third intervening layer of gelatinous material, the **mesoglea,** that varies in structure among the four classes of cnidarians. The outer **epidermis** layer covers the external surface of the body, and the inner **gastrodermis** layer lines a single internal body cavity. In the simplest cnidarians, Class Hydrozoa, the mesoglea is thin and gelatinous; in the Class Scyphozoa and the Class Cubozoa, the mesoglea is fibrous and contains amoeboid cells; and in the Class Anthozoa, the mesoglea has many amoeboid cells and thus resembles a true mesenchyme layer. Some scientists have suggested that the cellular components of certain cnidarians may represent a rudimentary or primitive third tissue layer, but most authors still consider the phylum to be essentially diploblastic.

A major distinguishing feature of the Cnidaria is the presence of cnidae (sing. cnida), tiny cell products formed by **cnidocytes,** specialized secretory cells that secrete the complex cnidae. Several kinds of cnidae have been described, but **nematocysts** are the most common and most numerous (see figure 8.1). Each nematocyst is a complex secretory product formed within developing cnidocytes. It consists of an outer protein capsule and a long, coiled tube often armed with spines or barbs. Toxins, enzymes, and other chemicals are also contained within the capsule of various kinds of nematocysts. Nematocysts are triggered by chemical and/or mechanical stimulation and explode with some force, literally turning inside out. Some types of nematocysts are able to penetrate the body of a prey organism, injecting their contents. Other types contain sticky substances that adhere to the prey and hinder its movement. No other phylum has members that produce nematocysts, but several animals in other phyla are predators of cnidarians, including some flatworms and some molluscs, and are able to incorporate nematocysts into their own tissues and make some use of them. Nematocysts are among the most complex cell products made by animal cells.

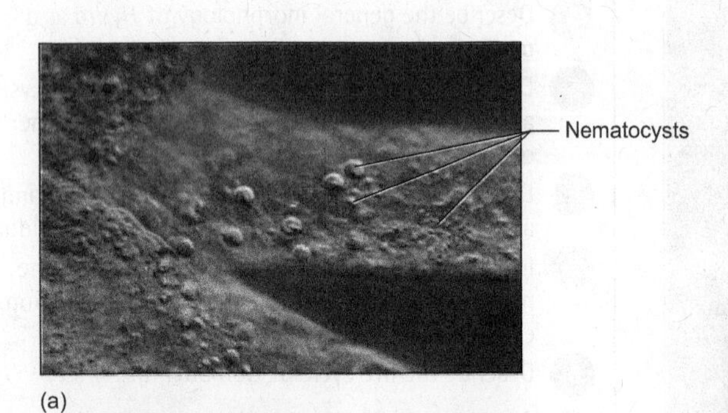

(a)

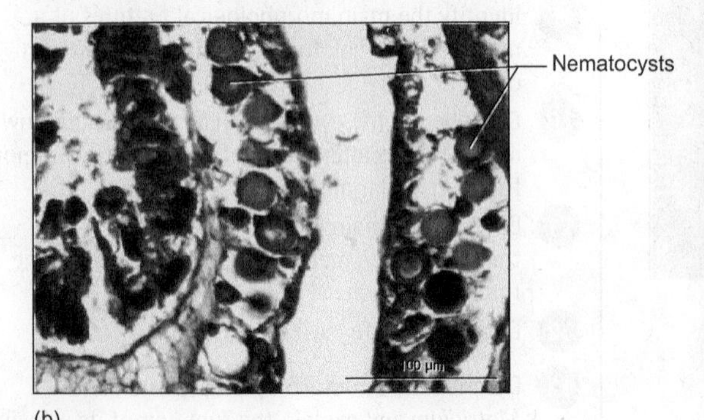

(b)

FIGURE 8.1 Nematocysts (*a*) Tentacle of living *Hydra.* (*b*) Cross section of tentacles of *Physalia.*

(a) Courtesy of Carolina Biological Supply Company, Burlington, NC.
(b) Photo by John Meyer.

The cnidarians are also noted for their prominent **radial symmetry.** Their body parts are arranged symmetrically around a central axis as in the medusa of *Polyorchis* shown in figure 8.2. Observe the cuplike shape of the animal that aids its swimming movements. Along with the Phylum **Ctenophora,** the Cnidaria constitute the **radiate phyla.** The Ctenophora (figure 8.3) is a relatively small group of marine animals that contains the comb jellies with soft, gelatinous bodies. Both phyla demonstrate **primary radial symmetry.** The Cnidaria and the Ctenophora are believed to represent groups evolved from protistan ancestors prior to the evolution of triploblastic animals. Scientists are currently using genetic and biochemical techniques to learn more about the early evolution of the Cnidaria, the Ctenophora, and other Metazoa.

In addition to the Cnidaria and the Ctenophora one other phylum, the Echinodermata, also exhibits radial symmetry, but only among adult animals. The radial symmetry of echinoderms has been shown to be a **secondarily derived feature** since larval stages of echinoderms are bilaterally symmetrical. Radial symmetry in echinoderms is believed to have evolved as a result of their sedentary lifestyle as bottom dwellers in the sea.

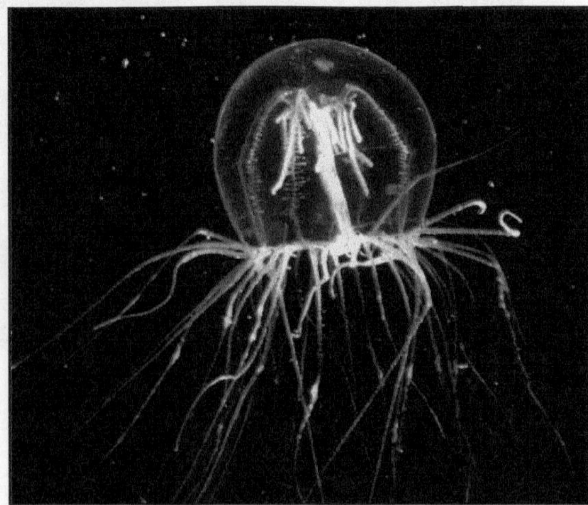

FIGURE 8.2 Medusa of *Polyorchis,* a small medusa found in shallow waters of the Pacific.

Ingram Publishing Age Fotostock.

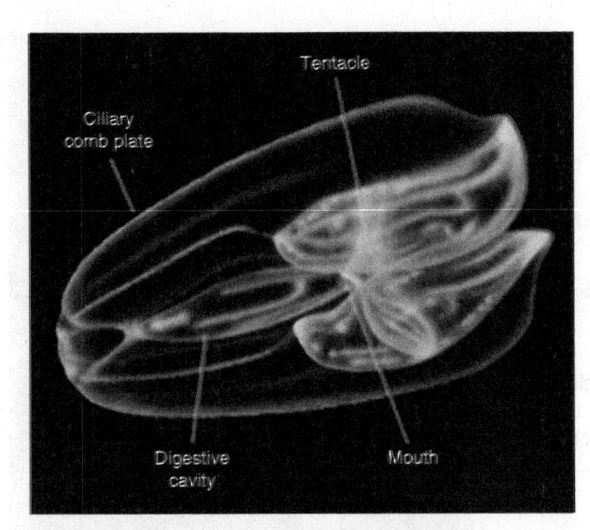

FIGURE 8.3 Drawing of a ctenophore. Note the longitudinal rows of ciliary comb plates used in swimming.
Courtesy of Carolina Biological Supply Company, Burlington, NC.

The former name of the Cnidaria, Coelenterata, refers to the large interior coelenteron, or **gastrovascular cavity.** The opening into this gastrovascular cavity is called a **mouth,** but it is a single opening that serves for both entrance and exit of materials. Tentacles are usually arranged around the mouth opening. This is another distinguishing characteristic of the phylum.

Two different **body forms** are exhibited in the life cycle of many cnidarians: an attached **polyp** stage and a free-swimming **medusa** stage. Many species have both polyp and medusa stages and their life cycles involve an alternation of these two forms. Although this type of life cycle is often called an **alternation of generations,** it really is an **alternation of two body forms** within a single generation. Other cnidarians have only a polyp form or only a medusa form in their life cycle, and the life cycle may also include one or more larval stages during development. We will study examples of several cnidarian life cycles in this exercise.

Another distinguishing feature of the Cnidaria is a diffuse network of nerve tissue rather than an organized central nervous system. Cnidarians lack an anterior concentration of nerve cells or anything resembling a brain as found in most other kinds of animals.

You will study members of three classes of cnidarians in this exercise. The freshwater *Hydra* is a member of Class Hydrozoa. Although not a very typical cnidarian, it is common in ponds and streams around the world, readily available, and makes a good model of a polyp. *Gonionemus*, also a hydrozoan, is studied as a representative medusa. *Obelia* and *Physalia* represent two different types of colonial Hydrozoa.

Class Scyphozoa, which includes the large oceanic medusae, is represented by *Aurelia*, and Class Anthozoa is represented by the sea anemone, *Metridium*, widely distributed in shallow waters along the Atlantic coast of the United States.

Classification

The phylum is divided into four classes:

Cladogram of Cnidaria

- Class **Anthozoa** (corals and sea anemones)
- Class **Hydrozoa** (hydroids and siphonophores)
- Class **Cubozoa** (sea wasps or box jellyfish)
- Class **Scyphozoa** (true jellyfish)

Class Hydrozoa (Hydroids and Siphonophores)

Most hydrozoa have a colonial polyp stage and a free swimming medusa stage. The medusae have a velum, a thin flap of tissue around the bottom of the bell, and bear gonads on their radial canals. The mesoglea is thin and gelatinous; there is a well-defined epidermis and gastrodermis. Most species are marine but a few live in fresh water. Hydrozoans often exhibit polymorphism with individuals specialized for certain functions. About 3,000 species. Examples: *Obelia, Hydra, Gonionemus, Polyorchis* (figure 8.2), and *Physalia* (Portuguese man-of-war).

Materials List

Living specimens
- *Hydra*
- *Daphnia* or *Artemia* larvae (food for *Hydra*)
- Sea anemones in aquarium (demonstration)

Preserved specimens
- *Gonionemus*, medusae
- *Metridium*
- *Physalia*
- *Aurelia*, medusae
- Ctenophores (demonstration)
- Assorted examples of dried corals
- Representative Hydrozoa, Anthozoa, and Scyphozoa (demonstrations)

Prepared microscope slides
- *Hydra*, cross section, longitudinal section
- *Obelia*, whole mount of polyp, whole mount of medusa
- Nematocysts, discharged (demonstration)
- *Gonionemus*, statocysts (demonstration)
- *Hydra*, whole mount, with buds (demonstration)
- *Hydra*, male with testes, whole mount (demonstration)
- *Hydra*, female with ovary, whole mount (demonstration)
- Green *Hydra* with symbiotic algae, whole mount (demonstration)
- Planula larva, whole mount (demonstration)

Continued–

–Continued

Aurelia marginal sense organs, scyphistoma, ephyra larva, strobila (demonstrations)

Chemicals

1% Acetic acid

0.01% Methylene blue solution

Pond water

Class Scyphozoa (True Jellyfish)

Most are large marine jellyfish with an abundant mesoglea, polyp stage reduced or absent, and velum lacking in the medusae. The mesoglea contains fibers and amoeboid cells. About 200 species, all marine. Examples: *Aurelia, Chrysaora* (sea nettle; see photo on page 109), and *Cyanea.*

Class Cubozoa (Sea Wasps or Box Jellyfish)

All species have small marine medusae with conical bells of four flattened sides and a velum or velumlike structure around the margin of the bell. Small polyps metamorphose directly into medusae. Common in all tropical seas, especially in the Indo-Pacific and Western Pacific. Their stings are dangerous to humans; the nematocysts contain some of the strongest toxins known. About 20 described species. Examples: *Carybdea, Chironex,* and *Chiropsalmus.*

Class Anthozoa (Sea Anemones and Corals)

Solitary or colonial animals with only a polyp stage (the medusa is absent), a pharynx or gullet is present, gastrovascular cavity partitioned by septa. Mesoglea with many amoeboid cells. About 6,000 species, all marine. Examples: *Metridium* (sea anemone), *Astrangia* (coral), *Gorgonia* (sea fan), and *Renilla* (sea pansy).

A Hydrozoan Polyp: *Hydra*

Class Hydrozoa

Hydra (figure 8.4) typifies the polyp form of a cnidarian. *Hydra* lives in freshwater streams, lakes, and ponds. It is usually attached to submerged sticks, stones, or vegetation and feeds on various small aquatic animals.

Hydra serves as a good example of the polyp stage of a cnidarian, but it differs from most other members of the Class Hydrozoa in several important ways. *Hydra* is solitary (not colonial), it lives in fresh water (not marine), and it has the ability to move around (not sessile). Also, *Hydra* has no separate medusa stage. The polyp develops gonads and releases gametes for sexual reproduction—events that occur during the medusa stage in other hydrozoans.

FIGURE 8.4 Living *Hydra.*

Courtesy of Carolina Biological Supply Company, Burlington, NC.

General Appearance and Morphology

- Examine a living specimen of *Hydra* in a dish of pond water. **Caution:** Be sure to use pond water and not tap water, since most tap water contains trace amounts of copper and other substances toxic to *Hydra.*

Locate the following structures on your specimen with the aid of figure 8.5: (1) the **pedal disc** at the lower end, which serves for attachment; (2) the cylindrical **column;** (3) a circle of **tentacles** at the free end (***how many?***); (4) the **hypostome,** an elevation between the bases of the radially arranged tentacles; (5) the **mouth** in the center of the hypostome; (6) **buds,** the products of asexual reproduction, may also be present; and (7) **ovaries** or **testes** are also present on the middle portion of the body in mature specimens (figure 8.10).

Behavior

- ***Does your specimen change in shape?*** Touch one of the tentacles with the tip of your dissecting needle. ***What is the reaction? What methods of locomotion are used by* Hydra?** Observe *Hydra* feeding in an aquarium, or add a few *Daphnia* or washed brine shrimp (*Artemia*) larvae to your dish near the specimen to observe feeding.

Cnidocytes and Nematocysts

After you have studied the basic form and behavior of your specimen, place it on a clean microscope slide in a drop of water and carefully add a coverslip. Take care not to crush the animal with the weight of the coverslip. Observe the numerous **cnidocytes,** which appear in clusters on the surface of the tentacles (figures 8.1 and 8.6). Each cnidocyte is a cell containing a **nematocyst,** or stinging capsule.

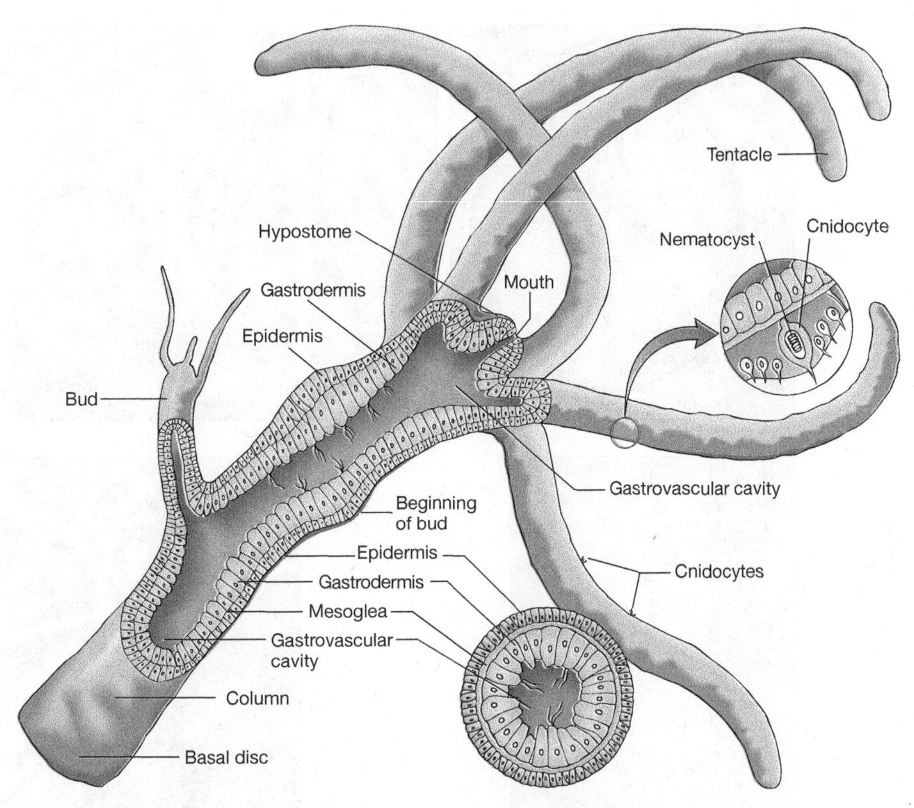

FIGURE 8.5 *Hydra*, body construction.

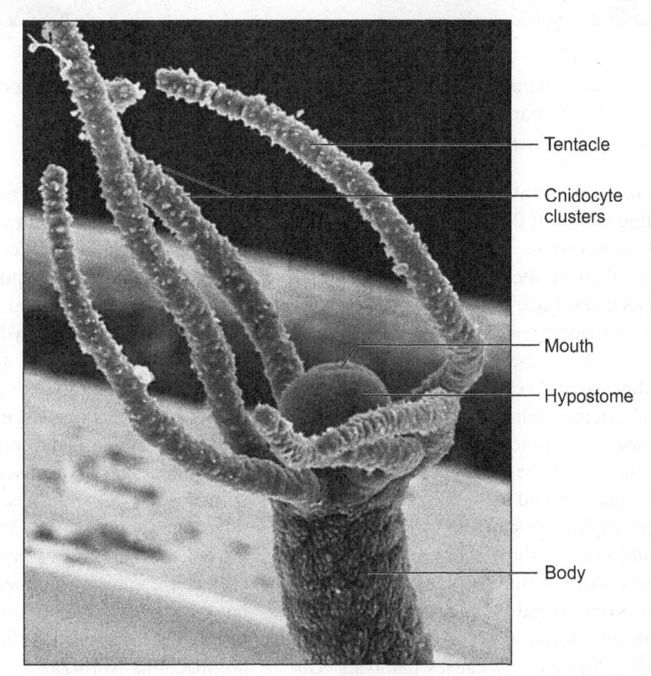

FIGURE 8.6 *Hydra*.
Scanning electron micrograph by Louis de Vos.

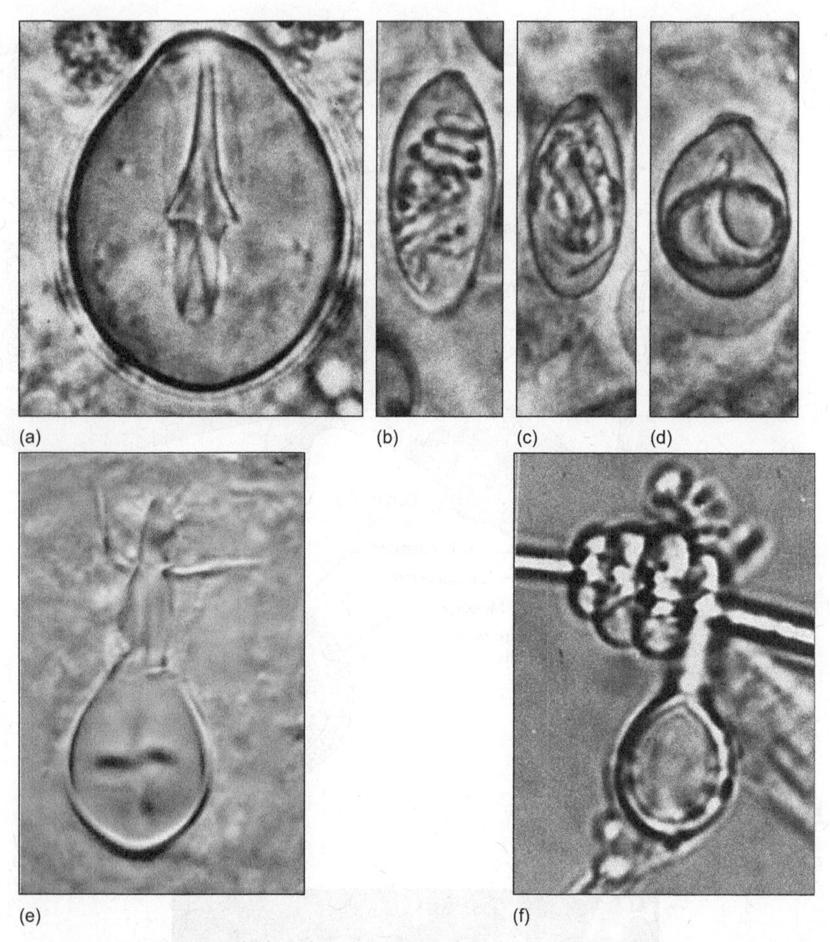

FIGURE 8.7 Nematocysts of *Hydra* (*a*) stenotele, (*b*) holotrichous isorhiza, (*c*) atrichous isorhiza, (*d*) desmoneme, (*e*) discharged stenotele, (*f*) discharged desmoneme.
Photographs courtesy of Carolina Biological Supply Company, Burlington, NC.

The cnidocytes of *Hydra* can be stained to aid in your observations by adding a drop of 0.01% methylene blue solution to the edge of the coverslip.

Nematocysts are complex **secretory products** formed within developing cnidocytes. Each nematocyst consists of an outer protein **capsule,** and a long, coiled **tube** often armed with **spines** or **barbs.** Toxins, enzymes, and other chemicals are also contained within the capsule of certain kinds of nematocysts.

Hydra has four different kinds of nematocysts, each with a distinctive structure and function (fig. 8.7). **Stenoteles** or penetrants are the largest of the four types. They play an important role in the capture and immobilization of prey. Discharged stenoteles (fig. 8.7e) show an open tubule, enlarged at its base, and bearing three sharp stylets that aid in penetrating the prey. A long thin tube extends from the enlarged base. Upon firing, these parts of the nematocyst are ejected with explosive force, penetrating the prey and injecting a potent toxin that quickly causes paralysis. Undischarged stenoteles (fig. 8.7a) are easily recognized by the large dart-like base of the shaft of the coiled tube.

The four types of nematocysts occur in numerous batteries on the tentacles and are arranged with one or two stenoteles in the center with the other types clustered around them. **Desmonemes** are much smaller and are slightly pear-shaped. They are the most abundant type of nematocyst in hydra. When discharged, their narrow tubes tend to form several tight coils that entangle bristles or other projections on the prey's body. **Holotrichous isorhizas** are larger than desmonemes and appear sausage-shaped. They contain a long coiled tube of uniform diameter. After discharge, the tube exhibits several spiral rows of tiny spines. These nematocysts, commonly thought to act as "sticky ropes" for prey capture, may also have some stinging or penetrating action. **Atrichous isorhizas** are similar in appearance to holotrichous isorhizas but are slightly smaller and lack spines on their discharged tubes. Their function is thought to be similar to that of the holotrichous isorhizas.

Tap on the coverslip of your wet mount of *Hydra* tentacles to induce a discharge of the nematocysts. When properly stimulated, nematocysts empty their contents with a rapid discharge as the coiled tube shoots out. In addition to tapping the coverslip to induce nematocyst discharge, you can also stimulate nematocyst discharge by adding a drop of dilute acetic acid (1% solution).

Study the discharged nematocysts under high power or with an oil immersion lens, if available. Observe the outer capsule, the long thread or tube, and the large spines or barbs at the base of the tube. ***How many kinds of nematocysts can you find on your slide with the aid of figure 8.7?***

Observe also the microscopic demonstration of discharged nematocysts.

Histological Structure

Hydra and other members of the Class Hydrozoa exhibit the simplest histological structure among the Cnidaria (figure 8.5).

- Examine stained slides of cross and longitudinal sections of *Hydra*, and note the following: (1) **epidermis,** the outer, thinner epithelial layer of cells; (2) **gastrodermis,** the inner layer of cells; (3) **mesoglea,** a very thin, noncellular layer between the epidermis and gastrodermis; and (4) **gastrovascular cavity,** or enteron, the internal cavity lined by the gastrodermis.

Cellular Structure

Hydra is significantly more complex than a sponge, not only in its general structure but also in the degree of differentiation of cellular structure and functions. Several different types of cells may be distinguished in the stained preparations. In the epidermis, under high power, try to distinguish the following cell types: (1) The large and abundant **epitheliomuscular cells,** which possess contractile processes or fibers at their base, all running lengthwise. ***What specific functions do these fibers perform?*** (2) The small **interstitial cells** at the bases of the epitheliomuscular cells. (3) The **cnidocytes.** (4) The mucus-secreting **gland cells** abundant on the pedal disc. Also among the epidermal cells are many small nerve cells (described in the next section). They can be seen only in specially stained slides.

In the inner gastrodermal layer, note the following cell types: (1) The abundant, large **digestive cells** with many vacuoles. These cells bear one or two flagella, and ingest food particles for intracellular digestion. They are also epitheliomuscular in character and possess contractile fibers that run transversely along their bases, thus providing a circular musculature. ***What is the function of the contractile fibers in the gastrodermal cells?*** Compare this function with the function of the contractile fibers in the epidermis. (2) The **gland cells,** which secrete either mucus or digestive enzymes. Mucus-secreting cells are abundant in the hypostome region. (3) The **interstitial cells** at the bases of the gastrodermal cells. Some other types of cells are also present but are difficult to observe except in specially stained preparations.

Some species of *Hydra* have symbiotic algae called zoochlorellae in the gastrodermal cells lining their gastrovascular cavity giving them a bright green color. This is a commonly cited example of symbiosis, and several research studies have been conducted on the symbiotic relationship between these two species. The symbiotic algae appear to contribute some of their nutrients to the hydra similar to the way that symbiotic golden brown zooanthellae contribute nutrients to corals in the ocean.

Nervous System

Specialized nerve cells, or **neurons,** are located among the cells of both the epidermal and gastrodermal layers of *Hydra.* These neurons may have two or more processes that connect via **synaptic junctions** with other neurons or with various types of receptor or effector cells.

Figure 8.8 shows an isolated multipolar neuron from the body wall of *Hydra.* Special techniques are required to see cnidarian neurons with either a light microscope or an electron microscope.

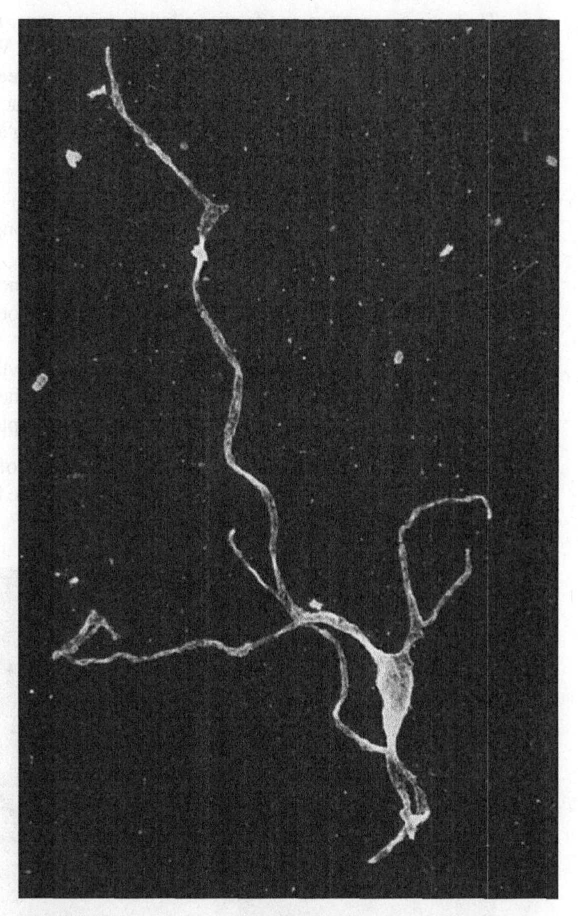

FIGURE 8.8 Isolated multipolar neuron from body column of *Hydra* with several branches or processes.

Scanning electron micrograph by J. A. Westfall and L. G. Epp from *Tissue and Cell* 17(2):165.

The interconnecting network of neurons in *Hydra* form a **nerve net** that lacks concentrations of neurons that form ganglia or a brain. This diffuse nerve net is characteristic of *Hydra* and other cnidarians. Nervous impulses tend to spread in a radiating pattern from the point of origin or stimulation because of the structure of synapses between adjacent neurons. Many of these synapses are symmetrical and are **nonpolarized,** thus allowing impulses to flow in both directions. Some polarized synapses also occur in cnidarians. Recent studies have revealed similar diffuse nerve nets in the nervous system of some higher animals, including the digestive systems of annelid worms and humans. The rhythmic peristaltic movements of your stomach and intestine after you eat are coordinated by such a nerve net.

Neurons also have processes connecting via synapses with epithelial muscular cells, gland cells, and nematocysts. Substantial evidence suggests that nematocyst discharge is at least partly under nervous control.

Feeding Behavior

Hydra is a carnivore and feeds on living crustaceans, rotifers, insect larvae, and other small animals (figure 8.9). When properly stimulated, *Hydra* exhibits a characteristic feeding response. You can observe the feeding behavior by placing a healthy, unfed *Hydra* in a clean watch glass containing about 10 ml of pond water (or *Hydra* culture solution).

- Add a few (6–12) washed brine shrimp (*Artemia*) larvae near the *Hydra* and record your observations. Note how the food organisms are captured; what happens to them during and after their capture; and the movements of the tentacles, hypostome, and other parts of the *Hydra.* Use the second hand on your watch to time various parts of this complex behavioral reaction of *Hydra.* Record your observations in the Notes and Sketches section at the end of this chapter.

Scientists have shown that the feeding reaction of *Hydra* is normally caused by body fluids oozing from the body of its prey, which has been pierced by nematocysts. Further experiments have shown that a feeding reaction can be elicited by a solution of a tripeptide, reduced glutathione.

FIGURE 8.9 *Hydra* capturing and ingesting *Daphnia.*
Courtesy of Carolina Biological Supply Company, Burlington, NC.

Reproduction

Hydra reproduces asexually by budding and sexually by the production of eggs and sperm (figure 8.10). The *Hydra* in most laboratory samples are asexual, and you can often find budding specimens among them. In nature, *Hydra* reproduce most of the year by budding, and at certain times form gonads—testes or ovaries—which appear as thickenings of the body wall (figure 8.10). Most species of *Hydra* are **dioecious** (sexes are separate), although a few species are **monoecious** (both sexes are in one individual). Observe the demonstration materials of male and female hydra with gonads.

Regeneration (Optional Exercise)

Hydra and other cnidarians have great powers of regeneration. Place a *Hydra* in a watch glass and cut across the middle of the body with a sharp scalpel or razor blade. Separate the two pieces into different watch glasses and observe their development during the next several days. Observe the formation of a new hypostome, mouth, and tentacles by the *basal half* of your *Hydra.* ***What happens to the other half?*** Keep a record of your observations in the Notes and Sketches section at the end of this chapter.

Demonstrations

1. *Hydra* with testes (microscope slide)
2. *Hydra* with ovaries (microscope slide)
3. *Hydra* budding (microscope slide)
4. Green *Hydra* with symbiotic algae (microscope slide)

A Hydromedusa: *Gonionemus*

Class Hydrozoa

Gonionemus is a small marine **hydromedusa,** or hydrozoan jellyfish, common in coastal waters in many parts of the world. The medusa of *Gonionemus* is the adult or sexually mature stage and serves to illustrate the typical structure of a hydrozoan medusa (figure 8.11).

- Examine a preserved specimen in a watch glass partly filled with water and note its umbrella-like form. The specimens are delicate and must be handled with care. Note the jellylike consistency of the medusa, due to the thick layer of **mesoglea** within the body. The bulk of the medusoid body is largely mesoglea.

The medusa usually swims with its convex **exumbrellar surface** upward and the concave **subumbrellar surface** downward. Observe the numerous **tentacles** around the margin of the bell. Note the small swollen **adhesive pads** and rings of nematocysts on the tentacles. *Gonionemus* lives in shallow waters and is often found on sea grass beds. It uses the adhesive pads to attach temporarily to the sea grass to

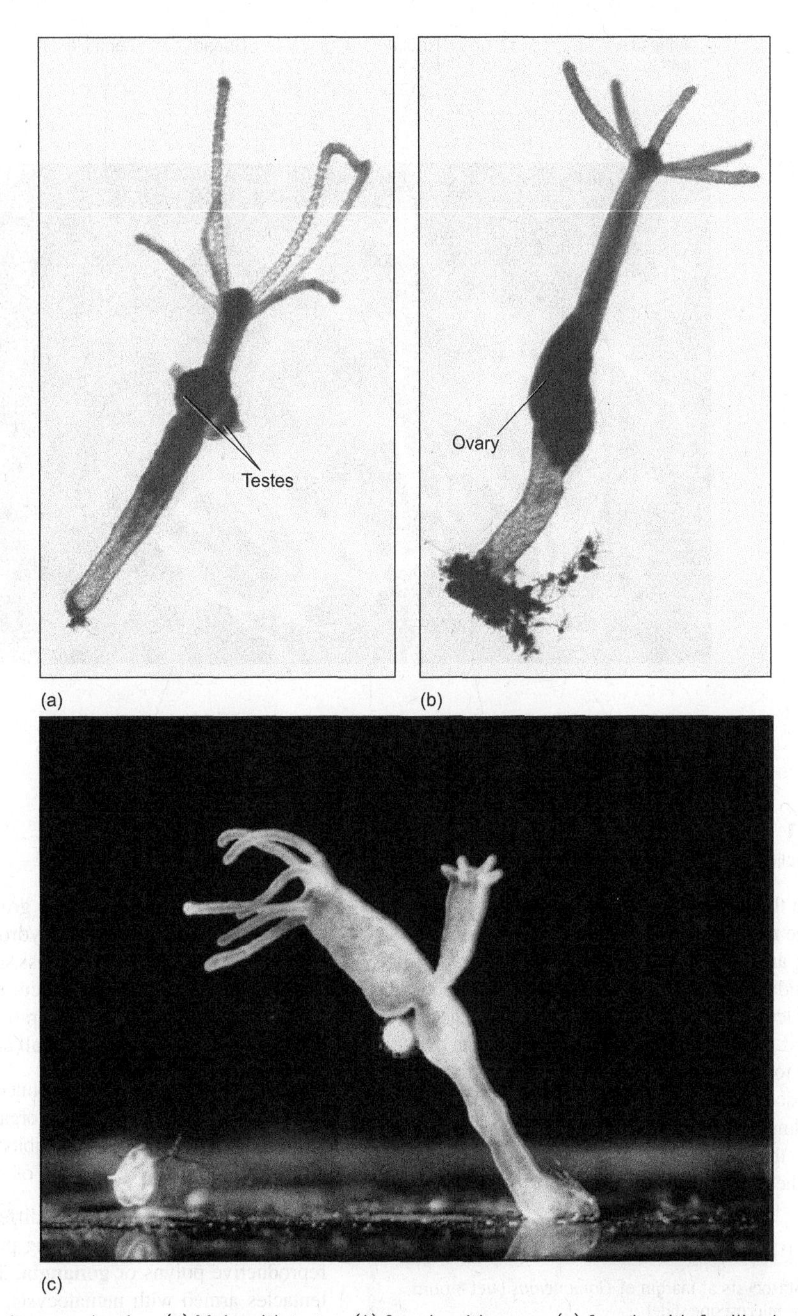

FIGURE 8.10 *Hydra* reproduction. (*a*) Male with testes, (*b*) female with ovary, (*c*) female with fertilized egg and a bud.
Courtesy of Carolina Biological Supply Company, Burlington, NC.

keep from being washed about. It is not an active swimmer. Extending downward from the center of the subumbrellar surface is the **manubrium,** with the **mouth** at its tip surrounded by four **oral lobes.** Around the inner margin of the bell, extending inward, is a thin circular flap of tissue, the **velum,** which is believed to aid in swimming. At the base of the manubrium is an expanded portion of the manubrium, the "stomach." Find the four **radial canals** extending from the stomach to the **circular canal** at the margin of the bell. The hollow tentacles connect with the circular canal, and their cavities represent a continuation of the gastrovascular cavity. Observe the numerous **nematocyst batteries** and the **adhesive pads** on the tentacles.

At the bases of the tentacles are round, pigmented structures that serve as light-sensitive photoreceptors. Between the tentacle bases are the **statocysts,** which serve as balancing organs.

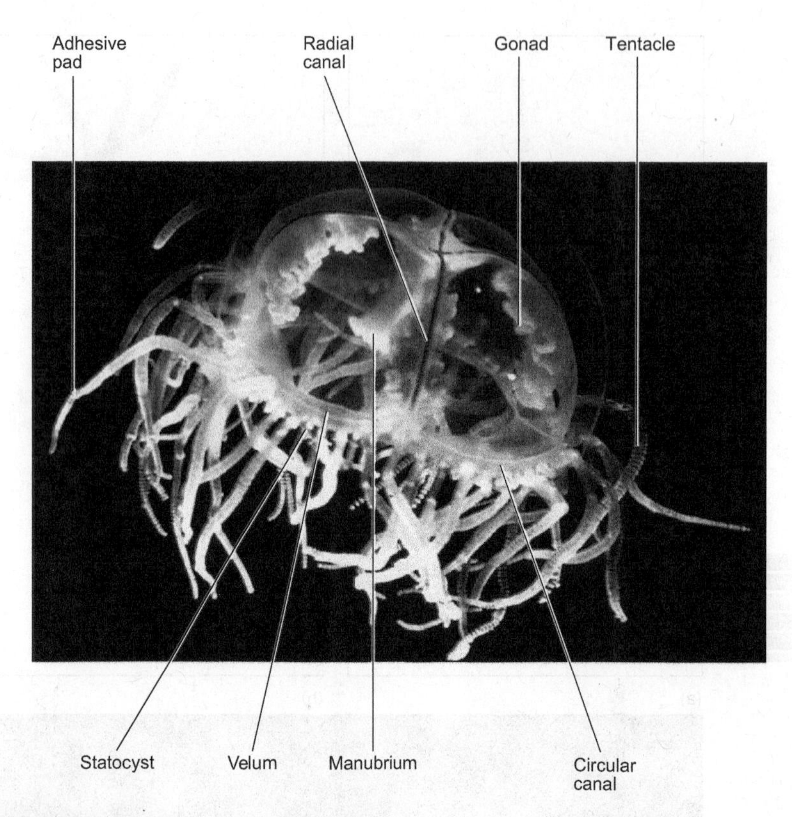

FIGURE 8.11 *Gonionemus* medusa.
Courtesy of Carolina Biological Supply Company, Burlington, NC.

Observe the folded gonads attached to the subumbrellar surface of the radial canals. The gonads on the specimen you are studying are either ovaries or testes, since *Gonionemus*, like most cnidarians, is dioecious. Gametes produced by the gonads are released into the sea, and the fertilized eggs develop into a ciliated **planula larva.** Later, the planula larva settles and attaches to a submerged object in the sea and transforms into a microscopic **polyp.** The polyp may reproduce asexually by budding and, under certain conditions, may form tiny medusa buds that detach and grow into mature medusae, thus completing the life cycle and the alternation of generations.

Demonstrations

1. Statocysts at margin of *Gonionemus* (wet mount or microscope slide)
2. Planula larva (microscope slide)

A Colonial Hydrozoan Polyp: *Obelia*

Class Hydrozoa

Obelia is a colonial marine cnidarian that illustrates the **complex life cycle** with alternating polyp and medusa stages found in many cnidarians (figure 8.12 and 8.13). Such species are members of a large group of common **colonial** organisms called **hydroids.** Hydroids are often found attached to shells, rocks, sea grass, and other objects in the intertidal zone but they also occur on floating algae and on the sea bottom. The polyp generation predominates in most hydroids, and the medusa is small and often short-lived.

- Examine a stained whole mount of the polyp or asexual stage of *Obelia* and study its organization. Like many colonial animals, *Obelia* exhibits **polymorphism**—the morphological specialization of its members.

You can distinguish two different kinds of individuals in an *Obelia* colony: feeding polyps or **hydranths** and reproductive polyps or **gonangia.** The feeding polyps bear tentacles armed with nematocysts, a mouth, a hypostome, and a delicate outer covering, the **hydrotheca.** Reproductive polyps consist of a central **blastostyle** on which **medusa buds** develop, and a thin outer covering, the **gonotheca.** At the distal end of the gonotheca is an opening, the **gonopore,** through which the newly liberated medusae escape. Note that the gonangia have no mouth or tentacles. ***How do they receive their nutrition?***

The hydranths and gonangia are attached to a main stem, or **hydrocaulus,** which consists of a cylindrical tube of living tissue, the **coenosarc,** and an outer secreted covering, the **perisarc.** In cross section, the coenosarc resembles

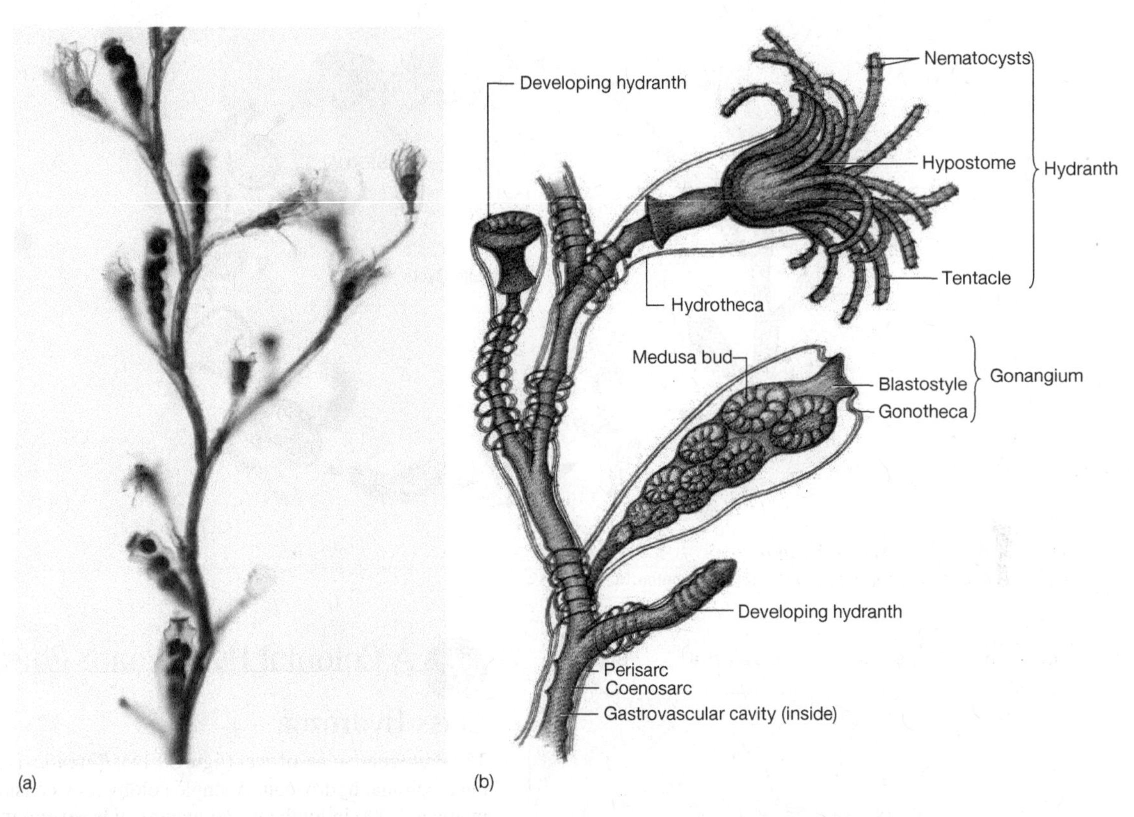

FIGURE 8.12 *Obelia.* (*a*) Polyp colony, whole mount. (*b*) Drawing of feeding individual and gonangium.
(a) Courtesy of Carolina Biological Supply Company, Burlington, NC.

a cross section of *Hydra* with an outer epidermis, an inner gastrodermis, and a thin intervening layer of mesoglea.

Alternation of Generations

The life cycle of *Obelia* is fundamentally similar to that of *Gonionemus* and illustrates the alternation of generations characteristic of the Phylum Cnidaria (figure 8.13).

The alternation of the asexual polyp generation and the sexual medusa generation is basically different from the alternation of generations in plants. Both the polyp and medusa generations of cnidarians are **diploid** (2n), while in plants the alternation is between a **haploid** (n) gametophyte generation and a **diploid** (2n) sporophyte generation.

In the life cycle of *Obelia*, the polyp generation produces medusa buds within its gonangia. The tiny, short-lived medusae escape to become part of oceanic plankton and produce either eggs or sperm. Fertilized eggs develop into ciliated **planula larvae,** which swim about in the sea for a time and settle to transform into a new polyp. Buds formed by asexual reproduction of the polyp do not detach, thus forming a colony.

Fact File

Cnidaria

- The phylum name Cnidaria comes from the Greek word for a nettle and refers to the presence of nematocysts, tiny harpoonlike spears. The firing of these nematocysts occurs at a velocity of 2 meters per second with 40,000 *g* acceleration. It is the fastest physical reaction in the animal kingdom.
- A nematocyst is not an organelle. It is manufactured by the Golgi apparatus within a cnidoblast cell and, therefore, qualifies as a secretory product. Indeed, it is the most complex secretory product known to exist.
- The word *polyp* means "many feet." It is derived from *poulpe*, the French word for octopus. Early French naturalists thought the tentacles of a hydroid resembled the "feet" (arms?) of an octopus.
- In most cnidarians, nerve impulses can be transmitted in either direction along a neuron or across a synapse. This is probably a distinct advantage for an organism with radial symmetry.

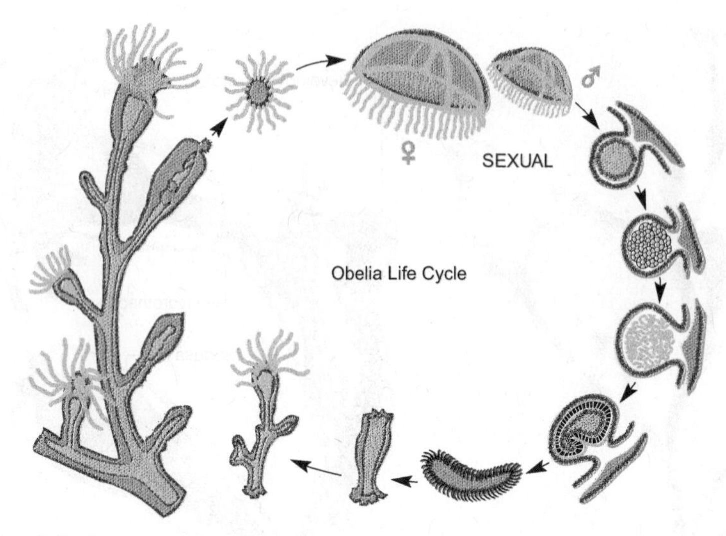

FIGURE 8.13 Life cycle of *Obelia*.
Courtesy of Carolina Biological Supply Company, Burlington, NC.

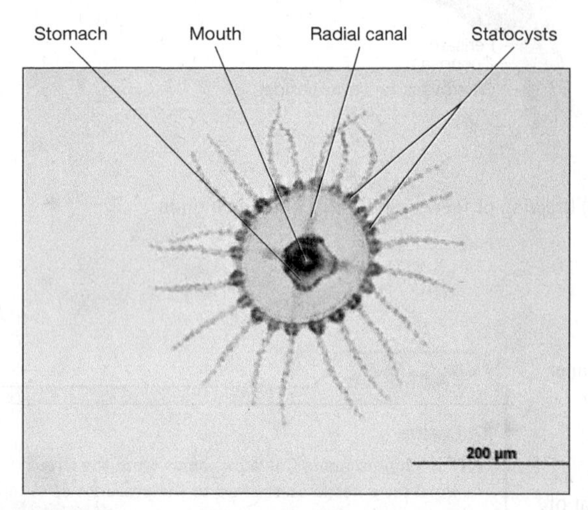

FIGURE 8.14 *Obelia* medusa, whole mount.
Photo by John Meyer.

The sexual stage of *Obelia* is a free-swimming medusa somewhat similar to *Gonionemus*, but much smaller in size (figure 8.14).

- Observe a stained whole mount of an *Obelia* medusa and observe its structure.

Demonstration

Preserved specimens of *Tubularia, Physalia,* and/or other hydrozoans

A Colonial Hydrozoan: *Physalia*

Class Hydrozoa

The Portuguese man-of-war (figure 8.15), *Physalia*, is a complex, colonial hydrozoan. A single colony may consist of as many as 1,000 individuals and may include several types of polypoid and medusoid forms. What looks like a single animal is in fact a superorganism made up of hundreds of feeding, reproductive, and defensive individuals closely joined in a colony. The familiar iridescent colonies of *Physalia* are commonly found along the beaches of Florida, the Gulf of Mexico, the South Atlantic, and sometimes as far north as Cape Cod.

The most prominent feature of *Physalia* is a gas-filled **float** above which is a sail-like **crest.** *Physalia* is transported by winds and oceanic currents, and its normal habitat is the open sea rather than the sandy beach where it is most often seen (and sometimes felt!) by bathers. The sting from the nematocysts on the tentacles of *Physalia* can be painful when touched but is rarely dangerous, except to infants, elderly persons, or highly sensitive individuals.

Below the float are numerous suspended **tentacles** and other structures made up of several kinds of modified polyps and medusae. Thus, *Physalia* is an unusual cnidarian—a single colony contains both polypoid and medusoid individuals of several types closely joined together, in contrast to separate polypoid and medusoid generations.

- Observe a preserved specimen of *Physalia* and identify the float, the crest, and the tentacles. See how many types of individuals you can identify among the structures attached to the float.

Velella and *Porpita* are two related colonial hydrozoans often seen on our Pacific coast and less often in the Gulf of

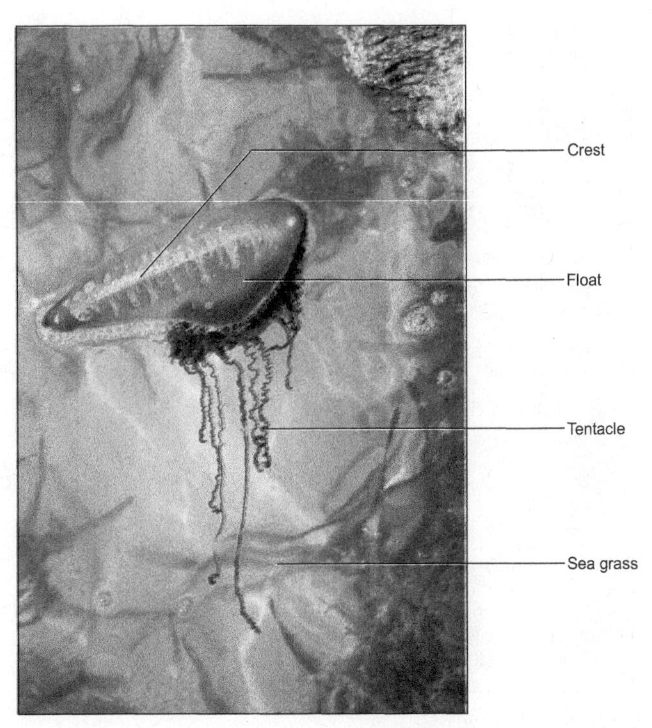

FIGURE 8.15 *Physalia.* Portuguese man-of-war. Approximately one-fourth life size.
Photo by C. F. Lytle.

Mexico and South Atlantic. Both exhibit polymorphism and habits similar to *Physalia.*

A Scyphozoan Medusa: *Aurelia*

Class Scyphozoa

Aurelia (figures 8.16 and 8.17) is a common, widely distributed marine medusa, or jellyfish. Large specimens may reach 30 centimeters (12 inches) in diameter. The polyp form, called a **scyphistoma,** is small, sessile, and lives attached to rocks and other submerged objects in shallow coastal waters.

- Study a preserved specimen of *Aurelia* to learn about the organization of a scyphozoan medusa. You will not need to dissect the specimen, since the transparent body readily shows the most important features. Handle the specimen with care and return it to the supply desk when you finish your study.

Observe the four-part **radial symmetry** and locate the four long **oral arms** arising from the corners of the square mouth. Along the arms, find the many short **oral tentacles** that help to capture food (small planktonic animals), which are then moved toward the mouth along the **ciliated groove** on the oral side of each arm. After passing through the mouth, the food enters the gastrovascular cavity. Internally, the gastrovascular cavity is divided into four **gastric pouches.** A ring of **gastric filaments** with many nematocysts within each gastric pouch immobilizes or kills any food organisms still alive.

FIGURE 8.16 Photograph of *Aurelia* medusa.
© Imagestate Media (John Foxx)/Imagestate.

Four horseshoe-shaped **gonads** surround the ring of gastric filaments within the four gastric pouches. Depressions on the subumbrellar surface of the bell beneath

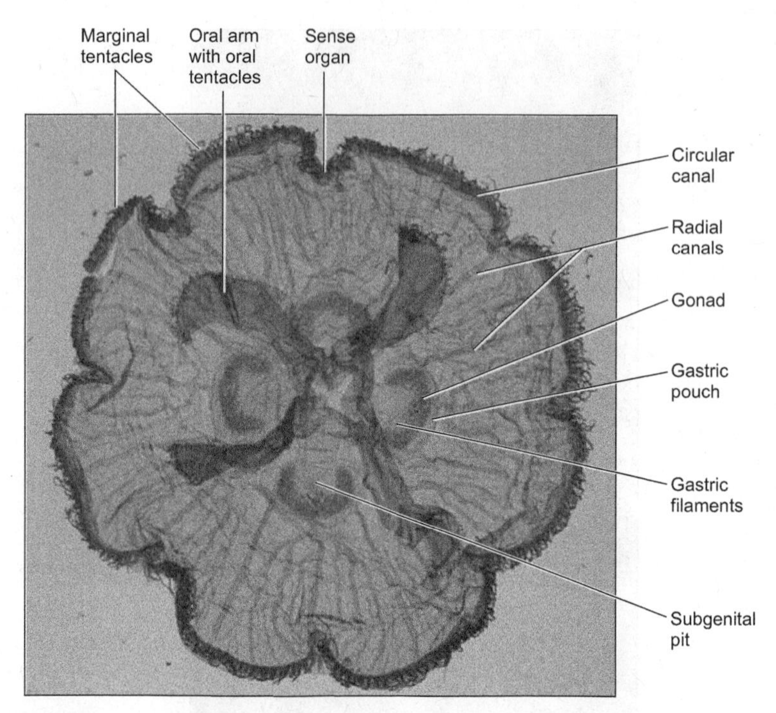

FIGURE 8.17 *Aurelia*, medusa, oral view.
Courtesy of Carolina Biological Supply Company, Burlington, NC.

the gonads are called **subgenital pits.** Their function is unknown.

Observe the complex branching system of **radial canals** that distribute food materials from the gastric pouches to other parts of the bell, and an outer **circular canal** around the margin of the bell. Also, around the margin of the bell, locate the eight marginal **sense organs.** These marginal organs are sensitive to touch and balance.

Reproduction and Life Cycle

Aurelia has a life cycle that involves both sexual and asexual reproduction with two different free-swimming larval stages plus free-swimming medusa and sessile polyp stages. **Asexual reproduction** (figure 8.18) starts with a small polyp called a **scyphistoma** attached to a rock, shell, or some other stable substrate. The scyphistoma grows and begins to bud off a series of disclike segments of its body below the tentacles. These disclike segments differentiate into small, free-swimming juvenile medusae called **ephyrae** (singular: ephyra) that swim about, grow, and metamorphose into small **medusae.**

Sexual reproduction starts with the medusa (figure 8.18). This is the sexually mature stage bearing gonads. **Female medusae** release eggs into the gastrovascular cavity from which they pass into the sea. **Male medusae** release spermatozoa from the testes into the gastrovascular cavity and then into the sea where **fertilization** takes place. Fertilized eggs develop into a ciliated sausage-shaped larva,

Fact File

Jellyfish

- Jellyfish have been on earth for more than 650 million years. They were here before dinosaurs and sharks.
- The lion's mane jellyfish, *Cyanea capillata*, is an arctic species that can grow up to 2 meters (6 feet) in diameter. Large individuals found in open water may have as many as 800 tentacles, with some being more than half the length of a football field.

Continued—

—Continued

- Some jellyfish are edible and are considered a culinary delicacy in some parts of the world. Dried and desalted, they are low in fat and calories but rich in minerals and nutrients.
- Australia's sea wasp, *Chironex fleckeri*, is probably the world's most dangerous jellyfish. The toxin of this cubomedusan is more potent than cobra venom and can kill a normal human in minutes.

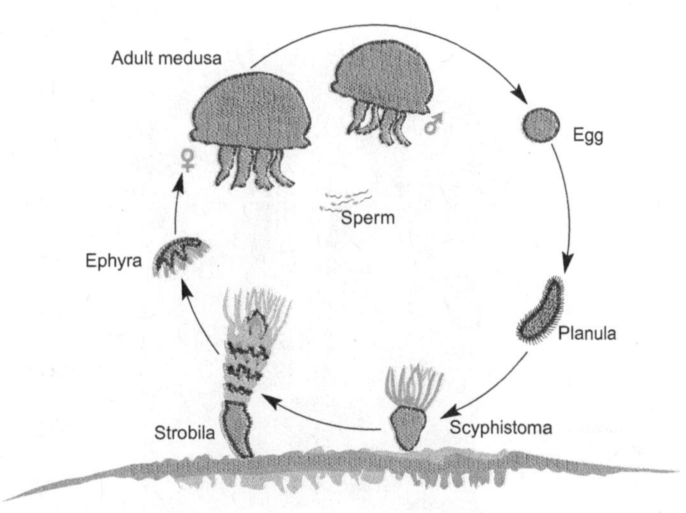

FIGURE 8.18 Life cycle of *Aurelia.*
Courtesy of Carolina Biological Supply Company, Burlington, NC.

the **planula.** After swimming about for a time, the planula settles, attaches to a substrate, and metamorphoses into a **scyphistoma.**

Demonstrations

1. Marginal sense organ of *Aurelia* medusa (microscope slide)
2. Scyphistoma of *Aurelia* or other scyphozoan (preserved or microscope slide)
3. Planula and ephyra larvae of *Aurelia* (microscope slide)
4. Strobila stage of *Aurelia* (microscope slide)

FIGURE 8.19 Photo of *Metridium*, a sea anemone.
Courtesy of Carolina Biological Supply Company, Burlington, NC.

An Anthozoan Polyp: *Metridium*

Class Anthozoa

Sea anemones are typically sessile cnidarians that attach to rocks, shells, pilings, and other hard substrates in the sea. Some species, however, burrow in soft sediments or are free-swimming. All nonetheless represent the polyp form of the cnidarians; the medusa generation is totally lacking in this class. The anemones and other Anthozoa represent the highest degree of specialization of the cnidarian polyp. The basic features of the anthozoan polyp are well illustrated by the common North Atlantic anemone, *Metridium* (figure 8.19). Similar or related genera also occur in the Pacific Ocean.

- Select a preserved specimen of *Metridium* and identify the **mouth** in the center of the **oral disc** surrounded by many short **oral tentacles,** and the **pedal disc,** which attaches to rocks or other hard substrates (figure 8.20).

Most of the internal anatomy can be observed in specimens that have been bisected longitudinally or horizontally. It is not necessary, therefore, to further dissect the specimens. Study precut preserved specimens and locate the following structures. Find the tubular **gullet** leading internally from the mouth to the large **gastrovascular cavity.** One or more **ciliated grooves** (siphonoglyphs) should be found along the edge of the gullet. Usually the cilia within these grooves beat inward to provide an oxygenated current of water into the gastrovascular cavity. Cilia on the remainder of the gullet wall beat outward to remove wastes and foreign particles from the gastrovascular cavity. During feeding, however, the beat of the cilia along the gullet wall is reversed and aids in moving food particles into the gastrovascular cavity. Observe the feeding of living anemones in an aquarium (if available) to better understand the interaction of the tentacles and ciliary currents on the oral disc and in the gullet during the feeding process.

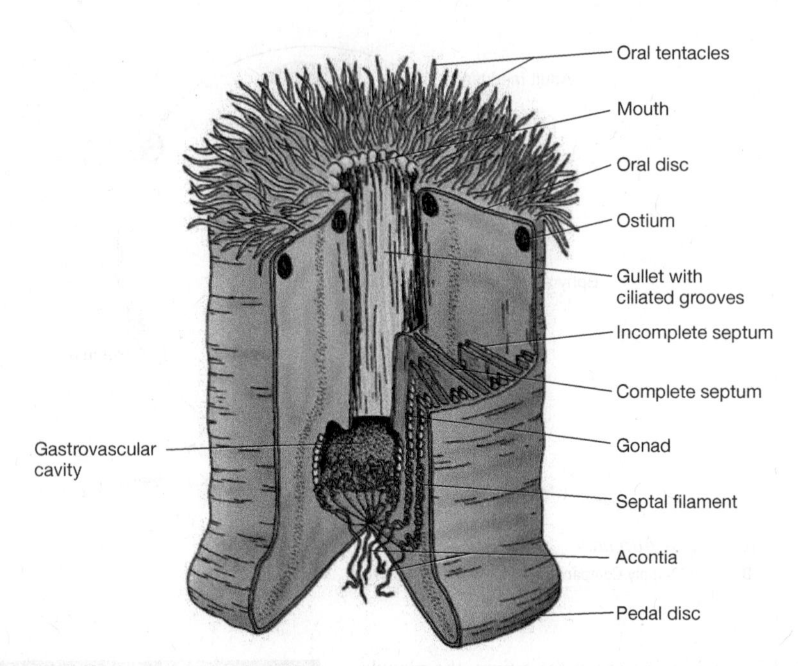

FIGURE 8.20 *Metridium*, partly dissected. Approximately life size.

A microscopic cross section of *Metridium* reveals more details of the internal anatomy, as shown in figure 8.21. Observe the thickened outer body wall with the outer epidermis, a thick layer of mesoglea, and the inner gastrodermis. The interior gastrovascular cavity is divided by numerous partitions called septa. Each septum is an extension of gastrodermis and mesoglea from the outer body wall. Some septa extend to the central pharynx and are called complete septa; others that do not extend to the pharynx are called incomplete septa. The inner free margin of incomplete septa bear from one to three thicker lobes, the **septal filaments,** that contain nematocysts, enzymatic gland cells, and phagocytic cells. The lower end of the septal filaments extend in thin, tentacle-like **acontia** and aid in the capture and digestion of prey. Gonads, when present, form as thickened ridges on the septa parallel and lateral to the septal filaments.

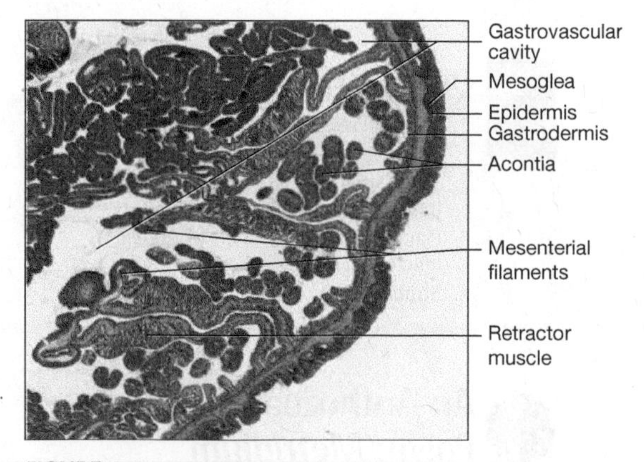

FIGURE 8.21 *Metridium*, cross section.
Photo courtesy of Betty Black.

Reproduction and Life Cycle

Reproduction in *Metridium* and most other anemones is both asexual and sexual. Some anemones reproduce asexually by splitting longitudinally **(longitudinal fission),** but the main asexual means of reproduction in *Metridium* is called **pedal laceration.** Bits of tissue from the pedal disc are split from the anemone as the animal moves along the substrate. These tissue pieces later regenerate an entire small anemone, literally in the footsteps of its parent.

Sexual reproduction occurs seasonally when gametes are released from the gonads on the partial septa into the gastrovascular cavity. The gametes are released and are fertilized in the sea. Fertilized eggs develop into free-swimming **planula larvae.** After a period spent as planktonic larvae, the planulae settle on a hard substrate and metamorphose into an anemone.

Corals

Class Anthozoa

Hard Corals, Subclass Hexacorallia (Zoantharia)

Reef-building (hermatypic) corals are anthozoans that usually live in colonies with many identical polyps interconnected by a thin living tissue covering the surface of the calcareous endoskeleton. They are commonly called hard corals because of their hard endoskeleton. The polyps exhibit six-part (hexamerous) **radial symmetry** (figure 8.22). Examine a specimen of hard coral and observe the symmetry of the skeleton. Like all anthozoans, corals have only a polyp form. There is no medusa stage in the life cycle. Individual polyps are small, usually only a few millimeters in diameter, and reside in a cylindrical cup secreted by a developing polyp on the surface of the endoskeleton. Radiating ridges **(calcareous septa)** project from the side and base of the cup covered by folds in the basal tissue of the polyp. These tissue-covered septa partially subdivide the polyp's gastrovascular cavity into several internal chambers.

Polyps deposit the mineral aragonite, a crystalline form of calcium carbonate, by taking up calcium and carbonate ions from seawater. Thus, a large coral reef can be formed by millions of individual polyps dividing asexually to form new polyps (figure 8.23a). Reef corals also reproduce sexually by releasing eggs and sperm into the sea. Spawning on a reef is usually synchronized by environmental factors that increase the probability of fertilization and subsequent distribution of offspring to populate new areas.

Corals can feed on small animals caught with the aid of their tentacles and nematocysts, but most hard corals get the majority of their nutrients from symbiotic algae. Many corals and some sea anemones contain intracellular algae, called zoochlorellae, in their gastrodermal cells. These corals require sunlight and live in shallow water less than 60 meters deep to allow sufficient light intensity for photosynthesis. These symbionts share some of their nutrients produced by photosynthesis with their anthozoan host, recycle nitrogenous wastes released by the hosts, and aid in deposition of the host's calcium carbonate exoskeleton Many interesting studies have been conducted on the physiological and biochemical interactions of these **symbiotic** partners.

Reef-building corals are very sensitive to temperature, light intensity, salinity, pH and other environmental factors. They are able to survive only within a narrow temperature range and require sufficient light intensity for growth of their zooanthellae. Coral reefs have an immense ecological impact by providing natural filtration of seawater, by mitigating the effects of wave action on adjacent islands and coastlines, and by providing habitats for a diverse community of marine organisms. Unfortunately, coral reefs are fragile communities that are susceptible to disease and predation and are increasingly endangered by human activities including oil spills, pesticide runoff from agricultural fields, sewage and sedimentation from coastal development, and global warming, which contributes to higher ocean temperatures and changes in sea level. Off the east coast of Australia, for example, large areas of the Great Barrier Reef have been damaged by the Crown of Thorns sea star, a marine predator that feeds on the soft tissues of the coral.

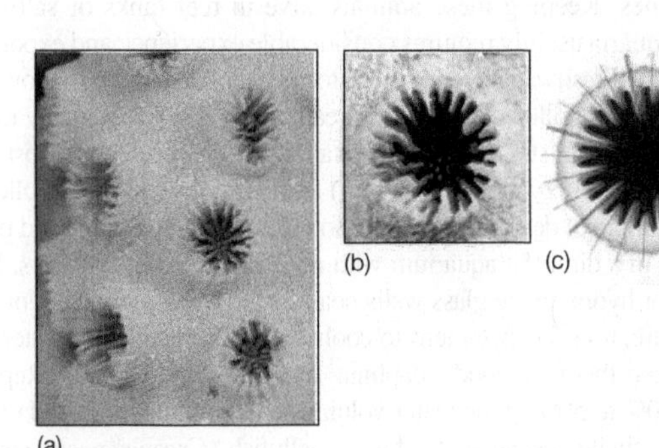

FIGURE 8.22 Hexacoral symmetry. (*a*) Photo of section of a hexacoral colony. (*b*) Photo of individual coral cup. (*c*) Drawing showing the planes of symmetry of a cup.
Photos (a) and (b) by John Meyer.

Many growth forms occur among the stony corals, and most species exhibit characteristic skeletons. A few species are solitary, such as *Fungia* (figure 8.23b) found in the Pacific Ocean. Other examples of a stony coral include *Astrangia* (figure 8.23c), the Atlantic star coral often found in shallow waters along the seacoast, and *Diploria* and *Meandrin*, brain corals with a structure resembling the ridges and grooves of a human brain (figure 8.23d).

Soft Corals, Subclass Octocorallia (Alcyonaria)

Octocorals are colonial forms whose polyps have **octomerous** (eight-part) **radial symmetry.** Study the photograph of a palm coral in figure 8.24a with its eight feather-like (pinnate) tentacles with many small side branches. Compare this octamerous symmetry to the hexamerous symmetry of the hexacorals shown in figure 8.22.

The polyps of soft corals are usually embedded in a soft matrix and are interconnected through their gastrovascular cavities, allowing exchange of water and nutrients. A notable exception is the pipe organ coral, *Tubipora* (figure 8.24c), which secretes a calcareous endoskeleton. It is prized by aquarists and other collectors for its bright red color and may be heading toward extinction.

Octocorals are especially prominent in tropical waters. Other examples are the precious red coral (*Corallium*), black corals (*Antipathes*), sea fans (*Gorgonia;* figure 8.24d), sea whips (*Ellisella*), sea pansies (*Renilla*), and sea pens (*Ptilosarcus*). Some of these species are quite colorful. They are often collected, dried, polished, and sold in souvenir stores as jewelry or decorative art.

- Study the demonstration materials illustrating several forms of coral. Compare this octamerous symmetry with the symmetry of the hexacorals shown in figure 8.22. Octocorals also produce a different kind of supporting endoskeleton (coenenchyma or mesoglea) containing a horny protein and/or microscopic spicules that may be separate or fused. Draw the different types of corals on your Notes and Sketches page. Identify each type that you draw.

FIGURE 8.23 Examples of Hexacorallia. (*a*) Elkhorn coral reef; note the yellowish-green color from symbiotic zooanthellae. (*b*) Fungia, a solitary coral. (*c*) *Astrangia danae*, small clusters common in shallow waters of the Atlantic. (*d*) Brain coral.

(a) © Elmer Frederick Fischer/Corbis. (b) © Brand X Pictures/PunchStock. (c) Courtesy of Carolina Biological Supply Company, Burlington, NC. (d) Ingram Publishing/SuperStock.

Demonstrations

1. Living sea anemones and/or corals in a marine aquarium
2. Representative preserved anemones
3. Assortment of preserved corals and dried coral skeletons

Collecting and Preserving Cnidaria

Marine hydroids and sea anemones are commonly found in coastal waters throughout the world. Look in tide pools along rocky shores and below the low waterline on wharfs and pilings. Most species are delicate and do not survive well in captivity. Reef-building corals grow only in tropical or subtropical oceans where the water temperature never falls below 20°C. A few temperate species form small colonies (usually 5–30 individuals) that encrust shells and rocks from Florida to Cape Cod and also along the Pacific coast. Corals and anemones can be purchased from vendors of saltwater aquarium supplies. Keeping these animals alive in reef tanks or saltwater aquaria usually requires considerable experience and expertise.

Freshwater hydra may be locally abundant in the shallow water of unpolluted ponds and streams where they are usually found clinging to stones, twigs, vegetation, or organic debris. Most species are very small (< 5 mm) and easily overlooked. Collect a sample of detritus and vegetation from the littoral zone and place it in a dimly lit aquarium with no aeration. After 24 hours, look for hydra on the glass walls near the air-water interface. For culture, transfer specimens to cool (< 20°C), clean pond water and feed them copepods, daphnia, or brine shrimp larvae. Replace 30% to 50% of the water volume weekly with fresh pond water. A single species of freshwater jellyfish (*Craspedacusta sowerbyi*) occurs in North America. Look for the medusa stage of this hydrozoan in small lakes or fish ponds between July and October.

For optimum preservation, Cnidaria should be fixed in 10% formalin because their soft bodies do not preserve well in

(a)

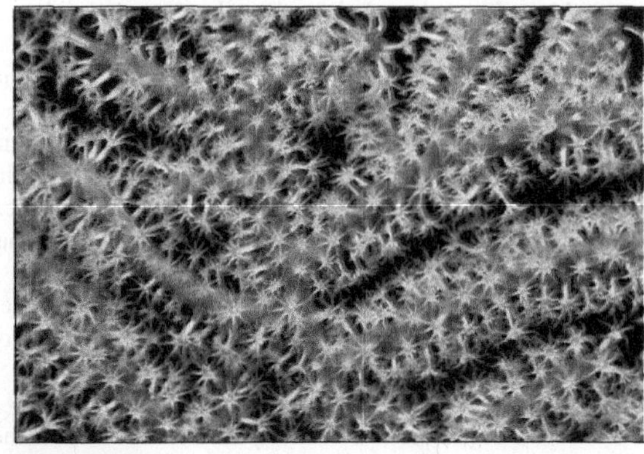
(b)

(c)

(d)

FIGURE 8.24 Examples of Octocorallia. (*a*) Palm coral showing octoradial symmetry and feathery tentacles. (*b*) Close-up of polyps on a sea fan. (*c*) Dried skeleton of *Tubipora* the pipe organ coral. (*d*) *Gorgonia*, sea fan.

(a) © Diane Nelson. (b) © ImageState/PunchStock. (c) Photo by John Meyer. (d) © Diane Nelson.

alcohol. CAUTION: Formalin is a toxic compound and must be used with care in a well-ventilated place. Specimens keep better in formalin, but if necessary, they can be stored in 70% ethanol after fixation in formalin for a few days. Small specimens can be dried and permanently mounted on glass microscope slides. Because of the skill and experience needed to produce satisfactory materials, it is usually better to purchase prepared specimens from a commercial vendor of biological supplies.

Key Terms

Alternation of generations the alternation of the sessile polyp and free-swimming medusa generations typical of the life cycle of the cnidarians.

Cnidocytes specialized cells of cnidarians that secrete and contain the stinging nematocysts.

Coenosarc the living portion of the tubular connecting portions of colonial cnidarians like *Obelia.* Consists of a simple cylinder of an outer epidermal tissue layer, an inner gastrodermal tissue layer, and an intermediate mesoglea surrounding a central gastrovascular cavity.

Colonial individual animals living in a group. Members of the group are usually produced asexually and often remain joined together by some physical connection. In simple colonies, members are often similar but some complex colonies exhibit polymorphism and/or specialization of individuals for certain functions.

Dioecious condition of an animal with both male and female sex organs borne in different individuals.

Diploblastic construction body with two layers of cells, typical of the cnidarians. Body has an outer epidermal tissue layer and an inner gastrodermal layer with an intervening noncellular mesoglea layer.

Epidermis outer tissue layer protecting the surface of an animal from its environment.

Gastrodermis inner tissue layer of animals bordering the digestive cavity.

Gastrovascular cavity a central cavity of an animal that serves both for digestion and circulation, with a single mouth opening that serves both as an entrance and an exit. A type of incomplete digestive system (without an anus).

Gonangium a type of reproductive individual in colonial Hydrozoa, such as *Obelia;* produces free-swimming medusae.

Hydranth feeding individual in a colonial hydrozoan, as in *Obelia.*

Medusa stage free-swimming stage in the life cycle of many cnidarians. Usually bears gonads and produces gametes.

Mesoglea the gelatinous layer between the epidermal and gastrodermal layers of cnidarians; simple and noncellular in the Hydrozoa, containing cells and/or fibers in the Scyphozoa and Anthozoa.

Monoecious condition of bearing both male and female sex organs in one individual. Usually not self-fertilizing.

Nematocysts the stinging capsules produced by the cnidocytes of cnidarians. A key characteristic of the phylum.

Nerve net the diffuse, interconnected network of nerve cells in the cnidarians. Lacks ganglia or other nervous centers. A very primitive type of nervous system.

Perisarc the nonliving outer covering secreted by the coenosarc of colonial hydrozoans. Surrounds the interconnecting coenosarc.

Planula larva a simple, ciliated, sausage-shaped larval form produced by many cnidarians. Develops from the zygote or fertilized egg.

Polymorphism exhibiting two or more distinct body forms in a single species, like the several types of individual polyps found in *Obelia, Physalia,* and many other cnidarians.

Polyp stage the sessile (attached) state in the life cycle of many cnidarians.

Radial symmetry body plan in which body parts are arranged symmetrically around a central axis, characteristic of the cnidarians and echinoderms.

Scyphistoma the inconspicuous polyp stage in the life cycle of certain scyphozoans.

Symbiosis an organism living in a close, continuing relationship with another organism of a different species, such as the algal cells living in the tissues of stony corals and other cnidarians.

Velum a thin, shelflike membrane found in most hydromedusae. It is located on the subumbrellar surface, just inside the perimeter of the bell.

Internet Resources

There are many valuable Internet sites with information about Zoology. Several sites containing pertinent zoological information for this chapter can be found on the McGraw-Hill Zoology web site at http://www.mhhe.com/zoology. Just click on this text's title.

Questions for Critical Thinking

1. Why do we say that cnidarians are the simplest animals with definite tissues? (Do you think the strict definition of "tissue" excludes other possibilities concerning cellular organization and cooperation?)
2. Speculate on the evolution of alternation of generations (polyp forms and medusa forms). Could the medusa form simply be an upside-down polyp? Or is the polyp a rightside-up medusa? If so, which form do you suppose evolved first?
3. Why do you suppose that radial symmetry did not become a big "evolutionary hit" among the animals? What drawbacks might such a symmetry have over bilateral symmetry?
4. Discuss the various nematocyst types and their functions. Do you think this variety on a common theme is an example of a basically good design being preadapted for many different possible functions?
5. Explain how the Portuguese man-of-war might be viewed as a "superorganism" (i.e., the polyps have taken on different functions to form a colonial animal).

Suggested Readings

Arai, M.N. 1997. *Functional Biology of Scyphozoa.* London: Chapman & Hall. 316 pp

Fautin, D.G. 2002. Reproduction of Cnidaria. *Can. J. Zool.* 80:1735–1754.

Mackie, G.O. 2002. What's new in cnidarian biology? *Can. J. Zool.* 80:1649–1653.

Moore, J. 2006. *An Introduction to the Invertebrates,* 2nd ed. New York: Cambridge University Press. 319 pp

Ruppert, E.E., Fox, R.S., and Barnes, R.D. 2004. *Invertebrate Zoology: A Functional Evolutionary Approach,* 7th ed. Belmont, CA: Thomson-Brooks/Cole. 1008 pp.

Vernon, J.E.N. 2000. *Corals of the World.* Townsville, Australia: Australian Institute of Marine Science. 3 vols. 1382 pp

Notes and Sketches

Chapter 9

Platyzoa: Platyhelminthes, Gastrotricha, and Rotifera

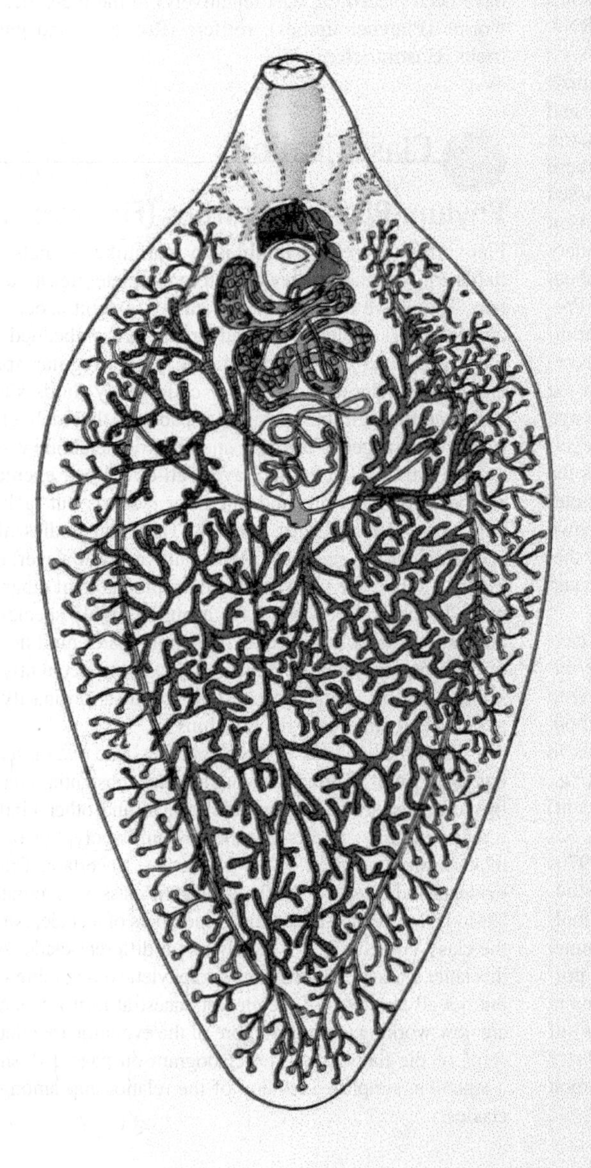

Objectives

After completing the laboratory work in this chapter, you should be able to perform the following tasks:

1. Briefly outline the characteristics of the Bilateria and explain how their structural organization differs from that of the Cnidaria and Ctenophora.
2. Discuss suggested evolutionary relationships among the Protostomia, identify the clades to which major phyla belong, and list the distinguishing characteristics of each clade.
3. List and briefly characterize each of the four classes of the Phylum Platyhelminthes.
4. Describe the behavior of a free-living flatworm (planarian) such as *Dugesia* and relate this behavior to the function of its sense organs.
5. Discuss the structure of the epidermis of a planarian.
6. Identify the major structures that can be seen in microscopic cross sections at various levels of the body of a planarian (for example, anterior, pharyngeal region, and posterior).
7. Explain the organization of the reproductive system of a planarian and identify its principal reproductive organs.
8. Discuss the general morphology of the trematode *Clonorchis* and compare it with a free-living turbellarian such as *Dugesia*.
9. Describe the life cycle of *Clonorchis* and identify the principal stages in microscopic preparations.
10. Describe the general morphology of the liver fluke *Fasciola* and identify its principal organs.
11. Describe the life cycle of a tapeworm such as *Taenia* or *Dipylidium*, and identify its principal stages in microscopic preparations.
12. Describe the anatomy of a gastrotrich, such as *Chaetonotus*, and explain how it differs in structure from a flatworm.
13. Describe the anatomy of a rotifer, such as *Philodina*, and locate its principal organs in a specimen or on a drawing.

Introduction

The vast majority of multicellular animals are bilaterally symmetrical—their body parts are arranged along a mid-sagittal plane such that the left and right sides are approximately mirror images of each other (see figure 5.3; page 71). Indeed, the only exceptions to this rule are the asymmetrical sponges and the radially symmetrical cnidarians and ctenoporans that you studied in Chapters 7 and 8. Adult echinoderms, which also exhibit radial symmetry, develop from bilaterally symmetrical larvae and the adult form is widely regarded as a secondary adaptation (see Chapter 14).

Most biologists have now come to accept the premise that all animals with **bilateral symmetry** (including echinoderms) are the descendants of a single common ancestor and therefore represent a monophyletic clade (the Bilateria) within the animal kingdom. Support for this hypothesis has come repeatedly from sequence analysis of DNA and RNA over a wide range of metazoan taxa.

Bilateral symmetry appears to be a hallmark of animals that have **triploblastic body construction**—tissues and organs that develop from three primary embryonic germ layers: endoderm, mesoderm, and ectoderm. Each of these layers develops into specific body components, as discussed in Chapter 5. Careful study of embryological development has prompted biologists to conclude that two evolutionary lines (clades) diverged from their bilaterian ancestor about 400 to 500 million years ago. One of these groups (the Protostomia) exhibits early consignment of cell fate (determinate development) and forms a mouth at the site (blastopore) where invagination first occurs in the blastula stage. In the other group, the Deuterostomia, cell fate within the embryo is determined at a relatively late stage (indeterminate development) and the site of the blastopore usually becomes the animal's anus. Deuterostomes include echinoderms (Chapter 14), the chordates (Chapter 15), and few other lesser-known phyla. The protostomes include many invertebrate phyla that have long frustrated zoologists' attempts at organization and classification.

The presence or absence of a coelom (central body cavity) was once regarded as a useful clue to protostome ancestry. Since flatworms (phylum Platyhelminthes) appear to lack a coelom, they were presumed to be an early offshoot of this lineage. Animals with a pseudocoelom (e.g., as in roundworms) or those with a well-developed coelom (e.g., annelids and mollusks) were regarded as later (more recent) branches of the phylogenetic tree.

This hypothesis persisted throughout much of the 1970s and 1980s, but it has not been well supported in recent studies of molecular genetics. Modern DNA sequencing technology, which allows rapid and economical analysis of animal genomes, is being paired with sophisticated computer programs that can match and compare homologous regions in the DNA of different animals. The number (and types) of genetic changes serves as a measure of the evolutionary "distance" two animals have "traveled" from their common ancestor.

One outcome of this research is a growing consensus that the protostome lineage diverged into at least two or three "super-phyla." One such group, the Ecdysozoa, is characterized by having an external cuticle that is periodically shed (molted) as the organisms grow. Members of this clade (including Nematoda and Arthropoda, Chapters 12 and 13) are linked by the inheritance of genes responsible for synthesis of the molting hormone ecdysone. A second protostome group, the Lophotrochozoa, gathers food with a ring of ciliated tentacles around the mouth (the lophophore) or has a distinctive trochophore larval stage. The molluscs (Chapter 10) and the annelids (Chapter 11) are the two most noteworthy members of this group. A third group, the Platyzoa, has recently (1998) been proposed as a sister group to the Lophotrochozoa. In the following pages of this chapter, we will examine members of the three largest phyla that have been placed (at least tentatively) in the Platyzoa: flatworms (Platyhelminthes), rotifers (Rotifera), and gastrotrichs (Gastrotricha).

Classification

Phylum Platyhelminthes (Flatworms)

Flatworms are soft, **triploblastic** wormlike animals with flattened, elongate bodies and an incomplete digestive system. They have no central body cavity (exhibit acoelomate construction). Instead, internal organs are embedded in a loosely packed parenchyma tissue with irregular spaces among the cells and clumps of cells. Body fluids simply percolate through these irregular spaces as they deliver nutrients and oxygen to the cells and remove metabolic wastes.

Free-living flatworms have well-developed excretory, reproductive, nervous, and muscular systems but lack circulatory and respiratory systems. Their flat bodies allow for sufficient gas exchange. Most flatworms, however, have evolved to become parasites or micro-predators of other animals. In the process, they have acquired highly specialized adaptations that often include reduction and/or modification of their organ systems. These modifications frequently obscure ancestral characteristics that scientists ordinarily use to deduce evolutionary relationships.

Classification of the Platyhelminthes, like that of numerous other animal groups, has undergone substantial changes in recent years due to new biochemical and other kinds of research. It now appears that the phylum is polyphyletic and relationships among the four classes are uncertain. Current evidence indicates that the three parasitic classes: (Trematoda, Cestoda, and Monogenea) are all members of a clade, but that the class Turbellaria is a member of a different clade. Also, this latter class is believed to be paraphyletic (containing some but not all members of the closest ancestral form). Scientists are now working to unravel more of the evolutionary relationships of the flatworms. The cladogram on page 131 shows a tentative, simplified version of the relationship among the classes.

Cladogram of Platyhelminthes

- Class **Turbellaria** (free-living flatworms, planaria)
- Class **Trematoda** (endoparasitic flukes)
- Class **Monogenea** (ectoparasitic flukes)
- Class **Cestoda** (tapeworms)

Class Turbellaria (Free-living Flatworms)

Mainly **free-living flatworms** with a dorsoventrally flattened body covered by a ciliated epidermis (figure 9.1). Mouth usually ventral, leading into a gastrovascular cavity with a single opening. Many marine species found on coral reefs and other substrates (figure 9.2), numerous freshwater species and a few terrestrial species. Together about 3,000 species. Examples: *Dugesia* (figure 9.1), *Bipalium* (figure 9.3), *Mesostoma, Polychoerus, Leptoplana.*

Class Trematoda (Endoparasitic Flukes)

Parasitic flatworms with a body covered by an external tegument (cuticle) secreted by underlying cells. Ovoid body typically with one anterior and one midventral sucker for attachment to the host. Internal parasites with complex life cycles, usually involving several successive larval stages and two or more hosts. About 9,000 species. Examples: *Clonorchis* (human liver fluke), *Fasciola* (sheep liver fluke), and *Schistosoma* (human blood fluke).

Class Monogenea (Ectoparasitic Flukes)

Small parasitic worms most commonly found on the gills and other exterior body parts of fish, but also occurring as external or internal parasites of amphibians. Monogeneans have a well-developed posterior attachment structure called an opisthaptor. They have a simple life cycle with a single host. About 400 species. Most are external parasites of fish.

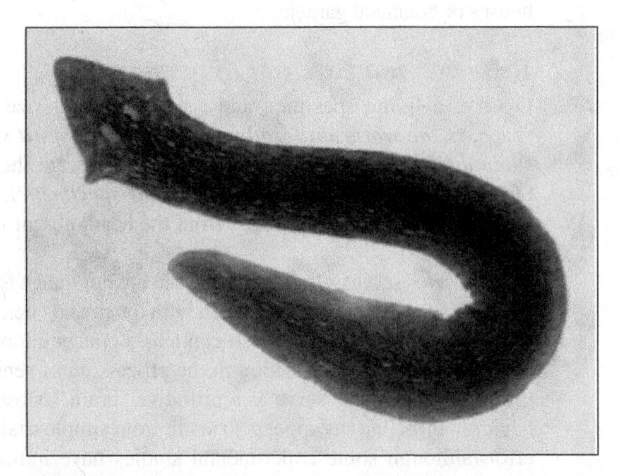

FIGURE 9.1 *Dugesia*, brown planaria, living.
Photo by John Meyer.

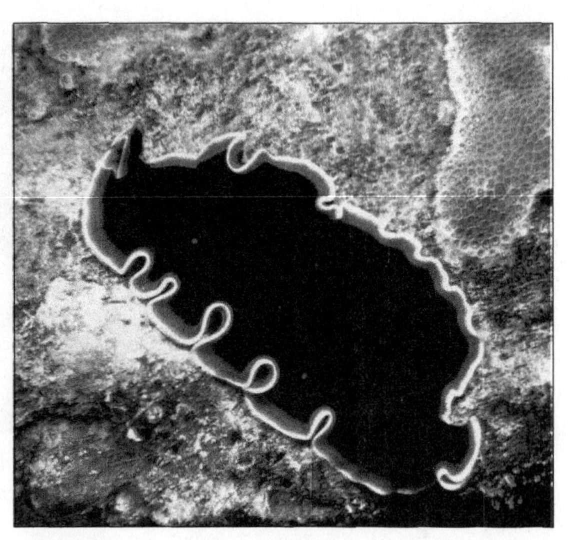

FIGURE 9.2 Ruffled flatworm, *Pseudobiceros*, on coral reef.
© Diane Nelson.

Examples: *Gyrodactylus, Polystoma,* and *Fundulotrema* (figure 9.13a).

Class Cestoda (Tapeworms)

Internal parasites with specialized scolex with hooks and/or suckers for attachment to host; body divided transversely into a series of similar proglottids; thick external tegument; mouth and digestive tract absent. Usually with a complex life cycle involving successive larval stages and alternate hosts. About 5,000 species. Examples: *Dipylidium caninum* (dog tapeworm), *Taenia pisiformis* (dog and cat tapeworm), and *Dibothriocephalus* (fish tapeworm).

Phylum Gastrotricha (Gastrotrichs)

Very small, aquatic animals that resemble flatworms but have a complete digestive system, a forked terminal appendage for attachment to the substrate, and spines, bristles, or scales covering the dorsal surface of the body. Marine species (mostly members of the order Macrodasyida) usually live in the capillary spaces between sand grains in bottom sediments. Freshwater species (mostly members of the order Chaetonotida) live on submerged vegetation, detritus, or in the sandy sediments of lakes and streams. About 750 species. Examples: *Chaetonotus, Macrodasys.*

Phylum Rotifera (Rotifers)

Microscopic animals found primarily in freshwater habitats. Body covered by a **syncytial epidermis** that secretes a flexible and telescoping **cuticle.** Complete digestive system with an anterior ciliated locomotory and feeding organ, the **corona,** and a specialized internal grinding organ, the

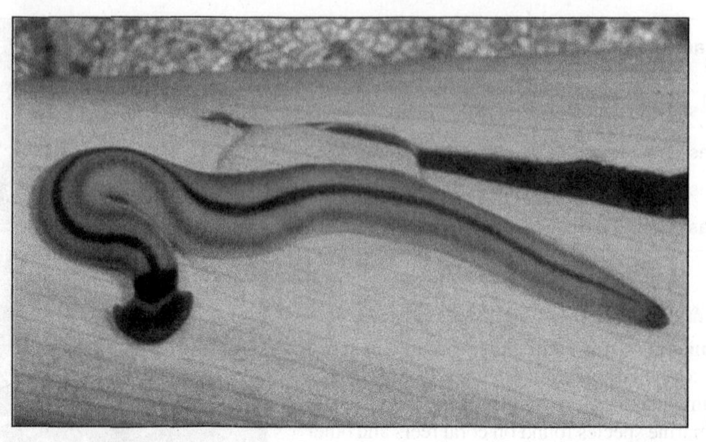

FIGURE 9.3 *Bipalium kewense*, a terrestrial flatworm found in tropical and mild temperate climates.
Photo by Paul M. Choate. Dept. Entomology and Nematology, University of Florida.

mastax. Many species have an adhesive **foot** for attachment to a substrate. About 2,000 species. Examples: *Philodina, Epiphanes, Asplanchna.*

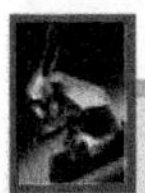

Materials List

Living specimens
- *Dugesia*
- *Stenostomum*
- Cercariae
- *Philodina*
- *Chaetonotus*

Preserved specimens
- Whole tapeworms

Prepared microscope slides
- *Dugesia*, whole mount, representative cross sections
- *Stenostomum*, whole mount
- *Clonorchis*, whole mount
- *Fasciola*, whole mount
- *Dipylidium*, whole mount
- *Taenia pisiformis*, scolex and representative sections
- *Taenia solium*, scolex and representative sections (demonstration)
- *Clonorchis*, miracidia, sporocysts, redia, cercaria, metacercaria (demonstrations)
- *Dibothriocephalus latus*, scolex and representative sections (demonstration)
- Onchosphere (six-hooked) larva (demonstration)
- Cysticercus larva (demonstration)

Plastic mounts
- *Dipylidium*, whole mount (demonstration)
- *Taenia pisiformis*, representative sections (demonstration)

Miscellaneous
- Methyl cellulose solution
- Beef liver
- Methylene blue solution (1%)

Free-living Flatworms: Class Turbellaria

A Planarian: *Dugesia*

Most members of the Class Turbellaria are marine. However, freshwater members of this class, commonly referred to as planarians, can be found in many springs, brooks, ponds, and lakes. These flatworms are typically grey, brown, or black and usually move about on the bottom or on submerged sticks, stones, or plants in shallow water. Many species are negatively phototactic and are found on the underside of stones or leaves. Common North American genera include *Dugesia, Phagocota,* and *Polycelis.*

The greenhouse planarian, *Bipalium kewense* (figure 9.3), is a large terrestrial species. It is usually 8-10 cm long, but occasionally reaches a length of 60 cm (nearly 2.5 ft!). Although this species is generally considered subtropical, it is often found farther north on moist walkways in or near greenhouses or botanical gardens.

Behavior and External Anatomy

Observe a living specimen and note its general size and shape. ***Is the worm uniformly pigmented, or does it show some distinctive pattern of pigmentation?*** Locate the anterior **head,** the **eyes** ***(how many on your specimen?),*** and the **auricles** (lateral projections from the head; absent from some species).

The eyes serve as light receptors but are not image forming. The auricles are well-equipped with touch and chemical receptors. The head region also contains a concentration of **nerve ganglia,** which function in the processing of sensory information and thus serve as a primitive "brain." Most behavior in turbellarians appears to result from simple trial and error, although some experimental studies have indicated that flatworm behavior can be modified to some extent by prior experience (i.e., they "learn").

Observe the smooth, gliding locomotion of the worm. This form of locomotion is due to the action of cilia on the ventral surface of the body, coordinated with rhythmic muscular contractions of the body. Note the behavior of the head and the auricles during locomotion. ***How is this behavior related to the function of sensory and nervous structures of the head?*** Gently touch the "head" of the worm with a clean dissecting needle. ***How does the worm react?*** Touch other body regions in a similar way and compare the reaction. Turn the worm over on its dorsal side. ***How does it react? Can you relate your observations of the worm's behavior to the structure of the head and the concentration of sense organs and nervous elements there? Can you speculate about the possible advantages to an animal of having an anterior head with a concentration of sense organs and nervous tissue?***

Internal Anatomy

- To study the anatomy further, obtain a microscope slide with a stained whole mount of a planarian (figure 9.4b).

Review the structures previously noted in the living specimen and also observe the **three-branched gastrovascular cavity.** Note the one anterior and two posterior branches of the cavity and the many smaller lateral branches, or diverticula.

- Study also a microscope slide with cross sections through the anterior, pharyngeal, and posterior portions of the body (figure 9.5). In the section from the anterior portion of the body, locate (1) the **epidermis,** the external layer of cells surrounding the body; (2) the large, vacuolated cells of the **gastrodermis** lining the digestive tract; (3) the layers of **longitudinal** and **circular muscles** lying just inside the epidermis; (4) the large, irregularly shaped cells of the **parenchyma** tissue, which fills most of the interior space of the body; and (5) the two large **ventral nerve cords.** The ventral nerve cords are connected at regular intervals by transverse nerves, giving it a ladderlike appearance (figure 9.4*c*).

Careful observation will reveal that the cells of the ventral epidermis are ciliated, but those of the dorsal epithelium are not. Mucus and other types of **gland cells** are present in the epidermis and in the underlying mesenchyme. Many of the epidermal cells contain densely staining **rhabdites.** Rhabdites are small, rodlike bodies whose function is not yet fully understood. There is some evidence that they are discharged if the worm is attacked and swell up to form a slimy coat for defense. Some of the gland cells have long ducts that extend to the surface. These gland cells produce mucus for lubrication and other sticky materials for adhesion, capturing prey, and other functions. Within the parenchyma, you should also find several small branches of the gastrovascular cavity surrounded by large vacuolated gastrodermal cells.

In the middle section through the buccal cavity of the planarian, find the large muscular **pharynx** lying within the **buccal cavity.** The pharynx has powerful muscles that allow it to be extended from the buccal cavity through the **ventral mouth** while feeding on small prey organisms (see figure 9.6). Within the pharynx, locate several types of muscles: inner and outer layers of **circular muscle,** inner and outer layers of **longitudinal muscle,** and bands of **radial muscle** that extend across the pharynx from the outside to the central lumen. Find the **ciliated epithelium** on the outside of the pharynx. Many **gland cells** that produce mucus and proteolytic enzymes can also be found within the pharynx.

- Study the section through the posterior portion of the body and note the several branches of the **gastrovascular cavity,** the ventral **nerve cords,** the circular and longitudinal **muscles,** and the **epidermis.**

Reproduction and Excretion

Many flatworms, including *Dugesia,* have male and female sex organs present together in the same individual. These animals are **monoecious**—the term is derived from Latin and means "one house." It appropriately describes the condition of **hermaphrodism,** in which ovaries and testes "live together" in the same body. Some other free-living flatworms are **dioecious** ("two houses") with male and female sex organs in different individuals.

In dioecious animals, sexual reproduction requires the involvement of two individuals of opposite sex. In monoecious animals, there is the potential for self-fertilization if eggs and sperm are produced simultaneously. Indeed, many scientists believe there is a strong tendency for parasitic species to become monoecious because it increases the opportunity for fertilization.

Although planarians are monoecious, they are not normally self-fertilizing. In sexual reproduction, sperm is transferred from the male system of one worm by the male copulatory organ, the **penis,** to the **seminal receptacle** of the partner. The sperm subsequently moves to the oviduct of the female reproductive system where fertilization occurs. The fertilized eggs are later deposited outside the body in cocoons (see figure 9.7) where they develop directly into young worms.

The reproductive system of *Dugesia* and other freshwater flatworms is small and difficult to observe except in special microscopic preparations. Some of the reproductive organs are shown in figure 9.4a, but most of them probably will not be visible in your slide.

In many species of planarians, however, the most common form of reproduction is **asexual.** A part of a worm breaks off and each part regenerates the missing structures. Planarians have great powers of regeneration, and even relatively small parts of a worm can develop into a complete animal. If time and materials permit, your instructor may be able to help you set up an experiment with regeneration in planarians. Several excellent resources for experimenting with planarian regeneration are available online. Try searching for "planarian regeneration".

Planarians have long been favorite animals for research on the process of regeneration. Recent research has focused on special cells called neoblasts that work like stem cells to give rise to other types of somatic cells and also to germline

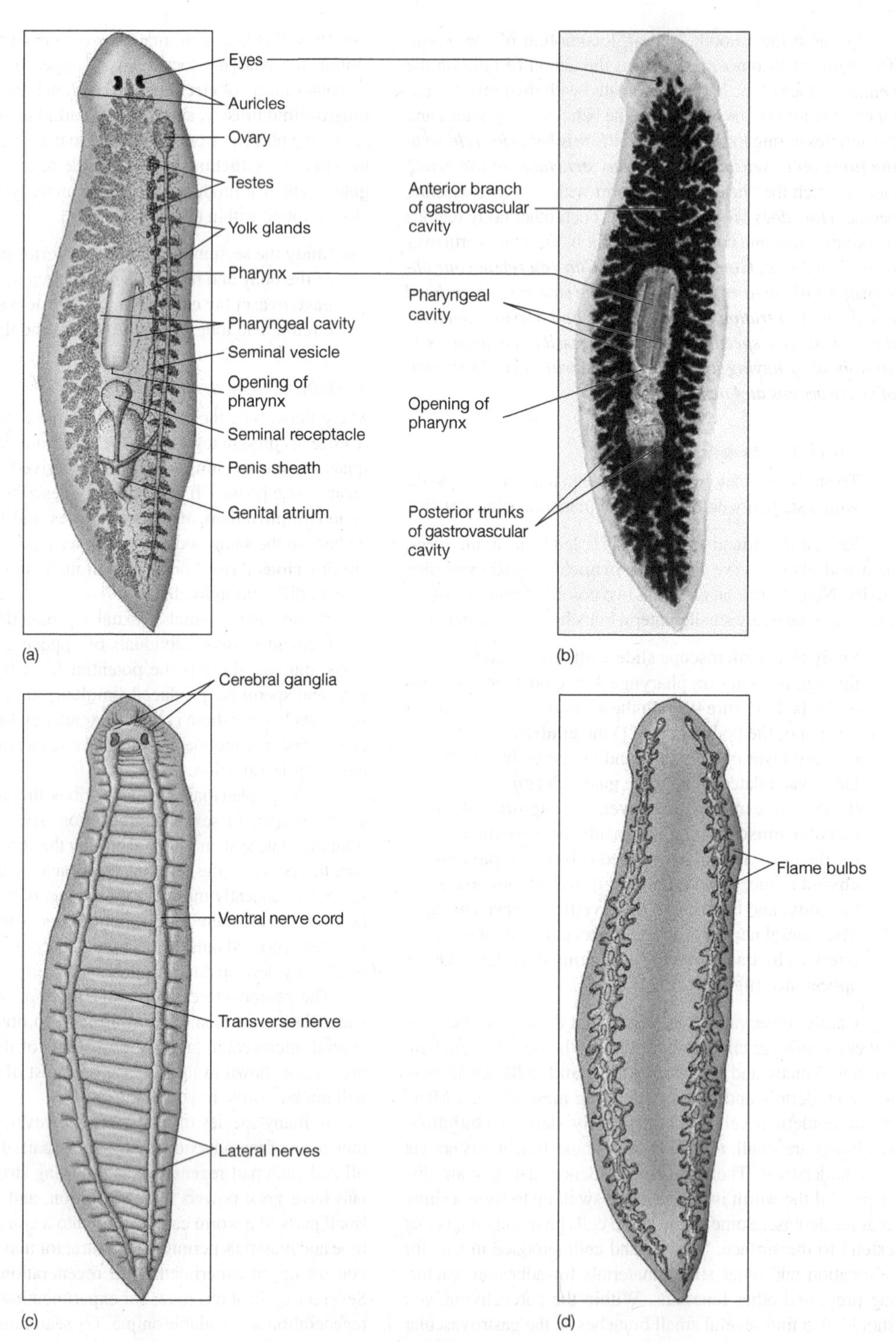

FIGURE 9.4 *Dugesia.* (*a*) Reproductive system. (*b*) Microscopic whole mount showing digestive system. (*c*) Nervous system. (*d*) Excretory system.

(b) Photograph courtesy of Carolina Biological Supply Company, Burlington, NC.

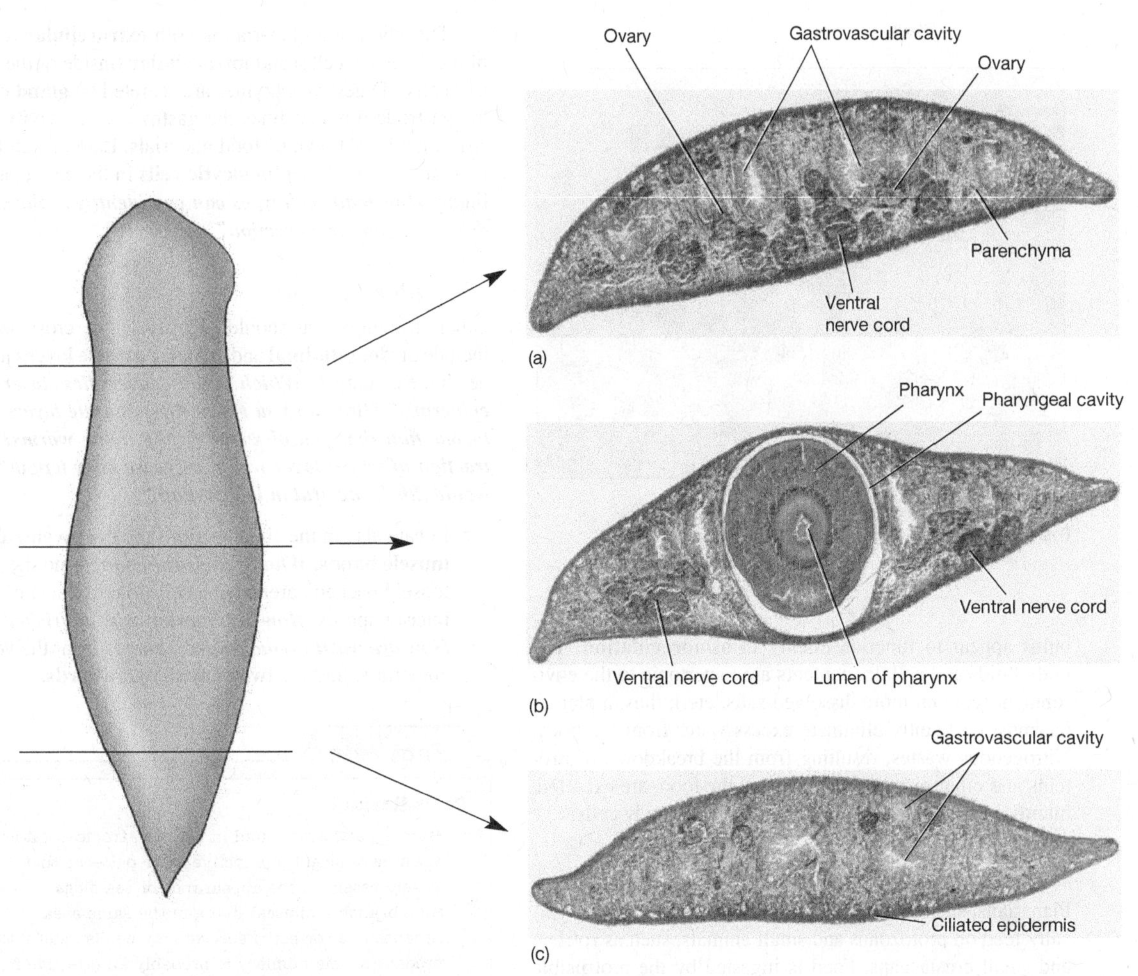

FIGURE 9.5 *Dugesia.* Cross sections through three body regions: (*a*) anterior region, (*b*) pharyngeal region, and (*c*) posterior region.
Photos (a), (b), and (c) courtesy of Betty Black.

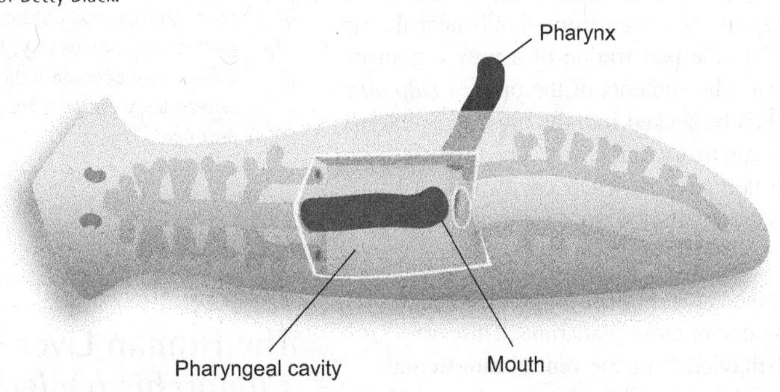

FIGURE 9.6 Drawing of *Dugesia* showing pharynx extended through ventral mouth.

cells. The complete genome of some species of planarians has now been sequenced and genes identified that regulate specific processes in regeneration.

The excretory/osmoregulatory system of *Dugesia* and other planarians is not well-developed and probably represents an ancestral condition. It consists of a system of **flame bulbs** (a type of protonephridia) interconnected by a system of collecting ducts (figure 9.4d) leading to a posterior excretory pore. The excretory structures are difficult to observe, except in special microscopic preparations. The flame

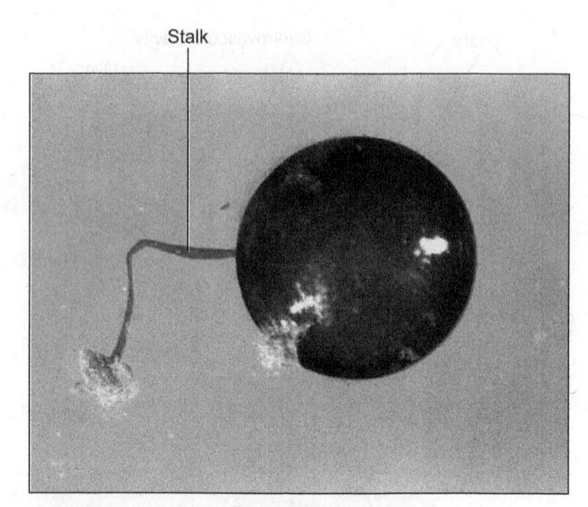

FIGURE 9.7 Cocoon of a planarian.
Courtesy of Carolina Biological Supply Company, Burlington, NC.

bulbs appear to function mainly in **osmoregulation.** The body fluids and cellular contents are hypertonic to the environment (contain more dissolved salts, etc.); thus, a planarian must constantly eliminate excess water from the body. Nitrogenous wastes, resulting from the breakdown of proteins and other nitrogenous matter in the food, are excreted, mainly as ammonia (NH_3), directly from the body cells.

Feeding and Digestion

Planarians, such as *Dugesia,* are chiefly carnivores and typically feed on protozoans and small animals, such as rotifers and small crustaceans. Food is ingested by the protrusible pharynx, which is extended through the midventral mouth while feeding (figure 9.6). Do not confuse the opening of the muscular pharynx with the actual midventral mouth opening. Proteolytic enzymes, secreted from glands near the tip of the pharynx, aid in the penetration of a prey organism, such as a crustacean. The contents of the prey (a *Daphnia,* for example) can then be sucked into the muscular pharynx and passed into the gastrovascular cavity. The digestive system of a planarian is a gastrovascular cavity with a single opening that serves as both the entrance for food and the exit for waste materials.

- Place a small piece of fresh beef or pork liver in a dish containing one or more planarians. Observe their feeding behavior. Note the ventral **mouth** and the extension of the protrusible **pharynx** through the mouth when food is located. ***How does the worm locate the food? What sensory structures may be involved?***

  How does this basic organization of the digestive system compare with that of a cnidarian or a higher animal, such as an earthworm or a frog? Which type of system would be more efficient? Why?

Digestion in a planarian is both **extracellular** (outside of the digestive cells) and **intracellular** (inside of the digestive cells). Digestive enzymes are secreted by **gland cells** in the gastrodermis that lines the gastrovascular cavity to assist in the breakdown of food materials. Later, small bits of food are engulfed by **phagocytic cells** in the gastrovascular lining. ***How many cell types can you identify in the gastrodermis in your cross section?***

Muscular System

Other structures you should identify in the cross sections include the **longitudinal** and **circular muscle layers** just beneath the epidermis. ***Which of these layers lies closer to the epidermis? How can you relate these muscle layers to the locomotion that you observed in the living worms? Contraction of which layer would increase body length? How would this be helpful in locomotion?***

- Locate also in the cross sections the **dorsoventral muscle bands.** ***What is their function?*** Find the loosely packed parenchyma cells that fill most of the interior spaces. ***How are interior cells nourished? How are wastes removed from them?*** Near the ventral epidermis, find the **two ventral nerve cords.**

Fact File

Turbellarians

- Marine flatworms found in shallow, tropical waters often have bright, distinctive color patterns that closely resemble the appearance of sea slugs (nudibranch molluscs) living in the same area. Because one (or both) species may be distasteful to predators, this mimicry is probably an effective form of defense.
- *Microstomum linare* acquires stinging capsules (cnidoblasts) from hydra that it consumes as food. Undigested cnidoblasts migrate out of the gastrovascular cavity, through the mesenchyme, and eventually become lodged in the flatworm's ectoderm where they remain functional and are used for defense.

The Flukes: Class Trematoda

The Human Liver Fluke: *Clonorchis (Opisthorchis) sinensis*

Members of the Class Trematoda are all **endoparasites** and have well-developed **suckers** for attachment—one located in the region of the mouth, and one located on the midventral surface. The outer covering of the trematode body is highly modified and lacks cilia. The outer layer, the **tegument** (formerly called the cuticle), is a syncytial extension of underlying cells embedded in the body wall. Electron microscope studies have revealed that the tegument has a complex structure. Tapeworms (Class Cestoda) also have a tegument with a similar structure.

The tegument serves an active role both in protecting trematodes from the digestive enzymes of the hosts and in the uptake of nutrients from the host gut. The tegument is an excellent example of morphological and physiological adaptation of a parasite for its very special mode of life.

Clonorchis sinensis (figure 9.8), formerly called *Opisthorchis sinensis,* is a common and important human parasite in parts of the world, particularly the Orient. Like many other trematodes, this species has a complex life cycle involving several hosts and a series of larval stages.

- Obtain a prepared microscope slide with a stained whole mount of an adult fluke and observe its size, shape, and general morphology under your stereoscopic microscope. Observe the **oral sucker** surrounding the mouth at the anterior end.

Behind the mouth is a muscular **pharynx,** a short **esophagus,** and two **intestinal caeca.** Note that the digestive tract has a single opening, the **mouth,** and thus represents an

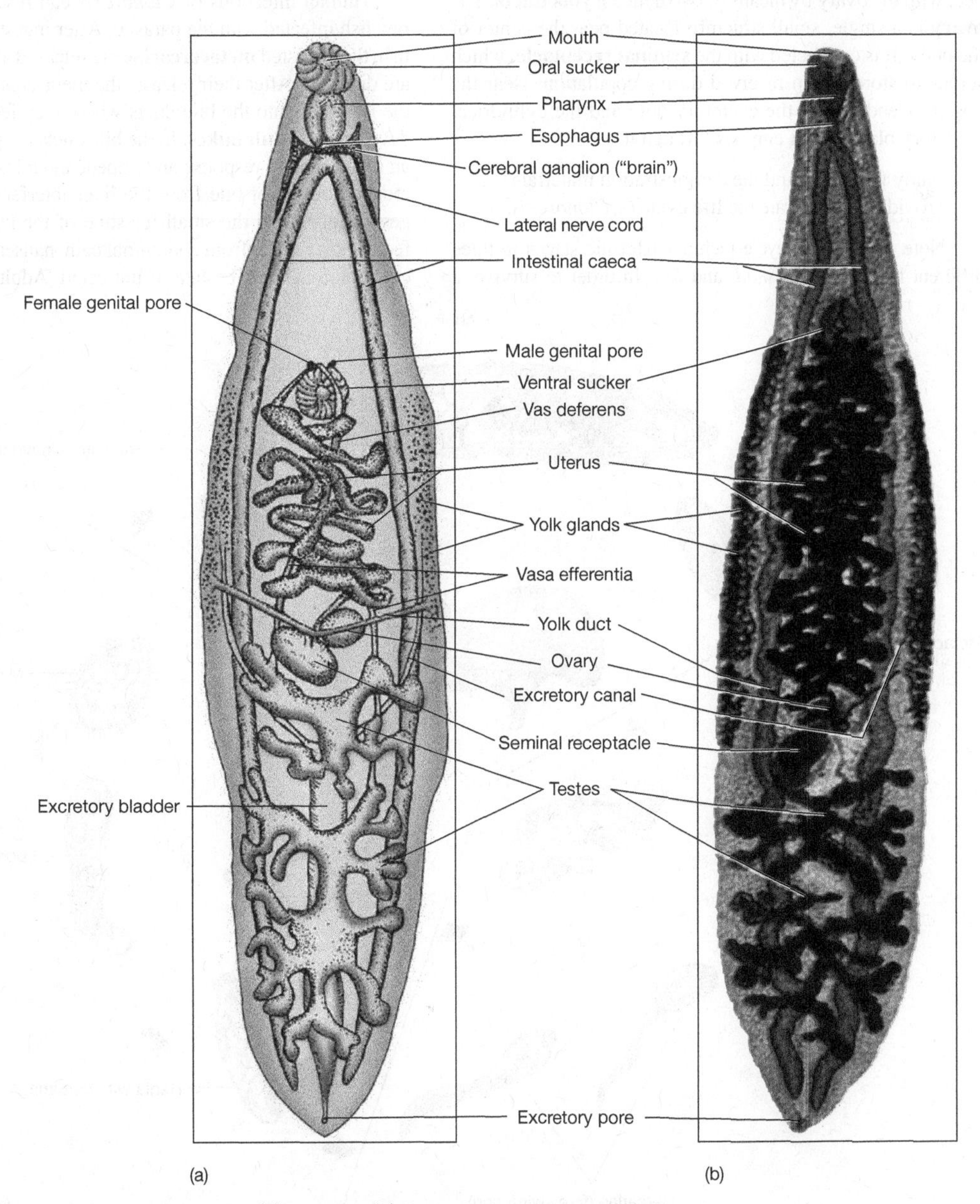

FIGURE 9.8 *Clonorchis,* liver fluke. (*a*) Drawing. (*b*) Stained whole mount.
(b) Photograph courtesy of Carolina Biological Supply Company, Burlington, NC.

incomplete digestive system. A bilobed **cerebral ganglion** ("brain") lies on the dorsal side of the pharynx (small and difficult to see in most slides). Near the branching of the intestinal caeca, note the ventral **sucker.**

Clonorchis has a well-developed reproductive system with both male and female sex organs (monoecious).

Find the long, coiled **uterus** filled with eggs just posterior to the large ventral sucker. Just posterior to the uterus lie the many-branched **testes** where the sperm are produced. Observe also the many small **yolk glands** along the lateral margin of the body in the midregion. The yolk glands connect with the ovary by means of two delicate **yolk ducts.** The **ovary** is a single, small structure located near the center of the body. It is connected with the **seminal receptacle,** which serves to store sperm received during copulation. Near the posterior end, locate the excretory pore and the cylindrical excretory bladder that empties through it.

- Study figure 9.9 and the demonstration materials provided to illustrate the life cycle of *Clonorchis.*

Note that the life cycle includes parasitic stages in three different hosts: *human, snail,* and *fish.* In order to survive, a parasite with such a complex life cycle including several hosts must have an effective means of transfer from one host to the next. Unless the proper host is available at the appropriate time, the life cycle cannot continue, and the parasite will die. This fact is used as the basis for the control of many parasitic diseases of humans and animals, such as malaria and schistosomiasis.

The adult liver fluke lives in the bile duct of humans or of other carnivorous animals. The host in which the adult (sexually mature) stage of a parasite resides is designated as the **definitive host.** All other hosts in the life cycle are termed **intermediate hosts.**

Human infections of *Clonorchis* can result from eating raw fish infected with the parasite. After ingestion of infected fish, the encysted **metacercariae** are released as the cyst walls are digested. After their release, the metacercariae migrate to the liver and into the bile ducts where they feed on bile and develop into **adult flukes.** In the bile ducts the parasites cause an inflammatory response and impede the release of bile. The reduced output of bile from the liver interferes with the digestion of fats in the small intestine of the human host. Infected hosts suffer from abdominal pain, nausea, diarrhea, and carcinoma that can be fatal if untreated. Adult flukes release

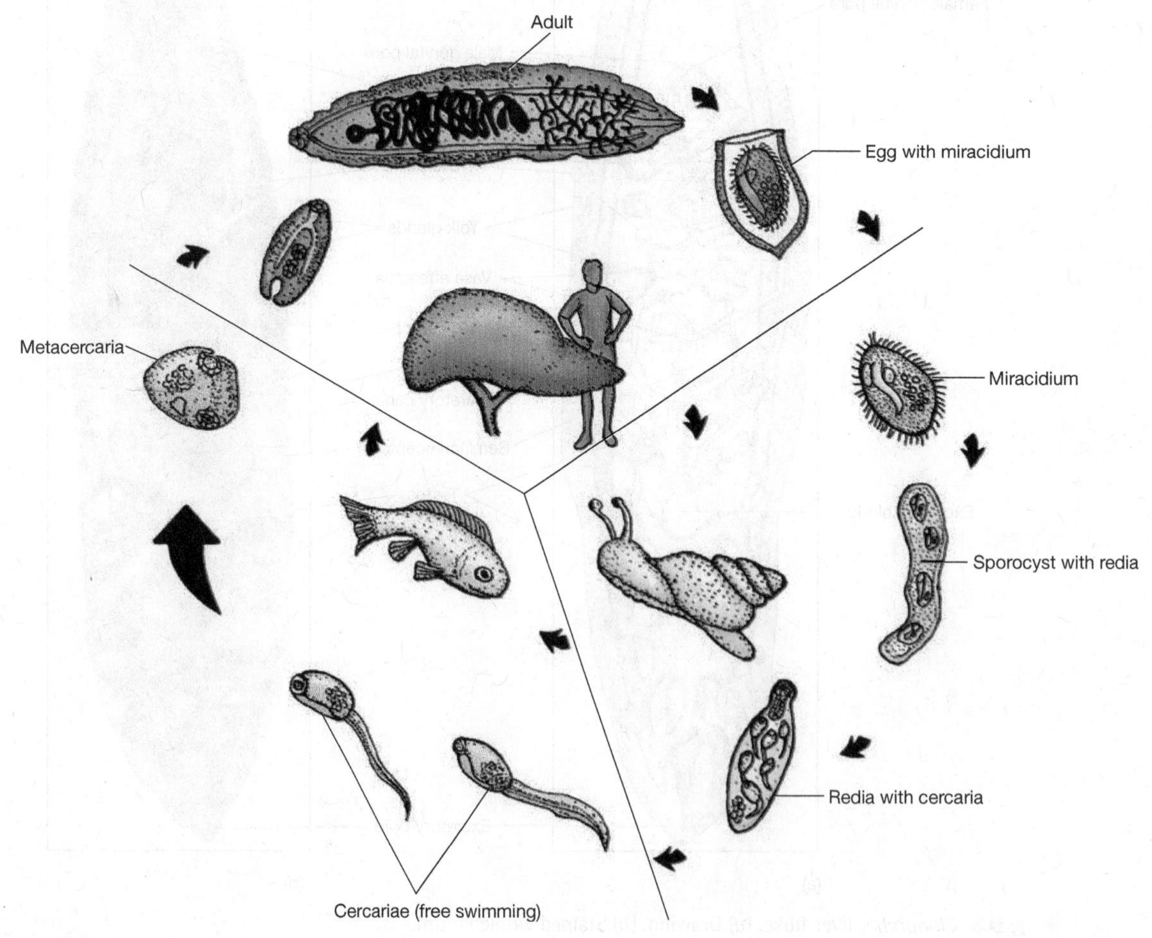

FIGURE 9.9 *Clonorchis,* life cycle.

fertilized eggs into the bile ducts that pass into the small intestine and are later voided in the feces of the human host.

The use of human waste for fertilizer in parts of the Orient leads to infection of the intermediate host, a snail, to continue the cycle. If these feces get into water, the eggs may be eaten by certain species of snails **(first intermediate host).** Indeed, the practice of using human waste as fertilizer is a major contributing factor to the survival of liver fluke populations in some parts of the Orient. Inside the digestive system of the snail, the egg hatches into a larval form called a **miracidium.** The miracidum lives in the tissues of the snail, passing through several other larval stages, (**sporocyst, redia,** and **cercaria**) and reproducing asexually to produce thousands of new larvae. The last larval stage, the cercaria, escapes from the snail and swims in the water until it contacts the **second intermediate host,** specific species of fish. When the fish is contacted, a cercaria burrows through the skin, sheds its tail, and encysts to form still another stage, the **metacercaria.** If raw or improperly cooked fish containing metacercariae is eaten by the human or another appropriate definitive host, the cyst walls are digested, and the metacercariae are released.

The Sheep Liver Fluke: *Fasciola hepatica*

This large trematode is similar in structure to *Clonorchis,* although it is considerably larger, and its reproductive system is more complex (figure 9.10). *Fasciola hepatica* is a

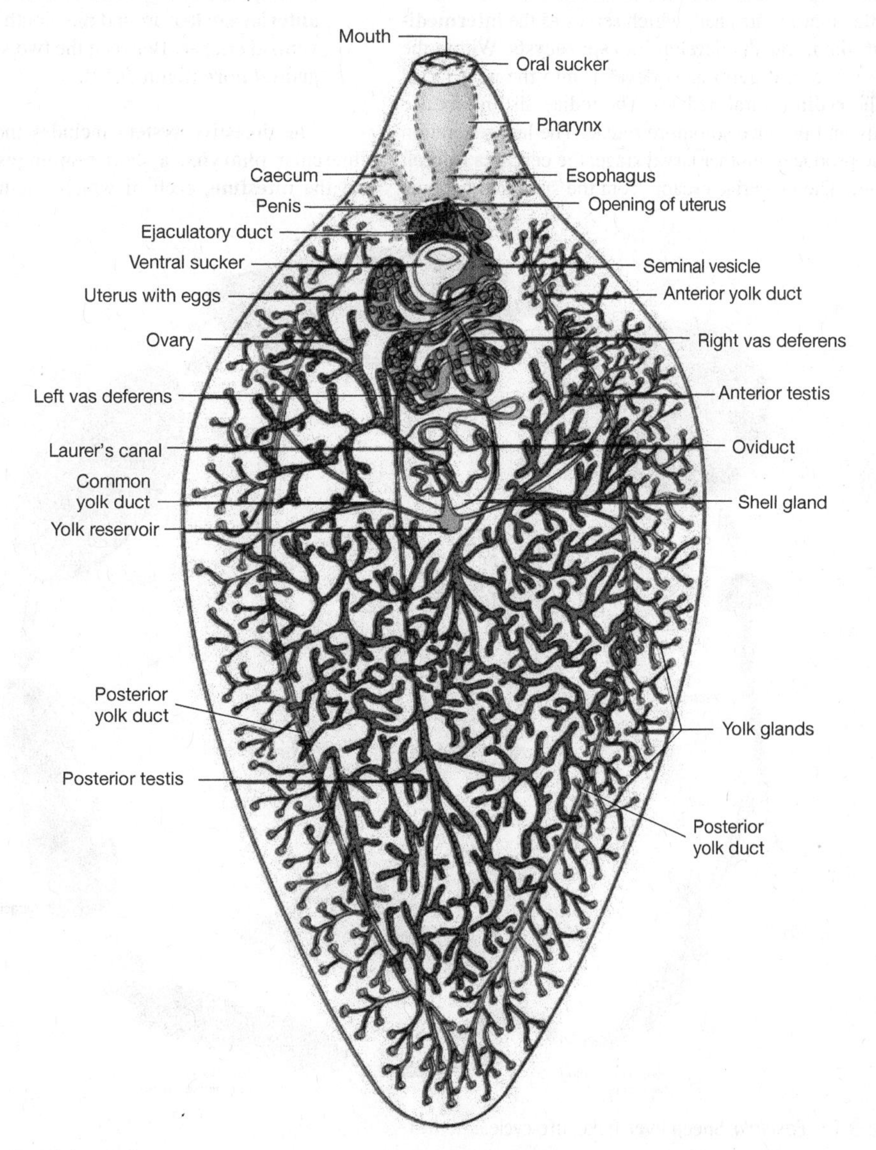

FIGURE 9.10 *Fasciola,* whole mount.

common parasite of sheep and cattle, but it also occasionally parasitizes other mammals, including humans. The adult flukes live mainly in the bile ducts of the mammalian host and cause the breakdown of the adjacent liver tissue, producing the disease called "liver rot."

The life cycle of *Fasciola hepatica* is of historic importance since it was the first to be worked out for a digenetic (two-host) trematode (figure 9.11). Adult worms live in the bile duct of a sheep and produce many **eggs** that pass from the bile duct to the intestine and are deposited with the feces. In water or warm, moist conditions, the **miracidia larvae** develop within the eggs and escape. These microscopic ciliated larvae must burrow into the appropriate species of snail within a few hours to continue their development. Unless they find an appropriate host, the miracidia die.

In the appropriate snail, which serves as the **intermediate host,** the miracidia develop into **sporocysts.** Within the sporocysts, several germ cells develop into the next larval stage, the **redia** (plural: rediae). The rediae also reproduce asexually and produce still more rediae. The last generation of rediae produces another larval stage, the **cercaria** (plural: cercariae). The cercariae escape from the snail and become free-swimming, until they reach some aquatic plants. Here they settle, lose their tails, and encyst as **metacercariae.** The encysted larvae can survive for several weeks on the plants, and if the plant and metacercariae are eaten by a sheep or other appropriate final or **definitive host,** the cyst wall is digested, and the larvae burrow through the intestinal wall to the body cavity and travel to the liver. During their migration through the liver, the young adults feed on the liver tissue.

- Observe the demonstrations of the various stages in the life cycle and try to understand the sequence of stages with the aid of figure 9.11.

  Obtain a slide with a stained whole mount of an adult *Fasciola* and note its general shape. ***How does its shape differ from that of Clonorchis?*** Locate the **anterior sucker** around the mouth and the nearby **ventral sucker.** Between the two suckers, find the **genital pore** (figure 9.12).

The digestive system includes the anterior **mouth,** a muscular **pharynx,** a short **esophagus,** and two branches of the **intestine,** each of which has many smaller lateral

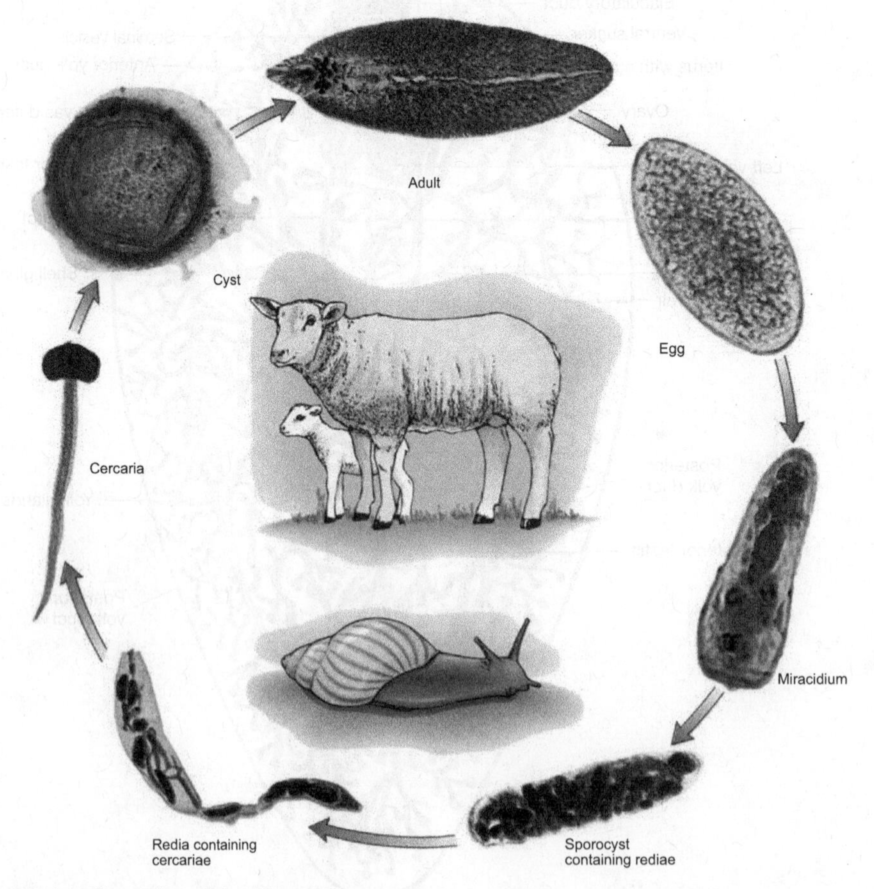

FIGURE 9.11 *Fasciola.* Sheep liver fluke, life cycle.

Photographs courtesy of Carolina Biological Supply Company, Burlington, NC.

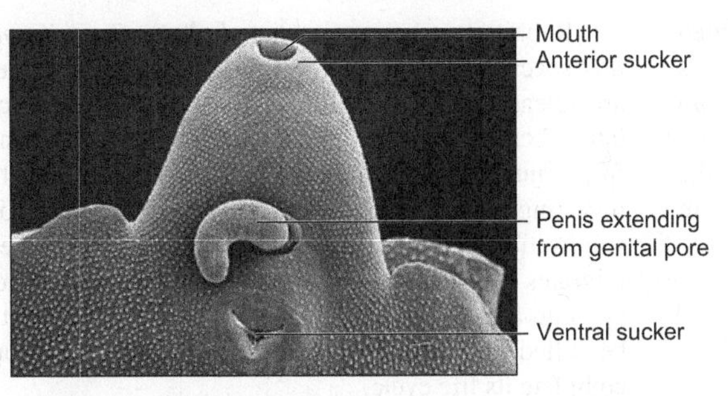

FIGURE 9.12 *Fasciola.* Ventral surface of anterior end. Note sculpturing on tegument.
Scanning electron micrograph by Louis de Vos.

branches, the **intestinal caeca** (singular: caecum). ***What would you expect to be the main purpose of the intestinal caeca?*** *Fasciola* is **monoecious** with both male and female sex organs in one individual. With the aid of figure 9.10, locate the principal organs of each system.

The female organs, found mainly in the anterior half of the body, include the following: **ovary, yolk glands, yolk ducts, yolk reservoir, shell gland, oviduct, uterus** with eggs, and the **opening of the uterus** just inside the genital pore.

The male organs include the following: **testes** (one anterior and one posterior), **vas deferens** ***(how many?),*** **seminal vesicle** ***(how many?),*** **ejaculatory duct,** and a muscular **penis.**

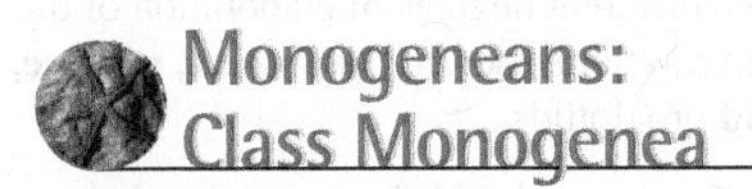

Monogeneans: Class Monogenea

Most monogeneans are ectoparasites that live on the outside of fishes. They are characterized by a distinctive posterior attachment structure called the **opisthaptor.** A few monogeneans parasitize turtles, amphibians, and some invertebrates. The opisthaptor is usually equipped with hooks, anchors, clamps, or suckers—a key feature for the identification of species. The anterior sucker (or prohaptor) is usually poorly developed. Most monogeneans have a simple life cycle with a single host. Fertilized eggs develop into a ciliated, hook-bearing larva that can attach to a new host and complete the life cycle. The two subclasses of the Monogenea are distinguished by the structure of their opisthaptor. The Monopisthocotylea (figure 9.13a) have a large sucker with hooks or a series of large and small hooks for attachment. The Polyopisthocotylea (figure 9.13b) bear a number of clamps rather than hooks.

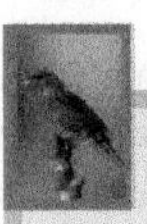

Demonstrations

1. Slides with adult stages of some other representative trematodes
2. Slides representing the stages in the life cycle of *Clonorchis* and/or *Fasciola*
3. Living cercaria

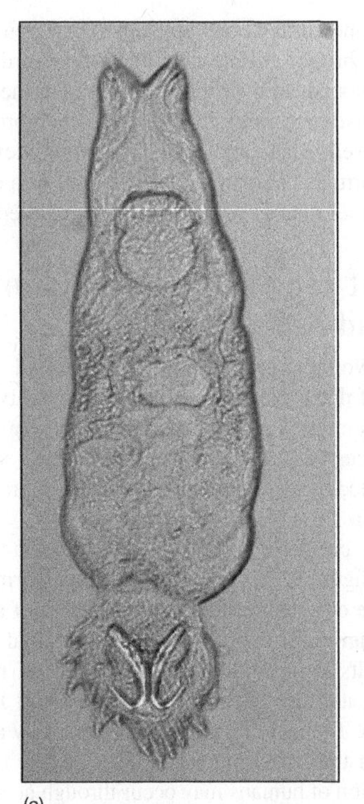
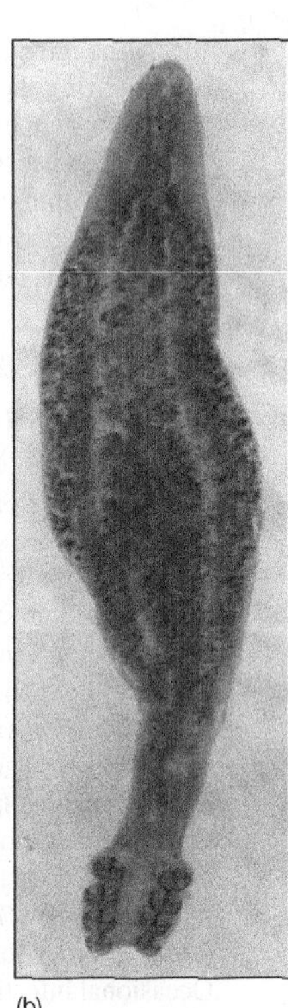

FIGURE 9.13 Representative monogeneans (*a*) *Fundulotrema,* a monopisthocotylean and (*b*) *Octomacrum,* a polyopisthocotylean.
(a) Photo courtesy of Stanley King of Dalhousie University, Halifax, CA; (b) Photo courtesy of John Forest of St. Marys University, Halifax, CA.

The Tapeworms: Class Cestoda

Tapeworms are internal parasites, highly specialized and well-adapted for their parasitic mode of life. Adult worms inhabit the intestines of various species of vertebrate animals while their larvae live in the tissues of some intermediate host. In general, the life cycles of tapeworms, or cestodes, are less complicated than those of the trematodes.

The flat, ribbonlike body of a tapeworm is typically divided into three major regions: an anterior scolex, a short, narrow neck, and a long, multisegmented strobila. The **scolex** is a specialized holdfast organ with suckers and/or hooks that attach the tapeworm to its host (figure 19.15a). The **neck** contains a budding zone where new reproductive segments, called **proglottids,** are produced asexually. The **strobila** is a long string of maturing proglottids, each slightly older than the one just in front of it. As the proglottids age, they undergo progressive development of reproductive structures, and eventually become packed with fertilized eggs. As these "gravid" proglottids reach the end of the strobila, they detach and pass out of the intestine with the feces to distribute their eggs.

Although a tapeworm's body appears to be segmented (metametric), the proglottids are not generally regarded as true body segments because of the way in which they are formed, and because each proglottid is a complete reproductive unit within itself. Many zoologists, therefore, view the body of a tapeworm as comparable to a colony, or a chain of individuals, rather than as being actually segmented.

Dog and Cat Tapeworms: *Dipylidium caninum* and *Taenia pisiformis*

Adults of these two tapeworm species may be found in the intestinal tracts of dogs, cats, wolves, and other carnivores. Both species are similar in structure and life cycle. Eggs are released in the feces of a primary host, as described above, but must be ingested by an intermediate host in order to continue development.

Dog fleas (or cat fleas) serve as the intermediate host for *D. caninum* (figure 9.14). Ingestion of a tapeworm egg by the larval stage of a flea results in development of a six-hooked **onchosphere larva** within the flea's intestinal wall. Once the flea molts to its adult stage, the tapeworm forms a bubblelike cyst and develops into a **cysticercoid larva** ("bladder worm"). The adult flea must be ingested by a dog or cat in order for the tapeworm to complete its life cycle. Occasional infection of humans may occur through accidental ingestion of an infested flea (most likely as a result of hand-to-mouth contact in young children who crawl around on the floor).

Fact File

Flukes and Tapeworms

- More than 200 million people worldwide are infected with the blood fluke (*Schistosoma*) that cause schistosomiasis (also called bilharzia). This disease is second only to malaria as the world's most prevalent infectious disease. Parts of Africa (especially Egypt), South America, and China are most severely affected.
- Tapeworms vary in length from less than an inch to more than 30 feet long. Some may live in the human body for more than 25 years.
- Three types of tapeworms live in the human digestive system: fish, beef, and pork. Incidence of the pork tapeworm, *Taenia solium*, is rare in the United States but common in Asia, eastern Europe, and Latin America. The fish tapeworm, *Diphyllobothrium latum*, native to Scandinavia and the Baltic area in northern Europe, is now worldwide in distribution. It is most common in Scandinavia, Japan, and parts of South America where raw fish is consumed. The beef tapeworm, *Taenia saginata*, occurs in California and New England but is more common in Africa, the Middle East, eastern Europe, Mexico, and South America.

Rabbits are the intermediate host for *T. pisiformis.* In this species, eggs develop into **onchospheres** before they are released from the proglottid. A tough shell encapsulates each onchosphere, protecting it for several months from injury and desiccation. A rabbit becomes infected by ingesting an onchosphere while foraging for food in the soil. Activated by digestive enzymes, the onchosphere larva migrates into the rabbit's liver or mesenteries and develops into a **cysticercoid** ("bladder worm"). The rabbit must be killed and eaten by a carnivore for the tapeworm to complete its life cycle.

- Obtain a prepared microscope slide or a plastic mount with the scolex and representative proglottids of *Dipylidium caninum* (figure 9.14) or *Taenia pisiformis* (figure 9.15). Identify the **scolex, neck,** and **strobila.** On the scolex, find the **rostellum** with its several rows of **hooks** and four lateral **suckers.** The hooks and suckers aid in attachment to the intestinal wall of the host. Note that there is no trace of a mouth or tubular digestive system. ***How does a tapeworm obtain its nourishment?*** Ancestral flatworms almost certainly did have a mouth and a digestive system. ***How would you explain their absence in tapeworms descended from ancestors that had such structures?***
- ***Where would you find the youngest proglottids in an intact tapeworm? The oldest?*** Study several proglottids in different stages of development and observe the different degrees of elaboration of the reproductive system. Identify **immature, mature,** and **gravid** proglottids.

Select a mature proglottid for more detailed study of its internal structure. Locate the female reproductive system and identify the **genital pore, vagina, seminal receptacle, ovary, oviduct, uterus,** and **yolk glands.** Among the male reproductive structures, locate the **testes, vas efferens, vas deferens,** and the copulatory organ, the **cirrus.** Along the lateral margin of each proglottid, identify the **longitudinal nerve cords** and **excretory canals.**

Study some of the gravid proglottids and observe the many ovarian capsules containing several eggs or embryos. ***Is the number of eggs per capsule always the same?***

Locate some of the transitional proglottids between the mature and **gravid** proglottids to observe the progressive **atrophy** of the reproductive organs. ***Which organs disappear first? Which persist the longest? How do these observations relate to this tapeworm's mode of reproduction?***

Adaptations for Parasitism

Tapeworms provide a good example of animals that have become adapted for a special mode of life. Tapeworms display several structural and physiological features that increase their chances of survival as internal parasites of vertebrates. To be effective parasites they must be able to successfully defend themselves against the various defense mechanisms of their host. Review what you have learned

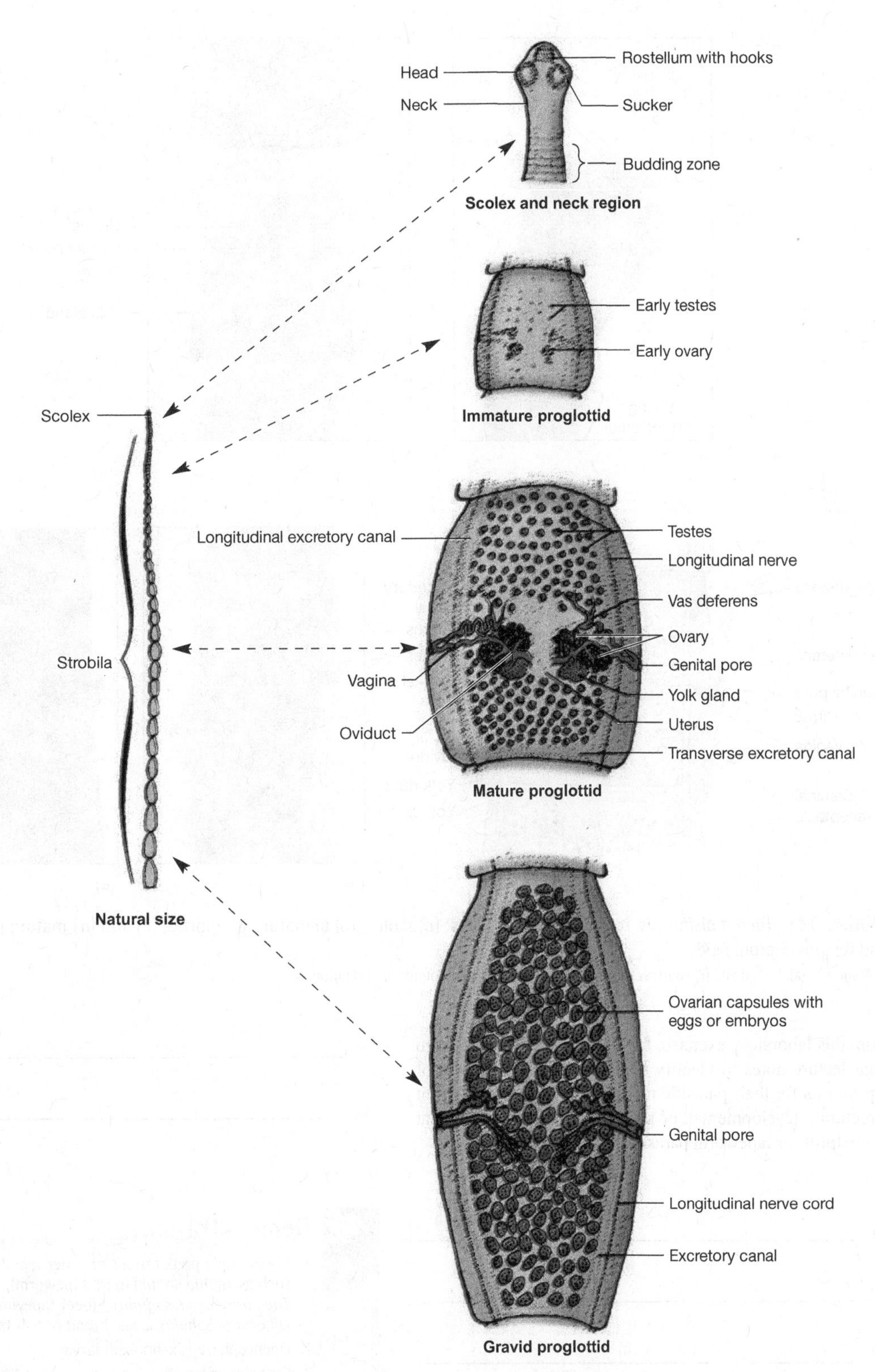

FIGURE 9.14 *Dipylidium*, representative sections of body.

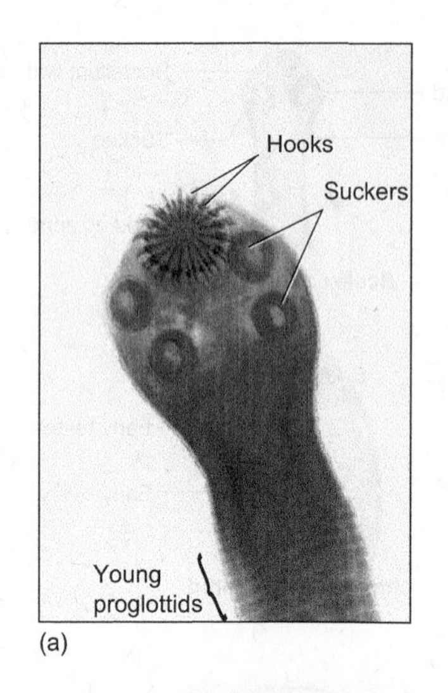

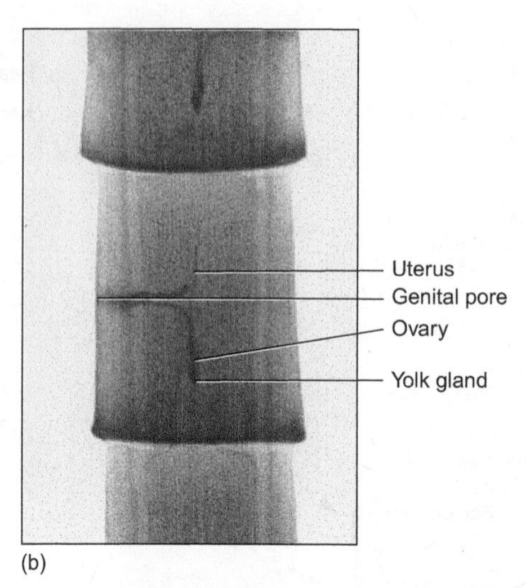

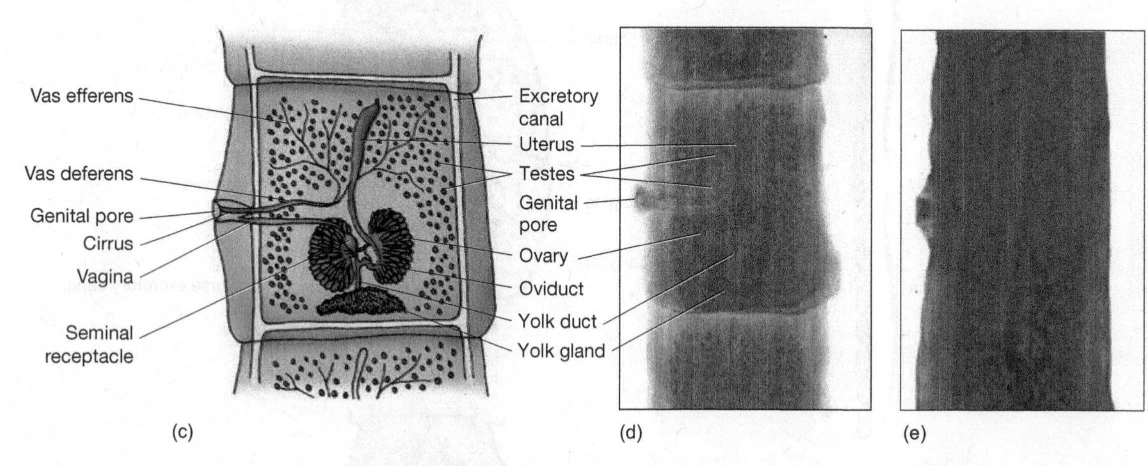

FIGURE 9.15 *Taenia pisiformis*, representative sections: (*a*) scolex, (*b*) immature proglottid, (*c*) and (*d*) mature proglottids, and (*e*) gravid proglottid.

Photographs (a), (b), (d) and (e) courtesy of Carolina Biological Supply Company, Burlington, NC.

from this laboratory exercise, from your textbook, and from your lecture notes to identify some of these adaptations of tapeworms for their parasitic mode of life. List six different structural, developmental, or physiological adaptations that are helpful for tapeworm parasitic existence.

1. ____________________

2. ____________________

3. ____________________

4. ____________________

5. ____________________

6. ____________________

Demonstrations

1. Microscopic preparations of other tapeworms such as *Taenia solium* (pork tapeworm), *Taeniarhynchus saginatus* (beef tapeworm), and *Dibothriocephalus latus* (broad or fish tapeworm)
2. Onchosphere (six-hooked) larva
3. Cysticercus larva
4. Preserved whole tapeworms

Collecting and Preserving Flatworms

Freshwater planarians (Class Turbellaria) are usually not difficult to find. Shake pond weeds into a white pan or look carefully on the underside of rocks in ponds and streams. Try putting a small pellet of dog food or egg yolk into an old nylon stocking. Secure this bag in a pond or stream bed overnight and in the morning look for flatworms on the surface of the bag. Planaria will survive well in an aerated aquarium or a flat bowl of pond water. Feed them small pieces of beef liver or egg yolk, but remove uneaten food after feeding and change the water weekly.

Marine flatworms live in all oceans of the world. They are commonly found on rocks or seaweed in shallow water. Hundreds of brightly colored species inhabit coral reefs and rocky beaches in tropical seas. Occasionally, they turn up as accidental introductions in marine aquaria where they may be mistaken for sea slugs (Phylum Mollusca). Most of the larger species are carnivorous and notoriously difficult to rear.

Parasitic flatworms (flukes and tapeworms) can sometimes be obtained from veterinarians. Stool (fecal) samples from various farm or domestic animals may contain eggs which can be isolated by flotation in a saturated solution of $ZnSO_4$ or $MgSO_4$ (epsom salts) and observed with 10× magnification. Premixed solutions and test kits with sample tubes can be purchased from vendors of veterinary supplies.

Cercariae, the free-living larval stage of trematodes, are also easy to obtain and study in the laboratory. Many aquatic snails serve as intermediate hosts of trematodes, and specimens collected in lakes, ponds, and streams can serve as a good source of cercaria larvae. Following collection, snails should be washed and placed in wide-mouth bottles. Examine the containers several times each day because the cercariae of some species are shed only at specific times. Hold the containers under a bright light and against a dark background. Cercariae will appear as small, white, swimming objects. Emerging cercariae stream out in a white cloud from heavily infected snails. Alternatively, infected snails (with directions for obtaining cercariae) can be purchased from a biological supply house.

- Capture cercariae with a capillary pipette and make a wet mount on a clean microscope slide after adding a drop of 1% methylene blue or other appropriate stain. Draw most of the water from under the coverslip with a piece of filter paper to compress the cercariae enough so that they are held in place but not crushed under the weight of the coverslip.

Phylum Gastrotricha

Gastrotrichs: *Chaetonotus*

You have probably never heard of gastrotrichs, yet they are among the most abundant small animals found in aquatic environments (figure 9.16). Members of this phylum have been largely overlooked by zoologists because they are so small—typically less than 0.5 mm in length. They can be found in both marine and freshwater habitats where they usually live as bottom-dwellers, attached to submerged vegetation and detritus or nestled in the tiny capillary spaces between individual grains of sand and gravel. Densities of 100,000 to 1 million individuals per square meter have been commonly reported from the sandbars of unpolluted streams and the littoral zones of shallow lakes and ponds.

- Make a wet mount of a living gastrotrich like *Chaetonotus* on a microscope slide and observe its appearance and behavior.

Most gastrotrichs have a translucent, wormlike body with distinctive bristles or scales scattered along much of the dorsal and lateral aspects of the head and trunk (figure 9.16). The ventral side is flat, covered by locomotor cilia that allow the animal to glide smoothly over objects in its path (much like a flatworm). Many freshwater species have bodies with a bowling pin shape—narrow behind the head and broader toward the rear where a forked tail, the **furca,** houses a pair of adhesive glands. These glands, and sometimes others near the head, secrete a glue-like substance for attachment to the substrate. Marine species generally have a more elongate body (often described as strap-shaped) with adhesive glands more widely distributed along the sides of the body.

Internal Anatomy

Gastrotrichs are **acoelomate** (no central body cavity) and have no circulatory or respiratory systems. Gas exchange occurs through the thin body wall. Most species have a pair of **protonephridia** (flame bulbs) near the midpoint of the trunk that probably serve both for excretion and osmoregulation as in flatworms (see page 135). Food consists mostly of bacteria, algae, small protozoans, and organic detritus collected by bristles or patches of cilia surrounding the mouth. The digestive system

FIGURE 9.16 Gastrotrich, drawing.

is complete, relatively unspecialized, and remarkably similar to that of nematodes (see page 194). From the mouth, food passes into a muscular pharynx and then through a simple tubular intestine that empties at a subterminal anus. The simple nervous system consists of a brain (formed by a pair of large ganglia in the head) and a pair of lateral nerves running the length of the body. Paired reproductive organs are located in the broad part of the trunk along with a mysterious caudal organ that appears to play some role in sexual reproduction.

Development and Life History

Gastrotrich embryos undergo a specific number of cell divisions and reach a determinant number of somatic cells by the time they hatch from the egg. Juveniles resemble adults—there is no larval stage. Growth occurs exclusively through an increase in the size of individual body cells over the span of only three or four days. This type of development, known as **eutely,** is otherwise found only in nematodes, rotifers, tardigrades, and dicyemid mesozoans (tiny parasites of cephalopods).

When gastrotrichs hatch, their reproductive organs already contain eggs that have begun **parthenogenetic** development. These eggs, rarely more than four in number, are laid singly over the course of three or four days while the animal is still growing in size. Most of these eggs hatch quickly into juveniles that are genetically identical to their parent. Occasionally, the last egg laid will be larger than the others and has a thicker shell that makes it resistant to desiccation and temperature extremes. This egg does not hatch immediately; it may remain dormant for up to two years, and must undergo an environmental stress (heat, cold, or dryness) before it will continue development. ***How would the population benefit from production of these dormant eggs?***

Following the production of parthenogenetic eggs, gastrotrichs enter a phase of sexual reproduction. Each individual becomes a hermaphrodite (capable of producing both eggs and sperm). Cross-fertilization between two individuals appears to be necessary because isolated individuals never produce sexual eggs. During this phase of life, the mysterious caudal organ grows in size. Its function has not yet been determined, but some people believe it plays a role in the exchange of gametes.

Alternation between asexual and sexual reproduction is not unusual among animals. ***Can you name other animals that do this?*** Gastrotrichs, however, are unique in allocating a fraction of each generation's reproductive output (the dormant eggs) as an insurance policy against the next catastrophic change in their environment. This is a good example of natural selection working at the population level to enhance survival despite frequent and unpredictable changes in the environment.

Collecting Gastrotrichs

Sandbars in unpolluted streams, submerged vegetation, and organic detritus near the banks of shallow lakes and ponds appear to be the most productive sites for finding gastrotrichs. Plant samples can be repeatedly washed with a 4% aqueous solution of $MgCl_2$, which acts as an anesthetic and helps detach the animals from their substrate. Concentrate specimens by straining the wash water through fine silk mesh, but exercise care because these animals are quite delicate. Gastrotrichs should be observed and processed when freshly collected because they do not survive long under laboratory conditions.

Fact File

- A gastrotrich's reproductive system apparently has no external opening. Eggs are laid through a small rupture in the body wall.

Phylum Rotifera

Rotifers are common and abundant freshwater animals. They are important constituents of the plankton of lakes, ponds, and streams, and are significant as a food source for many species of fish and other filter-feeding animals. They also live among mosses and lichens growing on rocks and trees, in soil and leaf litter and many other moist environments (figures 9.17 and 9.18). They are able to form **cysts** when temporary ponds or puddles dry up to survive unfavorable conditions and to be blown by winds to new places where the cysts can germinate when favorable conditions return. In fact, recent research suggests that such dispersal of cysts may have been important in the evolution of rotifers. Most rotifers live in fresh waters but some are found in brackish or seawater, and a few species are parasites. They may be free swimming in plankton, creeping forms among algae and other growth on the bottom of lakes and streams, or sessile, attached to the substrate and sometimes living in tubes they secrete or build.

Fact File

Rotifers

- Rotifers have a short life cycle and high reproductive potential. Under favorable conditions, 100 rotifers can grow to a population of 1 billion individuals in just 20 days.
- Rotifers are particularly abundant in shallow Antarctic lakes where few predators are able to survive.
- Rotifers in the genus *Conochilus* form spherical colonies and move together in a coordinated manner. They are so accustomed to living in a colony that, when separated as individuals, they swim in circles without any sense of direction.
- One rotifer, *Brachionus plicatilis*, serves as an important food source in aquaculture. Scientists in China and Japan have developed techniques for the continuous mass production of this species which is harvested and used as live food for the commercial production of white shrimp, crabs, and the larval stages of some marine fish.

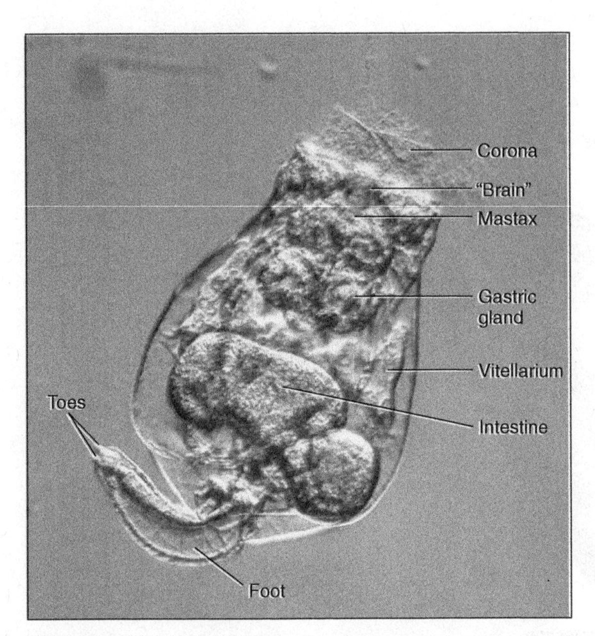

FIGURE 9.17 Living rotifer.
Photograph courtesy of Carolina Biological Supply Company, Burlington, NC.

Members of this phylum, the Rotifera, are bilaterally symmetrical, have a body cavity that is partially lined by mesoderm, and thus are **pseudocoelomate.** Their body is usually cylindrical and divided into a head, trunk, and foot. The body is covered by a **cuticle** which appears to be segmented, but there is no internal evidence of segmentation. The most characteristic feature of most rotifers is an anterior ciliated structure called the **corona** that surrounds the mouth. It serves to draw food particles to the mouth and in many species it is also important in swimming. Rotary movements of the cilia propel the rotifer through the water. A muscular **mastax** (pharynx) equipped with hard jaws for sucking and grinding food particles is often a prominent and characteristic feature because of its constant motion.

Several types of reproduction have been observed among rotifers. Rotifers are dioecious and may reproduce sexually or by parthenogenesis. Some species alternate between sexual reproduction and parthogenetic reproduction. Certain rotifers show sexual dimorphism in which the males are degenerate and lack a digestive system. They live just long enough to produce sperm for fertilization. Males are unknown in the Class Bdelloidea in which all known reproduction is parthenogenetic. Apparently bdelloidean rotifers have survived for millions of years without sex, contrary to conventional wisdom which holds that genetic diversity achieved through sexual reproduction is essential for long term survival of a species. Scientists have recently discovered some remarkable genetic mechanisms that enable rotifers to compensate for the lack of genetic diversity from sexual reproduction with other novel genetic mechanisms.

Rotifers have a complete digestive tract with a mouth and anus and usually feed on algae and other phytoplankton and dead and decomposing organic matter. Rotifers are very important ecologically as components in the aquatic food chain of many animals including copepods, fish, shrimp, and crabs. They have been used extensively in aquaculture as food for fish and other aquatic animals and have been proposed as candidates to be included in special aquaculture systems for portable ecosystems for future space applications.

Rotifers are also important as model animals for several types of research including studies of ageing, resistance to radiation effects, environmental and/or chemical induction of body form, and unusual gene function.

Classification

The phylum Rotifera is currently divided into three classes, Monogononta, Bdelloidea, and Seisonidea with about 2000 described species. The relationship of rotifers to other phyla is still largely unclear. For many years they were included along with several other groups of pseudocoelomate and presumed pseudocoelomate animals in the phylum Aschelminthes that has since been abandoned as a phylogenetic concept. There is good biochemical evidence of relationship to Acanthocephala, the spiny-headed worms, which is a small group (about 1100 species) of parasitic worms that parasitize many species of birds, fish and mammals. Acanthocephalans have a complex life cycle with insects and crustaceans as intermediate hosts. Formerly considered a separate phylum, Acanthocephala, they are now considered by some experts to be a class of rotifers.

Other evidence suggests that rotifers are related to flatworms, gastrotrichs, annelids, molluscs, and other members of the Lophotrochozoa.

Demonstration

Specimens of representative rotifers

A Rotifer: *Philodina*

Philodina (figure 9.18) is a common North American rotifer that illustrates many of the typical features of the Rotifera.

- Rotifers are easily observed with a compound microscope in a drop of water on a microscope slide under low power. Do not use a cover slip because rotifers tend to contract when disturbed. Rotifers can be picked up with a small pipette from a sample of pond water debris and transferred to a clean microscope slide. They can also be observed in a watch glass or other small glass container at higher power with a stereoscopic microscope.

Obtain a living specimen of *Philodina* and put it a water droplet to study its behavior and structure with the aid of the following description.

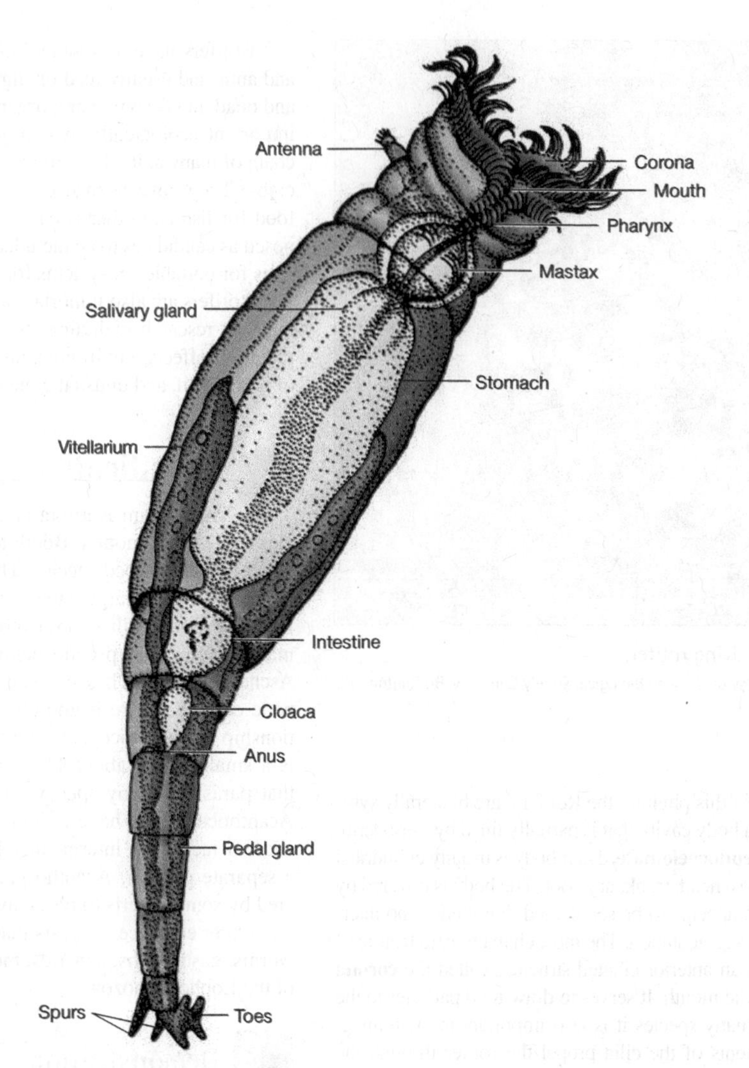

FIGURE 9.18 Rotifer, *Philodina.*

Living rotifers are superior to fixed and stained specimens for the study of general anatomy because of the serious contraction and distortion almost always encountered during their fixation and staining.

Observe the elongate cylindrical body, which can be divided into three general regions—the **head,** the **trunk,** and a posterior **foot** (figure 9.18). Superficially, the body is divided into several segments, usually about 16. The rotifers do not have true segmentation, but the cuticle covering the body is often divided into a number of superficial segments. Observe the telescoping of the segments when the animal retracts its head. Note also the large ciliated corona at the anterior end. Most rotifers have a conspicuous **corona,** which serves both for locomotion and for feeding. In *Philodina,* the corona consists of two large ciliated **trochal discs** and a posterior band of cilia, the **cingulum.** Special **retractor muscles** serve to retract the corona. Observe also, on the head, the dorsal fingerlike process, the **rostrum.** The rostrum is believed to be a sensory structure. Locate the long, tapering **foot** with two **spurs** and four retractable **toes** at the posterior end of the body.

The body wall is covered externally by a thin **cuticle** secreted by a syncytial **hypodermis.** Cell membranes are generally lacking in the organs and tissues of adult rotifers, and specific organs typically have a fixed number of nuclei. Although there are no definite muscle layers associated with the body wall of rotifers, several distinct **muscle bands** connect various parts of the body. The central body cavity of rotifers is a **pseudocoelom.**

The **mouth** is a funnel-shaped opening located at the base of the corona. It receives food particles swept down by ciliary currents created by the corona. Connecting with the mouth is a muscular **pharynx** which encloses a conspicuous grinding apparatus, the **mastax.** The mastax is a distinctive feature of the rotifers and can be easily identified in a living rotifer because of its active, almost constant movement. In *Philodina,* the mastax is specialized for grinding plankton, periphyton, and detritus, but other types of mastax are found in different species of rotifers, and their structure appears to be closely related to the food habits of the various species.

Posterior to the pharynx is a short **esophagus** surrounded by large **salivary glands.** The esophagus opens into a large, thick-walled **stomach** where digested food is absorbed by the gastrodermal lining. Behind the stomach is a short **intestine** leading into the **cloaca,** which in turn empties through the **anus.**

Most of the reproductive system of *Philodina* can be seen only in special preparations, but a pair of large, yolk-forming glands, the **vitellaria,** are readily visible in living specimens. Observe the two large vitellaria in the posterior part of the trunk region of your specimen. ***How many transparent nuclei can you identify in the vitellaria?*** The vitellaria are syncytial like most other rotifer organs and have a fixed number of nuclei. Other parts of the reproductive system include two small anterior **ovaries** (difficult to see), which connect via small ducts to the large vitellaria, and two tiny oviducts that lead from the vitellaria to the cloaca.

Males are unknown in the group of rotifers to which *Philodina* belongs (Order Bdelloidea), and the eggs produced develop exclusively by parthenogenesis. Other species of rotifers, however, do exhibit normal sexual reproduction.

Collecting Rotifers

Rotifers are easily collected in plankton and benthic (bottom) samples from ponds and lakes. Collect a plankton sample from a local pond or lake with a plankton net and examine portions of your sample in a cavity slide or in a watch glass under a stereomicroscope. Rotifers are best observed using transmitted or oblique lighting.

Bottom samples will yield other species of rotifers that live in association with the substrate rather than forming part of the plankton; some species of rotifers swim among the algae attached to stones, twigs, aquatic plants, and other submerged objects (figure 9.17). Other species of rotifers, called sessile rotifers, live attached to the substrate. Some of these sessile rotifers form delicate tubes from which they protrude while feeding and into which they can retract when disturbed or threatened.

By examining samples of plankton as well as material from bottom samples, you can observe some of the varied ways that rotifers have adapted for different modes of life in freshwater habitats.

Key Terms

Bilateral symmetry body form with parts arranged symmetrically along a midsagittal plane.

Cercaria (plural: cercariae) motile, free-swimming, tadpolelike larval stage in the life cycle of certain trematodes. Body resembles miniature fluke.

Cloaca a chamber at the posterior end of the digestive tract in certain types of animals that serves to receive wastes from the digestive tract, gametes from the reproductive tract, and/or wastes from the excretory system.

Corona a specialized organ at the anterior end of rotifers, usually ciliated, which typically serves for locomotion and food gathering. The circular motion of the coronal cilia in many species of rotifers provides the basis for the common name, "wheel animals," sometimes given to rotifers.

Cuticle a thick, non-cellular protective covering secreted by an underlying layer of epithelial cells, the hypodermis.

Cyst a resting or dormant stage of an organism usually with a protective covering and slowed metabolism formed by encystment. This process allows the organism to survive unfavorable environmental conditions and/or facilitates transport to new areas.

Definitive host final host in the life cycle of a parasite with more than one host; host in which the sexually mature parasite develops.

Dioecious having separate sexes; male and female sex organs occur in different individuals of the same species.

Eutely having a fixed and constant number of cells or nuclei in the adult stage for all members of the species.

Furca forked, terminal adhesive organ in gastrotrichs.

Hypodermis a type of tissue made up of epidermal cells that secrete an overlying cuticle

Intermediate host host in the life cycle of a parasite that bears a larval or immature stage of that parasite.

Mastax a specialized grinding organ found in the pharyngeal region of rotifers.

Miracidium ciliated larval stage in the life cycle of many trematodes. Develops from the fertilized egg and gives rise to the sporocyst stage.

Monoecious having both male and female sex organs in the same individual. Found in many flatworms, arthropods, and certain species of several other phyla.

Proglottid reproductive body division of a tapeworm. Since each proglottid contains a complete set of male and female organs, these units are often considered analogous to a complete individual, and thus the tapeworm as a colony.

Pseudocoelom a type of internal body cavity that lacks a mesodermal lining around the digestive tract. It arises embryonically as a remnant of the blastocoel.

Redia larval stage in the life cycle of certain trematodes. Formed by the sporocyst and produces many cercariae.

Scolex (plural: scoleces or scolices) specialized holdfast organ of the cestodes (tapeworms); with hooks and/or suckers for attachment.

Sporocyst saclike larval stage in the life cycle of some trematodes. Develops from a miracidium and produces several rediae.

Strobila the major portion of the body of a tapeworm excluding the neck and scolex. Includes all of the proglottids.

Tegument living syncytial protective outer covering of the cestode and trematode body; formerly called the cuticle; secreted by large underlying cells with numerous tubular cytoplasmic connections with the tegument.

Triploblastic construction body type with three distinct germ layers: endoderm, mesoderm, and ectoderm. Characteristic of flatworms and all higher metazoan phyla.

Internet Resources

There are many valuable Internet sites with information about zoology. Several sites containing pertinent zoological information for this chapter can be found on the McGraw-Hill Zoology web site at **http://www.mhhe.com/zoology**. Just click on this text's title.

Questions for Critical Thinking

1. Compare the Phylum Platyhelminthes structurally with the Phylum Cnidaria. What are the significant changes in symmetry, tissue formation, and life cycles from those found in the Cnidaria?
2. Discuss the importance of a triploblastic germ tissue construction. Of what evolutionary significance is the mesoderm?
3. Many parasitic flatworms have complex life cycles and multiple hosts. Speculate how these unusual life cycles might have evolved.
4. Why is encystment important to survival in many parasitic flatworms? Do you suppose that encystment in the different parasitic groups arose once or many times over the course of evolution? Support your answer.
5. Discuss the most important adaptations observed in flatworms. What adaptations are most important for endoparasites? Which are most important for ectoparasites?
6. Discuss the ecological importance of rotifers and gastrotrichs.

Suggested Readings

Combes, C. 2001. *Parasitism: The Ecology and Evolution of Intimate Interactions.* Chicago: University of Chicago Press. 552 pp

Moore, J. 2002. *Parasites and the Behavior of Animals.* New York: Oxford University Press. 338 pp

Moore, J. 2006. *An Introduction to the Invertebrates.* Cambridge: Cambridge University Press. 340 pp

Roberts, L, and J. Janovy Jr., 2008. *Foundations of Parasitology,* 8th ed. New York, McGraw-Hill. 728 pp

Thorp, J. H., and A. P. Covich, eds. 2010. *Ecology and Classification of North American Freshwater Invertebrates.* Boston: Academic Press. 1021 pp

Wallace, R.L. 2002. Rotifers: Exquisite Metazoans. *Integrated and Comparative Biology* 42:660–667.

Notes and Sketches

Chapter 10

Mollusca

Frank & Joyce Burek/Getty Images.

Objectives

After completing the laboratory work in this chapter, you should be able to perform the following tasks:

1. List and briefly describe the eight classes of living molluscs.
2. Identify the principal internal organs on a specimen of freshwater mussel or a marine clam and briefly explain the function of each organ.
3. Explain the pattern of water flow through a freshwater mussel and its importance.
4. Explain the reproduction and life cycle of a freshwater mussel.
5. Describe the organization of the garden snail *Helix* and identify its principal features on a dissected specimen.
6. Identify the chief external features of the squid and show four morphological features illustrating adaptation for its specific mode of life.
7. Briefly describe the main internal organs of the squid's digestive and circulatory systems. Contrast this circulatory system with that observed in a mussel or clam. Discuss which type of system appears to be more efficient in providing nutrition to organs and tissues and removing wastes.
8. Briefly discuss the current ideas about phylogeny of the molluscs and some of the lines of evidence that have influenced thinking about their phylogeny.
9. Explain the concept of adaptive radiation based on a simple body plan and how the molluscs illustrate this concept.

Introduction

Molluscs are a large and diverse group of animals that includes chitons, clams, oysters, snails, squid, and octopi (figure 10.1). This is the second largest phylum in

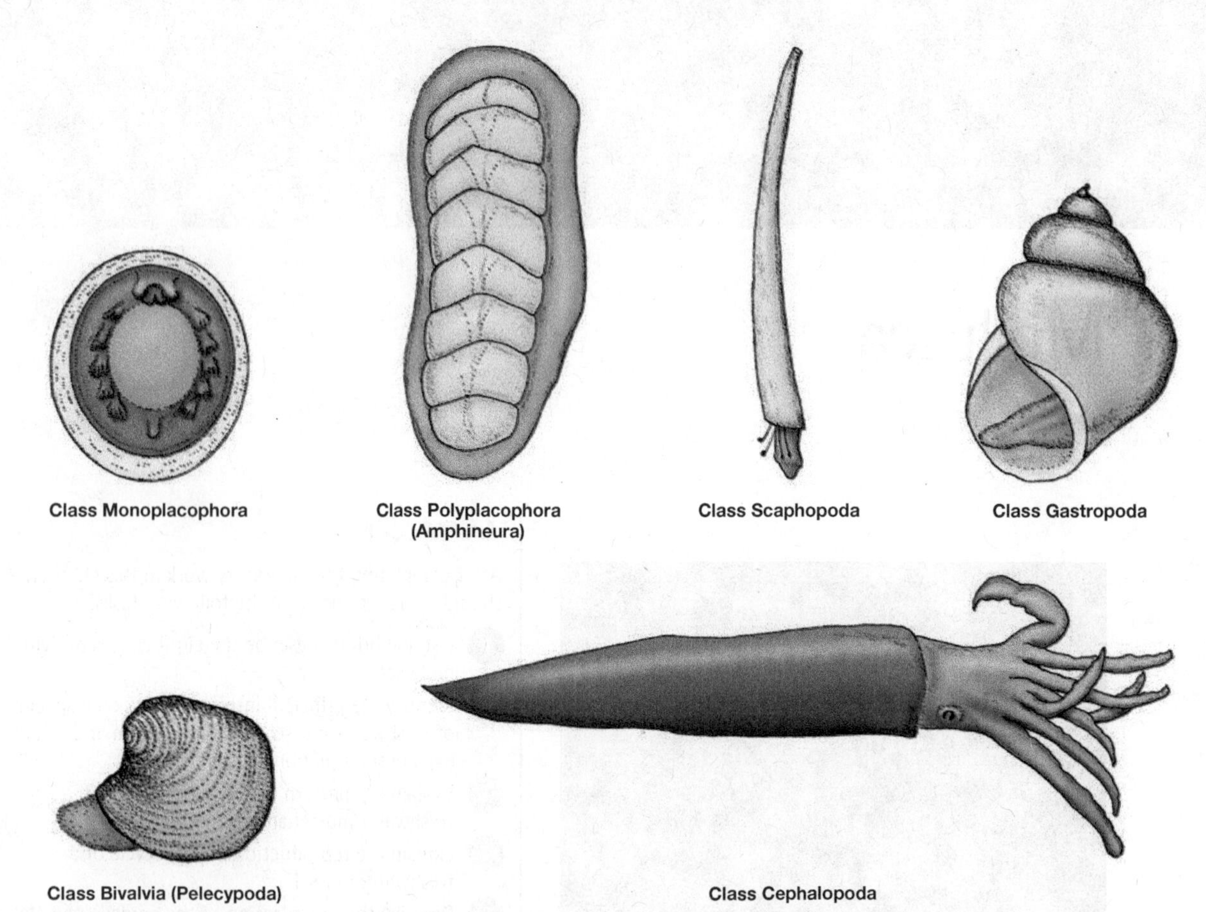

FIGURE 10.1 Six major classes of living molluscs. Two smaller classes are not shown.

terms of species, second only to the arthropods. It is also the largest marine phylum, with more marine species than any other phylum. Recent estimates of the number of scientifically accepted living species range from about 80,000–100,000 plus about 50,000–70,000 fossil species. Some 80 percent of all mollusc species are gastropods. Squid, octopi and other cephalopods have the most highly developed nervous systems and sense organs among the invertebrates. The recently described colossal squid may be the largest invertebrate in the world, measuring 10 meters long and weighing 500 kilograms (over ½ ton!).

Molluscs are an ancient group. They have an extensive fossil record dating back to the Cambrian era, 543–490 million years ago. Molluscs provide an excellent example of **adaptive radiation,** with a great diversity of forms that have become adapted for virtually every type of freshwater and marine habitat as well as many terrestrial habitats. Molluscs are one of the few animal phyla to successfully adapt to the rigors of the terrestrial environment.

Members of this phylum typically have soft, unsegmented bodies, which usually are enclosed, wholly or in part, by a thin fleshy layer derived from the dorsal body wall, the **mantle.** The folds of the mantle form a mantle cavity that typically encloses the gills. In most species, the mantle secretes a hard **shell.** In some of the more specialized molluscs, however, the shell has been lost or reduced, or has become embedded in the soft tissue.

Molluscs are **triploblastic coelomate** animals, as are the annelids and all other higher metazoan animals, including the chordates. The **coelom** is the principal internal body cavity in most of these groups and is completely lined by mesodermal tissue (muscle, connective tissue, etc.). The coelom arises during embryonic development as a cavity within the developing mesoderm (see Chapter 5). In the molluscs, the coelom is reduced, but the body wall is thick and muscular.

In most higher metazoans, the coelom expands greatly to become the principal internal cavity and serves many important functions. For example, it provides space for the development of internal organs, serves in the temporary storage of metabolic wastes, provides space for the temporary storage of gametes, and provides a hydrostatic (fluid) skeleton to facilitate the movements and burrowing activities of soft-bodied animals.

In most modern molluscs, however, the coelom is reduced to the cavities surrounding the heart, gonads, and excretory organs (nephridia). Abundant evidence indicates that molluscan ancestors had a more prominent and spacious coelom than that found on living forms.

The molluscan body typically consists of three major parts: an anterior **head,** a ventral **foot,** and a dorsal **visceral mass.** These basic parts are variously modified in different molluscs; clearly illustrating the remarkable diversity of form that can be achieved by alterations on a relatively simple body plan.

Because of the great diversity of form in the Phylum Mollusca, there is no "typical" mollusc. After you complete your study of the principal representative of this phylum (the freshwater mussel or a marine clam), it is important that you make a careful study of the demonstration material to gain a better appreciation of the body organization of other kinds of molluscs. We will study representatives of three of the classes of molluscs in more detail in this chapter.

Fact File

Molluscs

- The Shell Oil Company of today started as an importer of shells for making buttons. When the market for shell buttons began to collapse, the company was forced to diversify and got lucky with oil!
- Mollusca vs. Molluska. There has been a long discussion over the correct spelling of the name, but today Mollusca is generally accepted as the preferred American spelling and Molluska is accepted as the preferred British spelling.
- Molluscs are an ancient group of animals. Their ancestors can be traced back nearly 500 million years in the fossil record.
- More than 100,000 species have been described in the Phylum Mollusca—second in numbers only to the Arthropods. Many species have yet to be discovered and many recently discovered species are yet to be described and named.
- Paleontologists can use fossil shells to determine what the climate might have been like millions of years ago. This is done by comparing fossil shells with their closest living relatives of today and noting where they live, such as warm or cold and wet or dry climates.

Classification and Phylogeny

The great diversity of forms among the molluscs is clearly represented in the morphological diversity within the eight major groups of the phylum. These eight groups have traditionally been considered classes, and various theories have been advanced to account for the evolutionary relationships among the classes and the relationship of the molluscs with other animal phyla.

The origin of the molluscs and the relationships among the eight classes are still uncertain. Historically, scientists thought the molluscs were closely related to the annelids because of similarities in their early development and the occurrence of a trochophore-like larva in the life cycle of some molluscs. The 1952 discovery of *Neopilina,* the first living representative of the Monoplacophora, at first tended to support this idea. *Neopilina* showed evidence of segmentation and it was believed for a time that the Monoplacophora represented a "missing link," suggesting a common origin of the molluscs and the annelids. Subsequent studies, however, have failed to support this contention. Later, some scientists suggested that molluscs were derived from flatworms based largely on the adult morphology of the two groups. Thus, molluscs might have evolved from creeping, soft-bodied worms whose body differentiated into the classical molluscan head, foot, and visceral mass, and later developed a shell. Sufficient evidence has never been found to support this hypothesis. Recent biochemical evidence simply indicates that the molluscs belong in the large protostome group, the Lophotrochozoa, which includes both the annelids and the flatworms, but that they are not really closely related to either group.

The controversy over molluscan phylogeny arises from conflicting evidence from morphological and biochemical investigations. Some recent studies have concluded that the molluscs are not monophyletic; for instance, it appears that the scaphopods and the bivalves evolved separately from the other six molluscan classes. This provides an excellent illustration of the way science works; data and observations are collected from various sources using different methods to produce evidence sufficient to confirm or refute a conclusion. Every hypothesis or conclusion is subject to further investigation and is always contingent on new evidence that may be found to confirm or refute it. The cladogram on this page is just one interpretation of the possible phylogenetic relationship among the classes of molluscs without considering whether the phylum may be polyphyletic.

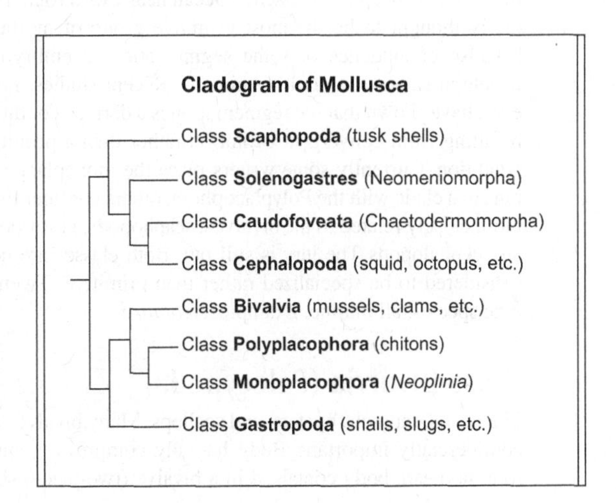

Class Solenogastres

The solenogasters (Neomeniomorpha). A small group of wormlike marine molluscs lacking both a radula and respiratory gills. The foot region is reduced to a narrow furrow, the pedal groove, and the head bears little more than a mouth opening. The body surface is covered by a cuticle imbedded with calcareous scales or spicules. Solenogasters usually live in deep water where they feed upon marine cnidarians. Most of ~250 known species are hermaphroditic; eggs are brooded in a cloacal pouch and may develop either into free-swimming larvae (trochophore stage) or directly into adults. Example: *Neomenia.*

Class Caudofoveata

Chaetodermomorpha. Wormlike animals, structurally similar to solenogasters, but often containing gills and a radula (sometimes reduced). Includes about 70 known species, all marine, that build vertical burrows in bottom sediments. Their food is mainly detritus and marine microorganisms. Sexes are separate, eggs are brooded in a cloacal pouch, and development typically includes a free-swimming trochophore larva. Members of Solenogastres and Caudofoveata were formerly grouped in a single class, the Aplacophora. Example: *Chaetoderma.*

Class Polyplacophora (Amphineura)

The chitons. Body oval shaped with eight dorsal calcareous plates and a large, flat foot. Algal feeders, mainly in the marine intertidal zone, but found at all depths of the ocean down to 4,000 meters. About 600 species. Examples: *Chaetopleura, Chiton* (figure 10.2a), *Cryptochiton.*

Class Monoplacophora

Molluscs with a conical, one-piece shell similar to the shell of a limpet. Known for many years only from fossil forms, until 1952 when the first living monoplacophoran was discovered in sediments deep in the Pacific Ocean near Costa Rica. Formerly thought to be the most primitive group of molluscs because of evidence of some segmentation in embryonic development and in adult specimens. Recent studies, however, have shown that the segmentation is a derived condition resulting from convergent evolution rather than a primitive condition. Currently some writers place the monoplacophorans in a clade with the Polyplacophora; others consider them more closely related to the bivalves, scaphopods, gastropods, and cephalopods The jury is still out. Both classes are now considered to be specialized rather than primitive. Twenty-nine species, all marine. Example: *Neopilina.*

Class Bivalvia (Pelecypoda)

The clams, mussels, oysters, and scallops. Many bivalves are commercially important. Body laterally compressed, small foot, no head, body contained in a bivalve (two-piece) shell hinged on the dorsal side. Most pelecypods are adapted for a sedentary life in marine or freshwater habitats. The mantle is composed of two large tissue folds; it encloses a large mantle cavity and secretes the hinged, bivalve shell. Typically they are filter-feeders and have specialized gills that serve to trap suspended food particles. About 15,000 species. Examples: *Anadonta* and *Unio* (freshwater mussels), *Mercenaria* (hard-shell clam), *Crassostrea* (an oyster), *Mytilus* (marine mussel), *Pecten* and *Aquipecten* (scallops).

Class Scaphopoda

The tooth shells or tusk shells. Body elongate dorsoventrally and encased in a tapered, tubular, one-piece shell open at both ends. The proboscis-like head sports slender tentacles. Burrowing marine molluscs found in sandy and muddy sea bottoms. Scaphopod shells were once used by several groups (or tribes) of Native Americans on the North American Pacific coast as money. About 900 species. Example: *Dentalium* (figure 10.2b).

Class Gastropoda

The snails, slugs, whelks, and limpets. Animals with a long, flat foot; a distinct head with eyes and tentacles; and a dorsal visceral mass usually housed in a spiral shell. Interestingly, several ancient groups of now-extinct gastropods were bilaterally symmetrical with coiling only in a single plane. The asymmetrical growth of the visceral mass and the overlying mantle is responsible for the spiral shape of the shell. Gastropods comprise the largest and most successful class of molluscs. About 15,000 fossil species and 35,000–40,000 extant species are known, making the gastropods the most diverse of all molluscan classes. Gastropoda is the second largest class of animals in number of species, second only to the Insecta. Ecologically, they are the most versatile molluscs with freshwater, marine, and terrestrial species. Some gastropods are carnivores, some are herbivores, and still others are parasites. Nudibranchs (commonly called sea slugs) are an order of colorful marine gastropods without a shell. They are often found on rocky shores, in seagrass beds, and on pilings. They are carnivores that feed on sponges, hydrozoans, and bryozoans. Their bright colors serve as warning coloration to repel predators. Examples: *Helix* (European garden snail), *Limax* (common slug), *Littorina* (periwinkle), *Busycon* (whelk), *Aplysia* (nudibranch), *Physa* and *Lymnea* (freshwater snails).

Class Cephalopoda

The squids, cuttlefish, nautili, and octopi. Cephalopods are the most advanced molluscs and possess a large head with conspicuous eyes; a complex nervous system; complex behavior; a mouth surrounded by 8 or 10 (or more) fleshy arms or tentacles; elongate body; shell often internal, reduced, or absent. They have the most highly organized nervous systems and sense organs among the invertebrate animals, including a complex, image-forming eye. There are three

FIGURE 10.2 Some representative molluscs. (*a*) *Chiton*, Polyplacophora; (*b*) *Dentalium elephantinum*, Scaphapoda; (*c*) *Nautius*, Cephalopoda; (*d*) *Sepia*, Cephalopoda.

(a) © Diane Nelson. (b) Photo from the Gladys Archerd Shell Collection, courtesy of Burton E. Vaughan. (c, d): © image100/PunchStock.

major groups: the nautiloids, the ammonoids, and the coleoids. Cephalopods range in size from 2 cm to nearly 15 meters in length. They are typically active marine animals, preying on various fish, molluscs, arthropods, and worms. About 650–700 species. Examples: *Loligo* (squid), *Octopus, Nautilus* (figure 10.2c) and *Sepia* (cuttlefish, figure 10.2d).

Materials List

Living specimens
- *Physa* or similar freshwater snail
- Representative molluscs (demonstration)

Preserved specimens
- Freshwater mussel
- *Loligo*
  - *Loligo*, liquid mount of dissection
- *Helix*
  - *Helix*, plastic or liquid mount of dissection
- Dextral (right-handed) and sinestral (left-handed) gastropod shells (demonstration)

Continued–

–Continued

- Dried pen of squid (demonstration)
- *Nautilus* shell (demonstration)
- Representative molluscs (demonstration)

Prepared microscope slides
- Mussel shell, cross section (demonstration)
- Glochidium larva (demonstration)
- Radula of gastropod (demonstration)

Chemicals
- Carmine or carbon powder (for demonstration of ciliary action)

Audiovisual materials
- Anatomy of the Freshwater Mussel

Demonstration

Living and preserved representatives of various classes of molluscs

A Freshwater Mussel

Class Bivalvia (Pelecypoda)

Freshwater mussels are sedentary animals that live in or on the bottoms of lakes, ponds, and streams. The particular species available for laboratory study vary from time to time and from place to place, and may include any of several large species in the genera *Anodonta, Lampsilis, Elliptio,* and *Quadrula.* The basic anatomy of the hardshell marine clam *Mercenaria* (formerly called *Venus*) is also very similar to the following description and can be used instead of the freshwater mussel. An illustration of a dissected *Mercentaria* is provided in figure 10.3.

Fact File

Bivalves

- Some oysters change their sex during their lives, first developing as males and later changing into females.
- A 1-acre "bed" of commercial mussels can produce as much as 10,000 pounds of mussel tissue per year. By comparison, 1 acre of pastureland produces only about 200 pounds of beef in a year.
- Many clams can burrow into sand, but some can burrow into hard substrates like wood or stone. Shipworms (*Teredo*), wormlike bivalves with a reduced shell used for boring, often damage wooden

Continued—

—Continued

ships and piers with their burrows. The clam (*Pholas*) can drill holes in soft types of rock.

- Giant clams (*Tridacna* spp.) harbor a type of symbiotic algae in their tissues. These photosynthetic algae produce carbohydrates that serve as nutrients for the clams. This symbiotic relationship supplements the clam's own filter feeding and allows it to survive in marine habitats with sparse plankton where other filter-feeders would starve.
- The zebra mussel (*Dreissena polymorpha*) is an invasive species that probably came to North America as an unintended passenger in the ballast water of a cargo ship. First discovered in the Detroit River in 1988, the species has now spread throughout most of the Great Lakes, the Mississippi, Ohio, and Tennessee rivers, and into many smaller lakes and streams.
- Byssal threads secreted by marine mussels have the tensile strength of steel. These threads are used to anchor the mussel firmly to its substrate.

- Place a mussel in a dissecting pan and note the hard **bivalve shell.**

The two valves of the bivalve shell are joined along the dorsal surfaces by a proteinaceous, elastic **hinge ligament.** The exterior surface of the valves is covered by a dark, horny material, the **periostracum.** Observe the concentric **lines of growth**

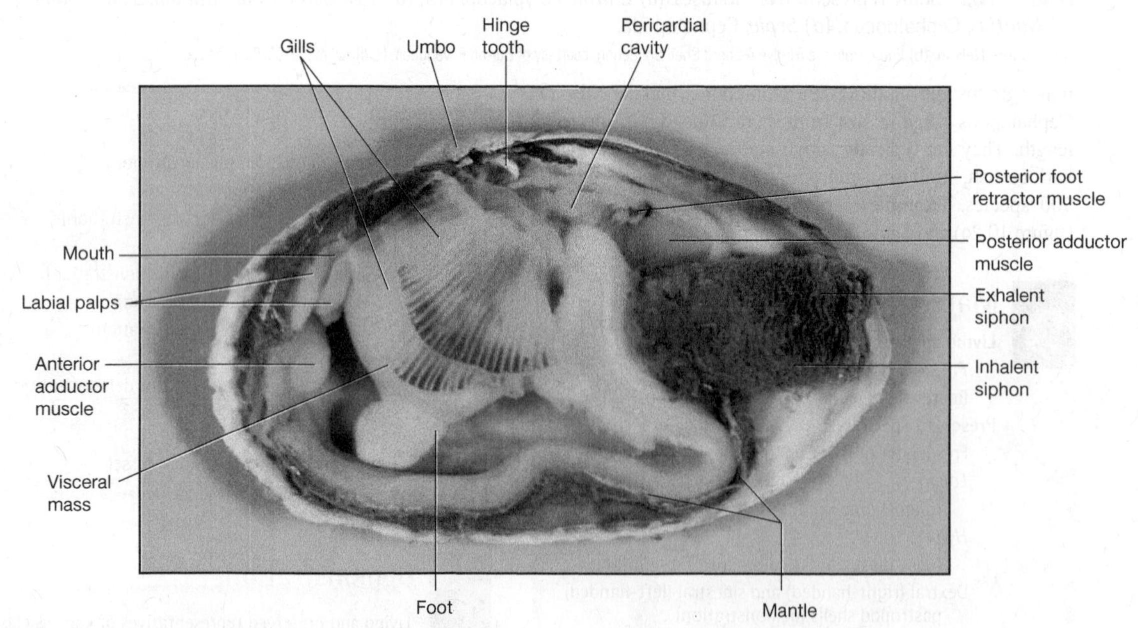

FIGURE 10.3 *Mercenaria,* marine clam, partly dissected.
Photo by John Meyer.

on the exterior surface of the shell; these lines are formed as the mantle secretes new material at the edge of the growing shell.

The shell actually consists of three layers, the exterior periostracum, a middle **prismatic layer,** and an inner **nacreous layer.** The exterior periostracum is made up of a structural protein, conchiolin, which retards dissolution of the shell by the slightly acidic waters in which freshwater mussels are typically found. The middle prismatic layer consists of crystalline calcium carbonate ($CaCO_3$) and provides strength. The inner nacreous layer ("mother-of-pearl") consists of numerous fine layers of $CaCO_3$ and is iridescent. Near the anterior end of each valve is a raised portion, the **umbo,** which represents the oldest part of the valve. At the edge of the shell, between the valves, you should be able to observe two openings between the edges of the mantle (figure 10.4): the (lower) **incurrent siphon** and the (upper) **excurrent siphon.**

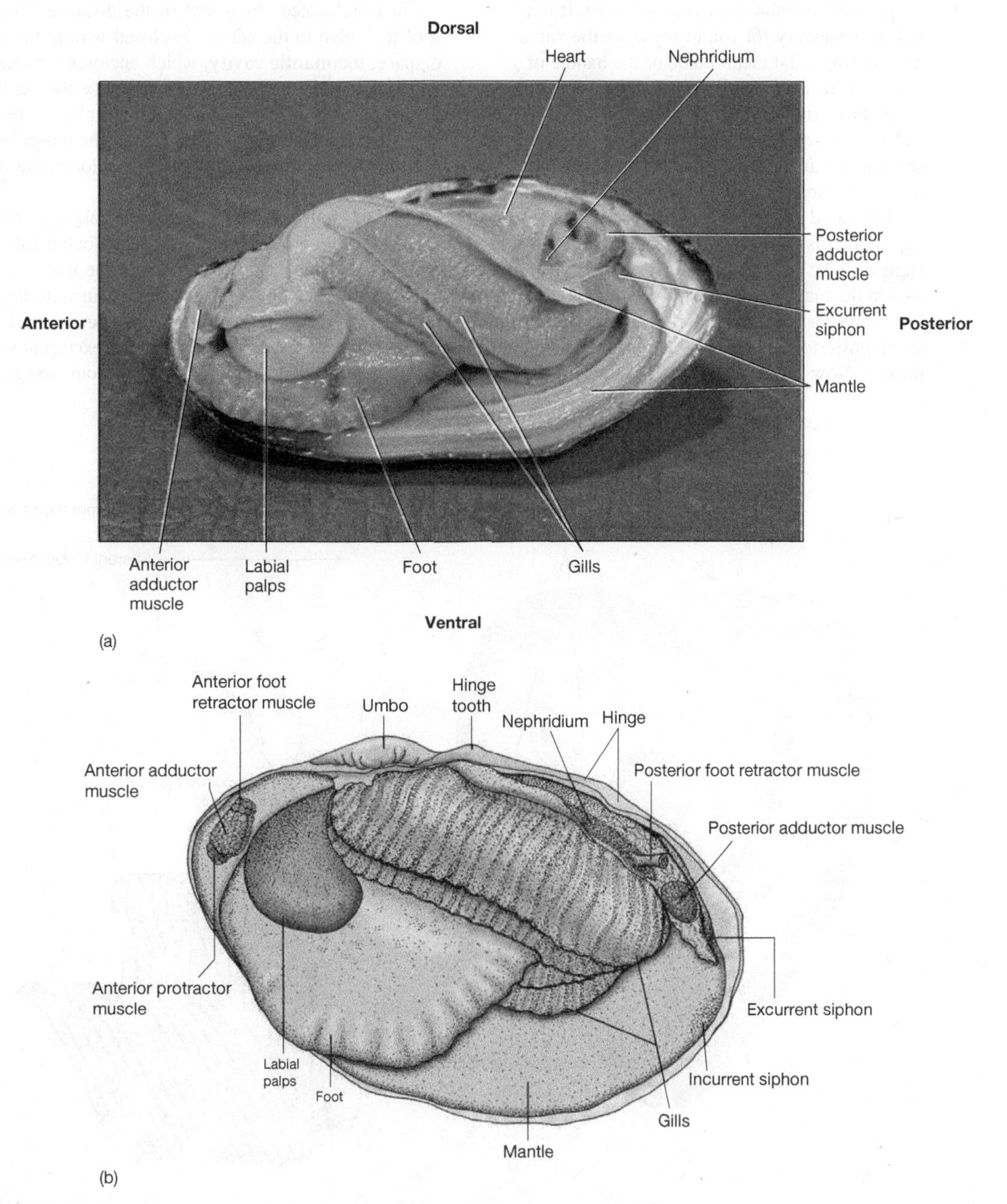

FIGURE 10.4 Freshwater mussel. (*a*) Part of mantle removed, lateral view. (*b*) Diagram.
(a) Courtesy of Carolina Biological Supply Company, Burlington, NC.

In figure 10.4, locate the two large muscles that hold the valves together, the **anterior adductor** and **posterior adductor muscles.** They work antagonistically to the **hinge ligament,** a strong chitinous structure on the dorsal edge of the valves that tends to keep them open. You must cut through the anterior and posterior adductor muscles (carefully) to access the interior organs.

- Check to see if your mussel has a wooden peg inserted to keep the valves spread apart. If not, it will be necessary for you to pry open the valves by inserting a flat metal blade or the handle of your forceps between the valves along the ventral edge. Twist the handle, and when the valves are sufficiently separated, place a wedge between them so that they are separated about 6–12 mm. Carefully insert your scalpel, blade first, into the space between the left valve and the closely adhering mantle in the region just below the anterior adductor muscle (figure 10.4). Keep the blade close against the shell, loosen the mantle from the valve, and cut through the large anterior adductor muscle. Repeat the procedure at the posterior end and cut the posterior adductor muscle. Now carefully lift the loosened left valve, separating the mantle from it as you lift. **Warning:** The heart is located in the pericardial cavity near the dorsal side of the shell; take care to avoid damage to this region.

Feeding, Digestion, and Respiration

The thin layer of tissue that lines each valve is the **mantle.** It attaches to the inside of each valve along the **pallial line,** which is located about half of the distance from the edge of the valve to the center. Enclosed within the mantle is a space, the **mantle cavity,** which encloses the other organs. Along the edge of the mantle, identify the ventral **incurrent siphon,** with **sensory papillae** lining its borders, and the dorsal **excurrent siphon.** Locate the muscular **foot,** the **gills,** the **labial palps,** and the **mouth** located at the base of the palps (figure 10.4).

The large gills play an important role in both respiration and feeding. Each gill consists of a double fold (**lamella**) of tissue (figure 10.5) suspended in the mantle cavity. The lamellae of each gill (left and right) join ventrally to form a **food groove** and dorsally to form an **excurrent chamber,** which carries water posteriorly to the excurrent siphon. The gills contain blood vessels and obtain some oxygen from the

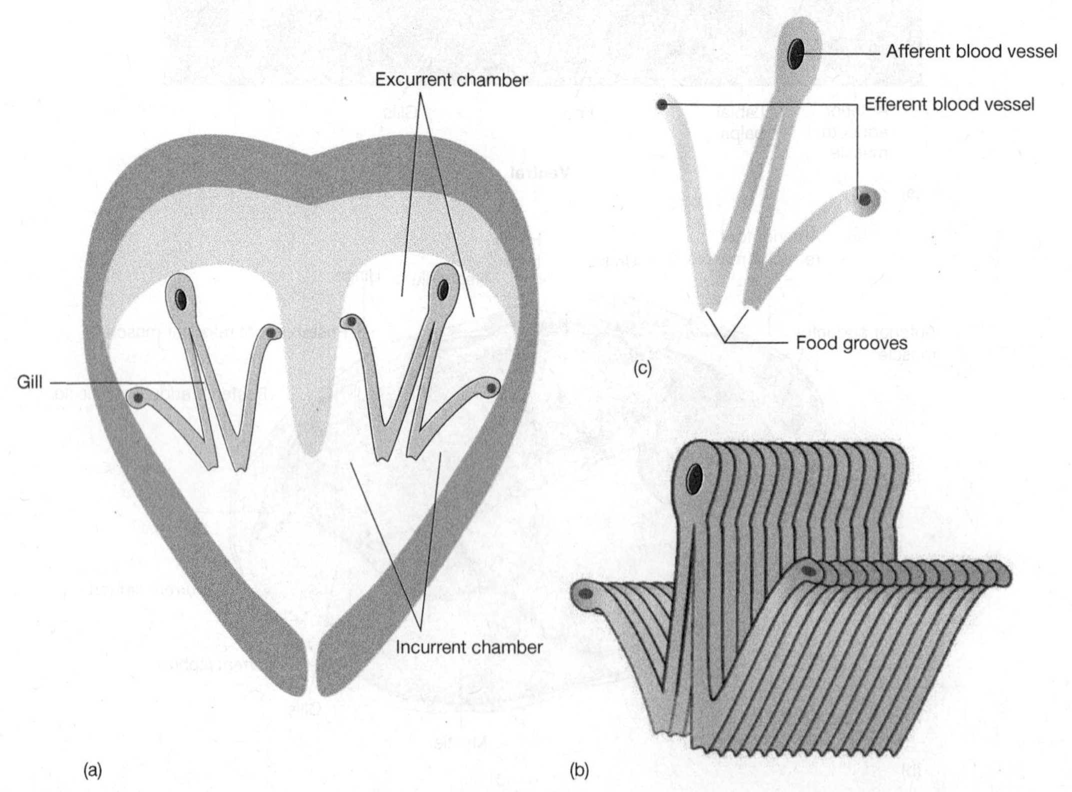

FIGURE 10.5 Freshwater mussel, cross section showing details of gill structure. (*a*) Clam cross section. (*b*) Gill bar showing relationship of gill filaments. (*c*) Single gill filament.

incoming water currents. The mantle is also highly vascularized and serves as another respiratory organ.

Cilia on the surface of the mantle and gills beat in one direction to create water currents in the mantle cavity. This ciliary action draws water into the incurrent siphon. Food particles contained in the incoming water are filtered from the water as it passes through the gills, and are trapped in mucus secreted by glands in the gill tissue. The trapped food particles are collected in the food grooves located on the ventral tips of the gills. The food particles are then transported anteriorly by coordinated ciliary movements to the labial palps and into the mouth (see figure 10.4). Along this route, nonfood particles are sorted out and eliminated.

- If live mussels or clams are available in the laboratory, you can observe the coordinated ciliary movements on the surface of the gills by placing a few particles of carbon powder (or carmine powder) on the moistened surface of a gill. Make certain that the surface of the gills is moist (not wet) before you apply the carbon, and take care to apply only a **few particles** of carbon as a marker. (Too much carbon will coat the whole surface of the gills and obscure action of the cilia.) When working with live mussels or clams, be sure to add water to your specimen frequently to keep the soft internal tissues moist and flexible. The soft tissues dry out rapidly when they are exposed to the air.

If the gills of your mussel appear fat or swollen, they may contain eggs or larvae. This is because the gills of female freshwater clams also serve as a brood pouch, or **marsupium** (see figure 10.7). Reproduction and the life history of the freshwater clam will be considered later in this exercise.

Most parts of the digestive system, including the **esophagus, stomach, digestive gland,** and **intestine,** are located within the foot and visceral mass (figure 10.6). To study the various structures enclosed within the **visceral mass,** you will need to cut along the ventral surface of the foot and bisect it into right and left halves. After you have bisected the foot, you should be able to locate the digestive structures mentioned previously and the yellowish **gonad** tissue that surrounds a portion of the intestine.

Find the mouth, which is behind the anterior adductor muscle. Food particles, collected on the gills and transported to the labial palps, pass through the mouth and esophagus and into the stomach, where the process of digestion begins. The stomach floor is folded into numerous ciliary grooves that aid in sorting food particles from sediment and other nondigestible particles. Such sorting begins when particulate matter is

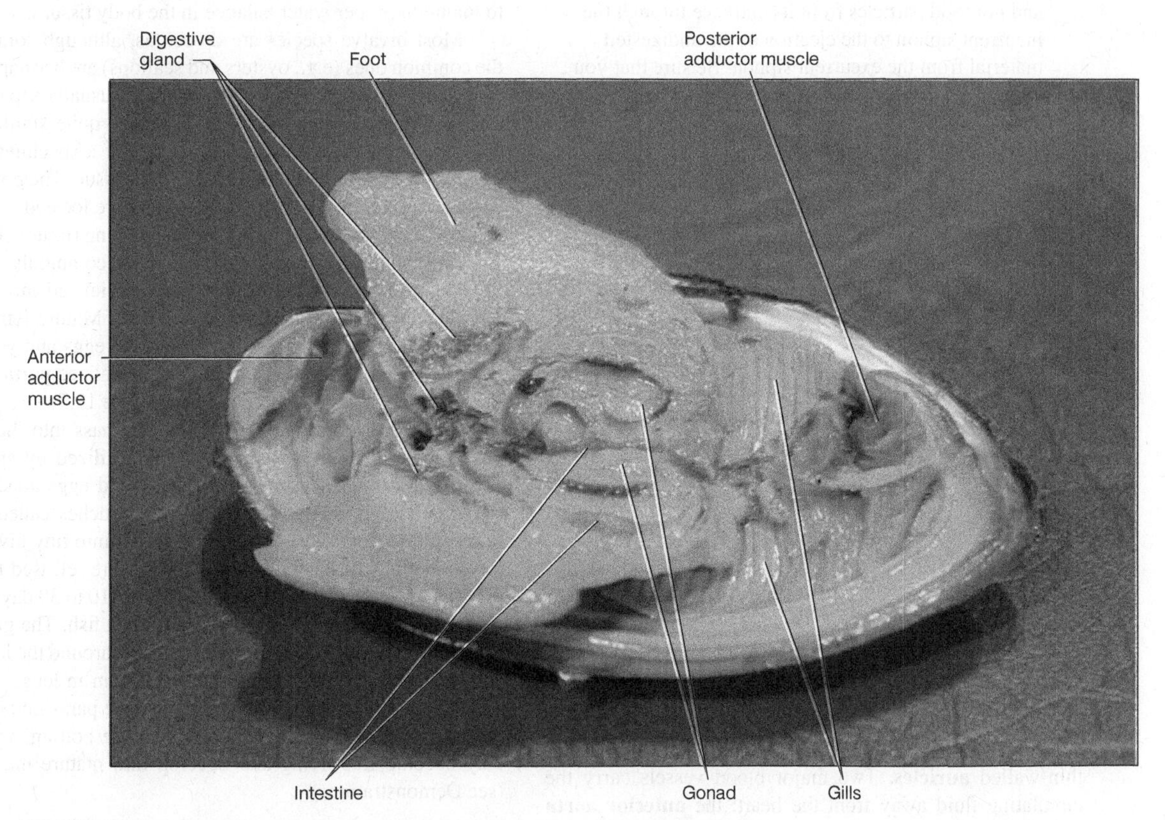

FIGURE 10.6 Freshwater mussel. Foot dissected to show internal structures.
Courtesy of Carolina Biological Supply Company, Burlington, NC.

trapped on the gills and it continues enroute to the stomach. Particles rejected in the stomach are passed into the intestine for elimination. The intestine traverses the entire length of the pericardial cavity (a remnant of the embryonic coelom). The surface area available for absorption of nutrients in the intestine is increased by the presence of a longitudinal infolding, the **typhlosole.**

Digestive enzymes are released into the stomach by the **crystalline style,** a gelatinous rod that extends into the stomach and releases enzymes as it is rotated by the action of certain stomach muscles. Other enzymes are secreted by the digestive gland. Digestion is both extracellular and intracellular. Small food particles are engulfed by phagocytic cells in the digestive gland, and digestion is completed intracellularly. The greenish pulpy tissue surrounding the stomach is the digestive gland. Undigested material passes through the intestine to the **rectum,** exits through the rectal tube or **anus** (which lies above the posterior adductor muscle), and is flushed from the mantle cavity via the excurrent siphon.

- Review the location and function of each part of the digestive system and observe how water currents and the gills play a central role in both digestion and respiration. To make sure you understand the processes of feeding and digestion, trace the path of some food and nonfood particles from its entrance through the incurrent siphon to the ejection of the undigested material from the excurrent siphon. Be sure that you know the role of each structure along this path.

Muscles

You previously located two large and important muscles, the anterior and posterior adductor muscles. Three other muscles facilitate movements of the foot. The **anterior protractor** aids in extending the foot, and the **anterior foot retractor** draws the foot into the shell. These two muscles are located near the mouth and the anterior adductor muscle. Often they are partially fused with the anterior adductor muscle and may be difficult to distinguish from it.

Another muscle, the **posterior foot retractor,** is found near the posterior end of the shell, slightly dorsal and anterior to the posterior adductor muscle. The posterior foot retractor draws the foot toward the posterior of the shell.

Circulation

The heart is located in the **pericardial cavity** found near the dorsal surface (figure 10.4a). The pericardial sinus is enclosed by a thin membrane, the **pericardium.** The pericardial cavity represents the reduced coelom of the mussel. Look within the pericardial cavity and locate the muscular **ventricle** surrounding the intestine. Attached to it are two thin-walled **auricles.** Two major blood vessels carry the circulating fluid away from the heart: the **anterior aorta** supplies the foot and most of the viscera, and the **posterior aorta** supplies the rectum and the mantle. The circulating fluid is called **hemolymph** and has functions similar to the blood and lymph of vertebrate animals. Although technically hemolymph differs from the blood of vertebrates, it is often called "blood" for convenience.

The circulatory system of the mussel is designated an **open system.** This means that the blood is not confined to a definite system of closed vessels, but that in the body tissues blood passes into large spaces or sinuses (figure 10.8). From the visceral organs, blood passes to the nephridia for the removal of metabolic wastes and then to the gills for gas exchange before returning to the heart via the veins. The mantle also serves as a respiratory organ, and oxygen-rich blood from the mantle returns directly to the heart. Dissolved in the nearly colorless blood is **hemocyanin,** a pale blue-green pigment that aids in carrying oxygen.

Excretion, Osmoregulation, and Reproduction

Ventral to the heart and embedded in the mantle tissue is a mass of brownish or greenish tissue, the **nephridia.** The nephridia are the excretory organs, or "kidney," of the mussel. They remove nitrogenous and other wastes from the blood. In freshwater mussels, the nephridia also serve an osmoregulatory function by excreting large amounts of water to maintain proper water balance in the body tissues.

Most bivalve species are dioecious, although some of the common ones (e.g., oysters and scallops) are hermaphroditic. In freshwater mussels, the sexes are usually separate. The male and female gonads are generally quite similar in appearance, so determination of the sex of a specimen requires microscopic examination of gonad tissue. The gonads consist of a yellowish mass of tissue and are located within the foot, surrounding a portion of the intestine (figure 10.6).

Freshwater mussels of the type most commonly used for laboratory dissection have a very specialized mode of reproduction not typical of other molluscs. Mature females can be identified easily by the presence of eggs and young larvae contained in the marsupium, a specialized portion of the gills that serves as a brood pouch (figure 10.7).

The ripe eggs are released, and they pass into the suprabranchial chamber, where they are fertilized by sperm brought in by water currents. The fertilized eggs attach to the gills, which enlarge to form brood pouches called the **marsupia** (figure 10.7). The eggs develop into tiny bivalve larvae, the **glochidia** (figure 10.9), which are released from the mussel to spend a portion of their lives (10 to 30 days) as external parasites on certain species of host fish. The growing glochidia induce the host to form a cyst around the larva. Mature glochidia measure only 0.5 to 5 mm in length, depending on the species of mussel. Later, the parasitic larvae detach from the host fish and move to the bottom, where they become free-living and develop into mature mussels (see Demonstrations).

- If you have a mature female with a marsupium, cut away a portion of the gill and remove a few of the eggs

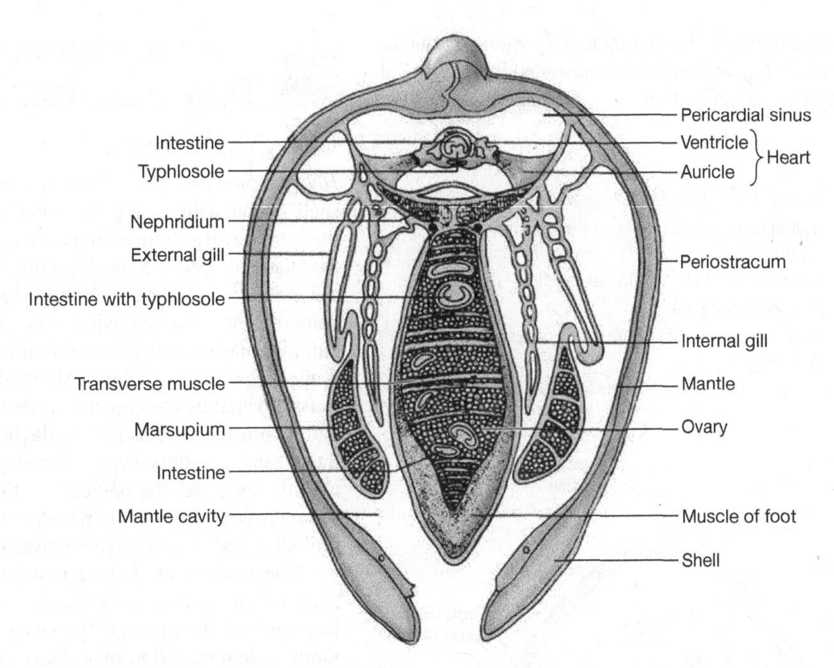

FIGURE 10.7 Freshwater mussel, cross section showing marsupium with eggs.

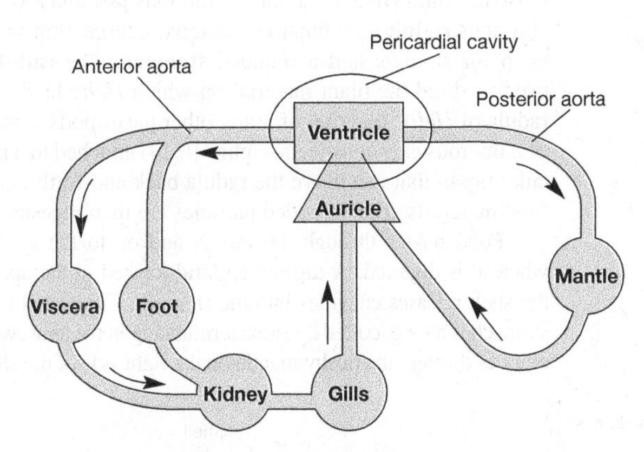

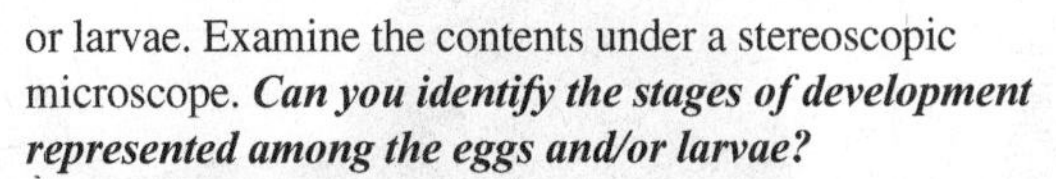
FIGURE 10.8 Pattern of open circulation in a mussel.

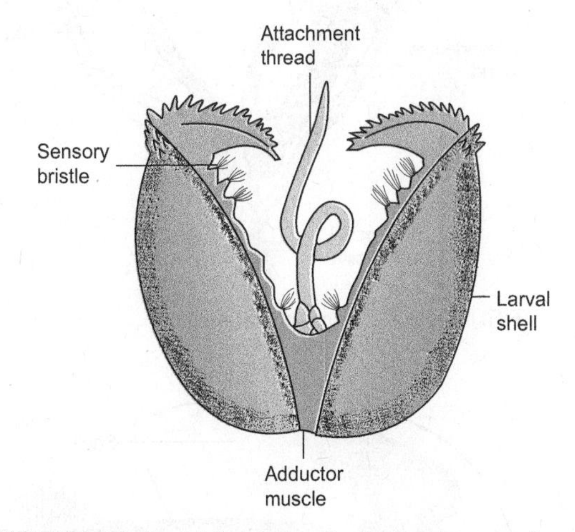

FIGURE 10.9 Glochidium larva of a freshwater mussel.
Courtesy of Carolina Biological Supply Company, Burlington, NC.

or larvae. Examine the contents under a stereoscopic microscope. ***Can you identify the stages of development represented among the eggs and/or larvae?***

Most marine pelecypods, including mussels, clams, and oysters, shed their gametes into the water. Here fertilization takes place, which leads to the formation and development of a free-swimming larva. A **trochophore larva** is typical of many marine molluscs and annelids (figure 10.10). Later in development the trochophore transforms into a **veliger larva** (figure 10.11) with a large, bilobed flap of ciliated tissue, the velum, which enables it to swim; a distinct head with tentacles and eyes; and a larval shell into which the head and velum can be retracted. Veliger larvae are found in the life cycle of many gastropod, pelecypod, and scaphopod molluscs, one indication of their close evolutionary ancestry.

Nervous Coordination

Although the mussels lack a differentiated head, the basic plan of their nervous system is similar to that of other molluscs and consists of three pairs of ganglia and their connecting nerves. A pair of **cerebropleural ganglia** (right and left) is located just posterior to the anterior adductor muscle; the two **visceral ganglia** lie just ventral to the posterior adductor muscle; and the **pedal ganglia** are embedded in the visceral tissue within the muscular foot.

Demonstrations

1. Living bivalves in aquaria to show action of siphons
2. Microscopic section of mussel shell
3. Glochidium (larva)
4. Living or preserved specimens of other common pelecypods

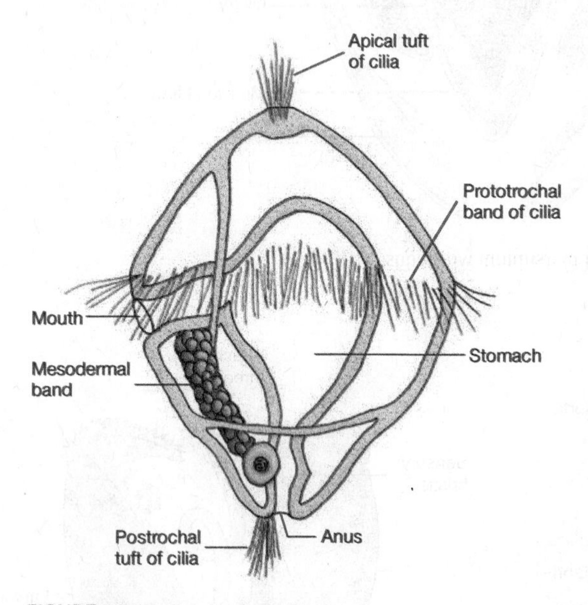

FIGURE 10.10 Trochophore larva.

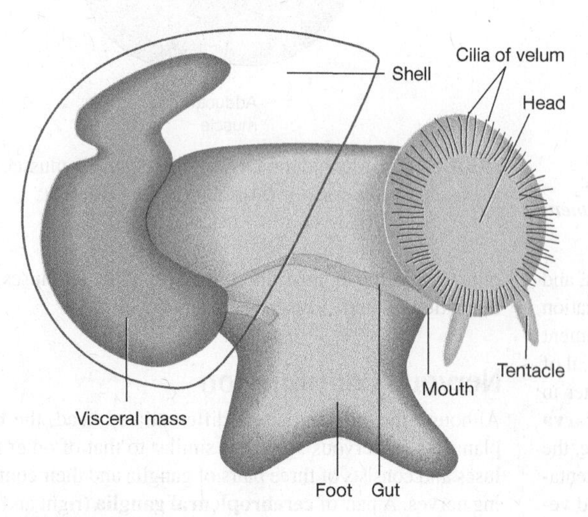

FIGURE 10.11 Drawing of veliger larva.

Helix: The Garden Snail (Demonstration)

Class Gastropoda

Helix (figure 10.12) is a large terrestrial **pulmonate** ("land snail") often called the garden snail or the edible snail. It is the snail (escargot) found in gourmet food shops and served in restaurants. Study a specimen that has been relaxed before preservation and identify its fleshy anterior **head** with two pairs of **tentacles,** one pair of **eyes** at the tip of the second pair of tentacles, and a ventral **mouth.** Posterior to the head is the large, muscular **foot.** Above the foot is the **visceral mass** covered by the coiled calcareous shell. When the animal is threatened or disturbed, the head and foot can be retracted into the dorsal shell. Some snails (not *Helix*) have a chitinous plate, the operculum, that covers the opening when the head and foot are withdrawn into the shell, thereby providing additional protection from predators such as birds.

The anatomy of *Helix* and most other gastropods is difficult to work out because the adult body is coiled and has lost much of the bilateral symmetry of its internal organs. Study a demonstration of a dissected snail to observe the principal internal organs (figure 10.13).

The digestive system consists of the ventral **mouth;** a muscular **pharynx** with a dorsal chitinous **jaw** and a ventral chitinous **radula;** a tubular **esophagus;** a large, thin-walled **crop** for storage; and a rounded **stomach.** The **radula** is used to shred the plant material on which *Helix* feeds. The radula of *Helix,* like that of many other gastropods, consists of numerous chitinous teeth (figure 10.14) attached to a muscular organ that can move the radula back and forth against food materials. The shredded particles are then ingested.

Food passes through the mouth and on to the stomach where it is digested. A **digestive gland** located in the apex of the shell releases enzymes into the stomach. Connected to the stomach is a long, coiled intestine terminating in the anus, which empties through the fleshy mantle on the right side of the shell.

FIGURE 10.12 *Helix,* garden snail, living.
Ingram Publishing/SuperStock.

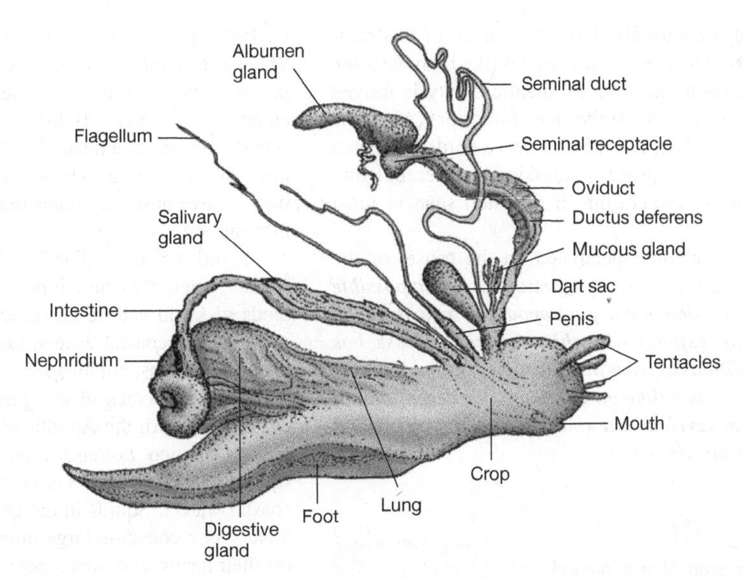

FIGURE 10.13 *Helix,* the garden snail, drawing of internal organs spread out for easier observation.

Respiration in *Helix* is accomplished by a highly vascularized portion of the mantle that serves as a **lung.** The circulatory system consists of a **heart** with one **auricle** and one **ventricle.** Blood is pumped from the ventricle through arteries to the principal organs, and then drains into **blood sinuses** (spaces) from which it flows to the vascularized lung portion of the mantle. A series of pulmonary vessels collect oxygenated blood from the lung and carry it to the heart.

Helix has no real brain, but the nervous system consists of four pairs of **ganglia** connected with each other by nerves and with major organs of the body. The principal sense organs are the two **eyes,** an **olfactory organ,** and two **statocysts,** which serve as balancing organs.

Helix is monoecious. A single, unpaired **gonad,** located near the apex of the shell, produces both eggs and sperm. Mating between two snails results in the exchange of sperm and cross fertilization. Later, each of the snails deposits eggs in a gelatinous mass. Development is direct into miniature snails without larval stages.

Fact File

Gastropods

- During mating, a pulmonate snail (e.g., *Helix* spp.) drives a calcareous dart through the body wall and deep into the internal organs of its mate. The dart apparently stimulates the release of gametes and reproductive hormones.
- Tyrian purple is a pigment that can be extracted from the hypobranchial gland of snails (*Murex* spp.). It was used as a textile dye as early as 1600 B.C. and its

Continued—

—Continued

production had become a large-scale chemical industry by the time of the Phoenicians (~900 B.C.). During the Greek and Roman empires, this "Royal" purple was highly valued as a colorant for garments worn by the rich and famous.

One of the unique features of gastropods is the process of **torsion,** a developmental process during the larval stages that involves a 180-degree rotation of the visceral

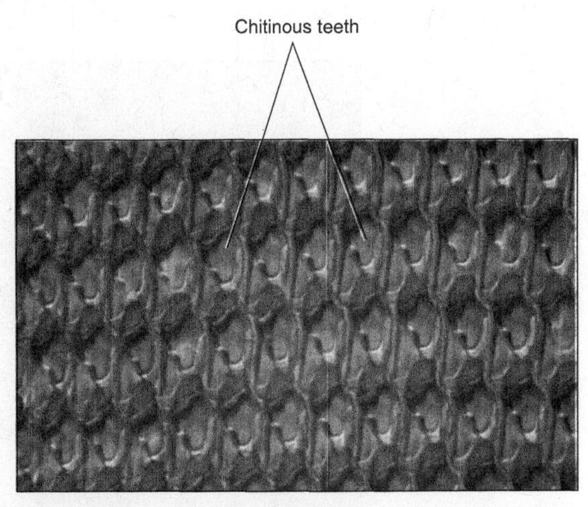

FIGURE 10.14 Radula of snail, stained preparation. Magnification 660×.
Courtesy of Carolina Biological Supply Company, Burlington, NC.

mass on top of the muscular foot. As a result of this rotation, organs originally on the left side of the body become located on the right side and the mantle cavity is moved from the back (posterior) to the front (anterior). Also, the nervous system with its two ventral nerve cords is twisted into a figure eight. The process of torsion is different from, and in addition to, the coiling of the shell seen in most gastropods.

Scientists have long speculated on the cause, or the possible advantages, of torsion in gastropods. ***What possible advantages of torsion for a gastropod can you think of that might help explain why this unusual process has persisted for millions of years in this group of animals? Note that some gastropods (like garden slugs) undergo torsion and then later in development undergo a secondary process of detorsion or untwisting.***

Demonstrations

1. Microscope slide of radula
2. Examples of dextral and sinestral shells
3. Examples of gastropods with reduced, internal, or no shell (slugs, nudibranchs, etc.)
4. Living or preserved specimens of other gastropods

A Squid: *Loligo*

Class Cephalopoda

Squid are large, swift-moving, and highly specialized molluscs (figure 10.15). They exhibit a well-developed nervous system, complex sense organs, and sophisticated patterns of behavior. They are good representatives of the Class Cephalopoda, the most highly evolved class of molluscs. The name *cephalopoda* means "head-foot" and is quite appropriate because the foot, one of the basic anatomical elements of molluscs, is highly modified in this class and actually serves as a head. The "head-foot" contains several large nerve ganglia, which coordinate nerve impulses, as well as several sense organs that provide important sensory information.

Squid are often abundant in the surface waters of the oceans, where they may form large schools. Dozens or hundreds of squid can be captured together in trawl nets cast overboard in coastal waters. Other species of squid inhabit the ocean depths, but all species are marine.

Several species of the genus *Loligo* are found in the coastal waters of the Atlantic and Pacific Oceans and in the Gulf of Mexico. *Loligo pealeii* is a common small Atlantic squid; *Loligo opalescens* is a small squid abundant in Pacific coastal waters. Squids in the genus *Loligo* are typically pelagic. They consume large quantities of fish by first biting off their heads, and then quickly chopping up the rest of the body with their sharp, parrotlike beaks.

External Anatomy

- Examine a preserved squid to identify its major external features. Living squid in an aquarium are interesting to observe for swimming, feeding, and other types of behavior, but are difficult to obtain and keep at inland locations.

On the preserved specimen observe the **head, tentacles,** and two large **eyes.** The eyes are equipped with a pupil, iris, cornea, lens, and retina and are capable of forming clear images; they are remarkably similar to vertebrate eyes in many respects. This is an interesting and important example

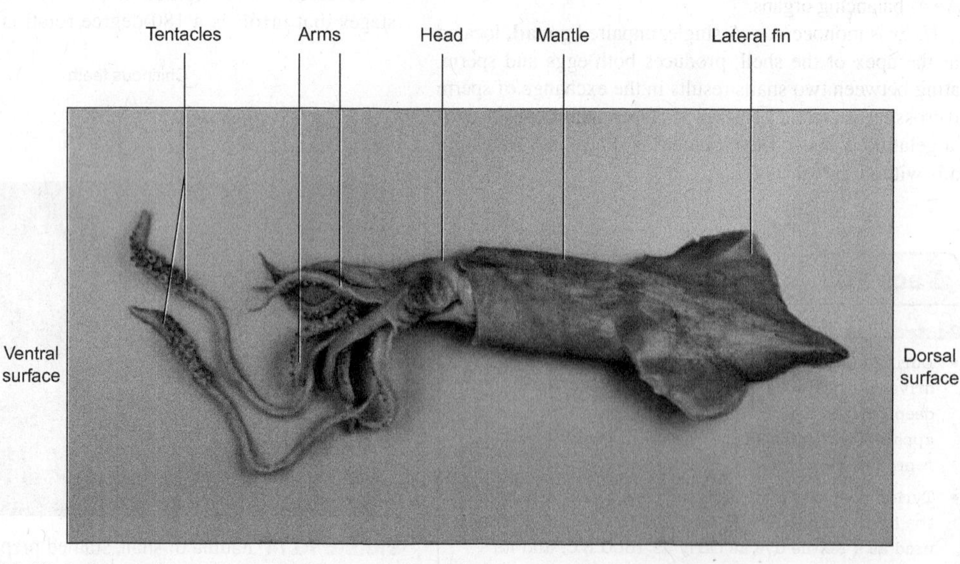

FIGURE 10.15 *Loligo,* squid, external anatomy, anterolateral view.
Photo by John R. Meyer.

of **convergent evolution,** in which two organisms with quite different evolutionary histories (squid and humans) have evolved quite similar structures. Recently, however, researchers discovered the Pax 6 gene, which regulates eye development in animals as diverse as flies, mice, and cephalopods. This suggests that eyes in virtually all animals may share a common genetic origin. Therefore, the eyes of invertebrates and vertebrates may be more homologous—at least genetically—than we formerly believed.

Note that the body of the squid is slender and tapered, which facilitates its swift movement through the water. The orientation of the squid is rather unusual, since morphologically the head and tentacles represent the **ventral surface** of the body, and the pointed end opposite the head represents the **dorsal surface** (figure 10.15). The funnel, or **siphon,** is located on the **posterior surface.** While swimming, a squid actually travels with its dorsal surface forward and its anterior surface up!

Squid swim by a type of jet propulsion in which water is squirted out of the funnel. *Loligo* alternately contracts its antagonistic circular and radial muscles to pump water in and out of the mantle cavity. Since the funnel is under muscular control, the direction of water ejection and thus the direction of movement can be swiftly changed. Movements of the arms and the two lateral fins also aid in steering the animal. The cephalopods are the fastest moving and most mobile of all molluscs.

Locate the four pairs of **arms** on the head of your specimen and one pair of elongated **tentacles.** The tentacles are retractile and play an important role in feeding. Study one arm and observe that it bears two rows of **suckers** (figure 10.16). ***Are the suckers all the same size? Which suckers are largest? Which are smallest?***

Each sucker is made up of a rounded cup attached to the arm by a stalk or pedicel. Remove a sucker from one of the arms and observe it under a stereoscopic microscope. Identify the **chitinous ring** with teeth, which supports the edge of the cup, and the small **piston** at the base of the cup. Note that the suckers on the two tentacles are found only near the distal ends.

A special modification of the suckers on one of the arms of mature males results in a reduction in the size of the suckers and an increase in the length of the pedicle supporting the terminal suckers (figure 10.16). This modified arm is called a **hectocotyl** and serves in the transfer of sperm bundles (**spermatophores**) to the female in mating. Such a modification of an anatomical structure by one sex (males in this case) and not the other is an example of **sexual dimorphism.**

Locate the muscular buccal membrane at the base of the arms and tentacles. In the center of the membrane is the **mouth** opening (figure 10.16). Inside the mouth, find the two chitinous **beaks,** which are used to seize and tear prey. Prey organisms are captured by swift movements of tentacles, which seize the prey, hold it fast with the suckers, and draw it toward the mouth aided by the other arms. Poison glands provide secretions that aid in subduing active prey.

The fleshy mantle surrounds the remainder of the body and encloses the internal organs. The ventral edge of the mantle adjacent to the eyes is called the **collar.** Locate the **funnel** extending from the posterior side of the collar. On the anterior side of the collar, you should find a projection of the mantle that represents the ventral tip of a translucent chitinous structure called the **pen.** In the squid, the skeleton is reduced to the pen plus several cartilages, (which protect the **cephalic ganglia** or "brain" within the head-foot), and other cartilages embedded in the collar region of the mantle and funnel.

Locate also the two large **lateral fins** extending from the mantle. As noted previously, the fins are movable and serve as steering aids during swimming. In the mantle epithelium are numerous **chromatophores,** specialized pigment-containing cells that allow the squid to change the color of its integument. Each chromatophore consists of a small sac of a single pigment (e.g., red, blue, yellow) surrounded by an elastic membrane attached to many small muscles. Contraction of these tiny muscles causes the pigment to expand and add to the general coloration of the squid. A chromatophore whose pigment is not expanded contributes little to the apparent color of the squid.

Since squid have several types of chromatophores, each with a different colored pigment, many shades of color can be achieved. This aids greatly in the camouflage and protective coloration of the squid. Live squid show almost constant changes of color because of the continuous activity of the chromatophores, which are under nervous (and muscular) control.

Internal Anatomy (Demonstration)

Study a demonstration of a dissected specimen to observe the principal organs (figures 10.17, 10.18 and 10.20). Posterior to the head is the **funnel** connected with two large **funnel retractor muscle** (figure 10.17). Dorsal to the funnel, locate the **rectum** and **ink sac.**

The internal organs are enclosed within the visceral mass and are covered with a thin layer of peritoneum tissue that is enclosed by the fleshy mantle. The squid has a complex digestive system with several specialized organs. The mouth is located on the ventral surface of the buccal mass, a round, hard, muscular bulb found in the head at the base of the tentacles. Projecting from the mouth, you may observe the two horny claws of the beak that are used to seize prey (figure 10.16). Food entering the mouth is further shredded by the radula within the buccal cavity inside the buccal mass. It then mixes with enzymes secreted by a pair of salivary glands. Further digestion and most nutrient absorption occur within a large **gastric caecum** that lies adjacent to the **stomach** (figures 10.17 and 10.18) and makes up most of the visceral mass. Digestive enzymes are passed through ducts from a large **liver** and a smaller **pancreas** and enter the **caecum** near its junction with the **stomach.** Waste materials collect in the **intestine** and are eliminated through the **rectum** partially surrounded by liver tissue (figure 10.18), and to the **anus** (located just dorsal to the head).

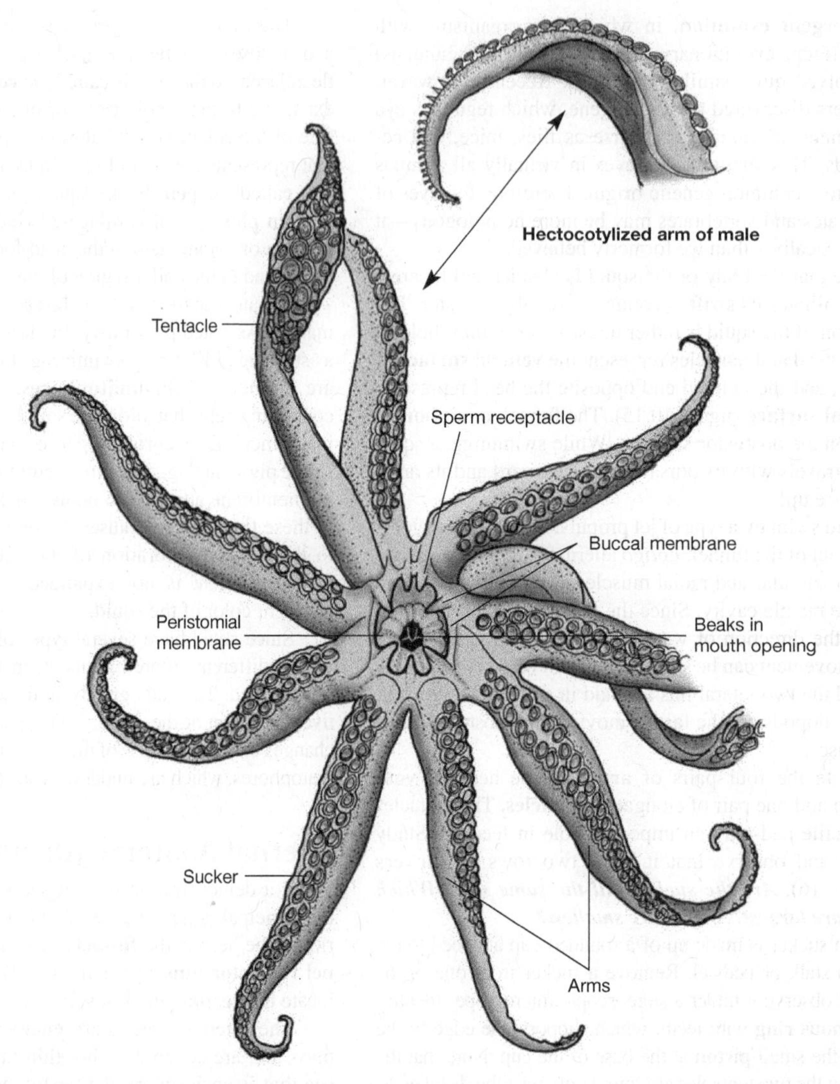

FIGURE 10.16 Squid, ventral view of mouth area, arms and tentacles.

Nervous System

A large brain lies between the eyes; it consists of five pairs of fused ganglia that encircle the esophagus. A giant axon is embedded in the mantle and controls movements of the tail, fins, and also some of the jet propulsion system. These neurons, which reach 1 mm in diameter, are unusually large and have often been used by neurophysiologists to study the biochemistry and biophysics of nerve transmission. Squid and other cephalopods have a highly developed nervous system and are capable of complex behavior. The highly developed cephalopod eyes are a classical example of convergent evolution because they have come to resemble the vertebrate eye, which evolved from very different ancestors.

Circulatory System

Squid and other cephalopods have a **closed circulatory system,** unlike the open circulatory system of other molluscs. Squid actually have **three hearts:** two branchial hearts and one systemic heart (figures 10.18 and 10.19). The branchial hearts pump blood to the left and right gills, which extend ventrally into the mantle cavity. Oxygenated blood from the gills returns to the separate systemic heart located in the center of the body between and just posterior to the branchial hearts. The systemic heart is three-chambered, with one ventricle and two auricles. It supplies blood to the mantle and the visceral organs.

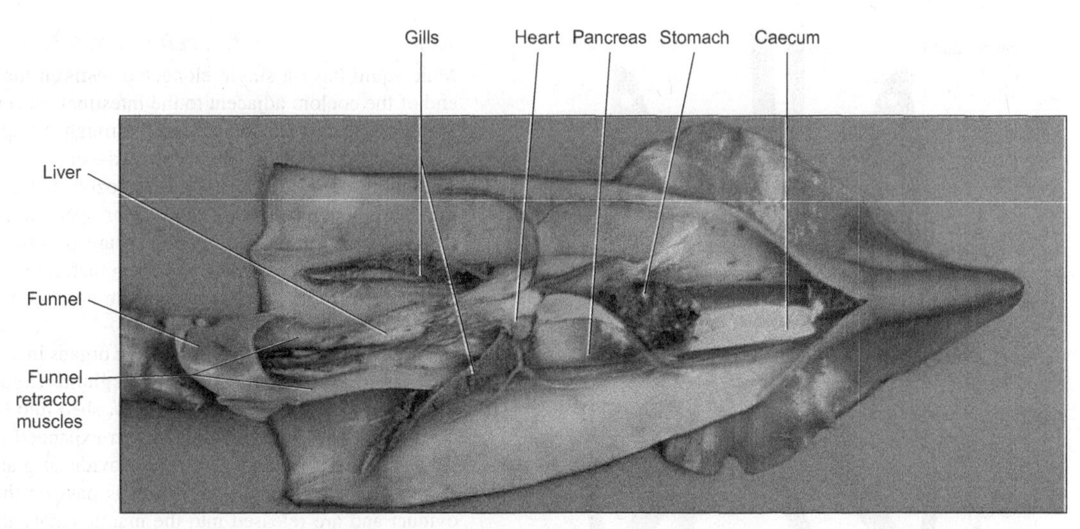

FIGURE 10.17 *Loligo, squid,* double-injected, dissected, showing arterial circulation (red), venous circulation (blue), and other internal organs.
Photo by John R. Meyer.

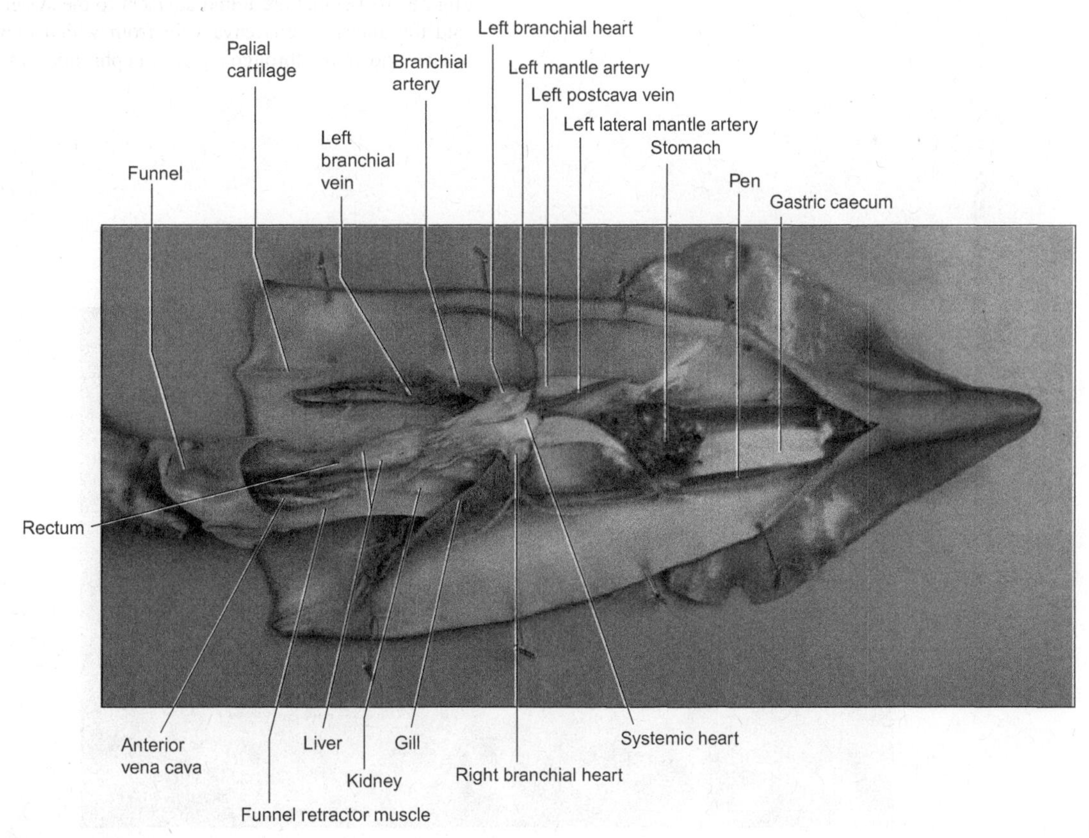

FIGURE 10.18 *Loligo*, squid, internal anatomy, posterior view.
Photo by John R. Meyer.

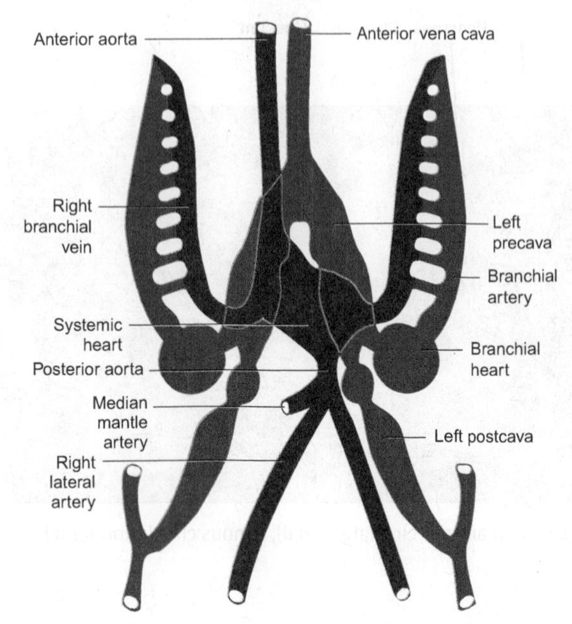

FIGURE 10.19 *Loligo*, squid, drawing showing three hearts and main blood vessels.

Reproductive and Excretory Systems

Male squid have a single elongated **testis** in the posterior end of the coelom adjacent to the intestinal caecum. Sperm are released into the coelom, pass through the sperm duct, and are bundled into **spermatophores**—complex structures containing many individual spermatozoa. Later, the spermatophores are stored in an anterior spermatophore sac. During copulation, spermatophores are picked up by the specialized **hectocotyl arm** and are transferred to the mantle cavity or to the seminal receptacle near the base of the arms of a female.

The most prominent reproductive organs in a mature female are two large, white **nidamental glands** (figure 10.20). Eggs are formed in a posterior **ovary,** shed into the mantle cavity, and pass into the oviduct. An expanded portion of the oviduct is modified to form an oviducal gland, which secretes the egg shell. Later, the eggs pass on through the oviduct and are released into the mantle cavity through an opening near the nidamental glands. These glands secrete gelatinous capsules that surround the eggs after they are released. Fertilized eggs develop within the egg capsule for several days until a ciliated juvenile squid is released into the plankton.

Squid have **paired kidneys,** triangular in shape, (figure 10.18) that are found adjacent to the systemic heart and the anterior vena cava vein from which they collect wastes and empty through a pair of nephridiopores.

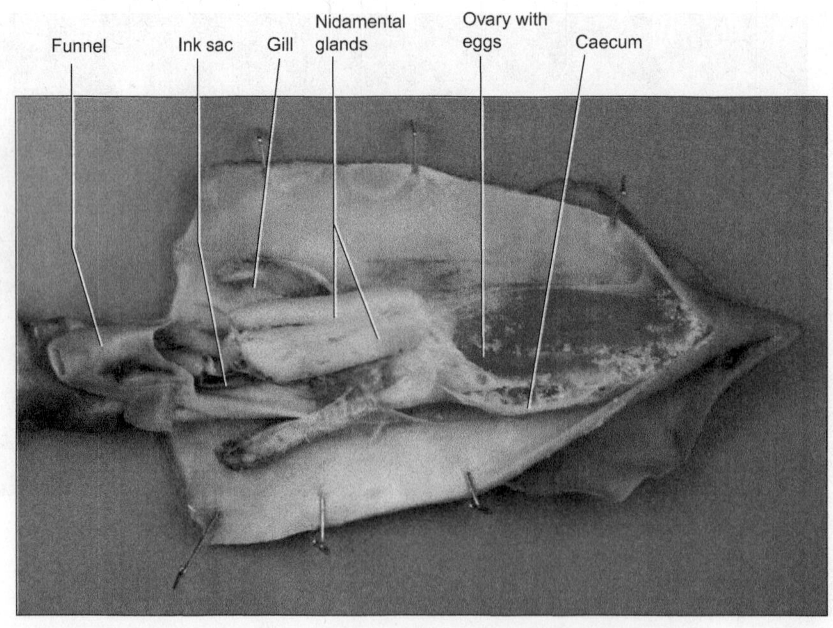

FIGURE 10.20 *Loligo*, squid, internal anatomy, posterior view, female reproductive system.
Photo by John R. Meyer.

Fact File

Cephalopods

- The colossal squid (*Mesonychoteuthis hamiltoni*) is claimed to be the world's largest invertebrate. A single juvenile specimen caught off the coast of New Zealand in 2007 had a total length (including its two long tentacles) of 10 meters (33 feet) and weighed 495 kilograms (1,091 pounds). This species lives deep in the ocean where it feeds on fish, large crustaceans, and other species of squid. Although rarely seen or collected, the discovery of these huge animals could lend credence to some of the legendary tales of giant sea monsters told by ancient mariners.
- The cuttlefish, a squidlike cephalopod (figure 10.2d), produces an internal shell within its mantle. This soft cuttlebone (mostly imported from Taiwan and the Philippines) is sold in pet stores and hung in bird cages to provide a natural source of calcium, magnesium, and other minerals.

Octopus (Demonstration)

Class Cephalopoda

The body of an octopus is globular and saclike with well-developed eyes and a complex brain (figure 10.21). Octopi usually crawl about the sea bottom using their eight flexible arms equipped with numerous suction cups used for locomotion and to grasp prey. When startled, they can also swim by jets of water forced through the funnel from the mantle cavity.

Octopi are carnivores and feed on molluscs by drilling holes in the shells with their radula and injecting a poison to kill their prey. They also have a powerful chitinous beak that is used to shred tissue or attack enemies.

FIGURE 10.21 *Octopus.*
Photographer's Choice/SuperStock.

Demonstrations

1. Dried pen of squid to illustrate internal skeleton
2. Preserved specimen or plastic mount of octopus
3. Shell of chambered nautilus

Collecting and Preserving Molluscs

Chitons

The rocky Pacific coast of North America is home to more than 100 species of chiton. They are much less common along the Atlantic coast where the seashore is more sandy. During the day, chitons hide in crevices and under rocks; at night they emerge and forage on coralline algae and diatoms. A few individuals can be kept successfully in a large, well-illuminated reef aquarium, but their radulae are hard enough to scratch the sides of an acrylic tank.

Chitons curl into a ball when pried from their substrate. For preservation, tie the body flat against a microscope slide or glass plate using nylon thread or fishing line. Store specimens in a solution of 9:1 ethanol (70%) and glycerine. Several changes of solution may be needed until body pigments stop leaching into the alcohol.

Gastropods

Slugs and land snails are most active at night or after a rain. They are commonly found in cool, moist habitats with an ample supply of dead organic matter (humus) covering the soil. In deciduous forests, look under rocks, logs, or the bark of dead trees. In residential areas, paving stones, foundation walls, and compost piles may provide suitable habitats. These animals can be kept in a moist terrarium with a shallow pool of fresh water. Provide vegetable scraps (e.g., lettuce, carrot, potato) sparingly as food and remove uneaten leftovers regularly to prevent growth of molds.

Freshwater snails are commonly found in lakes and ponds where aquatic vegetation provides a plentiful food supply. Keep snails and their host plants in a well-lit aquarium with a few seashells, crushed marble, or limestone as a source of calcium. Maintain water pH between 7 and 8. Below pH 7, calcium carbonate in the shell may begin to dissolve and weaken the shell.

Marine snails can be collected along rocky coastlines, in shallow tide pools, and on coral reefs. Cowries (*Cypraea* spp.), top snails (*Astraea* spp.), and other species that graze on algae can be kept successfully in a well-balanced saltwater aquarium. Predators and filter feeders are very difficult to maintain in captivity. Specimens are usually preserved in 70% ethanol, but it can be very difficult to kill them with the head and foot exposed outside the shell. Adding crystals of menthol or epsom salts to the water to anesthetise the specimen and slowly increasing the concentration of alcohol works for some species.

Whelk egg capsules are often found on beaches and in shallow water. They resemble a stack of poker chips linked together by chitinous strands. Each "chip" contains dozens of tiny whelks that are released into the sea when the tough cover of the egg case splits open. Egg cases can be dried and kept in the laboratory without any special treatment. If they still contain young whelks, they can be preserved and stored in 70% alcohol.

Nudibranchs (sea slugs) are colorful marine gastropods that lack a shell. Most species are nearly impossible to keep in captivity. As preserved specimens, they lose their color and shape. The best way to view them is in their natural habitat, and the best way to remember them is through photographs.

Bivalves

Clams (both marine and freshwater) use their foot to burrow into sand or mud bottoms. They can be collected by carefully digging or raking through the sediment. Oysters cement themselves to rock or other submerged objects and remain sessile throughout life. Marine mussels have byssal glands that secrete proteinaceous fibers (byssal threads) which attach them to hard substrates. Scallops live on top of the substrate and use their foot or "clapping" motions of the shells to move around. All of these animals are filter feeders. They require a specialized diet of phytoplankton and are very difficult to keep in captivity.

Laws regulating the collection of clams, oysters, and scallops vary from state to state. The laws are needed to protect endangered species and to prevent overharvesting and habitat destruction. In some states, a fishing license may be all that is needed to collect specimens. In other states, you will need a scientific collecting permit.

Key Terms

Adaptive radiation evolutionary diversification of a group of animals sharing common ancestry to fill numerous ecological niches.

Chromatophores specialized pigment-containing cells found in the integument of cephalopod molluscs and many crustaceans (Phylum Arthropoda).

Coelom central body cavity lined by mesodermal tissues; develops from spaces arising within the mesoderm during embryonic development. Characteristic of annelids, molluscs, and all higher metazoans, although modern molluscs have a secondarily reduced coelom.

Convergent evolution the evolution of two apparently similar structures within two or more groups of animals having different ancestral lines.

Foot in molluscs, the foot is a large ventral mass of muscular tissue; usually functions in locomotion; sometimes modified for other functions.

Glochidium larva (plural: glochidia) specialized larval form in the life cycle of certain freshwater bivalve molluscs; enclosed in a transparent bivalve shell, these larvae spend part of their life cycle as parasites on the gills of freshwater fish. Later they detach and develop into free-living adult mussels.

Hemocyanin a blue, copper-containing respiratory pigment in the blood of molluscs and certain other invertebrate animals. Carries oxygen to the body cells.

Mantle a thin, outer layer of tissue largely enveloping the bodies of most molluscs. Usually secretes a shell.

Marsupium a modified portion of the gills in females of certain freshwater clams and mussels; serves as a brood pouch for eggs and young larvae.

Pulmonate having lungs (vascularized portion of the mantle) for breathing air. In molluscs, this term describes certain gastropods that are adapted for terrestrial or freshwater habitats.

Radula a filelike, rasping organ found in most molluscs generally used for abrading food from the substrate.

Sexual dimorphism condition in which male and female members exhibit a different form.

Torsion a developmental event during the larval state of gastropods in which the visceral mass rotates 180 degrees counterclockwise in relation to the head and foot.

Trochophore larva the first developmental stage of most molluscs and some annelids. It is bilaterally symmetrical, free-swimming, and encircled with one or more rings of cilia.

Veliger larva typical larval form of many marine molluscs, with a prominent bilobed flap of ciliated tissue used in swimming; the flap can be retracted into the transparent shell. Also with a distinct head and tentacles.

Visceral mass the large, fleshy part of a mollusc's body that contains the digestive and reproductive organs.

Internet Resources

There are many valuable Internet sites with information about zoology. Several sites containing pertinent zoological information for this chapter can be found on the McGraw-Hill Zoology web site at http://www.mhhe.com/zoology. Just click on this text's title.

Questions for Critical Thinking

1. Discuss what is meant when one says that the molluscs provide an excellent example of adaptive radiation. First define *adaptive radiation* and then give specific molluscan examples to support your view.
2. What is the significance of knowing that molluscs are triploblastic coelomate animals? With which group(s) of animals does this fact suggest kinship and common evolutionary descent?
3. The coelom in molluscs is limited to cavities surrounding the heart, gonads, and excretory organs.

Does this indicate that the coelom has been reduced from that of an ancestral form that had a fully developed coelom, or does it mean that this coelom is primitive and never fully evolved? Give reasons for your point of view. Perhaps there is another viewpoint or two that might be considered?

4. Humans are dioecious and also exhibit sexual dimorphism. Discuss the various groups of molluscs and indicate which ones are dioecious and also exhibit sexual dimorphism. Do you think there is anything in common between those groups that are sexually dimorphic?
5. Prior to a few years ago, researchers believed that eyes evolved independently in the animal kingdom perhaps 12 different times. However, the discovery of the Pax 6 gene, which regulates eye development in all animals studied to date, from molluscs to primates, suggests that the eyes of all animals may share a common genetic origin. Find evidence to support the idea that the eyes of animals are not convergent (and therefore analogous) but are homologous and share a common ancestor.
6. Do you think there may be an evolutionary or physiological reason to explain why hemocyanin (a blue copper-containing compound) is the respiratory pigment of molluscs and other invertebrates instead of hemoglobin (an iron-based pigment) common in vertebrates?
7. Discuss the importance of the similarity of ciliated larvae found in many invertebrate marine groups. What does this imply—the common evolutionary origin of these larvae, or simply the fact that cilia are useful in marine environments? Do you believe the cilia borne by these larvae are analogous or homologous?

Suggested Readings

Abbott, R. T., and S. P. Dance. 1986. *Compendium of Seashells,* 3d printing, revised. Melbourne, FL, and Burlington, MA: American Malacologists, Inc.

Janus, Horst. 1989. *The Illustrated Guide to Molluscs.* London: Harold Starke Publishers.

Morton, J. E. 1979. *Molluscs: An Introduction to Their Form and Function.* Amhurst, NY: Prometheus Books.

Pfleger, V. 1998. *Molluscs.* Leicester, UK: Bookmart.

Wilbur, K. M., E. R. Trueman, and M. R. Clarke. 1988. *The Mollusca: Form and Function.* New York: Academic Press.

Notes and Sketches

Notes and Sketches

Chapter 11

Annelida

© Diane R. Nelson.

Objectives

After completing the laboratory work in this chapter, you should be able to perform the following tasks:

1. Briefly outline the characteristics of the Phylum Annelida and cite five significant differences observed in this phylum from the Phylum Platyhelminthes and the Phylum Mollusca.
2. List and briefly characterize each of the four classes of annelids. Cite an example of each class.
3. Identify the principal external features of the polychaete worm *Nereis* and explain the functions of these features.
4. Describe the internal structure of *Nereis* as a representative polychaete and identify the principal structures in a microscopic cross section or drawing.
5. Explain the difference in the coelom found in molluscs and that found in annelids. How do you think this difference has affected the evolution of these two groups?
6. Identify the parts of the digestive system in a dissected earthworm and give the function of each part.
7. Describe the basic pattern of circulation and identify the chief parts of the circulatory system of an earthworm. How does the pattern of circulation in the earthworm differ from that of a clam?
8. Identify the main reproductive organs and discuss reproduction in earthworms.
9. Describe the external anatomy of a leech and explain how it differs from that of an earthworm.

Introduction

The Phylum Annelida includes more than 15,000 species of segmented worms—animals whose cylindrical, elongate bodies are composed of a longitudinal series of segments (also called somites or metameres) (figure 11.1).

FIGURE 11.1 *Nereis*, clamworm.
Courtesy of Carolina Biological Supply Company, Burlington, NC.

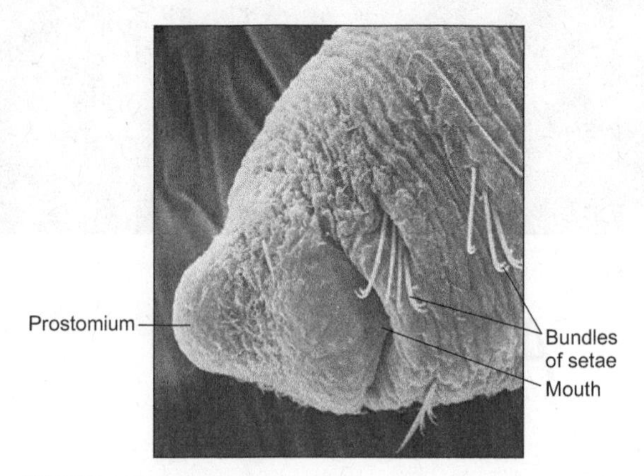

FIGURE 11.2 Mouth region of an aquatic oligochaete showing characteristic arrangement and structure of setae. Magnification 450×.
Scanning electron micrograph by Thomas Bouillon.

Annelids represent a classical example of metamerism, the serial repetition of body parts (see the discussion of metamerism in Chapter 5, page 73). For many years metamerism was considered good evidence that annelids and arthropods were closely related. Recent biochemical and other evidence, however, has refuted that idea and annelids are now thought to be more closely related to molluscs, brachiopods, and other phyla included in the Lophotrochozoa.

In addition to their conspicuous segmentation, the annelids also exhibit several other important structural features lacking in flatworms and roundworms, including a spacious coelom, a closed circulatory system, an efficient excretory system, a highly developed muscular system, a well-developed central nervous system, and an increased concentration of nerve centers (ganglia) and sensory organs at the anterior end of the animal (cephalization).

Annelids are the first animals we shall study that exhibit a well-developed coelom. Although molluscs are also coelomate animals, their coelom is reduced to cavities in which the heart (pericardial cavity), nephridia, and gonads are located. In contrast, annelids have a large coelomic cavity in which most of the internal organs are suspended. Annelid worms illustrate well the basic body organization of a coelomate animal and the relationship of the coelomic cavity, its peritoneal lining of mesodermal origin, and the internal organs.

Classification

During much of the 20th century, the phylum Annelida was divided into three classes: polychaetes, oligochaetes, and leeches. These classes were distinguished from one another by the pattern of **segmentation,** the type and distribution of setae, and the presence or absence of a clitellum. **Setae** (singular, seta) are thin, chitinous bristles or rods secreted by certain cells in the body wall (figure 11.2). Setae are used in locomotion and, for many annelids, also aid in feeding. The number, morphology, and distribution of setae is often important in the classification of annelids. The **clitellum** is a glandular swelling produced by certain segments. It may be permanent or formed only seasonally. It secretes a cocoon into which eggs are deposited for fertilization and later incubation.

Recently, this traditional view of annelid classification has been overturned by biochemical and genetic evidence that Echiura (a small group of marine worms known as spoon worms and formerly regarded as a separate phylum) are closely related to certain polychaetes. Similar research also suggests that leeches and oligochaetes are members of a single clade. In accord with these findings, the phylum Annelida is now thought to encompass two clades: one contains polychaetes and spoon worms, whereas the other includes oligochaetes and leeches, as illustrated in the following cladogram.

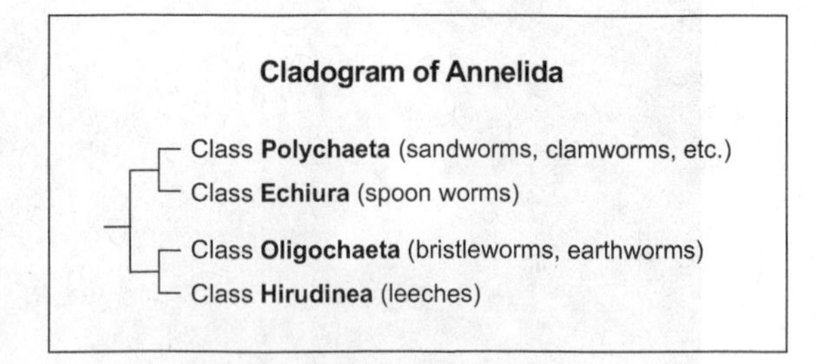

Phylogeny of the Annelida

The annelids have long been considered relatives of the arthropods because of their conspicuous metamerism and several embryological similarities. (See Chapter 5, page 73 for a discussion of metamerism.) Recently, biochemical studies have stimulated a revaluation of annelid affinities and many scientists now believe that the annelids are most closely related to the molluscs, brachiopods, and other phyla included in the Lophotrochozoa. In addition, the phylum has been expanded by the inclusion of two groups of marine worms formerly considered to be separate phyla: the deep-sea pogonophorans and the echiurans.

Inclusion of these two groups in the Annelida and its grouping within the Lophotrochozoa is supported by morphological, embryological, and genetic evidence, although some taxonomists still consider the echiurans to be a separate phyum. Another group of marine worms, the sipunculans, show numerous similarities to the annelids but are still considered by most scientists to represent a separate but related phylum. Classification of the annelids has attracted the attention of many scientists during the past few years and has undergone numerous changes. There are still some disagreements among the leading authorities on several issues concerning this group, so perhaps we could say that the classification of annelids is still a can of worms. Hopefully, with the accumulation of additional information on these interesting animals, we will gain a better understanding of the evolution and phylogenetic relations of the annelids.

Class Polychaeta (Polychaete Worms)

Polychaete worms typically show obvious segmentation, many lateral appendages (parapodia) with numerous setae, and a well-developed head region with appendages and sense organs. They lack a clitellum and permanent gonads. They are usually dioecious and most species are marine. Marine species often have a trochophore larva. This class now includes the archiannelids and the pogonophorans. More than 10,000 species. Examples: *Glycera* (bloodworm), *Bispira* (a feather duster worm), *Hermodice* (bearded fireworm), and *Riftia* (a giant tube worm found in deep sea trenches; see figure 11.3. We will study *Nereis,* the clam worm, more thoroughly in this unit as a representative polychaete worm.

Class Oligochaeta (Bristleworms)

This class includes the earthworms and many species of related freshwater worms that have many segments but few setae per segment, form a clitellum, but lack parapodia and a differentiated head region. They are generally monoecious, with permanent gonads and direct development. Oligochaetes are mostly found in the soil and in freshwater habitats. About 3,000 species. Examples: *Lumbricus* and *Allolobophora* (earthworms); *Eisenia* (dung worm); *Tubifex* (sewage worm, figure 11.4b); *Aeolosoma* (figure 11.4a); and *Chaetogaster.* The last three genera are aquatic and have many common freshwater representatives. We will study *Lumbricus,* an earthworm, as a representative oligochaete.

Class Hirudinea (Leeches)

This group includes the leeches, a relatively small class (about 500 species) of specialized annelids found mainly in freshwater habitats. A few species are also terrestrial. Leeches have dorsoventrally flattened bodies, anterior and posterior suckers, thick muscular bodies, no head or tentacles at their anterior ends, and lack parapodia and setae. The number of segments is limited (usually 34), and a seasonal clitellum is formed during the reproductive season. Fertilization is internal. Examples: *Hirudo, Glossiphonia, Haemadipsa.*

Class Echiura (Spoon Worms)

This class includes a small but widely distributed group of benthic marine worms previously considered as a separate phylum. They are found burrowing in sand or mud sediments and sometimes in gastropod shells or in crevices among rocks. Although adult echiurans lack metamerism, there is evidence of segmentation in the larval stages. Most species have an elongate scoop or spoon-shaped proboscis used for collecting food from the surrounding sediments; thus the common name of spoon worms for this group. About 150 species. Example: *Urechis* (the innkeeper worm). Although there are few species of echiurans, they are an important group of marine animals and are sometimes very abundant locally in shallow coastal waters. One arctic species has been used for human food.

Materials List

Living specimens
- Earthworm

Preserved specimens
- *Nereis*
- Leech
- Representative polychaetes (demonstration)
- Polychaete tubes (demonstration)
- Earthworm cocoons (demonstration)
- Representative oligochaetes (demonstration)
- Representative leeches (demonstration)
- Leech, dissected to show internal anatomy (demonstration)

Prepared microscope slides
- Earthworm, cross section
- Trochophore larva (demonstration)

Miscellaneous supplies
- Dissecting pans
- Pins

Audiovisual materials
- Chart of earthworm anatomy
- Anatomy of the Earthworm video

FIGURE 11.3 Some representative polychaetes. (*a*) *Glycera* bloodworm, a predatory worm common along the Atlantic coast of the United States and Canada often used as fish bait. (*b*) *Bispira*, a common and widespread feather duster worm. Note the parchment tubes from which the worms spread their feathery crowns. These are ciliated and serve to generate water currents, and to capture and sort particles from the seawater. Large particles are rejected, medium-sized particles are stored and used in the building of the parchment tubes, and small particles, including bacteria, other tiny microorganisms, and fine organic detritus, are moved to the mouth. The crown is also important in respiration. (*c*) *Hermodice*, bearded fireworm, on coral. Common on coral reefs in the tropical western Atlantic and among rocks and seagrass beds in shallow water up to 60 meters deep. Numerous setae along their sides are hollow and contain stinging venom, thus the name *fireworm.* (*d*) *Riftia*, the giant tube worm, lives in deep hydrothermal vent areas in the Pacific Ocean. Adult worms lack a functional digestive tract and get nutrition from symbiotic bacteria that use hydrogen sulfide for energy and synthesize organic compounds that are shared with the worms.
(a) NOAA Central Library. (b, c) © Diane R. Nelson. (d) OAR/National Undersea Research Program (NURP)/College of William & Mary/NOAA.

A Marine Annelid: *Nereis virens*

Class Polychaeta

The sandworm or clamworm, *Nereis virens* (figure 11.1), is a large marine annelid commonly found in many areas along the Atlantic coast of North America. *Nereis* usually lives in burrows in the sand or in mud bottoms of the shallow coastal waters during the day and emerges at night to search for food.

External Anatomy

- Select a preserved specimen of *Nereis* and observe the numerous body segments. Internally, the segments are separated by thin sheets of tissue called **septa.** Septa have some muscles, allowing for controlled body movement along the length of the coelom. Both septa and their mesenteries may be partially open, allowing for communication and material transfer between the coelomic compartments.

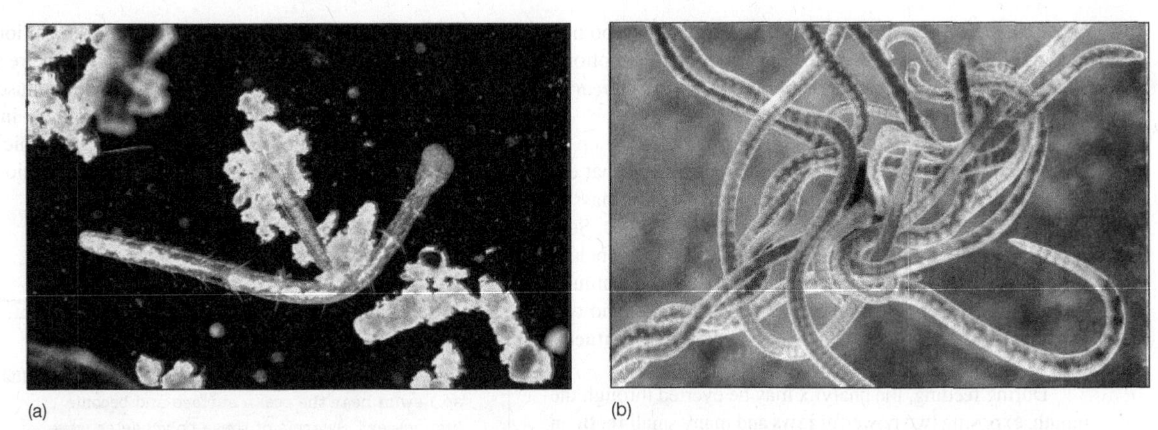

FIGURE 11.4 Some representative oligochaetes. (*a*) *Aeolosoma*, a small aquatic oligochaete, is common among algae and other vegetation on rocks and among debris in ponds and streams. The worms are suction feeders, moving along the substrate with cilia and creating a partial vacuum by arching their flattened head to suck up bacteria and small organic debris. Several species have colored pigment spots on their body surface. (*b*) *Tubifex*, sludge worms. These worms live in the bottom sediments of lakes and rivers worldwide and obtain nutrition by ingesting sediments, digesting bacteria and organic debris, and absorbing nutrients through their body wall. They are able to survive low oxygen conditions and polluted waters and often form thick clusters in nutrient-rich habitats.
(a) © Melba Photo Agency/PunchStock. (b) Photo by John R. Meyer.

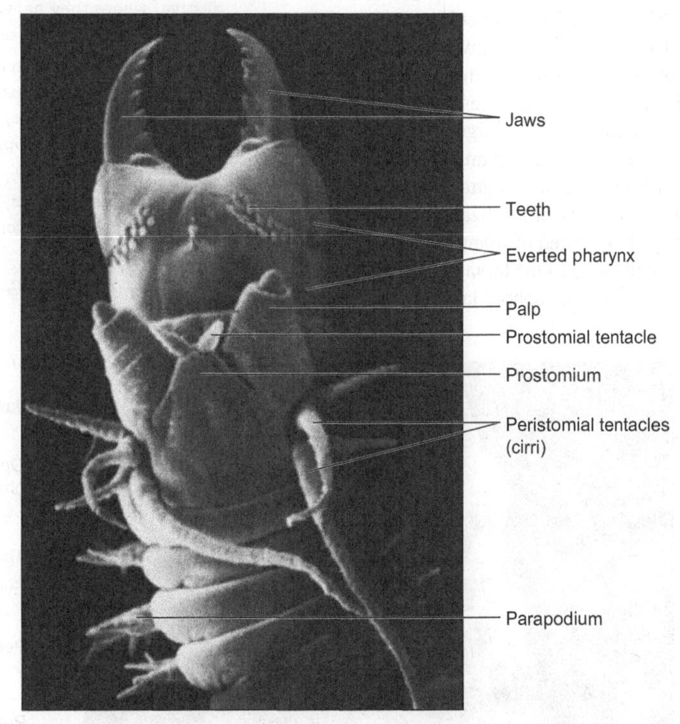

FIGURE 11.5 *Nereis*, head region, dorsal view.
Scanning electron micrograph by Betsy Brown.

The internal structure of *Nereis* is generally similar to that of the earthworm, which you will study later.

All the segments of *Nereis* except the first and the last bear a pair of lateral **parapodia** (figure 11.5). Observe the demonstration slide of a cross section of *Nereis* to study the structure of these appendages. Note that the parapodia are **biramous** (two-branched), with a dorsal **notopodium** and a ventral **neuropodium.** Each main branch bears bundles of chitinous bristles, or setae and also has several smaller side branches. The parapodia play an important role both in locomotion (swimming and creeping along the sea bottom) and in respiration. ***What aspects of their structure appear to be advantageous for these functions?***

Examine the well-developed head region made up of the **peristomium,** the first complete body segment that encircles the mouth, and the **prostomium,** a triangular mass of tissue located on top of the peristomium (figure 11.5). Several sensory structures are found on the prostomium, including two fleshy **palps** attached to the sides of the prostomium, two pairs of pigmented **eyes** on the dorsal surface, and two **prostomial tentacles.** Four pairs of **peristomial tentacles** (or cirri) are also found attached to the peristomium.

During feeding, the pharynx may be everted through the mouth, exposing two powerful **jaws** and many small **teeth** on the wall of the pharynx (figures 11.5 and 11.6). The jaws aid in seizing prey, and the pharyngeal teeth help hold the prey as the everted pharynx is withdrawn back into the mouth.

The sexes are separate in *Nereis* as in most polychaete worms, although there are no permanent gonads. The gametes (eggs and sperm) are formed in the body wall at certain times, pass into the coelom, and then are released into the sea. There the eggs may be fertilized and develop into a **trochophore larva** similar to the larvae of some molluscs (Chapter 10, page 161; figure 10.10). Other larval stages follow before metamorphosis into the adult worm.

Many species of polychaete worms show marked structural, physiological, and behavioral changes during the breeding season. Some species aggregate in large swarms and rise to the surface before releasing their gametes into the sea. This special behavior represents an important adaptation that serves to increase the probability of fertilization.

The trochophore larva formed by many marine polychaetes bears a close resemblance to the larvae of many marine molluscs. This similarity of trochophore larvae is one of the important pieces of evidence suggesting a close evolutionary relationship between these two phyla. Similarly, the resemblance of the trochophore larvae of annelids and molluscs to the marine larvae of several other invertebrate phyla, in addition to other similarities in their patterns of embryonic and larval development, suggests other important evolutionary relationships with groups such as the flatworms.

Fact File

Polychaetes

- During their spawning cycle, fireworms (*Odontosyllis* sp.) swim near the ocean surface and become luminescent. Swarms of these polychaetes were probably the mysterious lights that Christopher Columbus reported seeing on the night of October 11, 1492, as his ship approached the outer reefs of the Bahama Islands.
- Palolo worms (*Eunice* sp.) reproduce only once each year. In waters of the South Pacific, the annual spawn occurs exactly one week after the full moon in late October or early November. On some islands, the natives commemorate this event with a special festival where they harvest the worms and regard them as a gastronomic delicacy!
- *Aphrodita*, a polychaete worm whose body is covered with numerous fine setae giving it a superficial resemblance to a mouse, bears one type of setae that has unusual optical properties and is being used in research on fiber optics.
- Some polychaete worms, called feather-duster worms, are among the most colorful inhabitants of coral reefs.

Continued—

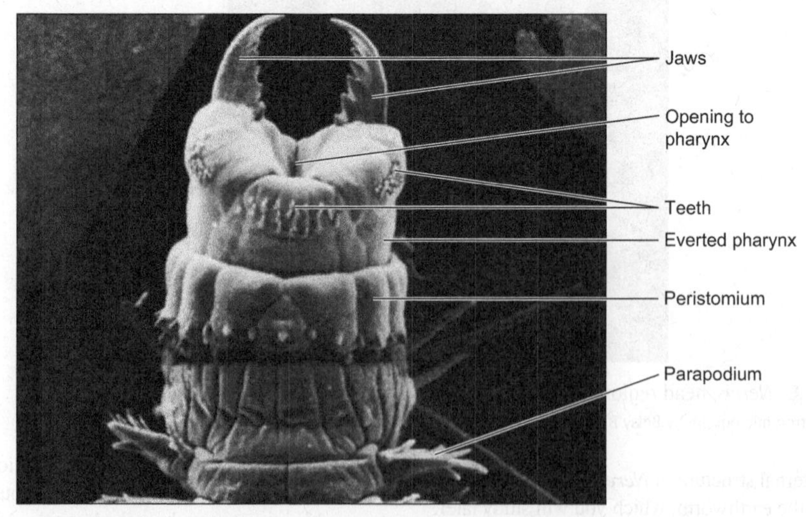

FIGURE 11.6 *Nereis*, head region, ventral view.
Scanning electron micrograph by Betsy Brown.

—*Continued*

- Some highly specialized polychaetes, called pogonophorans, or beard worms, live among the hydrothermal vents in deep-sea trenches thousands of meters below the surface where there is no light, no oxygen, high concentrations of H_2S, and water temperature is 2°C, except in the immediate vicinity of the vents whose discharge may reach temperatures above 400°C. These highly specialized worms lack a digestive tract and obtain their nutrition by absorbing nutrients directly from the seawater enriched by nutrients issued from the vents and from chemosynthetic bacteria embedded in the tissues of the worms.
- Lug worms (*Arenicola* spp.) live by burrowing through marine sediments and consuming the organic matter that passes through their digestive tract. In areas of the littoral zone where these worms are abundant, they may ingest as much as 1,900 tons of sand per acre each year. Over several years, the worms can redistribute virtually all of the sand in the littoral zone to a depth of about 60 cm.

Cross Section

Obtain a microscope slide of *Nereis* with a stained cross section of the midsection to help you understand the internal organization of a polychaete. First, locate the body wall with its external **cuticle** covering the surface of the body (figure 11.7). Beneath the cuticle is a layer of **hypodermal cells** that secrete the cuticle. Identify the **circular** and **longitudinal muscle** layers inside the hypodermis. These muscle layers enable the peristaltic movements of the body that assist the worm in creeping along the substrate. Between these muscle layers, find the thin **peritoneum** that lines the large central cavity, the **coelom.**

- ***What are the functions of the coelom? Why is it important to the evolution of higher animals?*** Review the discussion of the coelom and its functions in Chapter 5, page 72; figure 5.4.

Locate the large circular **intestine** in the middle of the coelom. Observe the large **gastrodermal cells** lining the inside of the intestine. These cells absorb nutrients from the food digested in the lumen of the intestine. Also identify the circular and longitudinal muscle layers and the outer thin peritoneum in the intestinal wall.

You should also be able to identify the **dorsal** and **ventral blood vessels** and the **ventral nerve cord** in your cross section. Depending on where your cross section was cut, you may be able to find some portions of the parapodia and fibers of the **oblique muscles** connecting the midventral line with the dorsal margin of the parapodia. These muscles aid in locomotion as well as other movements of the body. If your section was cut through the parapodia, you may also be able to identify the **setal sac** in which some of the parapodial setae are anchored and, adjacent to the setal sac, some **nephridial tissue.** The excretory organs of the polychaetes consist of many pairs of **nephridia** located in most body segments.

Later in this exercise you will have an opportunity to compare the internal structures of this polychaete to that of an oligochaete when you study cross sections of an earthworm.

Demonstrations

1. Representative genera of Polychaeta, such as *Amphitrite, Aphrodite, Arenicola, Chaetopterus, Hydroides, Sabella, Diopatra* and *Polygordius*
2. Examples of different kinds of tubes formed by polychaetes
3. Microscope slide of trochophore larva

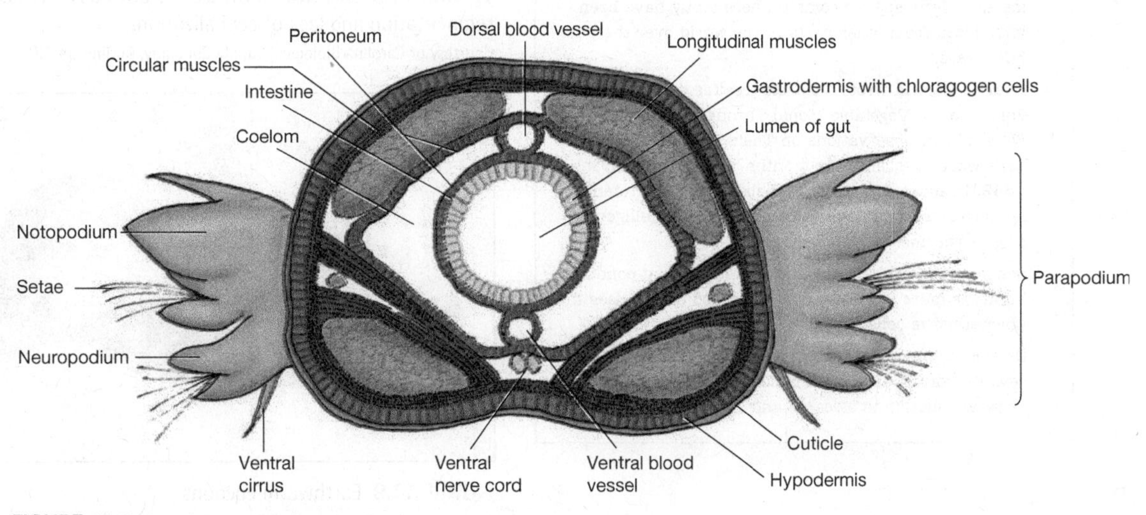

FIGURE 11.7 *Nereis,* cross section.

The Earthworm

Class Oligochaeta

Earthworms are terrestrial annelids that typically live in tunnels they burrow in the soil. Internally, they demonstrate many features typical of the Annelida, and they are the most commonly studied representatives of the phylum. The external features of earthworms, however, exhibit several modifications probably related to their burrowing habits and underground life. Most apparent of these external adaptations is the absence of a distinct head and the lack of parapodia (figure 11.8).

The traditional earthworm for classroom dissection is *Lumbricus terrestris,* upon which this and most other textbook descriptions are based. *Lumbricus terrestris* is a species common in Europe. It is known as the night crawler in northeastern United States and in eastern Canada where it has become more common in recent years due to its popularity as bait among fishermen and its widespread sale in bait shops.

Many of the earthworms supplied for dissection by North American supply houses in recent years, however, have been collected in other areas and are different species. These species often differ in the specific location of reproductive structures and pores and other minor anatomical details. In most respects, however, the anatomy of most species of earthworms should correspond to the following description of *Lumbricus terrestris.*

Fact File

Oligochaetes

- Most earthworm species native to North America were killed off nearly 15,000 years ago during the last ice age. Many species common here today have been introduced from other parts of the world over the past 300 years.
- Charles Darwin wrote a classic treatise entitled "The Formation of Vegetable Mould Through the Action of Worms With Observations on Their Habits." This 139-page work, published six months before his death in 1882, includes descriptions of elaborate experiments he performed to study the behavior and "intelligence" of earthworms.
- Some earthworms have giant nerve cells that conduct impulses more than 10 times the speed of impulses in normal nerve cells.
- Certain aquatic oligochaetes, called tubificids or sewage worms, are useful biological indicators of organic pollution in streams and rivers.

External Anatomy

- Obtain an anesthetized or freshly killed specimen of a large species of earthworm and examine the major external features. Preserved specimens may also be studied, but fresh specimens are far superior because the organs of preserved specimens are often fragile and their colors faded, making them difficult to identify.

Observe the long, cylindrical body, which tapers at each end (figure 11.8). The anterior end can be identified most readily by the **clitellum** a conspicuous swollen region including several segments located near the anterior end. The clitellum secretes material that forms the cocoon in which fertilized eggs are incubated (figure 11.9).

Note the absence of a distinct head in the earthworm and the apparent lack of external sense organs at the anterior end. ***How can you relate these adaptations to the burrowing habits of the earthworm?***

The mouth is located in the first anterior segment, and a small, rounded projection, the **prostomium** overhangs the mouth. The **anus** is located in the last segment.

Observe the darker and more rounded dorsal surface of the worm. Locate the setae, small chitinous bristles located

FIGURE 11.8 Earthworm. Notice the obvious external segmentation and lack of cephalization.
Courtesy of Carolina Biological Supply Company, Burlington, NC.

FIGURE 11.9 Earthworm cocoons.
Courtesy of Carolina Biological Supply Company, Burlington, NC.

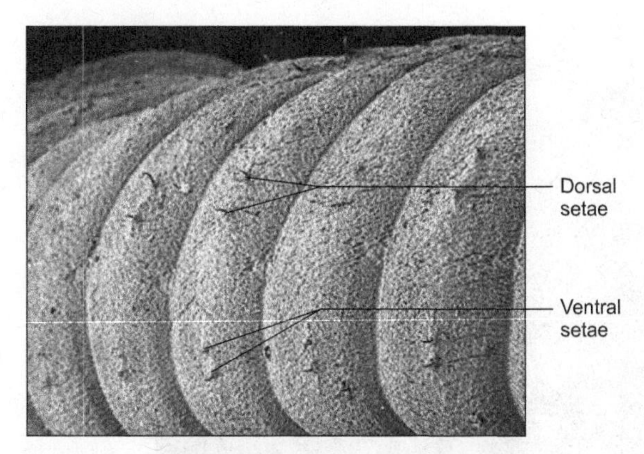

FIGURE 11.10 Earthworm, portion of body wall showing lateral rows of setae. Magnification 42×.
Scanning electron micrograph by Louis de Vos.

in each segment except the first and last. To find the setae, rub your finger lightly back and forth along the sides of the worm. Observe how the setae protrude slightly from the body surface and catch your finger—in one direction only! ***In which direction do they catch your finger? What does this tell you about the orientation of the setae? How would this aid in forward movements?***

The setae are located in four distinct rows. Two pairs of setae are located in rows on the ventral surface, and two pairs are located on the sides (figure 11.10).

Note the iridescent **cuticle,** a protective body covering that reduces water loss. The iridescence results from the many small striations in the surface of the cuticle. These striations cause the cuticle to act as a diffraction grating to separate incident light into separate wavelengths and to reinforce the wavelengths of those colors that produce the iridescence.

Also located on the surface of the body are several openings from the excretory and reproductive systems. A pair of small **excretory pores,** the nephridiopores, is found on the ventral surface of each segment except the first three or four and the last. Openings from the reproductive system include the following: a pair of **male pores** on the ventral surface of segment XV surrounded by swollen liplike structures, a small pair of **female pores** on the ventral surface of segment XIV, and openings to two pairs of **seminal receptacles** located laterally in the grooves between segments IX–X and X–XI. (Note: segments are usually designated by Roman numerals.) The use of a stereoscopic (dissecting) microscope will help to locate these tiny external openings to the excretory and reproductive systems of the earthworm.

Internal Anatomy

- Place your earthworm in a wax-bottom dissecting pan and carefully pin the specimen down, dorsal side up, with one pin through the prostomium and another near the posterior end. Study figure 11.15 and note the thin body wall and the closeness of the underlying parts before you begin your dissection. With the tips of your scissors, carefully puncture the body wall to one side of the dorsal midline and carefully make a longitudinal incision from behind the clitellum, continuing it anteriorly to the mouth. Cut to the side of the large dorsal blood vessel and avoid severing it. Take special care, also, to avoid damage to the brain, which is located near the mouth.

Figure 11.11 shows a dissected earthworm and its internal organs. Observe the large body cavity, the **coelom** that is divided into many distinct compartments by the **septa.** The septa are membranous, transverse partitions that divide the body of the earthworm internally into many segments.

Carefully separate the body wall from the internal structures and pin down the body wall on each side. Place the first pins on opposite sides of segment XV, easily located because of the presence of the large male pores. Insert the pins through the tissue and into the wax bottom of the pan at a 45-degree angle away from the worm to provide working space and a clear view of the internal organs. Place additional pins through each fifth segment (for example, segments V, X, XV, XX, XXV, XXX, and so on) to serve as convenient landmarks to identify the location of various structures. Cover the specimen with about 1 centimeter of water to keep the internal organs moist and flexible.

Digestive System

The digestive tract is a straight tube extending from the mouth to the anus. Examine your specimen and locate the **mouth** at the anterior end. The mouth opens into the small **buccal cavity** (figure 11.12). Immediately behind the buccal cavity is the muscular **pharynx.** The pharynx is attached to the body wall by numerous threadlike **dilator muscles** that can produce a sucking action to draw food materials through the mouth into the buccal cavity and pharynx. The tubular, thin-walled **esophagus** extends from segments VI to XIII and passes the food posteriorly to the **crop.**

Associated with the lateral wall of the esophagus of earthworms are two or more pairs of **calciferous glands.** These glands develop as evaginations of the esophageal wall and regulate the levels of calcium and carbonate ions (Ca^{++} and $CO_3^{=}$) and the pH of the blood.

Food is stored in the thin-walled, extensible crop before passing into the muscular **gizzard,** where it is ground into smaller pieces and passed to the **intestine** to be digested and absorbed. Undigested materials are passed to the **anus** for elimination.

Much of the intestine, as well as the dorsal blood vessel, is covered by a yellowish layer of **chloragogen tissue.** This tissue does not appear to be directly involved in digestion, but it does store glycogen and lipids and has other functions analogous to those of the liver in vertebrate animals. They deaminate proteins and release ammonia and urea, thereby serving an excretory function; they synthesize hemoglobin, the red oxygen-carrying pigment in the blood; and they provide some chemical defenses for the body.

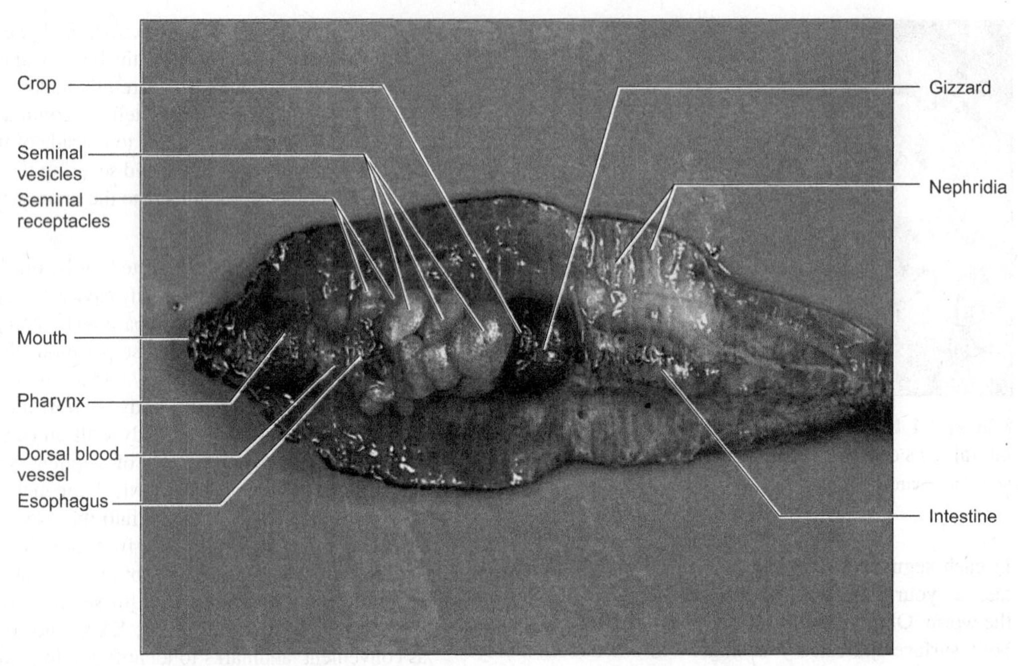

FIGURE 11.11 Earthworm, dissected, dorsal view.
Photograph by Ken Taylor.

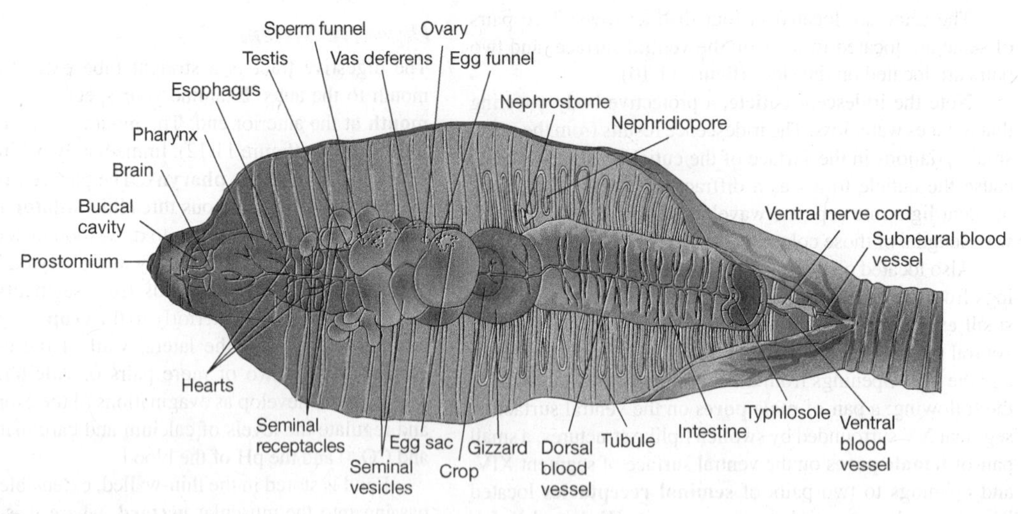

FIGURE 11.12 Earthworm, internal organs.

Circulatory System

The earthworm has a **closed circulatory system,** which contains red blood. The red color is due to the respiratory pigment, **hemoglobin,** dissolved in the **plasma,** or fluid portion, of the blood. In addition to the plasma, the blood also contains numerous colorless **blood cells,** or corpuscles.

Locate the large **dorsal blood vessel** lying just above the digestive tract, and the pair of **parietal vessels** extending laterally from the dorsal blood vessel in each segment. In segments VII to XI, inclusive, the parietal vessels are enlarged and much more muscular. These five pairs of enlarged **aortic arches** are often called "hearts." Their contractions aid in propelling the blood through the circulatory system.

The yellowish chloragogen tissue covers the surface of the large blood vessels. Some of the other major blood vessels include the **ventral vessel,** lying beneath the digestive tract; the **subneural** vessel (see figure 12.12), lying beneath the ventral nerve cord; and two **lateral neural vessels,** one on each side of the ventral nerve cord.

In an anesthetized worm, waves of muscular contraction can be readily observed along the dorsal vessel and the five pairs of "hearts." ***In which direction does the blood flow through these vessels?*** Valves in the dorsal vessel and in the hearts prevent backflow and aid in the efficient circulation of the blood.

Excretory System

Each segment, except the first three or four and the last, contains a pair of tubular excretory structures, the **nephridia.** Each nephridium is located partly in two adjacent segments. Consult figures 11.12 and 11.13. Locate the funnel-shaped opening into the coelom, the **nephrostome,** which projects through the septum into the next anterior segment. Also locate the **coiled tubular portion,** which empties through a ventral opening in the body wall, the **nephridiopore.**

The **calciferous glands,** formed as evaginations of the wall of the esophagus, also serve as excretory organs. They were described previously during our discussion of the digestive system.

Reproductive System

- Cut across the digestive tract about 12 mm behind the gizzard and carefully free it with the associated blood vessels from the underlying structures up to the level of segment IV.

The earthworm is monoecious: each individual has both male and female sex organs (figure 11.13). Despite the monoecious condition, however, cross fertilization between different individuals is necessary for sexual reproduction in earthworms (figure 11.14).

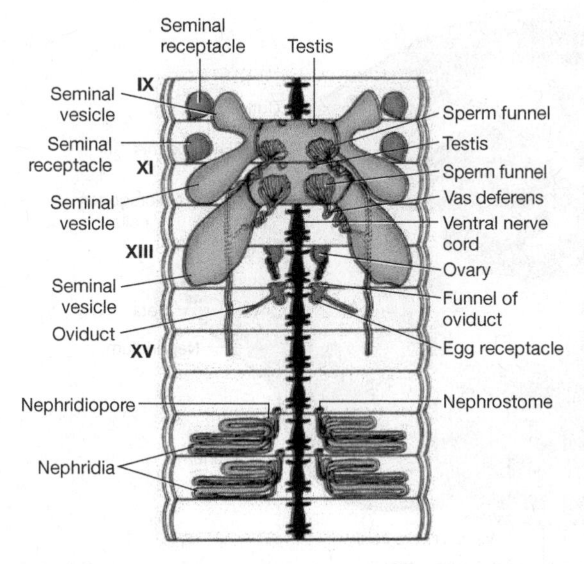

FIGURE 11.13 Earthworm, drawing of reproductive and excretory organs.

Several of the male reproductive organs are larger and more conspicuous than the female organs. Locate the following male reproductive structures and note the segments in which they are located: (1) three pairs of large **seminal vesicles** (storage sacs in which the sperm mature and are stored until copulation); (2) two pairs of small **testes** embedded within the tissues of the seminal vesicles; (3) two pairs of small ciliated **sperm funnels** (one pair on each side), which connect via short tubules to (4) two **vasa deferentia** (sperm ducts), which lead to the male genital pores on segment XV.

- Make a wet mount of a small piece of one of the seminal vesicles, gently squash the tissue, and examine under high power of your compound microscope. You should be able to observe many spermatozoa and perhaps specimens of the protozoan parasite *Monocystis,* a common parasite of annelids and arthropods. Many earthworms are infested with this apicomplexan parasite, which thrives in the tissues of the seminal vesicles.

The female reproductive structures are small and difficult to locate in most specimens. They include a pair of **ovaries** in segment XII, which discharge mature ova into the coelom. The ova are collected in two ciliated **egg funnels** and passed through the oviducts to the **female genital openings** on segment XIV. You should be able to locate two pairs of **seminal receptacles** in segments IX and X, where sperm is received during copulation. Sperm stored in the seminal receptacles is later used in fertilization of eggs released into the cocoon.

Nervous System

The nervous system of the earthworm consists of the following principal components: (1) an anterior pair of large **suprapharyngeal ganglia** (the "brain"); (2) two **circumpharyngeal connectives;** (3) a pair of smaller **subpharyngeal ganglia;** and (4) a **ventral nerve cord** with an

FIGURE 11.14 Earthworms mating. White material is sperm being exchanged.
NMPA/N. A. Callow.

enlarged ganglion in each segment (figure 11.13). Locate several nerves extending from the suprapharyngeal and subpharyngeal ganglia to the anterior segments. Three pairs of **lateral nerves**—two from the ganglion itself and one anterior to the ganglion—lead from the ventral nerve cord in each posterior segment.

So-called giant fiber cells are common but highly variable within the "brain" or the ganglia of the ventral nerve cord. These giant fibers are important for the rapid conduction of impulses used, for example, to generate startle reactions to predators.

Cross Sections

- Obtain a slide with a stained cross section of an earthworm and study its histological structure (figure 11.15). Observe the body wall and its composition. Locate the outer **cuticle** and the underlying **hypodermis.** ***What is the nature of the cuticle and how is it formed?*** Beneath the hypodermis is a thin layer of **circular muscles** and a thicker layer of **longitudinal muscles.**

Observe the thin layer of flattened cells, the **peritoneum** at the inner margin of the longitudinal muscle layer. Covering the outer surface of the intestine (and some of the larger blood vessels as noted earlier) is a specialized type of peritoneum, the **chloragogen.** The body cavity of the earthworm is a true **coelom.** ***What makes this a coelom?***

You may be able to observe within the coelom the **nephridia** lateral to the intestine. The nephridia in cross section may appear to be irregularly shaped since they are often cut through the coiled portions of the tubule (figure 11.16).

The wall of the intestine consists of several layers of cells, including an outer chloragogen, a thin layer of **longitudinal muscle** and **circular muscle,** and a layer of **endodermal epithelium** lining the cavity, or lumen, of the gut. Dorsally, observe the pronounced fold of tissue, the **typhlosole** which extends into the lumen and increases the surface area of the intestine available for absorption. Observe the layer of chloragogen tissue that surrounds the typhlosole.

Observe also the **dorsal** and **ventral blood vessels** and the **ventral nerve cord.** On each side of the nerve cord is a **lateral neural blood vessel,** and beneath the nerve cord is the **subneural blood vessel.**

Demonstrations

1. Charts and models to illustrate earthworm anatomy
2. Earthworm cocoons
3. Behavior of living earthworms (locomotion, reactions to stimuli, etc.)
4. Examples of other Oligochaetes, such as *Tubifex, Aeolosoma, Enchytraeus, Stylaria* and *Chaetogaster*

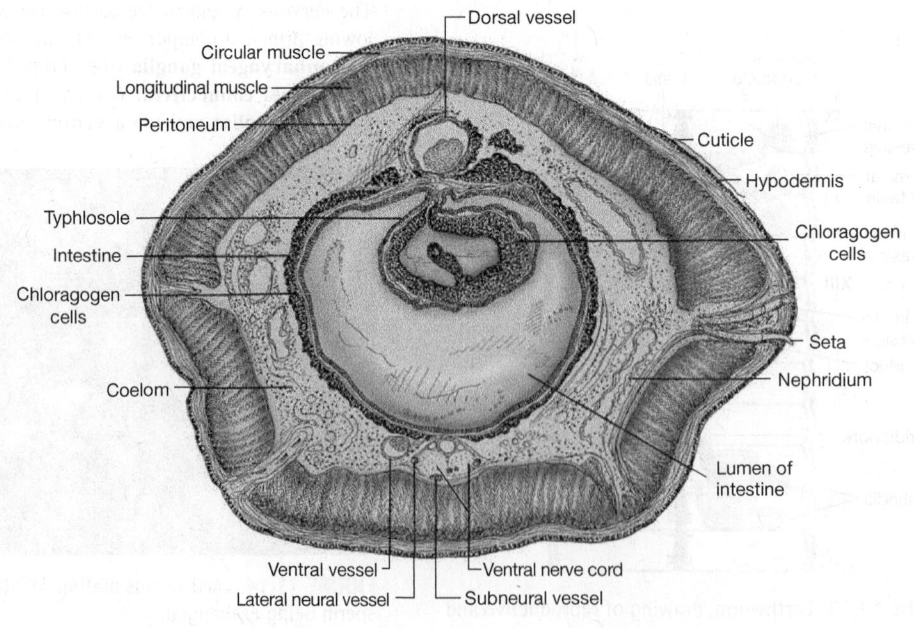

FIGURE 11.15 Earthworm, cross section.

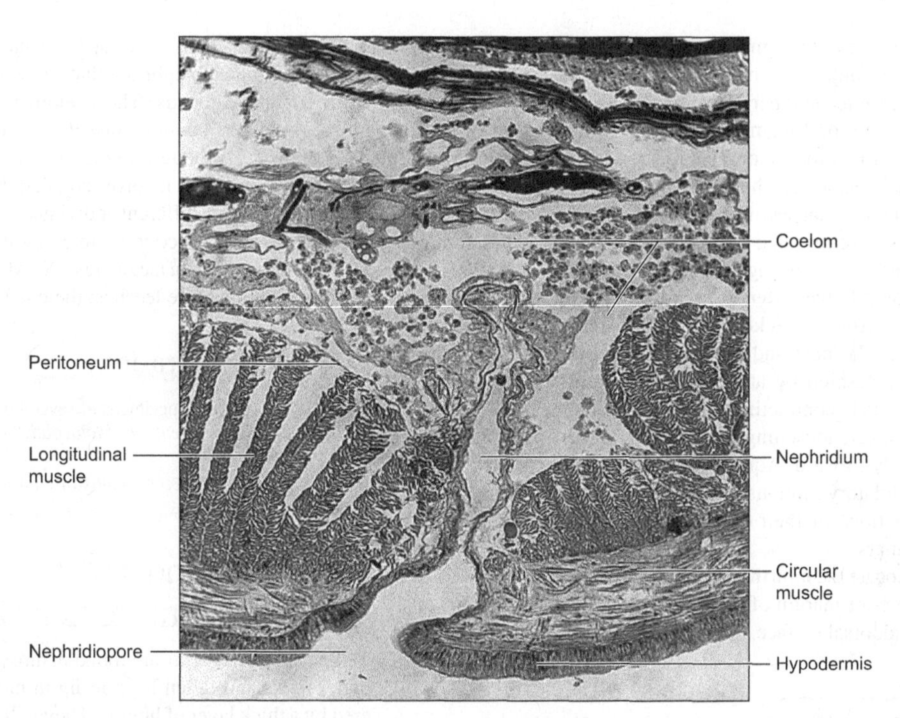

FIGURE 11.16 Earthworm, portion of cross section showing opening of nephridiopore through body wall.
Courtesy of Carolina Biological Supply Company, Burlington, NC.

Leeches

Class Hirudinea

Even though leeches are commonly called bloodsuckers, most leeches actually live in freshwater streams, ponds, or lakes as predators on small aquatic invertebrates or as scavengers rather than blood-sucking parasites. Some species, however, are external parasites on various vertebrate animals including fish, turtles, and mammals. There are also a few marine and terrestrial species.

Leeches are readily recognized as annelids because of the external appearance of segmentation. Another recognizable feature is the large posterior sucker that is used for attachment and also aids in locomotion (figure 11.17).

Leeches are relatively large annelids, and the adults of some species reach 25 cm in length. They **lack head appendages, parapodia** and have **no larval stages.** Their large interior coelom is not partitioned by septa and is largely filled with a loose mesohyl tissue. The leeches most commonly available for laboratory study in the United States are large aquatic species of the genera *Haemopis, Malacobdella,* or *Placobdella.*

External Anatomy

- Examine a preserved leech and observe the elongate, flattened body. The body is flattened dorsoventrally and usually tapers toward the anterior end. Note the absence of setae and lateral appendages. Locate the numerous rings or **annuli** that circle the body.

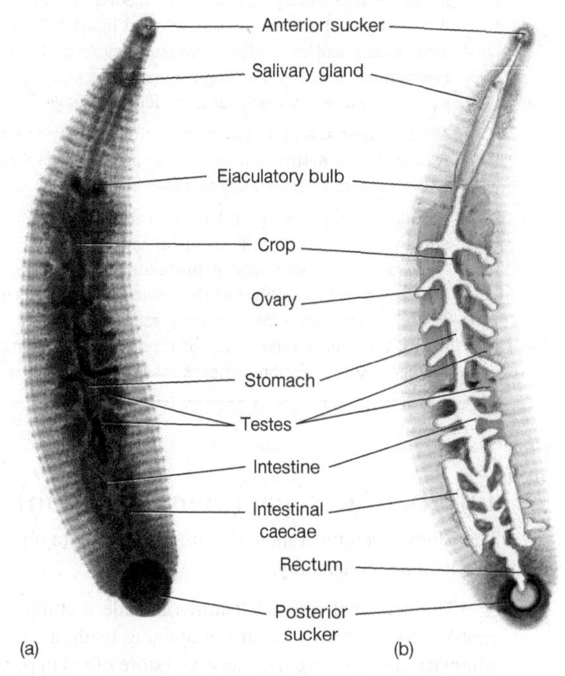

FIGURE 11.17 Leech. (*a*) Whole mount, stained. (*b*) Leech, drawing.
Photo courtesy of Carolina Biological Supply Company, Burlington, NC.

Leeches are segmented both internally and externally, however, unlike the oligochaetes and polychaetes, leeches have a limited number of internal body segments: 34, including two pre-oral segments and 32 post-oral segments. Also, the external grooves or annuli, do not all correspond to the internal segments. There may be from 1–5 external annuli for each internal segment in various species. Recent genetic studies have revealed how certain homeobox genes control the number and arrangement of segments in leeches.

Identify the **anterior sucker** around the mouth and the larger **posterior sucker.** The suckers serve both for attachment to the host and for locomotion. Leeches move in a looping fashion by attaching to the substrate by the anterior sucker, contracting the body, attaching by the posterior sucker, loosening the anterior sucker, and extending the body forward. Aquatic leeches can also swim by means of undulatory motions of the body resulting from rhythmic contractions of their muscular bodies. Some are graceful swimmers.

Locate the **mouth** in the center of the anterior sucker. At the anterior margin of the posterior sucker find the **anus** on the middorsal surface.

Fact File

Leeches

- A large European leech, *Hirudo medicinalis,* was used by physicians for "bloodletting" in the early days of medicine when many diseases and disorders were attributed to the accumulation of "bad blood." Today, leeches are still used after microsurgery (e.g., for the reattachment of severed fingers, ears, etc.) to reduce swelling caused by inadequate venous drainage.
- Most leeches are predators rather than parasites and feed on small arthropods and other small invertebrate animals. A few species are terrestrial.
- Hirudin is an anticoagulant peptide that occurs naturally in the salivary glands of the medicinal leech, *Hirudo medicinalis.* It prevents blood clotting by inhibiting the activity of thrombin, an enzyme that converts fibrinogen into the long, sticky threads of fibrin. The therapeutic value of hirudin is also being studied for prevention of heart attacks and strokes.

Internal Anatomy (Demonstration)

- Study a demonstration of a dissected leech to observe its internal anatomy.

The principal internal features include a complete digestive system with a **mouth** containing teeth, a muscular **pharynx,** a large **crop** (that serves to store blood in parasitic leeches), a small **stomach,** a narrow tubular **intestine,** and an enlarged **rectum** that opens dorsally via the **anus** located just anterior to the posterior sucker.

Leeches also have a closed circulatory system, several pairs of segmental nephridia that serve as excretory organs, and permanent gonads. The **coelom** is largely filled with loose tissue and contains one pair of ovaries and several pairs of testes. Although leeches are monoecious (hermaphroditic), reproduction involves copulation and reciprocal fertilization between different individuals. The fertilized eggs are deposited in a cocoon secreted by the clitellum formed by an enlargement of segments IX–XI. The eggs develop directly into miniature leeches; there are no larval stages.

Demonstrations

1. Preserved specimens of several different leeches, such as *Haemopis, Malacobdella, Placobdella,* and *Hirudo*
2. Dissected leech to illustrate internal anatomy

Collecting and Preserving Annelids

Terrestrial earthworms are found in most organic soils. Look under rocks and fallen logs or dig in moist soil that is covered by a thick layer of humus. They will thrive in containers of loose, loamy soil mixed with peat moss, shredded newspaper, dead leaves, or wood shavings. Add diced vegetable scraps, coffee grounds, or compost as food (no meat scraps). Additional information on "vermiculture" is readily available in the library or on the Internet.

Tubifex sp. worms are freshwater oligochaetes that live in mud on the bottom of lakes or slow-moving rivers. They may occur in high concentrations where the water is polluted with organic waste (sewage effluent). Collect specimens by dredging and transfer them to a deep pan or aquarium with several inches of organic mud and 4–6 inches of pond water. Provide aeration and change the water frequently. Feed sparingly with cornmeal, dried milk powder, or soymeal.

Aelosoma is a microscopic oligochaete common in wastewater treatment facilities, pond samples, and scrapings of algae from the bed of a lake or stream.

Leeches can often be found in freshwater ponds or streams that support healthy populations of fish, amphibians, and reptiles. Use a seine or dip net to collect them from aquatic vegetation or bait them with chunks of canned fish (salmon or tuna) inside a submerged burlap bag. They will survive for months in a well-balanced aquarium.

Nearly all polychaetes are marine. They occur in a wide range of habitats from estuaries and inshore waters to the open sea and hypersaline lagoons. Many species are found along intertidal areas, especially beyond the low waterline on sandy or muddy beaches. These worms can be recognized by their clearly segmented bodies, lateral parapodia bearing many setae, and head appendages. They can be divided into two groups, "sedentary" and "errant," based on their habits.

"Sedentary" polychaetes always live within burrows. Some make tubes of sand or parchment in the bottom sediments, whereas others build calcareous tubes attached to rocks, shells, or pilings. Common examples include:

- Plume worms, *Diopatra* spp. (figure 11.18), have iridescent bodies with bright red (blood-filled) gills. They build long, vertical tubes of parchmentlike material covered with sticks, stones, shells, and bits of sea grass. The tubes may be as much as a meter in length, with only 4–8 cm extending above the sediment.
- Feather-duster worms, *Hydroides* spp. (figure 11.19), have a feathery crown of brightly colored feeding appendages. They live in calcareous tubes secreted on the surface of shells or rocks.
- Parchment worms, *Chaetopterus* spp., are creamy yellow and glow in the dark. They construct U-shaped burrows in the sediment and feed by trapping plankton in a mucous net while circulating seawater through the burrow with specialized parapodia.
- Lug worms, *Arenicola* spp. (figure 11.20), lack parapodia and have a muscular head used for burrowing through marine sediments. They live on organic materials that pass through the digestive tract and play an important role in mixing sediments in shallow waters.

"Errant" polychaetes are wanderers—they do not have a permanent home or burrow. They may swim in open water or actively search for food along the bottom, among rocks and shells, or in sandy or muddy sediments. Common examples include:

- Clamworms, *Nereis* spp. (figure 11.1), are large marine predators that are commonly found in mussel beds and on mud flats where clams are numerous. They hide during the day and emerge at night to search for prey.
- Common blood worms, *Glycera dibranchiata* (figure 11.3a), are red to cream colored worms that have a long, eversible proboscis tipped with four hooks. The proboscis can be shot out rapidly to seize prey or to defend against an enemy. They are commonly sold as fish bait.

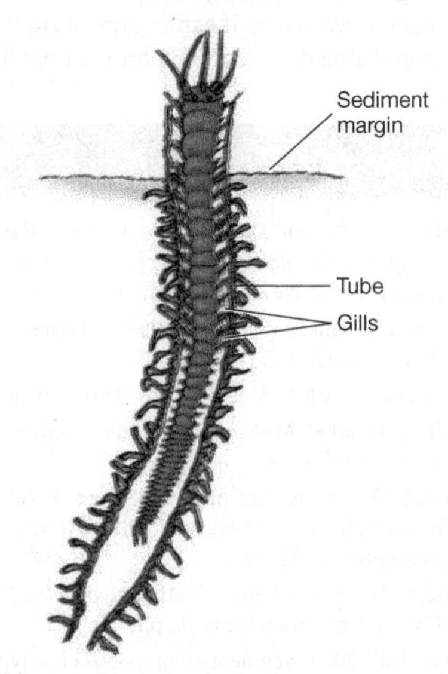

FIGURE 11.18 *Diopatra*, a plume worm.

Narcotizing agents such as carbonated water, magnesium sulfate (epsom salts), chloral hydrate, or chlorotone are often useful for relaxing annelids and reducing their movement under the microscope. When the animal is no longer responsive, it can be transferred to 10% formalin or 70–80% ethanol for long-term preservation. Adding several milliliters of glycerin to the solution will reduce specimen damage in case the preservative evaporates.

FIGURE 11.19 *Hydroides*, a tube-dwelling polychaete. These worms secrete a calcareous tube on stones and shells in shallow waters and have become distributed worldwide as an important fouling organism.
Photo by Charles Wyttenbach.

FIGURE 11.20 *Arenicola*, the lug worm. These sedentary polychaetes burrow in bottom sediments, feed on organic debris, and are important in mixing marine sediments in shallow waters.

Key Terms

Biramous having two branches or rami; as the parapodium of a polychaete worm with a dorsal notopodium and a ventral neuropodium.

Clitellum a thickened, glandular portion of certain midbody segments of many oligochaetes and hirudineans; characteristic of sexually mature individuals. Forms a cocoon in which eggs are deposited and incubated.

Coelom a central body cavity fully lined by peritoneum, an epithelial lining of mesodermal origin. It forms the principal body cavity of annelids and most higher phyla, although in some groups the coelom is secondarily reduced.

Nephridium (plural: nephridia) a tubular excretory organ found in annelids and other invertebrate phyla; may function both in osmoregulation and in the removal of nitrogenous and other body wastes.

Parapodium (plural: parapodia) a lateral appendage typical of polychaete worms; always occurs in pairs on opposite sides of the body; usually with many setae arranged in bundles.

Segmentation having a body consisting of many similar units or subdivisions arranged along the anterior-posterior axis. Each unit is called a segment, somite, or metamere. Also called metamerism.

Septum (plural: septa) a thin wall of tissue separating two adjacent segments or masses of tissue.

Seta (plural: setae) a chitinous bristle or rod secreted by certain epidermal cells of oligochaetes and polychaetes. Setae are also found in certain arthropods and other invertebrates.

Trochophore larva a marine larval form characteristic of many polychaete worms molluscs, and certain other invertebrate groups. Body is pear-shaped with one or more bands of cilia around the body, tufts of cilia at the anterior and posterior ends, and a complete digestive tract.

Internet Resources

There are many valuable Internet sites with information about zoology. Several sites containing pertinent zoological information for this chapter can be found on the McGraw-Hill Zoology web site at http://www.mhhe.com/zoology. Just click on this text's title.

Questions for Critical Thinking

1. Discuss several structural features found in the Phylum Annelida that are not found in the Phyla Platyhelminthes or Nematoda. All of these wormlike organisms have survived for millions of years. How do you think their body form has contributed to or limited their ability to invade new habitats? How do you measure organismal success? By the complexity of the organism? By the species diversity of the group? By the complexity of their life cycle? By the populations of its member species? Is one organism "more advanced" than another simply because it is more complex? Do complex species survive better than simpler species?
2. Compare the coelom of the molluscs with the coelom of the annelids. Of what advantage is an enterocoel when it is obvious that both groups are very successful? In fact, it may be argued that the mollusc's reduced coelom is an advantage and part of the reason for its diversity and adaptive radiation. What do you think?
3. Of what importance is increasing "cephalization"? Was this a significant evolutionary event (to have neural and sensory structures congregate in the anterior of the body)?
4. Segmentation really becomes dominant (some investigators say it appears for the first time) in the annelids. Of what importance is segmentation? Examine how segmentation might have contributed to the evolution of appendages and locomotion as well as the ultimate specialization of these appendages.
5. Parasites are generally considered to be organisms that live *within* or *on* another organism. Some investigators disagree that leeches are parasites, and instead call them "micropredators" because they obtain a blood meal and then fall off the host. How would you classify leeches?
6. Compare the trochophore larva with the larva of other marine invertebrate groups. Do you see any similarities? If so, what are they? Give an explanation for these similarities.
7. Discuss the ecological importance of oligochaetes in general. What do you suppose would happen in many ecosystems if earthworms were to suffer serious population declines? Support your predictions.

Suggested Readings

Brinkhurst, R.O., and D.G. Cook. eds. 1980. *International Symposium on Aquatic Oligochaete Biology (1st:1979: Sidney, B.C.)*. New York: Plenum. 529 pp.

Fauchald, K., and G.W. Rouse. 1997. Polychaete systematics. *Zool. Scripta.* 26:71–138.

McClintock, J. 2001. Blood suckers. *Discover.* 12(12):56–61.

McHugh, D. 2000. Molecular phylogeny of the Annelida. *Can. J. Zool.* 78:1873–1884.

Rouse, G.W., and K. Fauchald. 1998. Recent views on the status, delineation, and classification of the Annelida. *American Zoologist* 38:953–964.

Rouse, G.W., and F. Pleijel. 2001. *Polychaetes.* New York: Oxford University Press. 38 pp

Seaver, E.C. 2003. Segmentation: mono- or polyphyletic? *Int. J. Dev. Biol.* 47:583–595.

Chapter 12

Nematoda and Nematomorpha

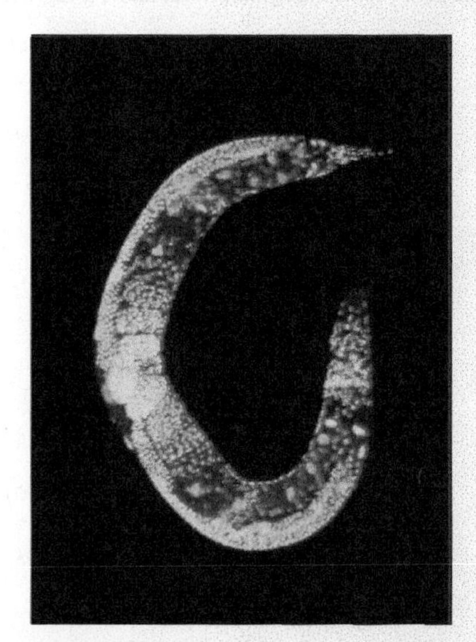

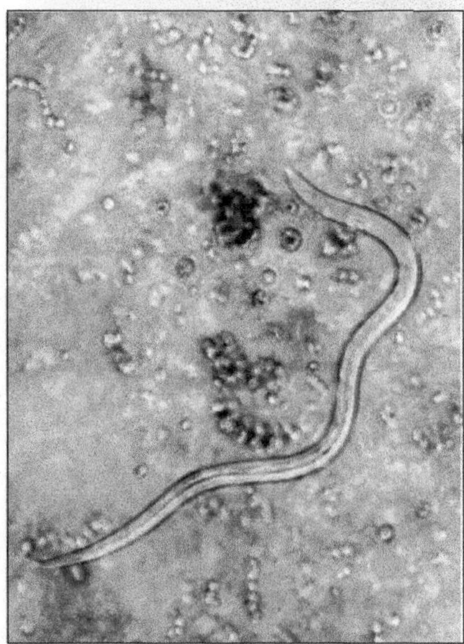

(Top) Leslie Saint-Julien, National Human Genome Research Institute.
(Bottom) © The McGraw-Hill Companies, Inc./Don Rubbelke, photographer.

Objectives

After completing the laboratory work in this chapter, you should be able to perform the following tasks:

1. Discuss suggested evolutionary relationships of the nematodes and nematomorphs and explain why they should or should not be grouped with other pseudocoelomate animals.
2. Distinguish between a pseudocoelom and a true coelom. Explain present views of the significance of the method of embryonic development of central body cavities of animals.
3. Describe the external morphology of the roundworm *Ascaris* and locate the chief external features on a preserved specimen. Distinguish between mature male and female specimens.
4. Describe the male and female reproductive systems of *Ascaris* and identify the principal reproductive organs in dissected specimens and on microscope slides.
5. Name three parasitic roundworms, explain their importance in human or animal health, and briefly describe their life cycles.
6. Describe the main morphological similarities and differences of the nematodes and nematomorphs.
7. Compare the body plan of a typical nematode to that of a typical flatworm and that of a typical annelid worm. Discuss the principal differences in body plans of the three animals.
8. Discuss the adaptations of a nematode and a Gordian worm for parasitism. How do these adaptations increase the chances of survival of the parasite?

Introduction

Historically, the nematodes (roundworms) and the Nematomorpha (Gordian worms) were considered to be **pseudocoelomate** and placed as classes within a phylum called Aschelminthes, along with several other groups of soft-bodied, wormlike animals. An important proponent of this grouping was the American scholar Libbie Hyman, who published a major series of volumes, *The Invertebrates,* between 1940 and 1967. This was a time when an animal's body form and its type of coelom formation were believed to offer important clues to its evolutionary relationships. Some of those groups included in the aschelminthes were later found not to be pseudocoelomate, and recent research has shown that body form is not always a good indicator of evolutionary relationships. Perhaps more importantly, recent biochemical and genetic studies have shown that some members of the former aschelminth group are not closely related. The term "ashelminthes" is therefore no longer capitalized since it is not now recognized as a legitimate taxonomic group. These more distantly related animals are now split among two or more higher taxonomic groups.

Although there is still considerable disagreement on the actual relationships among the Nematoda and related groups, genetic and biochemical evidence has led to assignment of many of the animal groups formerly included in the "ashelminthes" to other groups. Some of these groups are now included in the clades Lophotrocha or Placozoa as discussed in Chapter 9. Gastrotrichs and rotifers are now placed in the Platyzoa along with the flatworms.

Several other of these former "aschelminthes" groups are now included in the superphylum **Ecdysozoa,** including Kinorhyncha; Loricifera; Nematoda; Nematomorpha; and Priapulida. Other members of the Ecdysozoa include the Arthropoda, Onychophora, and Tardigrada. All of these

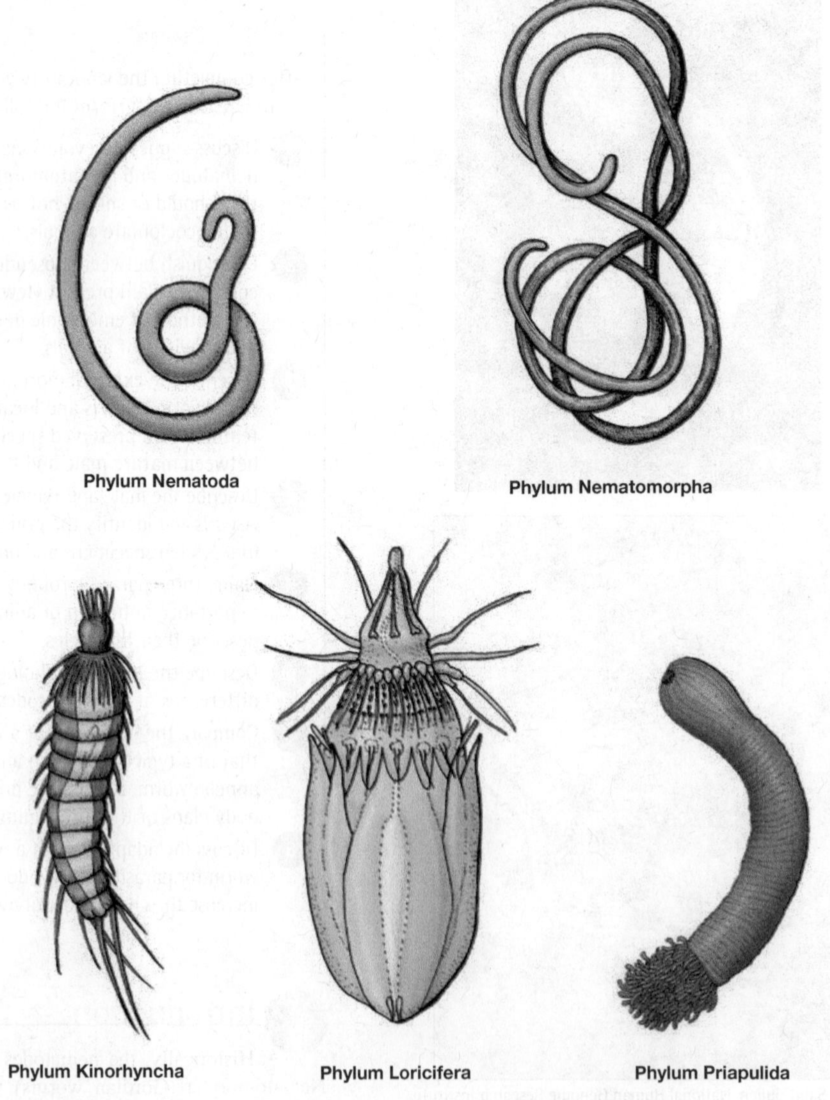

FIGURE 12.1 Nematoda and some other phyla formerly included in the aschelminthes.

animals have a cuticle, molt periodically, and are linked evolutionarily through the inheritance of genes responsible for synthesis of the molting hormone ecdysone.

This is an active area of research and the assignment of phyla to higher taxonomic groupings is subject to change as new evidence is obtained. Several of the other phyla seem to be more closely related to the Platyhelminthes and Nemertinea (ribbon worms). The Acanthocephala (spiny headed worms) are now believed to be closely related to Rotifera and are placed in superphylum **Lophotrochozoa** as discussed in Chapter 9.

Classification

Phylum Nematoda (Nemathelminthes or Roundworms)

Free-living and parasitic worms with an **elongate cylindrical body** tapered at both ends. The body is covered by a tough, flexible **cuticle** that is periodically shed as a concurrence of growth. Complete digestive system with a **triradiate pharynx** and a **specialized excretory system.** Cell number usually fixed and invariant during the adult stage, a condition called eutely, also found in rotifers. More than 15,000 described species. Examples: *Ascaris, Trichinella, Ancylostoma, Caenorhabditis, Anguillula.*

Materials List

- Living specimens
  - *Anguillula aceti*
  - *Caenorhabditis elegans*
  - *Philodina*
- Preserved specimens
  - *Ascaris*
- Microscope slides
  - *Ascaris* cross sections of male and female
  - *Ancylostoma* whole mount
  - *Enterobius* pinworm, whole mount
  - *Trichinella* whole mount (demonstration)
  - Representative Rotifera (demonstration)
- Plastic mounts
  - *Ascaris* male and female worms (demonstration)
  - *Ancylostoma* hookworm (demonstration)
- Miscellaneous
  - Nile blue solution

Phylum Nematomorpha (Horsehair or Gordian Worms)

A relatively small group of large worms found worldwide, but only about 11 species are found in the United States. The adults are long, thin, and may reach lengths up to several meters. The phylum is divided into two classes: the nectonematids, which parasitize marine crabs and shrimp; and the Gordiids, which parasitize insects and other small fresh water arthropods.

Phylum Nematoda

A Parasitic Roundworm: *Ascaris lumbricoides*

Ascaris lumbricoides (figure 12.2) is a large roundworm parasite found in humans. An estimated 25 percent of the world human population is infected with this intestinal parasite. Adult female worms can reach 18 inches in length (males are usually shorter). Living specimens are yellow or pink in color, but preserved specimens usually range from pink to brown. A related species, *Ascaris suum* is found in pigs. If a human ingests eggs of *Ascaris suum* the larvae hatch and move to the lungs and may cause a condition called "ascaris pneumonia" but the larvae do not develop into adults. Some parasitologists believe that these two nematodes are really one species that can infect both humans and pigs. The morphology of the two species is very similar and other scientists have suggested that they are two different species descended from a common ancestor.

External Anatomy

- Study a preserved specimen to determine some of the characteristic features of roundworms, members of the Phylum Nematoda.

Observe the elongate cylindrical body that tapers at both ends. The anterior end is more pointed than the posterior, which tends to be slightly blunt. The triangular **mouth** surrounded by three distinct **lips,** is located at the anterior end of the body (figure 12.3). Under your dissecting microscope, observe the three lips and the mouth. Also locate the **anus,** which appears as a transverse slit on the ventral side of the body near the posterior end. The ventral location of the anus is a convenient means of distinguishing the dorsal and ventral surfaces of the worm. You should be able to find four longitudinal lines extending lengthwise along the body wall: one dorsal, one ventral, and two lateral in position. These longitudinal lines are more apparent in living specimens than in preserved specimens, and their significance will be better understood after your study of the microscopic cross sections of the worm later in this exercise.

Further examination of your specimen should reveal its sex. *Ascaris,* like most nematodes, is **dioecious** (sexes separate) and the female worms are larger than the males (figure 12.4). The males are curved ventrally into a hood near the posterior end and bear a small pair of **copulatory spicules** protruding from the anus. Female worms have a ventral **genital pore** about one-third the distance from the anterior end. ***From your observation of the location of the copulatory spicules in the male, what can you conclude about the location of the male genital opening?*** Examine several specimens of *Ascaris lumbricoides* to determine whether you can find any other morphological differences between the sexes.

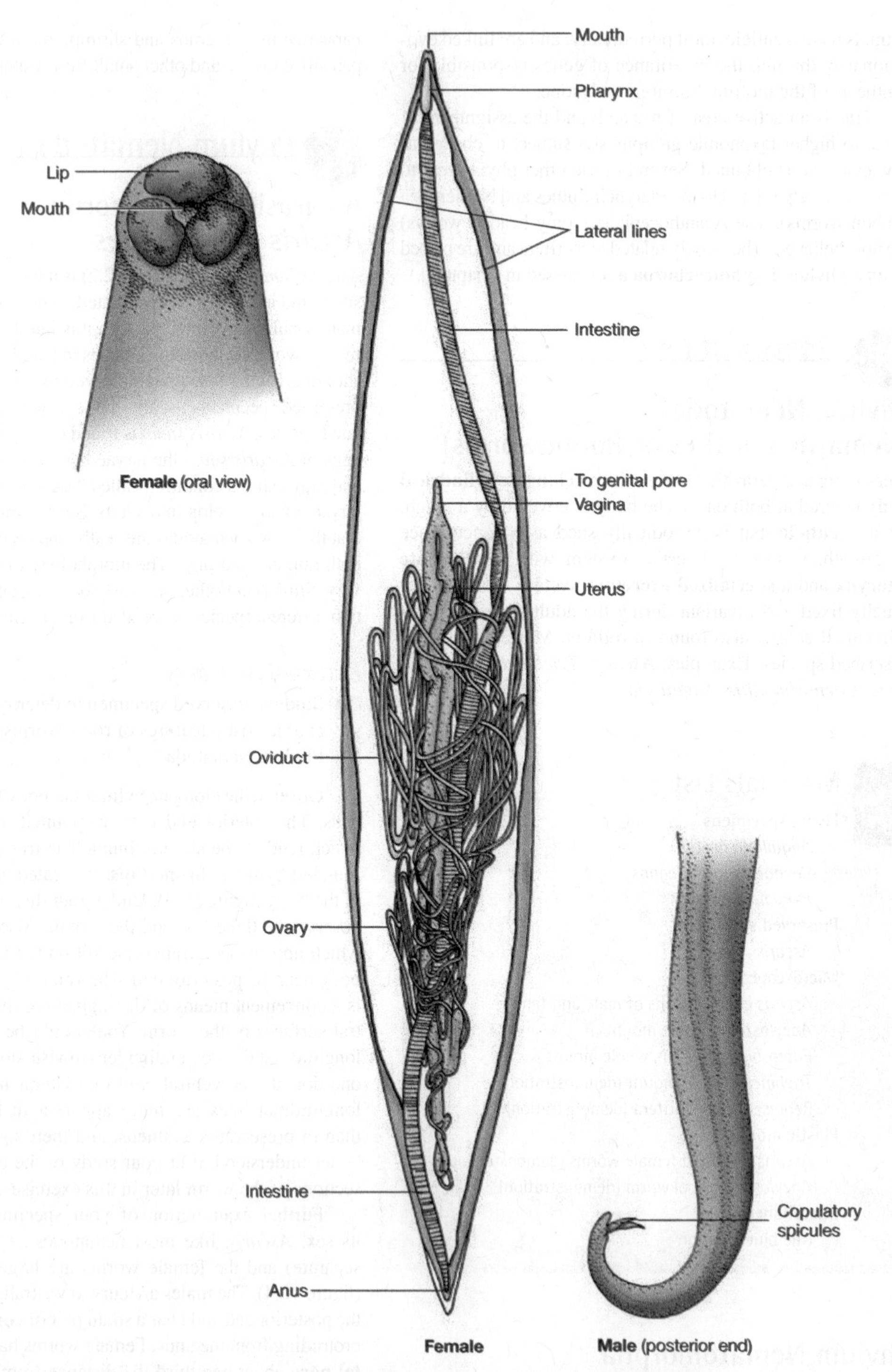

FIGURE 12.2 *Ascaris*, internal anatomy of dissected female and selected external features of female and male.

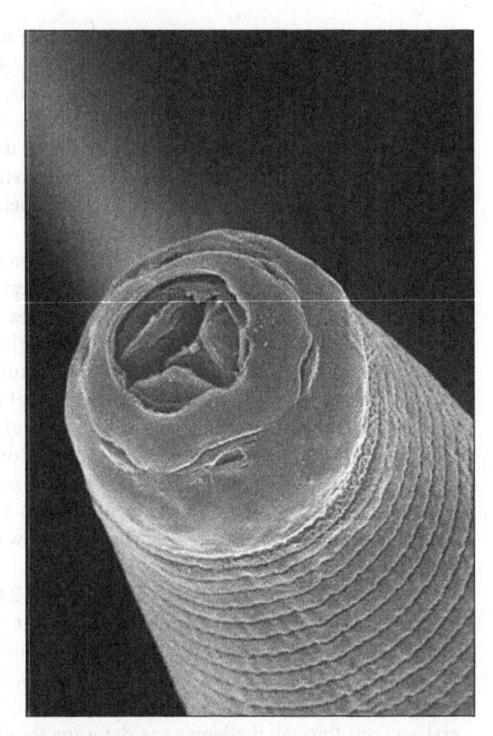

FIGURE 12.3 Anterior end of a nematode showing triangular mouth and lips. Although some nematodes have teeth, which aid in penetration of a host, *Ascaris* has none. Magnification 5,200×.
Scanning electron micrograph by Louis de Vos.

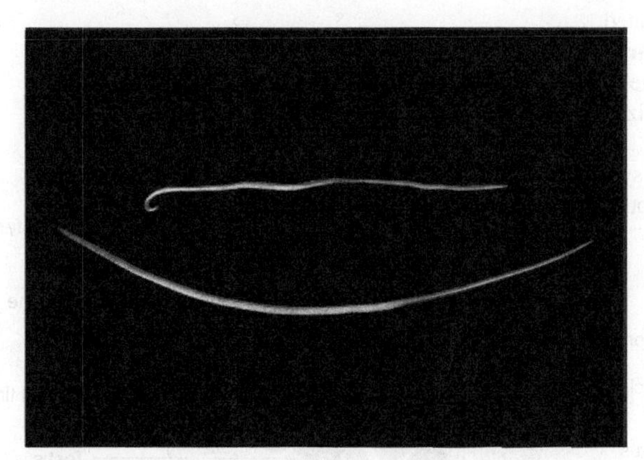

FIGURE 12.4 Male (top) and female *Ascaris.*
Courtesy of Carolina Biological Supply Company, Burlington, NC.

Internal Anatomy

- After you have completed your study of the external features of *Ascaris,* carefully dissect the worm by making a longitudinal slit down its dorsal side. Since the body wall is thin, the incision can easily be made with a pin or dissecting needle which lessens the risk of damaging organs. Pin down the body wall on each side of the worm in a wax-bottom dissecting pan to expose fully the internal structures, and cover your specimen with a little water to keep the internal organs moist and flexible.

> **CAUTION!**
>
> Although the *Ascaris* specimens you will receive in the laboratory have been preserved in formalin, the eggs found in the female specimens are enclosed in a very resistant "shell." They may not all be killed, even after several weeks in formalin. Although acquiring an infection in this way is a remote possibility, you should always follow good laboratory safety practices. Always wear dissecting gloves and avoid putting your hands in your mouth after handling live or preserved animals. You should wash your hands carefully after you complete your work.

Observe the long, ribbonlike **intestine** extending most of the length of the body and the numerous tangled threadlike coils surrounding the middle part of the intestine (figure 12.2). The female reproductive system is Y-shaped, and each branch of the Y consists of a long, threadlike **ovary**

> **Fact File**
>
> **Roundworms**
>
> - Most roundworms are very small—usually between 0.04 and 5.0 mm in length. A few species, however, are much larger. *Placentonema gigantisma,* found in the placenta of sperm whales, can grow up to 8 meters in length.
> - A remarkable feature of many nematodes is that the body of an individual has a genetically fixed and constant number of cells. Cell division stops at the end of embryogenesis and all subsequent growth involves only cell enlargement. For example, a fully developed embryo of ***C. elegans*** has 959 cells. As it matures, this nematode undergoes four juvenile molts, but the number of body cells never changes—each cell just grows larger.
> - *Wuchereria bancrofti* and *Brugia malayi* are parasitic nematodes that cause a serious human disease, lymphatic filariasis (also known as elephantiasis). Infective stages of these parasites are spread through the bite of mosquitoes. Over 40 million people in Africa, India, and the Americas are infected with these nematodes.
> - Nematode spermatozoa are unique in that they do not have a flagellum for locomotion. In some species, the sperm become amoeboid in order to move through the female's reproductive tract.
> - River blindness (onchocerciasis) is a parasitic disease caused by the nematode *Onchocerca volvulus.* Over 18 million people in tropical parts of Africa, the Arabian peninsula, and the Americas are infected by this parasite, which spreads through the bite of black flies.

connected with a larger **oviduct** and a still larger **uterus.** The distinction among these three organs is not apparent from a dissection, but is easy to observe in microscopic cross sections. The two uteri join to form a single midventral muscular tube, the **vagina,** which opens to the exterior via the female genital pore, or **vulva,** observed previously.

The male reproductive system is not branched, as in the female reproductive system, and it consists of a single, continuous tubular structure. This continuous structure includes a threadlike **testis** connected with a larger-diameter **vas deferens,** or sperm duct, a still larger diameter **seminal vesicle,** and a short, muscular **ejaculatory duct,** which empties into the **cloaca** at the posterior end of the digestive tract.

The digestive system consists of the ribbonlike **intestine,** which is attached anteriorly to a short muscular **pharynx** into which the mouth opens, and posteriorly to the **cloaca.**

Cross Sections

Prepared microscope slides with cross sections of male and female worms illustrate other aspects of nematode structure (figures 12.5 and 12.6).

- Examine a cross section of a female worm and identify the thick **cuticle** covering the body and the underlying layer of **hypodermis** which secretes the cuticle.

Observe the many **longitudinal muscle bands;** there are no circular muscles present in roundworms. This anatomical feature produces the peculiar whiplike movements often exhibited by living roundworms. Free-living nematodes are sometimes called "whipworms." Also associated with the body wall are the **dorsal** and **ventral nerve cords** and the two **lateral lines.** Located within each of the lateral lines is an **excretory canal,** part of the highly specialized excretory system of the nematodes.

Nematodes lack protonephridia but do have a unique system of gland cells connected to a canal system that empties through a ventral excretory pore. The function of this unique excretory system is not well understood, but *Ascaris* is known to excrete nitrogenous waste in two forms, ammonia and urea.

Observe the thin-walled intestine and note that the intestinal wall consists of a single layer of large **gastrodermal cells.** The body cavity of *Ascaris* is a **pseudocoelom,** since it lacks a mesodermal layer around the gut.

In the cross section of the worm, the female reproductive system should be represented by numerous round or oval structures of various sizes. The **ovaries** are small in diameter and bear some resemblance to tiny wagon wheels. A layer of columnar ovarian cells surrounds a central core, or **rachis.** The **oviducts** are slightly larger in diameter than the ovaries and have a central opening, or lumen, for the passage of eggs. If the section was cut through the region of the vagina, you could observe a single, large median structure, usually filled with eggs. Sections posterior to the vagina region usually show two large uteri, also filled with eggs.

Cross sections of male specimens (figure 12.6) exhibit features similar to female specimens except for the reproductive structures. The reproductive system is usually represented by several sections through the **testes,** which are small round structures containing many spermatogonia. Several sections through the larger **vas deferens** should also be present in the section. You should find numerous spermatocytes in the vas deferens of mature male specimens. A large **seminal vesicle** containing mature spermatozoa should also be present.

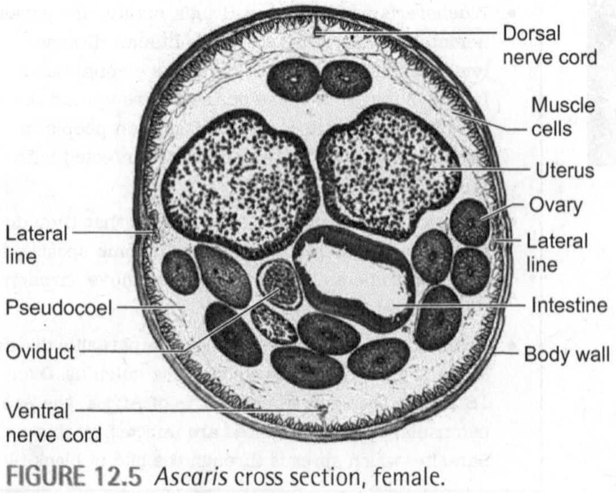

FIGURE 12.5 *Ascaris* cross section, female.
Courtesy of Carolina Biological Supply Company, Burlington, NC.

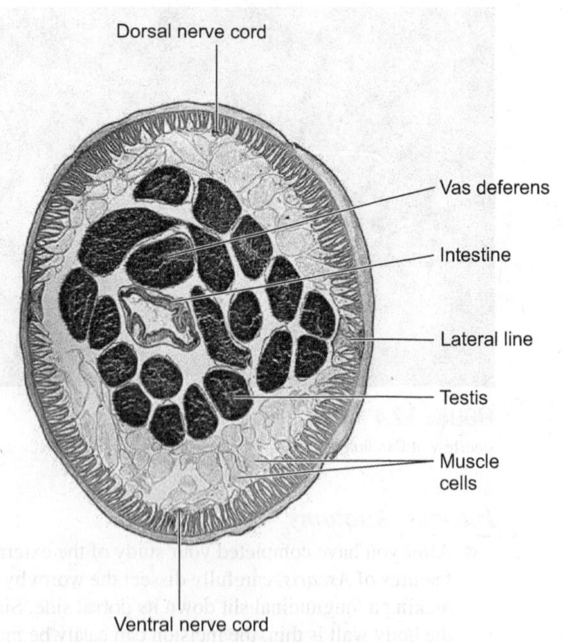

FIGURE 12.6 *Ascaris* cross section, male.
Courtesy of Carolina Biological Supply Company, Burlington, NC.

Life Cycle

Ascaris lumbricoides has a direct life cycle with no intermediate hosts. Mature female worms in the intestine of the host produce eggs that are expelled in the host feces. The fertilized eggs develop into larvae in warm, moist soil. If eggs containing larvae (figure 12.7) are ingested by an appropriate host, development may continue. Hatching occurs in the small intestine, and the larvae emerge, burrow through the intestinal wall, and are carried to the liver of the host via the hepatic portal vein. From the liver, the larvae migrate to the lungs. Later they move up through the bronchi and trachea and reach the oral cavity. From there, the worms are swallowed and are carried to the small intestine. Sexual maturity is reached within about two months, when fertilized eggs begin to be released. Fertilization occurs after copulation between male and female worms in the host's intestine.

A Parasitic Roundworm: *Trichinella spiralis*

Another common parasitic nematode is *Trichinella spiralis* the trichina worm. *Trichinella* is the causative agent of the disease **trichinosis** and occurs commonly in humans, swine, bears, rats, and other mammals. The disease is usually contracted by humans by eating raw or inadequately cooked pork or bear meat.

- Observe a microscope slide showing encysted worms in the skeletal muscles of a pig or a rat (figure 12.8).

Encysted worms in infected meat may be released through the digestion of the cyst walls by enzymes in the human digestive tract. The juvenile worms mature sexually in about two to three days and then mate. The females burrow into the wall of the small intestine and shortly after begin to produce living offspring. A single female may produce up to 1,500 young in her three-month life span. These juvenile worms enter the host's lymphatic system and are distributed through the body via the circulatory system. The young worms burrow mainly into the active skeletal muscles, where they grow to about 1 mm in length and become enclosed in cysts. Although such encysted worms cannot continue to develop unless the infected tissue is eaten by another suitable host, the encysted worms may remain alive within the cysts for several years.

Trichinella infections in humans often result in mild muscle aches that are mistaken for the flu. Such infections commonly go undiagnosed and sometimes are discovered only during an autopsy. Severe pain and serious damage can result, however, from the burrowing of adult and juvenile worms through the tissues of the body, and rare infections of the brain can result in death. Trichinosis has become less common in the US in recent years due to changes in raising pigs and improved hygiene.

Hookworms: *Ancylostoma duodenale* and *Necator americanus*

Ancylostoma duodenale and *Necator americanus* are the two most common and most serious human hookworms. *A. duodenale* is predominantly a northern species common in Europe, North Africa, northern China, and Japan, whereas *N. americanus* is primarily a tropical species occurring in many warmer areas of the world. At one time *N. americanus* was a very serious public health menace in the southeastern United States. Improved sanitation and control measures in the past few decades, however, have greatly reduced the severity of the hookworm problem.

Both species are principally human parasites, but are sometimes found in other animal species. Both species exhibit specialized **teeth** and/or **cutting plates** around the mouth (figure 12.9), which aid in the attachment of the adult worms to the wall of the small intestine. The body of *N. americanus* is smaller and more slender than that of *A. duodenale.*

- Study the demonstration slides of hookworms.

Life Cycle

The life cycles of the two species are basically similar. The adults live in the small intestine, where they attach to the

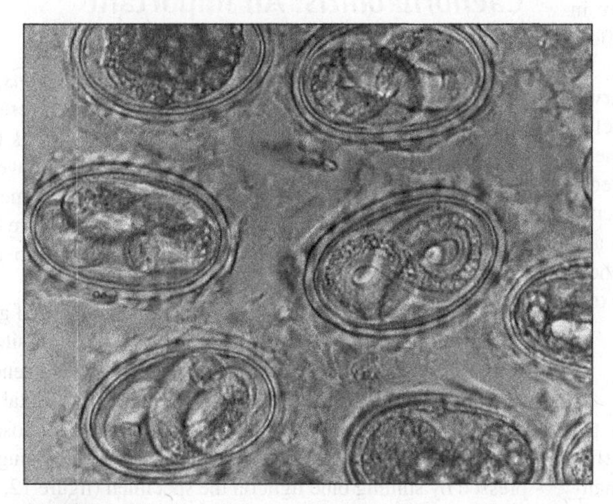

FIGURE 12.7 *Ascaris.* Eggs with developing larval worms.
Courtesy of Carolina Biological Supply Company, Burlington, NC.

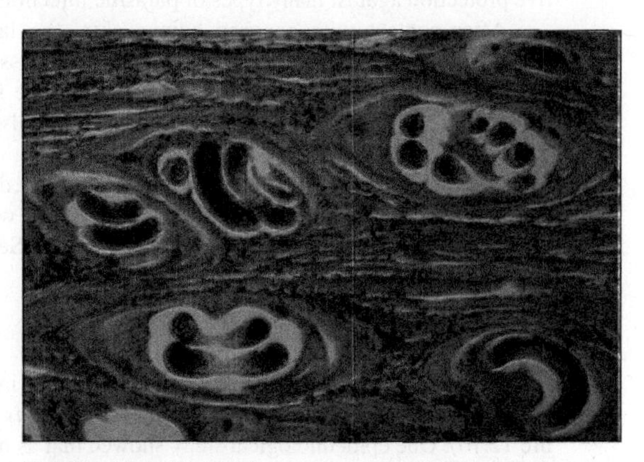

FIGURE 12.8 *Trichinella* larvae encysted in muscle tissue.
Courtesy of Carolina Biological Supply Company, Burlington, NC.

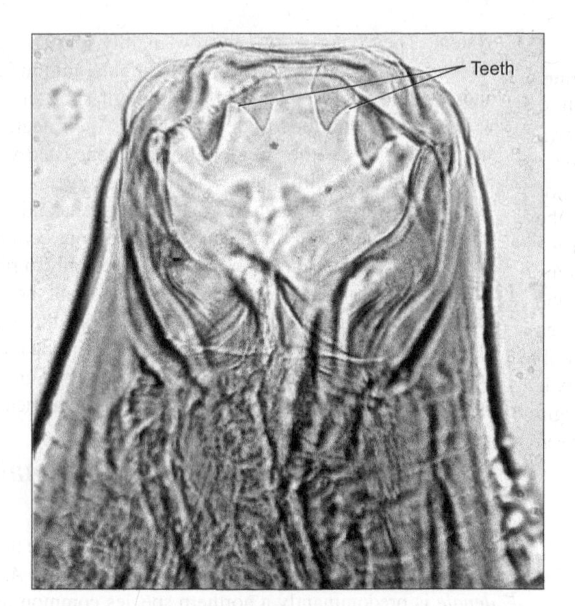

FIGURE 12.9 *Ancylostoma*, hookworm, mouth and buccal cavity showing teeth.
Courtesy of Carolina Biological Supply Company, Burlington, NC.

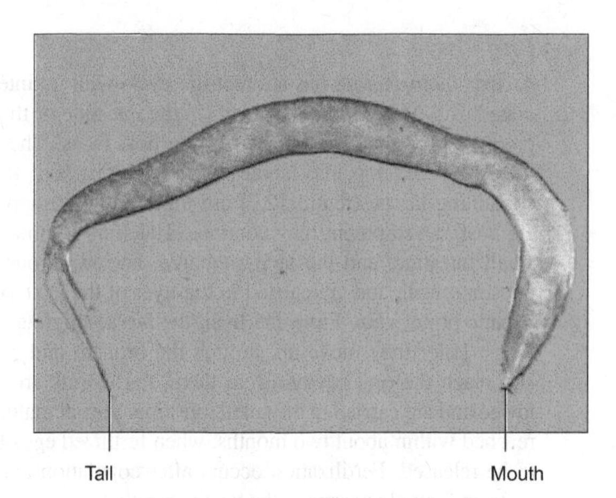

FIGURE 12.10 *Enterobius*, pinworm.
Courtesy of Carolina Biological Supply Company, Burlington, NC.

intestinal mucosa and feed on blood and fluids from tissues. Adult females produce large numbers of eggs each day (5,000–20,000). The eggs are released in the host feces, and the embryonated eggs develop and hatch in the dung. The young, free-living larvae feed on bacteria and other debris, grow, molt several times, and reach the infective stage in about five days. During this time, they live in the upper layer of soil and frequently migrate to the surface.

They are attracted to heat and are stimulated by physical contact. These stimuli facilitate contact and penetration of exposed human skin—such as that of a bare foot. Wearing shoes, therefore, instead of walking barefoot, is a very effective defense against hookworm infection. Often simple sanitation measures, like wearing shoes and taking care in the disposal of human feces, can be among the most effective protection against many types of parasitic infections.

After contact with exposed human skin, the larvae burrow through the skin until a blood or lymph vessel is reached. In the bloodstream, they are carried to the heart and then to the lungs. There they burrow into the alveoli. From the lungs, they are passed up the bronchi and trachea to the throat. From the throat, they may be swallowed and pass through the digestive tract to the small intestine, where they undergo a final molt and attach to the mucosa. Sexual maturity is reached in about six weeks.

Pinworm: *Enterobius vermicularis*

The most common nematode parasite of humans in the United States is the pinworm, *Enterobius vermicularis* (figure 12.10). One epidemiological study showed that as many as one-third of North American preschool children may be infected. Infections of adults are much less common.

- Obtain a microscope slide with a stained whole mount of an adult *Enterobius* and observe its size, shape, and internal organs.

Adult female pinworms are 8–13 mm long and have a pointed tail. They have enormous reproductive potential; a single female can lay more than 16,000 eggs in her lifetime.

Male and female adult worms live in the host intestine where they attach to the walls. Gravid females migrate posteriorly to the anus from which they often emerge at night to deposit their eggs on the surrounding skin. Infections and reinfections usually arise from ingestion of eggs transferred from contaminated bedding or from fingernails contaminated from scratching the anal area.

Free-living Nematodes

Caenorhabditis: An Important Model Research Animal

The free-living soil nematode, *Caenorhabditis elegans*, has become one of the most important animals used in research studies during the last 20 years (figure 12.11). It has been found to be useful for research studies in neural development, in the role of genes in development, and numerous other aspects of cell biology. To date, six scientists have been awarded a Nobel Prize for fundamental scientific discoveries made using *C. elegans* in their research.

One of those prizes was in recognition of the use of green fluorescent protein (GFP) for the precise location of sites of gene expression in living cells. After introducing the gene for production of GFP into the genome of an experimental animal, scientists can then visualize in a fluorescent microscope the precise location of cells in which this gene is being expressed by shining blue light on the specimen (figure 12.12).

C. elegans is a valuable animal for research studies because of its short life span, its transparent body, its simple

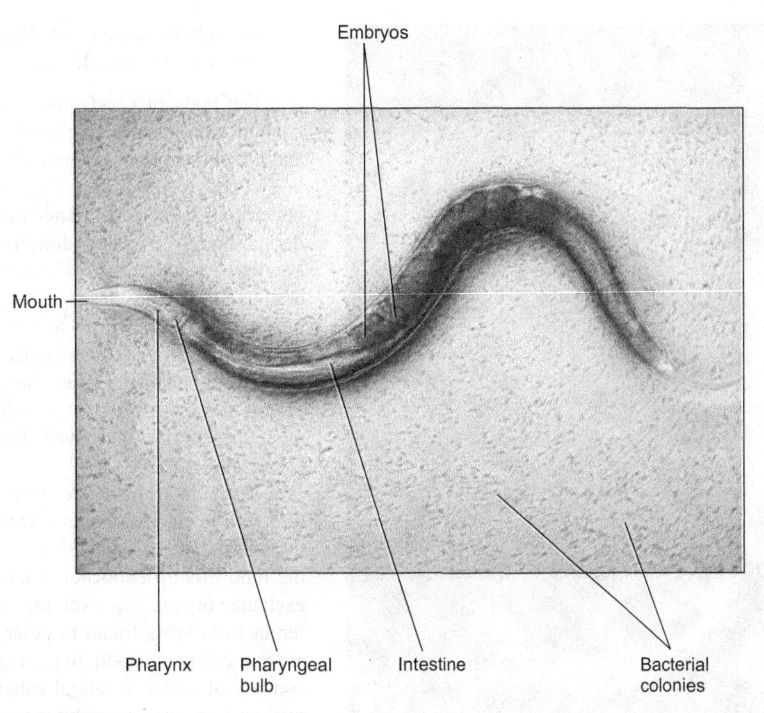

FIGURE 12.11 *Caenorhabditis elegans.*

Courtesy of Carolina Biological Supply Company, Burlington, NC.

anatomy, a fixed number of cells in the adult, and its ease of laboratory culture. The worms are easily cultured in petri dishes with a layer of *E. coli* on which they feed.

C. elegans naturally lives in temperate soils and reaches about 1 mm in length. It feeds on bacteria that grow on decaying plant material. Adult worms live about two weeks.

Like many other free-living nematodes, *C. elegans,* is predominantly hermaphroditic. Both males and females have been identified, but most populations consist primarily of hermaphrodites, typically with less than 0.05% males. Most individuals have two reflexed gonads that form a loop within the pseudocoel. Each tubular gonad consists of an **ovary** that extends anteriorly before looping back posteriorly and connecting with a larger **oviduct.** Locate the oviduct with large maturing **eggs** (and embryos) near the mid-ventral opening, the **vulva,** through which the embryos are released. **Sperm** are produced in the small **testis** also located in each tubular gonad. The sperm are stored in a spermatheca.

Eggs can be self-fertilized with sperm from the parent hermaphrodite or occasionally fertilized by sperm from a male. The eggs are fertilized internally in the hermaphrodite and laid a few hours later. After egg deposition and hatching, development takes about three days at 20°C during which the worms pass through four **larval (or juvenile) stages,** molting after each stage. The collagenous cuticle is shed and secreted anew after each larval stage.

Since males are produced only rarely in this species, self-fertilization in hermaphroditic individuals is the most common strategy of reproduction. Even this low frequency of males, however, is believed to be important in providing additional genetic variability to enhance the survival of the species.

Under certain conditions, crowding, starvation, etc., *C. elegans* can form a resistant stage called a **dauer larva** that can survive desiccation and other adverse conditions. This stage is similar to the infective stages of certain pathogenic nematodes.

Development has been studied so intensively that the embryonic development of *C. elegans* is more fully described than that of any other multicellular organism. For instance, a scientist can identify any specific cell at any time in development and know its complete developmental history. *C. elegans* has six chromosomes: five pairs of autosomes and either one or two X chromosomes. Males have a single X chromosome and hermaphrodites have two X chromosomes.

Similarly, its anatomy is known in great detail because of the small size and constancy among individuals. Scientists have been able to obtain more than 200,000 serial sections for electron microscopic studies to work out a detailed three-dimensional reconstruction of the entire worm. One example is the detailed structure of the entire nervous system, with its 302 neurons and their synaptic connections.

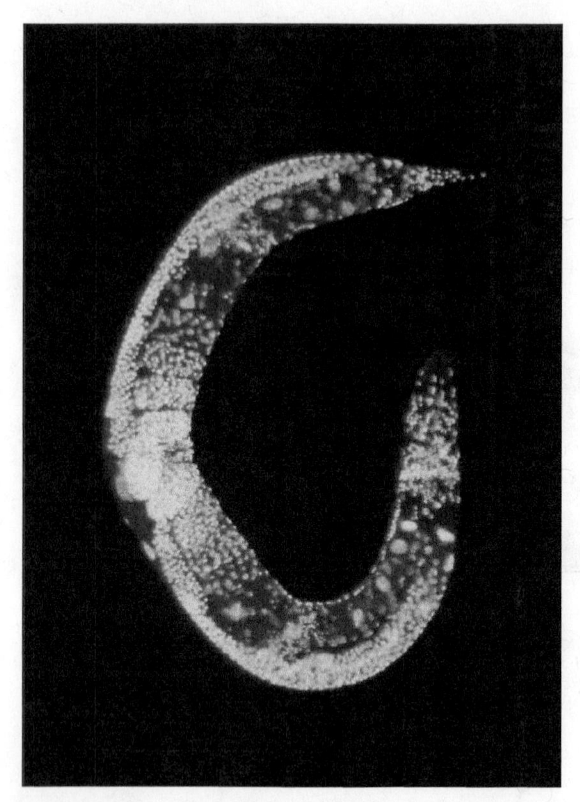

FIGURE 12.12 *Caenorhabditis elegans.* Fluorescent microscope photograph showing fluorescence of green fluorescent protein (GFP) due to activation of a specific gene at each site.

Leslie Saint-Julien, National Human Genome Research Institute.

Embryonic cleavage in *C. elegans* is precise and development is strongly determinate. Since the fate of each embryonic cell is fixed very early in development, the species is very useful in studies on the role of genes in development. The entire sequence of nucleotides in the genome of *C. elegans* has been determined and more than 500 genes have been identified and their roles investigated.

Many studies have followed the development of specific cells and the formation of their products, such as specific muscle proteins, in the larvae and adults of *C. elegans*.

If specimens of *C. elegans* are not available, *Rhabditis*, a related soil nematode similar in structure and readily available from most biological supply companies, can be substituted for this exercise.

- Prepare a wet mount with a few specimens of *C. elegans* and observe the slide under your compound microscope. Remember the modeling clay trick (see Chapter 6, page 85)! If the worms are too active for good observation, briefly hold the slide over a warm incandescent lamp or pass it quickly two or three times through the flame of an alcohol lamp or Bunsen burner to kill the worms. Be careful not to cook them!

The body of *C. elegans* resembles a flexible cylinder tapered at both ends; the anterior end is somewhat blunt, and the posterior end terminates in a pointed tail. The entire body is covered with a collagenous **cuticle** secreted by an underlying layer of **hypodermal cells.** Beneath the hypodermal layer is a layer of **longitudinal muscles** arranged in a series of muscle bands. There are no circular muscles. This arrangement of muscles results in the characteristic whiplike movements of the worms as they move along the substrate. The central cavity is a pseudocoel and is derived from the embryonic blastocoel (see Chapter 5). Within the pseudocoel is the digestive tract, which has no muscle layers or other mesodermally derived tissues separating it from the pseudocoelom.

The animals have a relatively simple histological structure with few types of specialized cells and tissues. Most of the internal organ systems are tubular and are suspended in the fluid-filled pseudocoel. There are no circulatory or gas exchange organs. The excretory system of nematodes is different from those found in other animals, and its functions are not well understood. In *C. elegans,* the **excretory system** consists of a pair of lateral canals that run along the body wall and are connected by a transverse canal that empties through a ventral excretory pore in the pharyngeal region.

The **digestive system** begins with an anterior mouth surrounded by three pairs of lips. The lips lead into the short **pharynx,** which terminates in a muscular **pharyngeal bulb.** Posterior to the esophagus is the long **intestine** which terminates in the **rectum.** The rectum empties through the ventral, subterminal **anus.**

The **nervous system** of *C. elegans* is simple, basically resembling that of *Ascaris* and consists of an **anterior nerve ring** that surrounds the pharynx, a **ventral nerve cord,** and a smaller **posterior nerve ring.** Several sensory and motor neurons connect with this central nervous system. Detailed studies of the nervous system of *C. elegans* have revealed that the entire nervous system of an adult of this species consists of only 302 neurons.

The Vinegar Eel: *Turbatrix aceti*

The vinegar eel (figure 12.13), *Turbatrix* is a tiny, free-living nematode often found worldwide in unpasteurized vinegar. Adult worms are about 1 mm long. Today, most vinegar sold in U.S. stores is pasteurized and has added preservatives to prevent the growth of vinegar eels and other undesirable flora and fauna. Vinegar eels are harmless, however, and would pose no health threat if accidentally ingested—they would just add a little protein to your diet!

Vinegar eels thrive in the acidic environment and feed on the yeast and bacteria growing in the sediments. This combination of yeast and bacteria is often called the "mother" of the vinegar. Most of the live worms are found near the surface of the liquid where the concentration of oxygen is higher.

From a biological point of view, "mother" of the vinegar is not a bad term for these sediments. ***Can you explain why this might be an appropriate name from what you know of the process of fermentation?*** Adult vinegar eels live about 10 months and become sexually mature in about four weeks. After maturity, females can produce about 45 young every 8 to 10 days, so a population can grow very rapidly.

Turbatrix has a long history in science. Vinegar eels were discovered in 1656 and were actually the first nematode described, 87 years before the first parasitic nematode was described in 1743. Interestingly, vinegar eels were used extensively as model organisms for research in embryology, physiology, ageing, toxicology, population growth, and other areas long before scientists turned to *Caenorhabditis elegans*. They are easy to culture and are readily available from biological supply companies.

- Examine a sample from the culture of the worms and observe their active, whiplike swimming movements. ***How can you relate their mode of locomotion to what you have learned about the anatomy of the nematodes? What kind of muscles are involved?***

  Select a few large worms for further study and place them in a small drop of vinegar on a clean microscope slide. If the worms are too active for study, they may be killed by briefly warming the slide over an incandescent lamp or a small flame. The specimens may then be observed unstained, or they may be lightly stained with hematoxylin or Nile blue sulphate in 70% alcohol.

Observe the blunt anterior end with its terminal mouth and the pointed posterior end. ***Where is the location of the anus?*** Identify the straight digestive tract consisting of the **mouth, pharynx, pharyngeal bulb** containing a three-part pharyngeal valve, an **intestine,** a thin-walled **rectum,** and a subterminal **anus.** In male specimens, the rectum opens into a **cloaca,** which in turn opens through the anus.

Select a large female worm for the study of the reproductive system. Identify the **female genital pore** on the ventral side of the body. This genital opening serves both for receiving sperm during copulation and for the release of the offspring, which are born live in this species. Extending posteriorly from the genital opening is a blind sac, the **seminal receptacle,** believed to serve for the temporary storage of sperm after copulation. Anterior to the genital pore is a single large **uterus,** which may contain several developing young. At its anterior end (about one-third the distance from the anterior tip of the body), the uterus is attached to a smaller **oviduct,** which bends posteriorly and connects with the threadlike **ovary,** where the eggs are produced.

Males may be recognized by their **copulatory spicules** and by the different structure of their reproductive system. Observe the opening of the **genital pore** as it enters the **cloaca,** the small spherical spermatozoa in the **vas deferens,** and the filamentous **testis,** which may extend almost to the middle third of the body, where it terminates in a slight enlargement. Observe a demonstration of a male worm that has been treated with strong formalin—this treatment causes the copulatory spicules to project from the anus.

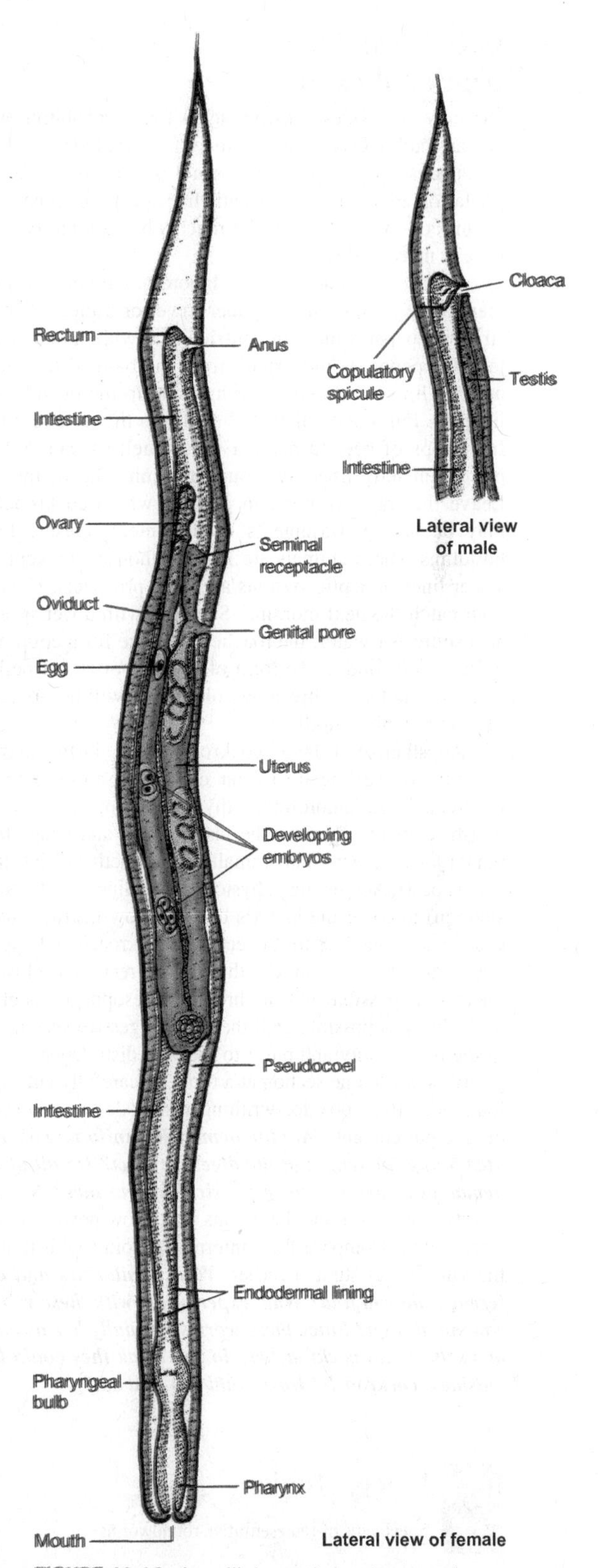

FIGURE 12.13 *Anguillula aceti*, the vinegar eel.

Cockroach Pinworms (Optional Exercise)

Virtually all species of cockroaches found cohabiting with humans harbor Oxyuroid nematodes in their hindguts. These nematodes (often called pinworms) belong to the family Thelastomatidae. They apparently live as endo-commensals, feeding on the contents of the roach's hindgut but causing little or no disability.

Use cockroaches from a laboratory colony or trap them alive by fastening a paper towel or a piece of construction paper around the outside of a wide-mouth glass jar. The paper should extend from the base of the jar to near its lip so a cockroach can climb up the outside and jump in. Put a slice of fresh bread into the jar and add a few drops of beer to make a yeast-smelling bait. Smear petroleum jelly liberally around the inner lip of the jar. Leave the trap overnight in an area where cockroaches may be active. Basements of apartment houses, farm buildings where animals are kept, outhouses, or vents to sewer lines or septic systems are good prospects. Harvest your catch the next morning. Supplied with a wet sponge as a source of water, the roaches will live for a couple of weeks. Add food in the form of dry dog or rabbit pellets and you can have a breeding colony that will be the envy of your neighborhood!

Anesthetize a large cockroach (late instar nymph or adult) by sealing it in a jar containing absorbent cotton treated with chloroform, ethyl acetate, or acetone (nail polish remover). Remove the insect's legs and wings then pin it (dorsal side up) into a small wax dissecting dish (black wax is best). Add enough physiological saline (0.75% NaCl solution) to cover the insect's body. Follow instructions in chapter 13, page 228 for dissecting a cockroach and exposing its digestive system. Cut through the rectum as close to the anus as possible and cut through the esophagus as close to the head as possible. Pull the entire digestive system out of the body cavity and pin it to the wax dish. Open the digestive system, one section at a time, by carefully cutting it longitudinally. Look for writhing nematodes as you tease out the gut contents. ***Are the nematodes uniformly distributed along the length of the digestive tract? Do all of the nematodes have the same physical appearance?*** Make a wet mount to examine the worms under low power in your microscope. Compare their internal anatomy with that of the vinegar eel studied earlier. ***What similarities and differences do you find? What experiments with these worms can you design? Since these worms normally live as commensals within cockroaches, do you think they could live outside a cockroach? How could you find out?***

Demonstration

Specimens of representative roundworms

Collecting Free-living Nematodes

Many species of free-living roundworms live in the soil, in decaying vegetation, and in many aquatic habitats, such as lakes, ponds, and streams. You can easily collect living aquatic nematodes for study in the laboratory by gathering small bits of decaying plant material from the edge or bottom of a pond or stream. Place the plant material in a small culture dish and gently tease it apart with dissecting needles and/or forceps.

You should easily identify the nematodes by their shape and characteristic whiplike movements. A stereoscopic microscope equipped with a substage mirror that allows viewing with transmitted light from the bottom works best with such material.

Soil-dwelling nematodes can also be obtained easily for laboratory study. Collect some samples of black or dark soil with ample organic material ("black dirt") and place small bits of the soil in water in a culture dish and examine in the same manner as described above.

Pieces of raw or boiled potato can also be used for bait to attract soil nematodes. Place the potato pieces in moist soil in the garden or a flower bed, under a rock or board, or beneath a log in the forest for a few days. After collecting the bait, you can immediately tease apart the potato pieces to search for nematodes, or you can incubate the pieces for a few days in a sterile petri dish or test tube in a warm place. The latter procedure may permit you to find eggs and developing larvae in the bait.

Phylum Nematomorpha

A Gordian Worm

Gordian worms, also called horsehair worms, are closely associated with water. They are called Gordian worms because of the tangles they often make with their very long and thin bodies, sometimes extending several meters in length (figure 12.14). The name comes from Greek mythology and the complex knots that King Gordius tied that no one else could untie. They are called horsehair worms also because of their long, thin bodies that resemble actual horsehairs. These worms are sometimes found in horse watering troughs as well as in ponds, streams, and at the edges of lakes. Among the common Gordian worms in the United States are species of the genera *Gordius* and *Polygordius*. Their anatomy and life history are similar although they have different hosts.

Adult worms are free living but the larvae are parasitic in crickets, cockroaches, locusts, and other insects. A few species are also found in spiders and terrestrial isopods (pill bugs or sow bugs). Although the biology of these worms is not well studied, their life cycle often begins with ingestion of tiny aquatic larvae by an appropriate host. Later, the larva penetrates the wall of the gut and enters the host body cavity. The larvae undergo one or more molts, mature in 4 to 20 weeks, and escape from the host in about two to

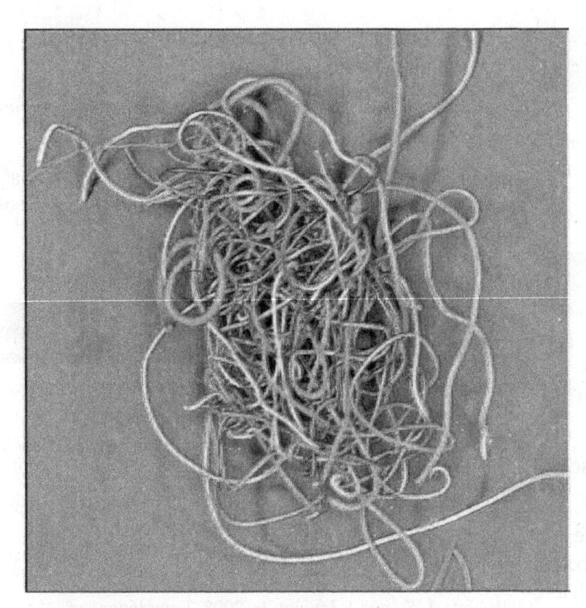

FIGURE 12.14 *Gordius*, adult worms, preserved.
Photo courtesy of Dr. Ben Hanelt and the Hairworm Biodiversity Project.

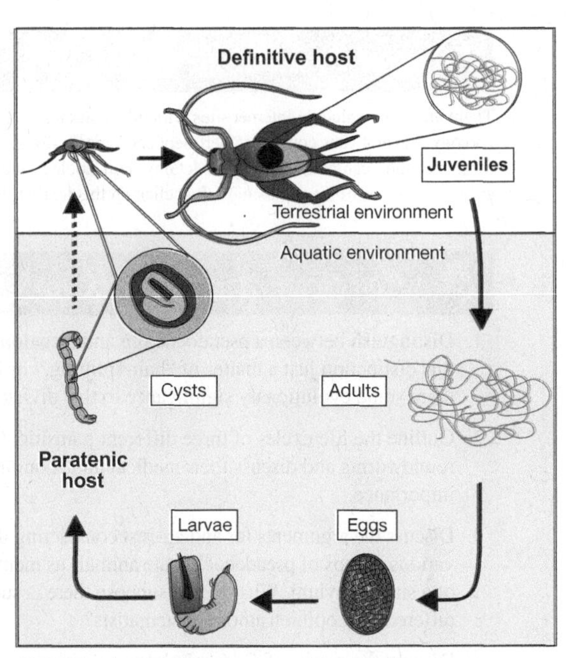

FIGURE 12.15 Life cycle of a Gordian worm.
Chart courtesy of Dr. Ben Hanelt and the Hairworm Biodiversity Project.

three months. It was long thought that the worms could only obtain nutrients while inside their cricket host because the digestive tract in the adults appears to be vestigial. Some recent studies, however, have provided evidence that the adult worms can also absorb some organic molecules through their vestigial gut and the body wall. Adult worms emerge after their host enters water. Research has shown that as it matures, the parasite modifies the behavior of its host and causes the host to enter the water. Once the worms sense the presence of water, they begin to emerge from the body of the host; this can occur in just a few seconds. The sexes are separate and the adult worms copulate in the water. Inseminated females deposit eggs in the water in long, gelatinous strings that hatch into small swimming larvae that may be ingested by a new host or, if not ingested, may form a cyst.

Another possible route for infection of an appropriate host and completion of the life cycle is believed to involve the infection of aquatic insect larvae in which the larval worms form cysts (figure 12.15). After the aquatic larvae emerge from the water and become flying insects, they become potential food for an appropriate host like a cricket. Crickets are known to be omnivores, and dead insects are a large part of their diet. This provides a possible route for infection of a suitable host for further development of the Gordian larva.

Internal anatomy is similar to that of nematodes in several respects, including structure of the body wall, cuticle, longitudinal muscles only in the body wall, and an anterior nerve ring around the pharynx. Differences include a single ventral nerve cord, a vestigial digestive tract, and the absence of circulatory, respiratory, and excretory systems.

Suggested Project on Gordian Worms

A group of international scientists are engaged in a study of the biodiversity of Nematomorpha. You can find information about the scientists and their research on these interesting but little-known parasitic worms on the following web site: http://www.nematomorpha.net/index.html

You can also assist in their research by contacting them through their Report-A-Worm program. Click on *Contact Us* on the group's home page.

Key Terms

Cloaca a chamber at the posterior end of the digestive tract in certain types of animals that serves to receive wastes from the digestive tract, gametes from the reproductive tract, and/or wastes from the excretory system.

Cuticle a thick, noncellular protective covering secreted by an underlying layer of epithelial cells, the hypodermis.

Hypodermis a type of tissue made up of epidermal cells that secrete an overlying cuticle.

Pseudocoelom a type of internal body cavity that lacks a mesodermal lining around the digestive tract. It arises embryonically as a remnant of the blastocoel.

Trichinosis a parasitic disease of humans and certain other carnivorous mammals caused by the nematode parasite *Trichinella spiralis.* Usually acquired by eating rare or uncooked pork or bear meat.

Internet Resources

There are many valuable Internet sites with information about zoology. Several sites containing pertinent zoological information for this chapter can be found on the McGraw-Hill Zoology web site at http://www.mhhe.com/zoology. Just click on this text's title.

Questions for Critical Thinking

1. Distinguish between a pseudocoelom and a coelom. Is this distinction just a matter of "hair-splitting," or is there some evolutionary significance to this division?
2. Outline the life cycles of three different parasitic roundworms and discuss their medical and economic importance.
3. Discuss the arguments for and against considering the various groups of pseudocoelomate animals as members of a single phylum. Why do you suppose there is such a difference of opinion among systematists?
4. Discuss the merits of *Caenorhabditis elegans* as a model laboratory animal for modern research. What disadvantages do you see with its use as a model laboratory animal?
5. Discuss the significance of completing the sequencing of the entire genome of *C. elegans*.
6. Why is it important that Gordian worms (as well as certain other internal parasites) alter the behavior of their hosts?

Suggested Readings

Brusca, R.C., and G.J. Brusca. 2002. *Invertebrates,* 2d ed. Sunderland, MA: Sinauer Associates. 936 pp

Pechenik, J. 2004. *Biology of the Invertebrates,* 5th ed. New York: McGraw-Hill. 590 pp

Polinar, C.O., Jr. 2010. Nematoda and Nematomorpha, pp. 237–277 in Thorp, J.H. and A.P. Covich. *Ecology and Classificatiion of North American Freshwater Invertebrates,* 3rd ed. (Aquatic Ecology).

Roberts, L.S., and J. Janovy, Jr. 2008. *Foundations of Parasitology*, 8th ed. Dubuque, IA: McGraw-Hill. 728 pp

Smith, D.G. 2001. *Pennak's Freshwater Invertebrates of the United States,* 4th ed. New York: John Wiley & Sons.

Notes and Sketches

Chapter 13

Arthropoda

NHPA/Robert Thompson.

Objectives

After completing the laboratory work in this chapter, you should be able to perform the following tasks:

1. Identify the physical characteristics that make Arthropoda distinctive from other invertebrate phyla.
2. Explain the difference between direct and indirect metamorphosis in arthropod development.
3. List the five subphyla of arthropods. Briefly characterize each subphylum and cite examples.
4. Identify the principal external features of the horseshoe crab, *Limulus;* briefly describe its method of feeding and locomotion.
5. Identify the chief external features of a spider on a preserved specimen.
6. Compare the basic morphological organization of a spider with that of the horseshoe crab, a crayfish, and a grasshopper.
7. Identify the chief external features of a crayfish. Compare and contrast the morphology of the crayfish with that of the horseshoe crab.
8. Discuss the concept of serial homology and give examples using the appendages of a crayfish.
9. Describe the respiratory system of a crayfish and explain which appendages are involved in ventilation or gas exchange.
10. Describe the circulatory system of a crayfish and compare it with that of an earthworm.
11. Describe the feeding mechanism of *Daphnia* and identify the chief structures involved.
12. Identify the principal external features of the cockroach *Periplaneta americana* using a specimen or a drawing.
13. Identify and explain the function of each of the principal internal organs of the cockroach *Periplaneta americana.*
14. Identify the major external features of the grasshopper *Romalea* and explain how they represent and illustrate the basic organization of an insect.
15. Describe the principal types of insect metamorphoses and development; identify the stages in complete metamorphosis.

Introduction

The arthropods represent the most diverse, abundant, and successful of the animal phyla. To date, over 1 million species of arthropods have been described. Recent estimates, however, suggest that this number may represent only ½ to ⅓ of the earth's total arthropod diversity. Members of this phylum dwell in almost every type of habitat—from the deep oceans to hot deserts and from coastal estuaries to mountain peaks. Their ecological roles include everything from herbivores to vertebrate parasites, benthic scavengers to filter-feeders, and zooplankton to pollinators of flowering plants. One class, the Insecta, is noteworthy as the only invertebrate taxon to have evolved the capacity for self-powered flight.

An arthropod's body is segmented, both internally and externally. In many cases, adjacent segments share a common function and may be organized into distinct body regions. These functional regions such as the prosoma and opisthosoma of a spider or the head, thorax, and abdomen of an insect are called tagmata (singular: tagma). Within a tagma, segments may be highly modified or fused together in various ways. Internally, the typical body plan includes a complete digestive system (mouth to anus), an open circulatory system with a dorsal heart that pumps blood forward toward the head, and a ventral nerve cord with segmental ganglia.

Exoskeleton

A chitinous **exoskeleton** is the hallmark of all arthropods. It serves not only as a protective covering for the body, but also as a surface for muscle attachment, a barrier to osmotic flow or evaporative loss of water, and a sensory interface with the environment. The exoskeleton is a laminated structure built from many layers of **cuticle,** a secretory product of underlying **epidermal cells.** Cuticle is a complex glycoprotein matrix containing various structural proteins bonded to microfibers of **chitin** (a large polysaccharide formed from N-acetylglucosamine monomers). Composition of the cuticle varies from place to place within the exoskeleton and is the main determinant of its physical properties. In some areas, adjacent protein molecules are cross-linked and stabilized with polyhydric phenols, quinones, or calcium compounds. These chemical reactions transform the cuticle into hard, rigid plates called **sclerites.** Elsewhere, the cuticle remains soft and pliable (as **membranes**) to allow movement.

Arthropods typically have a series of dorsal sclerites, **tergites** or **terga** (singular: tergum), and corresponding ventral sclerites, **sternites** or **sterna** (singular: sternum), connected laterally by flexible pleural membranes. Successive dorsal and ventral sclerites are joined by intersegmental membranes. This basic pattern of body architecture has been variously modified in different groups of arthropods. Sclerites often become modified through natural selection (fused together, enlarged, reduced, or vestigal). Adjacent segments may combine or fuse into functional body regions. Such flexibility in the exoskeleton and other body systems has certainly contributed to the great success of the arthropods.

Appendages

In early ancestors of the arthropods, each body segment bears at least one pair of lateral appendages. The basal portion of each appendage, the precoxa, arises from the lateral pleural membrane between the tergum and sternum. Further support for the appendage may be provided by development of additional pleural sclerites (pleurites) anterior and posterior to the point of attachment with the body. In marine arthropods, the appendages are often **biramous** (two-branched); they consist of a basal portion, the **protopodite,** and two apical branches, a lateral (outer) **exopodite** and a medial (inner) **endopodite.** Terrestrial (and most freshwater) arthropods have **uniramous** (unbranched) appendages.

Through natural selection and adaptive radiation, arthropod appendages have become variously modified for grasping, feeding, sensory perception, defense, swimming, walking, sperm transfer, and so forth. Because all of these appendages arise during development from similar embryonic tissue, they are regarded as homologous structures. The concept of **homology** is an important principle in zoology, and the study of structural homologies provides much valuable evidence about the evolutionary history and relationships of animals.

Molting and Metamorphosis

Despite its many advantages, the arthropod exoskeleton has limited ability to stretch or expand to accommodate an increase in body size. Therefore, an actively growing individual must periodically replace its exoskeleton with a new, larger version. This process is called **molting** (or ecdysis). It is a major event that occurs multiple times in the life history of all arthropods (and also in members of other phyla within the clade Ecdysozoa).

The molting process is triggered by hormones released as an arthropod's growth approaches the physical limits of its exoskeleton. Each molt represents the end of one growth stage **(instar)** and the beginning of another. In some species, the number of instars is constant but in others it may vary in response to temperature, food availability, or other environmental factors.

Molting provides a unique opportunity for postembryonic changes in body structure. These changes may include addition of body segments, modification of appendages, and in some cases, formation of wings. All such developmental changes in body form are known as **metamorphosis.** Since arthropods have become adapted to an extraordinary range of habitats and lifestyles, it is not surprising to find great diversity among their developmental life histories.

In some cases (spiders, centipedes, and crayfish, for example), eggs hatch into juveniles that are small versions of the adult—they change very little in appearance as they grow, molt, and mature into adults. This is known as **direct** (or epimorphic) development (figure 13.1*a*). In other arthropods (such as shrimp and barnacles), the egg hatches into a free-swimming, planktonic **larva** (called a nauplius) bearing little or no resemblance to the eventual adult. This is known as **indirect** (or anamorphic)

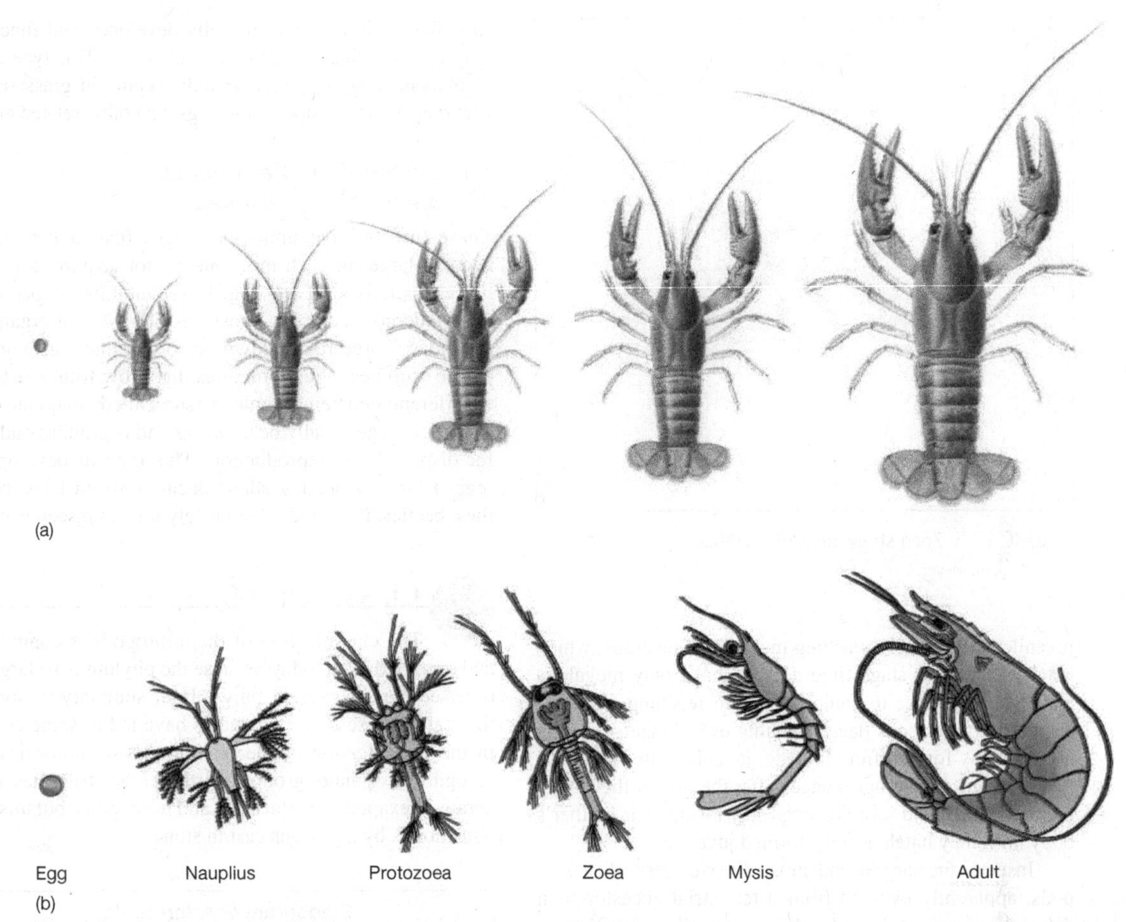

FIGURE 13.1 Two types of development in arthropods: (*a*) crayfish, direct development, (*b*) shrimp, indirect development.

development (figure 13.1*b*). The nauplius (figure 13.2) adds segments and appendages each time it molts, changing shape as it progresses through a series of growth stages. Brine shrimp, for example, molt about 15 times and pass through five distinctive stages before they reach maturity: nauplius, metanauplius, protozoea, zoea, and mysis.

Developmental life history appears to be related, at least in part, to the type of habitat and ecological role to which the animal has become adapted. Indirect development is most commonly found in marine crustaceans that spend the early part of their lives as zooplankton in the surface waters of the ocean. Direct development, on the other hand, is typical among arthropods that have become adapted for life on land or in fresh water. Some zoologists have suggested that indirect development represents the ancestral condition from which all arthropods evolved. As arthropods began to evolve specialized lifestyles, more and more stages of larval development occurred inside the egg before hatching. This is consistent with the observation that females of many crustaceans retain eggs in a brood sac for several weeks. Their larvae hatch at a later stage of development and pass through fewer

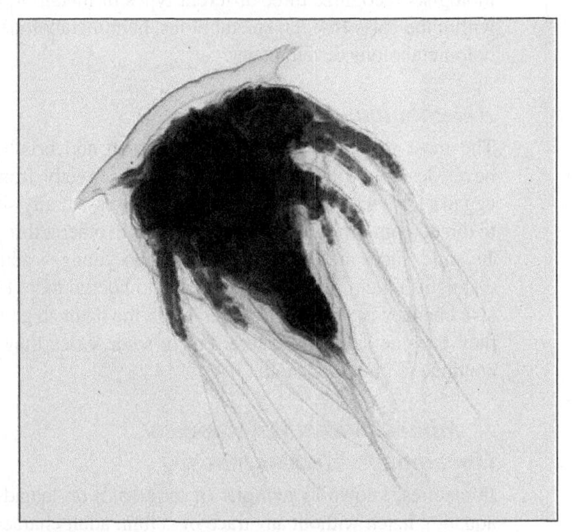

FIGURE 13.2 Nauplius stage larva of a barnacle.
Photo by John Meyer.

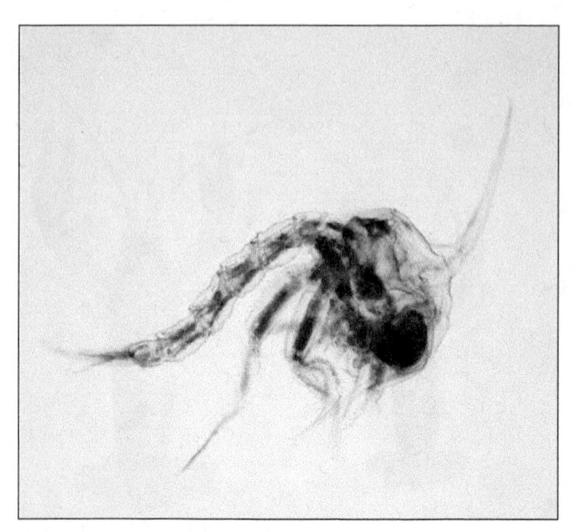

FIGURE 13.3 Zoea stage larva of a crab.
Photo by John Meyer.

juvenile stages before reaching maturity. Blue crabs, which hatch at the zoea stage (figure 13.3), have only megalopa and juvenile stages to complete before reaching maturity. Terrestrial arthropods delay hatching even longer—spiders emerge fully formed from the egg, juvenile mites and ticks add only one pair of appendages after the first of their three juvenile molts, and scorpion eggs stay inside their mother's body until they hatch as fully-formed juveniles.

Insects, the largest and most diverse class of arthropods, apparently evolved from a terrestrial ancestor with direct development. Eggs hatch into juvenile forms that, in some cases, undergo further metamorphosis as they develop wings or other structures found only in the adult stage. Entomologists recognize three different types of metamorphosis within the class Insecta: ametabolous, hemimetabolous, and holometabolous development.

Ametabolous Development

The most primitive insects, like silverfish and bristletails, never develop wings. These insects hatch directly from the egg in a form resembling miniature adults—essentially similar to direct (epimorphic) development found in other arthropods. Juvenile stages are called "young" (egg→ young→ adult) to distinguish them from other insects that undergo more extensive changes in form. Individuals reach the adult stage when they become sexually mature, but in some cases they may continue to grow and molt.

Hemimetabolous Development (Incomplete Metamorphosis)

Immatures, known as **nymphs** (if terrestrial) or "naiads" (if aquatic), hatch without any trace of certain adult characteristics. Wing buds and external genitalia begin to appear at the second or third molt, continue to grow and develop with each sucessive molt, and become fully developed and functional only after the final molt to the adult stage. This type of development (egg→ nymph→ adult) occurs in grasshoppers, cockroaches, leafhoppers, true bugs, and other related groups.

Holometabolous Development (Complete Metamorphosis)

These insects have immature stages (called larvae) that become larger at each molt but do not acquire any adult-like characteristics. Late-stage larvae molt into pupae where they undergo a complex transformation—larval organs and appendages are broken down (absorbed internally) and replaced with new adult structures that grow from clusters of undifferentiated (embryonic) tissue called imaginal discs. The adult stage usually bears wings and is primarily adapted for dispersal and reproduction. This type of development (egg→ larva→ pupa→ adult) occurs only in bees, butterflies, beetles, flies, and other closely related insect groups.

Classification

The classification of the arthropods is complicated and unsettled even today because the phylum is so large and diverse. We can present only a brief summary of some of the major classes. Recent studies have led to some changes in the classification of the phylum. Most authorities now recognize five major groups of arthropods—trilobites, chelicerates, hexapods, crustaceans, and myriapods, but this classification is by no means cast in stone.

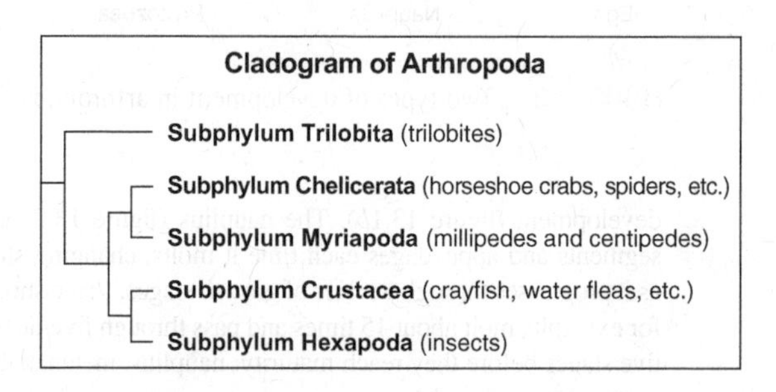

Subphylum Trilobita (Trilobitomorpha)

The Trilobites (figure 13.4). Primitive marine arthropods with a flattened, ovoid body divided into three parts, an anterior head (cephalon), a middle thorax: and a posterior pygidium ("tail"). Two longitudinal grooves divide the body into three lobes. One pair of simple antennae on the head; all other appendages biramous and similar.

Trilobites were abundant in marine habitats throughout the world during much of the Paleozoic Era. Populations declined during the upper Carboniferous Period and became extinct by the end of the Permian Period (230 million years ago). Over 1,500 genera have been described from the fossil record. They ranged in size from less than 10 mm to more than 60 cm in length.

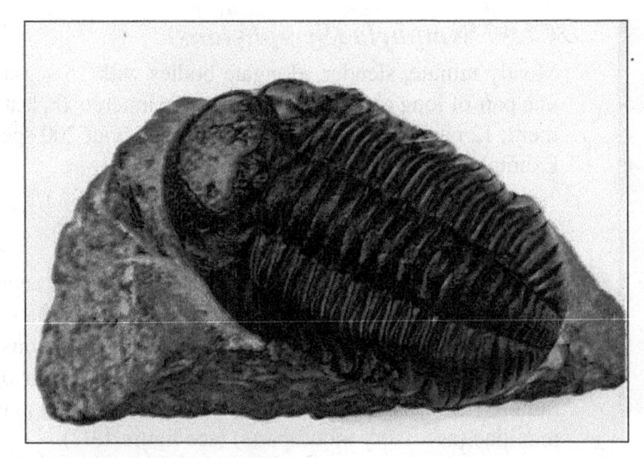

FIGURE 13.4 Fossil trilobite.
Photo courtesy of Charles Lytle.

Subphylum Chelicerata

The Chelicerates Body with distinct **prosoma** and **opisthosoma,** or prosoma and opisthosoma broadly fused. Antennae absent. Six pairs of uniramous appendages, including one pair of chelicerae (specialized pincerlike appendages modified for feeding). Chelicerates do *not* have mandibles.

Class Merostomata (Horseshoe Crabs)

Marine animals with abdominal book gills and one pair of lateral compound eyes. Well-developed cephalothorax and abdomen. Horseshoe crabs were a dominant marine life form until about 310 million years ago. Today only four species exist and are found only on the Atlantic coast of North America and on the Pacific coast of China and nearby areas. Example: *Limulus (Xiphosura)* (horseshoe crab).

Class Pycnogonida (Sea Spiders)

Small, marine, spiderlike chelicerates; well-developed cephalothorax and reduced abdomen. About 1,000 species. Example: *Pycnogonum.*

Class Arachnida (Scorpions and Relatives, Spiders, Harvestmen, Ticks, and Mites)

Mainly terrestrial animals; no marine species, although a few inhabit aquatic areas; with six pairs of appendages. From the anterior, the first pair of appendages are chelicerae with claw or fang; the second pair are pedipalps (variously modified for sensory perception, sperm transfer, grasping); the remaining four pairs are walking legs. Head and thoracic areas fused to form cephalothorax (= prosoma). Prosoma and opisthosoma distinct and narrowly connected by a pedicel, or else broadly fused. About 110,000 species. Examples: *Centruroides* (scorpion), *Argiope* (spider), *Dermacentor* (tick), *Trombicula* (mite) (see figure 13.5).

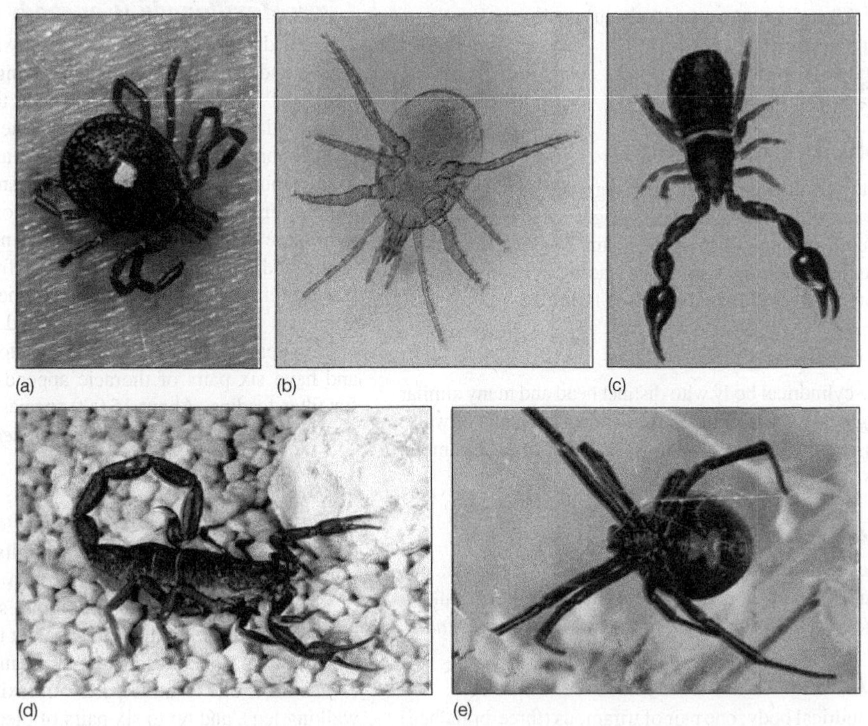

FIGURE 13.5 Some representative arachnids: (*a*) tick, (*b*) mite, (*c*) pseudoscorpion, (*d*) scorpion, (*e*) spider.
Courtesy of Carolina Biological Supply Company, Burlington, NC and John Meyer.

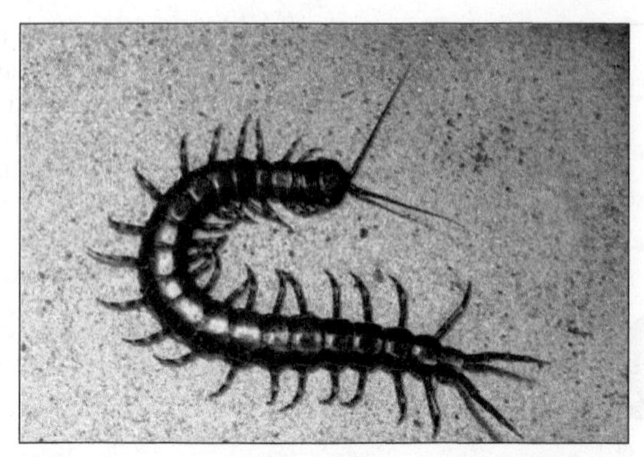
(a)

(b)

FIGURE 13.6 Common myriapods: (*a*) centipede and (*b*) millipede.
Photos (a) Clipart.com and (b) John Meyer.

Subphylum Myriapoda

Mainly terrestrial animals with two tagmata (head and trunk) and uniramous (unbranched) appendages; head with one pair of antennae, one pair of mandibles, and one or two pairs of maxillae. Respiration via spiracles leading to tracheal tubes; excretion by Malpighian tubules (see figure 13.6).

Class Diplopoda (Millipedes)

Elongate cylindrical body with distinct head and many similar trunk segments; one pair of short antennae; two pairs of walking legs per trunk segment. About 8,000 species. Example: *Narceus (Spirobolus).*

Class Chilopoda (Centipedes)

Elongate, flattened body with many similar segments; one pair of antennae; each trunk segment with one pair of walking legs. About 2,500 species. Examples: *Lithobius, Scolopendra.*

Class Pauropoda (Pauropods)

Tiny cylindrical body; one pair of triramous (three-branched) antennae; no eyes; trunk with nine or 10 pairs of walking legs. About 500 species. Example: *Pauropus.*

Class Symphyla (Symphylans)

Mostly minute, slender, elongate bodies with 15 segments; one pair of long antennae; one pair of spinnerets on last segment; 12 pairs of walking legs. No eyes. About 200 species. Example: *Scutigerella.*

Subphylum Crustacea

The Crustaceans Mainly aquatic animals with cephalothorax (fused head and thorax) usually covered by a dorsal carapace; hardened chitinous exoskeleton; biramous appendages. Head with one pair of antennae, one pair of biramous antennules, one pair of mandibles, and two pairs of maxillae (accessory mouthparts) (see figure 13.7).

Class Branchiopoda (Branchiopods)

Small crustaceans with trunk appendages flattened and leaflike, often modified for filter-feeding. Commonly live in temporary ponds or pools, mainly freshwater. About 800 species. Examples: *Daphnia* (water flea), *Artemia* (brine shrimp).

Class Ostracoda (Seed Shrimp)

Adult body is laterally compressed and protected by a clamshell-like carapace with a dorsal hinge. Usually with five pairs of appendages on the head, but only one to three pairs on a very short, unsegmented trunk. Common in both freshwater and marine habitats. About 13,000 species. Example: *Cypris.*

Class Maxillopoda (Copepods and Barnacles)

A highly diverse taxon characterized by a reduction of the abdomen and its appendages and by a single eye located in the middle of the the head. Copepods are tiny (usually <1 mm) but abundant in freshwater and marine habitats where they are a major component of zooplankton and serve as a primary food resource for other crustaceans, small fish, and baleen whales. Free-living forms swim with oarlike movements of their large antennae and thoracic appendages. Barnacles are larger and exclusively marine. The free-swimming larval stage resembles a copepod, but near the end of larval development it attaches to the substrate and loses its eye. Adults are sessile: they secrete calcareous plates over their carapace and have six pairs of thoracic appendages (cirri) modified for filter-feeding. About 15,000 species. Examples: *Cyclops* (copepod), *Balanus* (acorn barnacle), *Lepas* (goose barnacle).

Class Malacostraca

Large and diverse taxon containing over 22,000 species (more than half of all Crustacea). Adults usually have a head with one or two pairs of antennae, a 14-segmented trunk (8 thoracic somites and 6 abdominal somites), and an unsegmented telson. Each trunk segment typically bears a pair of biramous appendages including mandibles, two pairs of maxillae, one to three pairs of maxillipeds, five pairs of walking legs, and up to six pairs of pleopods (swimmerets). Most species are marine, but many have invaded freshwater habitats. Wood lice, also known as pill bugs or sow bugs

(a)

(b)

(c)

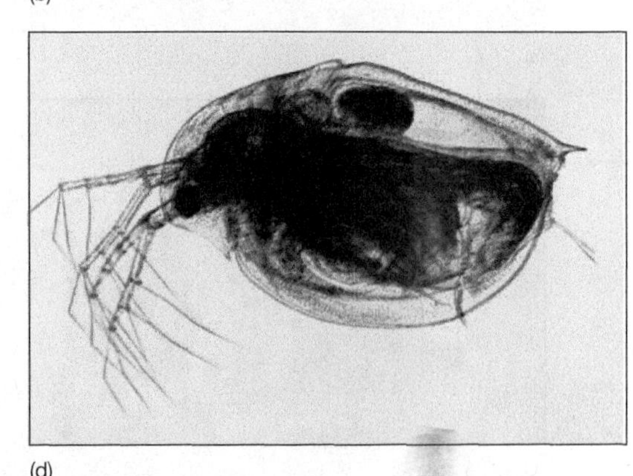
(d)

FIGURE 13.7 Representative crustaceans: (*a*) copepod, (*b*) amphipod or scud, (*c*) crab, (*d*) water flea.
Courtesy of Carolina Biological Supply Company, Burlington, NC and John Meyer.

(e.g., *Armadillidium* and *Porcellio*) are terrestrial, but restricted to moist habitats. Some members of the Order Decapoda are of considerable economic importance. These include *Homarus* (lobster), *Penaeus* (shrimp), *Procambarus* (crayfish), and *Callinectes* (blue crab). Equally significant are tiny marine forms in the Order Euphausiascea which, as larvae or adults, contribute to zooplankton and serve as a food resource for many other marine organisms. Example: *Euphausia* (krill).

Subphylum Hexapoda (Insects and Relatives)

Body with three tagmata (head, thorax, and abdomen), one pair of antennae, and three pairs of walking legs (see figure 13.8).

Class Insecta (Insects)

The largest and most diverse class on earth. More than 800,000 described species in about 30 orders occupying virtually every terrestrial and freshwater habitat. Includes scavengers, herbivores, carnivores, and parasites. Most species have one or two pairs of wings that are functional during the adult stage. Examples: *Romalea* (grasshopper), *Musca* (housefly), *Apis* (honey bee), *Pulex* (flea).

Materials List

Living specimens
- Crayfish (*Procambarus*, etc.)
- *Daphnia*
- Spider (*Argiope* or similar)

Preserved specimens
- *Limulus*
- *Romalea*
- *Argiope*
- Crayfish, female with eggs (demonstration)
- Representative crustaceans (demonstration)

Prepared microscope slides
- Crayfish, compound eye (demonstration)
- Crayfish, gill, cross section (demonstration)
- Insect cuticle (demonstration)
- *Argiope*, walking leg (demonstration)
- Mites and ticks

Plastic mounts
- *Argiope*
- *Peripatus*
- *Lepas*
- *Balanus*

Continued–

FIGURE 13.8 Common insects: (*a*) dragonfly, (*b*) beetle, (*c*) bee, (*d*) butterfly, (*e*) fly, (*f*) grasshopper.
Photos by John Meyer.

—Continued

Crayfish appendages
Millipede
Centipede
Honeybee
Insect life cycle

Audiovisual material

Anatomy of the Crayfish video

The Horseshoe Crab: *Limulus*

The horseshoe crab, *Limulus (Xiphosura) polyphemus* (figure 13.9), is common in the shallow waters and sandy shores of the Atlantic coast of North America from Canada to Mexico. It reaches a length at maturity of 0.5 meter. It is one of the few surviving members of an ancient group of

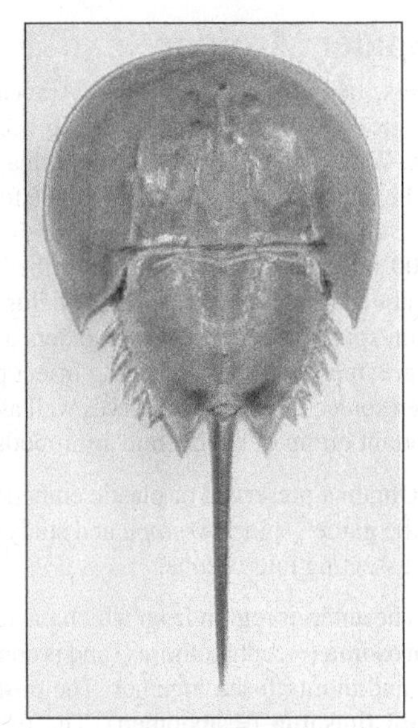

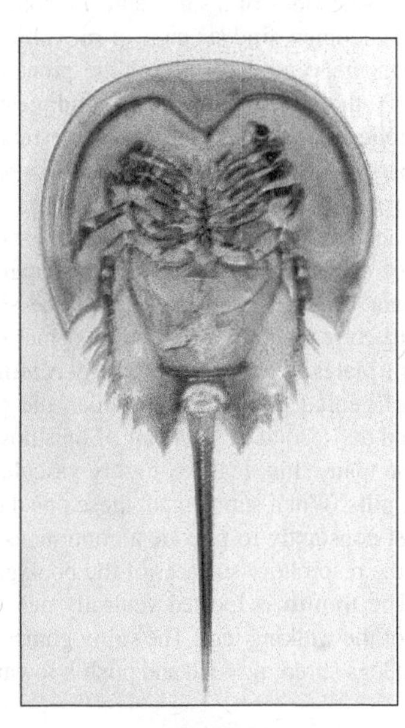

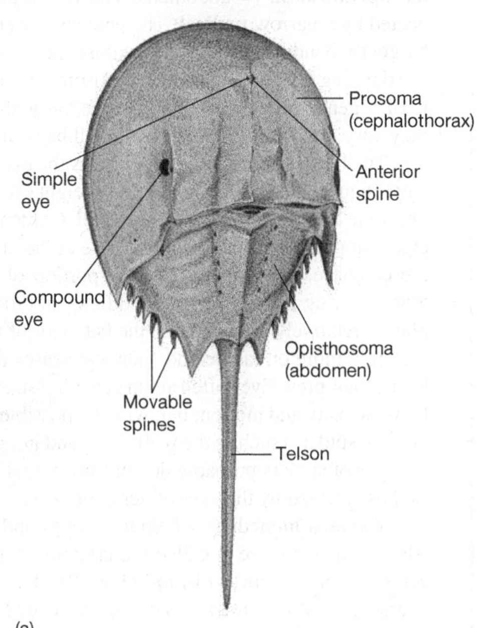

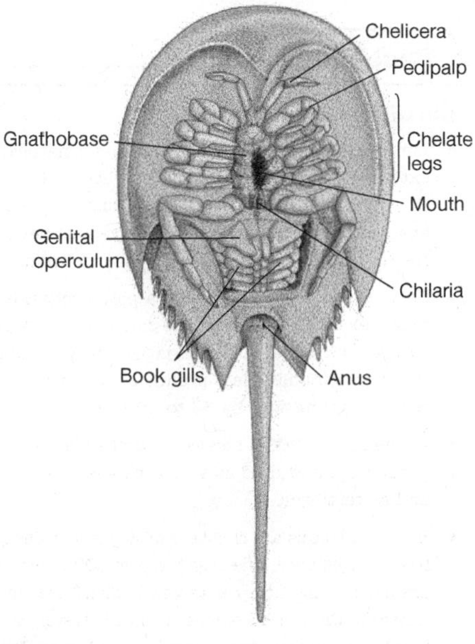

FIGURE 13.9 Horseshoe crab: (*a*) dorsal view, (*b*) ventral view.
Photo by John Meyer.

chelicerate arthropods whose ancestors date back at least 200 million years.

- Study a preserved or dried specimen of *Limulus* and observe the tough, leathery **carapace** that covers the **prosoma** (fused head and thorax) (figure 13.9*a*).

Behind the horseshoe-shaped cephalothorax is a tapering, hexagonal **opisthosoma** (abdomen) and a long posterior **telson.** On the dorsal surface of the carapace, identify the two lateral **compound eyes** and the two **simple eyes** located on opposite sides of a small anterior spine. Along the sides of the abdomen, find six pairs of **movable spines.**

On the ventral surface of the prosoma (figure 13.9*b*), identify the **seven pairs of appendages:** an anterior pair of **chelicerae** located in front of the mouth, four pairs of **chelate** (pincer) **legs,** and a longer sixth pair of appendages with specialized tips for cleaning the gills and pushing the carapace through the sea bottom sediments. Behind the sixth pair of appendages, locate a seventh pair of small, spiny **chilaria.** On the opisthosoma, find the six pairs of broad, flat structures, one pair per opisthosomal segment. The first pair of plates forms the **genital operculum,** which bears a pair of genital pores through which the gametes are shed. Behind the second to sixth pair of opisthosomal appendages are the many thin, leaflike, highly vascularized folds of the **book gills.** When submerged, these opisthosomal folds beat almost constantly to provide a continuous flow of seawater over the respiratory surfaces of the book gills.

The **mouth** is located ventrally near the center of the base of the walking legs. The spiny **gnathobases** at the base of the legs shred the food and push it toward the mouth.

Fact File

Horseshoe Crabs

- Limulus Amoebocyte Lysate (LAL) is a natural product extracted from the blood of horseshoe crabs. It is used throughout the pharmaceutical industry as a test to ensure that drugs and other intravenous devices are free of contamination by bacterial endotoxins.
- Chitin from horseshoe crabs, a polysaccharide found in the exoskeleton, is used to make nonallergenic sutures, wound-healing bandages, and skin dressings for burn victims. These products have been shown to reduce healing time by 35 to 50 percent.
- In the early 1900s, horseshoe crabs were dried, ground up, and used as a food supplement for poultry and as fertilizer.
- Horseshoe crabs are members of an ancient lineage that has changed very little over the past 500 million years. In prehistoric times, they were widely distributed in coastal waters of the world's oceans. Today, however, they are found only on the Atlantic coast of the United States and along the shores of China, Japan, and Indonesia.

Limulus feeds on worms, molluscs, and other small invertebrates in shallow ocean sediments. While *Limulus* burrows in the mud, it captures prey with its chelicerae. The crab plows through the sand by arching its back and pushing against the sand with the sixth pair of appendages and the telson. Awkward as it may sound, this crude form of locomotion and feeding has been effective for some 200 million years. *Limulus* is also able to swim short distances by vigorous beating of the abdominal platelike appendages.

A Spider: *Argiope*

Spiders, belonging to the Order Araneae, are another ancient group, with some fossils dating back over 300 million years. With two exceptions—Antarctica and the oceans below the intertidal zone—spiders can be found throughout the world. Of an estimated world total of 50,000 species, about 35,000 have been named and described. Spiders are often very abundant: for example, grassy fields may contain over 2 million spiders per hectare. All spiders are carnivorous, and thus are important for controlling insect pests. The common garden spider (*Argiope* sp.) serves well as an example of this important group of chelicerate arthropods (figure 13.10).

- Obtain a preserved or plastic embedded specimen of the garden spider *Argiope* and study it under your dissecting microscope.

The anterior region from which the legs project is called the **prosoma** (= cephalothorax) and is composed of the fused head and thorax fused together. The posterior part is called the **opisthosoma** (= abdomen). The two regions are connected by a narrow **pedicel.** The pedicel is seen most easily by gently bending the spider to expose the area between the overlapping opisthosoma and the prosoma. If your specimen is of a genus other than *Argiope,* the body shape and size may vary, but the anatomical parts will be similar.

The head region of the spider bears the eyes and mouthparts (figure 13.11). Most spiders have eight eyes (four pairs), but some have fewer or even none at all. Garden spiders have eight simple eyes called **ocelli,** visible at the anterior end of the carapace. The **carapace** is the portion of the exoskeleton covering the prosoma (fused head and thorax). Vision plays a relatively minor role in the behavior of most spiders; they rely more on tactile and chemical senses than vision to locate their prey. Eyes often aid in detecting subtle changes in light intensity and motion; but, with the possible exception of hunting spiders (such as the wolf, crab, and jumping spiders), the eyes of spiders probably do not form actual images, such as those formed by the eyes of vertebrates.

The area immediately below the eyes and down to the edge of the carapace is called the **clypeus.** A pair of chelicerae extend downward from below the clypeus. The chelicerae consist of a thick basal segment called the **paturon** and a small distal segment called the **fang** (figure 13.12). The fang articulates with the paturon. It is with the fang that a spider bites and poisons its prey.

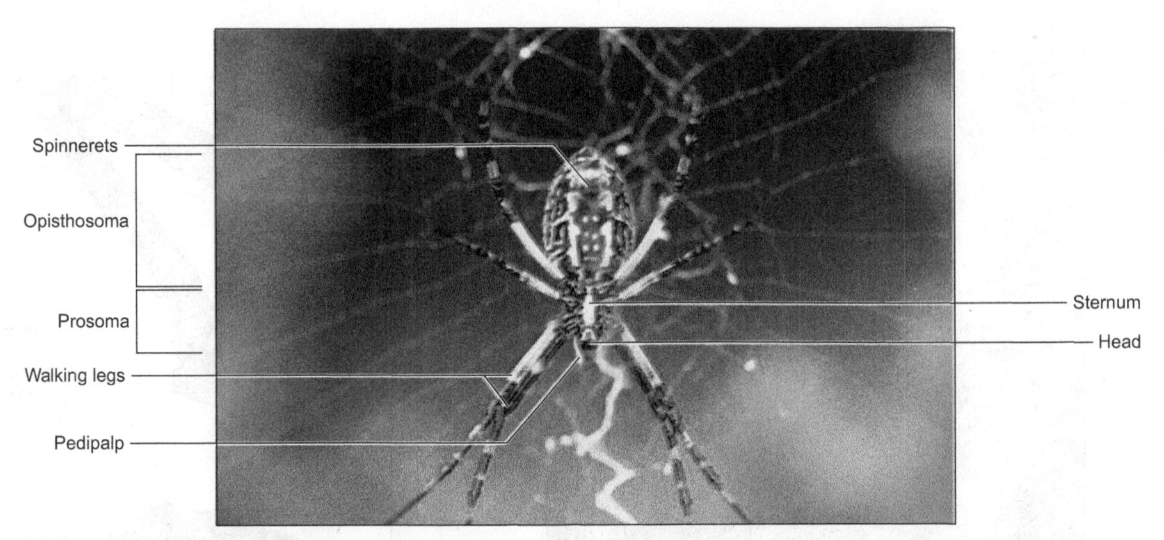

FIGURE 13.10 Garden spider, *Argiope*, ventral view.
Courtesy of Carolina Biological Supply Company, Burlington, NC.

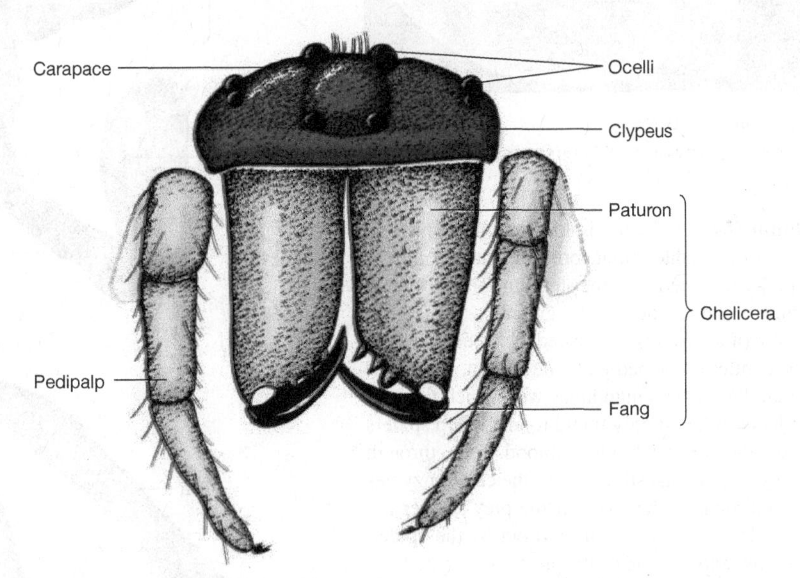

FIGURE 13.11 Garden spider, *Argiope*, anterior view of head.

Lateral to and slightly behind the chelicerae are a pair of **pedipalps** (= palps) that resemble small legs. The pedipalps are sexually dimorphic. In males, the distal end of the pedipalp is enlarged to form the **copulatory organ** (figure 13.13*b*). Now, turn the spider ventral side up and gently pry the legs away from the underside of the spider. Pedipalps consist of only six segments (figure 13.13*a* and *b*). The basal segment (coxa) of the pedipalp is expanded to form an accessory mouthpart called the **maxilla.** The maxilla often has a brush of hairs at its distal end called the **scopula.** The scopula aids in sponging up fluids from the prey as the spider feeds. The maxillae are manipulated to squeeze the fluids from the prey.

- Examine a walking leg from your specimen and in a prepared microscope slide. Identify the parts as shown in figure 13.13*c*. Compare the segments of male and female pedipalps (figure 13.13*a* and *b*) with those of a walking leg. ***How do these appendages illustrate adaptation? Which segments of the pedipalps show the highest degree of specialization?***

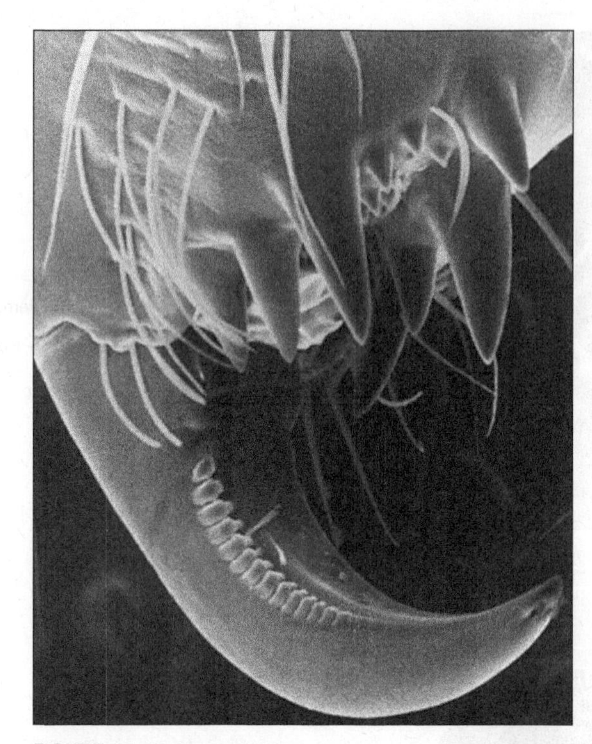
FIGURE 13.12 Fang of spider.
Scanning electron micrograph courtesy of North Carolina State University.

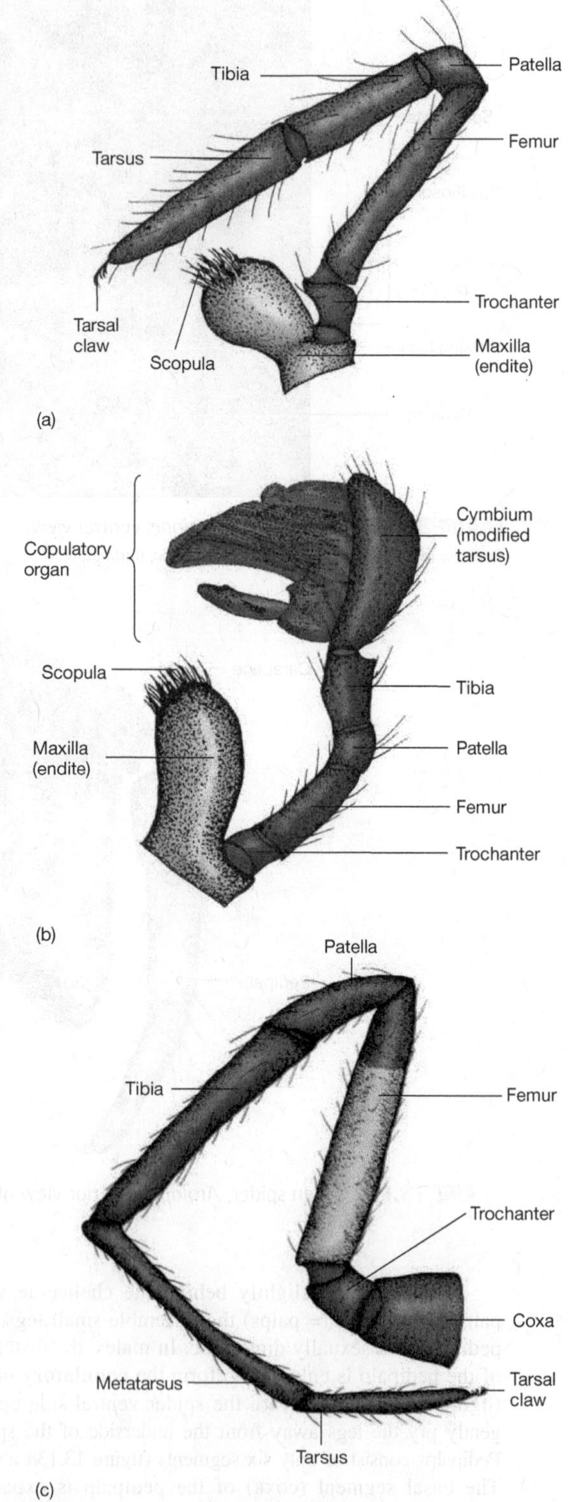

FIGURE 13.13 Garden spider, *Argiope*: (*a*) pedipalp of female, (*b*) pedipalp of male, (*c*) walking leg.

The **labium** (figure 13.14*b*) is located on the ventral surface between two endites (that form the anterior maxillae) and immediately anterior to the sternum. The labium is the ventral plate of the second fused segment.

The mouth (not visible) is located between the area formed by the endites of the pedipalp, the labium, and chelicerae. Note that there are no mandibles with which to chew the food. Spiders can ingest only liquid foods. When spiders "eat" their prey, they first suck out the blood. Then, through the wound, they pour in digestive juices (the chief enzymes are proteases and lipases) that convert the prey's inner tissues to a soup. The soup is then sucked out by the spider, leaving the hollow exoskeleton of the prey.

The **sternum** is the large plate on the ventral surface of the prosoma (figure 13.14*b*). It is surrounded laterally by the coxae of the four pairs of walking legs.

The pedicel connects the prosoma to the opisthosoma and represents the seventh body segment. Except in the very primitive spiders (Suborder Liphistiomorpha), none of the original opisthosomal segmentation is evident in spiders.

On the ventral, anterior portion of the opisthosoma, find the distinct groove that runs from one side to the other (figure 13.14*b*). This is the **epigastric furrow.** At each end of the epigastric furrow is a **lung slit,** marking the entrance to a **book lung.** Most of the air that larger spiders breathe enters through the lung slits. On the central, anterior edge of the epigastric furrow are the **gonopores.** If your specimen

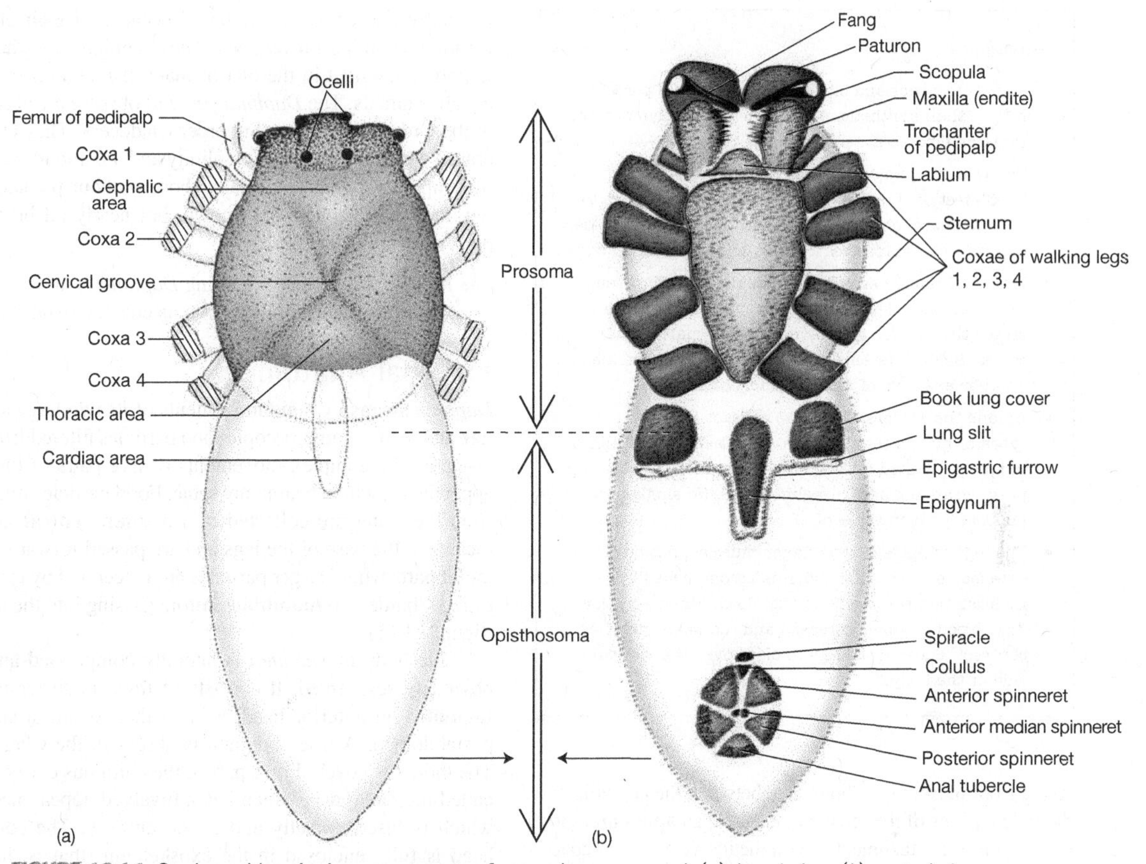

FIGURE 13.14 Garden spider, *Argiope*, most parts of appendages removed: (*a*) dorsal view, (*b*) ventral view.

is a female, there will be a large, intricate, sclerotized plate extending posteriorly, the **epigynum.** If your specimen is a male, a small plate may or may not be visible. Male spiders do not have a penis. Instead, the palpal organ of the pedipalp is enlarged and serves as the **copulatory organ** (penis analogue) (figure 13.13*b*). Some workers believe that the copulatory organ may have evolved from the tarsal claw. The female epigynum and male palps are very important taxonomic features.

Near the ventral, posterior end of the opisthosoma, locate the three pairs of **spinnerets** (figure 13.14*b*). Two pairs, the anteriors and posteriors, are larger and distinct; the anterior median pair is smaller and tucked in between the larger pairs. The single **anal tubercle** is located at the central posterior edge of the spinneret cluster. Between the anterior bases of the anterior spinnerets, find the small organ called the **colulus.** It is considered to be a small, vestigial spinneret. Immediately anterior to the spinnerets, in a groove, is an extremely small **spiracle,** which serves as the entrance to the tracheal system.

Turn the spider over and observe the dorsal side. On the central, slightly posterior part of the carapace is a depression that represents the **cervical groove** (figure 13.14*a*).

Fact File

Spiders

- During World War I, threads of silk from orb weaver spiders (*Araneus diadematus, Zilla atrica,* and *Argiope aurantia*) were used to make crosshairs for gunsights and other precision instruments.
- Young spiders may disperse on air currents by spinning out a long strand of silk that catches the wind like a parachute. This behavior, called ballooning, can carry the spider for long distances. In 1839, Charles Darwin observed ballooning spiders aboard the *H. M. S. Beagle* when he was more than 60 miles off the coast of South America.
- The weight of insects eaten every year by spiders is estimated to be greater than the total weight of the entire human population. Spiders can be very abundant, exceeding 2 million individuals per hectare in some grassland habitats.
- On an equal weight basis, spider silk is twice as strong as steel and more elastic than nylon. It is

Continued—

—*Continued*

also waterproof and biodegradable. It is 25 percent lighter than synthetic, petroleum-based polymers and remains flexible at temperatures as low as −40°C.

- The web of a typical orb weaver spider contains 20–60 feet of silk. The web takes about three hours to build and lasts only one night. Each morning, the spider eats its web, reprocesses the silk, and spins another web.
- Male and female spiders are usually very different in appearance. Females may be more than 10 times larger than males and have different markings and colors. Such differences between males and females is another example of **sexual dimorphism.**
- Among the spiders, there are at least seven different types of glands that produce silk. Each gland's silk is uniquely adapted for a special purpose (e.g., catching prey, wrapping eggs, dragline, etc.). No single species makes all seven types of silk.
- The legs of spiders have flexor muscles, but no extensor muscles. Extension is accomplished by hydrostatic pressure from the blood. When a spider dies, blood volume decreases and the flexor muscles contract, causing the legs to close over the ventral side of the body.

Aside from marking the boundary between the cephalic and thoracic regions of the prosoma, it is also an **apodeme,** serving as a muscle attachment site on the inside of the carapace.

There are no external, dorsal features worthy of note on the opisthosoma of spiders. However, in the central, dorsal anterior region is the cardiac area, beneath which lies the longitudinal **heart** that pumps blood anteriorly. Pale markings on the opisthosoma result from the deposition of the nucleotide guanine, an excretory product of spiders.

Demonstrations

1. Living *Argiope* or other spider in terrarium
2. Plastic mount of *Argiope*

Subphylum Crustacea, Class Branchiopoda

A Water Flea: *Daphnia*

The water flea, *Daphnia* (figure 13.15), is a common microscopic crustacean found in many bodies of fresh water. Its structure is simpler than that of the crayfish, and because of its transparency and small size, living *Daphnia* can be studied easily in the laboratory. *Daphnia* and other members of the Class Branchiopoda of the Subphylum Crustacea are popularly known as "water fleas" because of their characteristic swimming movements. These animals constitute an important element in the diet of many fish and other larger aquatic animals. The *Daphnia* you will observe are classified in the Order Diplostraca, Suborder Cladocera. They can tolerate a wide range of pH and salinity and are important fauna of temporary ponds, especially those without predators. A few marine Cladocera are known, but nearly all branchiopods are freshwater species.

- Make a wet mount of a living *Daphnia* on a microscope slide and observe its characteristic shape.

External Anatomy

Daphnia swims by rapid movements of its two large **antennae** and feeds on microscopic food particles filtered from the water by the complex movements of five pairs of thoracic appendages, which bear many setae. Food particles removed from the water are collected in a **median ventral groove** located at the base of the legs and are passed forward to the mouthparts where larger particles are macerated by the sclerotized (hardened) **mandibles** before passing into the mouth (figure 13.15).

The body of *Daphnia* is laterally compressed and not obviously segmented. It consists of three main regions or **tagmata:** an anterior head, a large thorax, and a smaller postabdomen. A true abdomen is absent in the Cladocera. The thorax is covered by a part of the chitinous exoskeleton called the **carapace,** which has a bivalved appearance, but which is fused dorsally and open ventrally. The compact head is fully enclosed in the exoskeleton (that is, it does not open ventrally) and is bent downward. Under certain conditions, some Cladocera exhibit seasonal changes in the morphology of the head by forming a dorsal elongation or "helmet." (Certain other parts of the body may be modified also.) This kind of morphological variation is called **cyclomorphosis,** and there is evidence that it is influenced by temperature, turbulence of the water, heredity, and other factors, such as the presence of predators.

The most conspicuous structure on the head is the single large compound eye with a central mass of pigmented granules surrounded by numerous hyaline lenses (figure 13.16). A single small *ocellus* (simple eye) lies posterior to the compound eye.

Locate the large **second antennae** attached near the posterior margin of the head. The **first antennae,** or antennules, are small and difficult to see. They are attached to the ventral side of the head just anterior to the attachment of the first antennae. The antennae bear **chemoreceptors** (specialized setae) that detect chemical changes in the surrounding water. The tapering projection of the head between the antennules is the **rostrum.**

The second antennae are very large structures that are used in swimming. Each second antenna consists of a stout, unjointed basal segment, and segmented dorsal and ventral rami. Each ramus consists of several segments and bears numerous **plumose** (featherlike) **setae.** The second antennae are

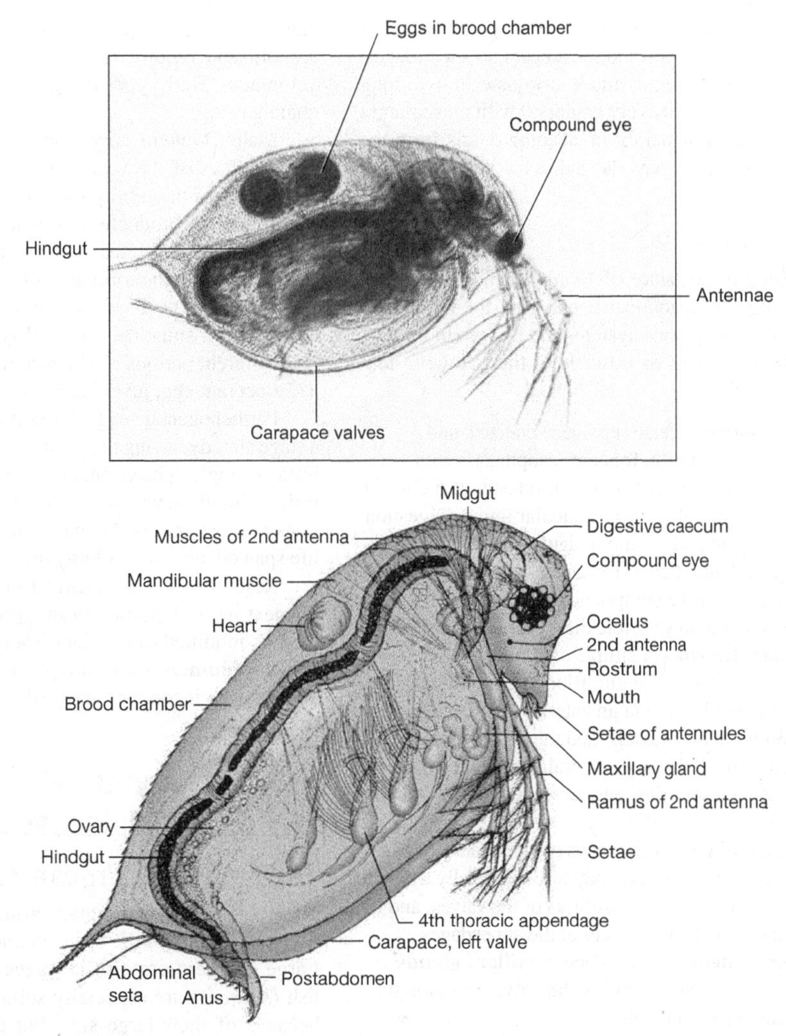

FIGURE 13.15 *Daphnia*, lateral view.
Photo courtesy of Carolina Biological Supply Company, Burlington, NC.

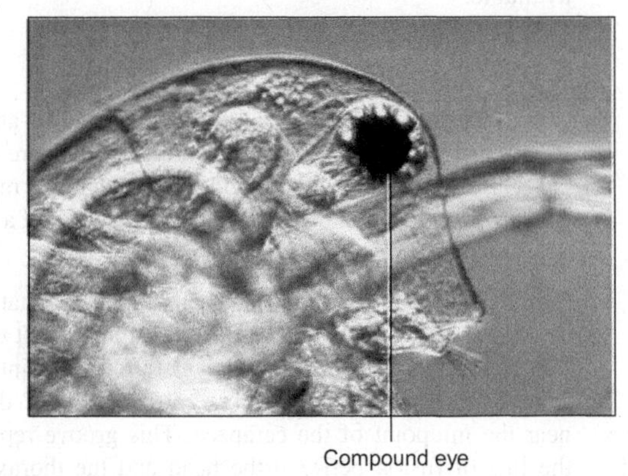

FIGURE 13.16 *Daphnia*, compound eye.
Courtesy of Carolina Biological Supply Company, Burlington, NC.

moved by a set of strong muscles that originate on the dorsal side of the carapace near the junction of the head and the thorax. Ventrally, near the same junction, the mouth may be found. It is surrounded by six mouthparts: (1) a dorsal median **labrum** (upper lip); (2) a pair of stout **mandibles;** (3) a pair of small pointed **maxillae,** which are used to push food between the mandibles; and (4) a ventral median **labium** (lower lip).

The **thorax** consists of six segments, and all but the fifth segment bear paired appendages. These five pairs of lobed, leaflike (foliaceous) appendages are basically biramous, but this condition is masked by their high degree of specialization. All of the thoracic appendages bear many hairs and setae used to strain food particles from the water.

As previously stated, a true abdomen is absent in *Daphnia,* but there is a definite **postabdomen** located at the posterior end of the body. It is usually held bent forward under the thorax so that the dorsal side of the postabdomen is directed downward. The postabdomen bears no appendages, but it has two rows of spines, and it terminates in two long abdominal setae. It is effectively equipped with musculature and appears to function mainly in cleaning debris from the thoracic legs, although it may also aid in locomotion.

Internal Anatomy

The complicated musculature of *Daphnia* tends to obscure some of the smaller anatomical details, but the major elements of most of the internal systems can be seen in a living specimen. Study as many as possible of the following features on your specimen.

1. **Digestive system**—Relatively unspecialized, and consisting of the **mouth, foregut** (esophagus) with a cuticular lining, **midgut** (stomach-intestine), and a **hindgut** (rectum), also with a cuticular lining. Digestion takes place principally in the midgut region, which does not have a cuticular lining. Located anteriorly (in the head region above the compound eyes) are two digestive caeca, which probably secrete digestive enzymes.
2. **Circulatory system**—A simple oval **heart** lies behind the head and dorsal to the gut. Blood enters through two lateral **ostia** and leaves via an anterior opening. There are no definite blood vessels, and the blood is roughly channeled through the large central **hemocoel** by a series of thin mesenteries, and powered by muscular contraction.
3. **Respiratory system**—Cutaneous or cuticular. Exchange of gases occurs through the body wall, especially through that portion lining the inner surfaces of the valves, and it also occurs through the surfaces of the appendages.
4. **Excretory system**—Two looped **maxillary glands** located near the anterior end of the valves are thought to be excretory in function.
5. **Nervous system**—Double **ventral nerve cord,** a few **ganglia,** paired **nerves,** and a **brain** located between the foregut and the compound eyes.
6. **Reproductive system**—Two elongated **ovaries** lie lateral to the midgut in the female *Daphnia.* Posteriorly, each ovary is connected to the **brood chamber** (a cavity dorsal to the body proper and located between the valves of the carapace; closed posteriorly by dorsal processes of the postabdomen) by a thin **oviduct.** In the male, two **testes** occupy the same location but are smaller in size and are continued posteriorly as **sperm ducts,** which follow the gut and open on the postabdomen near the anus. In some species of Cladocera, a part of the postabdomen of the male is modified to form a copulatory organ. The second antennae are large and inflated.

Under favorable conditions, a female *Daphnia* lays thin-shelled, diploid eggs, which contain little yolk and develop parthenogenetically. Under adverse conditions, however, the females produce thick-shelled, resistant eggs with large yolk reserves. The latter type of eggs are haploid and require fertilization by a male for their development. Both types of eggs are deposited in the brood chamber.

Male *Daphnia* appear in natural populations only at certain times of the year and often constitute only a small proportion of the total population. Recent research has indicted that the production of males in Daphnia populations is induced by the release of crustacean juvenile hormone which results in the formation of male eggs.

There are no larval stages in the development of the Cladocera (a situation not at all typical of Crustacea!), but four different periods can be recognized in the life history of a cladoceran: egg, juvenile, adolescent, and adult.

Parthenogenetic eggs develop in the brood chamber into a stage already having the adult form before their release (first instar young). Subsequent development continues through a series of molts in which the carapace and other chitinous portions are lost and the animal increases in size. The average life span covers about 17 **instars** (instar = intermolt stage).

Resistant eggs are retained in the brood chamber until the next molt. A portion of the brood chamber surrounding them is modified into a thick-walled protective case called an **ephippium.** At the next molt of the parent, the ephippium and the eggs it contains are shed.

Subphylum Crustacea, Class Malacostraca

The Crayfish: *Procambarus*

The crayfish is a large aquatic arthropod, which effectively illustrates many of the basic characteristics of the phylum. Large southern crayfish (*Procambarus*) or western crayfish (*Astacus*) are especially suitable for laboratory study because of their large size, but members of other common genera, such as *Cambarus* and *Orconectes,* can also be used. The American lobster *Homarus* is very similar in structure to the crayfish and may also be used when available.

External Anatomy

- Obtain an anesthetized or freshly killed crayfish and study the general organization of the body (figure 13.17). Observe that the body is divided into two major regions: an anterior **cephalothorax,** which is made up of a fused head and thorax, and a posterior **abdomen.**

The carapace is a rigid shield of exoskeleton that covers the dorsal and lateral sides of the cephalothorax. It extends forward between the compound eyes to form a pointed rostrum. The cervical groove forms an inverted "V" dorsally near the midpoint of the carapace. This groove represents the line of fusion between the head and the thorax. Both the carapace and the dorsal sclerites of the abdomen are hardened by the deposition of mineral salts (except in newly molted specimens). The crayfish body is made up of a total of nineteen segments. Segments 1–13 make up the cephalothorax (five in the head and eight in the thorax), and segments 14–19 constitute the abdomen.

Observe the two large **compound eyes** on movable eyestalks, two large **antennae** with a large scalelike exopodite (sometimes called second antennae), and two smaller biramous **antennules** (sometimes called first antennae). The **mouth** is ventral, surrounded and largely concealed by several pairs of modified appendages that serve as mouthparts (these will be studied later). Locate the five pairs of large **walking legs** and five pairs of small abdominal appendages or **swimmerets.** The first pair of walking legs bear large pincers or **chelae** and, thus, they are called **chelipeds** ("pincer legs"). ***How many of the other pairs of walking legs also bear chelae?*** On the last abdominal segment is a pair of large flattened lateral appendages, the **uropods.** Find the **anus** on the ventral side of the **telson,** a medial extension of the last abdominal segment (not an appendage).

Crayfish are dioecious, and each sex exhibits distinctive secondary sexual characteristics. Females (figure 13.18*a*) have a broad abdomen, rudimentary appendages (swimmerets) on the first abdominal segment, a **seminal receptacle** on the ventral surface of the exoskeleton between the fourth

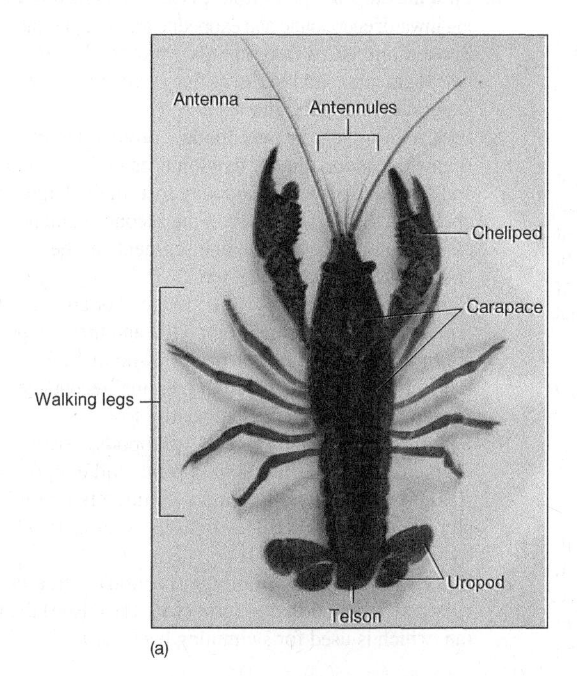

(a)

(b)

FIGURE 13.17 Crayfish: (*a*) dorsal view, (*b*) ventral view.
Photos by John Meyer.

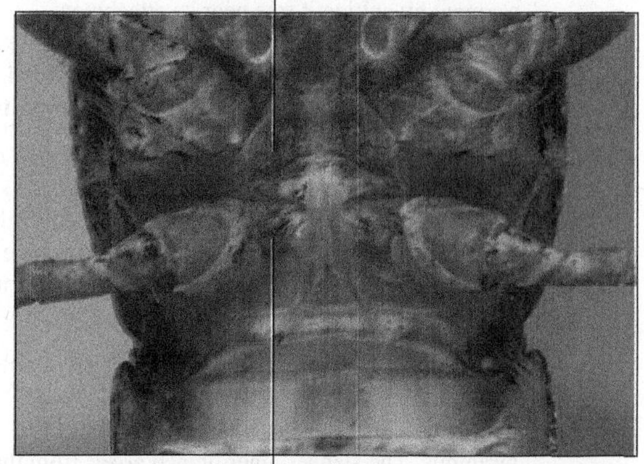

(a)

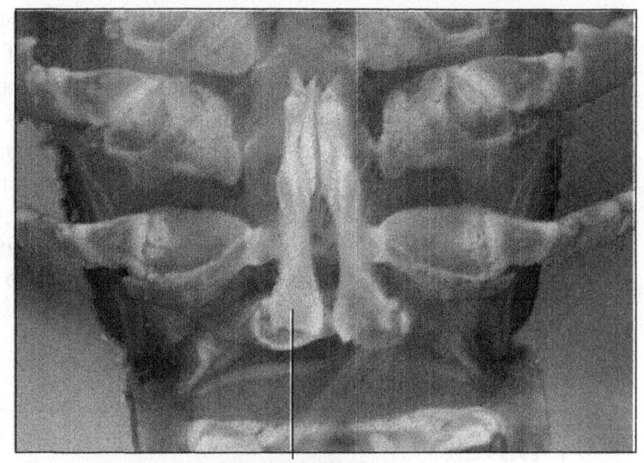

(b)

FIGURE 13.18 Crayfish, sexual dimorphism on ventral side: (*a*) female, (*b*) male.
Photos by John Meyer.

and fifth pairs of walking legs, and two female sex openings located at the bases of the third pair of walking legs. Males (figure 13.18*b*) have a narrower abdomen, enlarged swimmerets **(gonopods)** on the first abdominal segment (modified for the transfer of sperm to the seminal receptacle of the female), and openings of the vasa deferentia (male sex ducts) at the base of each fifth walking leg.

- Determine the sex of the specimen you are examining and also observe specimens of the opposite sex.

Appendages

The paired appendages of the crayfish provide an excellent illustration of **serial homology,** the adaptation of a longitudinal series of originally similar structures to carry on different functions. The appendages of primitive arthropods probably were **biramous** (two-branched). This is the simplest type of arthropod appendage and is most nearly approximated in the crayfish by the swimmerets of the abdominal segments.

- Remove an appendage from the third abdominal segment by cutting it free near its attachment, and observe that it consists of a basal portion, the **protopodite,** and two terminal branches, a lateral (outer) **exopodite** and a medial (inner) **endopodite.**

Carefully remove the appendages from one side of the body, starting with the first walking leg (cheliped). Next, remove the appendages forward of this cheliped, followed by those appendages behind the cheliped, and arrange them in order on a sheet of paper. Study each appendage with the aid of figure 13.19 and the following descriptions.

1. **Antennule** (first antenna) base composed of three segments, followed by two long, many-jointed filaments. This appendage is probably not truly biramous. The slender inner branch arises from the base of the outer thicker filament (expodite) and is, therefore, probably not homologous with an endopodite. The three basal segments are homologous with the protopodite. Within the basal segment of each antennule is a **statocyst,** a saclike sense organ for balance that opens dorsally via a small pore.
2. **Antenna** (second antenna) protopodite of two segments; endopodite long, many-jointed; exopodite a short scale.
3. **Mandible** basal segment of protopodite greatly enlarged as functional jaw; second segment small and forming the base of the palpus; endopodite, the two small distal segments of the palpus; exopodite secondarily lost.

NOTE:

The metastoma, or bifurcated "lower lip" between the mandible and the first maxilla, is a leaflike blade that fits closely over the convex surface of the mandible. It is not a true appendage.

4. **First maxilla** protopodite of two flat, leaflike segments; endopodite of two segments, the distal one narrow and pointed; exopodite secondarily lost.
5. **Second maxilla** protopodite of two flat, bilobed segments; endopodite, a small pointed segment; protopodite forming part of a large elongated plate, the **bailer.**
6. **First maxilliped** protopodite of two flat segments extending inward; endopodite and exopodite are both present.
7. **Second and third maxillipeds** protopodite of two segments; endopodite of five distinct segments; exopodite relatively thin.
8. Five walking legs or **pereopods** protopodite of two segments; endopodite of five joints as in the second and third maxillipeds; exopodite lost. In the large chelipeds (first walking legs), the second segment of the protopodite and the first segments of the endopodite are fused together.
9. **First swimmeret** (gonopod) reduced or absent in the female; in the male, the protopodite and the endopodite are fused, and the entire structure is modified for transferring spermatozoa to the seminal receptacle of the female; exopodite is secondarily lost.
10. **Second to fifth swimmerets** (pleopods) two-segmented protopodite; endopodite and exopodite flat and filamentous; second swimmerets might be slightly enlarged in males to assist sperm transfer by gonopods.
11. **Sixth abdominal appendage** (uropod) greatly enlarged and modified to form (with the telson) the tail fan, which is used for swimming backward.

Fact File

Crustacea

- It takes up to seven years for a lobster to grow to 1 pound, the smallest size legal for trapping. "Big George," a lobster caught off the coast of Cape Cod in 1974, weighed 37.4 pounds and was over 2 feet in length.
- Mantis shrimp (Stomatopoda) have the most sophisticated visual system in the animal world. Each stomatopod eye contains 16 different types of photoreceptors. It is able to perceive both ultraviolet and polarized light, and it can discriminate up to 100,000 different colors (about 10 times greater than the human eye).
- The European green crab, *Carcinus maenas,* and the Chinese mitten crab, *Eriocheir sinensis,* are invasive species that have become established on the west coast of the United States. They are regarded as a threat to coastal diversity and the commercial shellfish industry.
- "Sponge crabs" are females of the blue crab, *Callinectes sapidus,* that carry a spongy, external egg mass on the underside of their abdomens. A single large female may produce up to 8 million eggs during one spawning season.

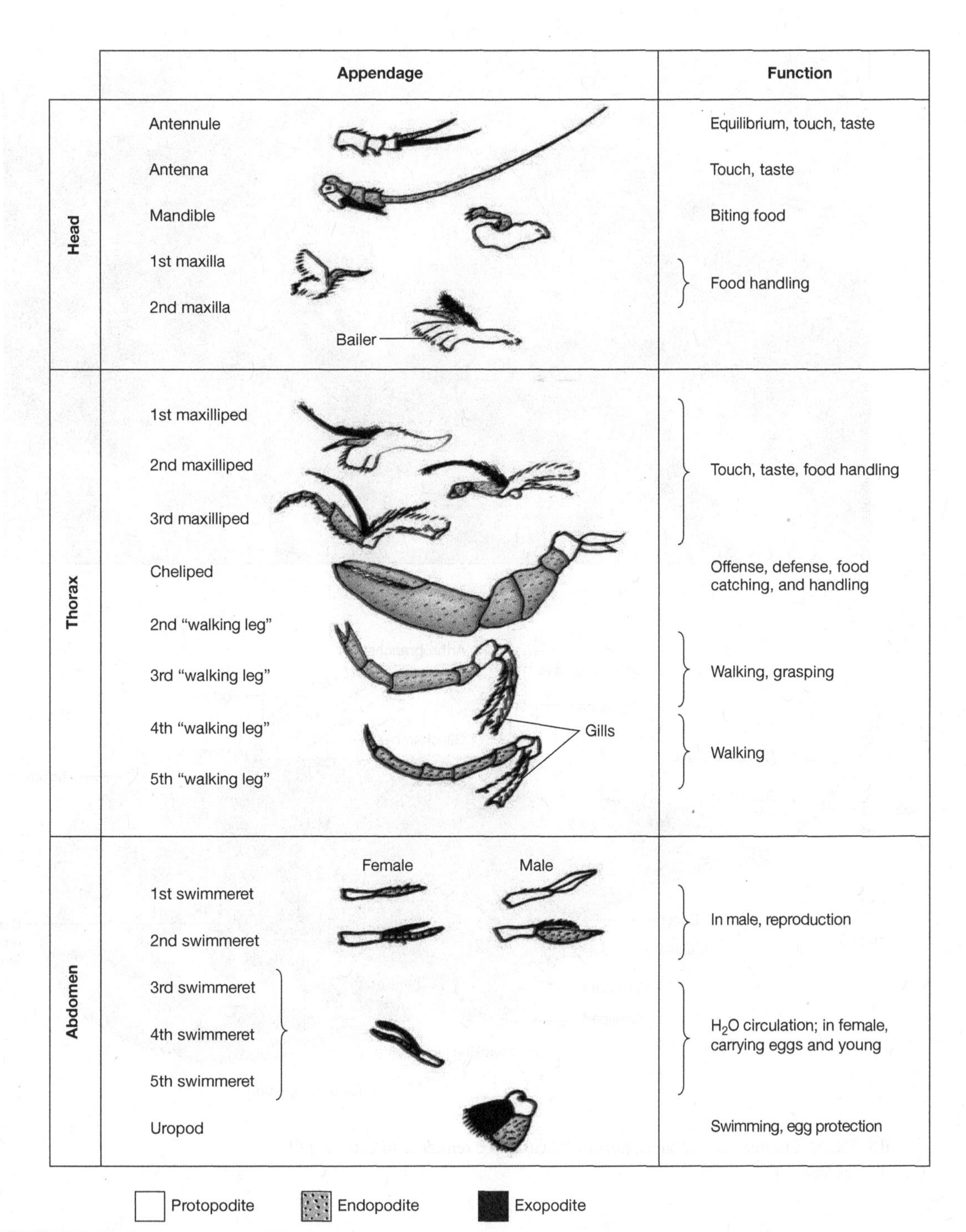

FIGURE 13.19 Functions of crayfish appendages.

Respiratory System

- Locate the gills within the **gill chambers** at each side of the carapace (figure 13.20). The lateral flaps of the carapace, covering the gills, are termed **branchiostegites.** Remove the left branchiostegite by cutting away the carapace with your scissors to expose the gills; make your cut carefully to avoid damage to the underlying gills.

The gills are feathery projections of the body wall that contain blood channels. Water is pumped through the gill

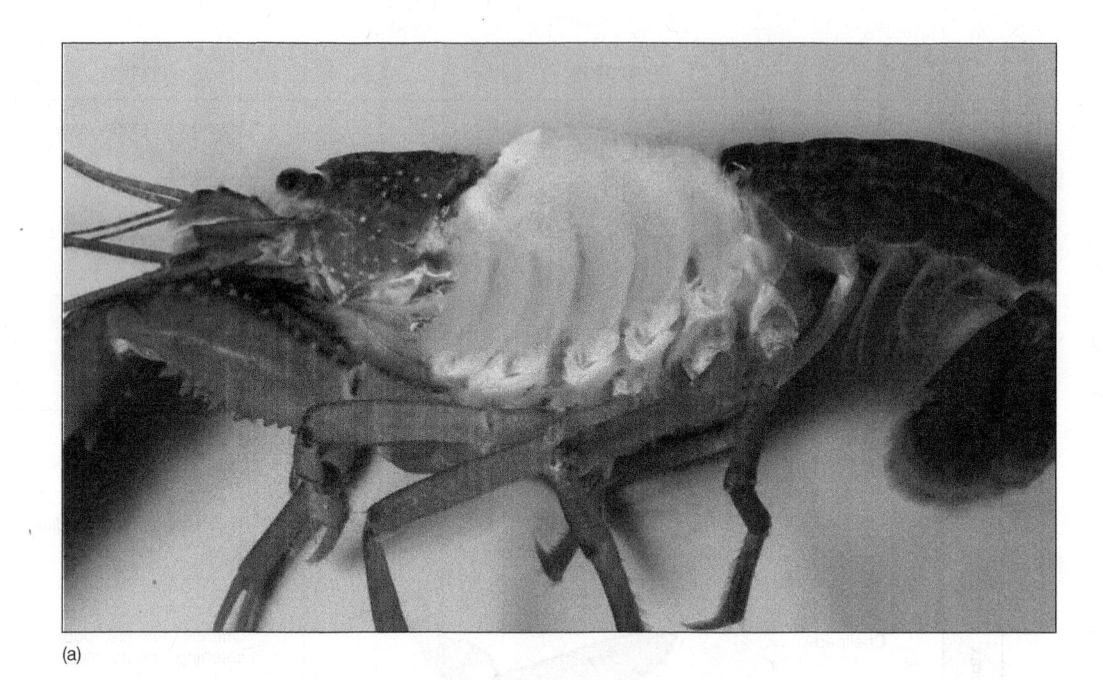

(a)

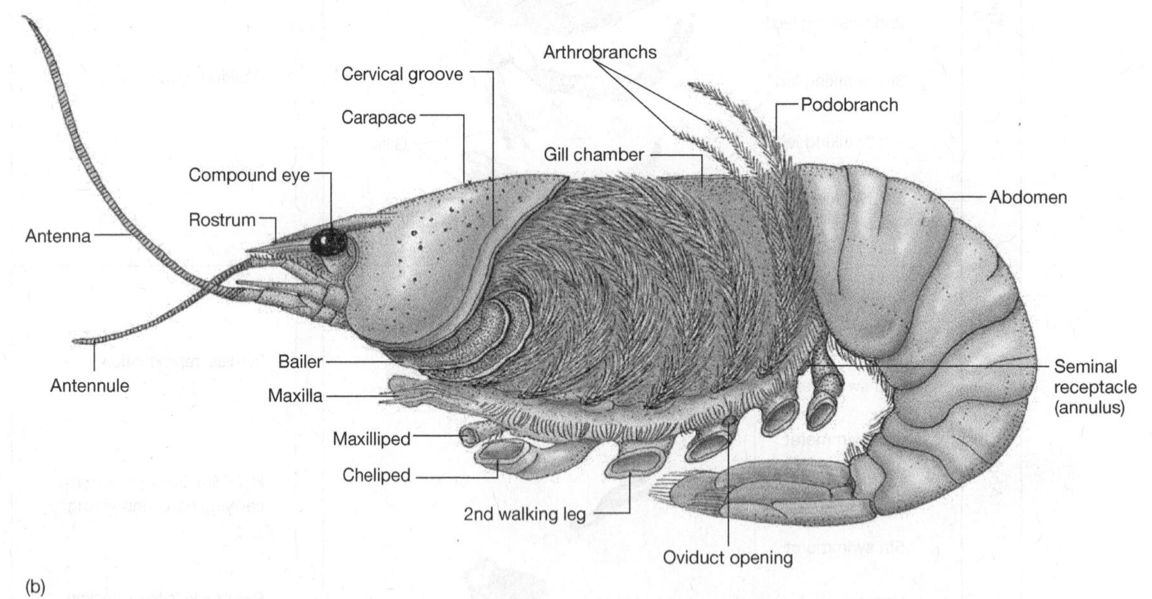

(b)

FIGURE 13.20 Crayfish, lateral view, portion of carapace removed to expose gills.
Photo by John Meyer.

chamber by the action of the **bailer,** a paddlelike projection of the second maxilla. Oxygen dissolved in the water diffuses across the thin walls of the gills and combines with the blood in the gills. Carbon dioxide diffuses across the walls of the gills in the opposite direction.

Observe that the gills occur in distinct longitudinal rows. ***How many rows of gills are there in your specimen? The outer row is attached to the base of certain of the appendages. Which ones? These outer gills are the podobranchs ("foot gills"). How many podobranchs do you find in your specimen?*** Separate the gills carefully with your probe or dissecting needle, and locate the inner row(s) of gills. These inner gills are the **arthrobranchs** ("joint gills") and are attached to the chitinous membrane which joins the appendages to the thorax (figure 13.20). ***How many rows of arthrobranchs do you find in your specimen?***

- Remove a gill with your scissors by cutting it free near its point of attachment and place it in a watch glass filled with water. Observe the numerous gill filaments arranged along a central axis. Also study a demonstration slide of a crayfish gill and note the afferent and efferent blood vessels in the central axis.

Internal Anatomy

- Remove the remainder of the carapace up to the rostrum by cutting forward from the rear of the carapace up to a level just behind the eyes. Leave the eyes, rostrum, and adjacent anterior portion of the carapace intact. With your scalpel, carefully separate the hard carapace from the thin, soft underlying layer of **hypodermis** as you cut away the carapace. The hypodermis is the tissue that secretes the carapace and other parts of the exoskeleton. Next, remove the remaining gills on the left side by cutting them near their points of attachment, and cut away the chitinous membrane underlying the gills to expose the internal organs as shown in figure 13.23.

Circulatory System Locate the small membranous **heart** just posterior to the stomach (figures 13.21 and 13.22). If you are dissecting a preserved specimen, you may find the heart filled with colored latex.

Identify the **pericardial sinus** (the space in which the heart lies), the **ostia** (openings in the wall of the heart), and the delicate **arteries** attached to the heart. Because of their small size, the arteries are often difficult to locate. Carefully search among the internal organs and try to locate the following arteries and determine which structures they supply with blood: (1) **ophthalmic artery,** (2) two **antennary arteries,** (3) two **hepatic arteries,** (4) **dorsal abdominal artery,** and (5) **sternal artery.** The circulatory system of the crayfish is an **open system;** blood is pumped from the heart through the major arteries that branch into smaller arteries and supply blood to the various organs of the body. From the arteries, the blood flows into open spaces, or **sinuses,** within and between the organs and then into a large **sternal sinus** along the floor of the thorax. Channels from the sternal sinus lead to the gills where the blood is oxygenated before returning to the pericardial sinus and into the heart via the ostia.

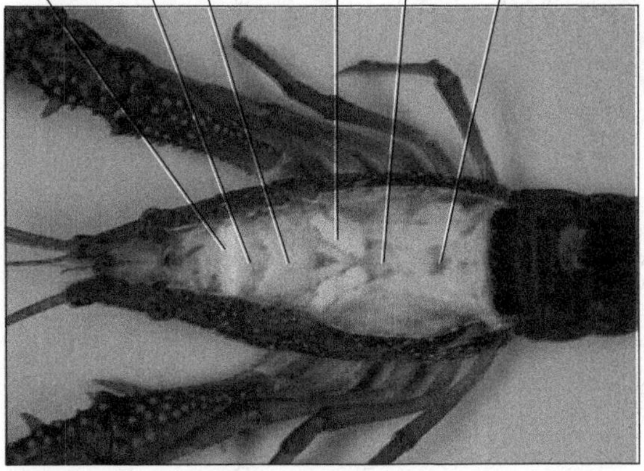

Crayfish blood consists of a nearly **colorless plasma** containing a dissolved respiratory pigment **hemocyanin** (a complex copper-containing protein) and numerous amoeboid blood cells.

FIGURE 13.21 Crayfish, dorsal view, upper portion of exoskeleton removed.
Photo by John Meyer.

Muscular System First locate some of the principal muscles of the crayfish. These include the two longitudinal bands of **abdominal extensor muscles** that run along the dorsal side of the thorax and extend back along the dorsal surface of the abdomen (figure 13.22). ***What is their function?*** Find the large **abdominal flexor muscles** in the abdomen lying below the extensor muscles. These large muscles serve to bend or flex the abdomen; hence, they provide the force for the quick backward thrust and propulsion of the animal when it is alarmed.

The **gastric muscles** attach the stomach to the inner wall of the carapace, and two large **mandibular** muscles **originate** (fixed end of the muscle) on the carapace and **insert** (movable end of the muscle) on the mandibles. Contraction of these muscles causes the crushing or grinding action of the mandibles. Each of the other appendages is also equipped with musculature: **flexor muscles** serve to draw the appendage closer to the body or to its point of attachment, and **extensor muscles** serve to extend, or straighten out, the appendage.

Female Reproductive System Locate the paired **ovaries** lying ventral and lateral to the pericardial sinus (figure 13.22) and trace the delicate **oviducts** to their openings at the base of the third walking legs. ***Are there eggs in the ovaries? What color are they?*** Locate again the **seminal receptacle** where sperm is deposited during copulation. Observe the demonstration of a female "in berry" with clusters of eggs attached to the bristles on the swimmerets.

Male Reproductive System Find the paired **testes** beneath the pericardial cavity, and trace the **vasa deferentia** to their openings. ***Where are the male openings located?*** Compare the first abdominal appendages of the male specimen to those of a female specimen (see figure 13.18).

Digestive System The **mouth** is obscured by the several oral appendages and can be located by moving these appendages aside with your forceps. Locate, behind the mouth, the tubular **esophagus** leading to the **stomach** just behind the rostrum. The stomach is divided into a large anterior **cardiac chamber** and a smaller posterior **pyloric chamber** (figures 13.22 and 13.23).

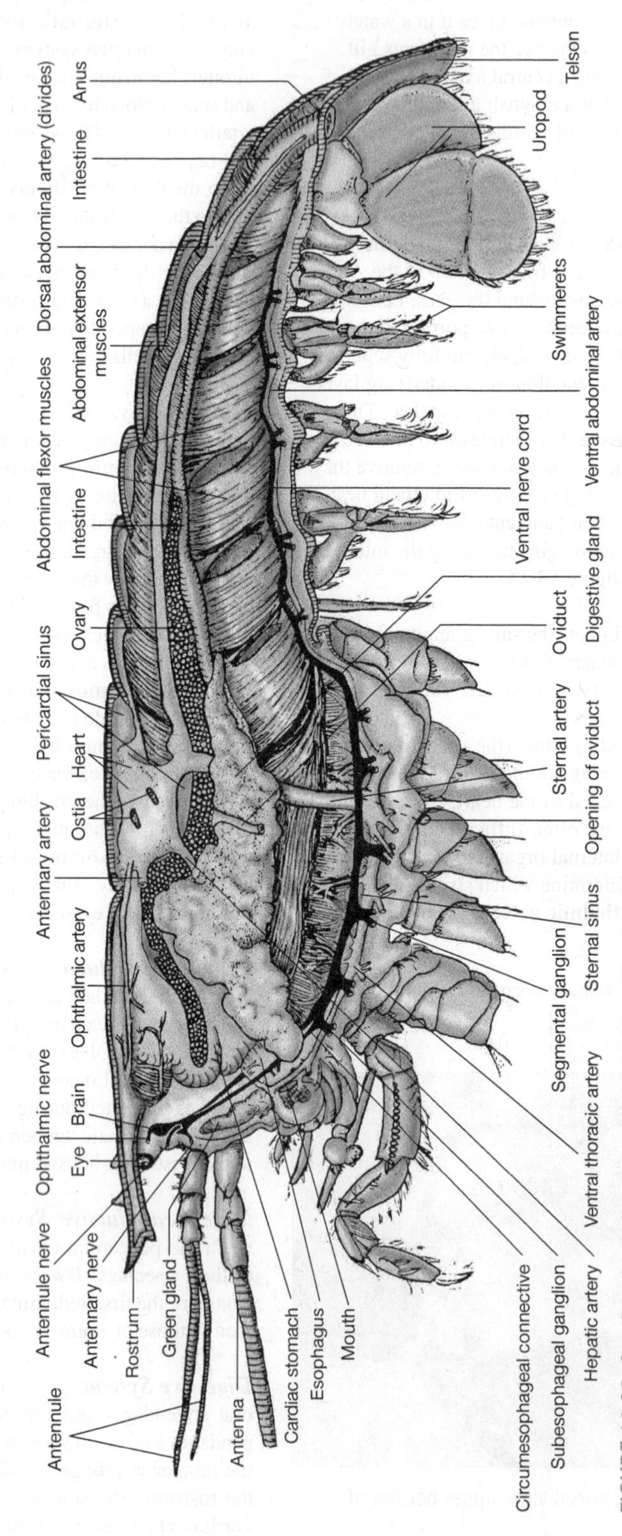

FIGURE 13.22 Crayfish, internal anatomy, female. Magnification 2X.

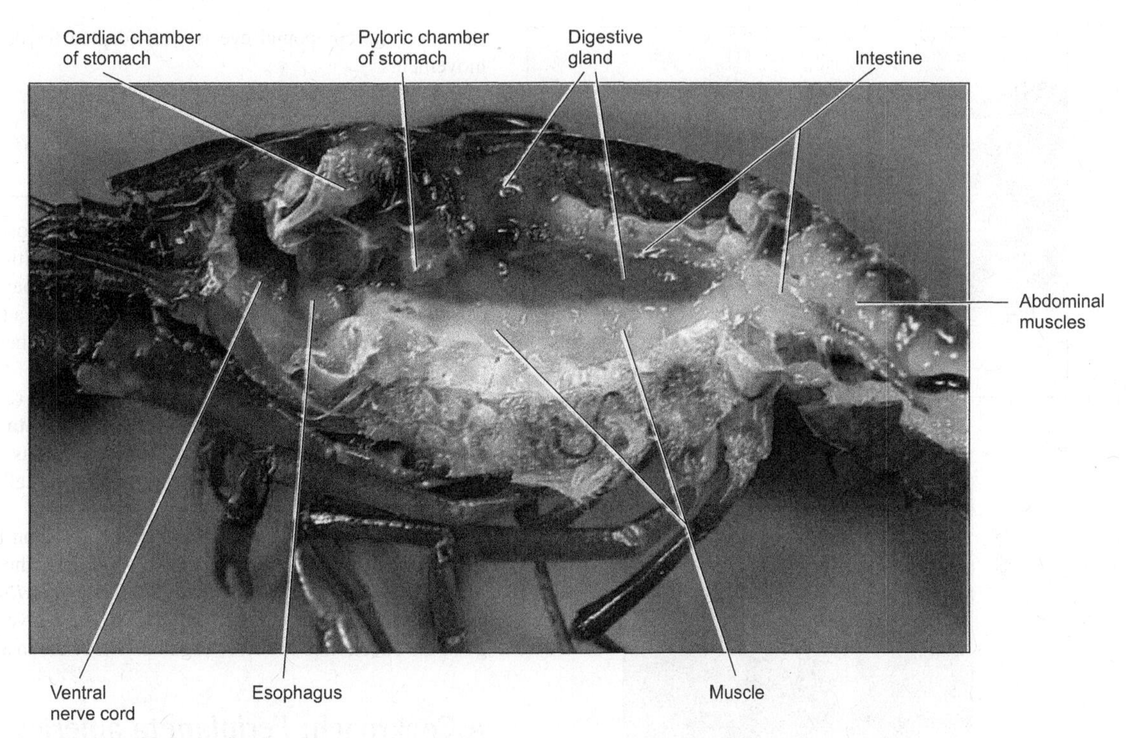

FIGURE 13.23 Crayfish, lateral view, gills removed to show internal organs. Magnification 2×.
Photo by Ken Taylor.

- Cut open the stomach and study the **gastric mill,** a grinding apparatus consisting of three chitinous teeth. Observe in the pyloric chamber the **chitinous bristles,** which serve as a strainer to prevent large chunks of food from passing unground into the intestine.

  You may also find calcareous **gastroliths** attached to the walls of the stomach. These calcareous bodies serve as stores of calcium salts, and they appear to play an important role in the calcification of the exoskeleton after molting.

  Locate the large digestive gland, or **hepatopancreas.** ***What is its function?*** Trace the **intestine,** behind the **stomach,** into the abdomen. Find the blind, saclike **intestinal caecum** in the abdomen and the terminal **anus.**

Excretory System The excretory structures of the crayfish are two large **green glands** located ventrally in the head region near the base of the antennae. These glands are usually dark red or brown (not green!) in preserved specimens.

- Find the two **excretory pores** at the bases of the antennae through which the nitrogenous wastes are discharged to the exterior.

Nervous System and Sense Organs

- Carefully remove the organs of the reproductive, digestive, and circulatory systems in the head and thoracic regions to expose the ventral surface of the cephalothorax. Also remove the muscles and intestine from the abdominal region, carefully cut away the skeletal plates that cover the anterior part of the nervous system, and locate the paired **ventral nerve cord** (figure 13.24*a*). Observe the **segmental ganglia** and their **paired lateral nerves.** Trace the nerve cord forward to the thorax and carefully cut away the skeleton that surrounds the nerve cord in this region. Follow the nerve cord anteriorly and locate the **supraesophageal ganglion** (the "brain") (figure 13.24b), the subesophageal ganglion, and the circumesophageal connectives, which run laterally and connect the supra- and subesophageal ganglia.

  Also find the several pairs of nerves that lead from the brain to the eyes, antennules, antennae, and mouthparts.

The most prominent sense organs of the crayfish are the **compound eyes** and the **statocysts,** both located previously. The crayfish also has numerous **sensory hairs** (tactile receptors and mechanoreceptors) found on various appendages that are sensitive to touch. Special **chemoreceptors** on the antennae, antennules, and mouthparts also provide senses of taste and smell.

The statocysts are a pair of small sacs located in the basal segments of the antennules. They have a cuticular lining bearing sensory hairs. A sand grain within the sac, the **statolith,** stimulates the sensory hairs and thus provides the animal with a sense of equilibrium. The lining of the statocysts and the statoliths are lost and replaced at each molt.

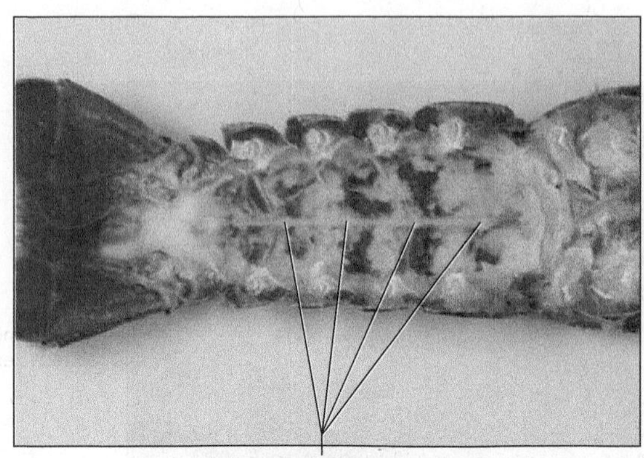

(a)

(b)

FIGURE 13.24 Crayfish nervous system: (*a*) ventral nerve cord, ventral view, (*b*) brain, dorsal view.
Photos by John Meyer.

Demonstrations

1. Compound eye (microscope slide)
2. Section of crayfish gill (microscope slide)
3. Female crayfish with eggs attached to swimmerets
4. Live crayfish in aquaria for study of behavior
5. Preserved specimens of several other types of freshwater and marine Crustacea

The large compound eyes are made up of many individual units, the **ommatidia.**

- Remove one of the compound eyes and examine its surface under your stereoscopic microscope. The surface of the eye is covered with a transparent **cornea** secreted by underlying cells. Observe that the cornea is divided into many sections, or **facets.**

Each facet of the cornea represents the outer end of an ommatidium. The compound eye probably forms a mosaic light-and-dark impression of its surroundings. Because of its structure, the compound eye is well-adapted for detecting movements.

Subphylum Hexapoda, Class Insecta

Insects constitute the largest, most diverse, and most widespread class of arthropods, and indeed of all the multicellular animals. More than one million species of insects have been described, and several thousand new species are added to the list each year. The insects are easily distinguished from the other classes of arthropods by a combination of the following characteristics: (1) **chitinous exoskeleton;** (2) **body divided into three distinct regions** called **tagmata** (head, thorax, and abdomen); (3) **one pair of uniramous antennae;** (4) **three pairs of jointed legs;** (5) and, usually, **two pairs of wings** in the adult stage.

In this exercise, we will examine two common insects as examples of this diverse group of animals: the cockroach *Periplaneta americana* and the grasshopper *Romalea microptera.* These two representatives should give you a good introduction to the basic organization of the insects.

A Cockroach: *Periplaneta americana*

Periplaneta americana is the largest of the three most common cockroaches in North America and Europe (figure 13.25). It is called the American cockroach even though the species appears to have originated in Africa. It lives in a close relationship with humans (not always harmonious!) and has accompanied our species as we dispersed into all parts of the world. *P. americana* is commonly found in restaurants, grocery stores, bakeries, homes, and other places where food is prepared or stored.

Fact File

Insects

- Traveling flea circuses were a popular form of entertainment in Europe during the 1800s. Mole fleas were used as performers because they are large (up to 5.5 mm in length) and cannot jump as far as most other fleas.
- The snowy tree cricket, *Oecanthus fultoni* (Family Gryllidae), is often called the temperature cricket. Adding 40 to the number of chirps it makes in 15 seconds will equal the ambient temperature in degrees Fahrenheit.
- Termites are usually the most dominant organisms in tropical forest environments. Their populations typically range from 2,000 to 4,000 individuals per square meter but may occasionally run as high as 10,000 individuals per square meter. Their biomass

Continued—

—*Continued*

(up to 22 g/m²) exceeds the combined biomass of all vertebrate species living in the same area.

- Beetles (Order Coleoptera) form the largest order in the animal kingdom. With about 500,000 different species, they include 40 percent of all insects and nearly 30 percent of all animal species. Social insects (ants and termites) make up another 20 percent of the total species diversity among animals.
- Flies (Order Diptera) are among the best aerialists in the insect world—they can hover, fly backward, turn in place, and even fly upside down to land on a ceiling.
- A lacewing's egg sits atop a slender stalk secreted by the female's reproductive system. For many years, biologists thought these eggs were the fruiting bodies of a fungus they called *Ascophora ovalis*. The true nature of these eggs was first discovered in 1737 by Rene Reaumur, a French physicist, biologist, and inventor.
- Although praying mantids usually feed on insect prey, they have been known to catch and eat small frogs, lizards, and even birds.
- Louse-borne disease is particularly common in wartime when soldiers are forced to live in crowded and unsanitary conditions. Trench fever was especially widespread during World War I, and was probably a major factor in the final collapse of the Russian army.
- Honeydew, an excretory product that is rich in sugars and amino acids, is produced by many species of Homoptera. Other animals use honeydew as a source of food. Honeydew from a mealybug, *Trabutina mannipara*, is regarded as the probable source of biblical manna.

Some authors place cockroaches in the Order Orthoptera along with the grasshoppers, crickets, mantids, and walking sticks. Other authors place them with the mantids in the Order Dictyoptera or in a separate Order Blattaria (or Blattodea). In this, as in many other cases, expert taxonomists do not always agree!

Mature specimens of *P. americana* range from 30–45 mm in length, and males tend to be slightly larger in size than females. Both sexes of this species have well-developed wings; males occasionally fly or glide through the air, although they are most often seen crawling along some surface. They tend to thrive in dark, moist places with an available food supply. They are negatively phototactic, avoid well-lit areas, and are most active at night. Cockroaches are omnivorous and have been reported to feed on fruit, fish, peanuts, flour, paste, paper, cloth, dead insects, and many other objects.

External Anatomy

Obtain an anesthetized or freshly killed cockroach and study its external anatomy. Preserved specimens may also be used for external anatomy, but fresh specimens are far superior for the study of internal structures. Specimens can either be anesthetized with carbon dioxide or immobilized by cold torpor simply by placing them in a closed container in a refrigerator for 20–30 minutes. Specimens killed by freezing are also suitable for dissection.

Observe the chitinous **exoskeleton** that covers the body. The exoskeleton is made up of many hardened plates, or sclerites, that are connected by thin membranes giving some flexibility to the exoskeleton. Observe that the body is divided into three regions: an anterior **head,** a middle **thorax,** and a posterior **abdomen** (figure 13.25). Each region consists of several segments, although many of the segments, particularly those making up the head, are fused together.

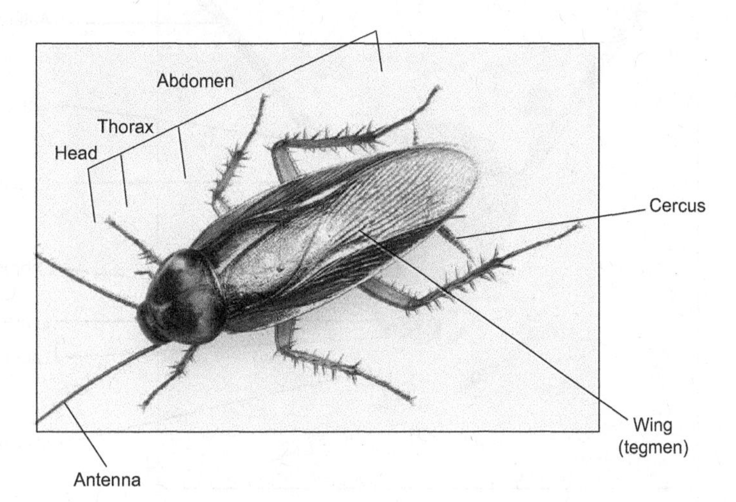

FIGURE 13.25 Cockroach, *Periplaneta*, Magnification 2×.
Courtesy of Carolina Biological Supply Company, Burlington, NC.

Identify the two large **compound eyes** and the two long, many-segmented **antennae** on the head. Adjacent to the base of each antenna is a white spot, believed to be the vestige of an **ocellus** or simple eye found in certain other insects. Viewed from the front, the head is roughly pear-shaped; the dorsal portion, which bears the kidney-shaped eyes and antennae, is broader than the ventral portion, which bears the mouthparts (figure 13.26).

The **mouthparts** are modified appendages that aid in grasping and macerating food. Locate the medial **labrum** (not a modified appendage), or front lip. Lateral and posterior to the labrum are the paired **mandibles.** Posterior to the mandibles are the paired **maxillae.** Each maxilla bears a maxillary palp that aids in manipulating food. The medial **labium,** or posterior lip, which consists of several fused parts, also has palps.

By lifting the labrum with your dissecting needle, locate the hardened mandibles that aid in chewing or macerating food for passage to the mouth. Between the mandibles is the **hypopharynx,** which extends upward into the preoral cavity; at its base is the mouth. A duct from the paired salivary glands empties at the base of the hypopharynx.

The head is joined to the thorax by a short, flexible **cervical region.** The thorax consists of three segments and bears two pairs of wings and three pairs of legs. The anterior thoracic segment (prothorax) never bears wings. It is larger than the next two segments and bears the pronotum (dorsal sclerite) that serve as a protective cover over the thorax. The **anterior wings** borne by the second thoracic segment (mesothorax) are thickened and help protect the membranous **second pair of wings** that arise from the third thoracic segment (metathorax).

Observe the numerous **veins** in the wings. The pattern of wing venation is specific and is characteristic of each species of cockroach; therefore, it often serves as an important taxonomic characteristic.

Two pairs of **spiracles**—small, slitlike openings into the tracheal system—are found in the membranes connecting the first and second and the second and third thoracic segments. Locate the spiracles on one side near the dorsal surface of the intersegmental membrane.

The abdomen consists of 11 segments, each with distinct dorsal and lateral plates called tergites and sternites, respectively. Each segment is joined to the next by a flexible **intersegmental membrane.** Several of the posterior segments have become fused so that some of the dorsal and ventral plates are not clearly identifiable. Each of the first eight segments bears a pair of oval spiracles. The **anus** opens in an unpigmented area on the last segment, dorsal to the external genitalia.

Cockroaches are sexually dimorphic. Sex can be determined in *P. americana* by observing the number of projections from the last abdominal segment. Males have four projections from this segment, two lateral **cerci** and two medial **styli** (figure 13.27). Females have two cerci but lack the styli.

Internal Anatomy

Start your dissection by removing the legs and wings near their bases with your dissecting scissors. Next, cut carefully through the terga of the thoracic and abdominal segments except the last abdominal segment. Make a shallow incision near the lateral margins of the terga. Be careful to avoid damaging internal organs as you cut through the terga. Do not cut into the head and last abdominal tergum. After you have completed these lateral incisions, pin the insect with its ventral surface down in a dissecting tray with a wax-covered bottom. Cover the insect with water or a saline solution (0.75% NaCl)

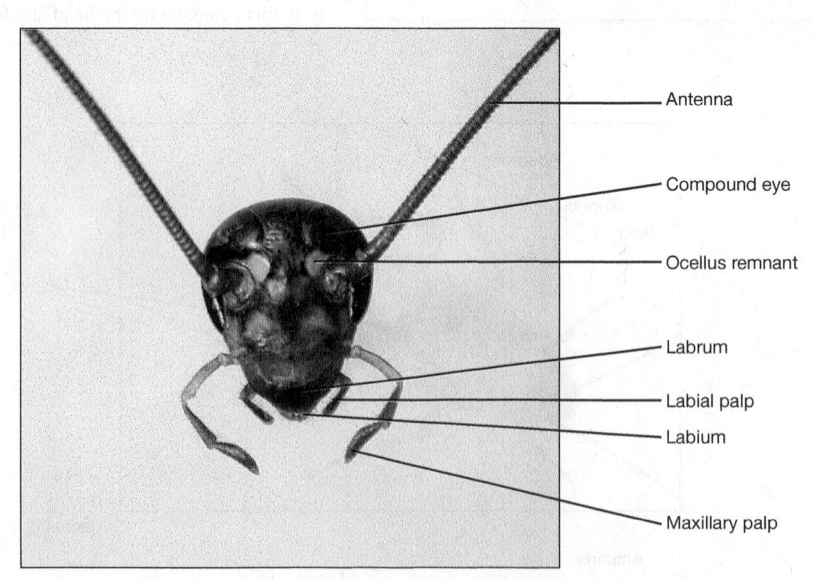

FIGURE 13.26 Cockroach, anterior view of head.
Photo by John Meyer.

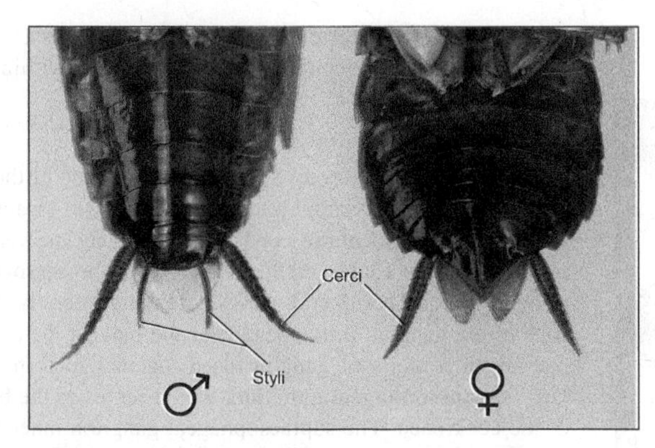

FIGURE 13.27 Posterior end of cockroach abdomen in male (left) and female.
Photo by John Meyer.

or 70% alcohol to keep its internal organs soft and pliable. Place the pins along the lateral margin of the insect beyond the incisions in the terga. Angle the heads of the pins outward to give you a clear view of the cockroach.

Beginning at the posterior end and working forward, carefully remove the thoracic and abdominal terga that have been cut. Use your forceps and a dissecting needle (or an insect pin) to separate the terga from the internal tissues. Take special care to separate the transparent dorsal blood vessel that runs along the dorsal midline just below the terga. Observe and cut the strong **dorsoventral muscles** attached to the thoracic terga. ***What do you think the function of these muscles might be? How might movements of the ventral body wall be important to the metabolism of an active insect?***

After removing the terga, locate the dorsal blood vessel again; the portion of this vessel in the abdominal region is called the **heart.** Blood enters the heart through paired **ostia** (slitlike valves) and flows anteriorly toward the head through the aorta. Blood passes from the aorta in the head and flows over organs in the **hemocoel** as it moves posteriorly through the body to the heart.

Running along the dorsal blood vessel are silvery **tracheae** and **air sacs** of the respiratory system. Much of the internal space of the abdomen is filled with a white, spongy **fat body.** The fat body is an important organ, somewhat analogous in function to the vertebrate liver. It is the major site of intermediary metabolism, and it also stores fats, amino acids, and carbohydrates.

Digestive System Carefully separate the heart, tracheae, fat body, and associated tissues to locate the digestive tract, which extends through the center of the abdomen. Adjacent to the digestive tract are the more transparent reproductive organs; be careful not to damage or remove them accidentally while removing parts of the fat body to locate the digestive tract.

Follow the digestive tract anteriorly and find the translucent **salivary glands** lateral to the digestive tract in the thorax. These glands secrete **amylase,** an enzyme that digests starch. The salivary glands empty through ducts leading to the base of the hypopharynx. Follow the digestive tract to its junction at the head and cut the digestive tract and the salivary glands at this junction. Free the digestive tract from the surrounding tracheae and membranes and pin the anterior end of the digestive tract at an angle with the body so that you can examine it in more detail (figure 13.28*a* and *b*).

As in other insects, the digestive tract of the cockroach is divided into three regions, a chitinized **foregut,** a **midgut,** and a chitinized **hindgut.** The foregut extends from the mouth via a narrow **esophagus** to an expanded **crop.** The crop serves mainly to store food. Posterior to the crop is a muscular bulb, the **proventriculus,** which works like a gizzard to break up solid food. A **stomoducal valve** at the posterior end of the proventriculus regulates the passage of food material into the midgut.

Digestion and absorption take place mainly in the midgut and in several fingerlike **gastric caeca** that extend out from the anterior end of the midgut. Several enzymes are secreted in the midgut and the caeca. The **pyloric valve** at the posterior end of the midgut controls the passage of materials into the hindgut.

The hindgut is composed of a large, dark-colored anterior **ileum;** a thin, light-colored **colon;** and a **rectum** found in the last abdominal segment (figure 13.28*b*). Many thin, translucent **Malpighian tubules** extend from the anterior end of the hindgut. These tubules are blind tubes that empty into the gut. They serve to remove metabolic waste materials and to regulate the concentration of various ions in body fluids. Undigested materials entering the hindgut from the midgut pass through the ileum and colon to the posterior rectum. The rectum is made up of six **rectal pads,** as well as other muscle and connective tissue, and serves to reabsorb water and to store waste materials temporarily before discharge through the anus.

Reproductive System Cut through the rectum and remove the digestive tract to study the reproductive system. If you have a gravid female, you will find large **ovaries** with developing **eggs** filling much of the abdominal space. A transparent tube, the **lateral oviduct,** extends posteriorly from each ovary and runs toward the ventral midline where it joins the other lateral oviduct to form the **common oviduct** (see figure 13.29).

The short common oviduct extends to the larger **vagina** or genital chamber near the posterior of the insect. Dorsal and posterior to the vagina are two white **accessory glands.** Anterior to the accessory glands is a small **spermatheca** (for holding sperm) and an elongate **spermathecal gland** (for nourishing the sperm).

In a male specimen, you will find a cluster of tubules near the base of the last tergite. These tubules are part of the **accessory glands,** which surround the paired **testes.** By carefully separating the tissues of the accessory glands, you may be able to locate the **testes** and the **seminal vesicles** in which the sperm are stored in bundles of **spermatophores** (figure 13.30).

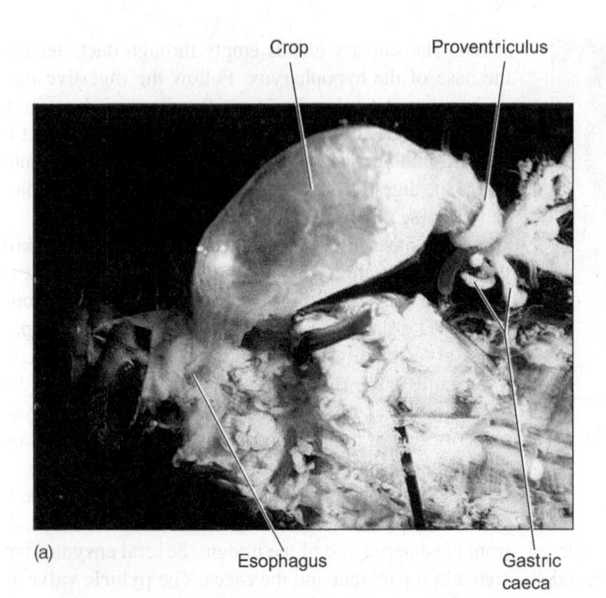

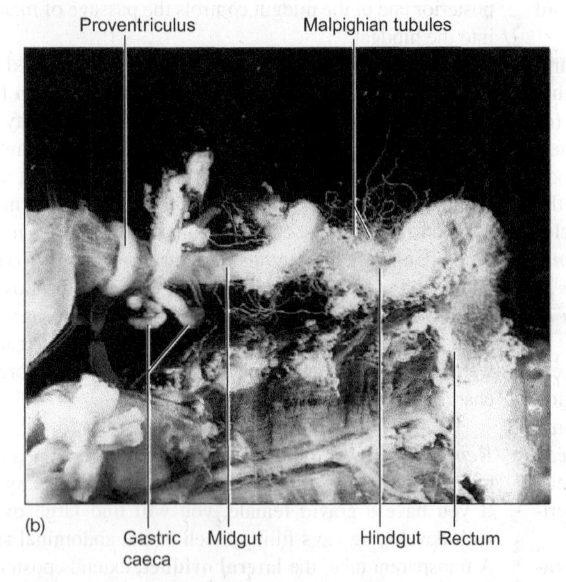

FIGURE 13.28 Cockroach, digestive system: (*a*) anterior portion, (*b*) posterior portion.
Courtesy of Carolina Biological Supply Company, Burlington, NC.

From the seminal vesicles, an **ejaculatory duct** runs posteriorly through a mass of muscles and sclerites near the posterior end of the insect. The ejaculatory duct terminates in a long, hooked **aedeagus,** which serves as a penis. Another **accessory gland** joins the ejaculatory duct near the base of the aedeagus.

Male *P. americana* have a complex copulatory apparatus consisting of several elongate processes equipped with hooks and barbs that aid in positioning the male and female genitalia during copulation. During mating, a spermatophore (discrete package of sperm and nutrients for the female) is transferred from the male and passes into the female genital chamber where it is stored within the spermatheca.

Nervous System Remove the remaining tissue of the fat body in the median ventral portion of the body to locate the translucent **ventral nerve cord** that runs along the ventral midline (figures 13.31 and 13.32). Observe the **segmental ganglia** and the peripheral nerves extending laterally from each of the ganglia. In the head near the base of the antennae are three pairs of ganglia fused together to form one large **supraesophageal ganglion,** which serves as the brain of the cockroach. The supraesophageal ganglion innervates the eyes, antennae, and other sense organs of the head. It connects to the anterior end of the ventral nerve cord by a pair of **circumesophageal connectives.** Ventral to the digestive tract at the anterior end of the ventral nerve cord is the **subesophageal ganglion,** also made up of three pairs of fused ganglia. This ganglion innervates the mouthparts of the cockroach.

A Grasshopper: *Romalea microptera*

Grasshoppers, along with katydids, locusts, and crickets, are members of a relatively unspecialized order of insects (Order Orthoptera). Specimens of the large, black, lubber grasshopper are usually provided for laboratory study (figure 13.33), but the following information generally applies to other common species of grasshoppers as well.

External Anatomy

- Obtain a preserved grasshopper and study its external features.

Externally, the body of a grasshopper is covered by a **chitinous exoskeleton** consisting of numerous hardened plates, or **sclerites,** separated by sutures, or thin membranous areas. The latter provide the flexibility necessary for movements of the body segments and the appendages. The body wall is made up of three principal layers—an outer **cuticle;** an underlying **hypodermis,** which secretes the cuticle; and a thin, noncellular **basement membrane.** The cuticle is a complex structure consisting of three layers: a very thin outer epicuticle, an exocuticle, and an innermost endocuticle lying directly upon the hypodermis. The endocuticle and exocuticle are composed largely of chitin, a nitrogenous polysaccharide (N-acetylglucosamine), while the epicuticle is nonchitinous and consists chiefly of a network of fibrous proteins impregnated with wax. Study the demonstration provided in the laboratory, showing the microscopic structure of insect cuticle.

The head of the grasshopper (figure 13.34) bears the eyes, antennae, and mouthparts. Two large **compound eyes** occupy prominent sites on opposite sides of the head, and three **ocelli** (simple eyes) form an inverted triangle on the front of the head between the compound eyes. The two

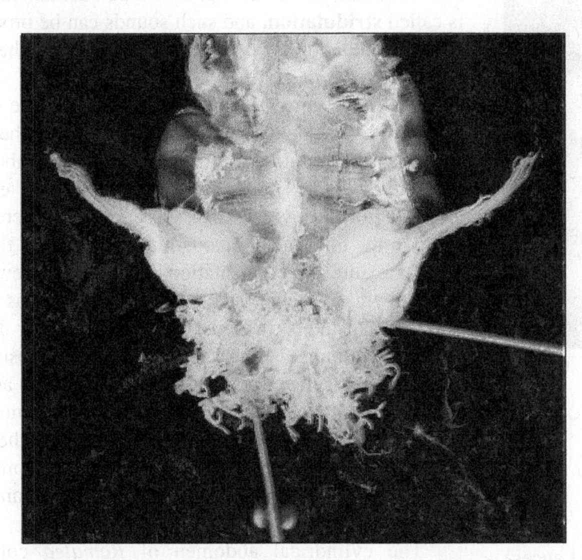

FIGURE 13.29 Cockroach, female reproductive system.
Photo by John Meyer.

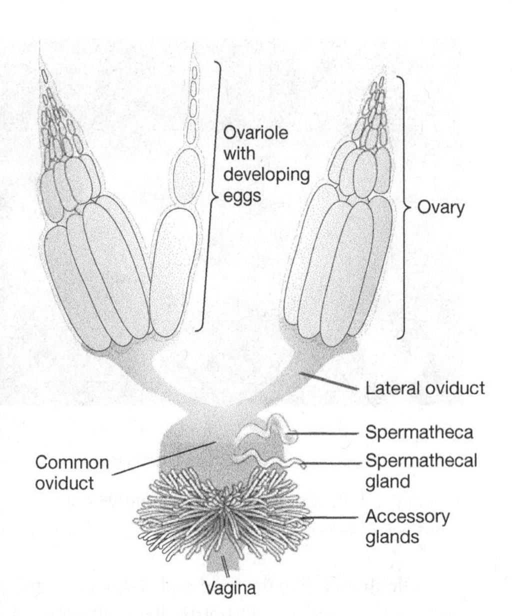

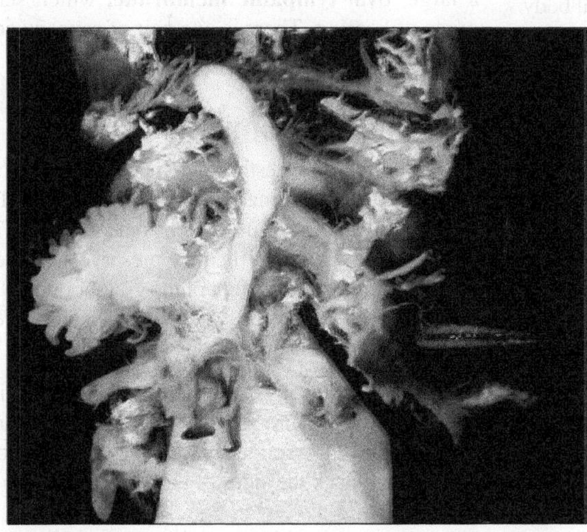

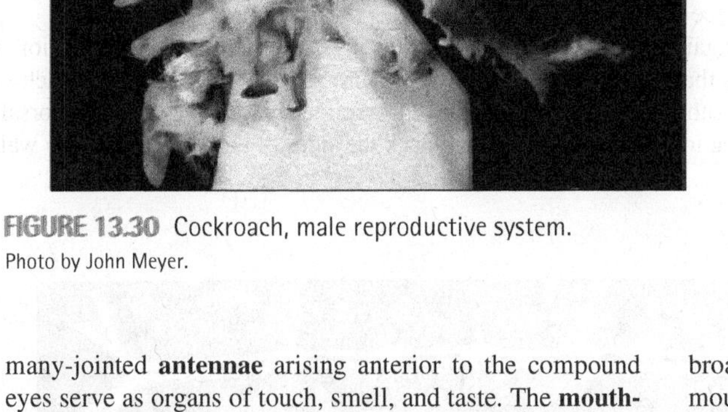

FIGURE 13.30 Cockroach, male reproductive system.
Photo by John Meyer.

many-jointed **antennae** arising anterior to the compound eyes serve as organs of touch, smell, and taste. The **mouthparts** of a grasshopper are relatively unspecialized in comparison with those of most other kinds of insects and are of the **chewing type.** They consist of seven parts: a dorsal **labrum** or upper lip; a pair of sclerotized (hardened) **mandibles;** a pair of **maxillae,** which manipulate the food; a broad median **labium** or lower lip; and, in the center of these mouthparts, a cylindrical **hypopharynx** at the base of which the salivary glands empty. The labium and the maxillae bear sensory palps, which project ventrally. The mouthparts can be most easily observed and studied by removing them carefully from the head with your forceps and studying them under a stereoscopic microscope.

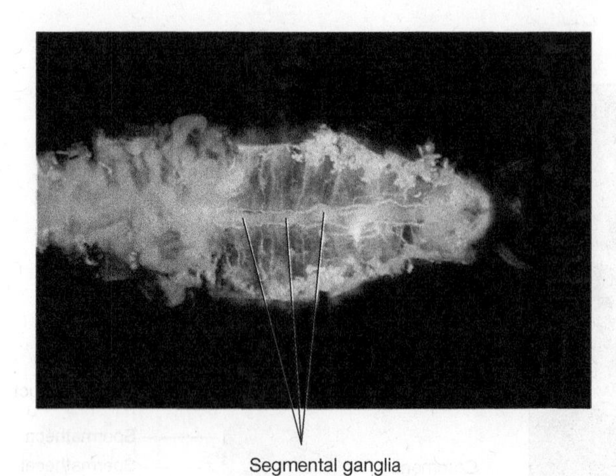

FIGURE 13.31 Cockroach, ventral nerve cord.
Photo by John Meyer.

The thorax, like the head and abdomen, consists of several fused segments, illustrating the multisegmented ancestry of the insects. In the thorax, there are three segments, a large anterior **prothorax,** followed by two smaller segments, the **mesothorax** and the **metathorax.** Each thoracic segment bears one pair of legs. Only the mesothorax and the metathorax bear a pair of wings.

The legs are clearly jointed and consist of five main segments: a short **coxa,** which is attached to the ventral body wall; a small **trochanter** (often difficult to see) fused with a stout **femur;** a slender, spiny **tibia;** and a distal **tarsus** subdivided into three **tarsomeres,** the last of which bears two terminal **claws.** Between the claws is a fleshy pad, the **arolium,** which provides a grip on smooth surfaces. All three pairs of legs are used by the grasshopper for walking and climbing, but the third pair (the metathoracic legs) is specially modified for leaping. They are equipped with an enlarged femur containing strong voluntary muscles and with an elongated tibia to provide additional leverage. Another modification of the metathoracic legs can be seen in male specimens of *Romalea* and its close relatives. The inner surface of the femur bears a row of small spines that is rubbed against the lower edge of the front wing to produce the sounds that comprise the characteristic song of the males. Certain other species of grasshoppers produce their song by rubbing together their forewing and hindwing. This type of sound production is called **stridulation,** and such sounds can be produced by insects of other orders by rubbing other parts of their bodies together.

The long, narrow **forewings** are borne on the mesothorax and serve as a covering for the hindwings when at rest. These leathery wings—found commonly in members of the orthopteroid orders (grasshoppers, cockroaches, etc.)—are called **tegmina** (figure 13.33). Observe the numerous **wing veins** in the thin, membranous **hindwings.** Note the difference in texture and pigmentation between the forewings and the hindwings. The grasshopper wing develops as a saclike outgrowth of the body wall; this outgrowth later flattens to form a thin double membrane, and the two opposing membranes fuse together enclosing tracheae, nerves, and blood sinuses. The hollow veins are formed by thickening of the cuticle around the **tracheae** (air tubes) or along the sinuses. The pattern of wing venation in insects is very precise and species-specific and therefore plays an important role in classification.

The cylindrical abdomen of *Romalea* consists of 11 segments, the last three of which are reduced in size and modified either for copulation or for egg laying (oviposition). Along the sides of certain thoracic and abdominal segments are 10 pairs of tiny respiratory openings, the **spiracles,** which open into the system of **tracheal tubules.** The first abdominal segment also bears on each side a large, oval **tympanic membrane,** which serves as an organ of hearing. There is one pair of **sensory cerci** on the eleventh segment.

Internal Anatomy

Freshly killed or anaesthetized specimens are most suitable for the study of internal anatomy, although well-preserved specimens can also be used with some success. The internal tissues and organs are difficult to preserve, however, because of the chitinous exoskeleton of the grasshopper, which limits penetration of the preservative fluids. Study figure 13.35 for orientation before starting your dissection.

- Remove the wings, and starting at the posterior end of the body, cut through the body wall along each side just above the spiracles. Remove the freed dorsal wall and pin back the sides of the remaining body wall in

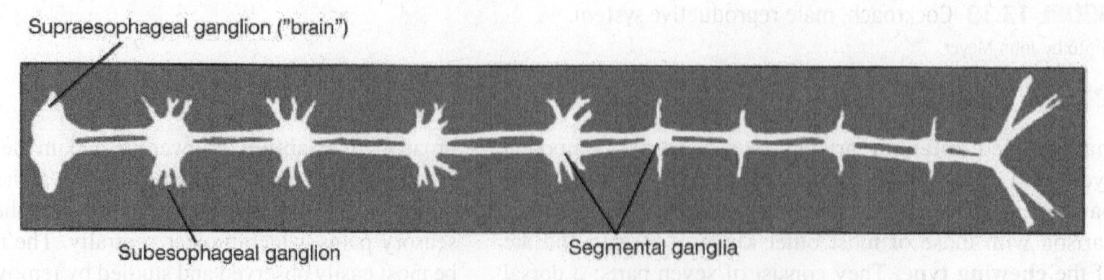

FIGURE 13.32 Cockroach, nervous system showing the brain, subesophageal ganglion and parallel ventral nerve cord with segmental ganglia.

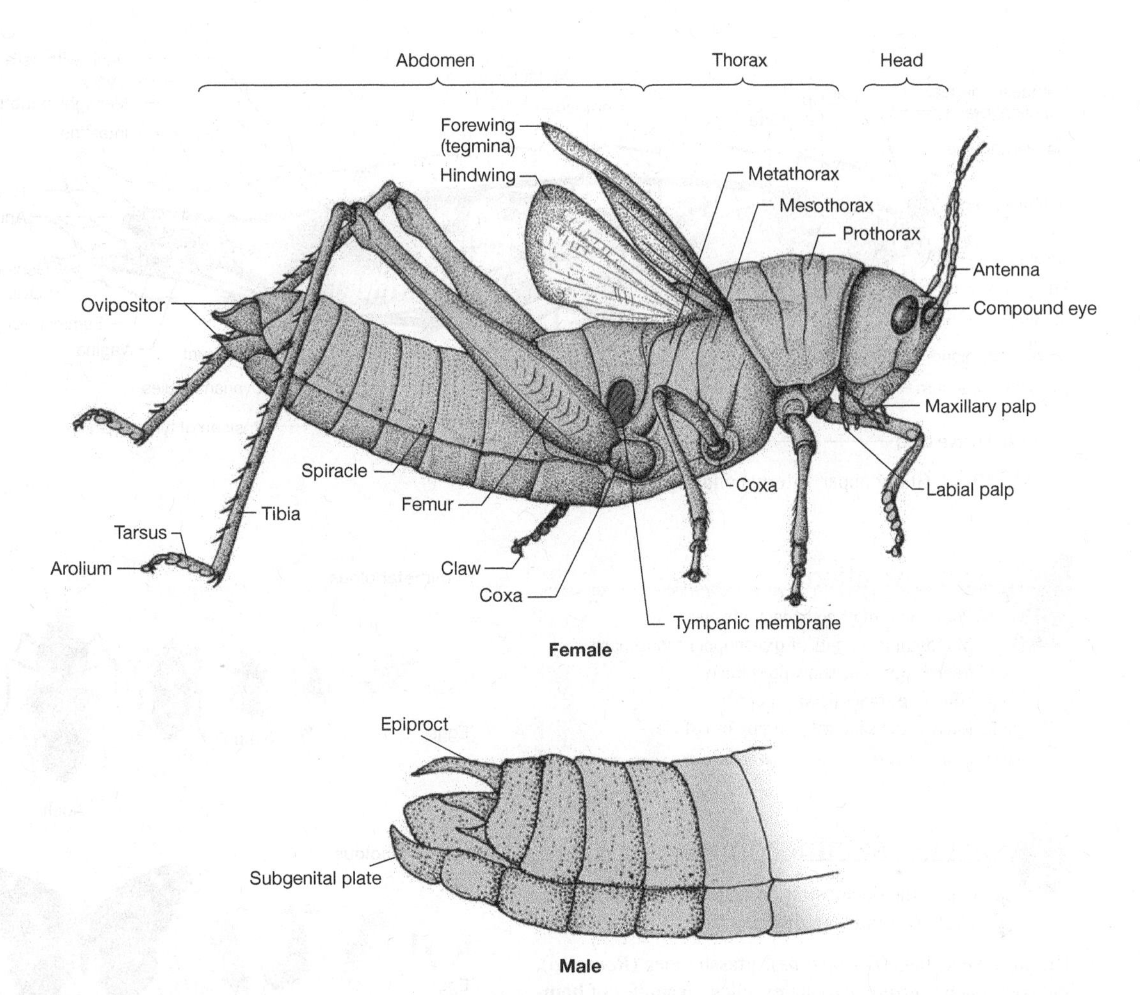

FIGURE 13.33 Grasshopper, *Romalea*, lateral view of female and posterior portion of male.

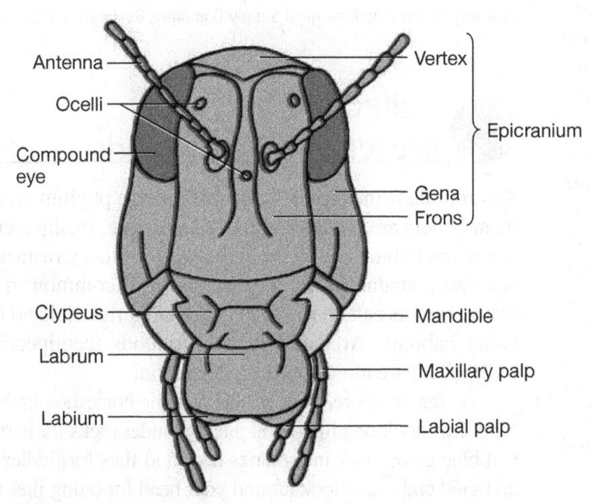

FIGURE 13.34 Grasshopper, head, frontal view.

your dissecting pan. Cover your dissection with water to keep the internal organs moist and pliable.

Observe the **hemocoel,** the tubular **heart** with **ostia,** and the various parts of the digestive tract. The digestive tract of insects consists of three main divisions: **foregut, midgut,** and **hindgut.** These divisions are often functionally subdivided for specific functions in particular insects.

The **Malpighian tubules** are outgrowths of the digestive tract that originate at the junction of the midgut and hindgut. They have an excretory function. Other internal organs include the **tracheal tubules** and the **gonads.** ***Is your specimen a male or female?*** Observe the texture and shape of the gonads and also the posterior segments of the abdomen to determine the sex. Along the ventral surface, find the **salivary glands** and the **nerve cord.** Note the **segmental ganglia** and the numerous **lateral nerves.** Observe the powerful muscles of the thorax associated with the legs and wings.

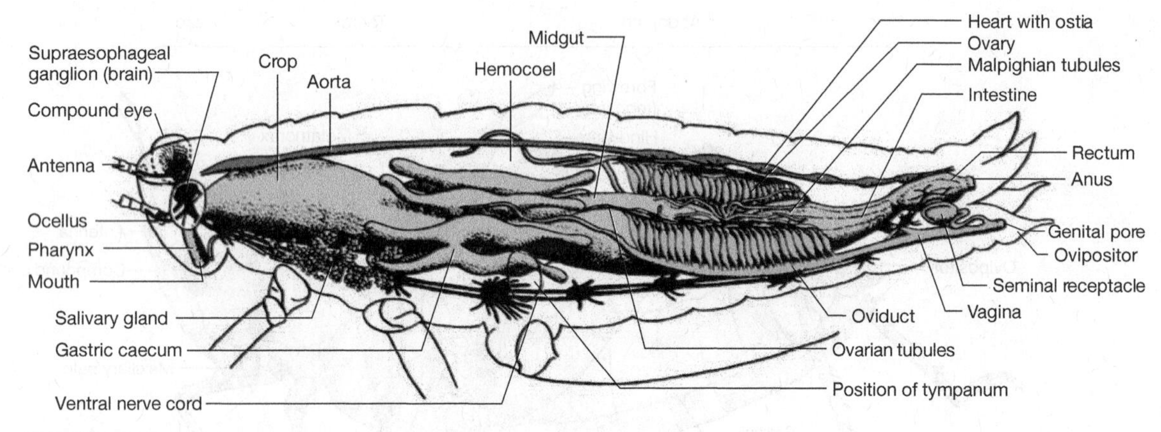

FIGURE 13.35 Grasshopper, internal anatomy, female.

Demonstrations

1. Plastic mount of cockroach life cycle
2. Microscopic mounts of grasshopper mouthparts
3. Microscope slide showing spiracle
4. Life cycle stages (larvae, pupae)
5. Microscope slide with section of cuticle

Insect Metamorphosis

- Study the demonstration materials on insect life cycles provided in the laboratory.

The milkweed bug (*Oncopeltus*), grasshopper (*Romalea*), and cockroach (*Periplaneta*) are excellent examples of **hemimetabolous development** (see figure 13.36*a*). Observe the eggs and several nymphal stages, and compare each of these with the adult. Note especially the changes in the head-to-body proportions and the gradual development of wing pads. Some hemimetabolous insects like dragonflies, stoneflies, and mayflies have nymphal instars (called **naiads**) that are adapted for aquatic life. These immatures usually have gills for underwater respiration. They eventually emerge from the water as winged adults.

Fruit flies (*Drosophila*), darkling beetles (*Tenebrio*), honey bees (*Apis*), and painted lady butterflies (*Vanessa*) are among the many insects that exhibit **holometabolous development** (see figure 13.36*b*). Note the striking difference in appearance between the larval and adult stages. Pupae may appear to be inactive, but radical changes take place internally. Most of the larval tissues are resorbed and adult organs are newly formed from special groups of cells called **imaginal discs** that remained undifferentiated throughout larval development. The winged adult that emerges from the **pupa** is primarily adapted for dispersal and reproduction.

- On figure 13.37, draw the life cycle for one of the insects studied.

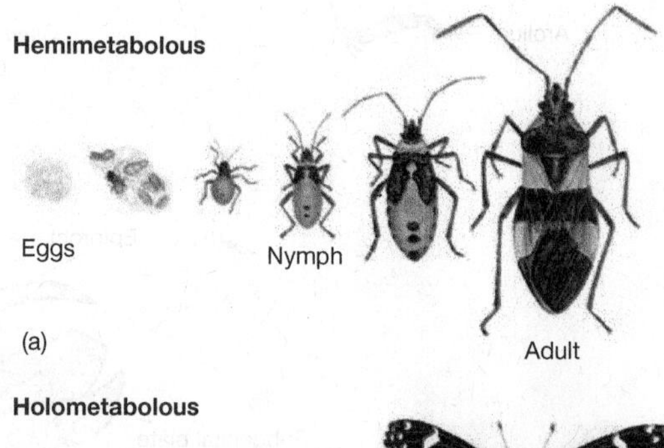

FIGURE 13.36 Stages in the life cycle of common insects: (*a*) incomplete metamorphosis and (*b*) complete metamorphosis.
Courtesy of Carolina Biological Supply Company, Burlington, NC.

Collecting and Preserving Arthropods

Arthropoda is the largest and most diverse phylum on earth; its members are commonly found in marine, freshwater, and terrestrial habitats throughout the world. Most Crustacea are aquatic: abundant in the oceans and also common in fresh water. Insects can be found in nearly all terrestrial and freshwater habitats. Arachnids and Myriapods (centipedes and millipedes) are almost entirely terrestrial.

At the seashore, look in the surf for horseshoe crabs and sand fleas, look on pilings and piers or under rocks for barnacles and blue crabs, look in estuaries and mud flats for fiddler crabs and sand crabs, and look around your head for biting flies (mosquitoes and deer flies)! Use a seine or dip net to collect dragonfly naiads, diving beetles, water bugs, crayfish, and other crustaceans

FIGURE 13.37 Student drawing of insect metamorphosis stages.

from freshwater ponds and streams. For the greatest diversity, sample in the mud and detritus of the littoral zone, among aquatic plants just offshore, or under rocks in shallow riffles.

Peel back the bark of dead trees and turn over rotting logs to find termites, carpenter ants, and a variety of wood-boring beetles. Ants, crickets, ground beetles, and terrestrial isopods (pill bugs and sowbugs) are easy to find in moist habitats, under bark or stones, near the edge of streams, and around building foundations. A diverse assortment of insects and spiders may be collected by sweeping through stands of clover, alfalfa, flowering plants, or tall weeds with an insect net. Some insects (especially moths) are attracted to lights at night. They may be collected on a bed sheet illuminated with white or UV light.

Mites, collembola (springtails), and other soil-dwelling arthropods can be separated from a large handful of leaf litter or ground cover by placing the sample on wire mesh (e.g., 1/4-inch hardware cloth) inside a large funnel. Suspend the funnel for several days over a beaker containing 70% ethanol. As the sample dries out, arthropods will move downward and eventually drop into the alcohol.

Winged insects (adult stage) are usually mounted on stainless steel insect pins and stored in an air-tight box with moth balls. Glue very small specimens to the tip of a paper "point." Wingless and immature insects, as well as most other arthropods, should be preserved in 70% ethanol. In order to prevent distortion and discoloration, caterpillars, maggots, and other soft-bodied individuals should be submerged in a fixative solution for several days before going into alcohol. Kahle's solution, a mixture of 95% ethanol, formaldehyde, glacial acetic acid, and water in a ratio of 15:6:2:30, is a good fixative for soft-bodied insects.

Key Terms

Biramous appendage a two-branched appendage characteristic of the crustaceans and certain other invertebrate groups. Consists of a basal portion (protopodite) and two distal branches or rami. The medial (interior) branch is the endopodite, and the lateral (exterior) branch is the exopodite.

Book gills respiratory organs consisting of many thin, highly vascularized sheets of tissue. Found in *Limulus.*

Cephalothorax fused head and thorax regions typical of the Crustacea and the Arachnida.

Chelicera (plural: chelicerae) pincerlike first appendage characteristic of the Chelicerata.

Cheliped one of the first pair of walking legs in the crayfish, lobster, and many other large crustacea. Equipped with a terminal pincerlike claw called a chela.

Chitin a complex structural carbohydrate that forms a chief component of the exoskeleton of many arthropods and other invertebrates. A polymer of N-acetylglucosamine.

Compound eye a photoreceptor composed of many individual units, the ommatidia. Typical of insects but also found in many crustaceans.

Endopodite the medial branch (ramus) of the distal portion of a biramous appendage.

Exopodite the lateral (exterior) branch (ramus) of the distal portion of a biramous appendage.

Exoskeleton an external supporting structure or support for the body. Found in arthropods and many other kinds of invertebrates.

Extensor muscle a muscle that extends or straightens out an appendage or other body part.

Flexor muscle a muscle that flexes or bends an appendage or other body part toward the body.

Green glands type of excretory organ found in the crayfish and many other crustacea. Located near the base of the antennae.

Hepatopancreas spongy internal organ of many arthropods that secretes digestive enzymes, absorbs food, and stores food reserves. Formed as an outgrowth of the midgut.

Homology two structures having the same embryonic origin although their functions may be the same or different.

Larva (plural: larvae) an independent, immature stage of an animal morphologically dissimilar to the adult.

Metamorphosis a change from one form to another in the developmental history of an animal.

Nymph an immature stage in the life history of insects exhibiting incomplete metamorphosis (hemimetabolism). Morphologically similar to the adult.

Ocellus (plural: ocelli) a simple eye found in many invertebrates.

Protopodite the basal portion of a biramous appendage in arthropods.

Pupa a metamorphic stage following the larva stage and preceding the adult stage in insects with holometabolism.

Serial homology the adaptation of a longitudinal series of originally similar organs to perform different functions, as in the appendages of the crayfish and many other arthropods.

Sexual dimorphism significant structural, physiological, and behavioral differences between the male and female members of a species.

Statocyst organ of equilibrium found in certain invertebrates, including crustaceans.

Statolith a sand grain or calcareous granule within a statocyst.

Internet Resources

There are many valuable Internet sites with information about zoology. Several sites containing pertinent zoological information for this chapter can be found on the McGraw-Hill Zoology web site at http://www.mhhe.com/zoology. Just click on this text's title.

Questions for Critical Thinking

1. Why do you suppose the arthropods are the most diverse and among the most abundant of all groups of organisms? Think in terms of cephalization, the specialization of the arthropodan appendages, segmentation, exoskeleton, generally high fecundity (fertility), and the range of food sources utilized.
2. Why do you suppose there are no truly marine insects? (It is true that some water striders are surface dwellers thousands of miles from shore, but there are no insects truly adapted to marine environments.) Explain your reasoning.
3. Examine the structural and chemical similarities of chitin and cellulose. (You may have to use an introductory biology textbook for this purpose.) Do you see any similarities? What are they? Is this surprising to you, or would you expect it based on evolutionary theory?
4. If crustaceans have modified legs for mouthpart appendages, this must mean that the original segmentation of the head region was composed of more than one segment. Count the mouthparts and estimate how many segments made up the "ancestral" anterior region of crustaceans (this also works for insects).
5. What is the significance of tagmatization (the amalgamation of body segments into distinct sections such as head, thorax, and abdomen)? Discuss the evolutionary importance of tagmatization.
6. Why is sclerotization so important to the exoskeleton of insects and crustaceans? What advantages does sclerotization provide?

Suggested Readings

Chapman, R.F. 1999. *The Insects: Structure and Function,* 4th ed. Cambridge: Cambridge University Press.

Fitzpatrick, J.F. Jr., J. Bamrick, E.T. Crawley, and W.G. Jaques. 1983. *How to Know the Freshwater Crustacea.* Dubuque, IA: William C. Brown Publishers.

Foelix, R.F. 1996. *Biology of Spiders.* New York: Oxford University Press.

Fortey, R.A., and R.H. Thomas (eds.). 1998. *Arthropod Relationships.* London: Chapman & Hall.

Gibb, T. 2006. *Arthropod Collection and Identification: Field and Laboratory Techniques.* Boston: Elsevier Academic Press.

Koenemann, S., and R.A. Jenner (eds.). 2005. *Crustacea and Arthropod Relationships.* Boca Raton, FL: Taylor & Francis.

Triplehorn, C.A., and N.F. Johnson. 2005. *Borror and DeLong's Introduction to the Study of Insects,* 7th ed. Belmont, CA: Thompson Brooks/Cole.

Notes and Sketches

Chapter 14

Echinodermata

© Diane R. Nelson.

Objectives

After completing the laboratory work in this chapter, you should be able to perform the following tasks:

1. List and briefly characterize the five classes of living echinoderms.
2. Cite five significant differences between the echinoderms and each of the following phyla studied previously: Arthropoda, Annelida, and Mollusca.
3. Identify the principal external features of a sea star. Explain the function of the madreporite, pedicellariae, dermal branchae, and tube feet.
4. Explain the role of the coelom in respiration and excretion in a starfish. How does this compare with the role of the coelom in annelids like an earthworm or a polychaete worm?
5. Identify the parts of the digestive tract of a sea star and explain the function of each part.
6. Identify the parts of the water vascular system of a sea star and identify a function for each part.
7. Identify the main external features of a sea urchin.
8. Compare the basic organization of a sea star, a sea urchin, and a sea cucumber.

Introduction

Unlike most other phyla discussed previously in this book, echinoderms are members of the "superphylum" Deuterostomia, a sister group to the Protostomia.

Deuterostomes exhibit some fundamental differences in embryological development from that seen in protostomes, particularly in the formation of the mouth, anus, and coelom. Refer to Chapter 5 for a more detailed explanation of differences between these two "superphyla."

The Phylum Echinodermata is a unique group of spiny-skinned marine animals, which includes the sea stars or starfish,

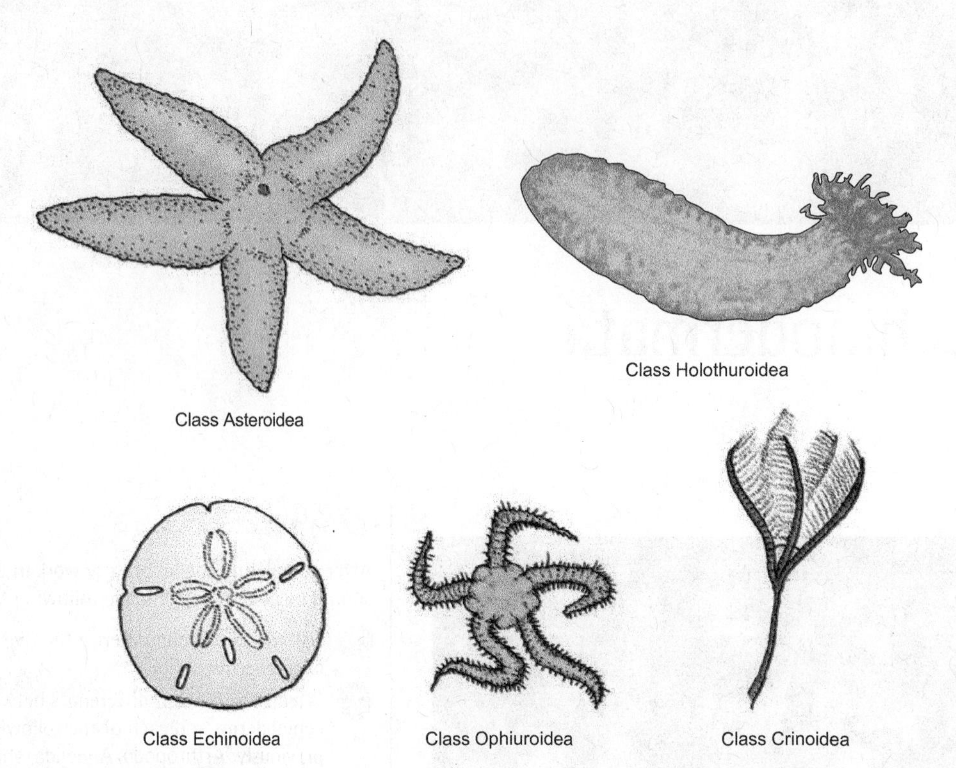

FIGURE 14.1 Classes of living echinoderms.

sea urchins, sand dollars, feather stars, sea cucumbers, brittle stars, and sea lilies (figure 14.1). Echinoderms are exclusively marine because they are unable to regulate the ionic composition of their body fluids and thus have not been able to colonize freshwater habitats. Nonetheless, echinoderms are widespread in the seas, occurring in shallow waters and in the depths of the sea in all parts of the world.

Echinoderms have inhabited the seas since the early Cambrian era, over 600 million years ago. They flourished in the seas at several times in the past, and many more fossil species have been described (some 13,000) than living species (about 7,000).

A prominent feature of living echinoderms is their five-part **radial symmetry** (pentaramous). The adult echinoderm body is typically divided into five parts radiating out from a central axis. These five radiating areas are called **ambulacra** and the areas between the ambulacra are called **interambulacra.** Echinoderm larvae, however, are **bilaterally symmetrical** and the radially symmetrical adult body results from a dramatic metamorphosis during development.

Another important feature of echinoderms is a calcareous **endoskeleton** made up of many small calcareous plates that may be fused into rigid **test** (as in sea urchins and sand dollars), or separated (as in sea cucumbers). It is called an endoskeleton because the surface of the skeleton is covered with a thin layer of living tissue. Echinoderms have a large enterocoelous **coelom** lined with a ciliated peritoneum, a portion of which forms the unique **water vascular system** during development.

The **water vascular system** is an interconnected system of fluid-filled tubes, reservoirs, and tube feet that serve several important functions in echinoderms. In addition to serving as an internal transport system for nutrients and waste products, it plays a vital role in gas exchange and defense. Hydraulically operated tube feet in various species are specialized for locomotion, food handling, burrowing, and sensory perception.

Many zoologists are especially interested in echinoderms because of their evolutionary relationships with the hemichordates and chordates. Recent genetic studies have revealed that sea urchins share as many as 7,000 genes with humans. Other evidence comes from embryological studies, including radial bilateral cleavage, enterocoelous coelom formation, and their endodermal skeleton.

Classification

Phylogeny and Evolution

Echinoderms are an ancient group of animals. Because of their hard exoskeleton, they have left an extensive fossil record. Despite this good paleontological record, there are some conflicting views regarding their evolution. The traditional view is that the ancestral echinoderm stock was free swimming and bilaterally symmetrical. This is suggested by the bilaterally symmetrical bipinneria larva found in many species. This evolutionary view also interprets the radial symmetry of adult echinoderms as a secondary feature acquired as an adaptation resulting from the sedentary, or sessile, habit of adult echinoderms.

The alternate, and less popular, view is that radial symmetry arose in free-swimming ancestors and that sessile groups arose several times independently from the free-swimming ancestors.

Other phyla of deuterostomes are the Hemichordata, Chordata, and the Xenoturbellida, a small group of little-known marine worms.

Ten classes of echinoderms are currently recognized by most scientists, five of which are known only from fossil remains. The five classes of living echinoderms are shown in the following cladogram.

Cladogram of Echinodermata

- Class **Crinoidea** (sea lilies and feather stars)
- Class **Ophiuroidea** (brittle stars and basket stars)
- Class **Asteroidea** (sea stars)
- Class **Echinoidea** (sea urchins and sand dollars)
- Class **Holothuroidea** (sea cucumbers)

Class Echinoidea

Sea urchins, heart urchins, and sand dollars (figure 14.2a and b). Echinoderms with a rigid test of fused skeletal plates. Body covered with movable spines. Five rows of tube feet (bearing suckers) around the test. There are about 950 species. Examples: *Arbacia, Lytechinus, Strongylocentrotus* (all sea urchins); *Echinarachnius, Mellita,* and *Dendraster* (sand dollars).

Class Holothuroidea

Sea cucumbers (figure 14.2c). Non-sessile soft-bodied animals having a flexible body wall with many tiny, embedded calcareous ossicles; no spines or arms. Body elongated in the oral-aboral axis to a cucumber or pickle-like form; a whorl of tentacles surround the mouth. Usually with five or more series of tube feet arranged in longitudinal rows. There are about 1,200 species. Examples: *Thyone, Cucumaria and Stichopus.*

Class Crinoidea

Sea lilies, feather stars (figure 14.3a, b, and c). Flowerlike echinoderms with a central calyx and five (or multiples of five) branching arms. Ciliated ambulacral groove along the

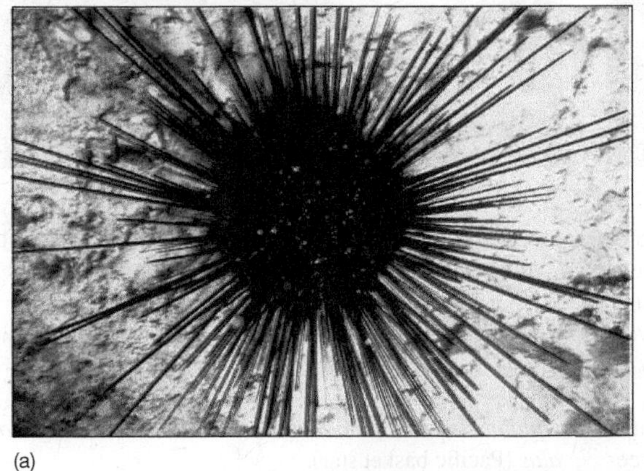
(a)

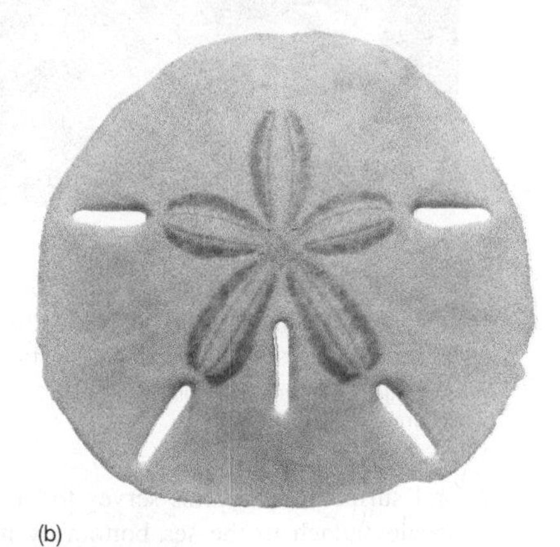
(b)

(c)

FIGURE 14.2 Representative echinoids and holothurians. (*a*) *Diadema*, sea urchin found in the western Atlantic in Florida and the West Indies. The sharp spines contain a toxin for defense. (*b*) *Mellita*, keyhole sand dollar found along the southern Atlantic coast, the Caribbean, and South America. (*c*) *Thyone*, sea cucumber, found in U.S. coastal waters from Maine to Texas.

FIGURE 14.3 Representative crinoids and asteroids. (*a*) Stalked crinoids at a Pacific Ocean deep sea volcanic site. (*b*) Fossil crinoid from a limestone bed. (*c*) Feather star, observe the pinnate arms extending from the central disc and note the absence of a stalk. (*d*) Necklace sea star, *Fromia*, New Guinea.

(a) Image courtesy of Submarine Ring of Fire 2006 Exploration, NOAA Vents Program. (b) Alan Morgan. (c) Comstock Images/Getty Images. (d) © Brand X Pictures/PunchStock.

oral surface of the arms serves for food gathering. Some species attach to the sea bottom by a stalk; others free-swimming. There are about 625 species. Example: *Cerocrinus* (crinoid), *Hathrometra* (=*Antedon*) (feather star).

Class Asteroidea

Starfish or sea stars. Animals with a star-shaped body, typically with five arms and a flexible skeleton with many calcareous plates (figure 14.3d). The arms are continuous with and not clearly distinct from the central disc. With ambulacral grooves on the oral side of each arm; tube feet bearing suckers. There are about 1,500 species. Example: *Asterias, Pisaster, Astropecten.*

Class Ophiuroidea

Brittle stars, basket stars (figure 14.4a and b). Star-shaped echinoderms with arms distinct from the central disc; no ambulacral grooves along the arms; tube feet absent or reduced to sensory organs. There are about 2,000 species. Example: *Amphipholis, Ophiothrix,* and *Ophioderma* (brittle stars); and *Gorgonocephalus* (Pacific basket star).

Materials List

Preserved specimens
- *Asterias*
- *Arbacia*
- *Cucumaria* or *Thyone*
- Fossil crinoid
- Aristotle's Lantern (demonstration)
- Sea star, dried to show ossicles (demonstration)
- Representative echinoderms (demonstration)

Prepared microscope slides
- Sea star, pedicellaria (demonstration)
- Sea star, dermal branchiae (demonstration)
- Sea star, bipinnaria larva (demonstration)
- Sea star, metamorphosis (demonstration)

Audiovisual material
- Anatomy of the Starfish video

(a)

(b)

FIGURE 14.4 Representative Ophiuroids. (*a*) Brittle star, observe the five flexible arms. (*b*) Basket star, *Astrophyton,* observe the five branched arms extending from the central disc.

(a) Courtesy of Carolina Biological Supply Company, Burlington, NC
(b) © Diane R. Nelson.

The Common Sea Star: *Asterias*

Class Asteroidea

Asterias forbesi is a common sea star (figure 14.5) found along the Atlantic coast of the United States. Obtain a living or preserved specimen of *Asterias* and observe its radial symmetry. The radial symmetry of adult echinoderms appears to be a secondary condition, since echinoderm larvae are bilaterally symmetrical. Observe a demonstration slide of the bipinnaria larva of the sea star. ***In what ways does the radial symmetry of the echinoderms differ from that of the cnidaria?***

The common sea star along the Pacific coast of the United States is *Pisaster,* whose anatomy is similar to that of *Asterias* and can be used in this exercise where available.

External Anatomy

Specimens preserved for laboratory study often are injected with a colored material to facilitate the study of the water vascular system, including the external tube feet. Keep your specimen wet by adding a little water to the dissecting pan. Note the **central disc** on the upper, or **aboral,** side; the five rays or **arms;** and the **madreporite**—a light-colored circular area near the edge of the disc at the junction of two rays. The madreporite is a calcareous disc that serves as the entrance to the water vascular system. The two arms adjacent to the madreporite comprise the **bivium;** the other three arms make up the **trivium.** The arm opposite the madreporite is referred to as the **anterior** arm.

Note the many spines scattered over the surface of the arms and the central disc. Among the spines there are also many small pincerlike structures located near the bases of the spines (see demonstration slide). These structures are the **pedicellariae** (figure 14.6).

- Carefully scrape a small area of the arm with your scalpel, add the scrapings to a drop of water, and observe the material under the low power of your microscope. ***Can you identify any pedicellariae among the scrapings? From their structure, what do you think the function of the pedicellariae might be?***

Also among the spines are numerous **dermal branchiae** (also called papulae), soft, hollow projections of the body wall used for gas exchange (see demonstration slide). Observe the skeletal calcareous plates, or **ossicles,** on a dried sea star specimen.

Note on the oral side of the sea star: the **mouth,** guarded by specialized **oral spines** and surrounded by a soft membrane, the **peristome;** the five **ambulacral grooves,** extending from the mouth and along the middle of each arm; and the numerous **tube feet** in the grooves, extending from the water vascular system. There is also a pigmented **eyespot** and a **sensory tentacle** at the tip of each ray, but these latter structures are usually difficult to find in preserved specimens.

Internal Anatomy

- Carefully examine figures 14.7 and 14.8 for orientation before starting your dissection. Cut off about one-half inch from the tip of the anterior arm. Then carefully cut along each side of the anterior arm to its junction with the central disc and across the top of the arm at the margin of the disc. Observe the hard, calcareous plates, or ossicles, of the skeleton as you cut. Remove the portion of the body wall covering the aboral surface of the arm and examine the large **coelom** within the arm, which houses the internal organs. The coelom is lined with a ciliated **peritoneum,** as mentioned previously, and is normally filled with coelomic fluid. This fluid carries oxygen and dissolved nutrients to various parts of the body. Next, remove the portion of the body wall covering the aboral side of the central disc, but leave the madreporite in place by carefully cutting around it.

FIGURE 14.5 *Asterias*, sea star. (*a*) Aboral view. (*b*) Oral view.
Photos by John Meyer.

Digestive System

The sea star has a complete digestive system. The mouth is on the oral surface and opens into a short esophagus. The esophagus leads to a large, two-chambered stomach that fills most of the central disc. The large, thin-walled cardiac chamber lies below a smaller, thick-walled pyloric chamber.

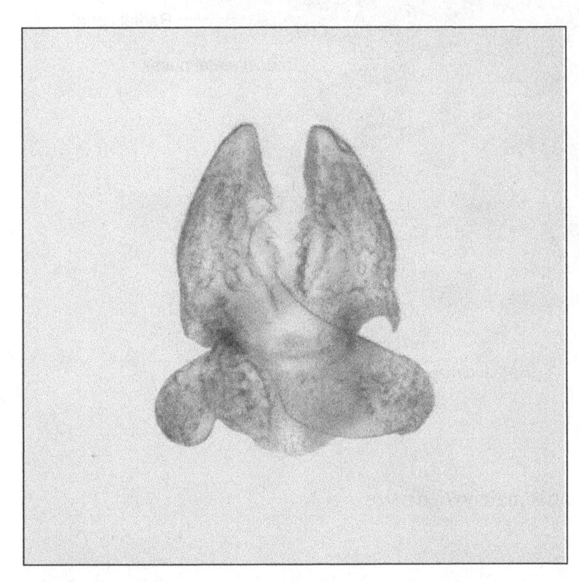

FIGURE 14.6 Pedicellaria, magnification about 300×.
Photo by John Meyer.

Emptying into the pyloric chamber through five pyloric ducts are five pairs of large, branched pyloric caeca (digestive glands). The pyloric caeca fill most of the interior space in the arms. Also leading from the pyloric chamber is a small, short intestine that empties through the aboral anus.

The cardiac chamber can be everted through the mouth to envelop its prey. Most sea stars are carnivores and feed on many kinds of invertebrates, including small clams, oysters, gastropods, crustaceans, polychaetes, other echinoderms, and small fish. Prey is often engulfed whole, including shell or exoskeleton, and the undigestible parts are later regurgitated. Typically, food is digested in the cardiac chamber of the stomach. Digestive enzymes are secreted principally by the pyloric caeca, which also play a major role in absorption and storage of food materials.

Water Vascular System

The **water vascular system** (figure 14.9) is an important feature of the echinoderms. It develops embryologically as a specialized part of the coelom and is lined with a **ciliated epithelium.** The water vascular system has several important functions, including locomotion, feeding, gas exchange, excretion, and sensory perception. The canals of this system are filled with a water vascular fluid that has a chemical composition resembling seawater plus some soluble proteins and a higher concentration of potassium ions. Suspended in the water vascular fluid are numerous **coelomocytes.** These are amoeboid cells that circulate

FIGURE 14.7 *Asterias,* sea star, oral view, partly dissected. Approximately life size.
Courtesy of Carolina Biological Supply Company, Burlington, NC.

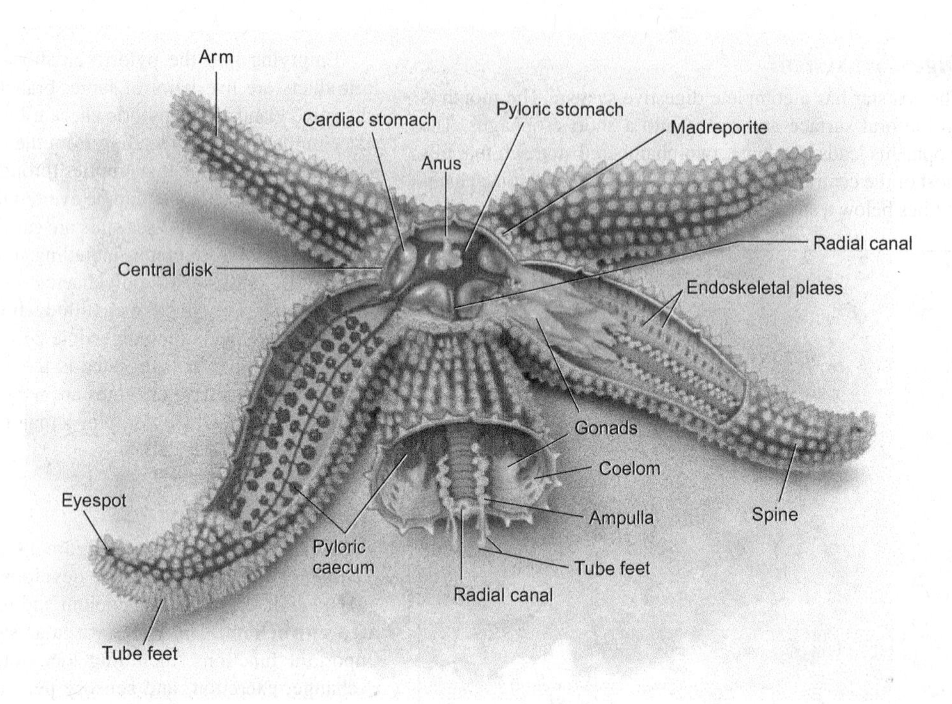

FIGURE 14.8 Sea star, internal anatomy, aboral view. Approximately 2/3 life size.

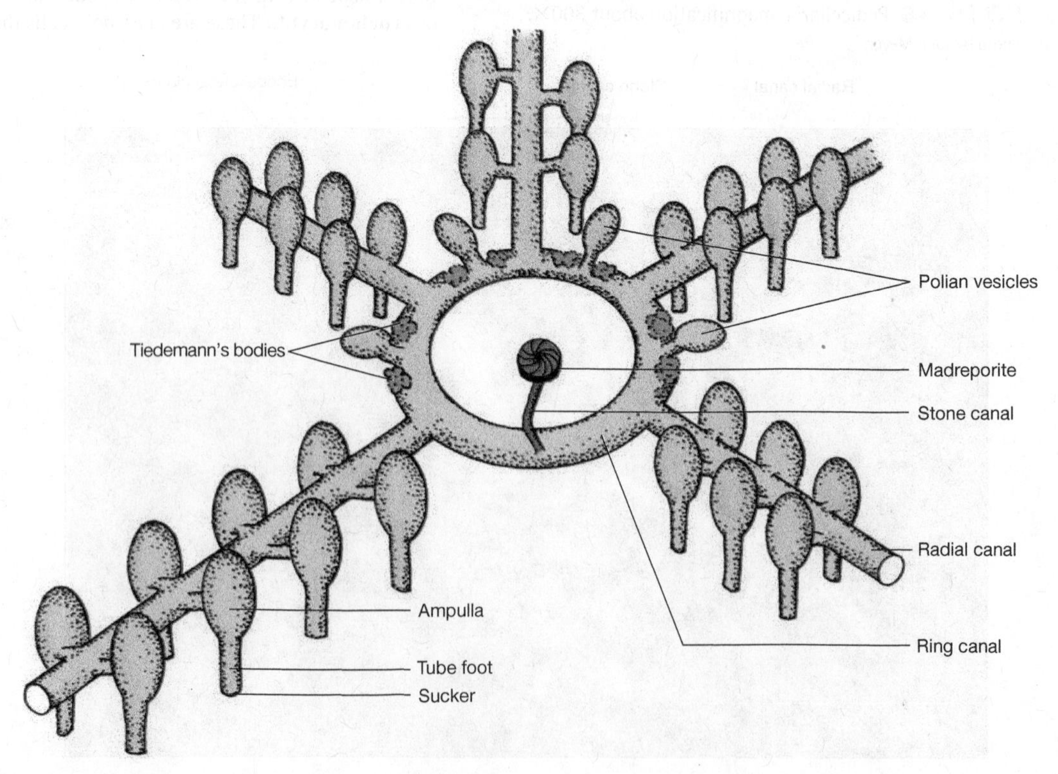

FIGURE 14.9 Water vascular system of a sea star.

through the canals and coelomic spaces. They ingest foreign materials and probably also play a role in excretion and nutrient transport.

Sea stars have a well-developed water vascular system. Among its principal parts are the external **madreporite,** seen previously on the aboral surface, which marks the entrance to a short **stone canal** (figure 14.9). The madreporite is covered by the thin, ciliated epidermis of the body surface, and is believed to function as a pressure regulator for the water vascular system. The stone canal gets its name from the calcareous deposits surrounding it. The stone canal leads to the **ring canal,** which encircles the mouth. Attached to the ring canal are one or more muscular sacs, the **Polian vesicles,** and four or five pairs of irregularly shaped **Tiedemann's bodies.** The Polian vesicles are believed to serve as fluid reservoirs and pressure regulators that maintain proper hydrostatic pressure within the water vascular system. The Tiedemann's bodies produce coelomocytes and may function in the chemical regulation of the coelomic fluid.

Fact File

Sea Stars and Brittle Stars

- Echinoderms have remarkable powers of regeneration. Missing arms, broken spines, and in some cases, even internal organs will grow back over a period of weeks or months. This explains why some arms of a sea star may be longer than others.
- Calcite crystals within the endoskeleton of some brittle stars (e.g., *Ophiocoma wendtii*) form a distributed array of microlenses on the aboral surface. Together with light-sensitive cells in the nervous system, these structures constitute a primitive "compound eye" that is capable of detecting nearby objects or the approach of a predator.

Five **radial canals** lead from the circular canal along the top of the **ambulacral groove** leading into each arm. Many short **lateral canals** connect the radial canals with each pair of **tube feet.** These tube feet are located in rows along the ambulacral grooves found on the oral side of each of the five arms. The Latin meaning of *ambulacrum* is literally, "a tree-lined walk"—a colorful and accurate description of the parallel rows of tube feet that border the radial canals extending outward along the arms of the sea star. Each tube foot consists of an internal bulblike **ampulla** attached to each suckerlike tube foot.

Within each lateral canal is a valve that closes to allow local pressure changes in the individual tube feet. Contraction of the circular muscles in the wall of the ampulla increases the internal pressure and extends the foot. Contraction of longitudinal muscles in the wall of the tube foot withdraws the foot. These pressure changes within the water vascular system (plus nervous control from the central nervous system) allow coordinated movements of the tube feet and, thus, coordinated locomotion of a sea star.

Nervous System

The sea star has a very simple nervous system. A circular **nerve ring** surrounds the mouth, and a **radial nerve** extends from the nerve ring into each arm. A simple light-sensitive **eyespot** at the tip of each arm is the only differentiated sense organ; however, sensory cells are scattered throughout the epidermis.

Coelom, Respiration, and Excretion

The coelom plays a role in several vital body functions in the echinoderms. The pouches of the larval coelom develop into both the body coelom and the water vascular system. The sea star has a large, fluid-filled coelom extending from the central disc into the arms. It is lined by a ciliated peritoneum that circulates the coelomic fluid through the body and into the dermal branchiae or papulae. These structures are thin-walled projections of the coelom that extend through the body surface in spaces between the ossicles. Oxygen and carbon dioxide are exchanged by diffusion across the thin walls, as are nitrogenous wastes, largely in the form of ammonia. The coelom also contains many amoebocytes that take up some waste materials and later migrate through the thin wall of the dermal branchiae and the tube feet. Echinoderms have no differentiated excretory organs. Their reduced hemal system appears to serve little function other than circulating digested nutrients.

Reproductive System

The reproductive system consists of a pair of **gonads** located near the base of each arm (see figure 14.8). The sexes are separate in sea stars, but cannot be distinguished except by microscopic examination of the gonad contents.

- Take a small amount of tissue from one of the gonads and macerate it in a drop of water on a clean microscopic slide. ***Can you identify any large oocytes (eggs) or any flagellated sperm?***

From each gonad a duct leads to an external pore on the aboral surface. Eggs and sperm are released by mature sea stars into the seawater, where fertilization occurs. The fertilized eggs develop into free-swimming bipinnaria larvae (figure 14.10). Observe that the bipinnaria larvae are bilaterally symmetrical rather than radially symmetrical like the adults. ***What does this suggest about the ancestry of the sea stars, and perhaps of the echinoderms?***

After a period of larval existence in the plankton, the larvae settle on some suitable substrate and undergo metamorphosis into miniature sea stars.

Fact File

- *Acanthaster planci* is commonly called the "crown of thorns" sea star. It is a pest on coral reefs where it feeds voraciously on the polyps. It has seriously damaged large portions of the Great Barrier Reef in Australia. Large spines on the aboral surface deliver a potent toxin but they also provide a safe haven for commensal shrimp and cardinalfish.
- Symbiotic bacteria living within some brittle stars produce substances that inhibit growth of harmful pathogens. These compounds are currently being investigated for their potential as antibiotic drugs.

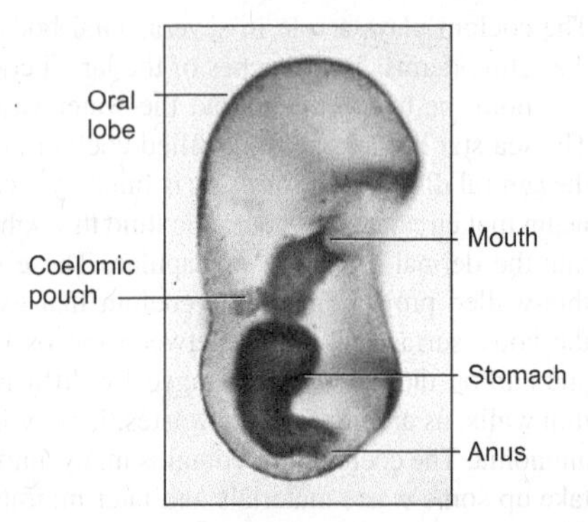

FIGURE 14.10 Bipinnaria larva of a sea star, lateral view.
Courtesy of Carolina Biological Supply Company, Burlington, NC.

A Sea Urchin

Class Echinoidea

Sea urchins are armless echinoderms with a rigid endoskeleton, or test. The endoskeleton is made up of ten radial areas of fused plates or ossicles: five **ambulacra** with holes bearing tube feet alternating with five **interambulacral plates** without holes. Sea urchins are usually globular or ovoid in shape and the aboral surface is covered with long, movable spines (figure 14.11). Among the spines are numerous tube feet and many **pedicellariae** which serve to keep the test clean of debris and fouling organisms.

- Study a living or preserved specimen of *Arbacia* or a similar sea urchin. Identify the rigid **test, spines, peristomial gills** and **pedicellariae.** Also locate the five series of **tube feet** along five symmetrically placed **ambulacral regions** around the globular body. In a living specimen, the tube feet can extend well beyond the spines.

On the oral side of the urchin (flattened surface), find the **mouth** surrounded by five **calcareous teeth** and a circular oral membrane, the **peristome.** Five pairs of **peristomial gills** are located around the peristome. The teeth are

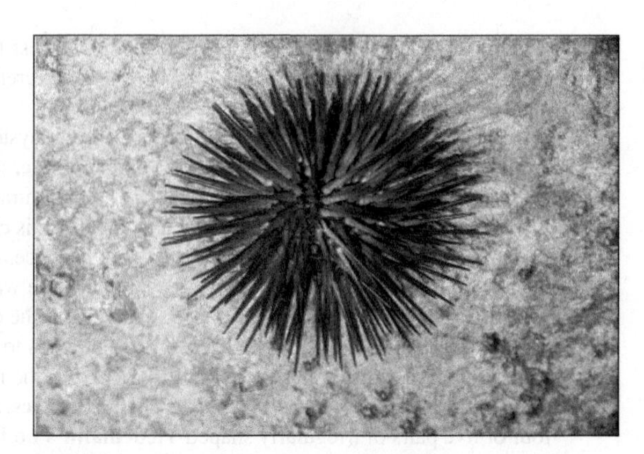

FIGURE 14.11 *Arbacia*, purple sea urchin.
Courtesy of Carolina Biological Supply Company, Burlington, NC.

connected internally to a complex chewing apparatus, called **Aristotle's Lantern,** which anchors the teeth and enables them to scrape algae from submerged rocks along the shore where they live. Although some sea urchins are scavengers, most species are principally herbivorous, and many species feed on algae from the sea floor; they are sometimes called the grazers of the sea. Their relatives, the sand dollars and heart urchins, are adapted for life on soft sea bottoms in which they burrow in search of food and shelter.

Internally, sea urchins have a short intestine that terminates in an anus near the center of the aboral surface. In a mature animal, much of the body cavity is filled with gonads (sexes are usually separate). At spawning time, eggs and sperm are released directly into the water. Before spawning, gravid individuals can often be induced to release gametes by electrical stimulation or by injecting potassium chloride (KCl) through the peristome and into the body cavity. Sperm remain "inactive" until they mix with seawater. Detailed instructions for collecting gametes and observing fertilization and embryonic development of sea urchins may be obtained from vendors of biological supplies. (See also Chapter 4, the section, sea star Embryology.)

Fact File

Sea Urchins and Sea Cucumbers

- A single sea urchin can produce as many as 20 million eggs. In some parts of the world, these eggs, called "roe," are regarded as a culinary delicacy.
- Some sea cucumbers excrete sticky threads (Cuvierian tubules) as a defense mechanism. These filaments are so sticky that they cannot be removed without shaving the skin. Natives of Palau, a chain of islands in the Western Pacific, smear this stickum on bleeding wounds as a bandage or apply it to their bare feet for protection against the sharp edges of broken shells on the beach.

Continued—

—*Continued*

- In a sea cucumber, self-defense may involve "autoevisceration"—a process that involves ejecting some or all of its internal organs. Predators are either scared away or appeased by this behavior; the sea cucumber may escape and soon regenerates its internal organs.
- In parts of Asia and the tropics, sea cucumbers are collected for food. After removing the viscera, people dry the skins and sell them as a base for soups and stews. Locally they are known as *trepang* or *beche-de-mer*. Like tofu, sea cucumber is flavorless but has the ability to soak up the taste of other foods and seasonings.
- Several species of pearlfish (Family Carapidae) live as commensals in the respiratory trees of some sea cucumbers.

A Sea Cucumber

Class Holothuroidea

Sea cucumbers are soft-bodied echinoderms with many tiny plates, or ossicles, embedded within a leathery body wall (figure 14.12). Most sea cucumbers are scavengers or filter feeders. The mouth is surrounded by oral tentacles (modified tube feet), which in life are in nearly constant motion. Food particles collect on the sticky surfaces of these tentacles and are passed with wavelike movements through the mouth into the pharynx.

Five **ambulacra** extend aborally from the mouth and are indicated externally by the rows of small tube feet. The areas between adjacent ambulacra are the **interambulacra.** Five longitudinal muscle bands lie internally beneath the ambulacra. The **anus** is at the aboral end of the body and opens from the cloaca. Study a demonstration of *Cucumaria* or other sea cucumber and identify its internal organs with the aid of figure 14.13 and the following description.

The sea cucumber has a large coelom that consists of several compartments. Most of the internal organs are contained in the largest of these compartmens, the **perivisceral coelom.** Observe the white calcareous **ring canal** at the base of the oral tentacles from which extend five **Polian vesicles** and a smaller **stone canal** that ends at the **madreporite.** Note that the madreporite lies within the coelom. ***How does this compare with the location of the madreporite in the sea star?*** The water vascular system consists of the ring canal with its attached Polian vesicles, stone canal, and five small radial canals that extend anteriorly, then loop posteriorly to run beneath the five longitudinal muscle bands that extend the length of the body. The radial canals extend through the body wall to connect with the rows of tube feet.

The long bundle of slender filaments extending posteriorly is the single gonad that divides into two parts. The digestive system leads from the mouth located in the center of the oral tentacles, with the pharynx passing through the ring canal and leading to the expanded stomach. Posterior to the stomach, the long intestine loops posteriorly, then anteriorly, before looping back to empty into the posterior cloaca. The anus is located at the posterior end of the cloaca.

Extending anteriorly from the cloaca are two respiratory trees that exchange respiratory gases with the coelomic fluid. Seawater is drawn into the cloaca by contraction of the numerous cloacal muscles that connect the cloaca with the outer body wall after closing the intestinal sphincter and opening the anal sphincter. Later, contraction of the circular muscles around the cloaca forces the seawater into the respiratory trees where gas exchange occurs with the surrounding coelomic fluid. Following gas exchange, contraction of

FIGURE 14.12 *Cucumaria*, sea cucumber. Note the elongate oral-aboral axis.
Courtesy of Carolina Biological Supply Company, Burlington, NC.

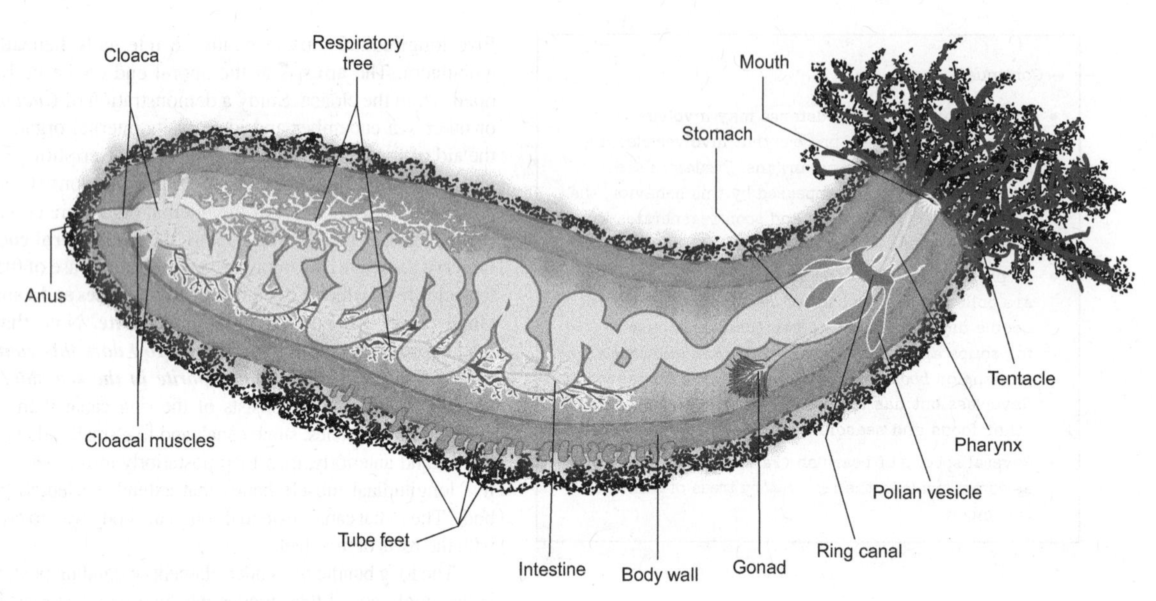

FIGURE 14.13 Sea cucumber, dissected.

muscles in the respiratory trees forces the seawater back into the cloaca, which relaxes its circular muscles and the anal sphincter to complete the process.

The hemal system includes a small hemal ring adjacent to the ring canal with five radial hemal canals that extend along the radial canals of the water vascular system in the ambulacra areas. A system of interconnected hemal canals lead from various areas of the intestine and connect with an intestinal vessel and a portal system of capillaries. The nervous system consists of an anterior nerve ring and five radial nerves that extend along the ambulacral areas with the radial canals.

Demonstrations

1. Pedicellariae (microscope slide)
2. Dermal branchiae (microscope slide)
3. Bipinnaria larva (microscope slide)
4. Sea star metamorphosis (microscope slide or color transparency series)
5. Preserved or dried Aristotle's Lantern from sea urchin
6. Dried sea star to show ossicles
7. Preserved specimens representing other types of echinoderms, such as crinoids, basket stars, and brittle stars

Collecting and Preserving Echinoderms

Echinoderms are widely distributed throughout the oceans and are commonly found as bottom-dwellers in both shallow and deep water. Look below the waterline on pier pilings, wharves, and intertidal rocks for sea stars and brittle stars. Many species are active and easier to find at night with a headlamp. Sea urchins often cling to rocks and shells in deeper water. Sand dollars and sea cucumbers usually burrow into sand or mud on soft parts of the seafloor. Search for them with a dredge net or carefully rake through loose sediments.

Most echinoderms can be transported in buckets of seawater or loosely wrapped in wet seaweed. They will survive for several days to weeks in a well-balanced saltwater aquarium. The animals should not be crowded. Sea urchins can be fed small pieces of beef or raw seafood; provide live snails, clams, or oysters as food for sea stars. Avoid overfeeding and fouling the water.

A solution containing 70% ethanol and glycerine in a 9:1 ratio is a good preservative for most echinoderms. Unfortunately, the natural colors fade quickly and cannot be preserved. Several changes of solution may be necessary until body pigments stop leaching into the alcohol.

Key Terms

Ambulacral groove one of the five longitudinal grooves on the oral side of the arms of a sea star that bear many tube feet.

Ambulacrum (plural: ambulacra) One of the five radial sections on the body surface of many echinoderms that extend from the central area and bear rows of tube feet. The radially symmetrical echinoderm body is usually divided into ten radial sections: the five ambulacra alternate with five interambulacra lacking tube feet; evident in sea stars, sea urchins and sea cucumbers.

Aristotle's lantern a rasping/chewing apparatus located within the mouth opening of sea urchins and used for scraping food particles from the substrate.
Dermal branchiae also called papulae, are small, thin-walled, fingerlike projections of the coelom that extend through the body surface in spaces between the ossicles of sea stars and other asteroids. Serve in gas exchange and waste removal.
Madreporite a sievelike opening to the water vascular system located on the dorsal (aboral) surface of many echinoderms.
Pedicellariae small, pincerlike structures found on the surface of many sea stars (Class Asteroidea), sea urchins (Class Echinoidea), and brittle stars (Class Ophiuroidea). Serve to keep the body surface free from debris and fouling organisms.
Radial symmetry type of body organization in which all parts of the body are arranged symmetrically around a central axis.
Test a rigid endoskeleton composed of many fused calcareous plates as exhibited by the sea urchins.
Water vascular system a unique organ system found only in the Phylum Echinodermata. Functions in feeding, attachment, locomotion, gas exchange, and sensory perception. Formed embryologically as a specialized portion of the coelom.

Internet Resources

There are many valuable Internet sites with information about zoology. Several sites containing pertinent zoological information for this chapter can be found on the McGraw-Hill Zoology web site at http://www.mhhe.com/zoology. Just click on this text's title.

Questions for Critical Thinking

1. Compare the structural similarities and differences among the Arthropoda, Annelida, and Mollusca. In what important ways do they differ from Echinodermata? What does this suggest about the ancestry of these phyla?
2. Explain how the water vascular system of echinoderms functions. Identify the madreporite, stone canal, ring canal, Tiedemann's bodies, dermal branchae, and tube feet and explain their interactions.
3. Echinoderms and chordates are both deuterostomes. Does this necessarily mean that one of these groups descended from the other? Explain why or why not.
4. Scientists refer to some types of animals as reduced or "degenerate" because they have lost certain ancestral characteristics, often as a result of some specialized lifestyle. Could you speculate that the echinoderms are a degenerate group as compared with the chordates? Explain why or why not.
5. Examine the variations in the form of different classes of echinoderms, keeping in mind the elongation of the oral-aboral surface in some groups (e.g., the Holothuroidea). Can you imagine how a bilateral larva could give rise to a radial creature, and then in the Holothuroidea evolve an almost elongated, biradial-bilateral body form?
6. Why do you suppose the echinoderms are so well-represented in the fossil record?

Suggested Readings

Calestani, C. 2006. Sea Urchin Genome Sequencing Conference. The genome of the sea urchin *Strongylocentrotus purpuratus*. *Science* 31(5801): 941–952.

Hendler, G. 1995. *Sea Stars, Sea Urchins, and Allies*. Washington, DC: Smithsonian Institute Press. 392 pp

Matranga, V. (ed.). 2005. *Echinodermata*. New York: Springer. 275 pp

Nichols, D. 1969. *Echinoderms*, 4th (revised) ed. London: Hutchinson University Library, 192 pp

Notes and Sketches

Notes and Sketches

Chapter 15

Chordata

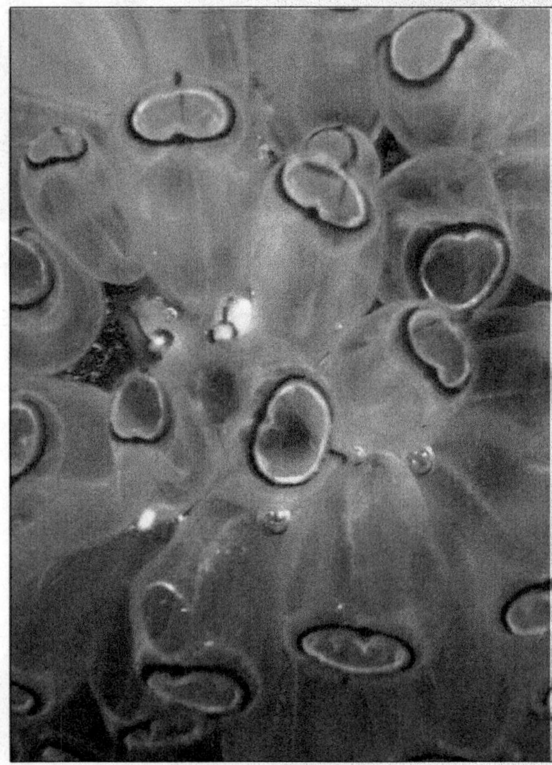
© Diane Nelson.

Objectives

After completing the laboratory work in this chapter, you should be able to perform the following tasks:

1. Identify five distinctive characteristics of the Phylum Chordata.
2. Describe the life cycle of a tunicate (Subphylum Urochordata) and identify the principal features of a tunicate tadpole larva. Explain the morphological and evolutionary significance of the larva.
3. Identify the main organs in an adult tunicate and discuss the adaptations of the adult for its sessile mode of life.
4. Explain the feeding mechanism of a tunicate and identify the principal structures involved.
5. Identify the main organs in a lancelet and discuss the significance of its basic chordate features.
6. Compare the organization of a tunicate and a lancelet and explain their main similarities and differences.

Introduction

The Phylum Chordata consists of more than 45,000 described species. These include a remarkably diverse range of life forms—from relatively simple marine animals, like sea squirts, to highly specialized birds and mammals. Superficially, these animals may appear to have little in common, but they are grouped into a single phylum because they all share, at some point in their life cyle, five highly distinctive characteristics: (1) dorsal, hollow nerve cord, (2) dorsal supporting rod (called the notochord), (3) paired pharyngeal gill slits, (4) endostyle, (5) post-anal tail. Each of these structures has made a unique contribution to the evolutionary development of this important animal group.

A chordate's central nervous system, which features a **dorsal, hollow nerve cord,** is different in both location and embryological origin from the nervous systems found in all other animals. All other animals with well-developed nervous systems have a solid nerve cord located in the ventral half of the body that forms when individual neurons grow outward from multiple sites of origin, called ganglia.

In chordates, however, the brain and nerve cord develop from embryonic tissue that folds inward or **invaginates** to form a longitudinal neural groove along the dorsal side of the embryo. This longitudinal groove running from head to tail later closes to form the **dorsal, hollow neural tube** characteristic of the chordates.

The **notochord** is a stiff but flexible rod that runs longitudinally just below the dorsal nerve cord. It contributes an element of support to the animal's body by resisting the tendency to shorten or "telescope" when longitudinal muscles contract. Its flexibility enables the musculature to produce undulating body movements, which, in combination with a postanal tail, improves locomotion in an aquatic environment. In vertebrates, such as birds and mammals, the notochord has been functionally replaced by a bony spinal column. Its remnants, however, still persist in developing embryos and eventually form the jellylike substance at the core of each intervertebral disc that serves as a cushion between bones in the adult spinal column.

Pharyngeal gill slits are perforations of the body wall that occur between the mouth and the esophagus. In primitive chordates, these slits are part of a filter-feeding apparatus: water enters the mouth, passes through the pharynx, and exits via the gill slits. Mucus produced by the **endostyle** (on the floor of the pharynx) helps trap ingested food particles before they exit. Gill slits are visible in the developing embryos of all chordates. Over the course of evolution, the primitive filter mechanism has undergone changes in structure and function. The endostyle has evolved into the thyroid gland, and the pharyngeal slits have become adapted as gills for gas exchange in aquatic vertebrates and as components of the middle ear in terrestrial vertebrates.

Living members of Chordata have successfully populated the land, the waters, and the air. They are commonly divided into three subphyla: Urochordata, Cephalochordata, and Vertebrata (figure 15.1). In this chapter, we will study representatives of the first two subphyla.

Cladogram of Chordata

- Subphylum **Cephalochordata** (lancelets)
- Subphylum **Urochordata** (tunicates)
- Subphylum **Vertebrata** (vertebrates)

Classification

The long-standing controversy over kinship among the three chordate subphyla was mostly settled in 2008 with publication of a complete genome sequence for the Florida lancelet, *Branchiostoma floridae*. By comparing

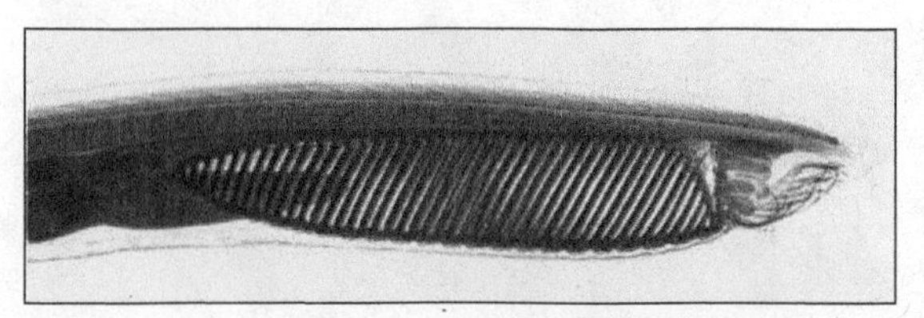

(a)

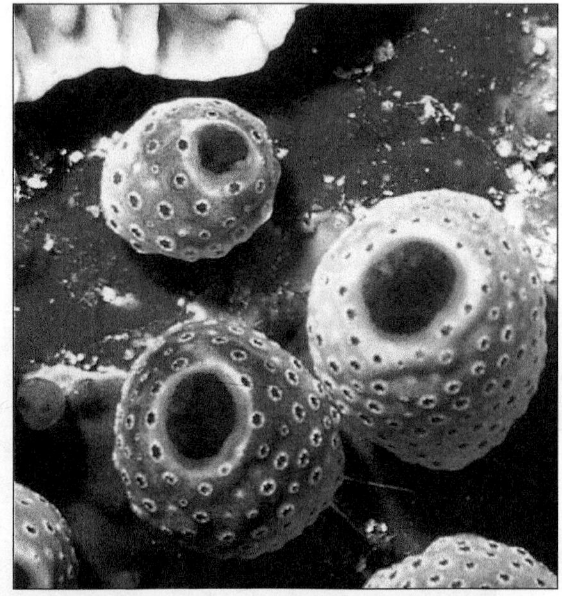

(b)

(c)

FIGURE 15.1 Representative chordates: (*a*) lancelets (Cephalochordata), (*b*) tunicates (Urochordata), (*c*) mammal (Vertebrata).

(a) © Diane Nelson. (b) Photo courtesy of Carolina Biological Supply Company, Burlington, NC (c) © 1996 PhotoDisc, Inc./Getty Images.

the number, structure, and location of lancelet genes with homologs in other Deuterostome genomes that have also been sequenced (including that of humans), biologists found strong evidence that the last common ancestor of lancelets and vertebrates lived about 550 million years ago, in the Cambrian Period of geological time. The DNA evidence further suggests that lancelets (Cephalochordata) are more distantly related to vertebrates than tunicates (Urochordata), despite the fact that tunicates tend to be less complex in physical structure. Unlike tunicates, which have lost many of their ancestral genes, vertebrates appear to have gained genes as a result of at least two "doubling" events over the past half-billion years. Each doubling of the genome created an extra copy of each gene. For the most part, these extra copies were useless (and eventually lost). But in special cases, particularly in genes that served regulatory or signaling functions, they became adapted to new roles and contributed to novel features of vertebrate biology. The story of our genetic heritage is just beginning to unfold. Only as we continue to sequence the genomes of other animals will we begin to fully appreciate the complexity and diversity of the vertebrate lineage.

Subphylum Urochordata (Tunicates or Sea Squirts)

Animals with a well-developed notochord and dorsal nerve cord in the free-swimming larva; specialized adults: sessile or planktonic, and lacking a notochord and dorsal nerve cord. About 1,600 species. Examples: *Molgula* (sea grape), *Styela, Amaroucium* (sea pork).

Subphylum Cephalochordata (Lancelets)

Elongate, fishlike chordates with a persistent notochord and dorsal nerve cord. About 20 species. Example: *Branchiostoma* (formerly called Amphioxus).

Subphylum Vertebrata (or Craniata)

Chordates with a distinct and characteristic head capsule (skull) of bone or cartilage that surrounds the brain. Vertebrata (*sensu strictu*) excludes the hagfish, which lacks a backbone.

Class Myxinia (Hagfish)

Eel-like marine scavengers with an oral sucker in place of upper and lower jaws. Lacking both a backbone and paired appendages. Disgustingly known for their copious production of slime (mucus) when handled. About 50 species. Example: *Myxine*.

Class Petromyzontida (Lampreys)

Similar in appearance to hagfish but with an abbreviated cartilaginous backbone. Many species breed in fresh water and migrate to the sea as adults. Formerly grouped with hagfish in the class Agnatha. About 40 species. Example: *Petromyzon*.

Class Chondrichthyes (Sharks, Skates, and Rays)

Cartilaginous fishes, usually with external placoid scales. About 550 species. Examples: *Squalus* (shark), *Raja* (skate), and *Chimaera* (ratfish).

Class Osteichthyes (Bony Fishes)

Fishes with skeleton primarily of bone, usually with swim bladder or lungs. About 21,000 species. Examples: *Amia* (bowfin), *Perca* (perch), *Lepistoseus* (gar), *Micropterus* (bass).

Class Amphibia (Salamanders, Frogs, and Toads)

Four-limbed vertebrates (tetrapods) with a soft, moist skin; three-chambered heart; eggs enclosed in gelatinous covering; fertilization and development usually restricted to fresh water. Adults usually aquatic or semiaquatic. About 3,100 species. Examples: *Rana* (frog), *Bufo* (toad), *Ambystoma* (salamander).

Class Reptilia (Lizards, Snakes, Turtles, and Alligators)

Four-limbed vertebrates with a dry, cornified skin; eggs enclosed in a protective shell resistant to drying (amniote egg); most with three-chambered heart. About 6,600 species. Examples: *Anolis* (lizard), *Chrysemys* (turtle), *Crotalus* (rattlesnake), *Alligator*.

Class Aves (Birds)

Winged vertebrates with feathers and constant high body temperature. About 9,600 species. Examples: *Sturnus* (starling), *Cyanocitta* (blue jay), *Corvus* (crow).

Class Mammalia (Mammals)

Warm-blooded vertebrates; body covered with hair; young nourished by milk produced by female. About 4,600 species. Examples: *Homo* (human), *Sus* (pig), *Equus* (horse), *Canis* (dog).

Materials List

Preserved specimens
- *Molgula*
- *Branchiostoma*
- Representative tunicates

Prepared microscope slides
- *Amaroucium*, tadpole larva
- *Branchiostoma*, whole mount
- *Branchiostoma*, cross section

Subphylum Urochordata, Tunicates or Sea Squirts

Urochordates are interesting, important, and unusual animals. They are interesting and important because they represent the simplest living chordates. They are unusual because adult urochordates became adapted to a special mode of life as sedentary, filter-feeding, marine animals, and, in the process, have lost two of the five basic chordate features. Larval urochordates, however, have retained all five chordate features and clearly establish the urochordates as legitimate ancestors of the higher chordates, including fishes, birds, mammals, and humans.

The Tunicate Larva

The larval stages of tunicates are of special importance because they clearly exhibit the five fundamental chordate characteristics: **dorsal hollow nerve cord, notochord, pharyngeal gill slits, endostyle,** and **postanal tail.** The first two of these features are lost in the adult tunicate, presumably because of the specialization of the adult due to its sessile mode of life.

Tunicate larvae are often called **"tadpole larvae"** because of their superficial resemblance to the tadpole larvae of amphibians.

- Study a prepared microscope slide with a whole mount of the larva of *Amaroucium,* or a similar tunicate larva, and observe its characteristic form (figure 15.2).

Note the **muscular tail** and thickened body. At the anterior end of the body, find the adhesive papillae with which the larva attaches to the substrate prior to metamorphosing into a sessile adult. Near the dorsal surface is a **sensory vesicle** with two conspicuously pigmented sense organs—a light-sensory **ocellus,** or eyespot, and a **statolith** that serves as a balancing organ.

Anterior to the darkly pigmented sensory vesicle, locate the **incurrent siphon** through which water is pumped into the pharynx. The **excurrent siphon** is found slightly posterior to the sensory vesicle, near the attachment of the tail. Locate also the **gill slits** in the wall of the pharynx.

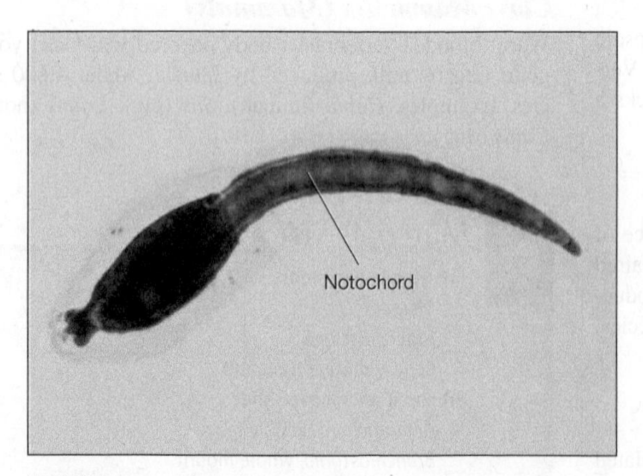

FIGURE 15.2 "Tadpole" larva of a tunicate.
Photo by John Meyer.

The **notochord,** the **dorsal nerve cord,** and conspicuous **muscle bands** are present in the tail. During metamorphosis, the larva settles to the bottom, attaches by secretions of glands in the adhesive papillae, and partly reabsorbs its tail. The notochord, nerve cord, and muscles are broken down, and several new organs appear in the body of the tunicate to complete the transformation into the adult. The life cycle of a tunicate is illustrated in figure 15.3.

The Adult Tunicate

Adult tunicates may be solitary or colonial, sessile or planktonic. The most familiar ones, however, are the sessile forms found attached to rocks, jetties, and pilings in shallow seas (figure 15.4).

The typical structure of an adult tunicate is well-illustrated by *Molgula,* the sea grape, found in many places along the Atlantic coast.

- Study the demonstration of a partially dissected specimen of *Molgula* and note the tough, fibrous outer covering called the **tunic** (see figure 15.3). This outer covering is unusual because it is composed partly of **cellulose,** a carbohydrate normally produced only by plants.

Projecting from the globular body are two siphons: the **incurrent siphon** is anterior, and its opening is divided into six lobes. The more posterior **excurrent siphon** has a square opening. Locate the numerous **gill slits** within the **pharynx.** Ventral to the pharynx are the short **esophagus,** the bulbous **stomach,** and the beginning of the **intestine.** The intestine continues dorsally and ends at the **anus** just below the excurrent siphon.

Molgula, like most other tunicates, is a specialized **filter-feeder.** Water is drawn in through the incurrent siphon by the action of cilia lining the internal chambers. The water is drawn into the large pharynx, and minute organisms suspended in the water (plankton) are trapped in the mucus on the pharyngeal walls as the water passes out through the many gill slits.

Special ciliary currents collect the food within the pharynx and pass it into the esophagus and on to the stomach. Water leaving the pharynx passes into the atrium, a large cavity surrounding the pharynx, and from the atrium the water flows out through the excurrent siphon.

The **endostyle,** a ciliated groove located on one side of the large pharynx, is of special importance. The endostyle secretes a stream of mucus that is carried along by the motion of the cilia to form a sheet that traps small food particles filtered from the incoming seawater. The mucous sheet with its captured food particles is then passed to the esophagus and into the stomach and through the digestive system. The endostyle has been found to be the forerunner of the vertebrate thyroid gland.

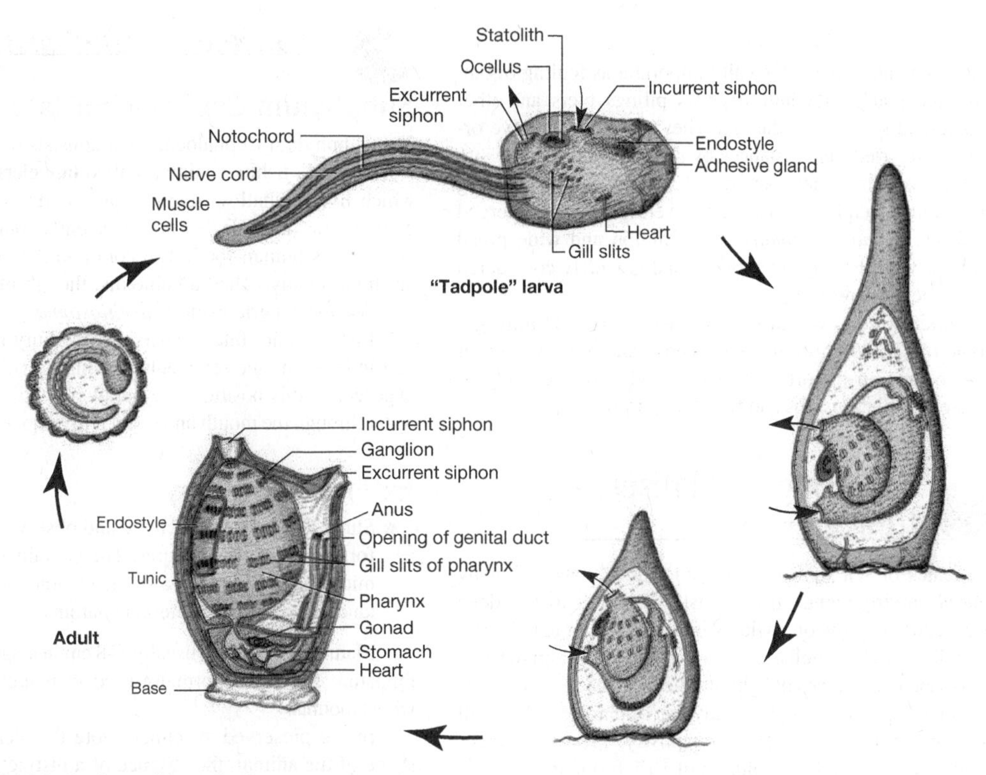

FIGURE 15.3 Tunicate, life cycle. Stages not to scale.

FIGURE 15.4 A colonial tunicate.
Courtesy of Carolina Biological Supply Company, Burlington, NC.

The circulatory system in *Molgula,* as in most tunicates, consists of a ventral **heart** located near the stomach. Attached to the heart are two large vessels that carry blood to other organs. The tunicate heart is unique in that it provides **two-way propulsion.** The heart first pumps blood in one direction, then reverses and pumps blood in the opposite direction.

In the adult tunicate, the dorsal hollow nerve cord and the notochord are absent. Their loss is generally believed to represent a specialization of the adult tunicate for its sessile mode of life.

Fact File

Sea Squirts

- Known for their habit of squirting out jets of water when disturbed at low tide and their physical appearance, various species of tunicates are also known as sea squirts, sea potatoes, sea lemons, sea peaches, sea grapes, sea pork, and dead men's fingers.
- Sea squirts are a popular seafood in the northeastern provinces of Japan, and in France, they are eaten raw with lemon juice or dried and eaten like potato chips.
- *Ecteinascidia turbinate,* a sea squirt found on coral reefs in the Caribbean Sea, is the natural source of Ecteinascidin, a powerful anticancer drug that has shown promising activity against human tumors in the breast, ovary, prostate gland, and connective tissues.
- The blood of some sea squirts contains high concentrations of vanadium or niobium, rare-earth elements that are scarce in the environment. These chemicals, once believed to be used in respiratory pigments, are now known to aid in defense by serving as antifouling and/or antipredator agents.

Importance of Tunicates

Some tunicates are ecologically important as fouling organisms that attach to the hulls of boats, pilings, piers, and other submerged surfaces in the sea. They are often invasive organisms carried long distances by ships and introduced into new areas where they can spread rapidly, increasing drag and adding weight to boats and covering other submerged surfaces. *Ciona intestinalis* is a common and widespread tunicate with these characteristics and is widely considered a nuisance organism.

Interestingly, within the past few years the entire genome of *Ciona intestinalis* has been sequenced and now it has become an important research model for the study of vertebrate relationships and for other purposes.

Collecting and Preserving Tunicates

Tunicates or sea squirts are common inhabitants of many marine environments, from coastal shallows to the deep ocean. Adults grow on a wide variety of submerged objects, including rocks, shells, corals, wharf pilings, mangrove roots, and even blades of kelp or other seaweeds. Sea squirts are filter-feeders; without a steady diet of plankton, they will survive for only a few days in captivity. Fresh specimens should be preserved in a solution of 10% formalin.

Demonstrations

1. Preserved specimens of solitary and colonial tunicates
2. Microscope slides of tunicate larvae

A Lancelet: *Branchiostoma*

Subphylum Cephalochordata

The Subphylum Cephalochordata consists of about 20 species of small, fishlike animals called **lancelets** (figure 15.5), which live in shallow seas in many parts of the world. In certain areas, the lancelets are sufficiently abundant that they are used as human food. The common American lancelets are traditionally called amphioxus, though most specialists now use the generic name *Branchiostoma*.

Lancelets are filter-feeders. Adults bury themselves tail first in loose marine sediments, leaving only the anterior end exposed. In this position, they draw a continuous stream of water through the mouth and strain it to remove food particles.

External Anatomy

- Study a preserved specimen and observe its general form and external features. Then obtain a prepared microscope slide with a stained whole mount of a lancelet to study its internal anatomy.

Mature lancelets are usually 5–8 cm in length, but smaller, immature forms are normally used in making microscopic whole mounts.

In the preserved specimen, note the slender, elongate shape of the animal, the absence of a distinct head, and the lack of paired fins or limbs.

- Handle the specimen with care and do not dissect it. Return it intact to the proper container when you have completed your study.

Refer to figure 15.5 and identify on your specimen the anterior **rostrum** and the **oral hood** bordered by a fringe of

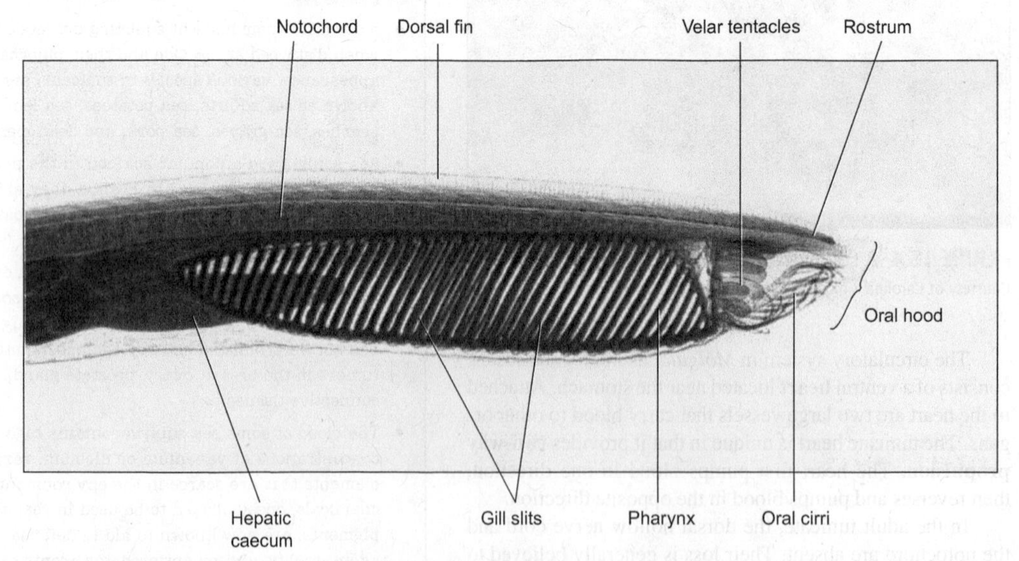

FIGURE 15.5 Lancelet, whole mount, carmine stain. Magnification 6×.
Courtesy of Carolina Biological Supply Company, Burlington, NC.

ciliated **oral cirri** enclosing the large **vestibule.** The **mouth** is an opening in a membrane, the **velum,** located at the rear of the vestibule. Surrounding the mouth are several **velar tentacles.** ***How many?*** Cilia arranged in bands along the walls of the vestibule form the "wheel organ," which generates water currents and carries seawater containing suspended food organisms into the mouth.

Internal Anatomy

Behind the mouth is a large **pharynx** with many diagonal **gill slits** on each side. Between the gill slits are **gill bars,** each supported by a thin, cartilaginous rod (figure 15.6). The pharynx plays an important role both in feeding and in gas exchange, but its primary function is in feeding. The gas exchange function is clearly secondary. Posterior to the pharynx is a straight **intestine,** which ends at the subterminal **anus.** A slender pocket, the **hepatic caecum** (believed by some workers to be homologous with the vertebrate liver), opens on the ventral side of the intestine near the junction of the pharynx and intestine and extends forward.

Locate the **dorsal, caudal** (tail), and **ventral fins.** Note the short **fin rays** composed of connective tissue within the fins. Observe the **atriopore,** a midventral opening located anterior to the ventral fin. Water taken into the pharynx passes out through the gill slits into the **atrium** and out of the atrium via the atriopore (figure 15.6).

Lancelets, like the tunicates studied earlier, are **filter-feeders.** Seawater containing planktonic organisms is drawn by ciliary currents through the mouth into the pharynx where food particles are trapped by mucous secretions. The mucus, containing trapped food particles, is swept posteriorly to the intestine, and the water passes out of the pharynx through the lateral gill slits. Gas exchange also occurs as the water passes through the gill slits and past the gill bars, which are highly vascularized with blood vessels.

Note the conspicuous V-shaped structures along each side of the body; these are the muscle segments, or **myomeres.** Contraction of these muscles produces a side-to-side lateral bending of the body, which aids the lancelet in swimming and burrowing in the bottom sediments where it commonly dwells.

Locate the dorsal **notochord,** which extends longitudinally just dorsal to the pharynx and intestine. The notochord is a thin, cartilaginous rod surrounded by a sheet of connective tissue. The contraction of the myomeres against the rigidity of the notochord produces the lateral swimming movements of the body that propel the lancelet forward. Find the **dorsal nerve cord** just ventral to the dorsal fin and dorsal to the notochord. At its anterior end is a slight enlargement, the **cerebral vesicle,** a very primitive sort of "brain."

The circulatory system of *Branchiostoma* consists of a network of elastic vessels similar to that found in the higher chordates, but it lacks a distinct heart. It is difficult to study the details of the circulatory system except by the dissection of specially prepared specimens, but portions of the circulatory system can be observed in the microscopic cross

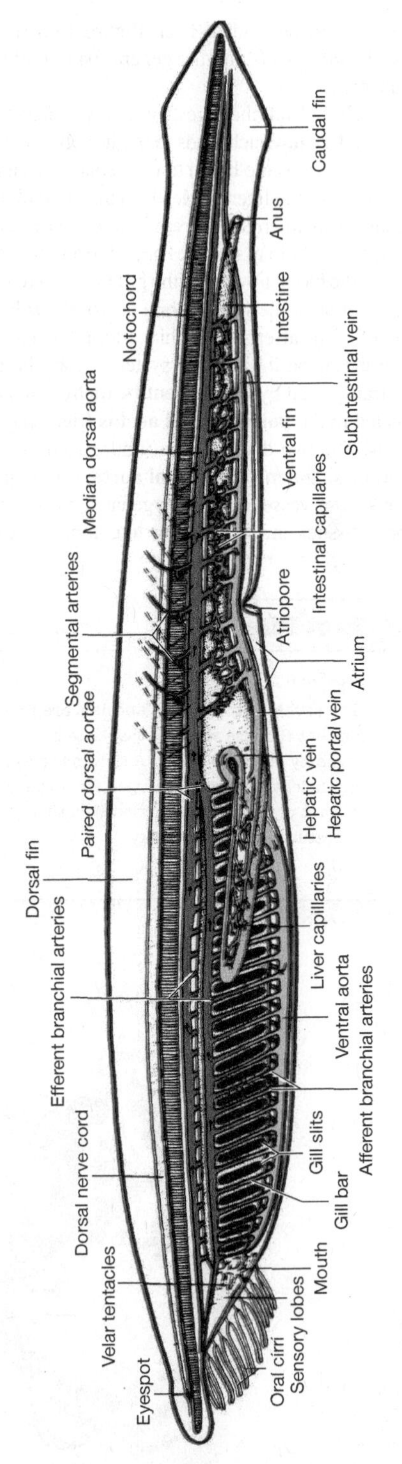

FIGURE 15.6 Lancelet, circulatory system. Small arrows in blood vessels show direction of blood flow.

sections to be studied later. Figure 15.6 shows the principal blood vessels and the general pattern of circulation in a lancelet.

Blood from the digestive tract is collected by the **subintestinal vein,** which leads to the **hepatic portal vein,** which, in turn, carries the blood to the hepatic caecum. The **hepatic vein** leaves the liver and leads to the **ventral aorta** below the pharynx. Numerous **afferent branchial arteries** (each with a contractile bulb at its base) branch from the ventral aorta and carry the blood upward to the gill bars where it is oxygenated. Pulsations of the ventral aorta and of the enlargements at the bases of the afferent branchial arteries appear to aid in pumping the blood through the system. From the gills, the blood is transported by the **efferent branchial arteries** and is collected in the **paired dorsal aortas** (right and left) above the gills. The two dorsal aortas (figure 15.6) join posteriorly to form a single **median dorsal aorta** just behind the pharynx. This latter vessel carries oxygenated blood posteriorly to the body tissues and the intestine to complete the circuit.

Fact File

Lancelets

- The earliest accounts of lancelets are found in ancient Chinese writings that date back to the T'ang Dynasty (618–907 A.D.). According to Chinese mythology, lancelets emerged from the dead body of a giant crocodile that carried Wen Chang (the God of literature) through the sky.

Continued—

—Continued

- Despite 500 million years of independent evolution, it is still possible to identify 17 gene linkage groups that are shared in both lancelets and humans. It is likely that these 17 groups represent parts of chromosomes from the most recent common ancestor of vertebrates and cephalochordates!
- In 1778, Piotr S. Pallas, a German zoologist, provided the first scientific description of the lancelet *Branchiostoma lanceolata* but he identified it as a sea slug. More than 50 years later, William Yarrell, an English biologist, studied the same species and concluded it was a primitive type of bony fish. It was not until the 1830s that Oronzio Gabriele Costa, an Italian naturalist, correctly reclassified the lancelet as a cephalochordate.

Cross Sections

Prepared microscopic cross sections will help greatly to supplement your observations of the preserved specimens and whole mounts, and will improve your understanding of the anatomy of *Branchiostoma.*

- Study several cross sections from different regions of the body and attempt to identify as many of the internal and external structures as possible.

The following description should aid you in identifying various structures in the cross section. It is based upon cross sections through different pharyngeal regions as illustrated in figure 15.7.

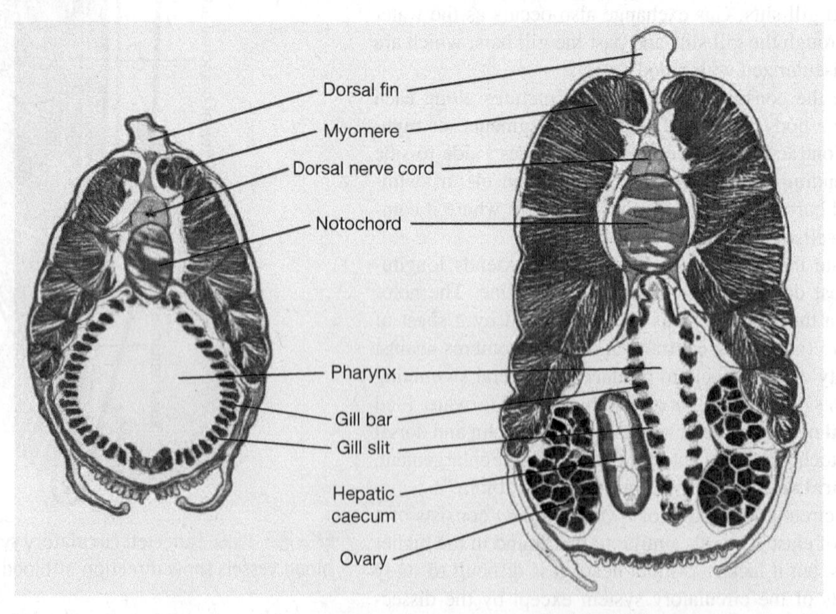

FIGURE 15.7 Lancelet, female, cross sections: (*a*) through anterior pharyngeal region, (*b*) through posterior pharyngeal region.
Photos courtesy of John Meyer and Carolina Biological Supply Company, Burlington, NC.

Observe the two-layered **skin,** with an outer **epidermis** consisting of a single layer of columnar epithelium, and an underlying **dermis,** a thin layer of gelatinous connective tissue. Find the **dorsal fin** and **fin ray,** the **myomeres** bounded by the myosepta of connective tissue, the **notochord,** and the **dorsal nerve cord** above it. Locate the **central canal** within the dorsal nerve cord. Observe also the nerve cells and fibers of the dorsal nerve cord and the **spinal nerves** (not present in every section because of variations in the plane of sectioning).

Observe the laterally compressed pharynx and the numerous **gill slits** between the **gill bars.** The ventral ciliated groove is the **endostyle,** or hypobranchial groove, and the dorsal ciliated groove is the **epibranchial groove.** Both play an important role in trapping and transporting food particles.

The chamber around the pharynx is the **atrium.** ***What tissue layer forms the lining of the atrium? What tissue layer lines the coelom?*** Note the **gonads** extending into the atrium and the section of the **hepatic caecum** also extending into it. Germ cells (eggs or sperm) released by the gonads pass into the atrium and out through the atriopore.

The paired cavities above and lateral to the pharynx are portions of the **coelom,** as is the small cavity found below the endostyle. Locate the **blood vessels** in your section, including the **paired dorsal aortae** above the pharynx and the **single median ventral aorta** below it. Also, find the **hepatic vein** or veins closely associated with the hepatic caecum. The **nephridia** are small ciliated ducts that connect the dorsal portions of the coelom with the atrium. The two **metapleural folds** at the two sides on the ventral surface of the body should be apparent.

Compare the structures observed in the cross section through the pharyngeal region with a cross section through a more posterior region, as illustrated in figure 15.7.

Collecting and Preserving Lancelets

Lancelets are found buried in sandy, off-shore sediments in most temperate and tropical regions of the world. Exceptionally high population densities are known to occur during late summer in Tampa Bay (Florida), Lagos Lagoon (Nigeria), Kingston Harbor (Jamaica), and Villa Cisneros (Western Sahara). They can be collected by gently straining sand through a wire screen. Like sea squirts and other filter-feeders, lancelets require a steady diet of phytoplankton and do not survive long in captivity.

Demonstrations

1. Preserved lancelets
2. Cross sections of lancelets through different regions

Key Terms

Endostyle a ciliated groove located on the ventral surface of the pharynx in tunicates and lancelets. Functions in the capture of food particles. Homologous with the thyroid gland of vertebrates.

Gill slits paired openings in the lateral walls of the pharynx in urochordates and cephalochordates. Important in food capture and secondarily in gas exchange.

Myomeres segmented muscle blocks arranged longitudinally along the dorsal portion of the lancelet body. Contraction of the myomeres, combined with the relatively stiff notochord, causes the lancelet body to flex, and produces effective lateral swimming and burrowing movements.

Notochord a stiff, supporting, cartilaginous rod of mesodermal origin found in the tunicate tadpole larva and in adult cephalochordates. Absent in adult tunicates.

Tadpole larva characteristic larval form of the tunicates with a notochord, dorsal tubular nerve cord, and paired gill slits. Superficially resembles an amphibian tadpole.

Internet Resources

There are many valuable Internet sites with information about zoology. Several sites containing pertinent zoological information for this chapter can be found on the McGraw-Hill Zoology web site at http://www.mhhe.com/zoology. Just click on this text's title.

Questions for Critical Thinking

1. Describe in detail the life cycle of the tunicate and explain why the larval form is believed to be of such morphological and evolutionary significance.
2. How does the adult tunicate vary in morphology from the larval form?
3. Compare the organization and morphology of a tunicate and a lancelet and explain their similarities and differences.
4. Discuss the advantages of filter feeding in a marine environment. Is this a common form of feeding? Identify the groups studied so far in which filter feeding has arisen.
5. Describe the main chordate features and explain the significance of these features.
6. Is the "tadpole" larva of tunicates really a tadpole? Why or why not? Do you think we should not use the word "tadpole" in this context? Support your reasoning.

Suggested Readings

Bone, Q. 1998. *The Biology of Pelagic Tunicates*. New York: Oxford University Press. 340 pp

Gee, H. 1996. *Before the Backbone: Views on the Origin of the Vertebrates*. London: Chapman & Hall. 346 pp

Putnam, N. H. et al. 2008. The amphioxns genome and the evolution of the chordate karyotype. *Nature* 453(19): 1064–1072.

Stokes, M.D and N. Holland. 1998. The lancelet. *American Scientist* 86:552–560.

Notes and Sketches

Chapter 16

Shark Anatomy

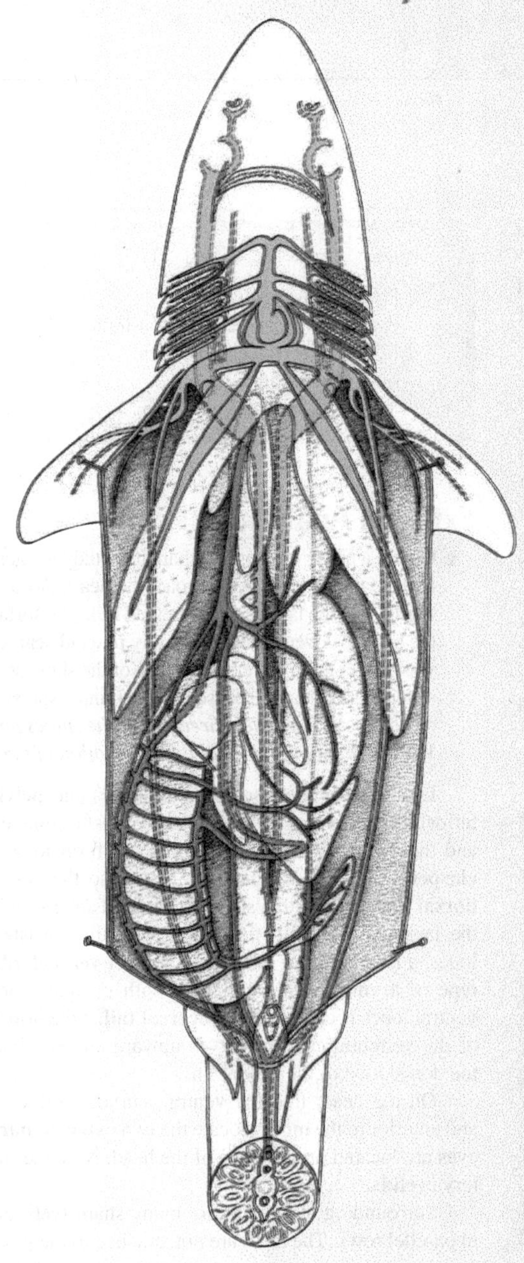

Objectives

After completing the laboratory work in this chapter, you should be able to perform the following tasks:

1. Identify the principal external features of the dogfish shark.
2. Locate the pelvic and pectoral girdles of the shark and explain their function.
3. Locate and identify the parts of the digestive system of the shark and explain the function of each part.
4. Locate the parts of the male and female reproductive systems of the shark and give the function of each part.
5. Locate the principal arteries and veins of the shark and explain the pattern of blood circulation.
6. Demonstrate the parts of the heart and explain the function of each part.
7. Describe the hepatic and renal portal circulatory systems and trace their paths in a specimen.
8. Describe the pattern of branchial circulation in a shark, explain its importance, and point out the chief blood vessels involved.
9. Identify the main parts of the shark brain.
10. Locate the eleven pairs of cranial nerves on a specimen and list their names.

The Dogfish Shark: *Squalus acanthias*

Sharks, skates, rays, and chimaeras are ancient fishes with a cartilaginous endoskeleton, biting jaws, paired appendages, and a tough, leathery skin that is usually covered with placoid scales. These fishes belong to the Subphylum Vertebrata, Class Chondrichthyes, Subclass Elasmobranchii and are commonly called elasmobranchs (because of their

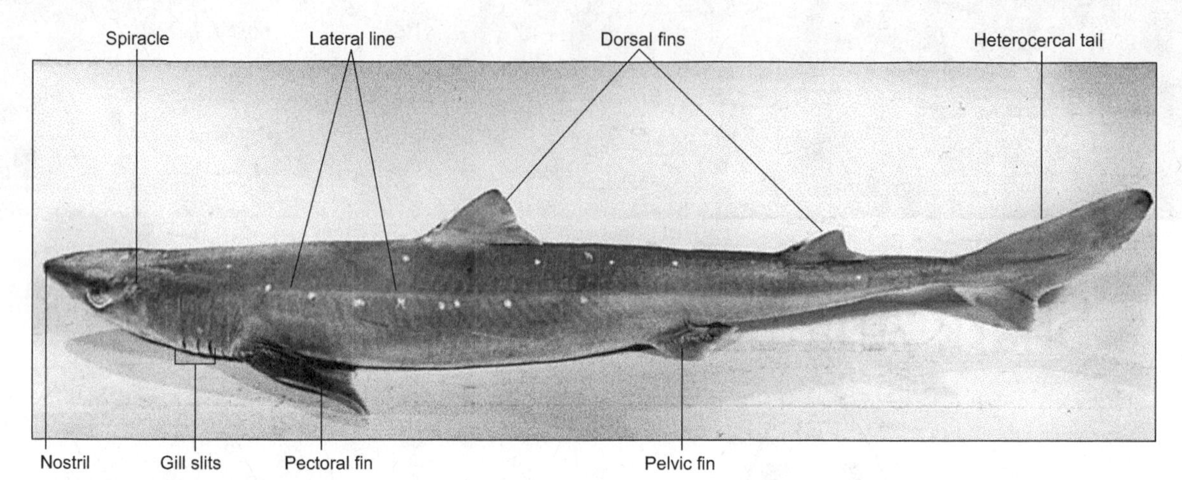

FIGURE 16.1 Shark, lateral view.
Courtesy of Carolina Biological Supply Company, Burlington, NC.

exposed gill openings) or cartilaginous fishes (because of the nature of their skeleton). Most of the members of this group are marine, although a few species have secondarily invaded fresh waters, as demonstrated by the freshwater sharks of Lake Nicaragua.

Paleontological studies of ancient shark's have revealed that this group evolved from ancestors with bony skeletons that inhabited fresh waters. Therefore, both the cartilaginous skeleton and the marine habitat of the elasmobranchs must be regarded as **specialized** (or secondarily derived) rather than as ancestral characteristics. Nonetheless, despite these specialized features, the elasmobranchs clearly illustrate many basic vertebrate features, and the shark has long been studied in zoology laboratories for this reason.

The spiny dogfish, *Squalus acanthias,* is a small shark (figure 16.1) common in shallow coastal waters on our Atlantic and Gulf coasts. A similar small shark from the Pacific coast, sometimes called *Squalus suckleyi* (which some authorities consider to be to the same species as the Atlantic form), is also commonly used for laboratory study. Mature specimens usually range from 1–3 meters in length, although smaller, immature specimens are normally used for laboratory study. Preserved specimens with the arterial and/or venous systems injected with latex are most satisfactory for dissection.

Materials List

Preserved specimens
- *Squalus acanthias,* mounted dissection
- *Squalus,* pregnant female with embryos in uterus (demonstration)
- *Squalus,* skeleton (demonstration)
- Bony fish skeleton (demonstration)
- Representative elasmobranchs (demonstration)

Continued–

–Continued

Prepared microscope slides
- Shark, placoid scales (demonstration)
- Shark, retina (demonstration)

Audiovisual material
- Anatomy of the Shark video

External Anatomy

- Obtain a preserved dogfish shark and study the general form of the body; note the broad, flat **head,** the tapered **trunk,** and the laterally compressed **tail.** The surface of the body is covered with many tiny **placoid scales.** Rub your finger lightly over the surface of the skin and feel the sandpaperlike texture caused by the minute spines borne on each scale. ***In which direction do the spines point? What is the significance of this orientation of the scales?***

Locate the paired **pectoral** (anterior) and **pelvic** (posterior) **fins.** In male sharks, the pelvic fins become enlarged and their inner borders are modified to form long, rodlike **claspers,** which aid in mating. Note also the two median **dorsal fins,** each with a sharp spine at its anterior edge, and the large **caudal** (tail) **fin.** The caudal fin consists of two lobes, a larger **dorsal lobe** and a smaller **ventral lobe.** This type of asymmetrical caudal fin with unequal dorsal and ventral lobes is called a **heterocercal tail.** Note how the end of the vertebral column curves upward and continues into the dorsal lobe of the caudal fin.

On the head, find the ventral, curved, slitlike **mouth,** and anterior to the mouth locate the two **external nares.** The **eyes** are located on the sides of the head. Note the rudimentary **eyelids.**

Surrounding the mouth are many sharp **teeth** arranged in parallel rows. The teeth are not attached to the jaw but are embedded in the tissue surrounding the mouth. Several rows of teeth grow behind the larger front rows and replace teeth that are broken or fall out. Sharks continue to grow new rows of teeth throughout their life. Scientists believe that the teeth were evolutionarily derived from the placoid scales embedded in the shark skin.

Shark teeth often become fossilized and are the most common fossil remains of sharks because their cartilaginous skeleton is less likely to become fossilized. Scientists estimate that it takes at least 10,000 years for a shark tooth to become a fossil. Sharks are a very old group of animals with earliest known species appearing some 450 million years ago.

Slightly behind and above the eyes, find the two **spiracles**—openings for water intake. The spiracles lead into the pharynx and represent a modified first pair of **gill slits.** Behind the mouth and near the ventral surface, find the five pairs of **external gill slits,** a fundamental characteristic of the chordates. Internally, the gill slits open into the **pharynx.**

Fact File

Sharks

- More than 400 species of shark live in today's oceans. They range in size from the spined pygmy shark, about 18 cm long (7 inches), to the whale shark, which can be up to 15 meters (50 feet).
- Shark cartilage (or compounds found in it) has been touted as a cure for cancer and arthritis, but its therapeutic value has not been proved. It is classified as a dietary food supplement, not a drug, by the Food and Drug Administration (FDA).
- Sharks may have up to 3,000 teeth at one time. Most sharks do not chew their food but gulp it down whole or in large chunks. The teeth are arranged in rows; when one tooth is damaged or lost, it is replaced by another. Most sharks have about five rows of teeth at any time. The front set is the largest and does most of the work.
- Cocos Island, off the coast of Costa Rica, boasts more sharks per cubic yard of water than any other place on earth. It is home to whitetip reef sharks, 40-foot whale sharks, and hammerheads that school by the hundreds.
- The great white shark, *Carcharodon carcharias,* is the world's largest marine predator. It can grow to 21 feet and weigh 4,000 pounds. Top speed in the water is about 25 mph. It is attracted by the odor of prey, and has the ability to detect one drop of blood in 25 gallons of water. It feeds primarily on bony fish, sea turtles, and marine mammals. Despite the reputation it earned from the movie *Jaws,* attacks on humans are relatively rare.
- Cookiecutter sharks, *Isistius* spp., use powerful suction and a twisting bite to suck "flesh-cookies" from the skin of large marine animals. Whales and dolphins are frequently marked by circular scars from previous encounters with cookiecutters.

Locate also the two light-colored **lateral lines** running posteriorly from the region of each spiracle to the tail on each side of the shark and the **cloacal aperature** located ventrally between the two pelvic fins. The lateral lines of the shark represent a special kind of sensory system found only in fishes and in some larval amphibians. It appears to function in the perception of water movements, pressure changes, and current changes, and thus aids in orientation and locomotion.

Feel the sandpaper-like texture caused by the minute spines borne on each scale. ***In which direction do the spines point? What might be the significance of this orientation of the scales?***

The tough skin of the shark is covered with these spines and research has shown that each of the scales causes a tiny vortex current in the seawater that reduces friction and aids the swift movements of the shark.

Internal Anatomy

- Take care in your dissection to avoid damaging structures important for subsequent study. Do not tear, cut, or pierce parts until you are sure that you know what you are doing. Never cut and discard parts until you are sure of their identity and know that you will have no further use for them.

  Be conservative in your cutting; often you will find that structures can be separated neatly and distinctly by teasing them free with a blunt instrument (a blunt probe, back of a scalpel blade, or even the handle of a scalpel) rather than by cutting. When cutting is necessary, make clean incisions with sharp instruments; dull scissors or a dull scalpel will tend to tear rather than to cut tissues and will lead to unsatisfactory results. Read the directions thoughtfully and follow them carefully *before* you begin your dissection.

Coelom and Visceral Organs

- Place your specimen on its back and carefully locate with your fingers the cartilaginous **pectoral** and **pelvic girdles,** which support the pectoral and pelvic fins, respectively (see figure 16.1). Then, using figure 16.2 as a guide, estimate the relative thickness of the body wall in your specimen. Carefully make an incision through the body wall along the midventral line from the pectoral girdle backward through the pelvic girdle, cutting to one side of the cloaca and ending your incision at a point just posterior to it.

If the interior of the body cavity appears to be oily, rinse it out carefully with **cold** water. Take care not to disturb the position of the internal organs during the washing.

- Put the waste water from your washing in the container designated by your laboratory instructor for toxic liquid waste since it contains some of the chemicals used to preserve the sharks.

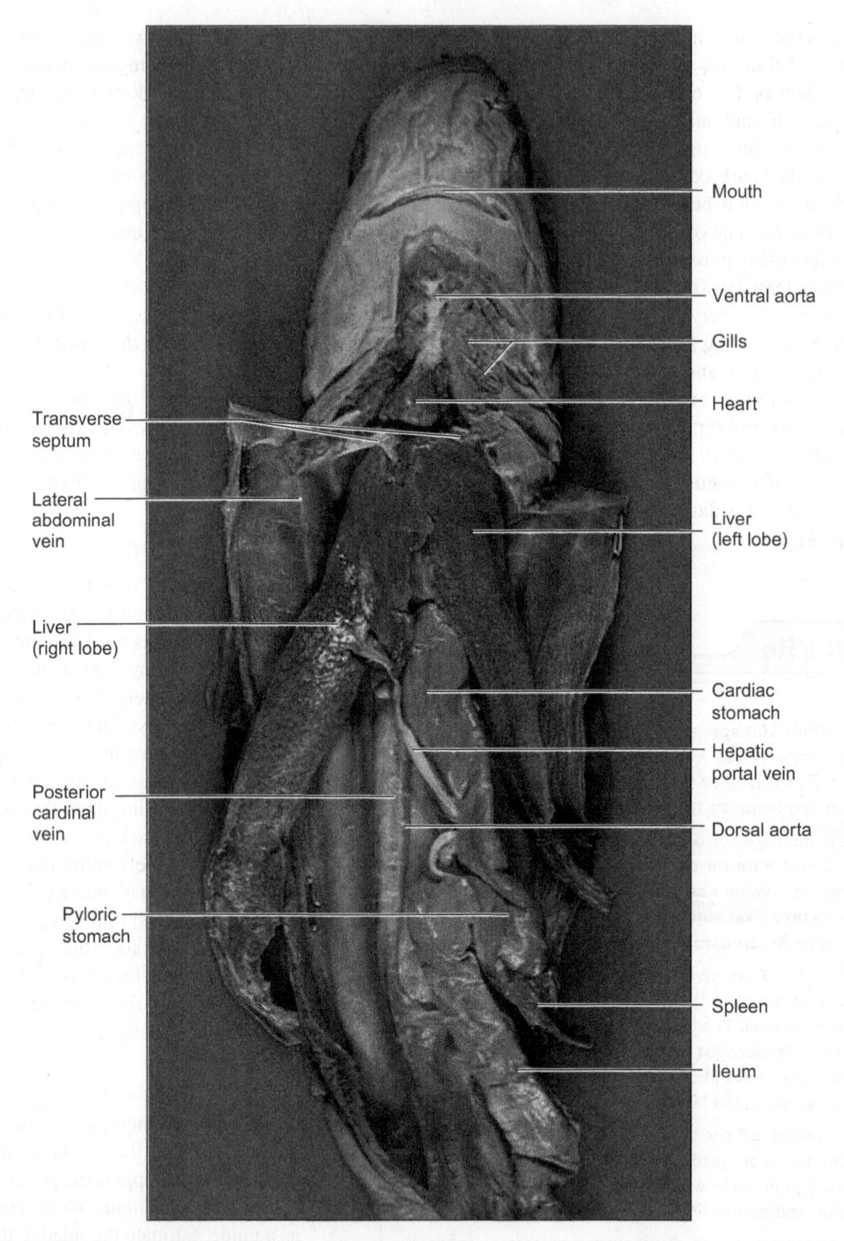

FIGURE 16.2 Shark, dissected, ventral view showing internal organs.
Photo by Ken Taylor.

Now make two transverse incisions through the body wall, one to the rear of the pectoral fins and one anterior to the pelvic fins, each about 2 inches in length. Pin or tie back the flaps of tissue to expose the large coelomic cavity containing the visceral organs. Carefully study figures 16.2, 16.3, and 16.4, and identify the major structures visible within the coelom. Note the location, relative size, shape, color, and texture of each structure. During your subsequent study, try to relate each structure with its principal function or functions.

The coelom of the shark is divided into two portions, the **pericardial cavity,** found anterior to the pectoral girdle, and the **pleuroperitoneal cavity,** found posterior to the pectoral girdle. These two portions of the coelom are separated by a thin partition, the **transverse septum.**

The smooth tissue lining the coelom is the **peritoneum,** which also covers the surface of the various organs suspended within the coelom. Dorsally, the peritoneum continues as a double epithelial membrane, the **dorsal mesentery** (figure 16.5), which supports the digestive tract within the coelom.

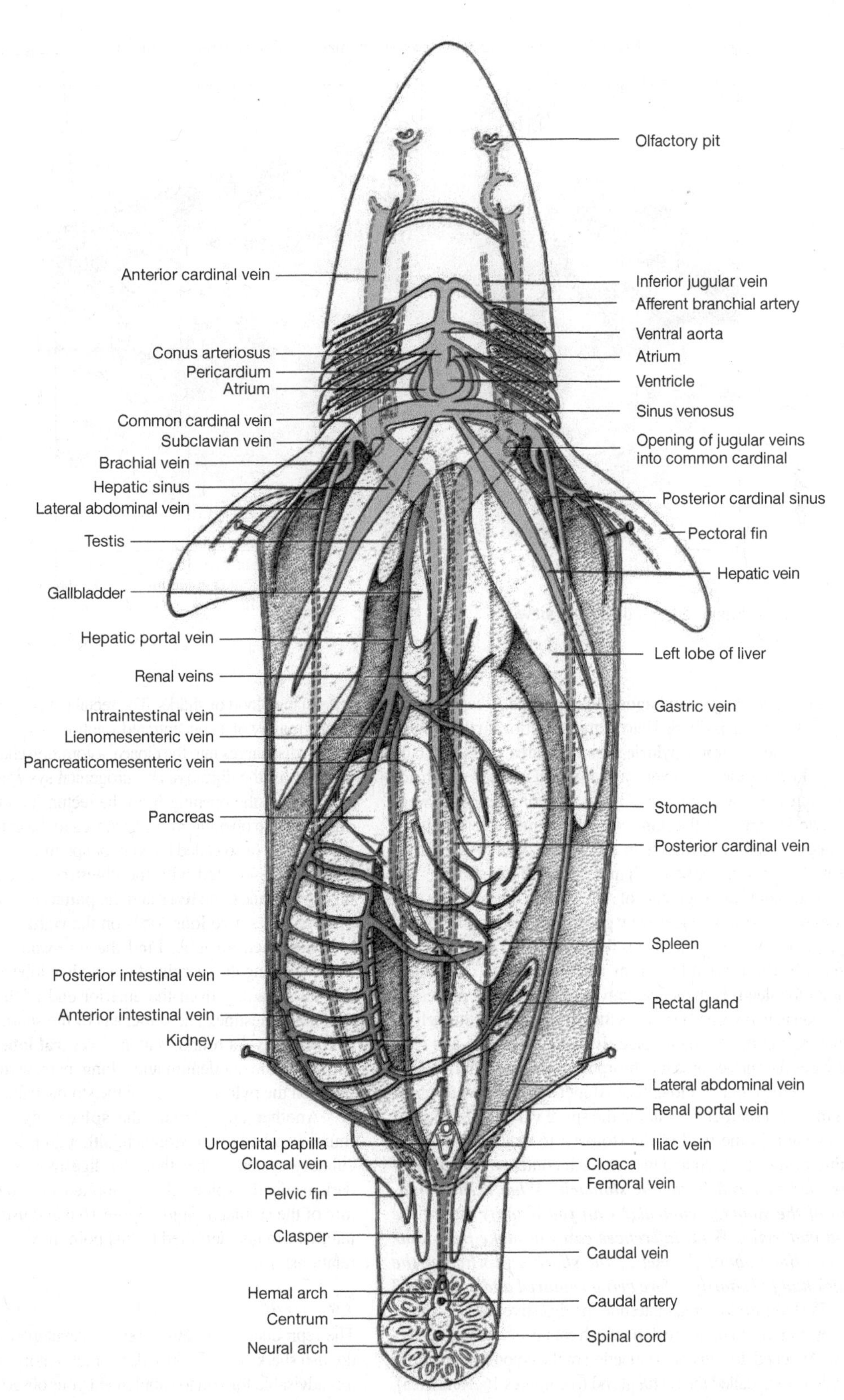

FIGURE 16.3 Shark, dissected, ventral view. Deoxygenated blood is blue and portal systems are orange.

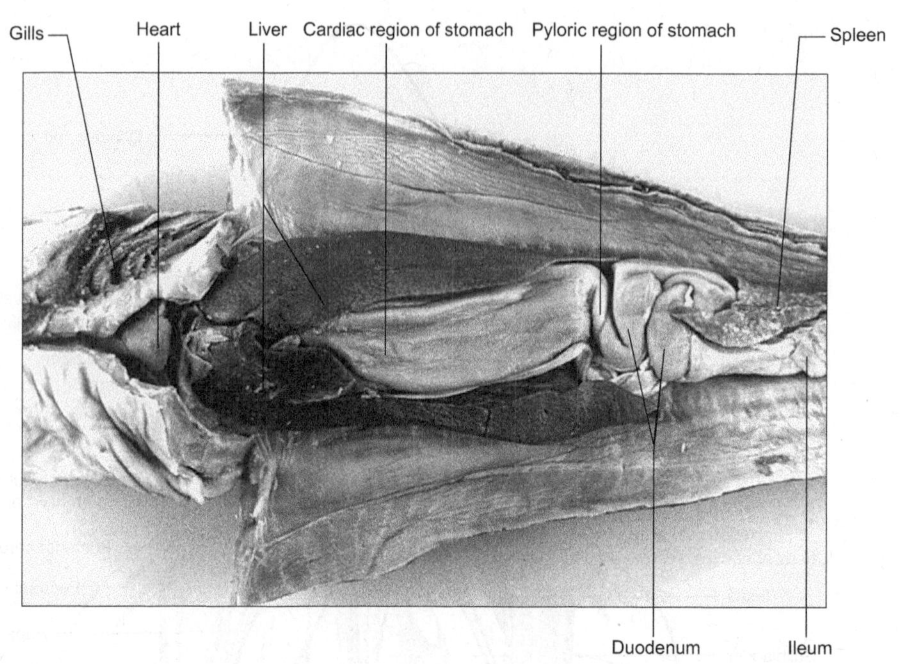

FIGURE 16.4 Shark, dissected, ventral view.
Photo by Carol Majors.

Locate the large **esophagus** anterior to the J-shaped **stomach.** The stomach is divided into a larger anterior **cardiac** region and a smaller posterior **pyloric** region (figures 16.4 and 16.6). The pyloric region lies beyond a sharp bend in the stomach and constitutes the lower portion of the J. The constriction between the pyloric region of the stomach and the small intestine is the **pylorus.** A sphincter muscle in this region controls the movement of food from the stomach into the **small intestine.**

The anterior segment of the small intestine is the **duodenum**—a short and narrow portion. The posterior, longer, segment of the small intestine is the **ileum.** Find the long **common bile duct** extending from the liver to the duodenum. Within the ileum is the **spiral valve,** which aids the process of digestion in two ways: (1) it slows the passage of food through the ileum, giving more time for digestive enzymes to act, and (2) it increases the surface area for absorption of released nutrients.

Carefully make a longitudinal incision in the wall of the ileum with a scalpel to examine the spiral valve. Make a similar incision in the wall of the stomach to examine the inner lining of the stomach and the stomach contents. ***Do you find any food material inside the stomach? What is the condition of the stomach contents? Can you identify any of the food materials? What inferences can you make from your observations about the diet of the shark and whether the shark had fed shortly before being captured and preserved?***

Posterior to the small intestine the digestive tract is continued by a short and narrow **colon,** which connects with a short **rectum.** Attached dorsally at the junction of the colon and rectum is a blind pocket called the **rectal gland** (see figures 16.3 and 16.5), which plays an important role in maintaining the proper salt balance in the blood of sharks. The rectal gland secretes a fluid consisting mainly of a concentrated solution of sodium chloride. The rectum discharges into the cloaca, a common chamber into which the ducts of the digestive and urogenital systems also terminate. Technically, the opening from the rectum into the cloaca is the **anus,** and the opening from the cloaca to the exterior is the **cloacal opening** (also called the vent or aperture).

Also associated with the digestive system are two large digestive glands, the **liver** and the **pancreas.** The liver consists of three lobes, two long lobes on the **right** and **left** sides, and a shorter **median** lobe. Find the thin-walled, greenish **gallbladder** along the margin of the median lobe and the common bile duct leading from the anterior end of the gallbladder to the small intestine. The **pancreas** of the shark consists of two distinct parts—a round, flattened **ventral lobe** attached to the surface of the duodenum and a long, narrow **dorsal lobe** lying between the pyloric portion of the stomach and the duodenum.

Another large organ, the **spleen** (figure 16.6), is also found attached to the stomach, although it is a part of the circulatory system rather than the digestive system. Locate the dark, triangular spleen closely applied around the outer curvature of the stomach. Study figure 16.6 and use it to review the internal organs identified to this point in your study and their relationships.

The Urogenital System

The reproductive system is poorly developed in the immature dogfish sharks usually provided for laboratory study. Therefore, it is advisable for you to supplement your observations of the reproductive system, particularly of the female shark, by viewing

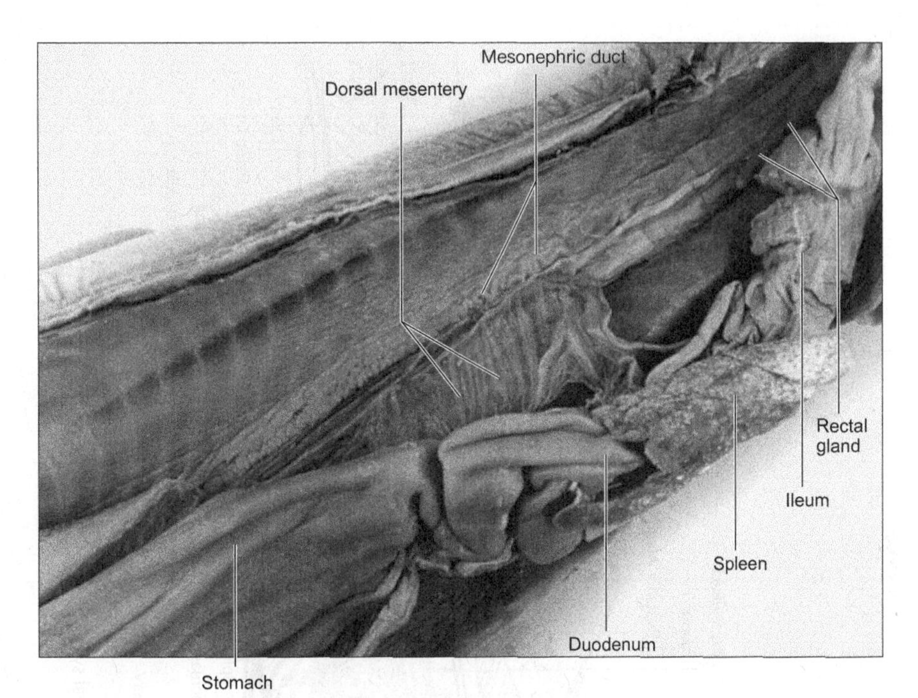

FIGURE 16.5 Shark, dissected, showing dorsal mesentery and abdominal organs of posterior region.
Photo by Carol Majors.

demonstrations of more mature specimens. Consult your instructor for information on the demonstrations available.

- Examine figure 16.7, which shows the urogenital systems of male and female sharks. **Do not remove** or **cut** any of the structures shown, but simply spread the abdominal organs apart to facilitate your observation. Note the long, flat **kidneys** closely applied to the body wall.

The kidneys of sharks, like those of all vertebrates, are actually **retroperitoneal;** that is, they are located **between** the parietal peritoneum and the body wall, and not suspended into the peritoneal cavity as are the other abdominal organs.

If you have a male specimen, observe the coiled **mesonephric ducts** as shown in figures 16.5 and 16.7 (also called the archinephric, pronephric, opisthonephric, or Wolffian ducts; sometimes incorrectly called ureters), which lie on the ventral surface of the kidneys and extend posteriorly to the cloaca. These ducts carry both urine from the kidneys and sperm from the testes. Locate the paired **testes**—a pair of elongated organs dorsal to the anterior end of the liver. Leading from each of the testes and emptying into the mesonephric duct are several small efferent ductules **(vasa efferentia).** The mesonephric ducts carry sperm to the paired **seminal vesicles** near the cloaca. Adjacent to the cloaca and emptying into it is the elongated **rectal gland.** The rectal gland secretes salt and aids in maintaining the osmotic balance of the shark's body fluids with the salinity of the surrounding seawater. Within the cloaca, the right and left mesonephric ducts join and empty through a common **urogenital pore.** The urogenital pore is located at the tip of the **urogenital papilla,** a small, fleshy projection from the dorsal wall of the cloaca.

If you have a female specimen, locate the **ovaries**—a pair of oblong, lobed bodies situated near the dorsal body wall above the anterior portion of the liver. From the ovaries, a pair of slender **oviducts** extend posteriorly along the length of the body cavity. The two oviducts originate as a common duct with a single opening, the **ostium tubae,** located anterior to the liver and ventral to the esophagus. (Search carefully in this area; the ostium tubae is often difficult to find.) From the ostium tubae, the **oviducts** loop anteriorly and laterally over the anterior end of the liver and pass back posteriorly along the dorsal body wall.

Ripe eggs discharged from the ovaries enter the body cavity, pass through the ostium tubae (see small arrow in figure 16.7), and enter the oviducts, where they may be fertilized. Most of the development of the embryos takes place in the **uterus,** an enlarged portion of each oviduct. The uteri open into the cloaca. Locate the enlarged **shell glands** (also called nidamental glands) attached to the oviducts. Note also the **mesonephric ducts** (you may require assistance from your instructor since these ducts are sometimes difficult to find in immature specimens) of the female shark lying along the ventral surface of the kidney. Trace one of the mesonephric ducts posteriorly to its entrance in the cloaca. Inside the cloaca, the mesonephric ducts empty through a **urinary pore** located on a **urinary papilla.** ***How do the male and female urogenital systems differ in this respect?*** Locate a student in your class who has a specimen of the opposite sex and compare the male

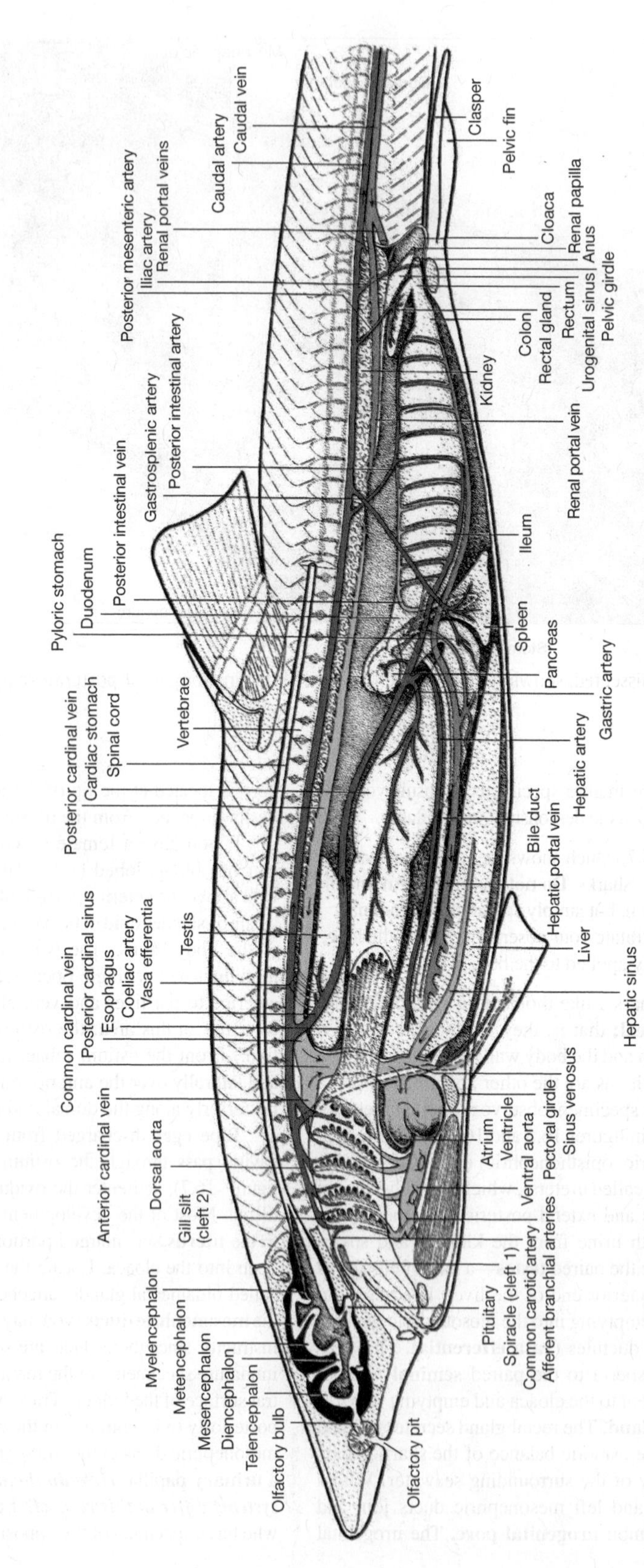

FIGURE 16.6 Shark, sagittal section. Oxygenated blood is red, deoxygenated blood is blue, and portal systems are orange.

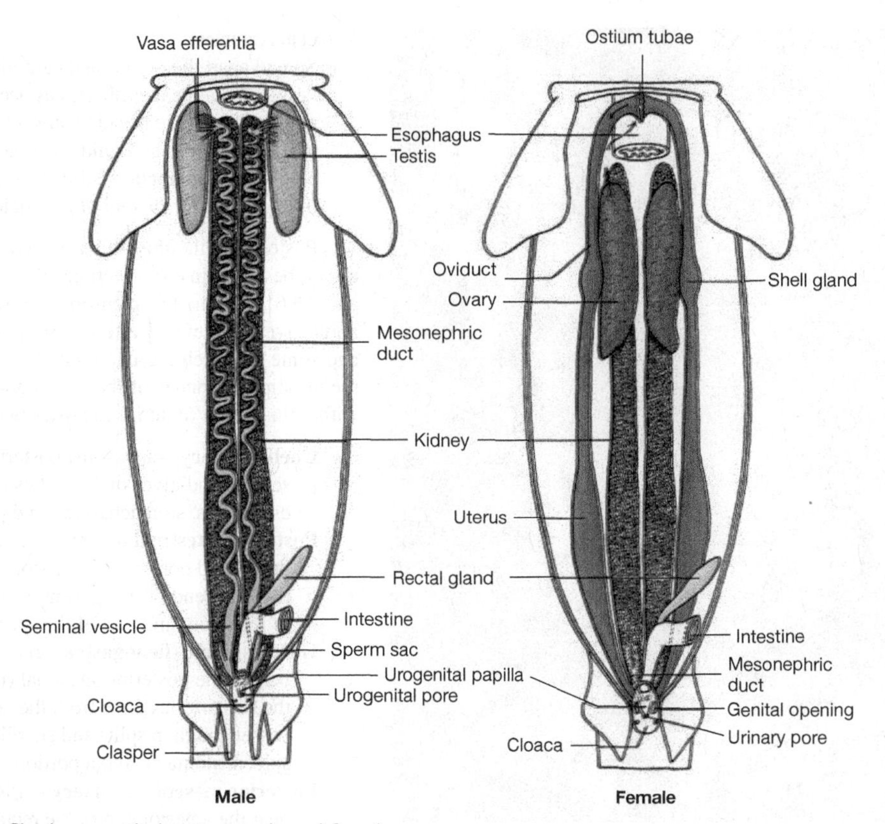

FIGURE 16.7 Shark, urogenital system, male and female.

and female reproductive systems or (alternatively) study the demonstration materials provided by your instructor.

Dogfish sharks (*Squalus acanthias*) give birth to living young, unlike some other sharks, which are egg-laying **(oviparous).** The dogfish shark is **ovoviviparous** since more-or-less typical eggs are produced but are retained within the reproductive system of the female until hatching. Many of the mature female sharks used for laboratory study are pregnant because of the unusually long gestation period of this species, which ranges from 20 to 24 months. *Note:* Many other species of sharks form a yolk sac placenta and give birth to living young without forming a shelled egg and thus are **viviparous.**

The Circulatory System

- Before you attempt to make a detailed study of the vascular system, you should study the general plan of the circulatory system and the direction of blood flow from the heart through the arteries to the capillaries of various organs and back to the heart by the veins (figures 16.6 and 16.9).

In certain cases, blood from the capillary beds in the tissues does not return directly to the heart but passes through an extra capillary bed (network) as well as an entire venous system en route back to the heart. The veins connecting the two beds of capillaries are called **portal veins.** Two **portal systems** exist in the shark, the **hepatic portal system** and the **renal portal system.** All portal systems begin and end in capillary beds. Study figures 16.3 and 16.9 for this preliminary survey and note the following principal structures: (1) the **heart,** which pumps blood to the gills where it is oxygenated and then flows to the dorsal aorta for distribution to various parts of the body; (2) the **hepatic portal system,** which returns blood chiefly from the digestive system to the liver and thence to the heart via the hepatic vein and sinus venosus; (3) the **renal portal system,** which returns blood from the posterior portion of the body to the kidneys, from which it goes to the heart via the postcardinal sinuses and the sinus venosus; and (4) the **anterior cardinal veins,** which collect blood from the head.

- From your study of the circulatory system, you should learn the following: (1) the names of the principal parts of the system, (2) the direction of blood flow in each part, (3) the organs served by the major blood vessels (both arteries and veins), and (4) the gains and losses from the blood as it flows through the capillaries of the various organs, particularly in regard to oxygen, carbon dioxide, nutrients, and nitrogenous wastes.

The Arteries The arteries of the shark can be conveniently studied in three groups: (1) the **visceral arteries,** consisting principally of the dorsal aorta and its branches

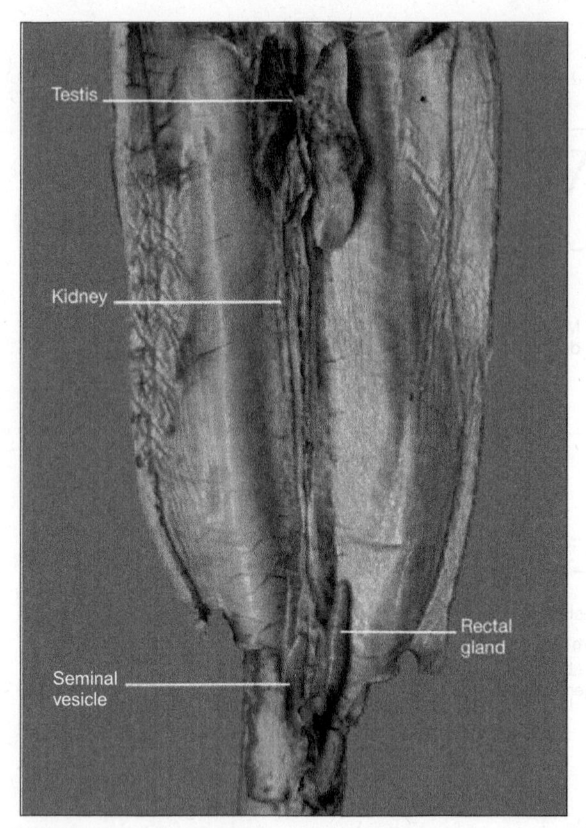

FIGURE 16.8 Shark, male urogenital system.
Photo by Toni Onks.

(figures 16.6 and 16.9); (2) the **afferent branchial arteries** and their branches, which carry blood to the gills; and (3) the **efferent branchial arteries,** which carry blood away from the gills and connect with the dorsal aorta.

Visceral Arteries:

- Spread apart the organs and carefully separate the blood vessels from the mesenteries as necessary to locate the arteries. Study the principal arteries in your specimen, using figure 16.6 as a guide. Locate first the large median **dorsal aorta,** visible through the peritoneal lining of the dorsal wall of the coelom.

Posteriorly, the dorsal aorta is continued as the caudal artery, best seen in cross sections of the tail (see figures 16.3 and 16.6). Within the pleuroperitoneal cavity, the dorsal aorta gives off several arteries, some of which are paired and some of which are unpaired. The following is a list of the principal branches of the dorsal aorta that can be found within this cavity (anterior to posterior):

Coeliac artery—arises just posterior to the transverse septum and gives off branches to the gonads, esophagus, stomach, liver, and pancreas.

Posterior intestinal artery—arises from the aorta dorsal and posterior to the stomach and near the posterior end of the mesentery and supplies one side of the ileum and part of the spiral valve.

Gastrosplenic (lienogastric) artery—arises just behind the posterior intestinal (in some specimens the two arteries arise from the aorta as a single vessel and then split) and supplies blood to the spleen, stomach, and a portion of the pancreas.

Posterior mesenteric artery—arises from the aorta near the anterior end of the rectal gland, which it supplies.

Iliac arteries (2)—a pair of large arteries arising just anterior to the cloaca. Supply the pelvic fin and the posterior portion of the body wall.

The Heart, Afferent Branchial Arteries, and Gills:

- Slice off the skin and muscles from the ventral surface of the head posterior to the mouth. Carefully cut away the muscles just anterior to the pectoral girdle until you

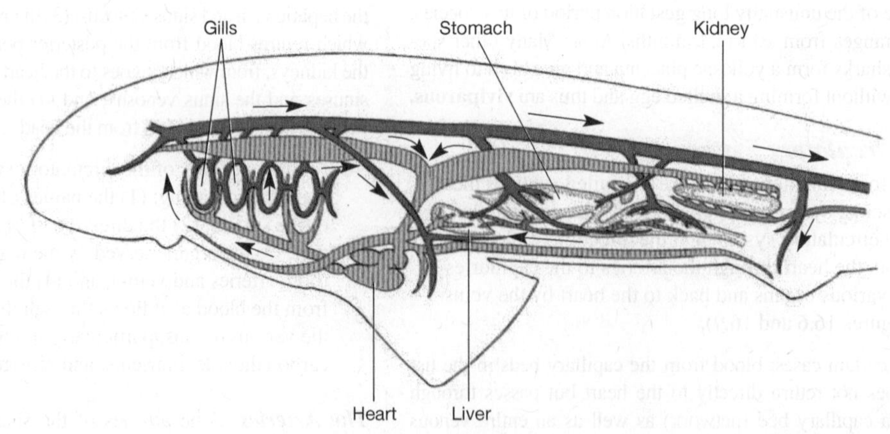

FIGURE 16.9 Shark, diagram of circulatory system. Oxygenated blood is red, deoxygenated blood is blue, and portal systems are orange.

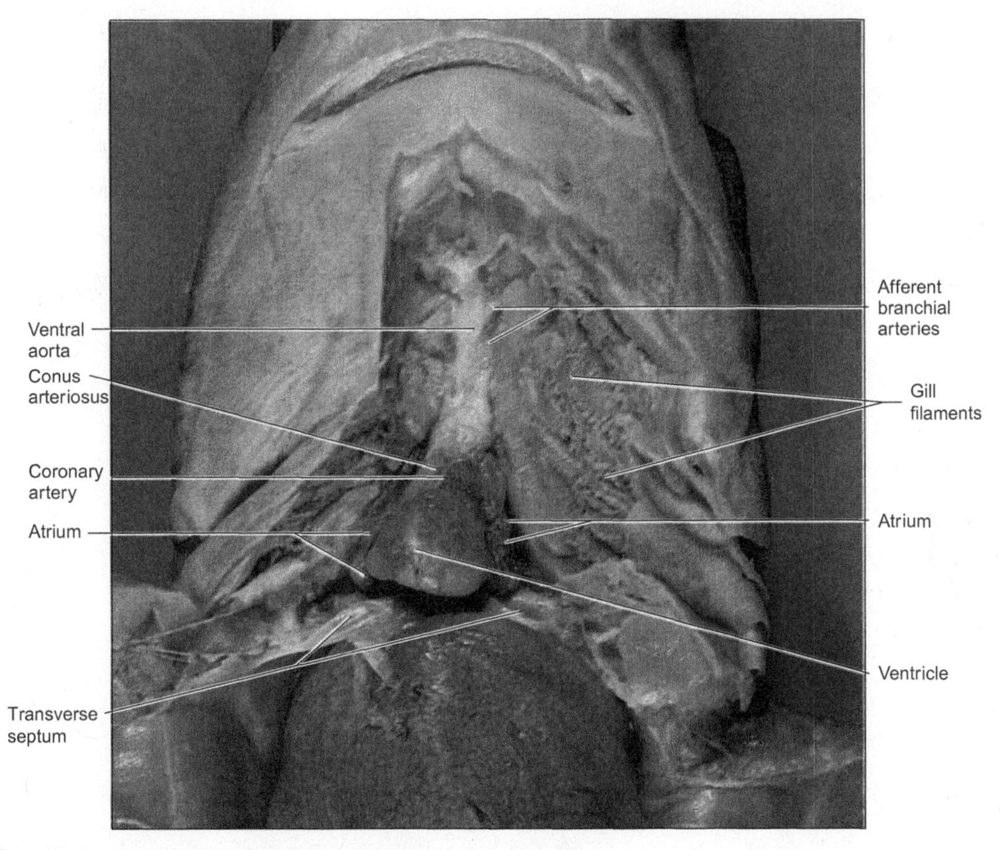

FIGURE 16.10 Shark, heart and afferent branchial arteries, ventral view.
Photo by Ken Taylor.

reach the membrane enclosing the pericardial cavity. Cut through the membrane to expose the heart within its cavity, and carefully cut through the pectoral girdle and remove a section of the girdle (about 3 mm wide) to further expose the heart. Consult figures 16.10, 16.11, and 16.12, and identify the parts of the heart and the surrounding blood vessels. Note that the shark has a **two-chambered heart** with a thick-walled, muscular **ventricle** and a thin-walled **atrium** (auricle). Observe that the large atrium is broad and extends beyond the ventricle on both sides. Identify also the **sinus venosus**, a flattened, thin-walled sac closely applied to the posterior surface of the ventricle and lying between the ventricle and the **transverse septum.** Lift up the posterior end of the heart to facilitate your viewing of the sinus venosus. Note that the transverse septum separates the **pericardial cavity** from the **pleuroperitoneal cavity,** thus dividing the **coelom** into two distinct parts. Locate the muscular **conus arteriosus** extending anteriorly from the ventricle.

The shark is commonly said to have a two-chambered heart; however, the conus arteriosus and the sinus venosus also have cardiac muscle. Technically, therefore, the shark can be said to have a four-chambered heart!

Blood enters the heart through the sinus venosus, passes into the atrium, then on to the muscular ventricle, and is pumped out through the conus arteriosus into the ventral **aorta.** The ventral aorta extends anteriorly from the heart and gives rise to five pairs of **afferent branchial arteries** (figure 16.12). Trace the ventral aorta forward, and on one side locate the afferent branchial arteries that carry blood to the gills for oxygenation. Cut away the lower portion of one of the gills and note the **cartilaginous bars** within the **gill arches:** these support the numerous **gill filaments.** Find also the **gill rakers,** cartilage-supported, fingerlike projections from the gill arches that guard the internal gill slits and prevent large food particles from entering the **gill chamber.** Count the number of gills and gill slits on your specimen. The **spiracle,** although nonfunctional as a gill in the shark, is customarily still designated as the first gill slit (figures 16.6 and 16.13). Note the many small branches of the afferent branchial arteries, which carry blood into the gill filaments where **gas exchange** occurs.

Efferent Branchial, Subclavian, and Carotid Arteries:

- Cut through the angle of the jaw on the left side of the head and continue the incision posteriorly through the pharynx, esophagus, and the ventral body wall to a

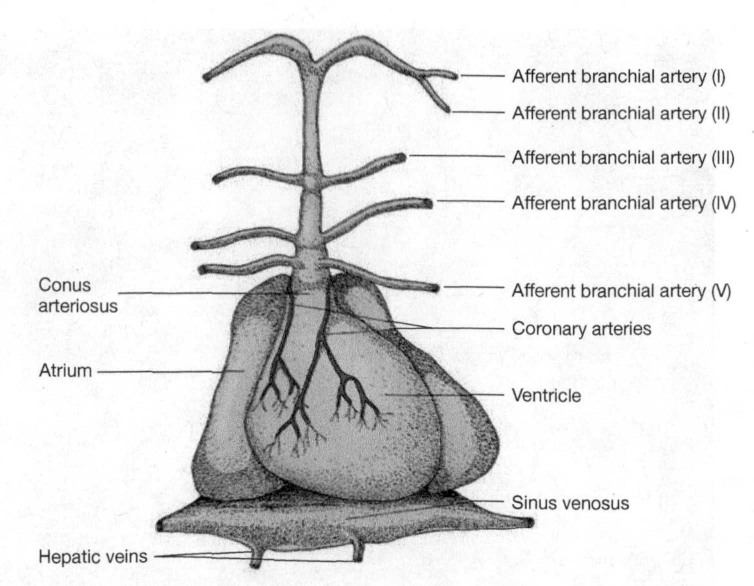

FIGURE 16.11 Shark heart, ventral view.

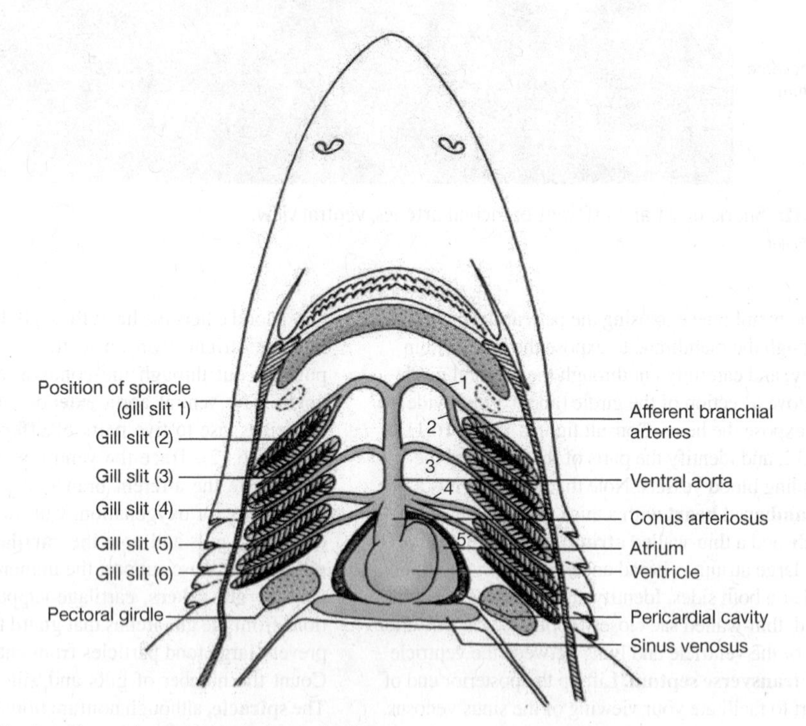

FIGURE 16.12 Shark, afferent branchial arteries and gills.

point just beyond the pectoral girdle. Make a similar cut from the angle of the jaw on the right side through the pectoral girdle. Fold back the lower jaw and the parts immediately behind it to expose the roof of the mouth and pharynx. With your forceps and scalpel, carefully remove the membrane from the roof of the mouth and pharynx to expose the **efferent branchial arteries.**

Study figure 16.13 and continue your dissection until you have exposed and identified all of the efferent branchial

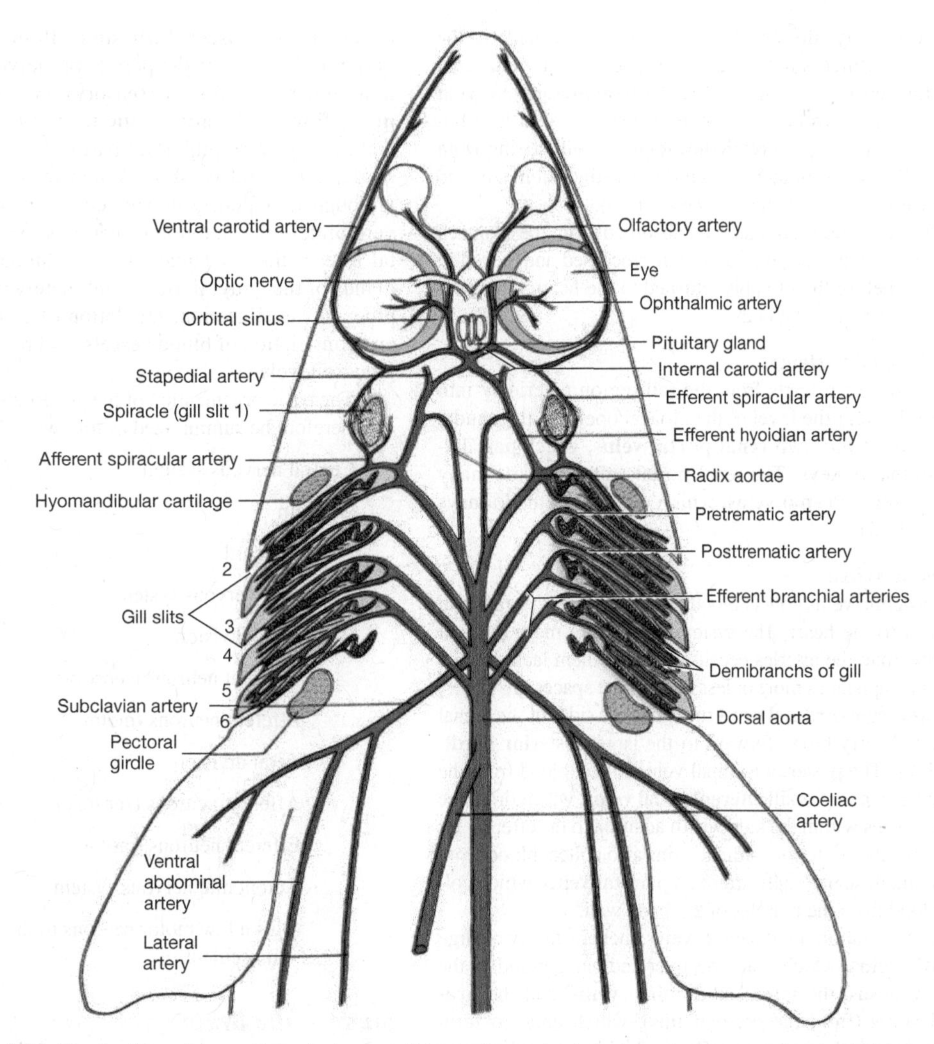

FIGURE 16.13 Shark, efferent branchial arteries, carotid arteries, and subclavian arteries are red and orbital sinus is blue.

arteries and other major blood vessels in this region. Note that each efferent branchial artery is formed by the union of two smaller arteries, a **pretrematic artery** and a **posttrematic artery,** which collect blood from the demibranchs ("half-gills") on the anterior and posterior sides of each gill slit. The pre- and post-trematic arteries are joined at both the dorsal and ventral ends of the gill slit, thus forming an **arterial collecting loop** (figure 16.13), which encircles the spiracle.

Find also the **dorsal aorta** and the paired **subclavian arteries** running toward the two lateral pectoral fins. Arising from each subclavian artery are two smaller arteries that carry blood posteriorly. These branches are the **ventral abdominal arteries** (larger and more lateral in location) and the **lateral arteries** (smaller and more medial in location).

The Veins The venous circulation (figure 16.3) of the shark includes three systems: (1) the **hepatic portal system,** which carries blood from the digestive tract to the liver; (2) the **renal portal system,** which carries blood from the tail region to the kidneys; and (3) the **systemic veins,** which collect blood from the other tissues and organs of the body. All three venous systems eventually empty into the two large common cardinal veins and, finally, into the sinus venosus.

Hepatic Portal System:

Study figures 16.3 and 16.6, and identify on your specimen the **hepatic portal vein** extending anteriorly from the stomach and intestine toward the liver. Note that the hepatic portal vein runs along the bile duct and enters the liver. Inside the liver, it branches several times and terminates in several capillary beds.

Posteriorly, the hepatic portal vein is formed by the junction of three smaller veins: (1) the **gastric vein** from the stomach (left tributary), (2) the **lienomesenteric vein** with tributaries from the intestine and spleen (central tributary), and (3) the **pancreaticomesenteric vein** arising from the confluence of branches from the intestine, stomach, and pancreas (right tributary) (see figure 16.3).

Some of the food materials absorbed from the stomach and intestine are removed and then processed and stored in the liver before the blood is returned to the heart and circulated to the rest of the body.

Renal Portal System:
The caudal vein leads from the tail region anteriorly into the trunk. Near the level of the cloacal opening, the caudal vein divides into two **renal portal veins,** which pass lateral to the kidneys. The renal portal veins give off many small **afferent renal veins,** which empty into the kidneys (figure 16.3).

Systemic Veins:
The systemic veins collect blood from the various organs and return it to the heart. The systemic veins are more difficult to study than the arteries because most of them lack definite walls and appear as more or less open tissue spaces or sinuses. The **posterior cardinal veins** run along the sides of the dorsal aorta and carry blood forward to the large **posterior cardinal sinus.** The posterior cardinal veins collect blood from the kidneys via many small **efferent renal veins,** which drain the renal sinuses within the kidneys. In addition to the efferent renal veins, the posterior cardinal veins also collect blood from many small, segmentally arranged **parietal veins,** which collect blood from the muscles of the body wall.

Other important systemic veins include the two large **hepatic veins,** which drain the liver and carry blood to the sinus venosus, the **lateral abdominal veins,** and the **brachial veins** (from the pectoral fins), which unite to form the short **subclavian veins.** On each side, the subclavian vein joins with the **posterior cardinal sinus** and the **anterior cardinal vein** to form the short **common cardinal vein** (duct of Cuvier), which empties into the **sinus venosus.** The openings of the anterior cardinal veins can be found by slitting open one of the common cardinal veins and using your probe to locate the aperture of the anterior cardinal vein.

Nervous System and Sense Organs

The nervous system of the shark, like that of other vertebrates, consists of two principal components: (1) the **central nervous system,** including the brain and the spinal cord, and (2) the **peripheral nervous system,** including the various nerves that connect the brain and spinal cord with various other parts of the body. The nerve fibers of the peripheral nervous system can be divided anatomically into those that innervate the outer portions of the body, such as the skin and voluntary muscles (the **somatic division**), and those that innervate the internal organs or viscera (the **visceral division**). Both the somatic and visceral divisions of the peripheral nervous system contain numerous **afferent** (sensory) and **efferent** (motor) **nerve fibers.** The **autonomic nervous system** is a specialized and very important part of the visceral division of the peripheral nervous system in higher vertebrates, although it is poorly developed in the shark. In higher vertebrates, the autonomic nervous system plays a vital role in the coordination of the smooth muscles and glands of the body; thus, it is intimately involved in such processes as digestion, regulation of heart rate, dilation and constriction of blood vessels, and regulation of blood glucose levels.

The basic organization of the vertebrate nervous system can therefore be summarized as follows:

- Central nervous system
  - Brain
  - Spinal cord
- Peripheral nervous system
  - Somatic division
    - Afferent neurons (sensory)
    - Efferent neurons (motor)
  - Visceral division
    - Afferent neurons (sensory)
    - Efferent neurons (motor)
    - Autonomic nervous system
    - (plus a few motor neurons to the branchial musculature)

Study of the Brain

Your study of the nervous system of the shark will be limited to a brief survey of the brain and cranial nerves since these parts will suffice to introduce you to the pattern of organization of the vertebrate nervous system. Some further aspects of the peripheral nervous system of vertebrate animals will be studied in some of the later exercises in this manual.

The brain and cranial nerves of the shark are illustrated in figures 16.14 and 16.15. Study these figures carefully before undertaking the dissection of the nervous system. The successful dissection and study of the nervous system requires patience, care, and attention to detail. Haste and carelessness most often lead to disappointing results and confusion.

- After you have carefully studied figures 16.14 and 16.15, take the shark head saved from your earlier dissection and carefully remove the skin from the dorsal side. The brain is encased in a cartilaginous braincase called the **chondrocranium** that protects it. After you have removed all of the skin from

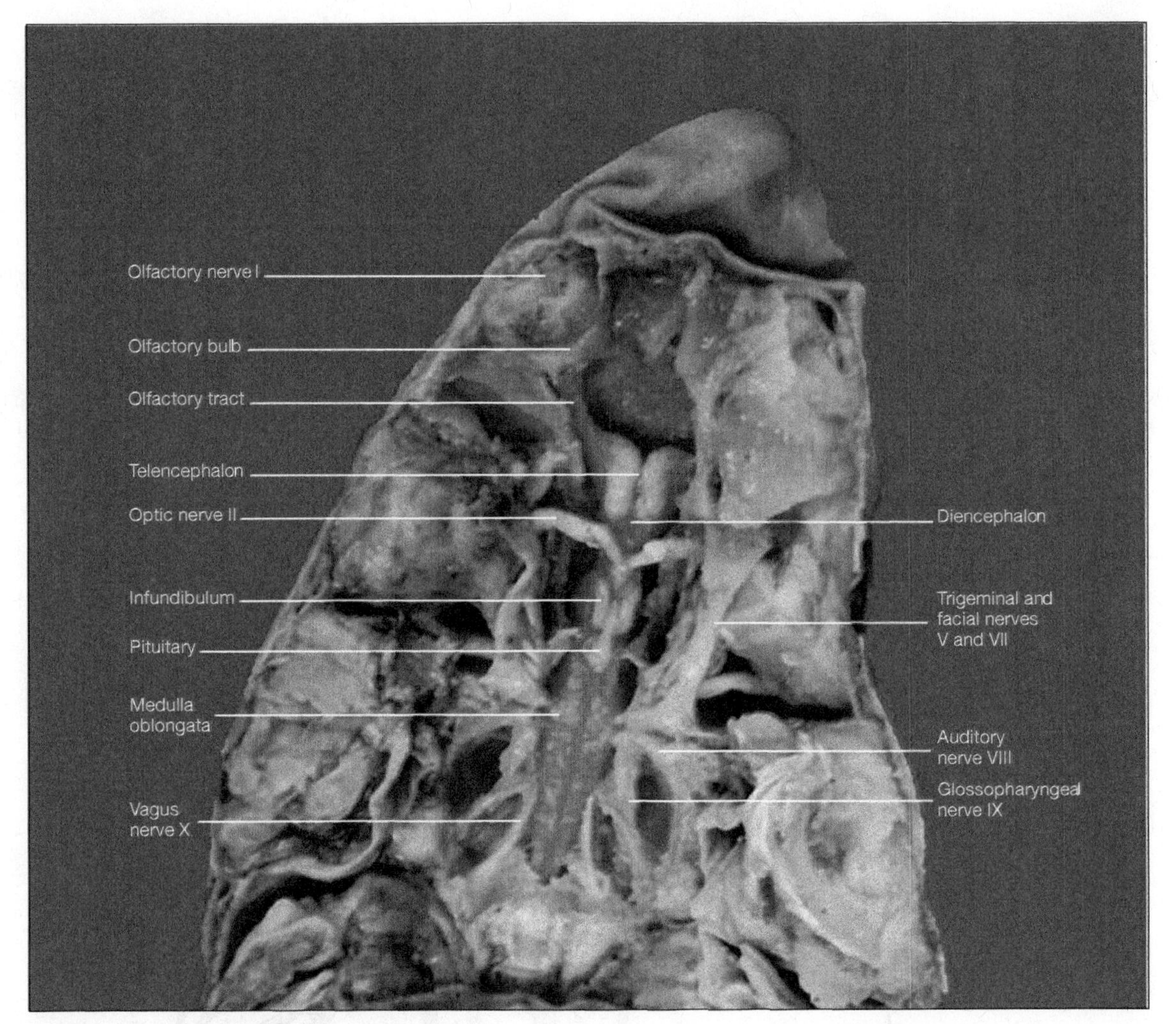

FIGURE 16.14 Shark brain, ventral view.
Courtesy of Betty Black.

the dorsal surface, slice away thin sections of the chondrocranium until the dorsal and lateral surfaces of the brain are exposed. Be careful not to cut or tear away the delicate nerves that pass through the several small openings (**foramina**) of the skull.

On the sides of the skull, near its posterior margin, are the two fused **otic capsules,** which enclose the **semicircular canals** of the inner ear. Carefully shave away the cartilage from the otic capsule and locate the semicircular canals, which are important organs of equilibrium.

During the early stages of embryonic development, the brain of vertebrates becomes divided first into **three distinct vesicles.** These three lobes are identified as the **forebrain** (prosencephalon), the **midbrain** (mesencephalon), and the **hindbrain** (rhombencephalon). This early embryonic differentiation of the three primary brain divisions (which together form the **brain stem**) reflects the evolutionary history of the vertebrate brain, since the brain of the earliest vertebrates also consisted of three divisions, each associated closely with one of three major sense organs: the nose, the eye, and the ear (including the lateral line).

The brain of adult sharks and other recent vertebrates, however, consists of five major divisions rather than three. Two of the three primary brain divisions (the forebrain and hindbrain) divide again during later stages of embryonic development. The brain of **adult sharks** therefore consists of **five major divisions:** (1) the **telencephalon,** (2) the **diencephalon,** (3) the **mesencephalon,** (4) the **metencephalon,** and (5) the **myelencephalon.** Locate each of

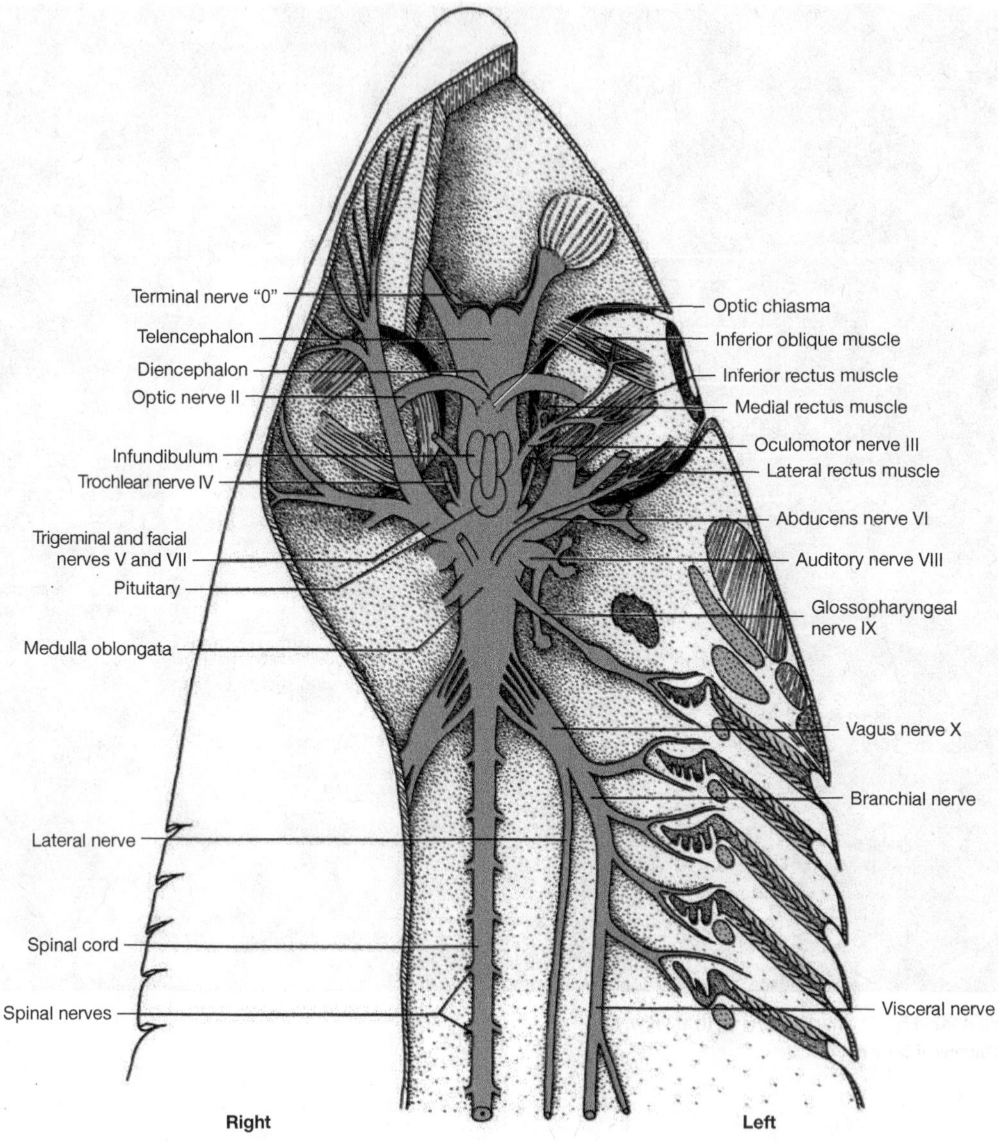

FIGURE 16.15 Shark head showing brain, cranial nerves, and eye muscles, ventral view.

these five divisions of the shark brain on your specimen. These divisions and their relationships are summarized in table 16.1.

Survey of the Brain

1. The anteriormost **telencephalon** is represented by the **olfactory bulbs,** the **olfactory tracts,** and the **olfactory lobes.** Just behind the two olfactory lobes are the less prominent **cerebral hemispheres.** They can be identified as slight swellings behind the olfactory lobes and are separated from the olfactory lobes by a shallow groove.
2. The **diencephalon** lies behind the telencephalon and appears on the dorsal surface of the brain as a narrow depressed area just anterior to the optic lobes. The diencephalon bears the **epiphysis** or pineal gland—a slender stalk extending anteriorly up through an opening in the roof of the skull. The pineal or "third eye" has long been an enigmatic organ of vertebrates; its function in the shark is not well understood. Ventrally the diencephalon bears the **infundibulum** and the **hypophysis,** which together comprise the **pituitary gland** of the shark. These structures will be studied later.
3. Behind the diencephalon are the two large lateral swellings, the **optic lobes**—important brain centers that

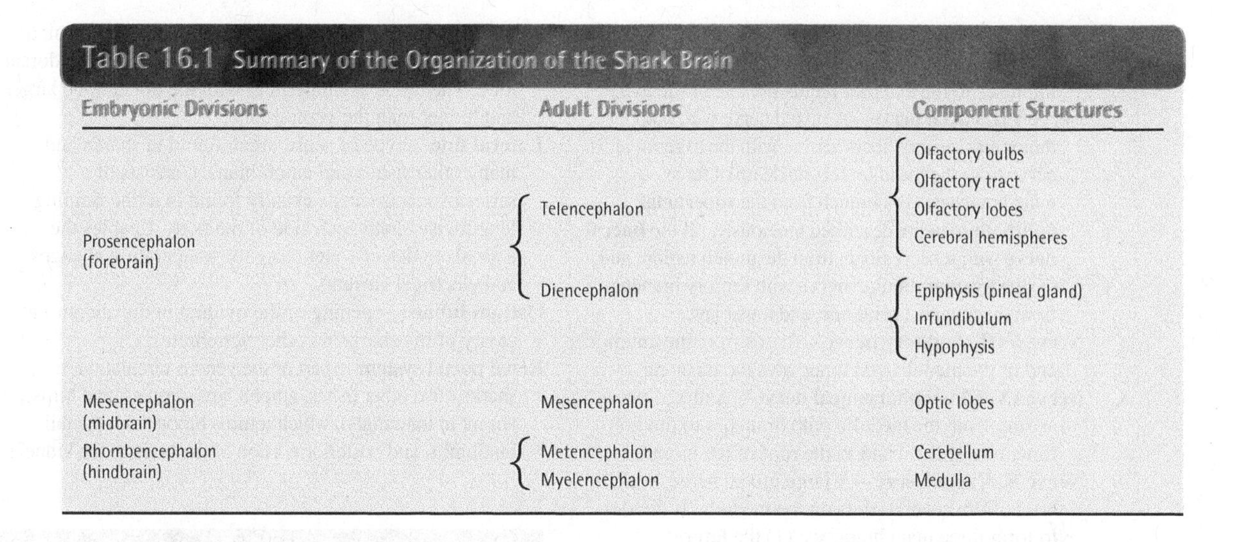

Table 16.1 Summary of the Organization of the Shark Brain

Embryonic Divisions	Adult Divisions	Component Structures
Prosencephalon (forebrain)	Telencephalon	Olfactory bulbs Olfactory tract Olfactory lobes Cerebral hemispheres
	Diencephalon	Epiphysis (pineal gland) Infundibulum Hypophysis
Mesencephalon (midbrain)	Mesencephalon	Optic lobes
Rhombencephalon (hindbrain)	Metencephalon	Cerebellum
	Myelencephalon	Medulla

play an essential role in vision; here the optic nerves terminate. The optic lobes represent the **mesencephalon.**

4. The **metencephalon** consists mainly of the **cerebellum,** a single large median lobe lying behind the paired optic lobes.
5. The **myelencephalon** consists principally of two parts—the triangular **medulla oblongata,** the most posterior portion of the brain, which tapers into the connecting spinal cord, and the two lateral **auricular lobes.** The latter extend forward from the medulla and can be seen adjacent to the posterior portion of the cerebellum. The auricular lobes serve as important centers of equilibrium.

Cranial Nerves

Sharks have eleven pairs of **cranial nerves.** Each of these nerves connects with a specific portion of the brain and connects that part of the brain with a specific organ or organs of the body. Some of the cranial nerves consist wholly of **sensory neurons** (carry impulses to the brain), and other cranial nerves are entirely made up of **motor neurons** (carry impulses from the brain), and about half of them contain mixtures of sensory and motor elements.

Cranial nerves were first studied seriously in humans and were given names and numbers (I to XII) based on their location and function in humans. (It is customary to use Roman numerals to designate the cranial nerves.) Subsequent study of other vertebrates, however, has revealed that the human pattern of cranial nerves does not hold for all vertebrates. In sharks, for example, only 10 of the cranial nerves correspond to those of humans (nerves I through X). The eleventh cranial nerve in the shark is a small anterior nerve, the terminal nerve, not present in humans. It is designated as nerve 0.

Students often use mnemonic devices to help them remember the names of cranial nerves and other anatomical parts. In the case of the cranial nerves of the shark, one such saying is, "**T**here, **o**n **o**ld **O**lympus' **t**op, **t**he **a**rmy **f**avorites **a**re **g**reat **v**ictories." See if you can make up another such saying in which the words start with the same letters as the 11 cranial nerves of the shark.

Locate each of the cranial nerves on your specimen with the aid of figures 16.14 and 16.15 and the list provided below.

Nerve 0. Terminal nerve—A delicate sensory nerve attached to the olfactory tract and arising from the median surface of the olfactory lobe and extending into the nasal region. (It is numbered "0" because it was discovered after the numbering system for cranial nerves became established.)

Nerve I. Olfactory nerve—Carries sensory impulses from the olfactory epithelium within the olfactory sac to the olfactory bulb (difficult to find).

Nerve II. Optic nerve—Arises in the retina of the eye and runs to the ventral surface of the diencephalon where it crosses the brain and carries impulses to the optic lobe on the opposite side of the brain. The prominent crossing of the optic nerves on the ventral surface of the diencephalon is called the **optic chiasma** (to be studied later).

Nerve III. Oculomotor nerve—Originates in the mesencephalon and divides into four branches, which carry motor impulses to the muscles of the eye.

Nerve IV. Trochlear nerve—Originates from the dorsal surface of the mesencephalon and passes through the chondrocranium to carry impulses to the superior oblique muscle of the eye.

Nerve V. Trigeminal nerve—A mixed nerve (with both motor and sensory fibers), which arises from the anterior part of the medulla and enters the orbit. It is the largest of the cranial nerves in the shark and divides into four main branches: (1) the superficial ophthalmic nerve to the skin of the head, (2) the deep ophthalmic nerve to the skin of the snout, (3) the infraorbital nerve to the region of the mouth and ventral surface of the snout, and (4) the mandibular nerve to the jaw muscles and skin of the lower jaw.

Nerve VI. Abducens nerve—A motor nerve arising from the ventral surface of the medulla and carrying impulses to the external rectus muscle of the eye.

Nerve VII. Facial nerve—A mixed nerve with both motor and sensory fibers arising with the trigeminal nerve from the medulla. It is made up of three main branches: (1) a branch from the **superficial ophthalmic nerve** described previously, (2) the **buccal nerve** with sensory fibers from the mouth region, and (3) the **hyomandibular nerve** with sensory branches from the tongue, lateral line, and lower jaw.

Nerve VIII. Auditory nerve—Arises from the anterior end of the medulla and innervates the inner ear.

Nerve IX. Glossopharyngeal nerve—A mixed nerve arising from the medulla with branches to the first functional gill slit and to the roof of the mouth.

Nerve X. Vagus nerve—A large mixed nerve, which arises from several roots on the medulla. It divides to form three main branches: (1) the **lateral line trunk** to the lateral line, (2) the **branchial trunk** to the gills (except the first), and (3) the **visceral trunk** to the heart and abdominal organs.

Ventral Surface of the Brain:

- After you have completed your study of the dorsal parts of the brain and the cranial nerves, study figure 16.14 and carefully dissect away the anterior portion of the roof of the mouth and the cartilage underlying the brain. Study the ventral surface of the brain and locate the **optic chiasma** where the large optic nerves cross, the two lobes of the **infundibulum,** the **hypophysis,** and the origin of the **sixth cranial nerve** (abducens) described previously.

Demonstrations

1. Dogfish skeleton, preserved or plastic mount
2. Microscope slide of placoid scale
3. Pregnant female shark showing embryos in uterus
4. Preserved mount of dissected triply injected shark
5. Preserved mount of shark brain
6. Preserved representatives of Superclass Elasmobranchiomorphii: skates, rays, chimaeras

Key Terms

Chondrocranium The cartilaginous braincase found in the sharks and other elasmobranchs.

Hepatic portal system portion of the venous circulation in sharks and higher vertebrates that collects blood from the stomach and intestine, and returns it to the liver where many food materials absorbed from the gut are removed and processed for storage.

Heterocercal tail type of tail found in sharks in which the caudal vertebrae are deflected upward into the dorsal lobe. The tail is asymmetrical with the dorsal lobe larger and longer than the ventral lobe.

Lateral line type of sense organ found in sharks and many other fishes and amphibians. Consists of a series of sensory cells usually found in a line running lengthwise along each side of the body. Enables the animals to detect water currents, temperature changes, and electrical currents.

Ostium tubae opening of the oviduct in the abdominal cavity of the shark and other vertebrates.

Renal portal system part of the venous circulation in sharks (also other fishes, amphibians, reptiles, and birds; absent in mammals), which returns blood from the tail, hindlimbs, and posterior portion of the body to the kidneys.

Internet Resources

There are many valuable Internet sites with information about zoology. Several sites containing pertinent zoological information for this chapter can be found on the McGraw-Hill Zoology web site at **http://www.mhhe.com/zoology**. Just click on this text's title.

Questions for Critical Thinking

1. Although the elasmobranchiomorphs are an ancient group of fishes, systematists now regard them as having specialized rather than "primitive" characteristics. Why might they be considered primitive?
2. Discuss the significance of an entirely closed circulatory system coupled with a heart that beats regularly (rather than irregularly as those found in most invertebrate groups).
3. What is the significance of dividing the coelom into a pericardial cavity and pleuroperitoneal cavity? Is this a sort of internal vertebrate "tagmatization" because it compartmentalizes distinct functions such as the gas exchange and digestive systems?
4. What is the physiological significance of the two portal systems (the hepatic portal system and the renal portal system)? What advantages would several capillary beds offer that a single capillary bed cannot?
5. Describe and discuss the increased complexity shown by the evolution of the shark brain and nervous system over those of invertebrates.
6. How does the lateral line function? Do you suppose it would be easier to approach a shark by swimming downward toward its dorsal surface or by approaching it from the sides? Explain your reasoning.

Suggested Readings

Ashley, L.M., and R.B. Chaisson. 1988. *Laboratory Anatomy of the Shark,* 5th ed. Dubuque, IA: McGraw-Hill Higher Education. 80 pp

Carrier, J.C., J.A. Musick, and M.R. Heithaus. 2004. *Biology of Sharks and Their Relatives.* Boca Raton, FL: CRC Press. 608 pp

Compagno, L.V., M. Dando, and S. Fowler. 2005. *Sharks of the World.* Princeton, Princeton University Press. 480 pp

Ferrari, A., A. Ferrari, and A. Bennett. 2002. *Sharks.* Richmond Hill, Ontario, Canada: Firefly Books. 258 pp

Gans, Carl, and T.S. Parsons. 1988. *A Photographic Atlas of Shark Anatomy: The Gross Morphology of Squalus acanthias.* Chicago: University of Chicago Press. 106 pp

Skomal, G., and N. Caloyiahis. 2008. *The Shark Handbook: The Essential Guide for Understanding the Sharks of the World.* Kennebunkport, ME: Cider Hill Press. 280 pp

Notes and Sketches

Notes and Sketches

Perch Anatomy

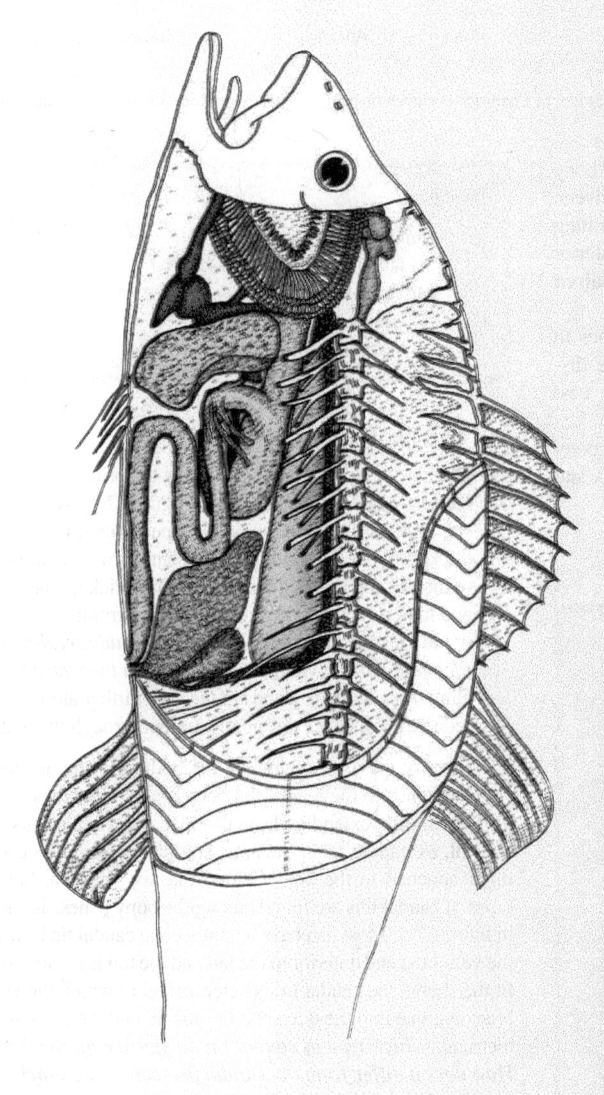

Objectives

After completing the laboratory work in this chapter, you should be able to perform the following tasks:

1. Locate and name the main external features of a perch.
2. Locate the axial skeleton, appendicular skeleton, and visceral skeleton on a prepared perch skeleton and explain the function of each. Describe and point out the main parts of a trunk vertebra.
3. Describe and locate the main divisions of the perch musculature.
4. Locate the gills, show the main parts of a gill, and explain the function of each part.
5. Identify the parts of the digestive system.
6. Describe the heart of a perch and locate its main parts.
7. Describe the basic pattern of circulation in a perch.
8. List the five major divisions of the perch brain and demonstrate their location on a specimen.
9. Discuss the principal morphological similarities and differences between the perch and the shark.

The Perch *Perca* sp.

The yellow perch, *Perca flavescens,* is a common bony fish found in lakes and streams throughout most of the United States; it is native to the central United States and southern Canada, and is widely stocked elsewhere. A similar species, *Perca fluviatilis,* the European perch, is common in Europe. Although less commonly studied in zoology and comparative anatomy laboratories than the shark, the perch provides an excellent representative of the more common modern bony fishes.

The yellow perch is a member of the Class Osteichthyes, the bony fishes, the largest class of living vertebrates. Although this and other directions for laboratory dissection are based on the yellow perch, similar species, like the European

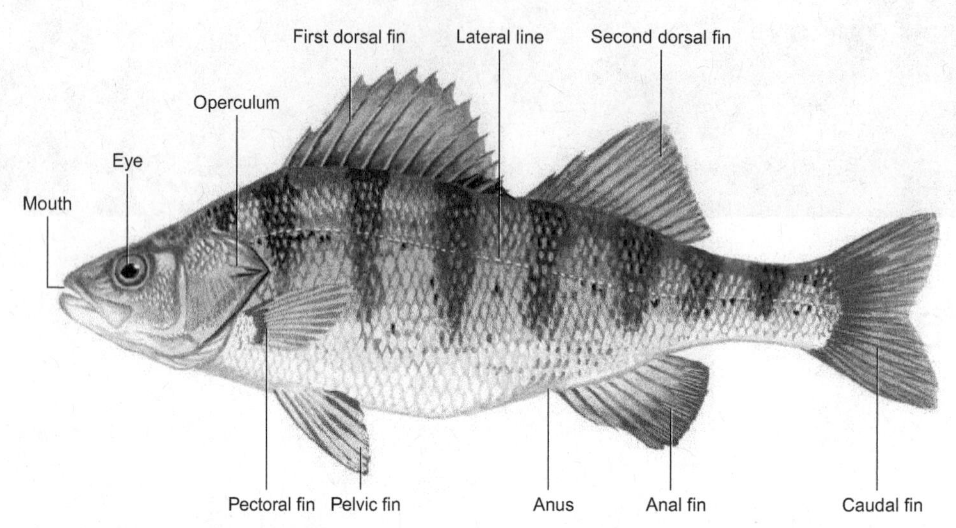

FIGURE 17.1 Perch, external anatomy, lateral view.

Painting from Fisherman's Guide: Fishes of the Southeastern United States, by Charles S. Manooch III and Duane Raver, Jr. North Carolina State Museum of Natural History, Raleigh, NC, 1984. Used by permission.

perch, may be used as available. Over 30,000 species of bony fishes have been described from the lakes, streams, rivers, and oceans of the world. Among the principal distinguishing features of the Osteichthyes are a **bony skeleton,** an anterior **terminal mouth, dermal scales,** a **homocercal tail, paired nostrils,** and ears with **three semicircular canals.**

Mature specimens range from 6 to about 12 inches in length. Preserved or freshly killed specimens may be dissected, but preserved, latex-injected specimens are best suited for study of the circulatory system.

You should also study a goldfish or other small bony fish in an aquarium to observe swimming movements and other aspects of fish behavior.

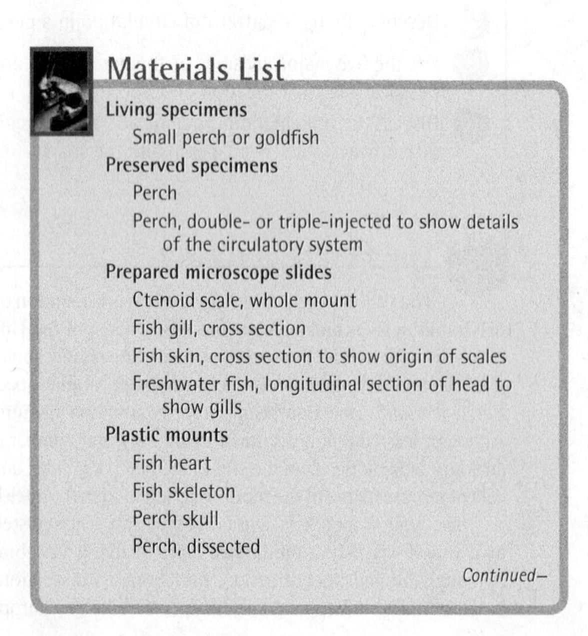

Materials List

Living specimens
- Small perch or goldfish

Preserved specimens
- Perch
- Perch, double- or triple-injected to show details of the circulatory system

Prepared microscope slides
- Ctenoid scale, whole mount
- Fish gill, cross section
- Fish skin, cross section to show origin of scales
- Freshwater fish, longitudinal section of head to show gills

Plastic mounts
- Fish heart
- Fish skeleton
- Perch skull
- Perch, dissected

Continued–

–Continued

Audiovisual material
- Anatomy of the Perch CD

Miscellaneous
- Aquarium
- Model of dissected perch

External Anatomy

- Obtain a preserved perch and study the principal features of its external anatomy (figure 17.1). Note the streamlined **fusiform** (spindle-shaped) **body,** that is thickest about one-third of the distance from the mouth to the tail and tapers in both directions. Pick up the fish and look directly at the mouth from the front; observe the ovoid cross section of the fish. ***How would this shape facilitate movement of the fish through the water?*** Numerous mucous glands in the skin further aid in reducing resistance during movement through the water.

Identify the three regions of the body: the anterior **head,** which extends to the rear of the bony operculum covering the gills; the **trunk,** extending from the operculum to the anus; and the **tail,** extending from the anus and posteriorly. The caudal fin is attached to the body behind the last vertebra. Several types of caudal fins are found among the bony fishes, as shown in figure 17.2. Most fish have a homocercal caudal fin in which the vertebrae end anterior to the tail and the tail is symmetrical. In the sharks the caudal fin is heterocercal in which the vertebrae continue into the dorsal part of the fin and the fin is asymmetrical. ***Which type of caudal fin do you see on the perch? How does it differ from the caudal fin seen in the shark?***

On the head, find the two sets of double **nostrils,** two large **eyes** (no eyelids), and the large **mouth** equipped with **teeth.**

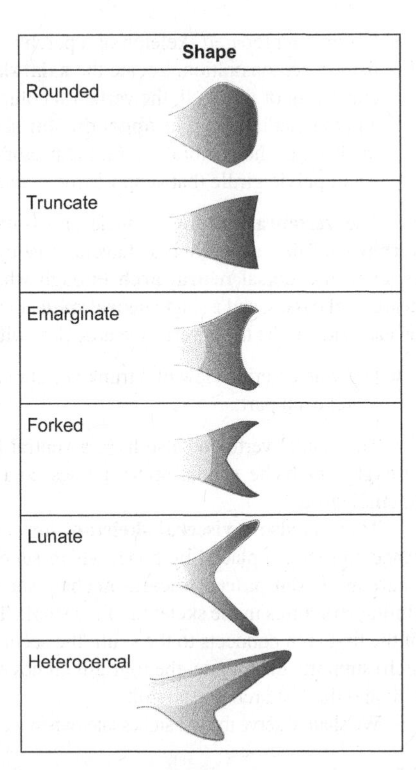

FIGURE 17.2 Types of caudal fins found in fishes. The first five types shown are homocercal; only the last type is heterocercal.

Where are the teeth found? A bony **operculum** covers the gills on each side of the head; under each operculum are four **gills.** The operculum is attached at the front and on the dorsal side, but is open behind and ventrally, providing for the release of water.

On the ventral surface, just anterior to the tail, locate the **anus** and the **urogenital opening(s).** Female perch have a single urogenital opening anterior to the anus; males have a separate genital pore and also a urinary pore located on a small urinary papilla. (see figure 17.14).

Observe the several fins attached to the body: four unpaired **median fins** (two dorsal fins, one anal fin, and one caudal fin) and two sets of **paired fins** (two pectoral fins and two pelvic fins). All fins are membranous extensions of the skin supported by numerous bony **fin rays.** The fins aid in stabilizing the fish and in directing its movements through the water.

On each side of the fish, extending from the operculum behind the eye to the base of the tail, is a **lateral line.** The lateral line is a specialized sensory organ system that detects vibrations and current directions in the water. It appears to aid fishes in orientation, in avoiding obstacles in the water, and in escaping predators.

The exterior surface of the perch is covered by a tough **skin** that contains many **mucous glands.** The skin produces the **scales,** which protect the surface of the body and are arranged in a well-ordered pattern of longitudinal and diagonal rows. Note how the posterior portion of each scale overlaps the anterior portion of the next scale. Each scale is formed in an epidermal pocket and extends posteriorly from the pocket. The scales grow continuously during the life of the fish and are not regenerated if lost.

Fish form several kinds of scales, and the shape and construction of the scales can help in the identification of different species and are useful indicators in other ways. Four common types of fish scales are illustrated in figure 17.3. Fish scales are homologous with the teeth and hair of mammals and we now know that some of the genes involved in tooth and hair development in mammals are involved in the development of scales. Scales are produced from the mesoderm layer of the skin, so they are different from the scales of reptiles that are formed in a different way.

Scales become larger as the fish grows by depositing new material around the margin; thus, a scientist can determine the age of a fish by microscopic examination of the "rings" in its scales (figure 17.4).

Growth of a fish depends on temperature and the abundance of food. Fish are **poikilothermic,** so their body temperature and metabolic rate vary with environmental temperature. They grow more slowly during the winter. Scales develop a wider layer of material during warm months when fish grow more rapidly. Scientists can determine the age and growth rate of the fish at various times by studying the growth rings on their scales. Similar annual depositions are found in certain fish bones, for example, the **otoliths** (ear bones) found in the ear.

Fact File

Perch

- Yellow perch (*Perca flavescens*) are also known as American perch, bandit fish, calico bass, convict, coon perch, coontail, Eisenhower, jack perch, lake perch, racoon perch, red perch, redfin, redfin trout, ringtail perch, ringed perch, river perch, sand perch, or striped perch. This is a good example of why biologists prefer scientific names over common names!
- The largest yellow perch on record weighed 4 lbs 3 oz. It was caught in 1865 from Cross Wicks Creek, New Jersey. The average weight of a fully grown adult is about 2 lb.
- Perch spawn at night in early spring when water temperature rises into the mid-40s (°F). Females release 5,000 to 50,000 eggs in a long, convoluted string of gelatinous material that would be 3–7 feet long, if straightened out. Upon contact with water, the egg string swells to several inches in diameter. It usually drifts with the current until it becomes snagged on submerged vegetation. Eggs hatch in 10–30 days depending on temperature (27 days at 47°F).
- Yellow perch are strictly carnivorous. They are primarily bottom-feeders with a slow, deliberate bite. They will eat almost anything, but prefer minnows, insect larvae, crayfish, plankton, and worms. They are visually attracted by movement of prey and feed only during daylight hours. Early mornings and late afternoons are peak feeding times and, consequently, these are the best hours for fishing.

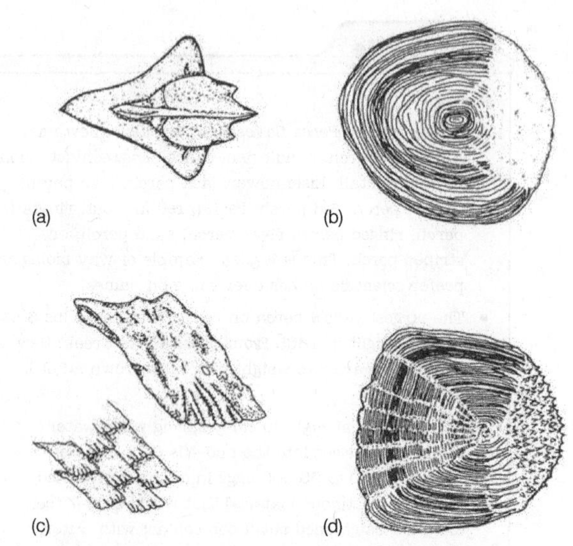

FIGURE 17.3 Some types of fish scales. (*a*) placoid, (*b*) cycloid, (*c*) ganoid, (*d*) ctenoid. From *Fishes of Tennessee* by David A. Etnier and Wayne C. Starnes, 1993. Courtesy of the University of Tennessee Press.

- Remove a scale from your specimen, make a wet mount on a microscope slide, and observe it under low power. Note the numerous concentric ridges **(annuli)** on the scale and the many fine **teeth** on the posterior portion of the scale. This type of scale is called a **ctenoid** ("comb") **scale** because of the presence of these teeth.

Internal Anatomy

Skeletal System

The scales, fin rays, and some of the bones of the skull of the perch represent elements of a **dermal exoskeleton,** but the chief supporting structure of the body consists of a **bony endoskeleton.**

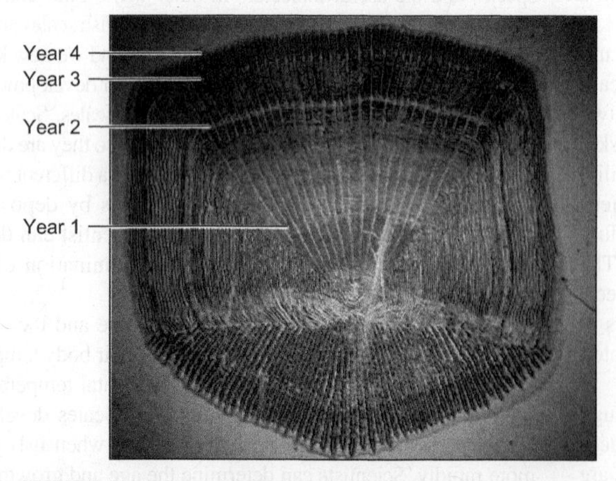

FIGURE 17.4 Ctenoid fish scale showing growth rings. Photo courtesy of the DE Division of Fish & Wildlife, 4876 Hay Point Landing Rd, Smyrna, DE 19977.

- Observe a prepared skeleton of a perch or other bony fish on demonstration. Locate the **axial skeleton** consisting of the skull, the vertebral column, the ribs, and the medial fins. The **appendicular skeleton** is made up of the pectoral girdle, the pectoral fins, and a small pelvic girdle that supports the pelvic fins.

The **vertebral column** is made up of many individual **vertebrae.** The trunk vertebrae have a large cylindrical **centrum** with a dorsal **neural arch** through which the dorsal nerve cord passes, and a single **neural spine.** Lateral processes on each side of the trunk vertebrae articulate with the ribs.

- Draw an anterior view of a trunk vertebra in figure 17.5. Label each part.

The caudal vertebrae also have a ventral **haemal arch,** through which the caudal artery passes, and a supporting **haemal spine.**

There is also a **visceral skeleton,** first formed of cartilage and later replaced by bone, which supports the gills. There are **seven paired visceral arches** that correspond to similar structures in the skeleton of the shark. The upper part of the first arch connects to the skull; the second arch (hyoid arch) supports the tongue; the four gill arches each support a gill; and the last arch has no gill.

We shall observe the gill arches later when we study the gills.

Muscular System

The muscular system of the perch is relatively simple compared to that of terrestrial vertebrates. Most of the body musculature consists of **segmental muscles** (myotomes) (figure 17.6). Contractions of these myotomes result in flexing of the body, which aids in swimming. Adjacent myotomes are separated by a **myoseptum** of connective tissue. The myotomes are also separated into dorsal and ventral portions by a **transverse septum.** The muscle segments dorsal to the transverse septum are called **epaxial muscles,** and the muscle segments ventral to the transverse septum are called **hypaxial muscles.**

FIGURE 17.5 Student drawing of a trunk vertebra.

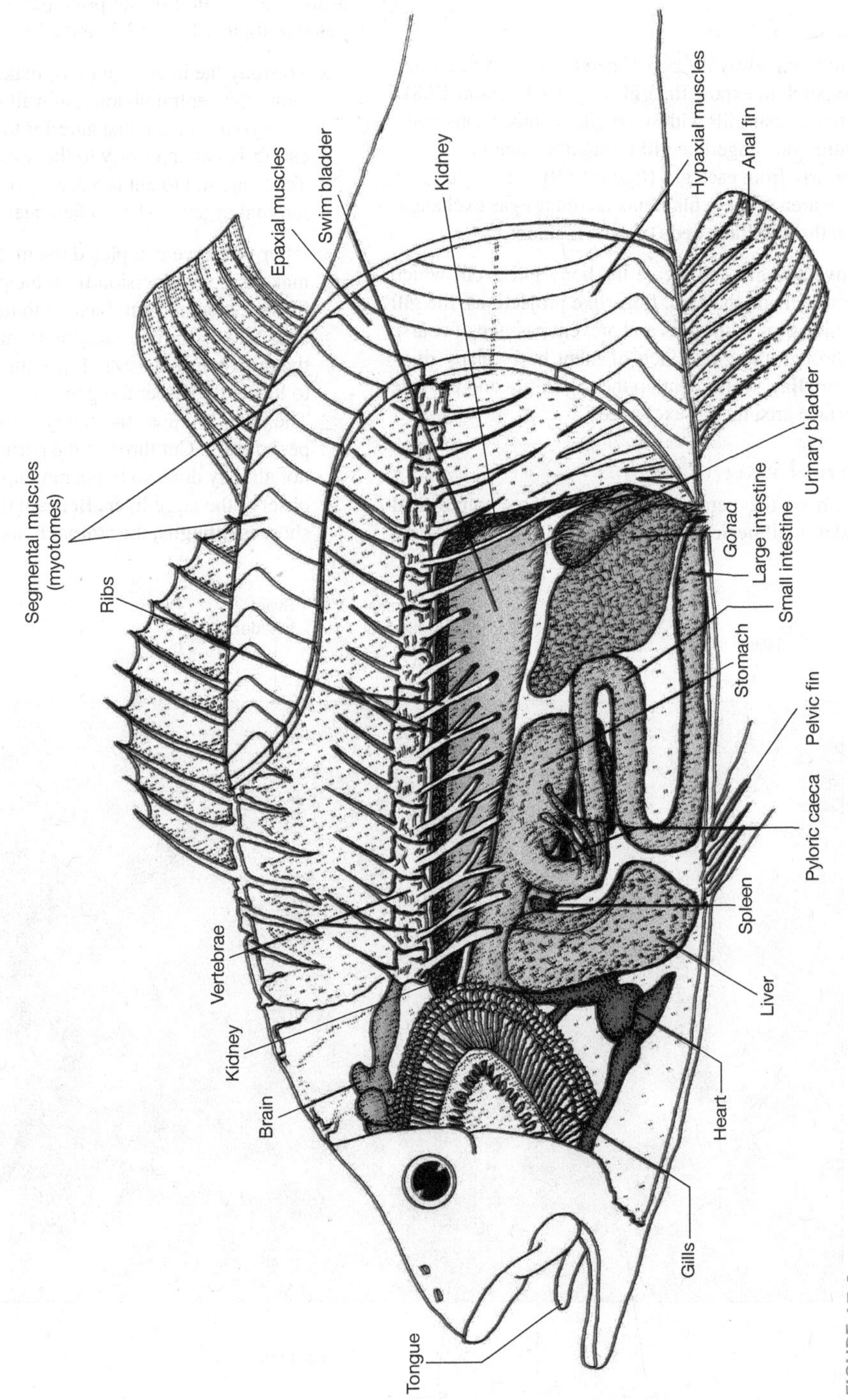

FIGURE 17.6 Perch, partly dissected to show muscles and internal organs, lateral view.

More specialized muscles in the head region serve to move the lateral fins, mouthparts, jaws, gill opercula, gill arches, and associated parts.

Respiratory System

- Carefully cut away the bony operculum from one side of the perch to expose the **gills** (figures 17.7 and 17.8). Locate the four gills within the gill chamber. Observe the numerous fingerlike **gill filaments** extending posteriorly from each gill (figure 17.9). The large surface area of these filaments facilitates gas exchange within the **capillary beds** of each filament.

Remove one gill and locate the bony **gill arch,** which supports the gill and the hard, fingerlike projections, the **gill rakers,** which protect the gills and prevent passage of coarse material through the gills. Each filament bears many disc-like thin **lamellae,** which contain the capillaries and provide a large surface area for gas exchange.

Coelom and Visceral Organs

The **coelom** of the perch consists of a large **peritoneal cavity** and a small **pericardial cavity.** The peritoneal cavity contains the stomach, liver, and other digestive organs, swim bladder, and other visceral organs. The pericardial cavity is located anterior to the peritoneal cavity and encloses the heart. The location of the principal internal organs is illustrated in figures 17.6, 17.7, and 17.8.

- To study the internal organs, make a longitudinal cut along the ventral abdominal wall with your scalpel. Start your incision just anterior to the anus and carefully cut anteriorly to the level of the pelvic girdle. Take care not to cut too deeply or you may damage internal organs to be studied later.
- After you have completed the midventral incision, make a second incision from the posterior end of the first incision and cut dorsally to the level of the lateral line. Make a similar incision from the anterior end of the midventral incision. Raise this portion of body wall to locate the visceral organs in the peritoneal cavity. The outer lining of the cavity is the parietal ("wall") peritoneum. Cut through the peritoneum if you have not already done so in opening up the body wall and observe the large **liver.** Beneath the liver, locate the short **esophagus,** the **stomach,** and the **small intestine.**

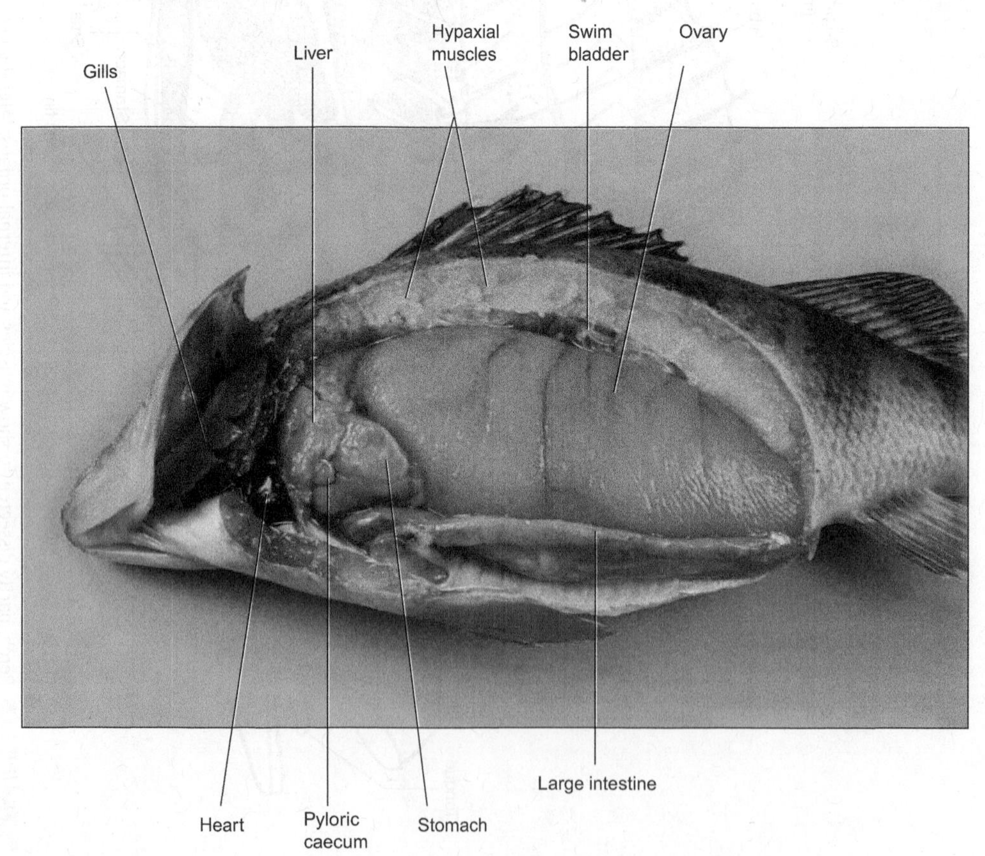

FIGURE 17.7 Perch, female internal anatomy.
Photo courtesy of John R. Meyer.

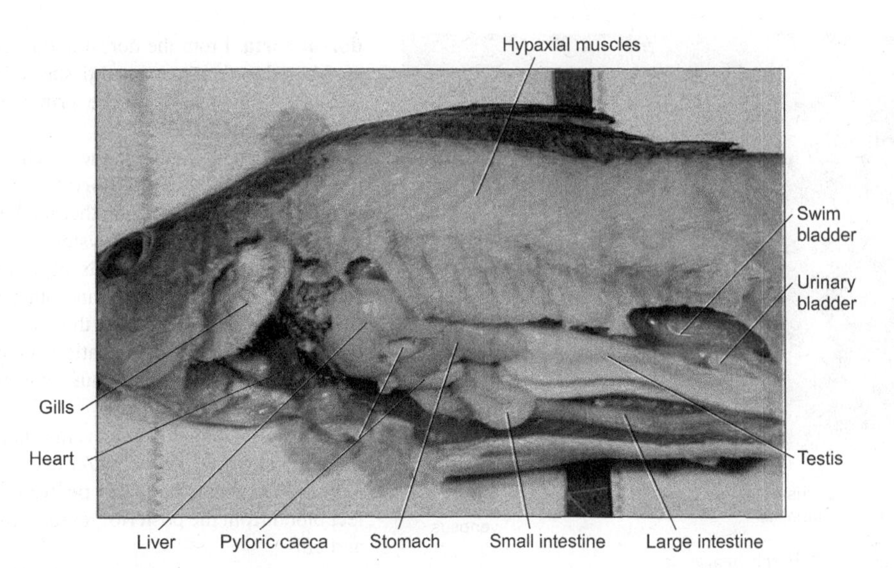

FIGURE 17.8 Perch, male internal anatomy.
Photo courtesy of Betty Black.

Extending from the anterior part of the small intestine are three short sacs—the **pyloric caeca.** Posteriorly, the small intestine empties into the **large intestine,** (figure 17.7), which terminates at the anus.

Dorsal to the digestive tract, find the large **swim bladder** or air bladder. (It is often deflated in preserved specimens.) The swim bladder is a hollow, gas- or air-filled sac that serves as a buoyancy organ. Altering the volume of gas within the swim bladder assists the perch in compensating for the differences in the specific gravity between its body and that of the surrounding water while swimming at various depths.

Above the swim bladder and underneath the vertebrae are two long, dark **kidneys.** As in the shark, the kidneys are retroperitoneal, located outside the parietal peritoneum adjacent to the dorsal body wall. Other organs in the peritoneal cavity include the **spleen,** an elongate organ lying along the posterior surface of the stomach; the **pancreas,** on the ventral surface of the intestine (often difficult to find); the **gonads,** posterior to the stomach and dorsal to the intestine; and the **urinary bladder,** found posterior to the gonads.

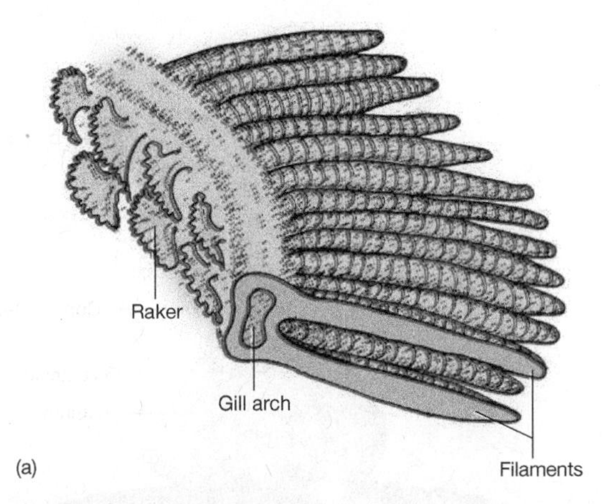

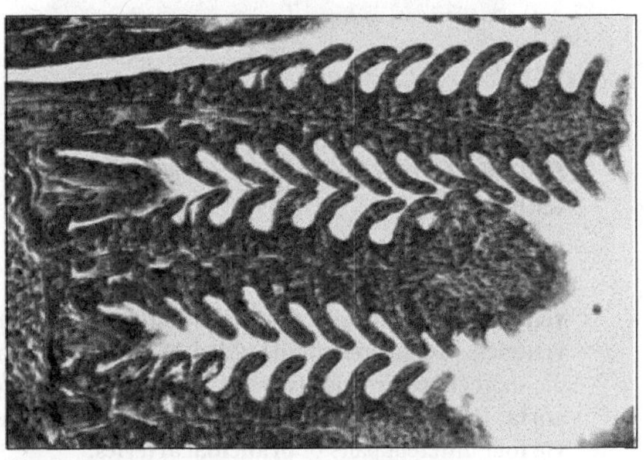

FIGURE 17.9 Fish gill. (*a*) Portion of a gill. (*b*) Microscopic cross section of a gill bar.
(b) Courtesy of Carolina Biological Supply Company, Burlington, NC.

Circulatory System

- The **heart** is located in the pericardial cavity, which lies ventral to the gills and anterior to the pelvic fins (figure 17.10). Carefully cut through the pectoral girdle and the muscles anterior to the girdle and cut away part of the lateral body wall to expose the heart and major blood vessels.

Like other fishes, the perch has a **two-chambered heart** (figure 17.11). Find the thin-walled **atrium** and the thick-walled, muscular **ventricle.** Blood passes from the **sinus**

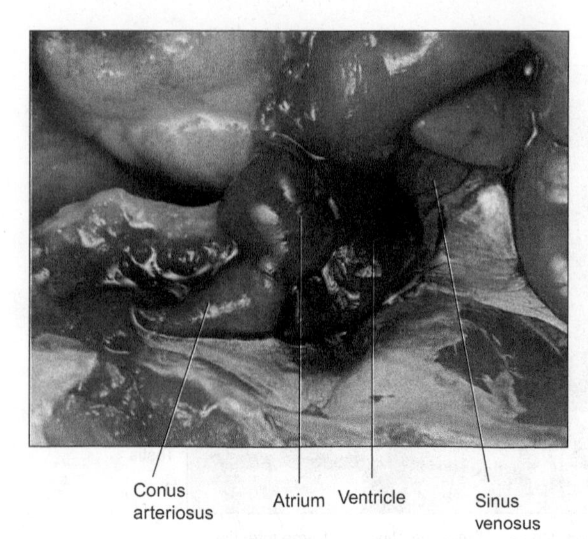

FIGURE 17.10 Perch heart.
Photo courtesy of John R. Meyer.

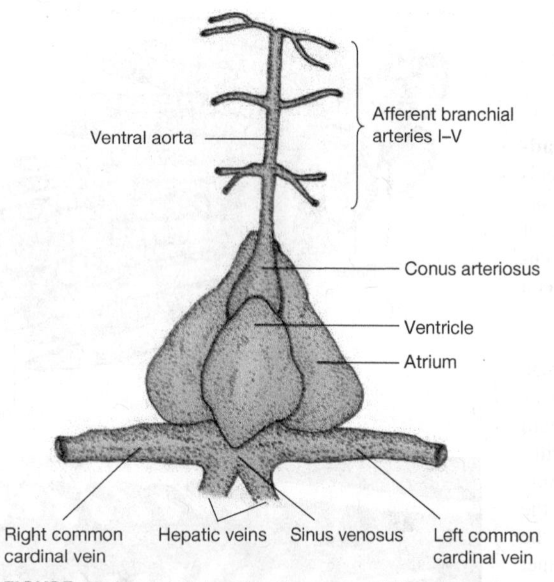

FIGURE 17.11 Perch heart and associated blood vessels, ventral view.

venosus to the **atrium** and from the atrium to the muscular **ventricle.** Contraction of the ventricle forces the blood into the short **conus arteriosus** and out through the short **ventral aorta.** From the ventral aorta, the blood passes to the gills via four different pairs of **branchial arteries.**

The **afferent branchial arteries** lead to extensive capillary beds in the lamellae of the gills, where the blood is oxygenated (figure 17.12). The oxygenated blood is collected by the **efferent branchial arteries,** which empty into the **dorsal aorta.** From the dorsal aorta, arteries carry oxygenated blood to the organs and tissues of the head, trunk, and caudal regions. Some of the principal blood vessels are shown in figure 17.13.

The venous system of the perch consists of two main divisions: (1) the **hepatic portal system,** and (2) the **systemic veins.** The perch and other modern bony fishes lack a well-developed renal portal system as seen in the shark. The hepatic portal system consists of veins that collect blood from the stomach, intestine, and other visceral organs, and carry it to capillary beds in the liver. From the liver, the blood is collected by the hepatic vein and is carried via the posterior cardinals to the sinus venosus and the heart for recirculation.

The principal systemic veins include a pair of large **anterior cardinal veins** that collect blood returning from the head region and a pair of **posterior cardinals** that collect blood from the posterior regions, including the kidneys and tail.

Urogenital System

The **kidneys** are two long, slender organs lying dorsal to the swim bladder and ventral to the vertebral column. They filter nitrogenous wastes from the blood and empty posteriorly through the **archinephric** (or Wolffian) **ducts,** which lead to the **urinary bladder.** From the bladder, urine passes into the **urogenital sinus** and out through the **urogenital pore** (figure 17.14). In the male, the urinary pore and the genital openings are separate. In the female, there is a common urogenital pore through which both systems empty.

The reproductive system of the **male** includes two long-lobed **testes** lying posterior to the stomach and ventral to the swim bladder (figure 17.14). Two **vasa deferentia** carry the sperm to a common **genital sinus,** which opens via the **genital pore.**

In the **female** is a single, fused **ovary** (figure 17.7), a large, saclike structure that releases the eggs through a short **oviduct** to the **urogenital pore.**

Nervous System

Like the nervous system of the shark, that of the perch consists of two main divisions: the **central nervous system** (brain and spinal cord) and the **peripheral nervous system** (nerves connecting the brain and spinal cord with other parts of the body).

We shall confine our brief study of the nervous system of the perch largely to the **brain** (figures 17.15a and 17.15b). The brain of the adult perch consists of five major divisions: (1) the **telencephalon,** (2) the **diencephalon,** (3) the **mesencephalon,** (4) the **metencephalon,** and (5) the **myelencephalon.** The specific parts of the brain making up these five divisions are summarized in table 17.1.

- To expose the brain for study, you must remove the skin from the dorsal surface of the skull behind the

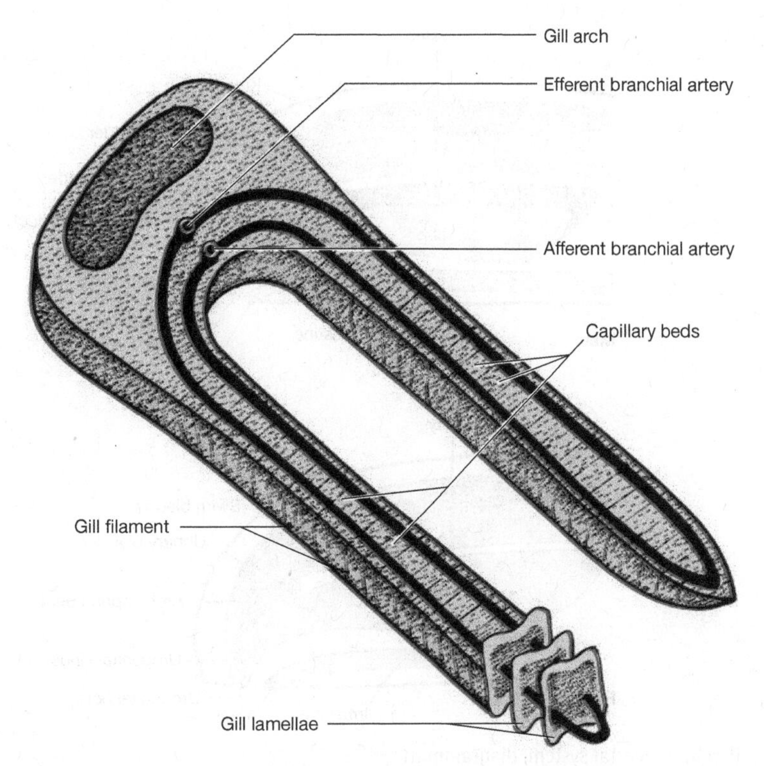

FIGURE 17.12 Perch gill, pattern of circulation within a filament.

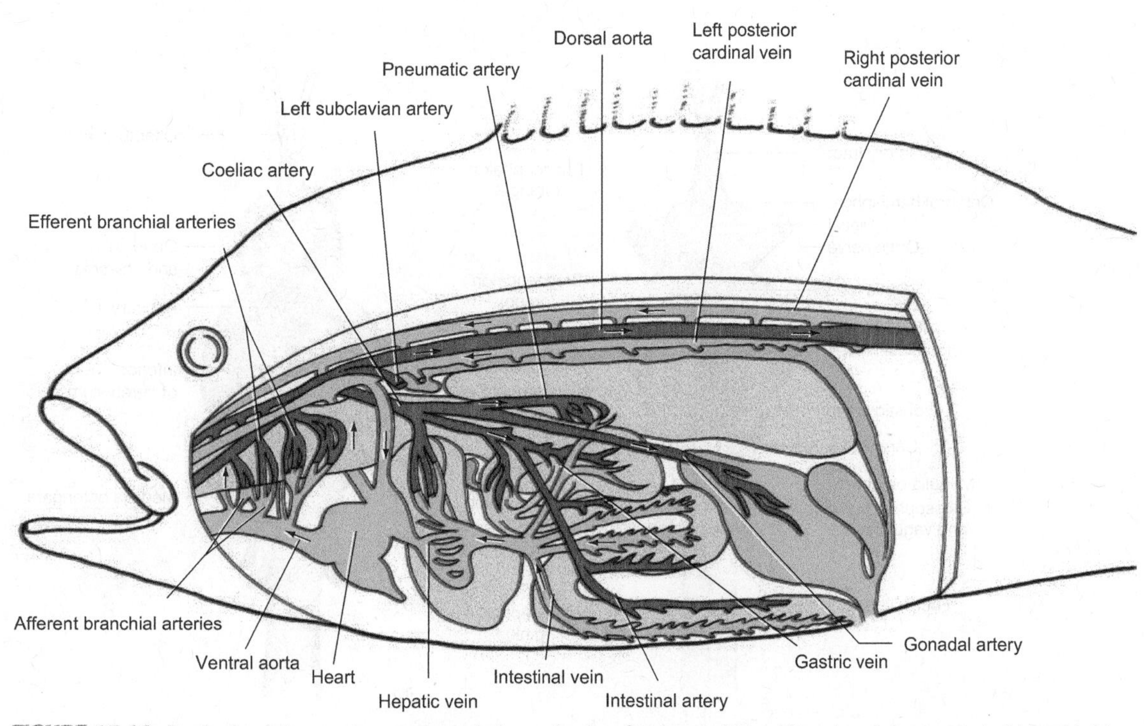

FIGURE 17.13 Perch circulatory system, major arteries and veins. Oxygenated blood is red, and deoxygenated blood is blue.

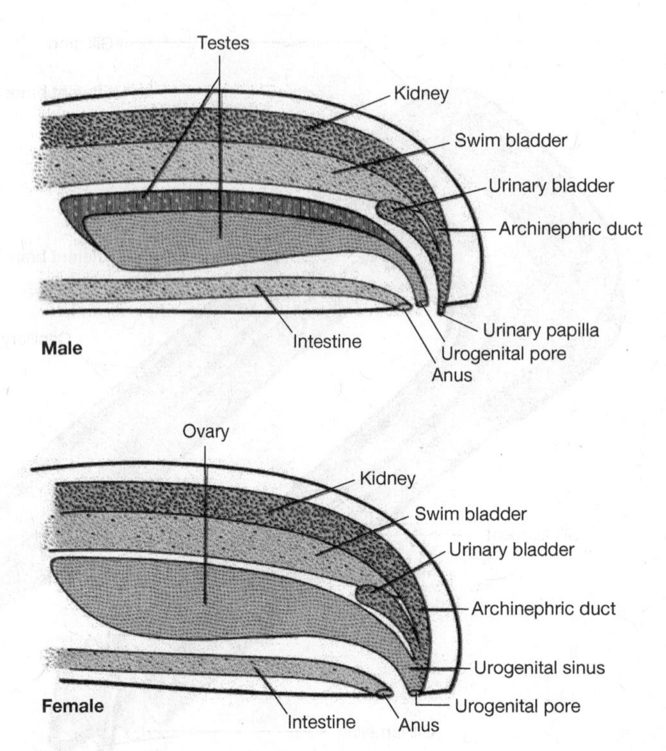

FIGURE 17.14 Perch urogenital system, diagrammatic.

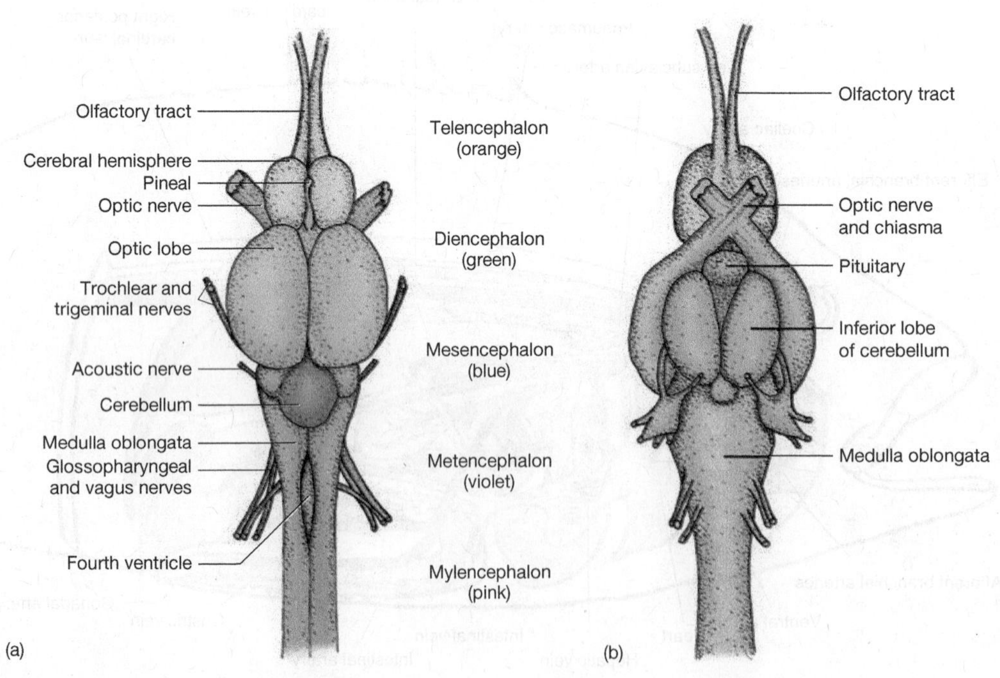

FIGURE 17.15 Perch brain. (*a*) Dorsal view. (*b*) Ventral view.

Table 17.1 Components of the Perch Brain

Division	Chief Structures
Telencephalon	Olfactory lobes, cerebral hemispheres
Diencephalon	Thalamus, hypothalamus, pineal, pituitary
Mesencephalon	Optic lobes
Metencephalon	Cerebellum
Myelencephalon	Medulla oblongata

eyes (figure 17.16). Carefully shave away the bony roof of the skull above the brain, taking care not to damage the delicate tissues beneath. You will find the brain enclosed in a gelatinous mass that must be removed to expose the brain. Also around the brain is a pigmented membrane. Carefully remove the membrane and identify the principal structures of the brain with the aid of figures 17.15 and 17.16.

Associated with the brain of the perch and other bony fishes are **10 pairs of cranial nerves** (table 17.2). Nerve 0, the terminal nerve of the shark, is absent in bony fishes.

The spinal cord leads posteriorly from the brain to the tail and passes through the neural arches of the vertebrae. One pair of **spinal nerves** arises from the spinal cord for each vertebra.

Table 17.2 Cranial Nerves of the Perch

Nerve	Function
I. Olfactory	Olfaction (smell)
II. Optic	Vision
III. Oculomotor	Eye movements
IV. Trochlear	Superior oblique muscle of eye
V. Trigeminal	Jaw muscles, touch
VI. Abducens	Lateral rectus muscle of eye
VII. Facial	Taste, lateral line, skin of head
VIII. Acoustic	Inner ear and lateral line
IX. Glossopharyngeal	Gill muscles and lateral line
X. Vagus	Gills, heart, anterior part of digestive tract, lateral line

Demonstrations

1. Microscope slide of ctenoid scale
2. Microscope slide of fish skin showing origin of scale
3. Mounted or plastic-embedded fish skeleton
4. Model of dissected perch
5. Fish heart, plastic mount
6. Microscope slide of fish gill, sectioned through filaments
7. Living fish in aquarium

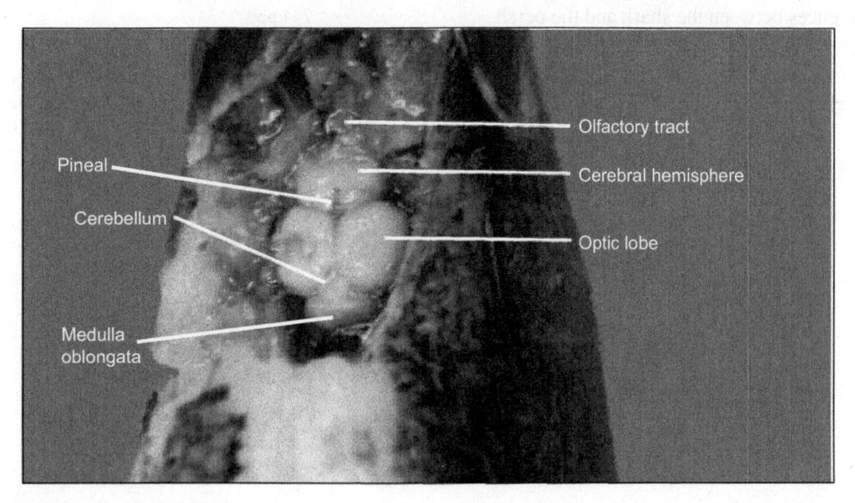

FIGURE 17.16 Perch brain, dissected, dorsal view.
Photo courtesy of Betty Black.

Key Terms

Appendicular skeleton portion of the endoskeleton that supports the appendages. Consists of the anterior pectoral girdle and the posterior pelvic girdle.

Archinephric duct a primitive type of kidney duct that collects urine from a series of segmentally arranged nephrons; also called a Wolffian duct.

Axial skeleton portion of the endoskeleton that supports the longitudinal axis of the body. Consists of the skull, the vertebral column, the ribs, and the medial fins in bony fishes.

Ctenoid scale type of dermal scale made of a thin sheet of bonelike material, circular or oval in shape, and bearing spines ("teeth") on the posterior margin.

Homocercal tail type of tail with upper and lower parts of the tail symmetrical and with the vertebral column ending at the center of the base; found in most bony fishes.

Swim bladder a gas- or air-filled sac found in the abdominal cavity of most bony fishes; serves mainly as a hydrostatic organ assisting with buoyancy in modern fishes.

Visceral skeleton portion of the endoskeleton that supports the gills. Consists of seven paired gill arches.

Internet Resources

There are many valuable Internet sites with information about zoology. Several sites containing pertinent zoological information for this chapter can be found on the McGraw-Hill Zoology web site at http://www.mhhe.com/zoology. Just click on this text's title.

Questions for Critical Thinking

1. Discuss the components and functions of the axial skeleton, the appendicular skeleton, and the visceral skeleton. What advantages (if any) does the rigid endoskeleton have over a cartilaginous endoskeleton of the shark?
2. Describe the basic morphological similarities and differences between the shark and the perch.
3. Speculate on the advantages and disadvantages of the heterocercal tail of the shark versus the homocercal tail of the perch. Think about maneuverability, thrust, steering ability, etc.
4. Describe in some detail the operation of the gas exchange system. When water warms, it loses its free oxygen. How would this affect the gas exchange system of a bony fish? Could the fish literally "drown" if the water temperature got too high? Explain.

Suggested Readings

Barr, B.M. L.M., Page, J.P. Sherrod, E.C. Beckham, and J. Tomelleri. 2011. *Peterson's Guide to the Freshwater Fishes,* 5th ed. Boston: Houghton Mifflin Harcourt. 688 pp

Caillet, G.M., M.S. Love, A.W. Love, and A.W. Eberling. 1996. *Fishes: A Field and Laboratory Manual on their Structure, Identification, and Natural History.* Long Grove, IL: Waveland Press. 202 pp

Chaisson, R.B., and W.J. Radke. 1991. *Laboratory Anatomy of the Perch,* 4th ed. Dubuque, IA: McGraw-Hill Higher Education. 67 pp

Fitzgerald, D.G. et al. 2001. Application of otolith analyses to investigate broad size distributions of young yellow perch in temperate lakes. *J. Fish Biol.* 58(1): 248. 263 pp

Moyle, B.G., and J.J. Cech Jr. 2004. *Fishes: An Introduction to Ichthyology.* Upper Saddle River, NJ: Pearson Prentice Hall. 744 pp

Notes and Sketches

Chapter 18

Frog Anatomy

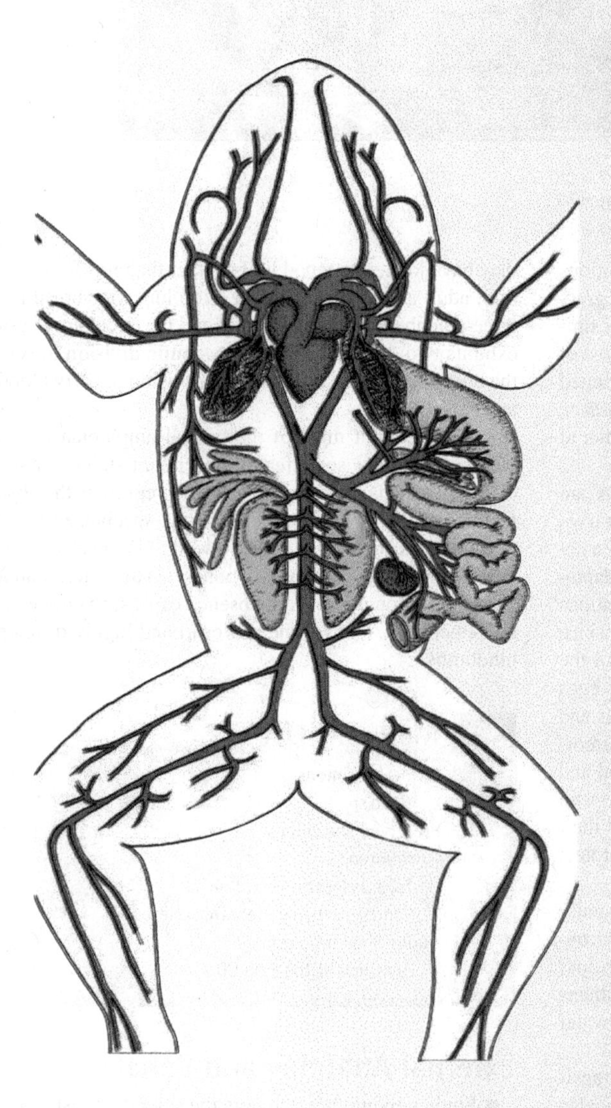

Objectives

After completing the laboratory work in this chapter, you should be able to perform the following tasks:

1. Identify the main structures within the oral cavity of the frog and explain the function of each.
2. Describe the external features (secondary sexual characteristics) that distinguish mature male and female frogs.
3. Describe the major divisions of the frog skeleton and explain the chief function(s) of each division.
4. Describe the principal parts of the digestive system of the frog and explain their functions.
5. Describe the structure of the frog heart and identify its major parts.
6. Describe the respiratory system of the frog and identify the main structures involved in gas exchange.
7. Describe the basic pattern of blood circulation in the frog and identify the principal blood vessels involved.
8. Identify the important parts of the urogenital system of both male and female frogs and explain their functions.
9. Identify the five main regions of the frog brain and the components of each. Briefly explain the main function of each part.
10. Describe the stages of frog metamorphosis.

Rana pipiens or *R. catesbeiana*

Frogs are the most commonly studied representatives of the Class Amphibia, Subphylum Vertebrata, Phylum Chordata (figure 18.1). Although there is no "typical vertebrate" any more than there is a "typical person," a study of frog anatomy does serve to illustrate effectively the basic

FIGURE 18.1 Leopard frog, *Rana pipiens.*
Photograph courtesy of Carolina Biological Supply Company, Burlington, NC.

body organization of a vertebrate animal. The descriptions in this chapter are based primarily on *Rana pipiens* (the grass or leopard frog), but *Rana catesbeiana* (the bullfrog) or similar species may be used for these studies. Actually, most of the frogs now sold by biological supply houses in the United States are a mixture of *Rana forreri* and *Rana berlandieri.* Most of the anatomy of these species is essentially similar to that described in this chapter.

It is important to remember that the amphibians are transitional animals that typically live a portion of their lives in water and another portion on land. Thus, they exhibit a peculiar mixture of characteristics—some representing adaptations for terrestrial life and some representing adaptations for life in water. Mating nearly always occurs in water, since the eggs lack the protective outer coverings that permit the terrestrial existence of birds and reptiles. Amphibian eggs hatch into **tadpoles,** an immature, larval form with gills and a muscular tail for swimming. Later the tadpoles metamorphose into four-legged adults, which are semiterrestrial and become sexually mature. Some amphibians, however, spend all of their lives in the water, and a very few spend all their lives on land, having evolved special mechanisms to protect their eggs from desiccation.

The skin of adult amphibians is smooth and usually moist. Generally, the adults live in wet or moist environments because they are very susceptible to water loss through the skin. Toads are rather exceptional amphibians that have developed a tough, horny skin that reduces water loss and thus allows them to spend more time on land.

Since the amphibians represent an evolutionary transition between the fishes and the terrestrial animals, they also exhibit numerous adaptations for their amphibious life in and out of water. The skull is broad, flat, and light weight; they have jointed tetrapod limbs rather than fins for locomotion; adult amphibians often develop lungs for breathing; a three-chambered heart is formed; and the circulatory system exhibits two distinct circuits: a **systemic division** to supply the body organs and a **pulmonary division** to carry blood to and from the lungs.

Frogs exhibit most of the typical amphibian features, but they also show some features peculiar to their own mode of life. The salamanders more nearly represent the typical features of the Amphibia. Some of the specialized features of the adult frog include the following: (1) the absence of a tail, (2) the loss of certain skull bones, (3) the anterior attachment of the tongue, (4) the absence of ribs, (5) the lack of a distinct neck, and (6) the powerful and highly developed hindlimbs.

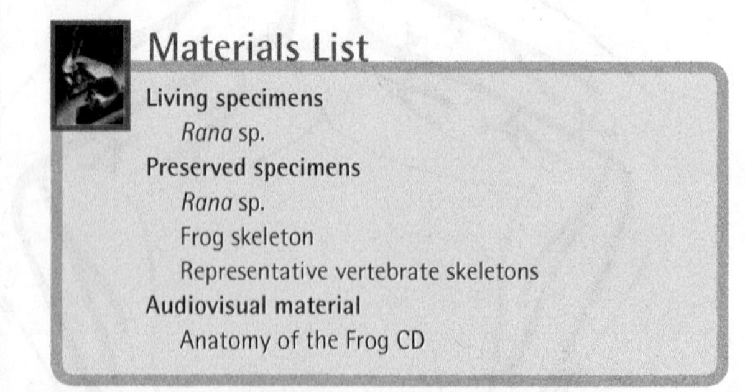

Materials List

Living specimens
Rana sp.
Preserved specimens
Rana sp.
Frog skeleton
Representative vertebrate skeletons
Audiovisual material
Anatomy of the Frog CD

External Anatomy and Behavior

- Study a living frog and note the smooth, moist, and pliable skin. Observe the pattern of coloration of the dorsal and ventral surfaces of the body.

How does the coloration differ on the two surfaces? Compare the color pattern of your specimen with those of others in the laboratory. ***How much variation in color patterns do you observe?*** The specific pattern of spots has been shown to be genetically determined.

Note the broad, flat head with the large mouth, the nostrils or **external nares,** two conspicuous **eyes,** and the circular **tympanic membranes** located behind the eyes. Bordering the eye is a fleshy lower eyelid and a less-prominent upper eyelid. A third transparent inner eyelid, the **nictitating membrane,** helps keep the eye moist while the frog is on land and also helps protect the eye under water from abrasion.

- Observe demonstrations of living frogs in an aquarium and in a terrarium. Study a frog in the aquarium and observe its swimming behavior.

How does the frog propel itself through the water? Which limbs are most prominent in swimming? What adaptations of the limbs are most important for swimming? What is the posture of the frog while it is floating? Compare the behavior of frogs in an aquarium filled with water at room temperature and the behavior of frogs in an aquarium filled with cold water. ***Which frogs are more active? How can you explain the difference in their activity? Which frogs remain submerged for longer periods of time: those in the cold aquarium or those in the warm (room temperature) aquarium?*** Frogs are **poikilothermic,** which means that their body temperature fluctuates with the thermal environment around them. ***How does this fact help you to explain the difference in behavior between the frogs in the cold aquarium and those in the warm aquarium?***

- Observe also the behavior of frogs in a terrarium.

When exposed to the air, frogs breathe by drawing air through the external nares into the oral cavity. Later, air is forced into the lungs by closing the internal nares and raising the floor of the mouth. Some gas exchange with the blood also occurs across the moist epithelium of the oral cavity and through the skin covering the exterior of the body.

- Compare the locomotion of frogs in the water with frogs in a terrarium or on a laboratory table. ***How are the limbs used in locomotion on land?***

To observe the feeding behavior of a frog, place some living house flies, fruit flies, or other small insects in a small aquarium with a frog. ***How does the frog react to the food organisms? How are the insects captured?***

Fact File

Frogs

- When a frog swallows food, its eyes close and the bulgy eyeballs sink into the head applying internal pressure that helps push food down into the throat.

Skeletal System

The skeleton of the frog (figures 18.2 and 18.3) consists chiefly of bone and cartilage. It supports the various parts of the body, protects delicate organs such as the brain and spinal cord, and provides surfaces for attachment of the skeletal muscles.

- Study a prepared skeleton of a bullfrog (plastic-embedded specimens are excellent for this purpose) and compare the general organization of the frog skeleton to that of other vertebrates on demonstration.

The skeleton of **vertebrates** consists of the **somatic skeleton** (skeleton of the body wall and appendages) and a small **visceral skeleton** (skeleton of the pharyngeal wall—prominent in fishes as support for the gills and as a part of the jaws, but much reduced in higher vertebrates). In the frog, the visceral skeleton is represented principally by the **hyoid apparatus,** a small bone and cartilage structure that helps support the floor of the mouth, the base of the tongue, and parts of the jaws and larynx.

The major components of the frog skeleton and some of their relationships are shown in table 18.1.

The **skull** consists of three main parts (figure 18.2): (1) the narrow braincase, or **cranium;** (2) the paired **sensory capsules** of the ears, nose, and the large eye sockets, or orbits, for the eyes; and (3) the **visceral skeleton** (consisting of parts of the jaws, hyoid apparatus, and the laryngeal cartilages).

- Study the figures and locate the various bones of the skull on your specimen.

Table 18.1 Components of the Frog Skeleton

Visceral Skeleton
- Hyoid apparatus

Somatic Skeleton
- Axial skeleton
  - Vertebral column
  - Sternum
  - Skull
- Appendicular skeleton
  - Pectoral girdle
  - Forelimbs
    - Humerus
    - Radio-ulna
    - Carpals
    - Metacarpals
    - Phalanges
  - Pelvic girdle
  - Hindlimbs
    - Femur
    - Tibiofibula
    - Tarsals
    - Metatarsals
    - Phalanges

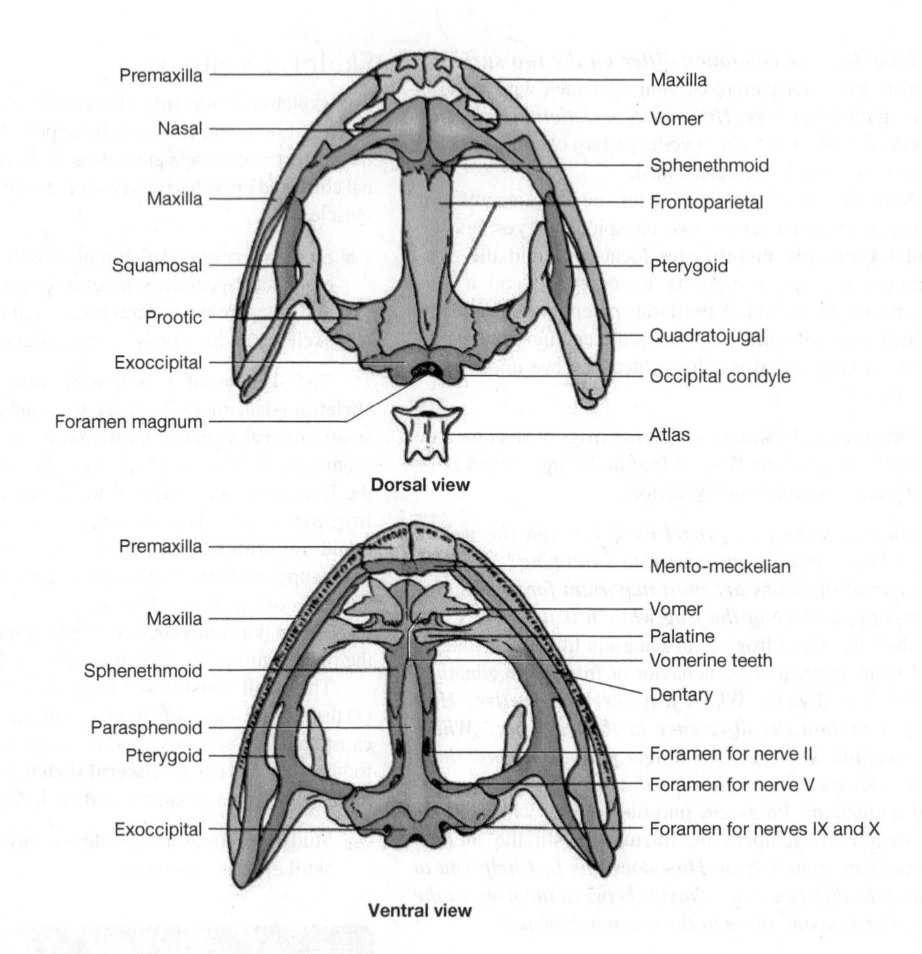

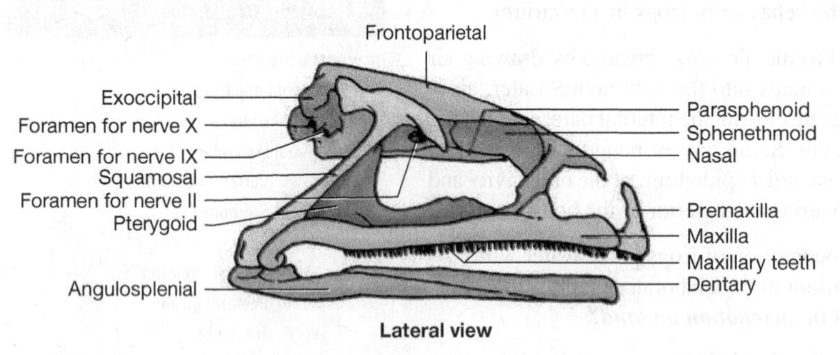

FIGURE 18.2 Frog skull.

The **vertebral column** is made up of 10 vertebrae. The first vertebra, the **atlas,** articulates with the base of the skull (figure 18.2). It has no transverse processes and is the only **cervical** (neck) **vertebra** in the frog. The next seven vertebrae are the **abdominal vertebrae** (figure 18.3). Caudal to the abdominal vertebrae is a large **sacrum** with two strong transverse processes that join with the **ileum.** The last vertebra is the long **urostyle** (figure 18.3).

- Study the parts of an abdominal vertebra with the aid of figure 18.3.

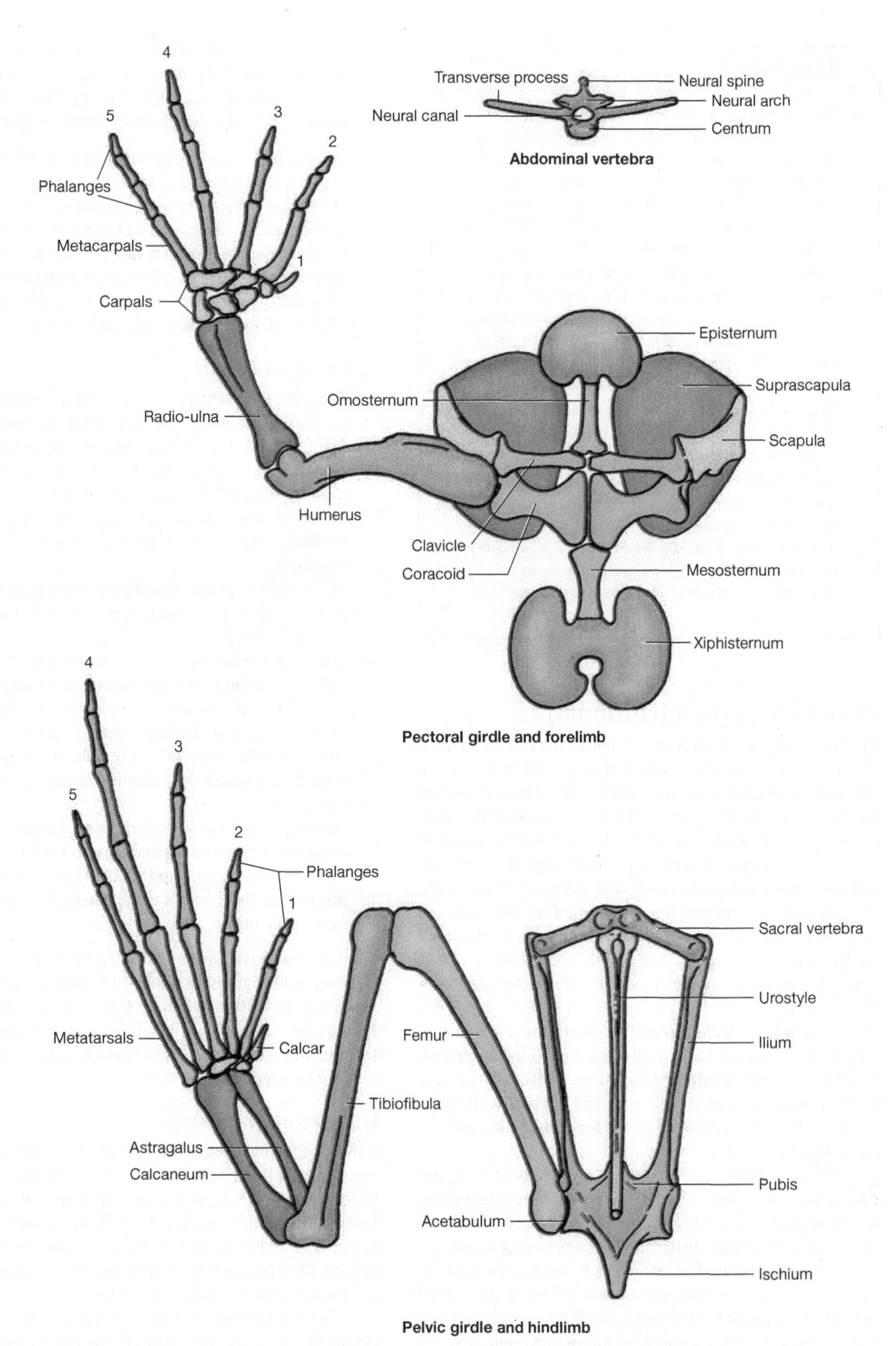

FIGURE 18.3 Frog skeleton.

Fact File

- Most frogs and toads have a long, sticky tongue attached near the front of the jaw. A quick flick of this tongue is used to capture prey. But *Xenopus*, the African clawed frog commonly studied in biology laboratories, does not have a tongue. It grabs prey with its front legs, which are equipped with special claws—hence the name "clawed frog."
- Many frogs, including the poison dart frog, have neurotoxins in the mucus coating their skin. Scientists are studying the pharmacological effects of these compounds and some of them are currently being evaluated for use as pain killers and antidepressants for humans.
- Frogs first appeared in the fossil record during the Jurassic period, about 190 million years ago. Scientists believe their jumping legs may have evolved as an adaptation to avoid being eaten by dinosaurs.
- Occasionally throughout history there have been tales of frogs raining down from the sky. We now know this can actually happen when a cyclonic windstorm passes over a lake or pond, sucks up water containing frogs, and dumps them elsewhere as the storm dissipates!

Muscular System (Optional)

Effective study of the muscular system of the frog takes time, patience, and good dissection technique. There are some 200 muscles in the frog, so a complete study of its musculature would take a good deal of time. In this exercise, we shall concentrate only on a few muscles to illustrate the organization of the muscular system and to learn something about the relationship between muscles, bones, and other parts of the body.

Some of the principal muscles of the frog are illustrated in figures 18.4 and 18.5. The skeletal muscles of the adult are well-adapted for swimming and for locomotion on land. Note the large, highly developed muscles associated with the hindlimbs.

A skeletal muscle typically consists of a fixed end, called the **origin,** and a movable end, called the **insertion.** The fleshy middle portion is called the **belly.** Most of the skeletal muscles taper at their ends and connect with tough white cords of connective tissue, the **tendons,** which serve to attach them to bones.

The movement or effect caused by a muscle is its **action.** Thus, each skeletal muscle has a characteristic origin, insertion, and action. Most muscles are arranged in pairs or groups that are **antagonistic:** they have opposing actions.

Muscles are classified according to their actions. Some general classes of skeletal muscles are as follows: **extensors**—muscles that straighten or extend a part; **flexors**—muscles that bend one part toward another part; **adductors**—muscles that pull a part back toward the axis of the body; and **abductors**—muscles that draw a part away from the axis of the body.

- For a detailed study of the muscles of the frog, you must first remove the skin from a specimen. Specimens especially preserved in alcohol are best for this purpose although formalin-preserved specimens may also be used.
- The chemicals used for preservation can be irritating to the skin and eyes. You should wear plastic or rubber gloves and protective eye gear when dissecting preserved specimens. See the section on laboratory safety following the preface of this book for further information about the properties of chemicals used for preservation and for other safety precautions. If you follow a few simple safety rules, work in the biology lab can be safe and rewarding.

Skinning the Frog

1. Place the frog ventral surface up in a dissecting pan. Carefully cut through the skin along the midventral line from the anterior tip of the jaw to the region of the cloaca. Pull the skin up and away from the underlying muscles to avoid damage to the muscles.
2. Make two transverse cuts through the skin around the body anterior to the forelimbs and anterior to the hindlimbs.
3. Make additional cuts around the opening of the cloaca, and around the eyes and tympanic membranes on each side of the head.
4. Carefully loosen the skin from the underlying muscles with a ***blunt (not sharp) instrument***. Observe the thin, white sheets of connective tissue between the skin and the muscles. Note the large areas where the skin is not attached to the muscles. These areas are **subcutaneous lymph sacs** where this colorless fluid collects in a living frog.
5. Gently peel off the skin to expose the muscles of the abdomen, head, and appendages. Loosen the skin at the base of each limb and carefully pull from the base toward the distal end of the appendage, turning the skin inside out.

After you have finished skinning the frog, compare your specimen with figures 18.4 and 18.5. Identify first the principal muscles of the abdomen. Carefully free and separate the muscles with a blunt instrument as you work. Follow the direction of the muscle fibers as you gently separate one muscle from another.

Muscles of the Hindlimb

A study of the muscles of the well-developed hindlimbs provides an excellent introduction to the principles of muscle anatomy. Table 18.2 lists the principal muscles of the hindlimb. Start with the muscles of the dorsal surface of the thigh (upper leg) and identify the five principal muscles on this surface. Observe how these muscles provide control of the various movements of the leg and foot.

Continue your study with the muscles of the ventral surface of the thigh and then identify the muscles of the shank (lower leg). ***How many pairs of antagonistic muscles can you identify from the hindlimb?***

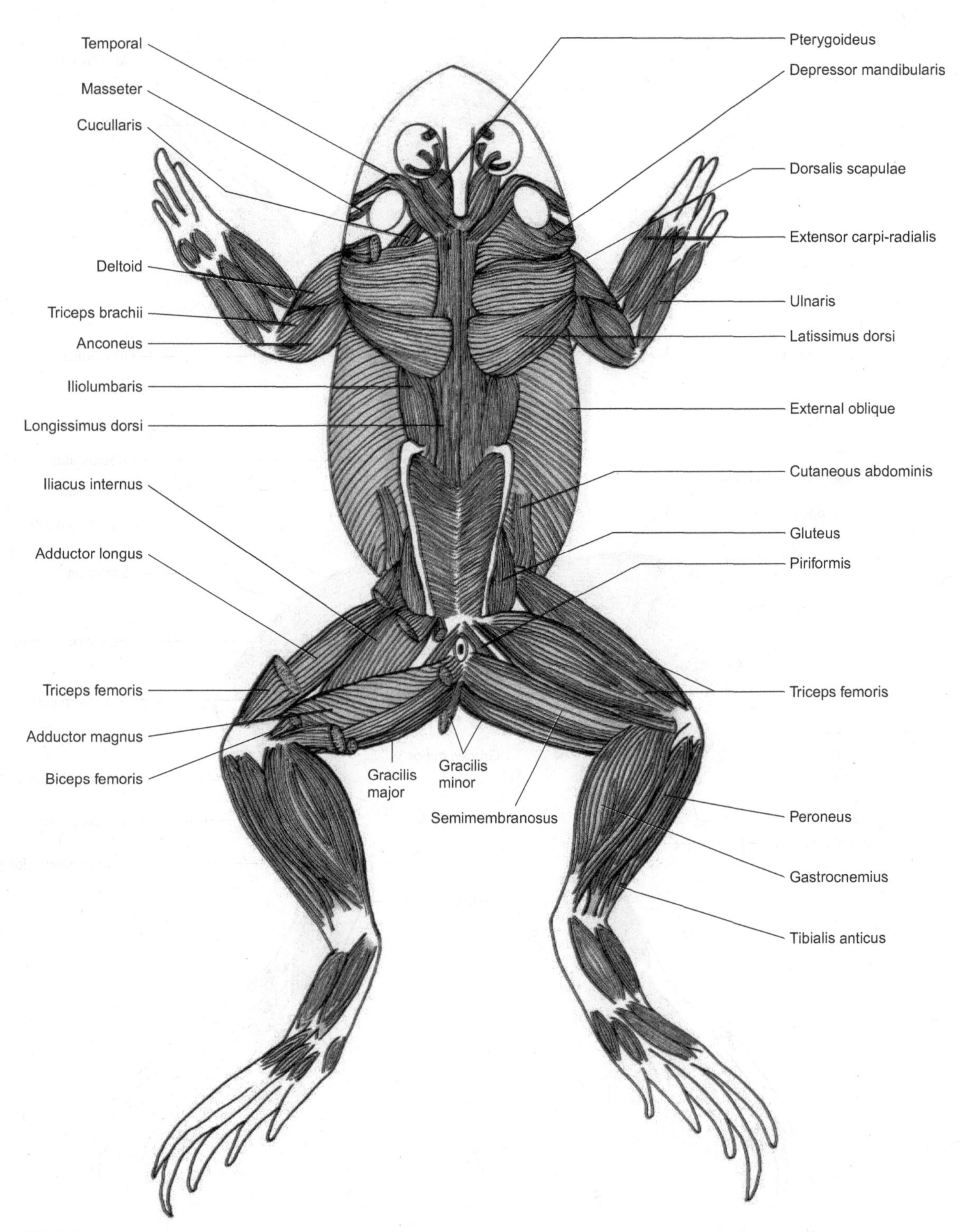

FIGURE 18.4 Frog, muscular system, dorsal view.

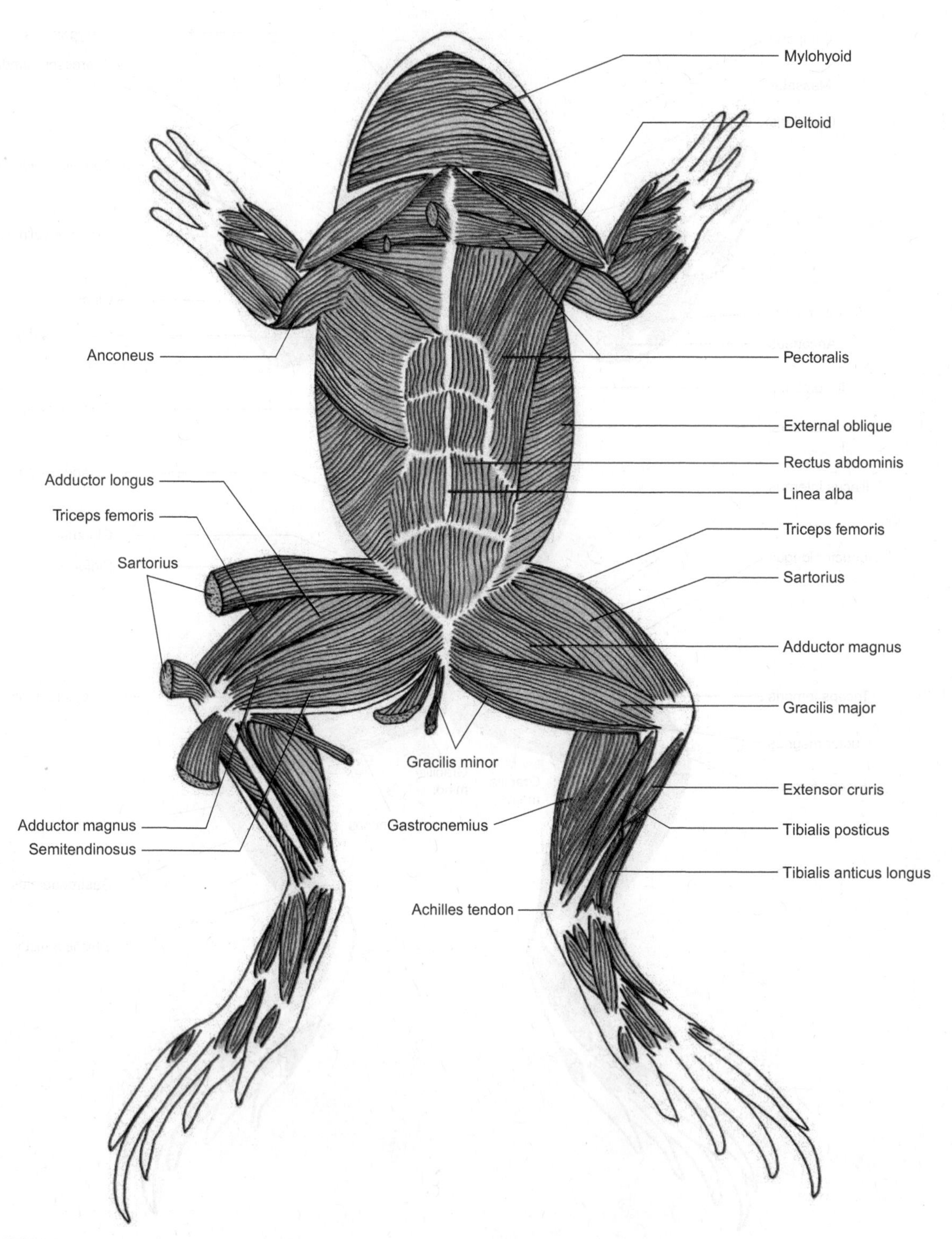

FIGURE 18.5 Frog, muscular system, ventral view.

Table 18.2 Chief Muscles of the Thigh and Shank

Muscle	Location	Origin	Insertion	Action
Dorsal Muscles of the Thigh				
Triceps femoris (three parts)	Lateral surface of thigh	Ilium, acetabulum	Tibiofibula	Flexes thigh and extends shank
Rectus anterior femoris		Ilium	Fascia attached to tibiofibula	Flexes thigh and extends shank
Vastus internus		Acetabulum	Tibiofibula	Flexes thigh and extends shank
Vastus externus		Ilium	Vastus internus	Flexes thigh and extends shank
Gluteus	Medial and anterior to triceps femoris	Ilium	Femur	Rotates thigh
Semimembranosus	Medial to triceps femoris	Ischium and pubis	Tibiofibula	Extends thigh and flexes shank
Biceps femoris (iliofibularis)	Between triceps femoris and semimembranosus	Ilium	Tibiofibula and femur	Extends and adducts thigh; flexes shank
Piriformis	Small muscle near cloacal opening; between biceps femoris and semimembranosus	Urostyle	Femur	Extends and rotates thigh
Ventral Muscles of the Thigh				
Sartorius	Large, flat midventral muscle	Pubis	Tibiofibula	Flexes thigh and shank
Gracilis major (rectus internus major)	Medial to sartorius	Pubis	Tibiofibula	Extends thigh and flexes shank
Gracilis minor (rectus internus minor)	Medial to gracilis major	Pubis	Tibiofibula	Extends thigh and flexes shank
Adductor magnus	Medial and partly beneath sartorius	Ischium and pubis	Femur	Adducts thigh
Adductor longus	Beneath sartorius	Pubis	Femur	Adducts thigh
Semitendinosus	Under and between gracilis minor and adductor magnus	Ischium	Tibiofibula	Extends and adducts thigh, flexes knee
Muscles of the Shank				
Gastrocnemius	Medial surface of shank	Femur	Achilles tendon	Flexes shank and foot
Peroneus	Lateral to gastrocnemius	Femur	Distal end of tibiofibula	Extends shank and foot
Tibialis anticus longus (tibialis anterior longus)	Anterior to tibiofibula	Femur	Tarsal bones	Lifts foot and flexes ankle
Tibialis posticus (tibialis posterior)	Posterior to tibiofibula	Tibiofibula	Tarsal bones	Flexes foot
Extensor cruris	Anterior to and partly beneath tibialis anticus longus	Femur	Tibiofibula	Extends shank

Muscles of the Shoulder and Forelimb

At least 10 principal muscles of the shoulder and forelimb aid in control movements of the **humerus,** the large bone of the forelimb. On the dorsal side of the shoulder find the large **dorsalis scapulae** that originates on the **scapula** and inserts on the upper end of the **humerus** (figure 18.4). It serves to extend the **forelimb.** Just caudal (posterior) to the dorsalis is the **latissimus dorsi** that originates on the fascia of the back caudal to the dorsalis and inserts on the upper end of the humerus. The latissimus dorsi retracts the forelimb. Several smaller muscles including the **anconeus,** the **biceps brachii** (not shown), and the **triceps brachii** control the movements of the radio-ulna and the lower portion of the forearm.

On the ventral surface of the shoulder region, the two principal muscles controlling movements of the humerus and the upper portion of the forelimb are the **pectoralis,** a large muscle that covers most of the chest, and the **deltoideus,** found on the front of the shoulder. The pectoralis is divided into several parts that originate on the abdominal muscles and sternum and insert on the upper end of the humerus. Contractions of anterior portions of these muscles extend the forelimb while contractions of the posterior portions of the same muscles retract it. Several smaller muscles in the lower forearm serve as extensors and flexors to control movements of the "hand" and digits.

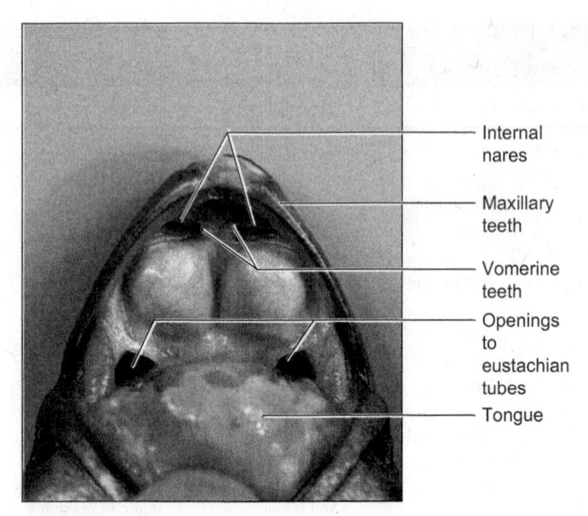

FIGURE 18.6 Frog, oral cavity, ventral view.
Courtesy of Carolina Biological Supply Company, Burlington, NC.

Oral Cavity

- Obtain a preserved or anesthetized frog from your instructor to study the anatomy of the **oral cavity.**

Frogs can be anesthetized by immersion in a solution of tricaine (ethyl m-aminobenzoate methanesulfonate) or urethane. You and your instructor should make sure that your frog is properly anesthetized before you proceed.

Place the frog on its back and make a small cut with your scissors at each corner of the mouth so the jaws can be opened widely. Rinse out the oral cavity with cold water if there is an accumulation of mucus. Consult figure 18.6 and locate the following structures in the oral cavity of your specimen: the **maxillary teeth** on the margin of the upper jaw, the **vomerine teeth** on the palate (these teeth are used for holding food rather than for chewing since the food is swallowed whole); the **internal nares** (probe through them from the **external nares**); and the openings into the **eustachian tubes.** Observe also the **tongue** and its attachment.

- ***What advantage does this peculiar attachment of the tongue have for the frog?***

Posterior to the oral cavity (behind the tongue) is the **pharynx.** At the rear of the pharynx is the opening into the esophagus; on its ventral surface find the **glottis**—a slitlike opening into the **larynx.** A pair of short **bronchi** connect the larynx to the two lungs.

Males of many species of frog have openings into two **vocal sacs** located at the rear of the mouth just anterior to the openings to the eustachian tubes. Inflation of the vocal sacs serves to amplify the croaking sounds. Vocal sacs are absent in female frogs.

Internal Anatomy

- Before continuing with your study of internal anatomy of the frog, you should review the general instructions for dissection provided in the front matter starting on page xxi.

If you are dissecting an **anesthetized** specimen, take care to minimize the cutting of blood vessels to prevent the blood from obscuring your view of the internal organs. Keep your specimen covered with cold water during dissection. Fasten the frog, ventral side up, to the bottom of the dissection pan with pins through the tip of the jaw and all the limbs.

If you have a **preserved** specimen for dissection, it may be saturated with the preservation fluid or it may simply be moist and come in a plastic bag. In either case, the tissues should be moist and pliable.

- Study the general location of internal organs of the frog in figures 18.7 and 18.8 to make certain that you understand the approximate location of the internal organs before you begin your dissection. Proceed carefully in your dissection to avoid unnecessary damage to body parts that you may need to study later.

Lift the skin from the body with your forceps and make a **longitudinal cut** slightly to one side of the midventral line and forward from the pelvis to the tip of the lower jaw.

Observe the subcutaneous lymph spaces and the attachment of the skin to the body as you free the skin from the ventral body surface. Make short transverse cuts in the skin at the anterior and posterior ends of the trunk and pin back the flaps of skin on both sides.

Consult figure 18.5 and locate the following three large muscles before continuing with your dissection: the **pectoralis,** the **rectus abdominis,** and the **external oblique.** Note also the whitish **linea alba** (the midventral connective tissue joining the lateral muscles) along the midventral line. The ventral abdominal vein lies beneath the muscular body wall along the midventral line.

Now lift the muscles of the abdomen with your forceps and make a **longitudinal incision** through the body wall with your scissors a little to one side of the linea alba. Cut through the bony **sternum,** which supports the forelimbs (see figure 18.3), and up to the posterior end of the lower jaw, taking care not to cut the ventral abdominal vein. Avoid damage to the internal organs within the coelom. Trace the ventral abdominal vein to the liver and make two short transverse incisions through the body wall just in front of the liver so the muscles can be pinned back to expose the internal organs. Remove a 10–12 mm section from the middle portion of the sternum and pin the forelimbs back to provide access to the internal organs in this region as shown in figure 18.7.

Survey of Internal Organs

Now that you have completed the preliminary phase of your dissection, make a brief survey of the internal anatomy, referring to the figures cited in the following description.

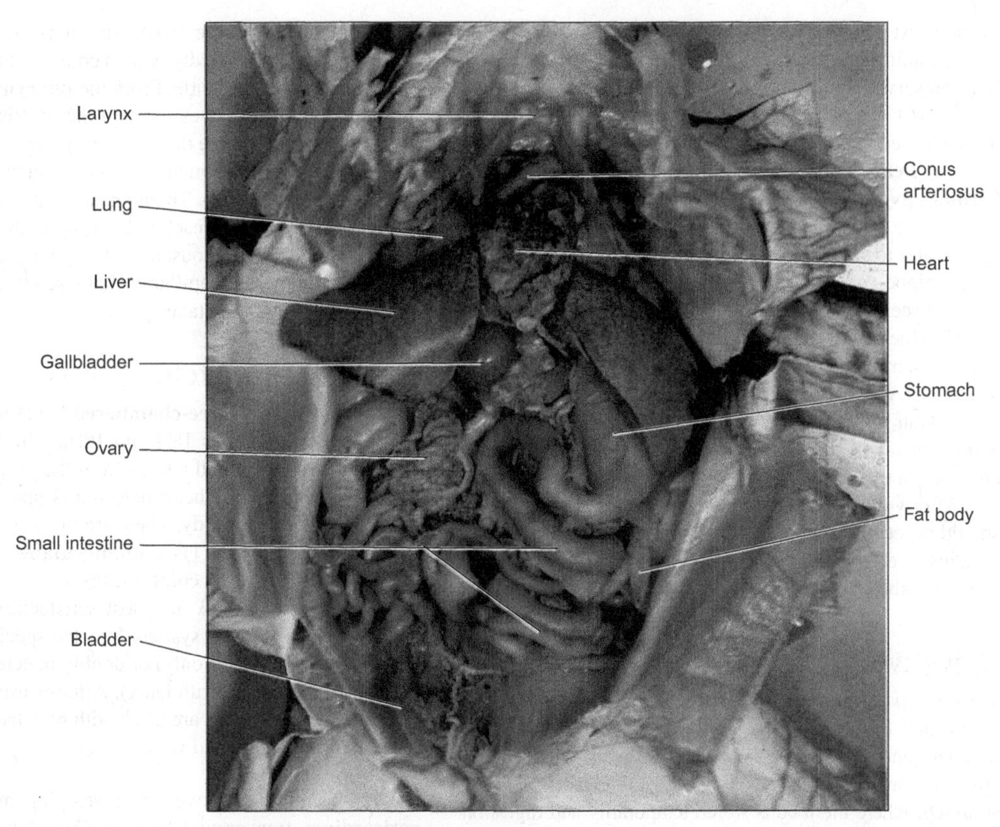

FIGURE 18.7 Frog, internal organs, female, ventral view.
Photo courtesy of Betty Black.

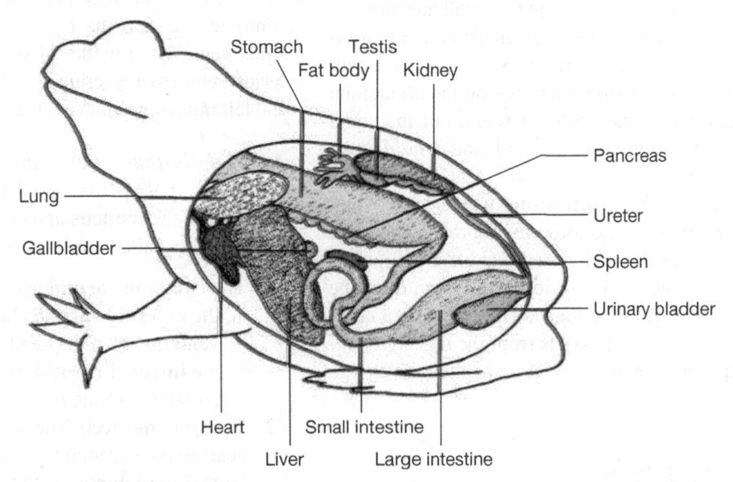

FIGURE 18.8 Frog, internal organs, lateral view.

Note the position, size, shape, color, and texture of each organ. Consider also the relationship of each organ to other organs and attempt to determine the function(s) of each organ as you work. If you have a female specimen, and the ovaries and oviducts are filled with eggs and greatly enlarged, consult figure 18.14 and carefully remove the ovary and oviduct from one side of the body to facilitate your study of the other parts.

The location of the principal internal organs is illustrated in figures 18.7 and 18.8. The large, three-lobed **liver,** located just posterior to the **pectoral girdle,** is one of the most prominent internal organs. The liver is dark in color in both preserved and living specimens. Adjacent to the liver on the right side of the frog (viewed from the ventral surface as in a dissecting pan) is the curved, tubular **stomach.**

The **heart** is enclosed in a membranous sac, the **pericardial cavity,** which lies anterior to the liver and partly beneath the pectoral girdle. You will open the pericardial cavity later to study the heart and its relationship with its attached blood vessels. The two **lungs** are located on the sides of the visceral cavity posterior and lateral to the heart and dorsal to the liver. In a living or freshly killed frog, the lungs will be semitransparent, pink, and inflated. In preserved specimens, they will often be deflated.

If your specimen is a mature female, much of the visceral cavity will be filled with many black and white **eggs** enclosed in the large, membranous **ovaries.**

After you have located these major interior organs, you should proceed with a more detailed study of specific organ systems as directed by your laboratory instructor and as described in the following sections.

Digestive System

Trace the pathway of the digestive system starting with the mouth and the **oral cavity.** Behind the **tongue,** locate the **pharynx** and the opening into the **esophagus.** The esophagus is a short, cylindrical tube that passes food material to the **stomach,** where the food is stored temporarily and digestion begins. The muscular stomach also provides a thorough mixing of the food mass by its contractions. At the posterior end of the stomach, find the **pyloric valve,** a sphincter that controls the release of the stomach contents into the small intestine.

The anterior segment of the small intestine is the **duodenum,** which receives secretions from the liver and pancreas through the common bile duct. Behind the duodenum is the convoluted **ileum,** the posterior section of the small intestine where digestion is completed and where most absorption of nutrients into the bloodstream occurs. The ileum empties into the **large intestine,** where most water as well as certain vitamins and ions are absorbed. The large intestine is also where the undigested residue is temporarily stored as fecal material. Posteriorly, the large intestine narrows and empties into the **cloaca.** The cloaca is a common chamber that collects materials from the digestive, excretory, and reproductive systems prior to their discharge through the cloacal opening.

Respiratory System

Adult frogs accomplish the necessary exchange of gases by using three different parts of their body: the **skin,** the **lungs,** and the **lining** of the **oral cavity** and **pharynx.** Locate the following parts of the respiratory system of the frog, which supplement the gas exchange accomplished through the skin.

Find the **external nares** located on the anterior part of the head (see figure 18.1) and the **internal nares** on the roof of the oral cavity (figure 18.6). Air enters via the nares and passes into the **nasal cavity,** which connects the internal and external nares on each side. From the **pharynx,** posterior to the oral cavity, air passes through the **glottis** into the **larynx,** a cartilaginous tube that divides posteriorly to form two **bronchial tubes** or **bronchi.** The two bronchial tubes connect with the two **lungs.** The lungs are located dorsal to the heart and liver, one on each side of the body (figure 18.7). Inside each lung are thousands of tiny **air sacs,** each surrounded by numerous capillaries. In these sacs, the exchange of gases with the blood takes place.

Circulatory System

Amphibians have a **three-chambered heart** with two **atria** and a **ventricle** (figures 18.7 and 18.11). In the adult frog, part of the venous blood moves from the heart to the lungs and then returns to the heart before it is pumped out to the various organs of the body. There are two major divisions of the circulatory system: (1) the **pulmocutaneous circulation** and (2) the **systemic circulation.** Preserved frogs that have been injected with latex are most satisfactory for detailed study of the circulatory system. Injected specimens may be singly injected (arteries only) or doubly injected (both arteries and veins injected with latex). Arteries usually are filled with red latex and veins are filled with blue latex, except for the pulmonary artery and vein.

Heart Carefully remove the enveloping membrane, the **pericardium,** from around the heart. Consult figures 18.7 and 18.9, and identify the thin-walled right and left **atria,** the thick-walled **ventricle,** and the **conus arteriosus,** which divides into two branches, the right and left **truncus arteriosus.** Lift the ventricle and find the dark-colored, thin-walled, triangular **sinus venosus** on its dorsal side (see figure 18.12), the **posterior vena cava** entering it at the posterior end, and the right and left **anterior vena cavae** entering it at the anterior end.

Arterial System Study the arterial system with the aid of figures 18.9, 18.10, and 18.11. Note the left and right branches of the **truncus arteriosus,** each giving rise to three large arteries called aortic arches.

1. The **common carotid artery.** This divides into
   a. the **external carotid** (lingual artery) leading to the ventral part of the head and to the tongue;
   b. the **internal carotid** to the dorsal part of the head and brain. (Note the carotid body.)
2. The **systemic arch.** The two systemic arches encircle the heart to pass around the pharynx and unite to form the large **dorsal aorta.** Note the arteries leading from each systemic arch:
   a. the **short occipitovertebral artery** dividing into the **occipital artery** leading to the skull and the **vertebral artery** leading to the vertebral column;
   b. the **subclavian artery,** giving off branches to the shoulder region and extending into the arm where it becomes the **brachial artery;**

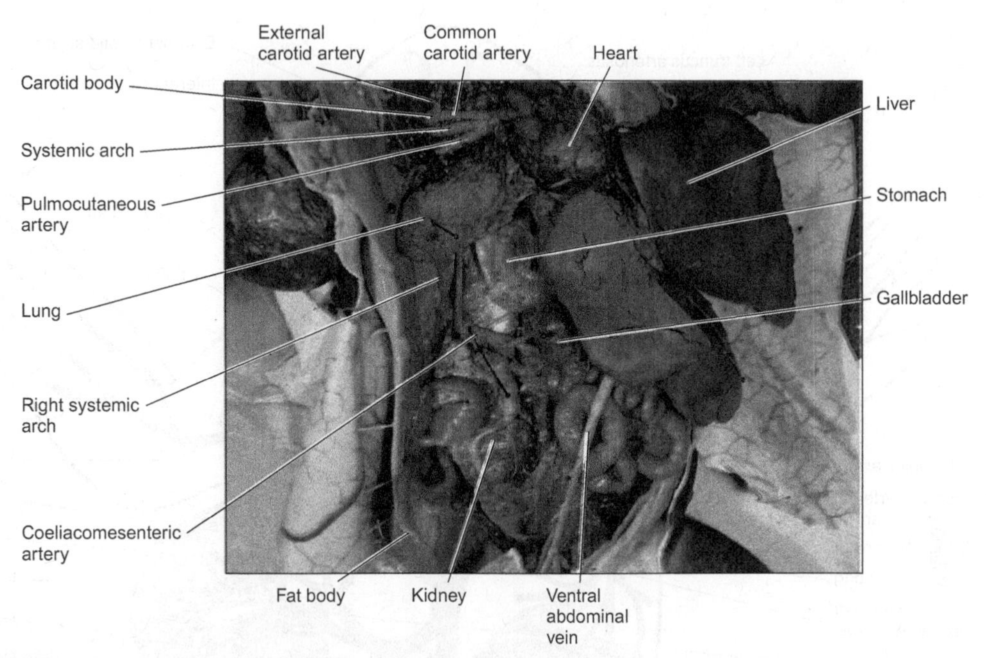

FIGURE 18.9 Frog, dissected, heart and major arteries, ventral view.
Photo courtesy of Betty Black.

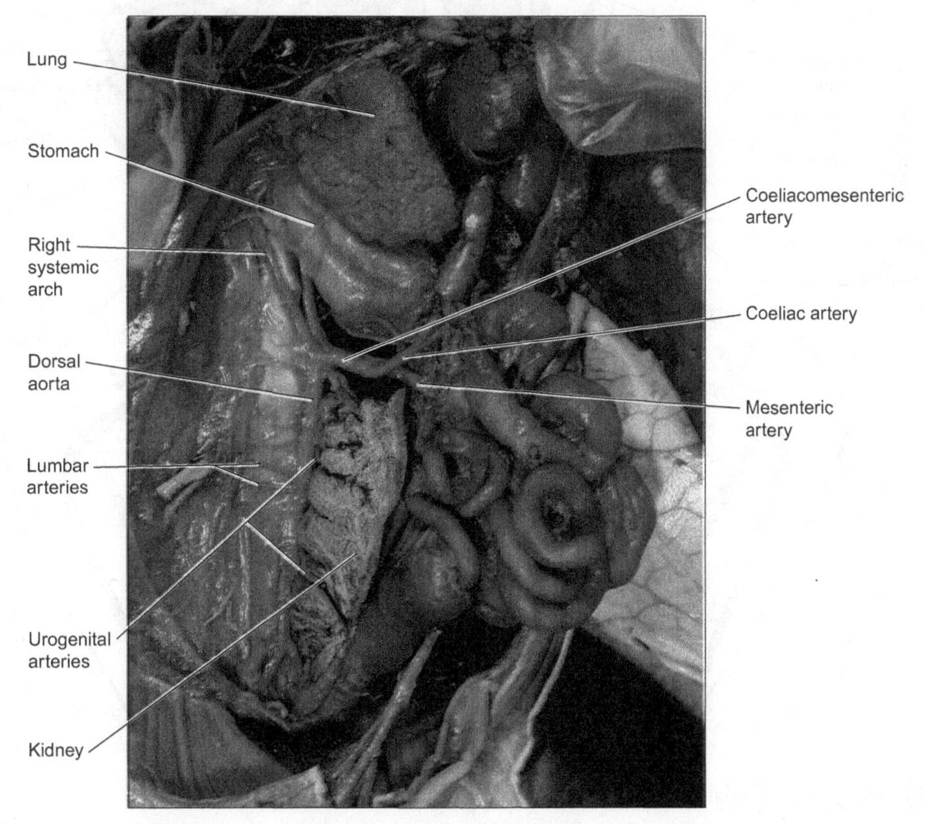

FIGURE 18.10 Frog, dissected, dorsal aorta and major abdominal arteries.
Photo courtesy of Betty Black.

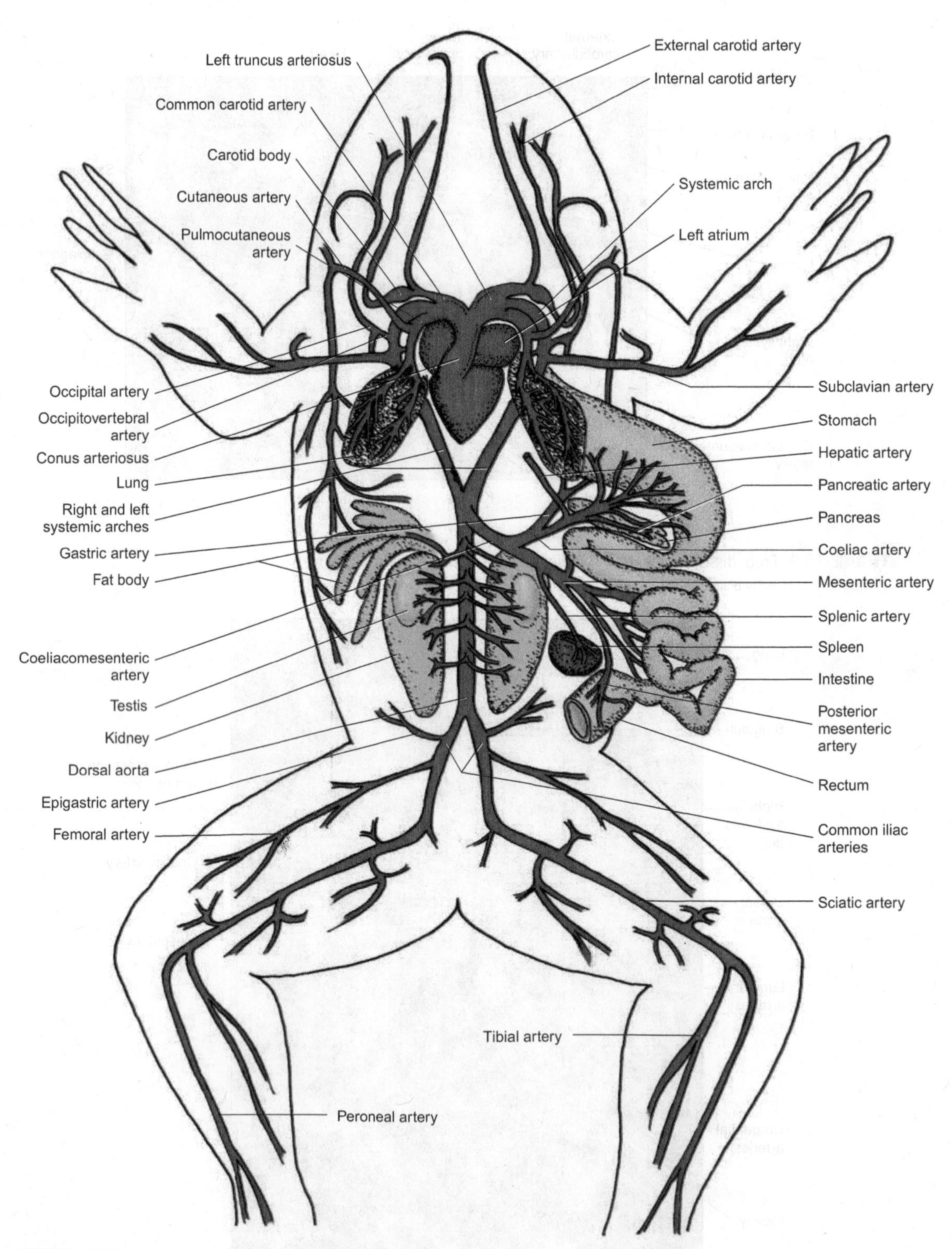

FIGURE 18.11 Frog, arterial system, ventral view.

c. the **dorsal aorta;**
d. the **coeliacomesenteric,** which divides into the **coeliac** and the **mesenteric;**
e. the **right** and **left gastric arteries** to the stomach, the **pancreatic artery** to the pancreas, and the **hepatic artery** to the liver (these latter arteries are all branches of the coeliac);
f. the **mesenteric artery** to the intestine, with branches to the spleen (the **splenic artery**) and to the rectum and large intestine **(posterior mesenteric artery);**
g. the **urogenital arteries** to the kidneys, gonads, and fat bodies (**renal artery** to kidneys and **genital artery** to gonads). (The urogenital arteries vary considerably in number and arrangement in different individuals. Sometimes they do not inject well because of their small size and may be difficult to find. ***How many are there in your specimen?***); the **lumbar arteries** to the dorsal body wall (several pairs of small arteries arising from the dorsal surface of the aorta—not shown in the figure, also small and often hard to find);
h. the **common iliac arteries,** divisions of the dorsal aorta;
i. the **epigastric artery** to the bladder and body wall of that region;
j. the **femoral artery** in the thigh;
k. the **sciatic artery,** a continuation of the external iliac artery furnishing branches to most of the muscles of the leg; and
l. the **peroneal** and **tibial arteries,** divisions of the sciatic to lower parts of the leg as shown in figure 18.11.

3. The **pulmocutaneous artery.** This divides into
   a. the **pulmonary artery** going to the lung and
   b. the **cutaneous artery** going to the skin. Note that this artery and its branches carry **deoxygenated** blood.

Venous System Veins are blood vessels that carry blood **toward** the heart. Study the venous system of your specimen with the aid of figure 18.12 and locate the following veins.

1. The two (right and left) **anterior venae cavae** into which enter: (a) the **external jugular vein** with its branches, the **lingual vein** from the tongue and floor of the mouth and **maxillary vein** or mandibular from the jaw; (b) the **innominate vein** into which flows the **internal jugular vein** from the deeper parts of the head and the **subscapular vein** from the shoulder; and (c) the **subclavian vein** formed by the union of the **musculocutaneous vein** from the muscles and skin of the side and back, and the **brachial vein** from the forelimb.
2. The **posterior vena cava** into which enter: (a) the **hepatic veins** from the liver; (b) the **renal veins** from the kidneys; and (c) the **genital veins** from the gonads.
3. The **hepatic portal system** consists of: (a) the **abdominal vein,** which enters the liver and is formed by the union of the two **pelvic veins;** (b) the **hepatic portal vein,** which carries blood from the stomach **(gastric vein),** the intestine (**mesenteric vein**), and the spleen (**splenic vein**).
4. The **renal portal system.** The **renal portal veins** carry the blood to the kidneys. They receive the blood from (a) the **dorsolumbar vein;** (b) a branch of the **femoral vein** from the hindlimb; and (c) the **sciatic vein** from the thigh. Note that the blood from the femoral vein may go to the kidney by way of the renal portal vein or to the liver by way of the pelvic and abdominal veins.
5. The **pulmonary veins.** These veins carry **oxygenated** blood from the lungs and unite to enter the left auricle.

The Urogenital System

The excretory and reproductive organs are closely associated, and together comprise the **urogenital system.** If your specimen is a female with large ovaries concealing the other visceral organs, consult figures 18.13 and 18.14, and carefully remove one of the ovaries. The excretory structures are similar in the two sexes of the frog. With the help of figure 18.13, again note the two **kidneys,** each with the **adrenal gland** on its ventral surface; the **ureter** leading from the posterior border of the kidney to the cloaca; the **urinary bladder,** which empties into the cloaca on the ventral side; and the **fat bodies.**

The kidneys are the major excretory organs of the frog. They are responsible for the removal of most of the wastes from the body, although some wastes are also lost through the skin. The kidneys remove most of the nitrogenous wastes from the blood, which are excreted in soluble form as **urea** and **ammonia.** These important organs also play a major role in **homeostasis,** the maintenance of a constant internal environment by removing excess ions and other substances from the blood and by conserving other substances of limited availability.

Specimens should be distributed in the class so that both male and female specimens are available at each table. After you have completed the study of the urogenital system on your own specimen, find a specimen of the opposite sex and make a comparative study so that you will be familiar with the organization of the urogenital system in both the male and female.

In a female specimen (figures 18.12 and 18.14), locate on one side the **ovary** attached dorsally and suspended into the coelom by a **mesentery;** the coiled **oviduct** with a funnel-shaped opening, the **ostium,** at its anterior end; and a **uterus,** an enlargement near the cloaca. Cut open the cloaca and use your probe to find the openings of the oviducts, the ureters, and the urinary bladder.

Near the time of their maturation, the eggs are released from the ovaries into the coelom and pass through the ostia

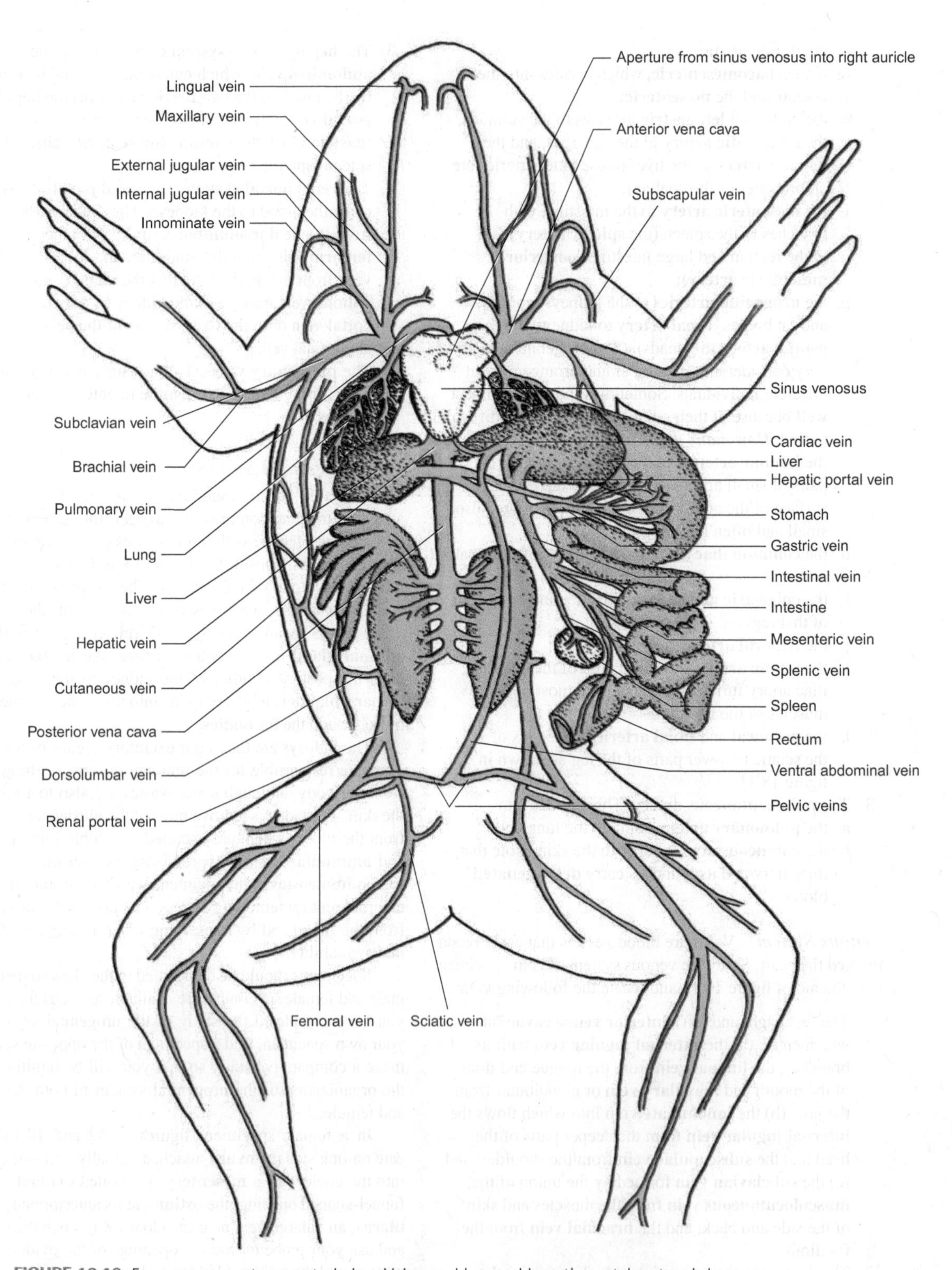

FIGURE 18.12 Frog, venous system, ventral view. Veins are blue, and hepatic portal system is brown.

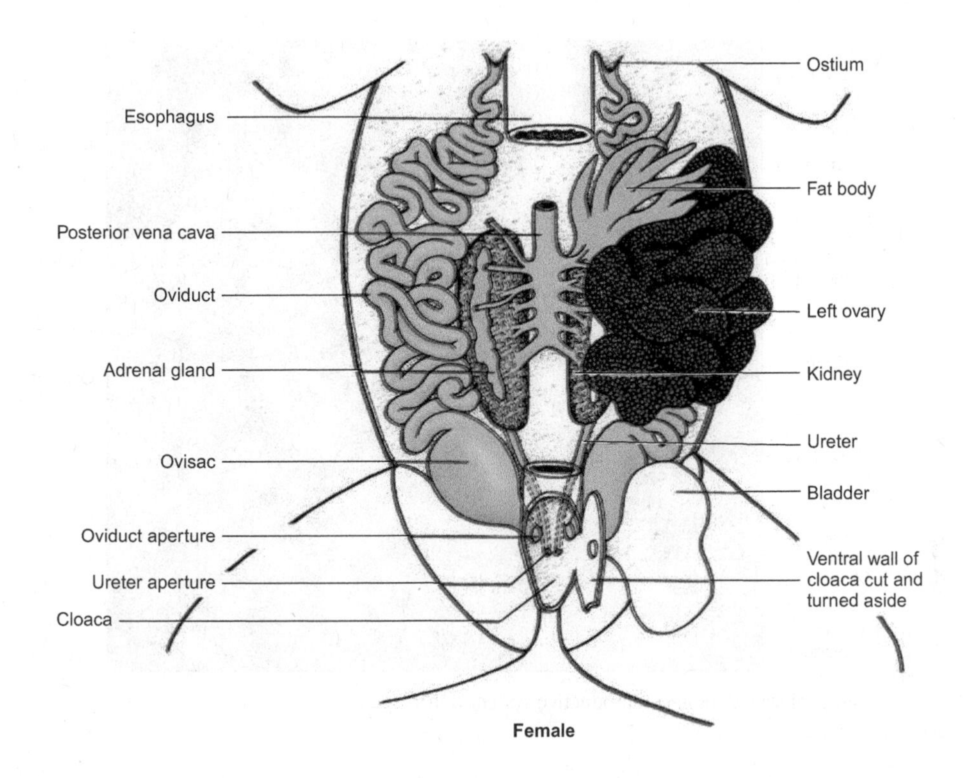

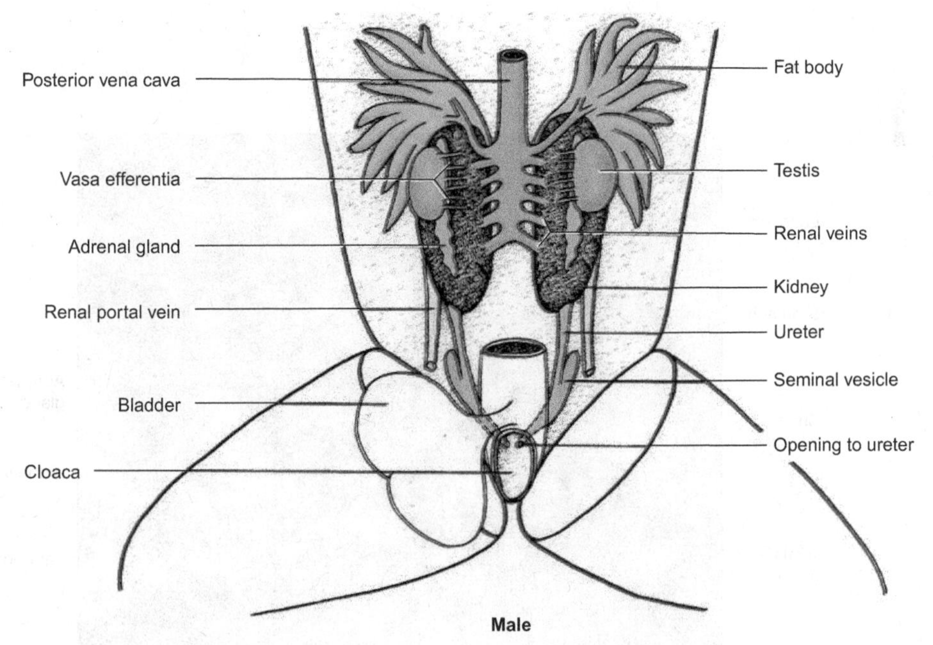

FIGURE 18.13 Frog, urogenital system, male and female, ventral view. Systemic veins are blue and hepatic portal system veins are yellow.

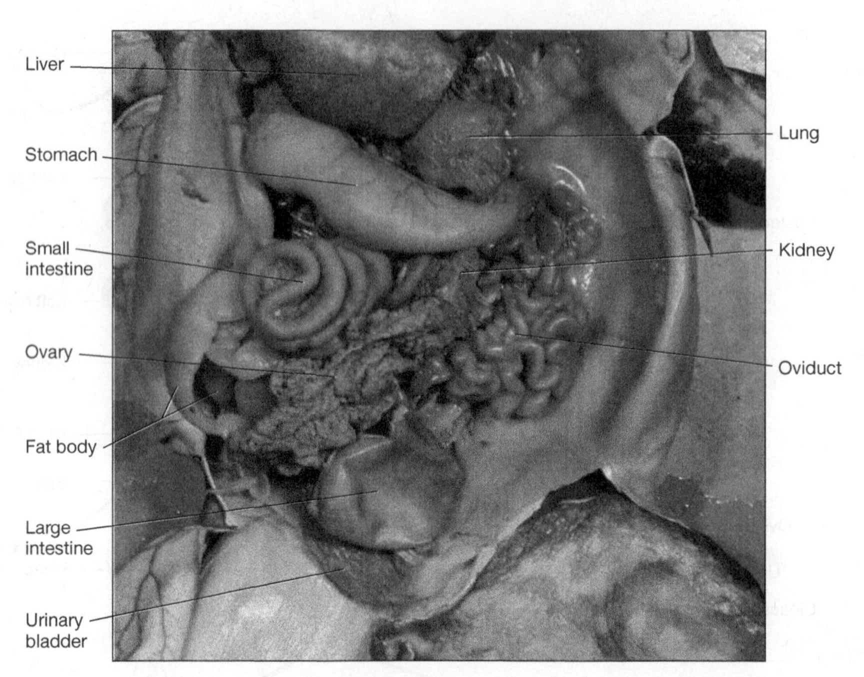

FIGURE 18.14 Frog, dissected, female reproductive system, ventral view.
Photo courtesy of Betty Black.

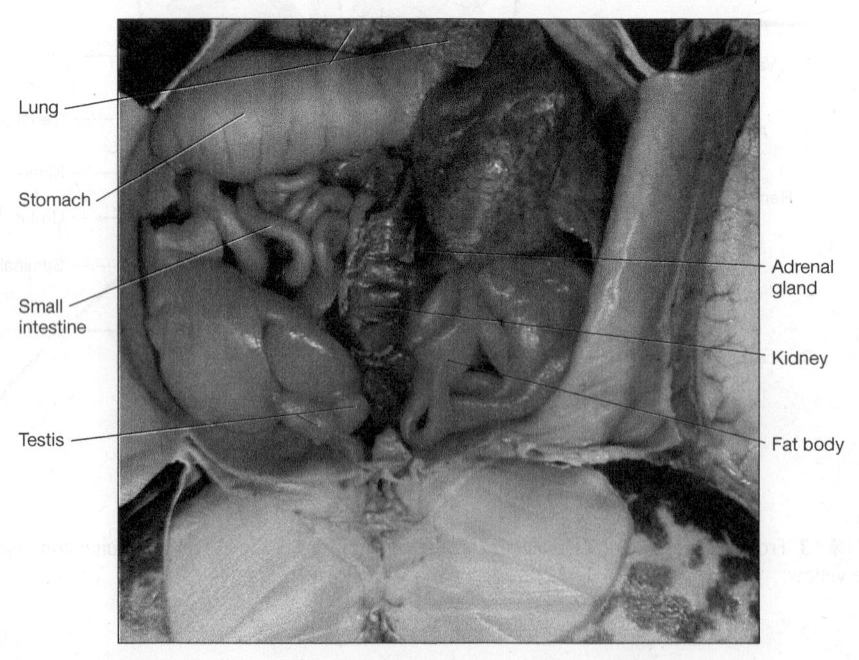

FIGURE 18.15 Frog, dissected, male reproductive system, ventral view.
Photo courtesy of Betty Black.

into the oviducts. As the eggs pass down the oviduct, they are covered with several layers of jellylike material secreted by glands in the walls of the oviducts. The eggs collect in the uteri before they are discharged through the cloaca and into the water. Fertilization is external: the male mounts the female and discharges sperm on the eggs as they are discharged. This mating posture is called **amplexus.**

In a male specimen (figures 18.13 and 18.15), locate the two **testes,** each suspended from the dorsal abdominal wall by a mesentery, and the several **vasa efferentia,** the small ducts that carry sperm from the testes to the kidneys. The sperm are carried from the kidneys to the cloaca via the **ureters,** which serve as genital ducts in the male. Male leopard frogs also frequently have vestigial oviducts (mesonephric ducts) alongside the kidneys.

Nervous System

The nervous system of vertebrates consists of three main parts: (1) the **central nervous system,** which includes the brain and the spinal cord; (2) the **peripheral nervous system,** which includes the nerves extending from the central nervous system; and (3) the **autonomic nervous system,** a specialized portion of the peripheral nervous system that regulates the glands of the body and the visceral organs.

- Successful study of the nervous system of the frog requires careful dissection. Do not rush the dissection, because you may damage important structures necessary for your study. Carefully remove the skin from the dorsal surface of the head between the eyes and along the vertebral column. Clear away the muscles and connective tissue underlying the skin to expose the skull and vertebral column. With your scalpel, carefully shave thin sections of bone from the skull until you expose the brain. Then carefully pick away additional small pieces of bone with your forceps until the brain is completely exposed. Continue the same procedure to expose the vertebral column and the spinal cord, but leave the exposed nervous system in place.

The Brain Consult figures 18.16, 18.17 and 18.18 and identify the five main regions of the brain: the **telencephalon,** the **diencephalon,** the **mesencephalon,** the **metencephalon,** and the **myelencephalon.** The anteriormost telencephalon bears four distinct lobes: two **olfactory lobes** and, posterior to them, two **cerebral hemispheres.** Extending anteriorly from the olfactory lobes are the **olfactory nerves,** which carry impulses from the olfactory epithelium to the brain.

Posterior to the telencephalon is the diencephalon, a diamond-shaped depressed area directly behind and somewhat between the posterior portion of the cerebral hemispheres. The **pineal gland** (epiphysis) is a small, inconspicuous body attached to the dorsal wall of the diencephalon,

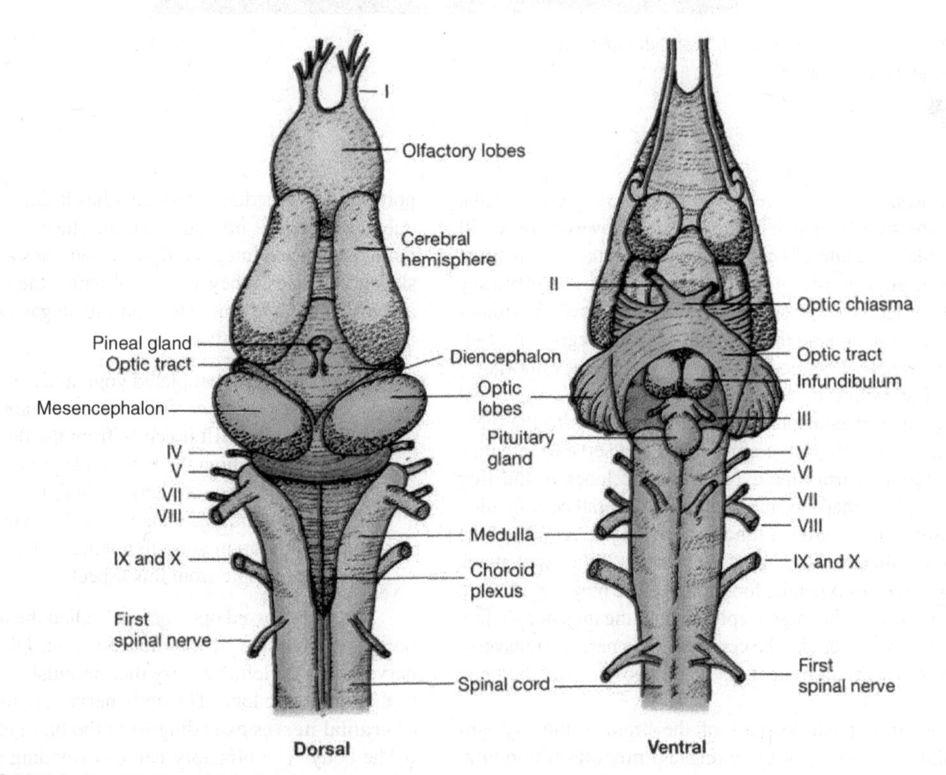

FIGURE 18.16 Frog brain and cranial nerves.

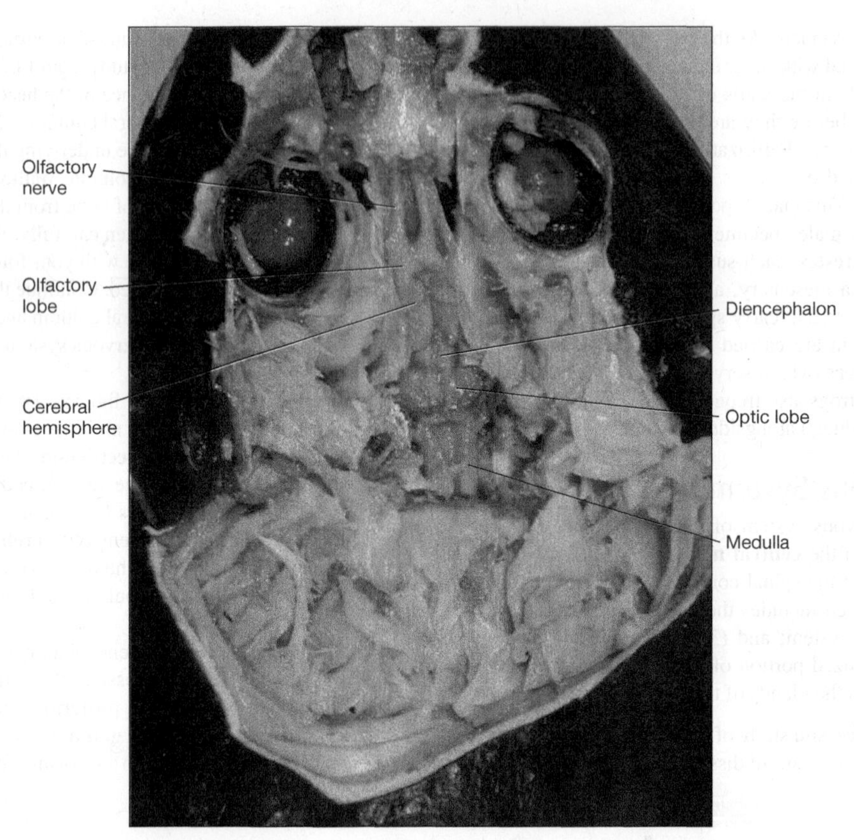

FIGURE 18.17 Frog, dissected, brain, dorsal view.
Photo courtesy of Betty Black.

which probably will be removed in your dissection of the skull. The stalk by which it was attached, however, may still be visible. The lateral walls of the diencephalon constitute the **thalamus.** Below the diencephalon lies the **pituitary gland,** which you will study later. The principal function of the telencephalon and diencephalon is the integration of olfactory signals. Chemical perception is of great importance to the frog, both for feeding and defense.

The mesencephalon, immediately posterior to the diencephalon, bears two large **optic lobes** that serve to integrate nerve impulses from the eyes. The optic lobes in the frog also provide perhaps the most important overall coordination of sensory information, a function carried out mainly by the cerebral hemispheres in higher vertebrates. Frogs will attack an insect seen as potential food long before they can smell it.

Posterior to the mesencephalon find the metencephalon, which is represented by the **cerebellum,** a narrow transverse portion of the brain lying immediately posterior to the optic lobes.

The most posterior part of the brain is the myelencephalon, consisting of the elongated **medulla oblongata,** which tapers gradually into the **spinal cord.** The anterior portion of the medulla oblongata has a thin roof that is frequently removed during dissection. The cerebellum and medulla receive and integrate signals from the ears and from the skeletal muscles. They serve to monitor the general state of activity in the body and also coordinate gas exchange, feeding, and muscular activity.

- When you have completed your study of the dorsal side of the brain, ***carefully*** detach the anterior portion of the brain and lift it gently from the floor of the skull. Continue detaching the brain and spinal cord from its ventral connections and gradually free the entire nervous system. Place the isolated nervous system ventral-side-up in your dissecting pan and study the structures visible from this aspect.

Note the crossed optic nerves, called the **optic chiasma,** on the ventral side of the diencephalon. Observe how the nerves from the left eye carry their impulses across the brain to the right optic lobe. The optic nerves are the **second pair** of **cranial nerves** extending from the brain to various parts of the body. The olfactory nerves extending anterior to the olfactory lobes are the **first pair** of **cranial nerves.**

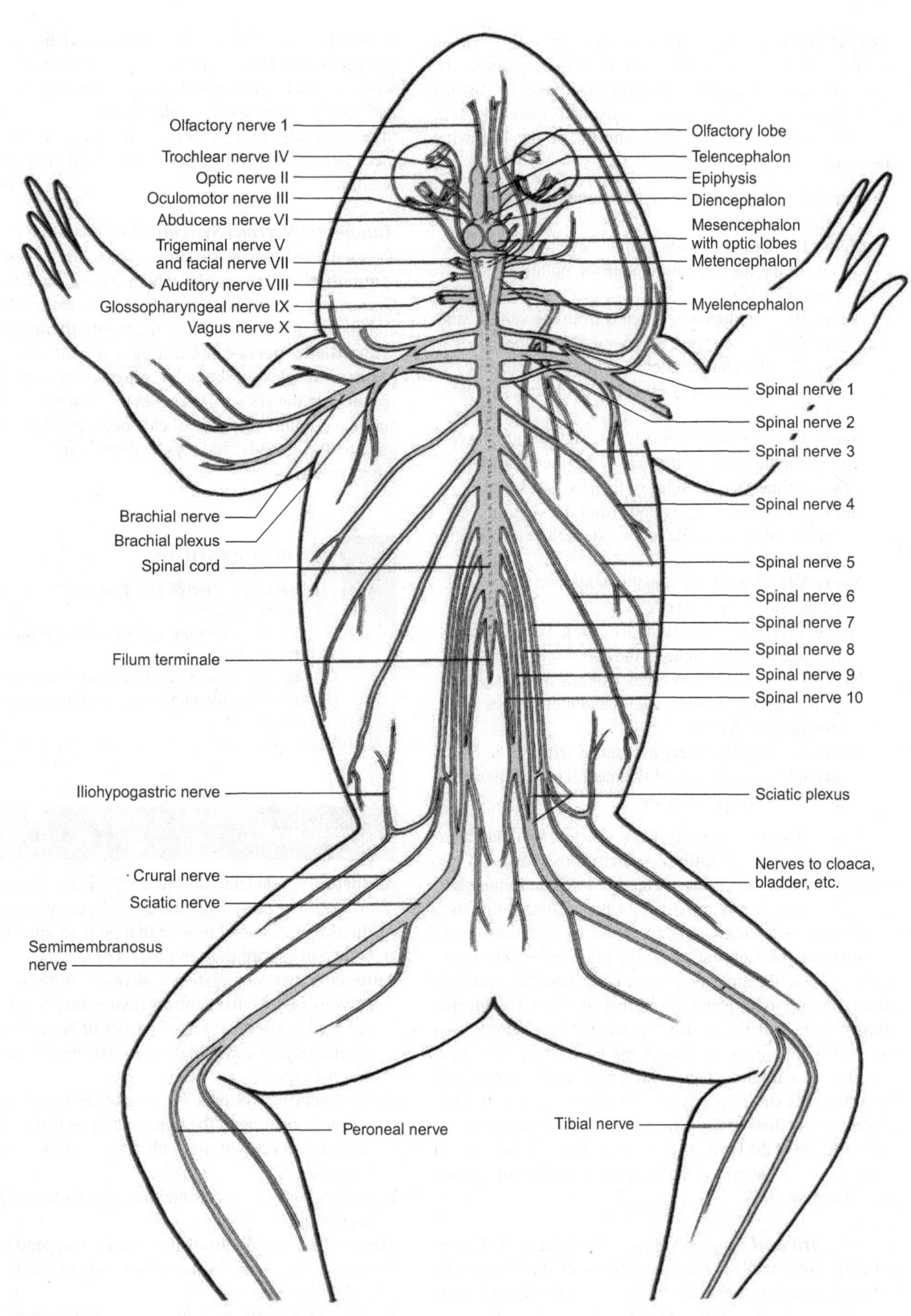

FIGURE 18.18 Frog, nervous system, dorsal view.

Cranial Nerves There are 10 pairs of cranial nerves in the frog, extending from the brain to various parts of the body. (Fishes and amphibians generally have 10 distinct pairs of cranial nerves; reptiles, birds, and mammals have 12 pairs.) Refer to figure 18.18 and locate the following cranial nerves.

Nerve I Olfactory, extends from the naris to the olfactory lobe.
Nerve II Optic, extends from the eye to the ventral side of the diencephalon. Note the optic chiasma (figure 18.16).
Nerve III Oculomotor, extends from the ventral side of the mesencephalon to the muscles of the eye.
Nerve IV Trochlear, extends from the dorsal side of the mesencephalon to the superior oblique muscle of the eye.
Nerve V Trigeminal, closely associated with Nerve VII, the facial, from the side of the medulla to the skin of the face, muscles of the jaw, and tongue.
Nerve VI Abducens, extends from the ventral surface of the medulla to the external rectus muscle of the eye.
Nerve VII Facial, previously mentioned in connection with Nerve V.
Nerve VIII Auditory (statoacoustic), extends from the side of the medulla to the ear.
Nerve IX Glossopharyngeal, extends from the side of the medulla to the muscles and membranes of the tongue and pharynx.
Nerve X Vagus, closely associated with Nerve IX, extends from the side of the medulla to the heart, lung, and digestive organs.

Next, observe the ventral outgrowth from the diencephalon, the **infundibulum,** which corresponds to the posterior lobe of the pituitary gland in birds and mammals. Located posterior and proximal to the infundibulum is a dorsal outgrowth originating from the roof of the mouth, the **anterior lobe of the pituitary** (also called the hypophysis). Together these two lobes constitute the **pituitary gland,** an important gland of internal secretion **(endocrine gland).** Note that the adult pituitary gland develops from two different sources—a ventral outgrowth from the diencephalon (the infundibulum) and a dorsal outgrowth from the oral cavity (the hypophysis). The hypophysis frequently adheres to the floor of the skull and may be detached during your removal of the brain. If you do not find it on the isolated nervous system, search on the floor of the skull in the appropriate location.

Spinal Cord and Spinal Nerves Observe now the spinal cord. Note that the terminal portion of the spinal cord in the adult frog is threadlike and is called the **filum terminale.** Ten pairs of **spinal nerves** are attached to the spinal cord (figure 18.18). Each nerve has two roots: the **dorsal root** and the **ventral root.** Be careful not to disturb the sympathetic nerves while tracing the spinal nerves with which they unite. Note the large size of the **second spinal nerve.** Find the **brachial plexus** and the **sciatic plexus** where several of the spinal nerves leading to the forelimbs and hindlimbs extend from the spinal cord separately, then unite, and later rebranch. Trace the large **sciatic nerve** to the sciatic plexus and follow its principal branches into the hindlimb.

Autonomic Nervous System Examine figure 18.19, which shows the autonomic system in solid black. Note the two **sympathetic nerve trunks,** which originate anteriorly as the Gasserian ganglia and extend backward along the systemic arches and dorsal aorta. Find the **sympathetic ganglia;** the **sympathetic nerves** connecting with the spinal nerves; the **cardiac plexus;** the **solar plexus;** the large peripheral **splanchnic nerves** with branches to the stomach, intestines, and other organs; and the delicate **peripheral nerves,** which pass to the gonads, kidneys, spleen, adrenal glands, and other organs.

Demonstrations

1. Bullfrogs—injected and dissected to show the various systems
2. Various vertebrate skeletons for comparison with the bullfrog
3. Beating of the frog's heart and circulation of the blood through capillaries of the web of the frog's foot

Key Terms

Abductor a skeletal muscle that pulls an appendage or body part away from the central axis of the body.
Adductor a skeletal muscle that pulls an appendage or body part toward the central axis of the body.
Autonomic nervous system division of the vertebrate nervous system that controls involuntary functions, such as breathing or heart rate. Consists of branches from several cranial and spinal nerves leading to various internal organs.
Central nervous system portion of the vertebrate nervous system consisting of the brain and the spinal cord.
Cranium portion of the skull that encloses the brain; the braincase.
Extensor a skeletal muscle that extends or straightens an appendage.
Flexor a skeletal muscle that bends an appendage.
Homeostasis maintenance of a steady internal physiological state.
Peripheral nervous system portion of the nervous system outside the central nervous system made up of the sensory and motor nerves connecting the organs and tissues with the central nervous system.

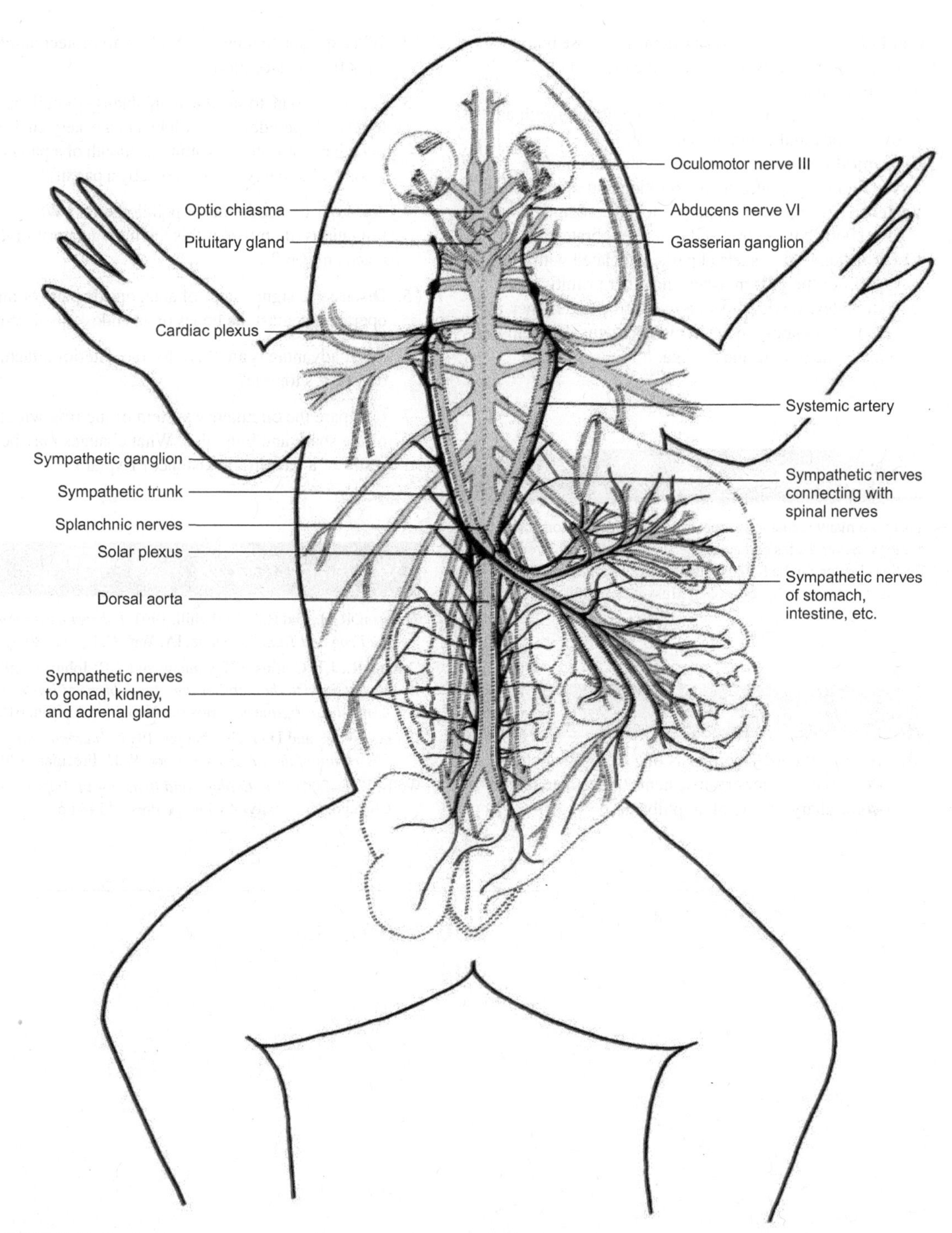

FIGURE 18.19 Frog, nervous system with autonomic system in black superimposed on other parts of the nervous system, ventral view.

Poikilothermal refers to an organism whose body temperature varies with that of the thermal environment.

Tadpole larval form (typical of amphibians) with an ovoid body and a thin muscular tail.

Urogenital system combined organ system of vertebrates specialized for excretion and reproduction.

Vertebra unit of the spinal (vertebral) column or "backbone" of the frog and other vertebrates.

Visceral skeleton skeletal parts associated with the support of the gills in fishes and other primitive vertebrates; vestiges remain in the frog and higher vertebrates mainly in the hyoid apparatus that helps support the tongue and trachea.

Internet Resources

There are many valuable Internet sites with information about zoology. Several sites containing pertinent zoological information for this chapter can be found on the McGraw-Hill Zoology web site at http://www.mhhe.com/zoology. Just click on this text's title.

Questions for Critical Thinking

1. Describe the metamorphosis of a frog. What changes occur in the major organ systems? Speculate on the evolutionary history of amphibians.
2. What special adaptations evolved to protect amphibian eggs from desiccation?
3. Is it appropriate to say that amphibians—for all their terrestrial aptitudes—are still tied to a watery environment (whether that water be within the mouth of a parent frog, a pond, or a foamy nest excreted by a parent)?
4. Are frogs homeotherms or poikilotherms? What limitations (if any) does this thermoregulatory strategy place on them?
5. Discuss the significance of antagonistic pairs of muscles operating against the bones of an endoskeletal system.
6. What advantages are there for the anterior attachment of a frog's tongue?
7. Compare the circulatory system of the frog with that of the shark and bony fish. What changes may be the result of a terrestrial existence? Explain.

Suggested Readings

Chaisson, R. B., and R.A. Underhill. 1993. *Laboratory Anatomy of the Frog and Toad.* Dubuque, IA: Wm. C. Brown. 80 pp

Conant, R., J.T. Collins, I.H. Conant, and T.R. Johnson. 1998. *A Field Guide to the Reptiles and Amphibians of Eastern and Central North America.* Boston: Houghton Mifflin. 615 pp

Walker, W. F., and D.G. Homberger. 1997. *Anatomy and Dissection of the Frog.* New York: W.H. Freeman. 120 pp

Wells, K. D. 2007. *The Ecology and Behavior of Amphibians.* Chicago: University of Chicago Press. 1148 pp

Notes and Sketches

Fetal Pig Anatomy

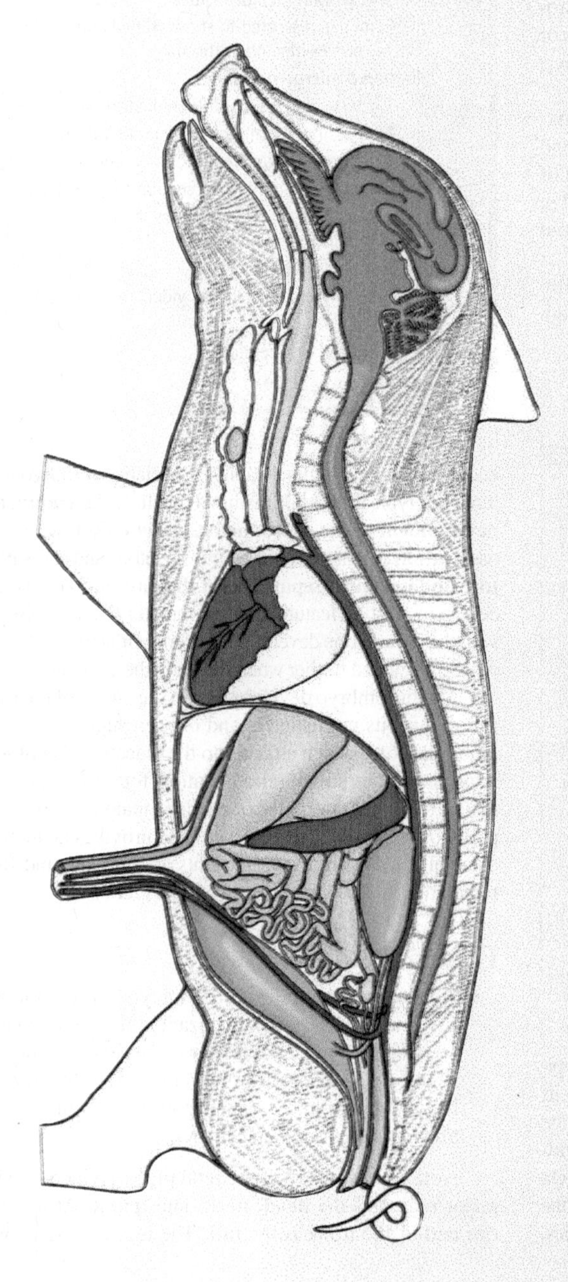

Objectives

After completing the laboratory work in this chapter, you should be able to perform the following tasks:

1. Locate and identify the principal external features of a fetal pig.
2. Identify the chief components of the axial and appendicular skeletons of the fetal pig.
3. Locate five major skeletal muscles of the fetal pig and explain their origins, insertions, and actions. Distinguish between extension, flexion, adduction, and abduction.
4. Demonstrate on a specimen the divisions of the coelom and the five principal mesenteries that support the abdominal organs.
5. Locate and identify the organs of the digestive system of the fetal pig.
6. Find and explain the function of the parts of the urogenital system of the fetal pig. Demonstrate the differences between male and female specimens.
7. Trace the main circulatory pathway on a specimen. Explain the main changes in circulation that occur at birth.
8. Point out the similarities and differences among the hearts of a fish, a frog, and a fetal pig or other mammal. Discuss the significance of the differences.
9. Locate the main arteries and veins on a dissected specimen.
10. Identify the five divisions of the mammalian brain on a pig or sheep brain, and point out the brain parts that make up these divisions.

The Fetal Pig: *Sus scrofa*

The fetal pig is often chosen for study as a representative mammal because it contains the organs and organ systems typical of most mammals and, thus, also those of humans. The principal distinguishing characteristics of the Class Mammalia include the following: a body surface covered with hair; an integument (skin) with several types of glands; a skull with two occipital condyles; seven cervical (neck) vertebrae; teeth borne on bony jaws; movable eyelids and fleshy external ears (pinnae); a four-chambered heart; a persistent left aorta; and a muscular diaphragm separating the thoracic and abdominal cavities. Mammals also are "warm-blooded," or **homeothermic,** which means they maintain a constant and elevated body temperature. The young develop within the uterus of the female, have a placental attachment for fetal nourishment, and are enveloped by special fetal membranes (amnion, chorion, and allantois). Milk to nourish the young after birth is produced by mammary glands. Other common representatives of this group include moles, bats, whales, mice, deer, monkeys, horses, cattle, and humans.

The fetal pig serves well as an example of mammalian anatomy because of its convenient size, availability, and relatively low cost. Fetal pigs are removed from the uteri of pregnant sows sold to meat packers. The gestation period in swine is 16 to 17 weeks; full-term fetal pigs measure about 30 cm (12 in) in length and weigh 1–1.5 kg (2–3 lb). At 14 weeks, pig embryos average about 23 cm (9 in) in length. The number of pigs in a litter averages about 8–10, although occasionally the litter may go as high as 18.

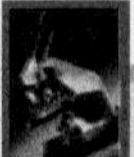

Materials List

Living specimens
- Mammalian spermatozoa (demonstration)

Preserved specimens
- Fetal pig
- Cat skeleton (demonstration)
- Mammalian placenta (demonstration)
- Mammalian lung (demonstration)
- Fetal pig, injected and dissected to illustrate circulatory system (demonstration)
- Mammalian heart, dissected (demonstration)
- Sheep brain (demonstration)
- Fetal pig, dissected to show cranial and spinal nerves (demonstration)

Prepared microscope slides
- Pig testis, cross section (demonstration)
- Mammalian spermatozoa (demonstration)
- Mammalian lung, cross section (demonstration)
- Mammalian ovary, cross section (demonstration)
- Mammalian spinal cord, cross section (demonstration)

Audiovisual materials
- Anatomy of the Fetal Pig video

Other
- Mounted cat skeleton
- String

Fact File

Pigs

- During the War of 1812, a New York pork packer named Uncle Sam Wilson shipped a boatload of several hundred barrels of pork to U.S. troops. Each barrel was stamped "U.S." on the docks, and it was quickly said that "U.S." stood for "Uncle Sam," whose large shipment seemed to be enough to feed the entire army. This is how "Uncle Sam" came to represent the U.S. government.
- Many anatomical and physiological features of the pig are nearly identical to humans. As a result, they are sometimes used as a source of valves for human heart surgery and insulin for treating diabetes.
- Pigs have an extremely sensitive sense of smell. In Europe, they are trained to locate truffles, an edible mushroom that grows underground. In World War I, pigs were used as "sniffers" to detect land mines on the battlefield.

Many of the anatomical features of the fetal pig are typical of mammals, although some special features related to its embryonic condition will also be apparent during your study. Among the typical features of mammals are the **four appendages, mammary glands, lungs, four-chambered heart, persistent left aortic arch, muscular diaphragm** that divides the coelomic cavity (into anterior pericardial and pleural cavities, and a posterior abdominal cavity). Specialized embryonic features not typical of adult mammals include the **rudimentary development of the reproductive system,** the **low degree of ossification** in many bones of the skeleton, and the **soft texture and indistinct separation of the muscles.** There are also certain interesting features of the circulatory system associated with the intrauterine development of most mammals. The latter will be discussed further when we study the circulatory system.

The pig embryo develops within the uterus of the mother and obtains its nourishment and oxygen supply through the umbilical cord, which attaches to the placenta. The **placenta** is an important structural adaptation found in most mammals. It is made up partly of uterine (maternal) tissues and partly of embryonic (fetal) tissues and provides an important channel for the supply of nutrients and oxygen and for the removal of metabolic wastes from the embryo.

External Anatomy

- Obtain a preserved fetal pig from your instructor and first study the general organization of the mammalian body and the principal features of its external anatomy (figure 19.1). Fresh hearts and brains can sometimes be obtained from the meat department of small, local grocery stores and butcher shops.

Note that the body of the fetal pig is divided into three major regions—the **head, neck,** and **trunk.** At the posterior end of the trunk is the **tail.** The trunk consists of two

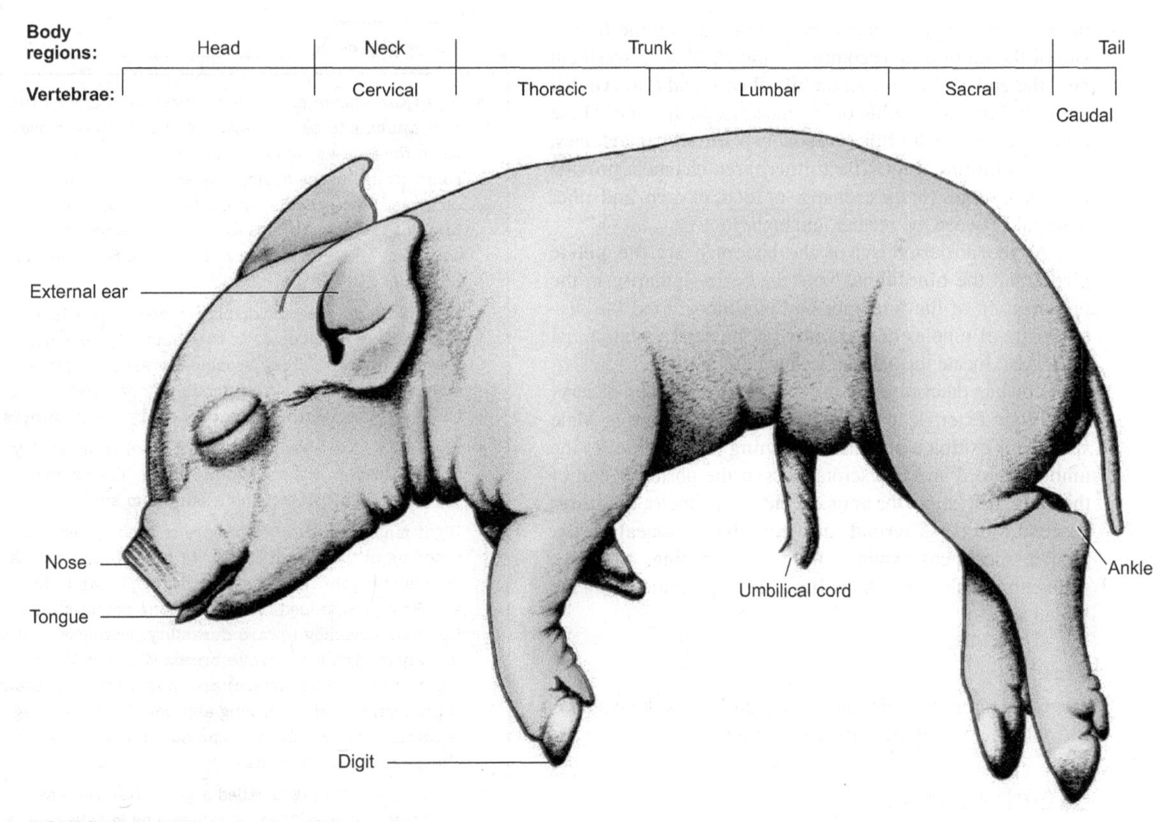

FIGURE 19.1 Fetal pig, external features.

principal subdivisions, the anterior **thorax** and the posterior **abdomen,** separated internally by the muscular diaphragm. Observe the two pairs of appendages, one pair of **forelimbs** and one pair of **hindlimbs.** Note that the limbs of the pig are directed ventrally rather than laterally as in the case of the frog studied earlier. This difference in orientation of the appendages represents an important advance of higher vertebrates and is closely linked with the evolution of the rapid and efficient locomotion characteristic of most mammals.

Head

Principal external features of the head include the large **mouth,** the **eyes,** and the **ears.** Note the elongated **snout** with a pair of **external nares** at the anterior end. On the snout, find the stiff sensory hairs, or **vibrissae.** Observe also the strong lower jaws below the mouth and the hard, rounded dorsal cap between the ears. This bony "cap" represents a portion of the braincase, or **cranium,** which protects the soft brain.

Neck

Connecting the head with the trunk is a short, stout **neck.** The neck is equipped with powerful dorsal muscles. Squeeze the dorsal portion of the neck with your fingers to feel these muscles. ***What is the significance of these muscles in the characteristic rooting of pigs?*** Locate these neck muscles in figure 19.6.

Thorax

The thoracic region comprises the anterior half of the trunk and includes the **pectoral girdle;** the **forelimbs;** and the **ribcage,** which encloses the lungs. Observe carefully the forelimbs and the feet. Note that the feet consist of **four digits.** The pentadactyl ("five-fingered") structure of the primitive vertebrate limb has been secondarily lost. The third and fourth digits are large and make up the two halves of the "cloven hoof" of the pig. The second and fifth digits are present but are reduced in size and are directed posteriorly. Feel the ribs and sternum in the thoracic region. These important skeletal parts provide protection for the lungs, heart, and major blood vessels within the pleural cavity.

Abdomen

The abdominal region lies posterior to the thorax and contains the **peritoneal cavity** within which the abdominal organs are suspended. Observe the ventral **umbilical cord.** The point of attachment of the umbilical cord to the abdominal wall of the fetus is called the **umbilicus;** after birth, the scar (on the young pig's abdomen), marking the former site of the umbilicus, becomes the **navel.** Make a fresh cut near the end of the severed umbilical cord and observe that there are **four large tubes** or channels within the tube. These tubes include an **umbilical vein,** two **umbilical arteries,** and an **allantoic duct.** The former three channels provide important routes for the exchange of food, oxygen, and other material between the mother and embryo.

At the posterior end of the abdomen are the **pelvic girdle** and the **hindlimbs.** Note the basic similarity in the organization of the forelimbs and hindlimbs. Find the several pairs of nipples or **mammae** on the ventral abdominal wall. Also locate the anus under the base of the tail.

You can determine the sex of your specimen by observing certain external features in the abdominal region. Male specimens exhibit a **urogenital opening** just posterior to the umbilical cord and two scrotal sacs at the posterior end of the body, just below the **anus.** In older specimens, the **testes** descend into these **scrotal sacs** from the peritoneal cavity. Female specimens exhibit a urogenital opening, or **vulva,** just ventral to the anus. A **genital papilla** protrudes from the vulval area.

Tail

The **tail** is short, flexible, and usually curled. Its skeletal axis is a continuation of the vertebral column.

Skeletal System

The general organization of the mammalian body can best be understood after study of the skeleton, the underlying supporting framework for the body. Unfortunately, the skeleton of the fetal pig is not very suitable for such study because in fetal pigs the skeleton is incompletely ossified and, consequently, some of the bones are not distinct. Therefore, you should refer to a mounted skeleton of a cat or similar vertebrate on demonstration for the purpose of general orientation. Figure 19.2 is provided to assist you in identifying various parts of the cat skeleton; compare it with figure 19.3, which illustrates the skeleton of a late-term pig embryo.

The skeleton of the pig and other mammals is made up of two divisions: the **axial skeleton** (skeleton of the body axis including the skull, vertebral column, ribs, and sternum) and the **appendicular skeleton.** The appendicular skeleton provides articulation for the four limbs with the axial skeleton. The skull is a complex structure consisting of many fused bones that enclose and protect the brain and other organs of the head. The cat skull is more convenient to study than the skull of the pig and is more typical of the adult condition. Both skulls are made up of the same bones, although their shapes do vary. Figure 19.4 will help you identify the principal bones of the skull.

The principal bones of the cat skeleton are listed in the following outline. The vertebral column of the pig and other mammals exhibit some differences in the specific numbers of the vertebrae in some regions.

Fact File

- Pigs (also known as swine or hogs) were among the first animals to be domesticated. The Chinese raised them for food as early as 7000 B.C. They were common in Europe during the Greek and Roman empires. Hernando de Soto, a Spanish explorer, brought the first 13 pigs to the North American continent when he landed near Tampa Bay, Florida, in 1539.
- Originally intended for display in zoos, potbellied pigs were first imported to the United States from Vietnam in 1985. They became a craze, and prices skyrocketed. Today, their popularity has waned, but they are still regarded as ideal pets by many owners.
- Pigs don't have sweat glands. On a hot summer day, they will roll in the mud to keep cool. The layer of mud also helps protect their skin from sunburn.
- Eight major breeds of swine are commonly used for breeding in the United States. In general, the five dark breeds (Berkshire, Duroc, Hampshire, Poland China, and Spot) are valued for their strong potential and for their tendency toward durability, leanness, and meatiness. The three white breeds (Chester White, Landrance, and Yorkshire) are sought after for their reproductive and mothering abilities. Crossbreeding is common to exploit the combination of desirable characteristic of two lines.
- A young female pig is called a gilt. After she has produced her first litter of babies, she is called a sow. An adult male is known as a boar, or if he has been neutered, a barrow. The offspring of a boar and a sow are called piglets. There are usually 8–10 piglets in a litter, each weighing 2½–3 pounds at birth.

Components of the Skeleton

A. **Axial skeleton**
   1. Skull
   2. Mandible (lower jaw)
   3. Hyoid apparatus (several small bones that support the tongue and larynx; remnant of the visceral skeleton of fishes and amphibians)
   4. Vertebral column
      - Cervical vertebrae (7)
      - Thoracic vertebrae (13)
      - Lumbar vertebrae (7)
      - Sacral vertebrae (3)
      - Caudal vertebrae (20–23)
   5. Ribs
   6. Sternum

B. **Appendicular skeleton**
   1. Pectoral girdle
      - Scapula

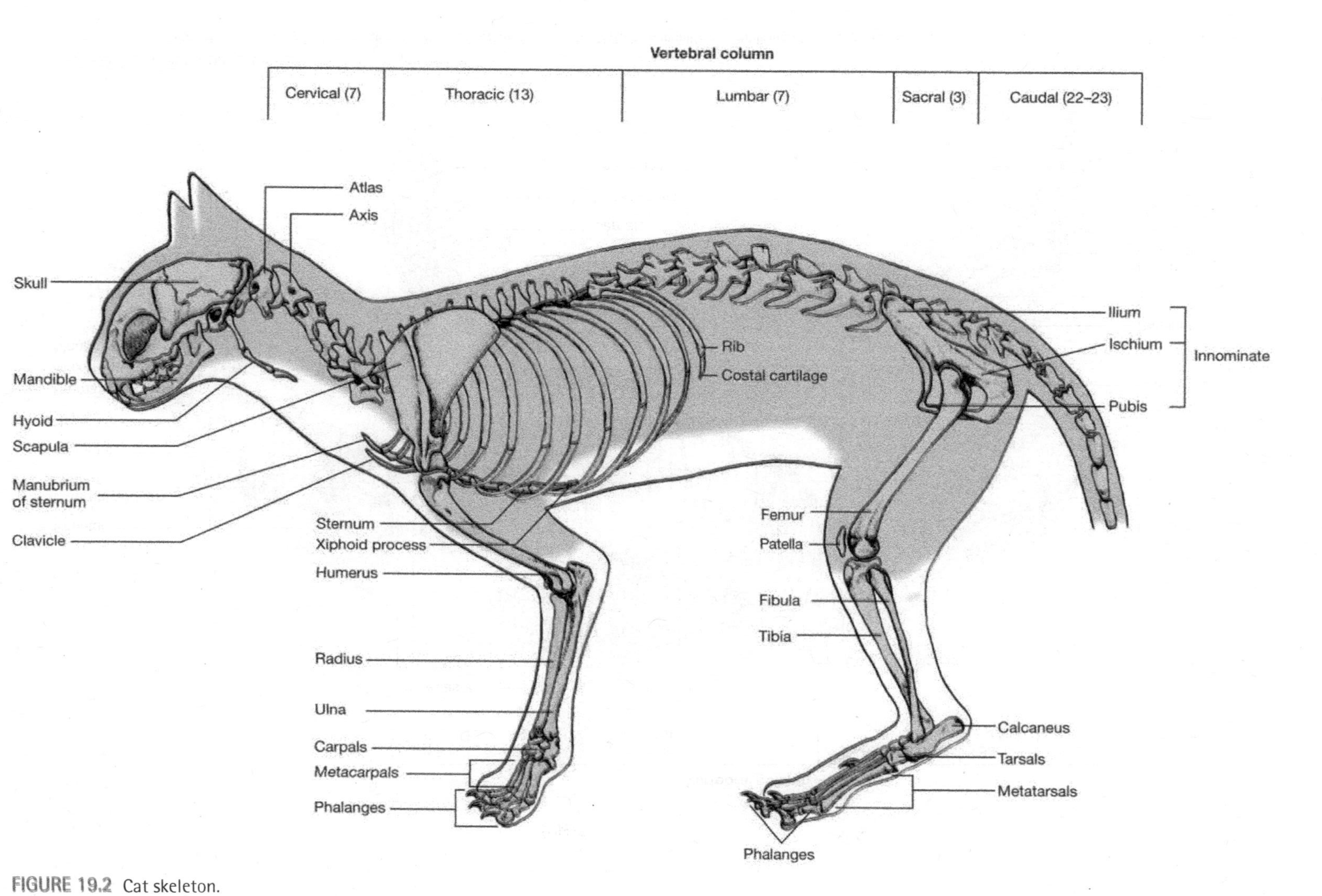

FIGURE 19.2 Cat skeleton.
From Kendall/Hunt Publishing Company Biology Plate Series Part I, *Zoology*. Copyright © 1975 by Kendall/Hunt Publishing Company. Reprinted with permission.

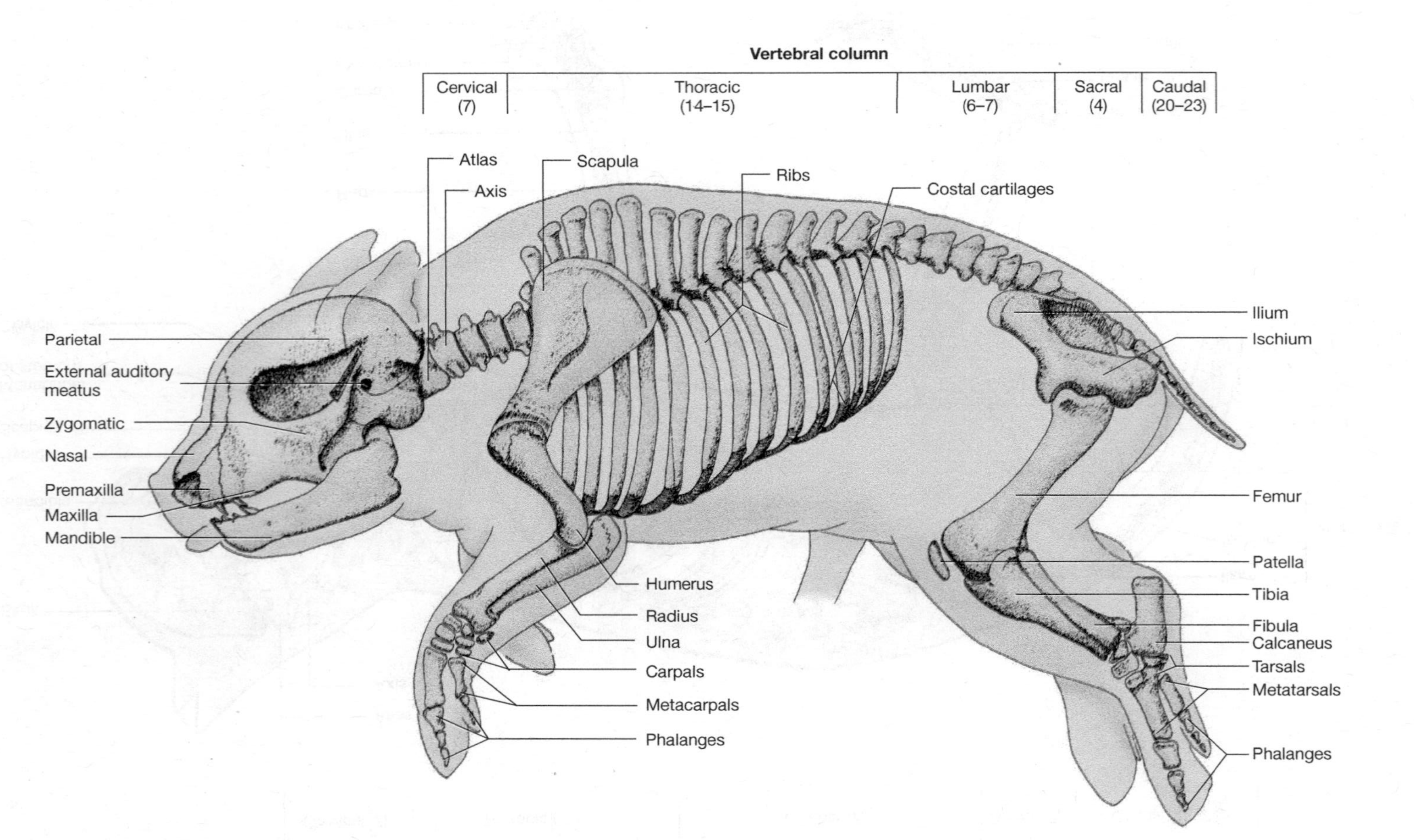

FIGURE 19.3 Fetal pig skeleton.

From Kendall/Hunt Publishing Company Biology Plate Series Part I, *Zoology*. Copyright © 1975 by Kendall/Hunt Publishing Company. Reprinted with permission.

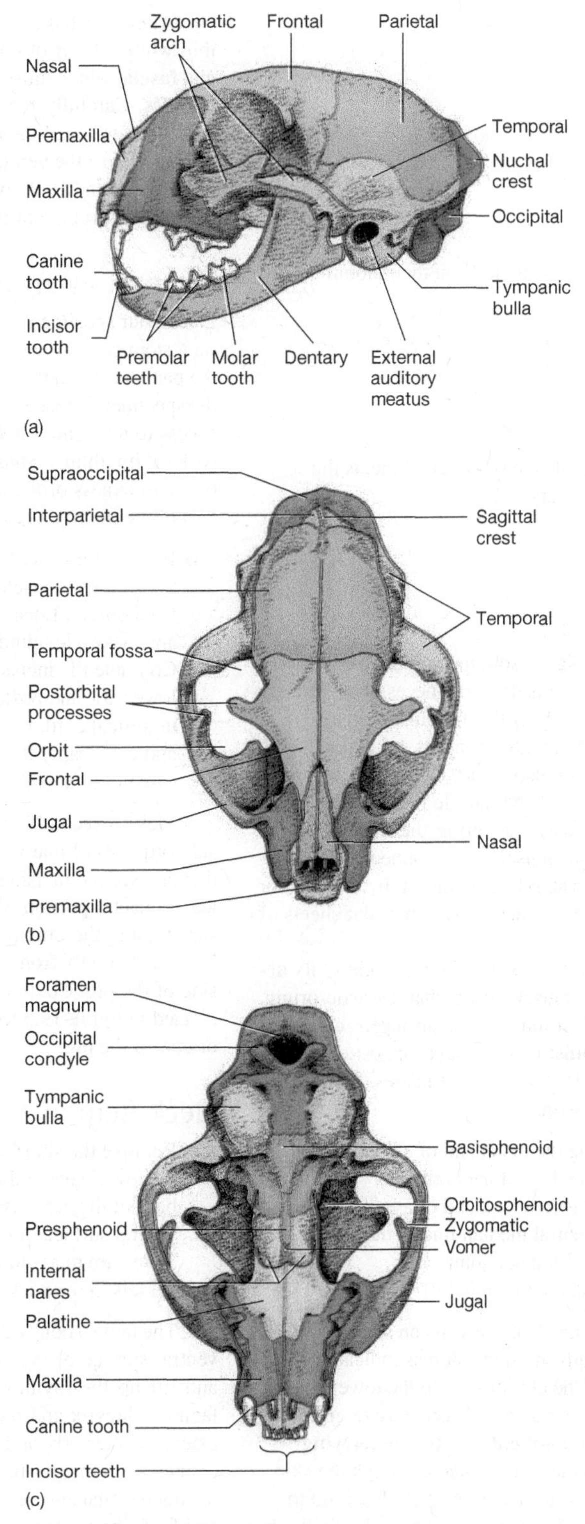

FIGURE 19.4 Cat skull: (*a*) lateral view, (*b*) dorsal view, (*c*) ventral view.

2. Forelimb
    - Humerus
    - Radius
    - Ulna
    - Carpals
    - Metacarpals
    - Phalanges
3. Pelvic girdle
    - Ilium, Ischium, Pubis } fused to form the innominate
4. Hindlimb
    - Femur
    - Patella
    - Tibia
    - Fibula
    - Tarsals (The calcaneus, or heel bone, is the largest of the tarsals.)
    - Metatarsals
    - Phalanges

Muscular System

The muscles of the fetal pig are soft, incompletely developed, and easily torn. Special care is therefore necessary if you are to be successful in your study of the muscular system.

A skeletal muscle typically consists of a fixed end, called the **origin,** and a movable end, called the **insertion.** The fleshy middle portion of the muscle is called the **belly.** The whitish connective tissue covering the muscles is the **deep fascia,** and skeletal muscles are connected to cartilage, bone, ligaments, or skin by this fascia, by tendons, or by aponeuroses (singular: aponeurosis)—thin, flat sheets of tough connective tissue.

The movement or effect caused by a muscle is its **action;** thus, each skeletal muscle has a characteristic origin, insertion, and action. Most muscles are arranged in pairs or groups that are **antagonistic:** they have opposite actions. Several actions of muscles have special names. Some common actions are the following:

Adduction moving the distal end of a bone closer to the ventral median line of the body
Abduction moving the distal end of a bone farther away from the ventral median line of the body
Flexion bending a limb at a joint
Extension straightening a limb

- The pig must be skinned before you can study its muscles. Make a midventral incision as indicated in figure 19.5. Extend the cut forward to the lower jaw and backward to the hindlimbs. **Take care to cut only through the skin; do not cut into the underlying musculature.** Next make an incision through the skin on the medial (inner) surface of the right forelimb to the hoof. Make a similar incision on the right hindlimb. Remove the skin from the right side of the body, and from the right forelimb and hindlimb.

After you have removed the skin, you will observe a thin whitish layer of connective tissue. This is the **superficial fascia,** which must be removed to expose the underlying muscles. Carefully remove the superficial fascia and study first the muscles of the dorsal side (figure 19.6). Then study the muscles of the ventral surface with the aid of figure 19.7. Tables 19.1 and 19.2 give the origins, insertions, and actions of the most prominent muscles of the pig.

General Internal Anatomy

Place your specimen ventral-side-up in a dissecting pan and tie a stout cord to one of the forelimbs. Pass the cord under the pan and attach the cord to the other leg in order to secure the specimen in place, to spread the legs apart, and to provide access to the ventral body surface. Repeat the tying process with the hindlimbs. Study figure 19.8 to obtain a general idea of the thickness of the body wall and the approximate location of the internal organs before starting your dissection.

- Begin your dissection by making an incision through the skin and muscles of the ventral body wall overlying the sternum. Locate the sternum between the forelimbs and make a longitudinal incision as shown in figure 19.5. Continue the incision anteriorly to the level of the lower jaw and posteriorly to a point just anterior to the umbilical cord. Use a sharp scalpel for your dissection and cut cleanly through the skin and muscles. Take care not to damage any underlying organs.

Make a second incision from the region of the umbilical cord toward one side of the body and then connect the first and second incisions by cutting around the umbilicus to leave intact a portion of the ventral body wall immediately surrounding the umbilical cord. The fourth incision should be made laterally from the umbilical region toward the other side of the pig. Continue your incisions in the sequence indicated in figure 19.5 to provide good access to the internal organs of the pig.

Neck Region

- Remove the skin from the ventral surface of the neck, the lower jaw, and the left side of the head up to the base of the ear. Also, cut through the muscles on the ventral surface of the neck to expose the thymus gland. Take care to avoid damage to the several large blood vessels in the neck.

The large, spongy tissue of the **thymus gland** covers the ventral surface of the larynx and trachea. Carefully loosen and lift up the thymus gland and identify the large cartilaginous **larynx** and the smaller **trachea** (figure 19.8) that extends posteriorly and splits into two smaller **bronchi** that connect with the two lungs. Locate the cartilaginous rings of the trachea that ensure a continuous passage for air to pass to and from the lungs. Just behind the larynx, locate the small bilobed **thyroid gland** on the ventral surface of the trachea just posterior to the larynx. The thyroid gland is darker in

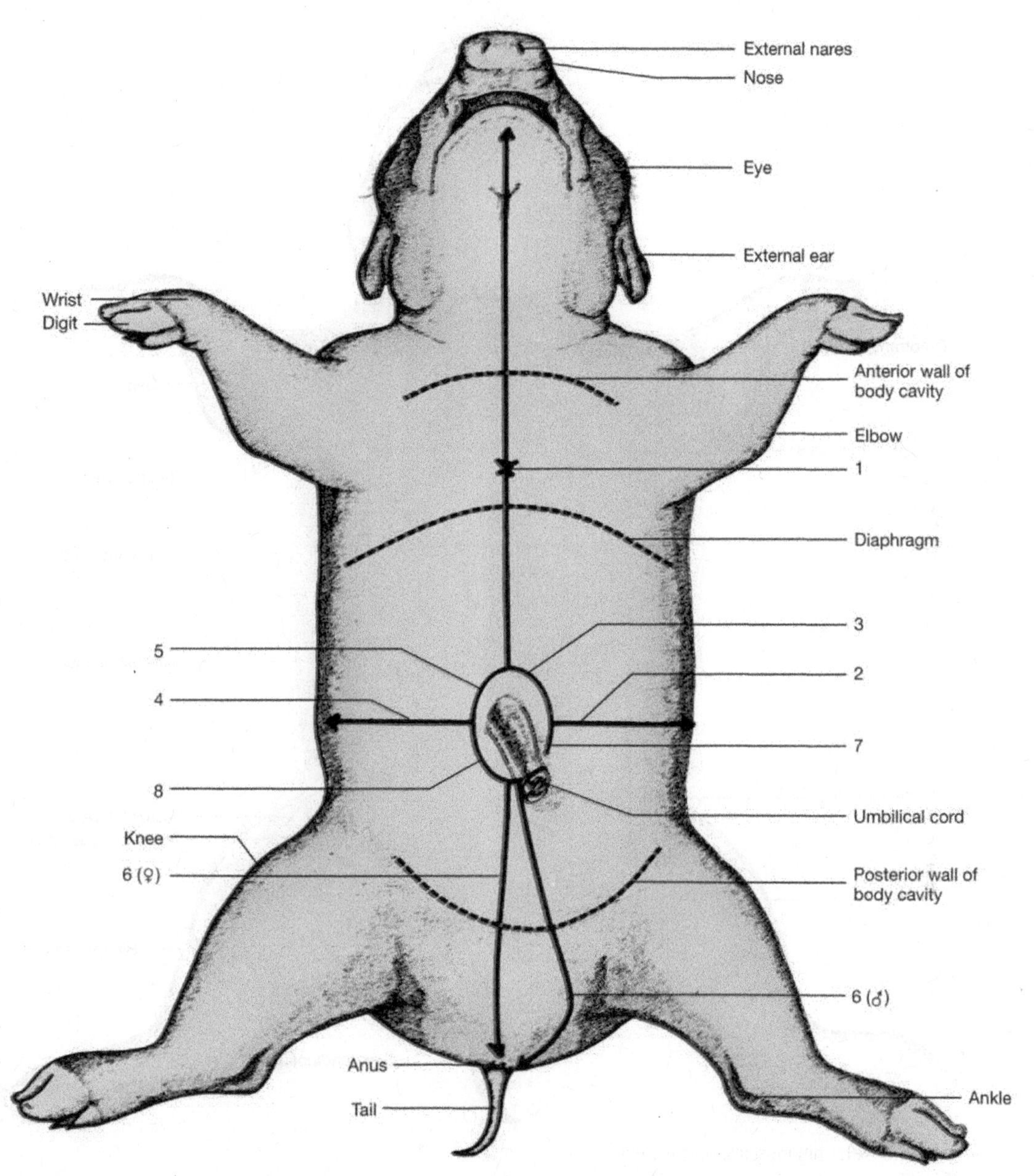

FIGURE 19.5 Fetal pig, ventral view with lines to indicate incisions for dissection. Numbers show sequence of recommended incisions.

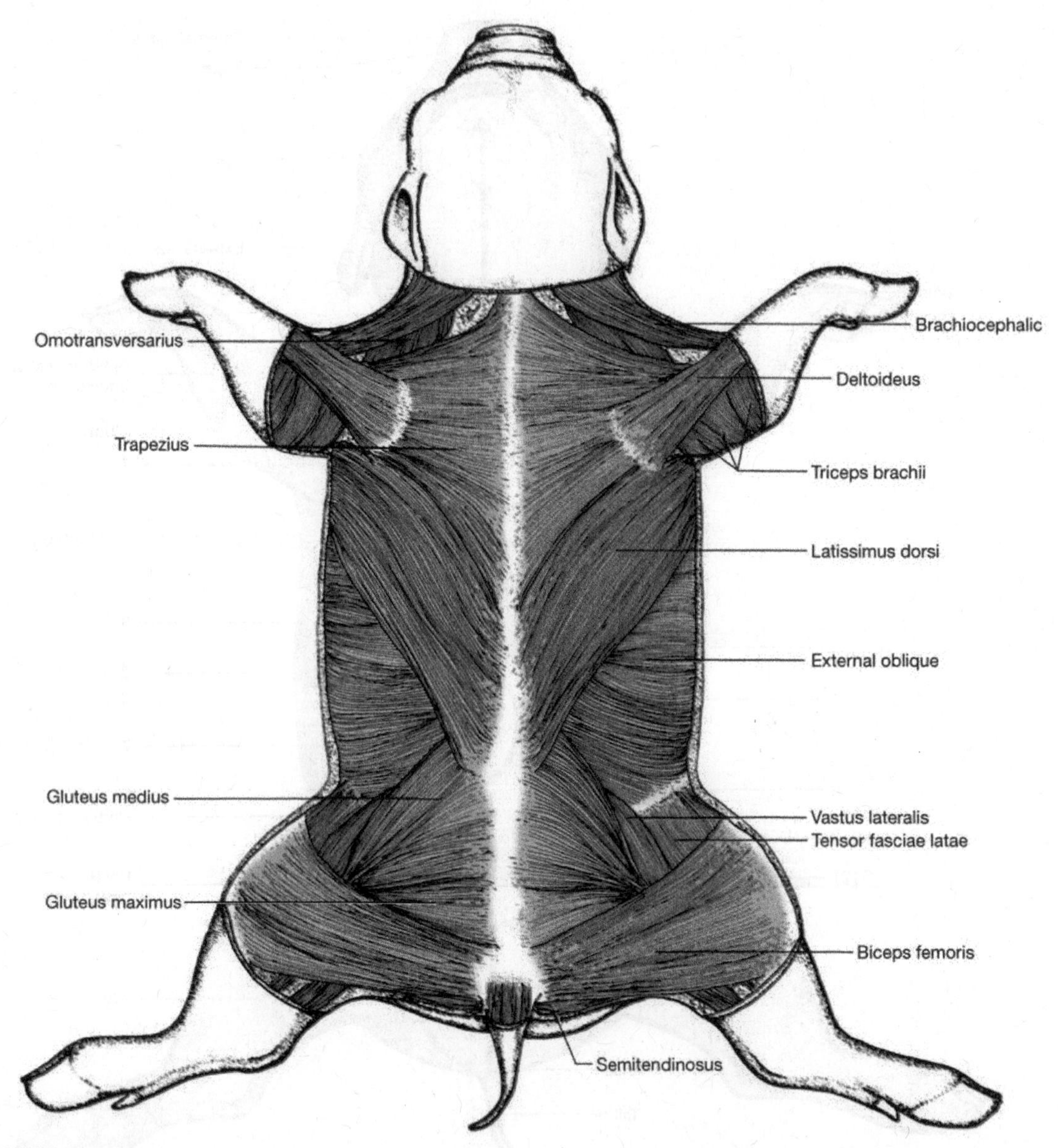

FIGURE 19.6 Fetal pig muscles, dorsal view.

From Kendall/Hunt Publishing Company Biology Plate Series Part I, *Zoology*. Copyright © 1975 by Kendall/Hunt Publishing Company. Reprinted with permission.

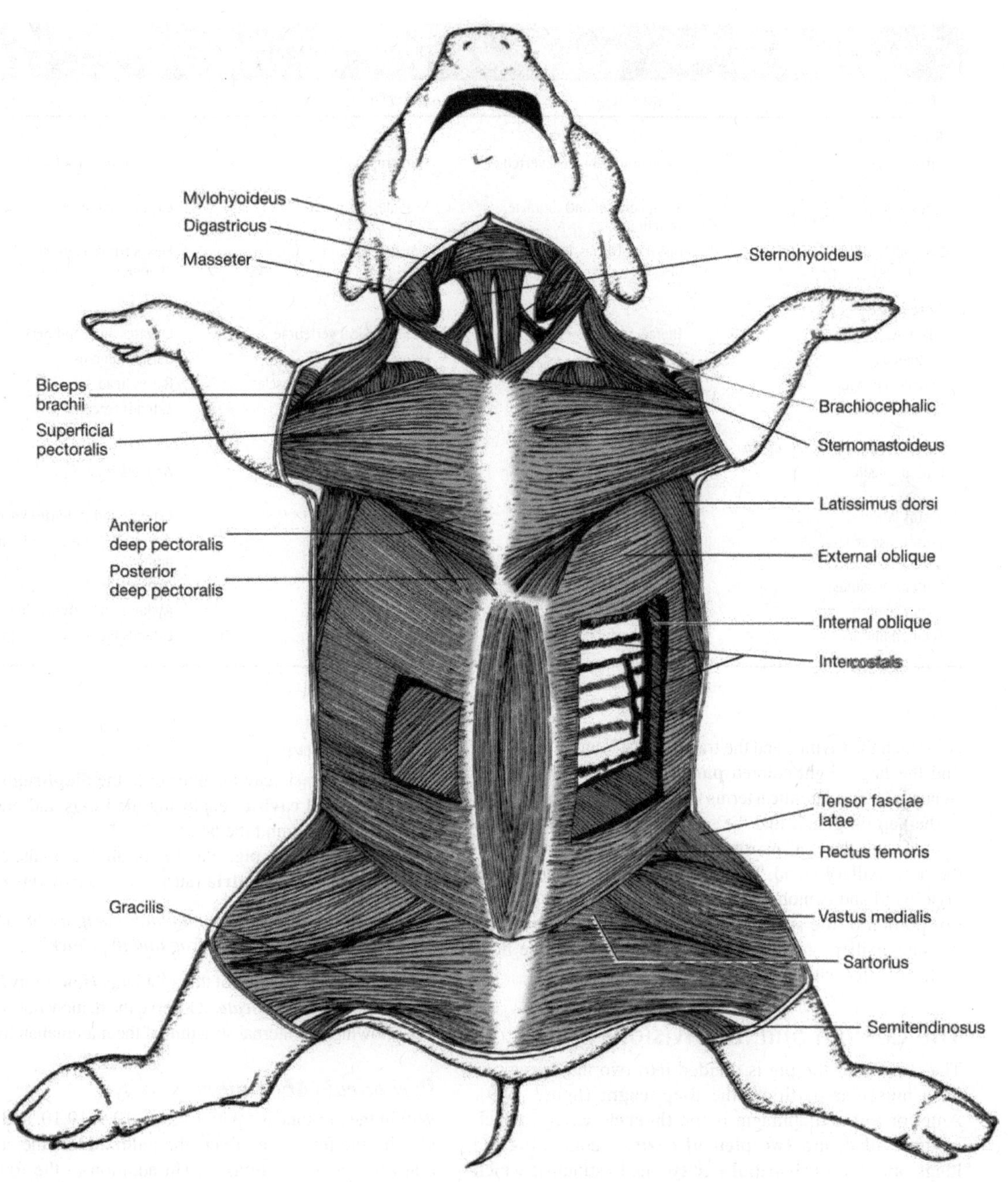

FIGURE 19.7 Fetal pig muscles, ventral view.

Table 19.1 Muscles of the Dorsal Region

Muscle	Origin	Insertion	Action
Back and Side			
Latissimus dorsi	Lumbar and thoracic vertebrae	Humerus	Draws humerus upward and backward
Trapezius	Skull; cervical and thoracic vertebrae	Scapula	Elevates shoulder
External oblique	Last 9 or 10 ribs, lumbodorsal fascia	Linea alba, ilium	Flexes trunk, constricts abdomen
Forelimb			
Splenius	Thoracic vertebrae	Skull, cervical vertebrae	Elevates head and neck
Deltoideus	Scapula	Humerus	Raises humerus
Brachiocephalic	Skull	Humerus, shoulder muscles	Raises head
Triceps brachii	Scapula, humerus	Ulna	Extends forearm
Pelvic Region and Hindlimb			
Gluteus medius	Longissimus dorsi muscle, ilium, sacroiliac	Femur	Abducts thigh
Vastus lateralis	Femur	Patella and tibia	Extends and stabilizes knee
Tensor fasciae latae	Ilium	Patella and tibia	Flexes hip joint, extends knee joint
Gluteus maximus	Sacral and caudal vertebrae	Fascia lata	Abducts thigh
Biceps femoris	Ischium, sacrum	Patella, leg, thigh	Abducts and extends limb
Semitendinosus	Caudal vertebrae, ischium	Tibia, calcaneus	Extends hip, flexes knee joint

color than the thymus and the trachea. Just ventral to the ear, find the large, light-colored **parotid gland** (figure 19.20). Beneath (or in anatomical terms "deep") and slightly anterior to the parotid gland, find the smaller **submaxillary gland.** Also, locate the flat, narrow **sublingual gland** anterior to the submaxillary gland. The sublingual gland lies under the mylohyoid and geniohyoid muscles and surrounds the anterior portion of the submaxillary duct. The ducts from both the submaxillary and sublingual glands open into the floor of the mouth cavity.

The Coelom and Its Divisions

The coelom of the pig is divided into two major regions by a muscular partition, the **diaphragm** (figure 19.9). Anterior to the diaphragm is the **thoracic cavity,** which is subdivided into **two pleural cavities** containing the lungs, and the **pericardial cavity** (mediastinum), which contains the heart and the large blood vessels connected with the heart. The thin epithelial lining of the pleural cavity is the **pleura;** that of the pericardial cavity is the **pericardium.** Posterior to the diaphragm is the large **peritoneal** (abdominal) **cavity** within which the abdominal organs are suspended. Each of the abdominal organs is enclosed in a thin layer of mesodermal epithelium **(visceral peritoneum);** a similar layer also covers the inner surface of the body wall **(parietal peritoneum).** The double sheets of peritoneum that support the abdominal organs are called **mesenteries.**

Thoracic Cavity

Within the thoracic cavity, anterior to the **diaphragm,** there are two **pleural cavities** enclosing the **lungs** and one **pericardial cavity** around the heart.

The heart of the pig, like that of all mammals, consists of four chambers: **two atria** (auricles) and **two ventricles.**

- ***How does the heart of mammals compare in this regard with that of the frog and the shark?***

Locate the four lobes of the right lung. ***How many lobes of the left lung can you locate?*** Observe the demonstration materials showing the internal structure of the mammalian lung.

Peritoneal (Abdominal) Cavity

Within the peritoneal cavity (figures 19.9, 19.10, and 19.11) identify the four-lobed **liver,** the **gallbladder,** the common **bile duct,** and the **stomach.** (In addition to the four main lobes of the liver, there is also a small caudate lobe attached to the posterior portion of the right lateral lobe.) Find the **esophagus** where it connects to the stomach.

The upper portion of the stomach, into which the esophagus empties, is called the **cardiac region.** A circular muscle, the **cardiac sphincter,** lies at the junction of the esophagus and the cardiac stomach and regulates the passage of material into the stomach. It empties into the large **fundus,** seen externally as a large bulge on the left side of the stomach. The lower part of the stomach, which empties into the small intestine, is the **pyloric region.**

Table 19.2 Muscles of the Ventral Region

Muscle	Origin	Insertion	Action
Head and Neck Region			
Mylohyoideus	Mandible	Hyoid bone	Raises floor of mouth, tongue, and hyoid bone
Digastricus	Mastoid process of skull	Mandible	Depresses mandible
Masseter	Zygomatic arch of skull	Mandible	Elevates jaw, closes mouth
Sternohyoideus	Sternum	Hyoid bone	Retracts and depresses hyoid and base of tongue
Forelimb and Pectoral Girdle			
Brachiocephalic	Nuchal crest, mastoid process of skull	Humerus and shoulder	Inclines or extends head, draws forelimb forward
Biceps brachii	Scapula	Radius, ulna	Flexes forearm
Superficial pectoralis	Sternum	Humerus	Adducts humerus
Anterior deep pectoralis	Sternum	Scapula, supraspinatus muscle	Adducts and retracts forelimb
Posterior deep pectoralis	Sternum, 4th to 9th ribs	Humerus	Retracts and adducts forelimb
Side and Belly			
Latissimus dorsi	Lumbar and thoracic vertebrae	Humerus	Draws humerus upward and backward
External oblique	Last 9 or 10 ribs	Linea alba, ilium	Constricts abdomen, flexes trunk
Internal oblique	Lumbodorsal fascia	Linea alba	Constricts abdomen, flexes trunk
Intercostals	Ribs	Adjacent rib	Pulls ribs together
Rectus abdominis	Pubic symphysis	Sternum	Constricts abdomen, bends vertebral column
Pelvic Girdle and Hindlimb			
Tensor fasciae latae	Ilium	Patella and tibia	Flexes hip joint, extends knee joint
Rectus femoris	Ilium	Patella and tibia	Extends knee, flexes hip
Vastus medialis	Femur	Patella and tibia	Extends leg
Semitendinosus	Ischium	Tibia	Extends hip, flexes limb
Sartorius	Iliac fascia and tendon of psoas minor muscle	Patella and tibia	Abducts hindlimb, flexes hip joint
Gracilis	Pubis	Patella and tibia	Adducts hindlimb

The constriction between the stomach and small intestine is the **pylorus.** Passage of material from the stomach into the small intestine is regulated by a specialized circular muscle, the **pyloric sphincter.** Identify the **greater curvature** of the stomach (the longer, convex side), the **lesser curvature** (the smaller, concave side), and the large body of the stomach that lies between the two.

Locate the elongated, dark-red **spleen** along the greater curvature of the stomach and the light-colored granular **pancreas** found posterior to the stomach in the curve of the duodenum. Also observe the loosely coiled **small intestine** on the right and the tightly coiled **large intestine** on the left. The large intestine consists of four segments: cecum, colon, rectum, and anal canal. The anterior portion is the colon. Locate the **caecum,** a blind sac attached near the anterior end of the **colon.** Posterior to the colon is the **rectum,** followed by the anal canal leading to the **anus.** The main function of the colon is resorption of water, but some vitamins are synthesized here and the vitamins and some fatty acids are absorbed. Most absorption of nutrients occurs in the small intestine. At the junction of the small and large intestines is the **ileocolic valve,** which regulates passage of materials into the large intestine. Posteriorly, the large intestine is modified to form the rectum.

Several mesenteries that support specific organs can be identified: (1) the **falciform ligament,** found along the umbilical vein inside the peritoneal cavity, which connects the liver to the ventral body wall; (2) the **coronary**

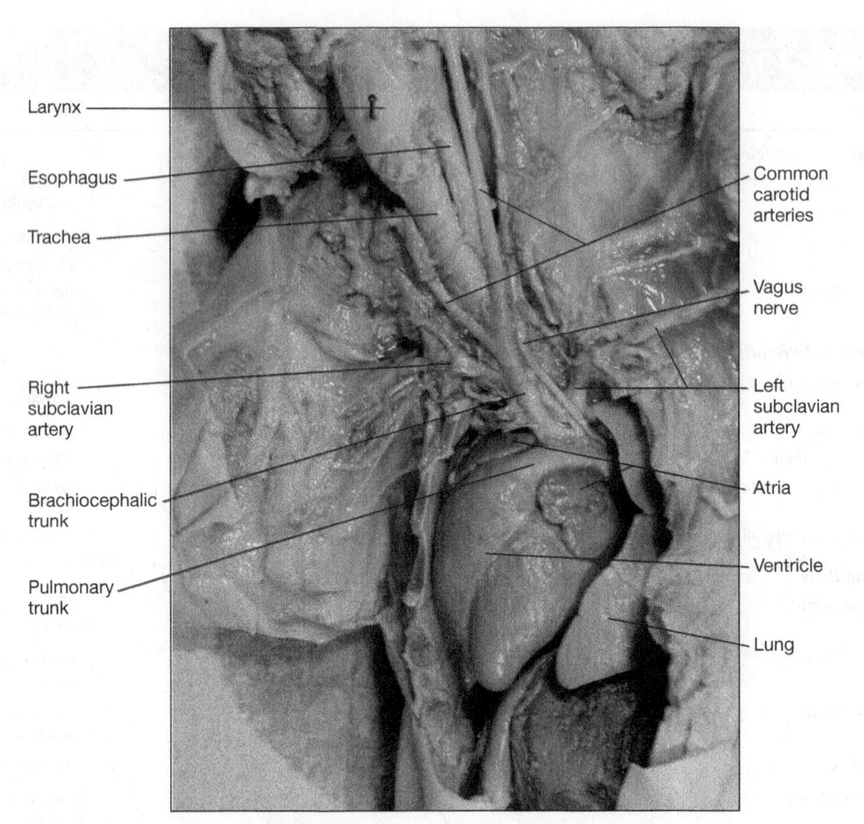

FIGURE 19.8 Fetal pig, dissected, neck and thoracic region, ventrolateral view.
Photo courtesy of Betty Black.

ligament, which attaches the anterior surface of the liver to the diaphragm; (3) the **greater omentum,** which attaches the greater curvature of the stomach to the colon, spleen, and most of the small intestine; (4) the **lesser omentum,** which attaches the lesser curvature of the stomach to the liver, and connects the anterior portion of the duodenum; and (5) the **mesentery proper,** which suspends the small intestine from the dorsal midline. Other smaller mesenteries support the colon, the rectum, and the gonads from the dorsal body wall.

Find the chief mesenteries that support the abdominal organs: **falciform ligament, coronary ligament, greater omentum, lesser omentum,** and **mesentery proper.**

Locate the **urinary bladder** located along the ventral body wall between the two umbilical arteries (figure 19.9), which extends dorsally and posteriorly from the umbilicus. On either side of the bladder find the **two umbilical arteries** (figure 19.9).

Observe the large blood vessels that supply the intestines along their length (figures 19.15 and 19.19) as you remove this portion of the digestive tract. Examine the interior of the large and small intestines and of the caecum. ***How do their interior walls differ in structure?*** Notice also the numerous, light-colored lymph nodes found between the two layers of the mesentery. ***Why do you think lymph nodes would be found in this area?***

The Urogenital System

The excretory and reproductive (genital) systems are best studied together because they are closely related in development and because both systems use common ducts.

The Urinary System

Refer to figures 19.12 and 19.13, and locate the kidneys on your specimen. A **ureter** leads from each kidney to the large **urinary bladder** (an enlargement of the embryonic allantois). Find the **allantoic duct,** which leads from the bladder through the umbilical cord. Note also the **renal arteries** (figure 19.16), which enter the kidneys and the large renal veins (figure 19.19) through which blood leaves the **kidneys.** Locate also the two large umbilical arteries (figure 19.9), which enter the **umbilical cord.** Along the medial border of each kidney, find the long, narrow, whitish **adrenal glands.** The adrenals are important endocrine glands.

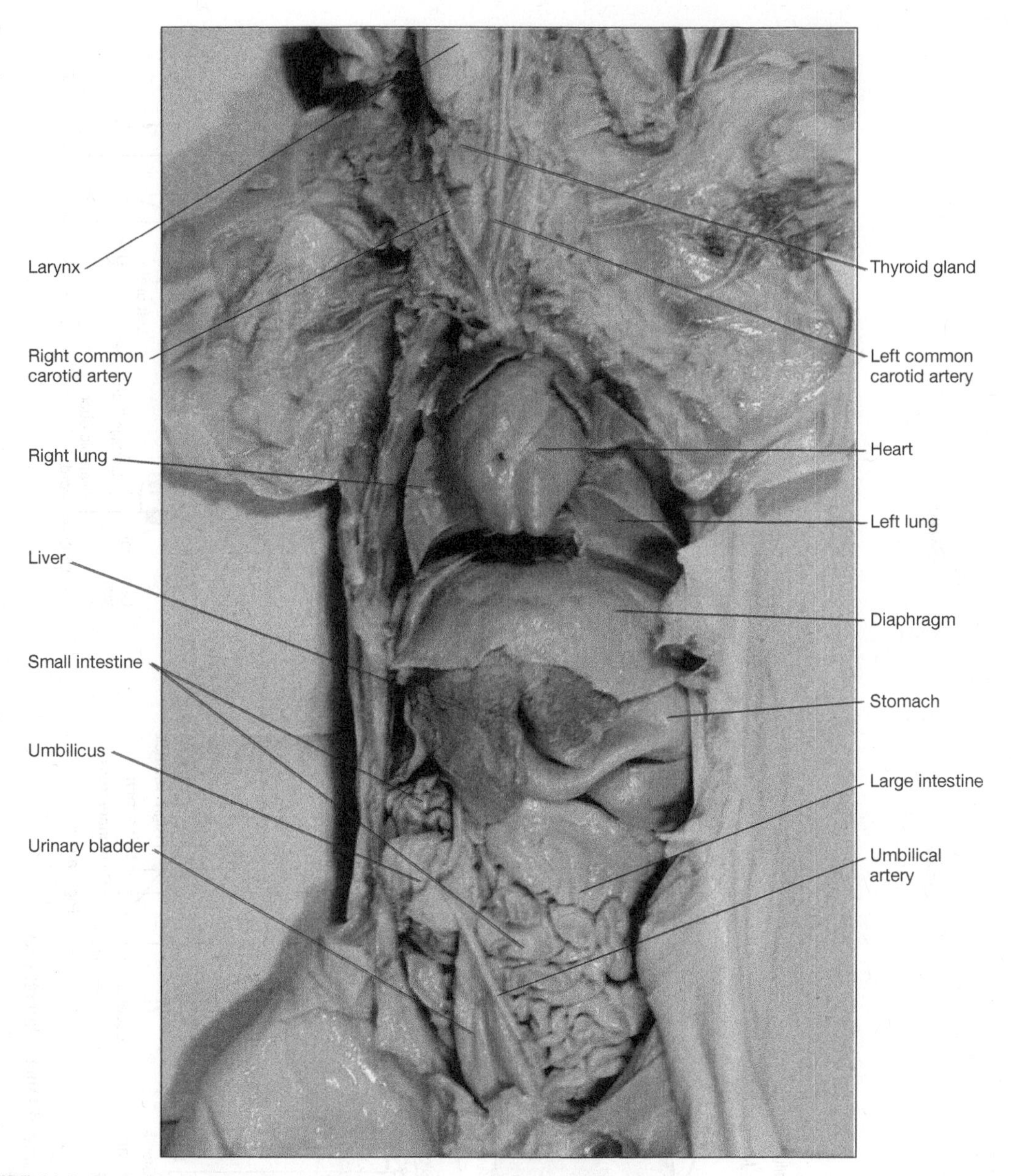

FIGURE 19.9 Fetal pig, dissected, ventral view.
Photo courtesy of Betty Black.

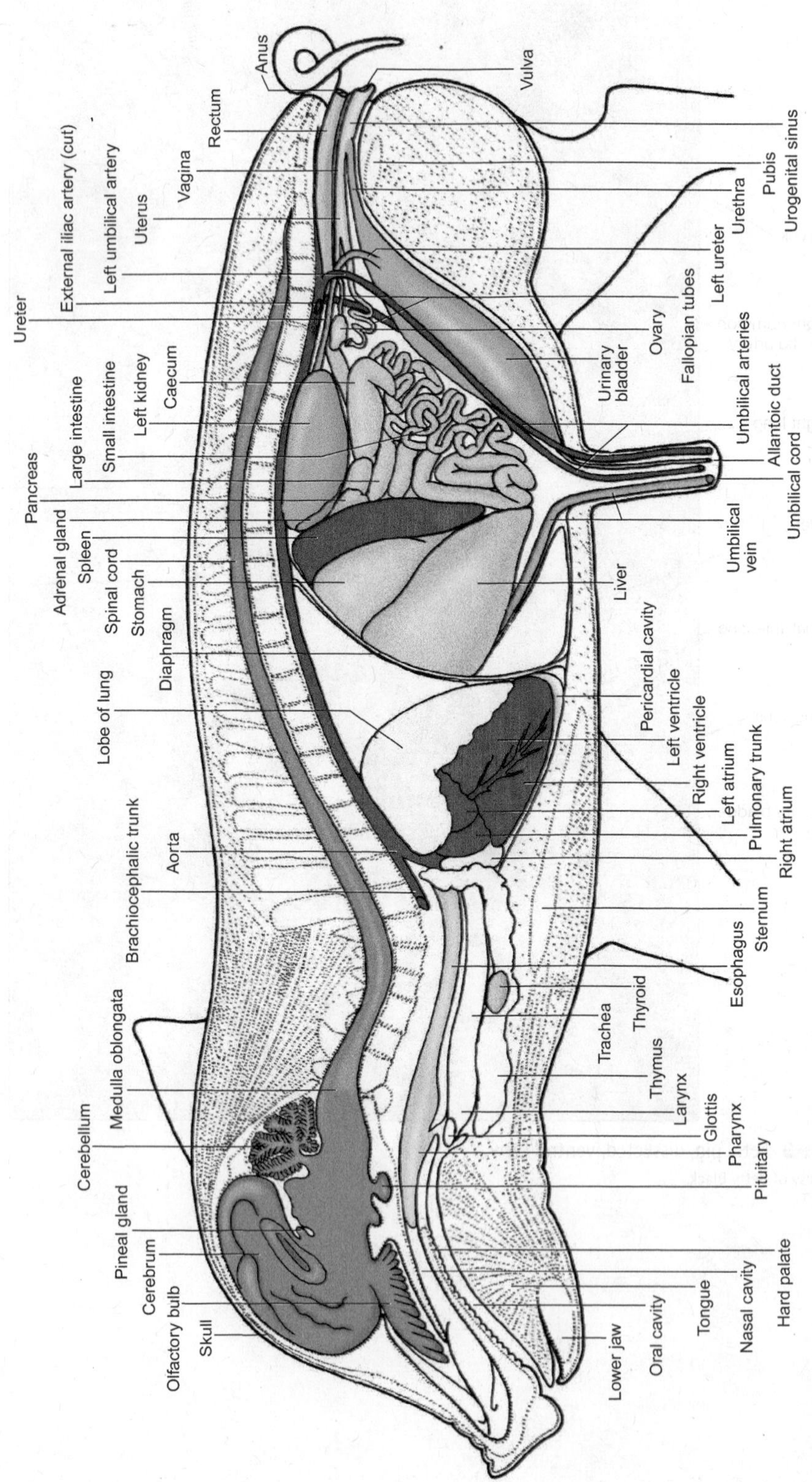

FIGURE 19.10 Fetal pig, internal anatomy of female, lateral view.

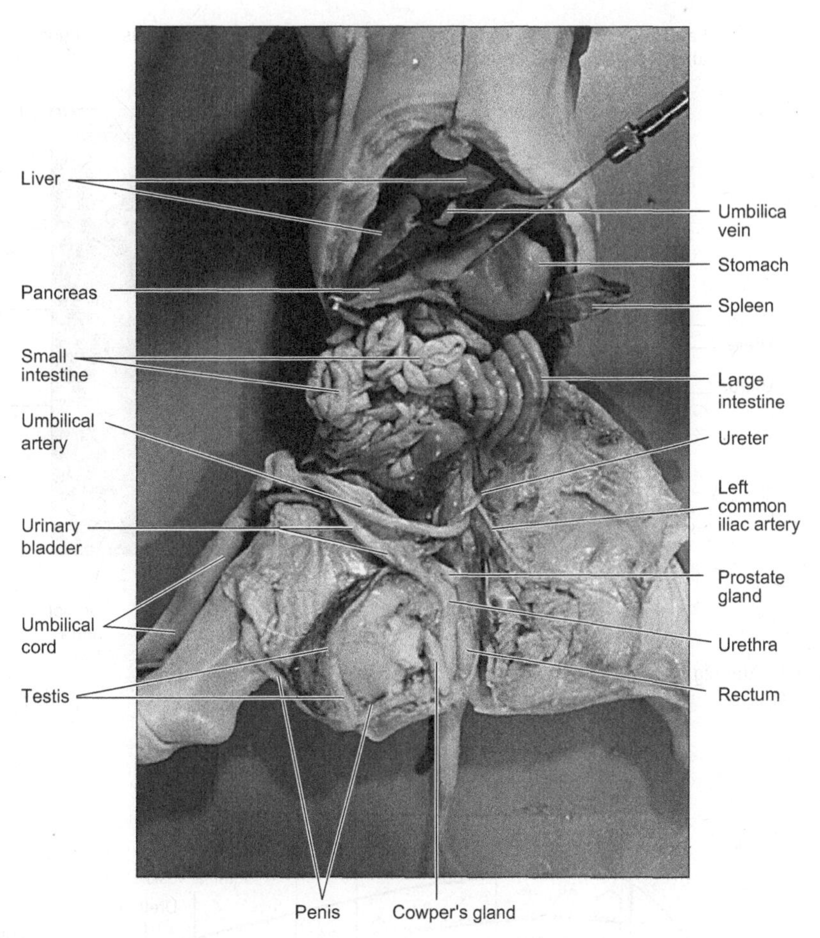

FIGURE 19.11 Fetal pig, abdominal organs, male, ventral view.
Photograph by John Vercoe.

- The specimens provided for each laboratory section should consist of both male and female pigs. Study the genital (reproductive) system of your specimen according to the following directions and then study a specimen of the opposite sex that was assigned to one of your classmates.

The Female Genital System

- Study figures 19.10 and 19.12, and locate the principal female reproductive organs. Then cut through the skin, muscles, and pelvic bone along the midventral line of your specimen and pull the limbs wide apart to locate these parts on your specimen. Find the two small, bean-shaped **ovaries** near the posterior end of the peritoneal cavity.

Attached to the dorsal surface of each ovary is a small convoluted duct, the **fallopian tube** (oviduct). Locate the wide, ciliated funnel, called the **infundibulum,** at the terminus of each fallopian tube. The opening into the end of the fallopian tube is the **abdominal ostium.** Trace the fallopian tubes to the larger **horns of the uterus** (where embryonic development takes place), which connect with the small median **body of the uterus.** The uterus continues posteriorly as a thick, muscular tube, the **vagina.** The vagina and the urethra both empty into a common chamber, the **urogenital sinus** (vestibule). Carefully cut open the urogenital sinus by making an incision along one side, and use a blunt probe to locate the openings of the urethra and the vagina. Also, search on the ventral floor of the urogenital sinus for the **clitoris,** a small, rounded papilla. *Do not* confuse the clitoris (homologous with the penis of the male) with the larger externally obvious **genital papilla.** Just ventral to the anus is the **urogenital opening.**

Study also the demonstrations of microscopic sections of a mammalian ovary to see germ cells in various stages of development and of pig embryos at various stages.

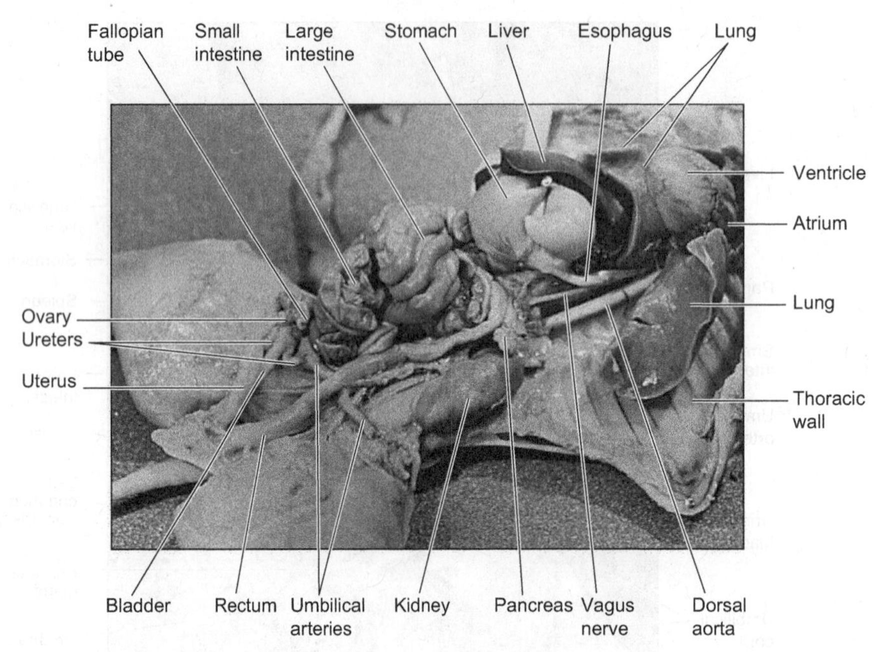

FIGURE 19.12 Fetal pig, thoracic and abdominal region organs, female, ventral view.
Photograph by John Vercoe.

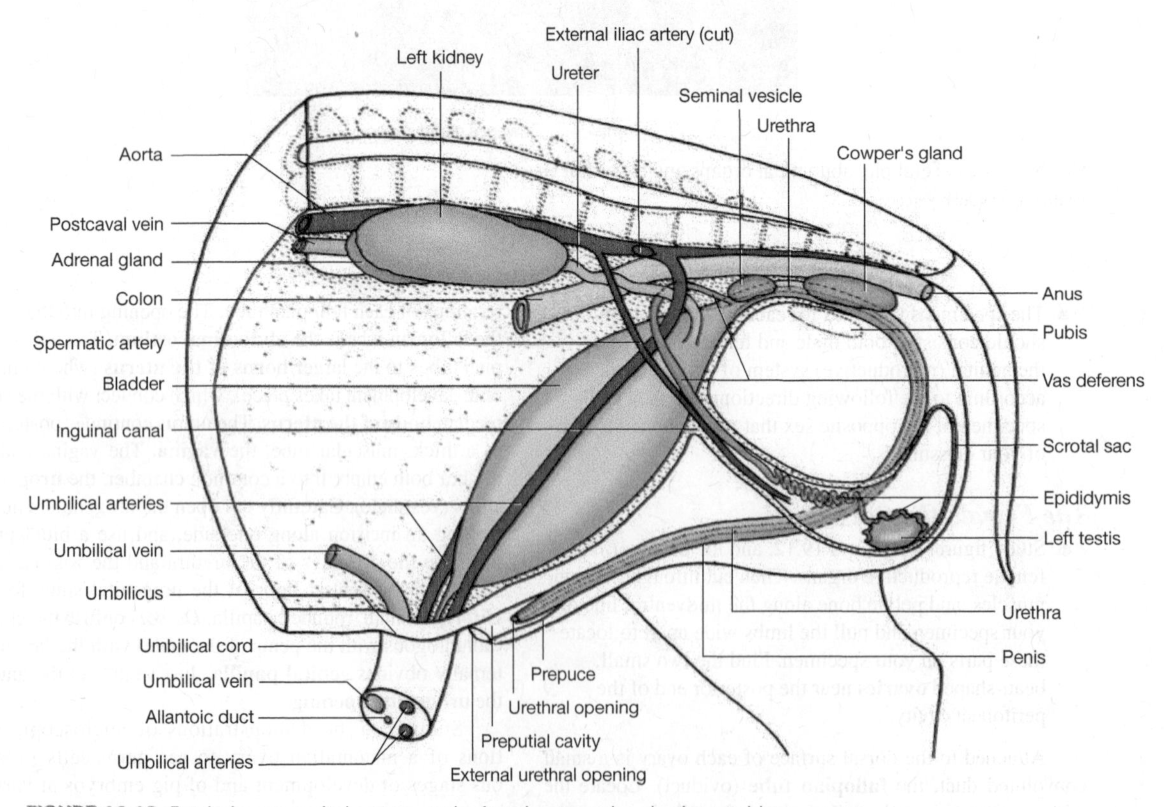

FIGURE 19.13 Fetal pig, urogenital system, male. Arteries are red and veins are blue.

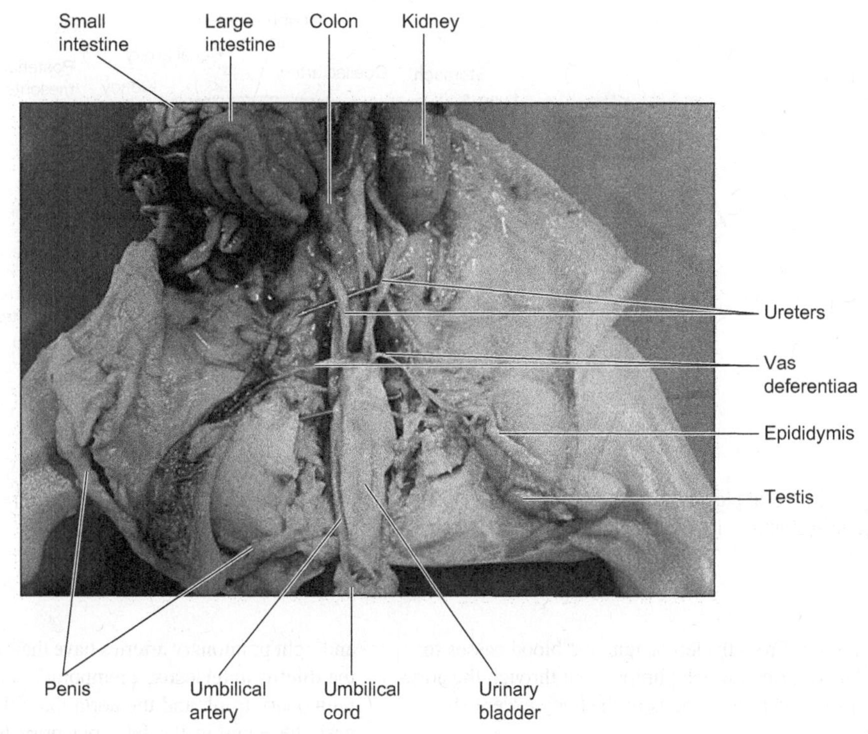

FIGURE 19.14 Fetal pig, urogenital system, male, ventral view.
Photograph by John Vercoe.

The Male Genital System

- Study figures 19.13 and 19.14, and note the relative position of the principal male reproductive organs. Cut through the scrotal sac and down through the pubis about 3 mm to one side of the midventral line, pull the hindlimbs carefully apart, and identify the principal male organs.

First locate the two **testes** (singular: testis). In larger and older specimens, the testes will be located within two **scrotal sacs** (collectively referred to as the scrotum), but in younger specimens, the testes may not yet have descended completely from the peritoneal (abdominal) cavity where they initially grow and differentiate. As the male pig grows larger, the developing testes descend from the peritoneal cavity through the **inguinal canal** into the scrotal sacs. In younger specimens, the testes may be found anywhere along this descending path.

Now locate the **epididymis** (figure 19.14), a mass of coiled tubules lying along one side of each testis. The epididymis connects with the vas **deferens,** which passes through the inguinal canal along the **spermatic artery** and vein, and crosses over the ureter to enter the **urethra** (figures 19.11 and 19.14).

Locate the **penis,** a long, thin cylinder lying beneath the skin just posterior to the umbilical cord. Cut through the skin overlying the penis to free the penis and trace it posteriorly to its junction with the urethra. Dorsal to the urethra, near the entrance of the **vasa deferentia,** find the two small **seminal vesicles.** Adjacent to the seminal vesicles is the **prostate gland** (poorly developed in younger specimens). Also note the two **Cowper's glands** (bulbourethral glands) lying alongside the urethra near its junction with the penis.

The tissue overlying the penis on the ventral abdominal wall is called the **prepuce,** and the cavity enclosed within the prepuce is the **preputial cavity.** Opening from the preputial cavity to the exterior is the male **urogenital opening.**

Observe the demonstration of the pig testis showing **germ cells** in various stages of maturation. See also the demonstrations of mammalian **spermatozoa.**

Circulatory System

To assure your understanding of circulation in mammals, you should review the basic organization of the mammalian circulatory system prior to undertaking a dissection and study of the circulatory system of the pig (figure 19.15). Among these essential features are the following:

1. A four-chambered heart, which provides an efficient separation of the systemic and pulmonary divisions of the circulatory system.
2. Oxygenated blood from the lungs is carried to the heart by the **pulmonary veins,** which empty into the

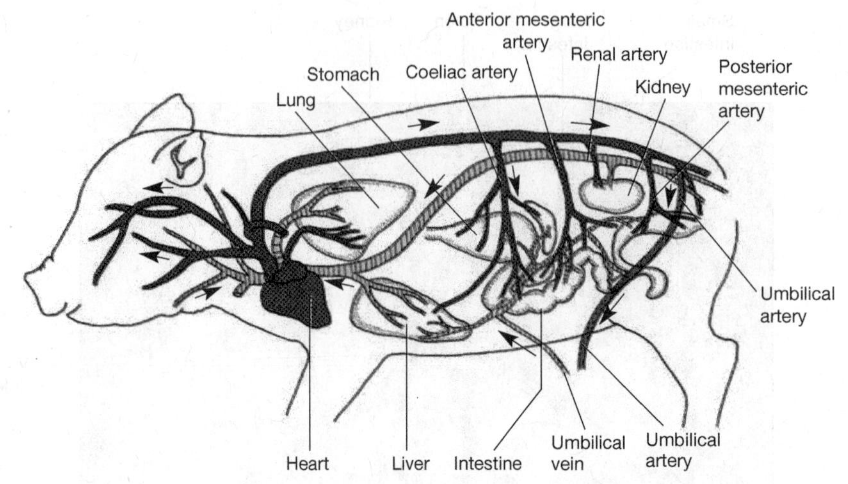

FIGURE 19.15 Fetal pig, pattern of circulation. The heart, systemic arteries, and pulmonary veins are red, pulmonary arteries and systemic veins are blue, and the hepatic portal system is yellow.

left atrium. From the left atrium, the blood passes to the left ventricle, which pumps it out through the aorta and its branches to all parts of the body **except the lungs.**

3. Blood from the organs of the body (except the lungs) is returned to the right atrium by the large precaval and postcaval veins. From the right atrium, it goes to the right ventricle and then back via the **pulmonary arteries** to the lungs to be oxygenated.
4. Mammals have a **hepatic portal system,** but do not have a **renal portal system,** characteristic of many lower vertebrates such as the frog and the shark. The latter has been lost during the evolution of the mammals.

Heart and Arterial System

Completely remove the thymus gland to observe the heart. Identify its four chambers, the muscular ventricles, and the two thin-walled atria. Blood from the systemic veins enters the vena cava and then flows into the **right atrium,** followed by the **right ventricle.** From the right ventricle, the blood is pumped through the pulmonary arteries to the lungs. From the lungs, the blood returns to the left atrium through the pulmonary veins. From the **left atrium,** the blood passes to the **left ventricle,** where it is pumped into the aorta and then to the various organs of the body. We will study the internal structure of the heart later.

Refer to figures 19.8, 19.9, 19.16, and 19.17 for the following study of the large arteries attached to the heart. On the top (anterior surface) of the heart, locate the arch of the **aorta,** the most dorsal structure, and the more ventral **pulmonary trunk,** which exits the right ventricle and splits into the left and right pulmonary arteries. Note also that the upper part of the pulmonary trunk (where the left and right pulmonary arteries have their origin) connects with the **ductus arteriosus,** a temporary connection between the pulmonary trunk and the aorta that allows the blood to bypass the lungs in the fetal pig prior to birth. Study figure 19.16 and locate the major systemic arteries described in the following paragraphs.

> **SUGGESTION**
>
> **It is recommended that at this point you study the veins located anterior to the heart before continuing with the remaining arteries. The veins in this area lie ventral to the arteries, so they are likely to be damaged as you try to separate and trace the arteries. Skip to the section entitled Venous System on page 341.**

The **brachiocephalic trunk** is the first large branch from the arch of the aorta and leads toward the head; it branches into the **right subclavian artery,** the external branch of which supplies the right forelimb, the right anterior body wall, and the right side of the neck (figures 19.16 and 19.17). Just anterior to the origin of the right subclavian, the **brachiocephalic trunk** branches into the left and right **common carotid arteries,** the external branches of which supply the skull, brain, tongue, cheek, and face. The internal branches supply the brain and the skull. The **left subclavian artery** arises independently from the aorta and just to the left of the brachiocephalic trunk. Locate the series of arteries branching from the left and right subclavian arteries: (1) the large **brachial artery** leading into the forelimb, (2) the **thyro-cervical artery** leading into the thyroid and parotid glands, and (3) the **sternal**

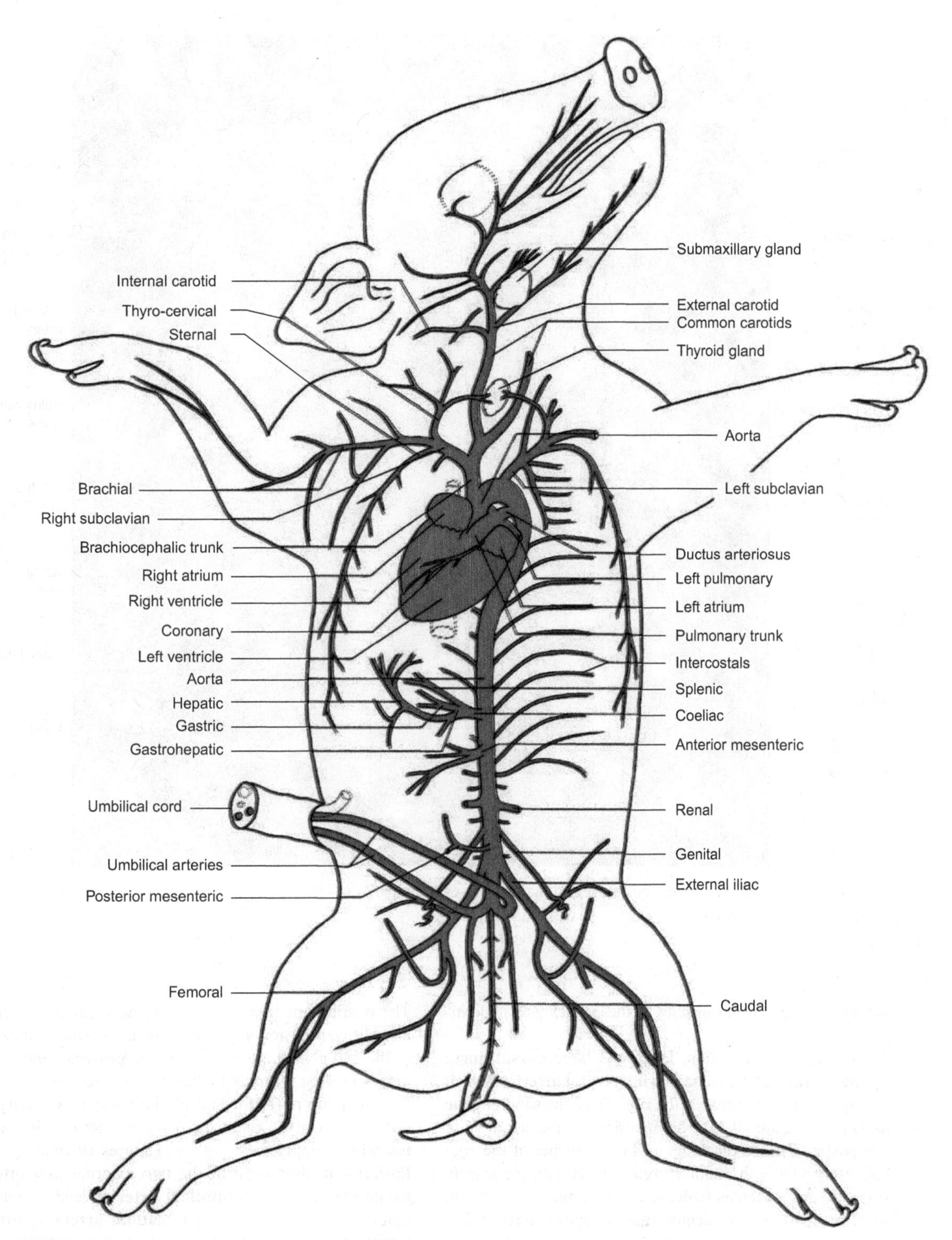

FIGURE 19.16 Fetal pig, arterial system, ventral view.

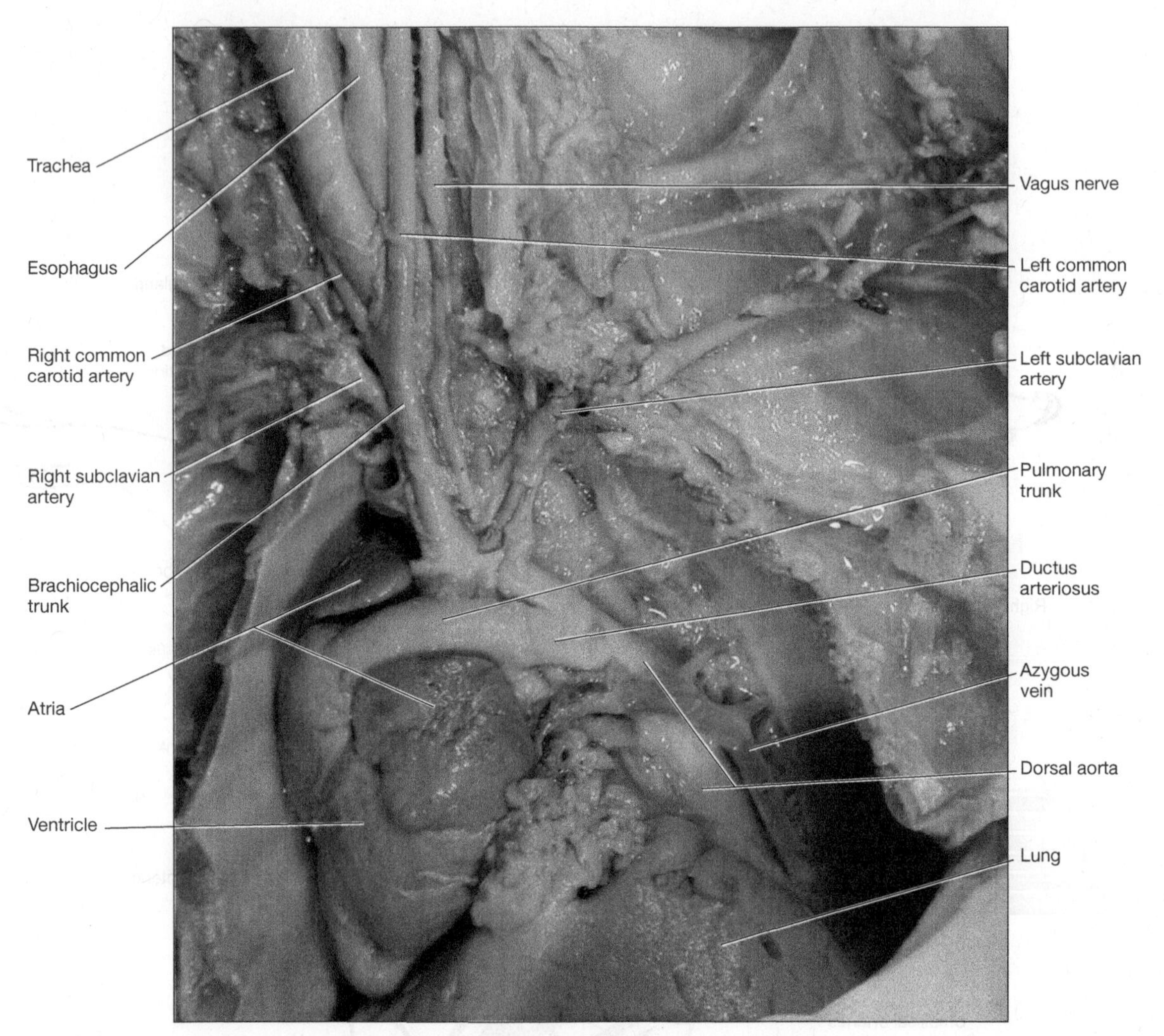

FIGURE 19.17 Fetal pig, dissected, heart and major arteries attached.
Photo courtesy of Betty Black.

artery leading into the muscles of the thoracic and abdominal walls.

In the thoracic region, locate the large **dorsal aorta** (figure 19.18) and the numerous **intercostal arteries,** which supply the muscles between the ribs. Trace the short **coeliac artery** (see figure 19.15) from its origin on the dorsal aorta just posterior to the diaphragm. Three branches of the coeliac artery supply the stomach **(gastric artery);** the spleen, stomach, and pancreas **(splenic artery);** and the stomach, liver, pancreas, and duodenum **(gastrohepatic artery).** Locate also the **anterior mesenteric artery** leading from the dorsal aorta to the pancreas, small intestine, and large intestine. It originates just posterior to the coeliac artery. Two **renal arteries** lead to the kidneys, and two **genital** arteries (ovarian in female, spermatic in male) lead to the gonads. The ovarian artery travels laterally whereas the spermatic artery travels posteriorly to the gonad. Posterior to the origin of the two genital arteries, locate the **posterior mesenteric artery** leading ventrally to the colon and rectum.

Near the posterior end of the abdominal cavity, the dorsal aorta gives rise to two large external **iliac arteries,** which supply the hindlimbs (figures 19.16 and 19.19). Posterior to the origin of the two external iliac arteries, locate the two large **umbilical arteries** leading into the umbilical cord and the single **caudal artery** continuing into the tail region. The umbilical arteries are fetal vessels that carry blood to the fetal portion of the placenta. These arteries degenerate after birth, and only those branches of the umbilical artery supplying the bladder persist in the adult pig.

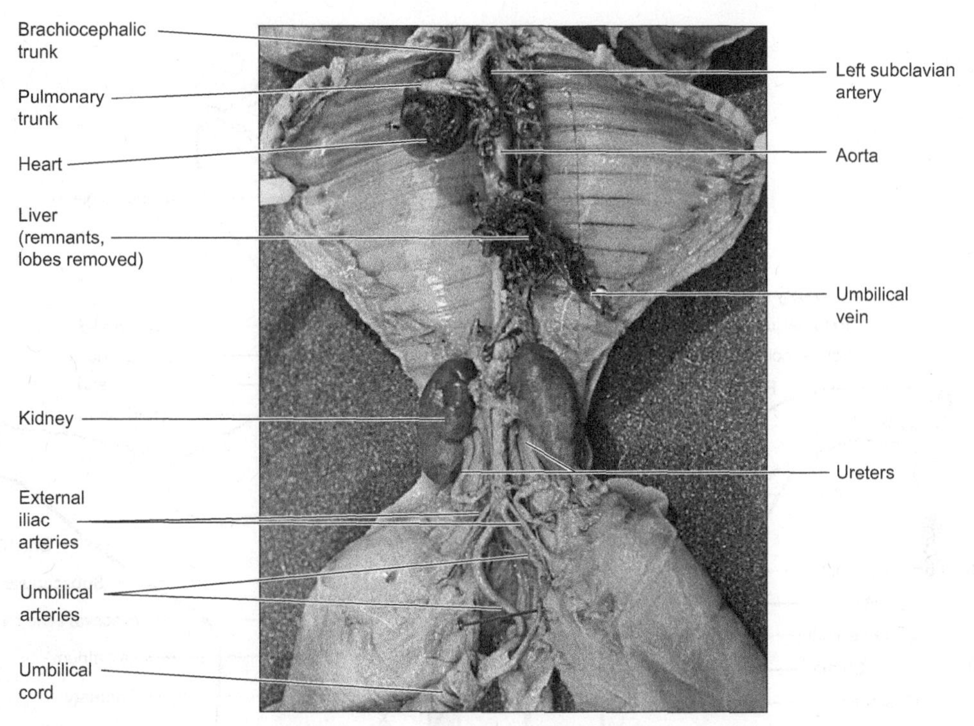

FIGURE 19.18 Fetal pig, thoracic and abdominal regions, principal blood vessels.
Photograph by John Vercoe.

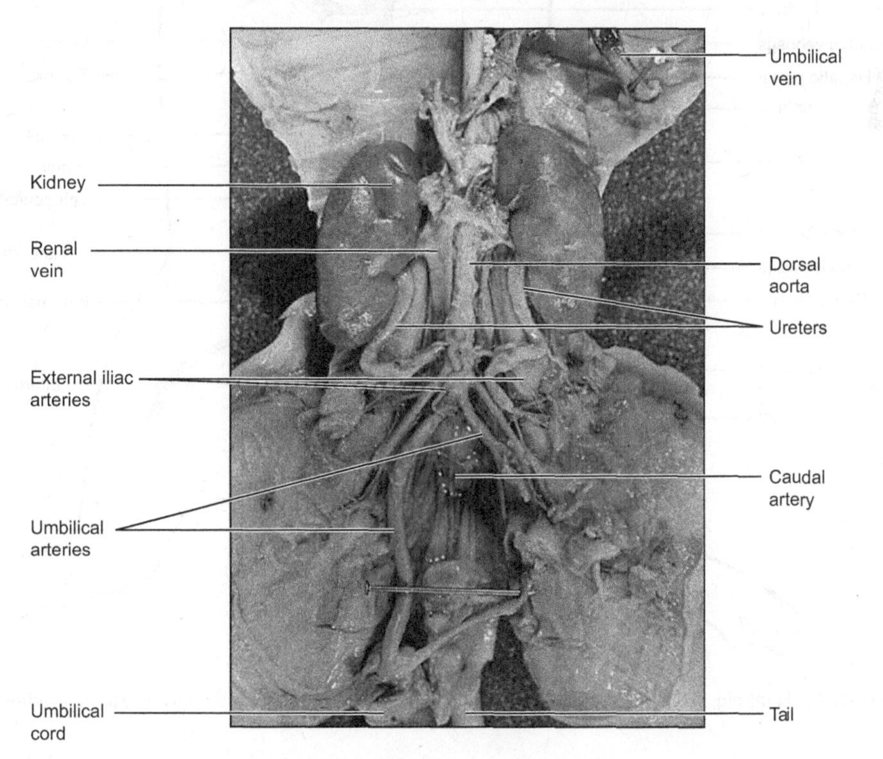

FIGURE 19.19 Fetal pig, abdominal pelvic regions, principal blood vessels.
Photograph by John Vercoe.

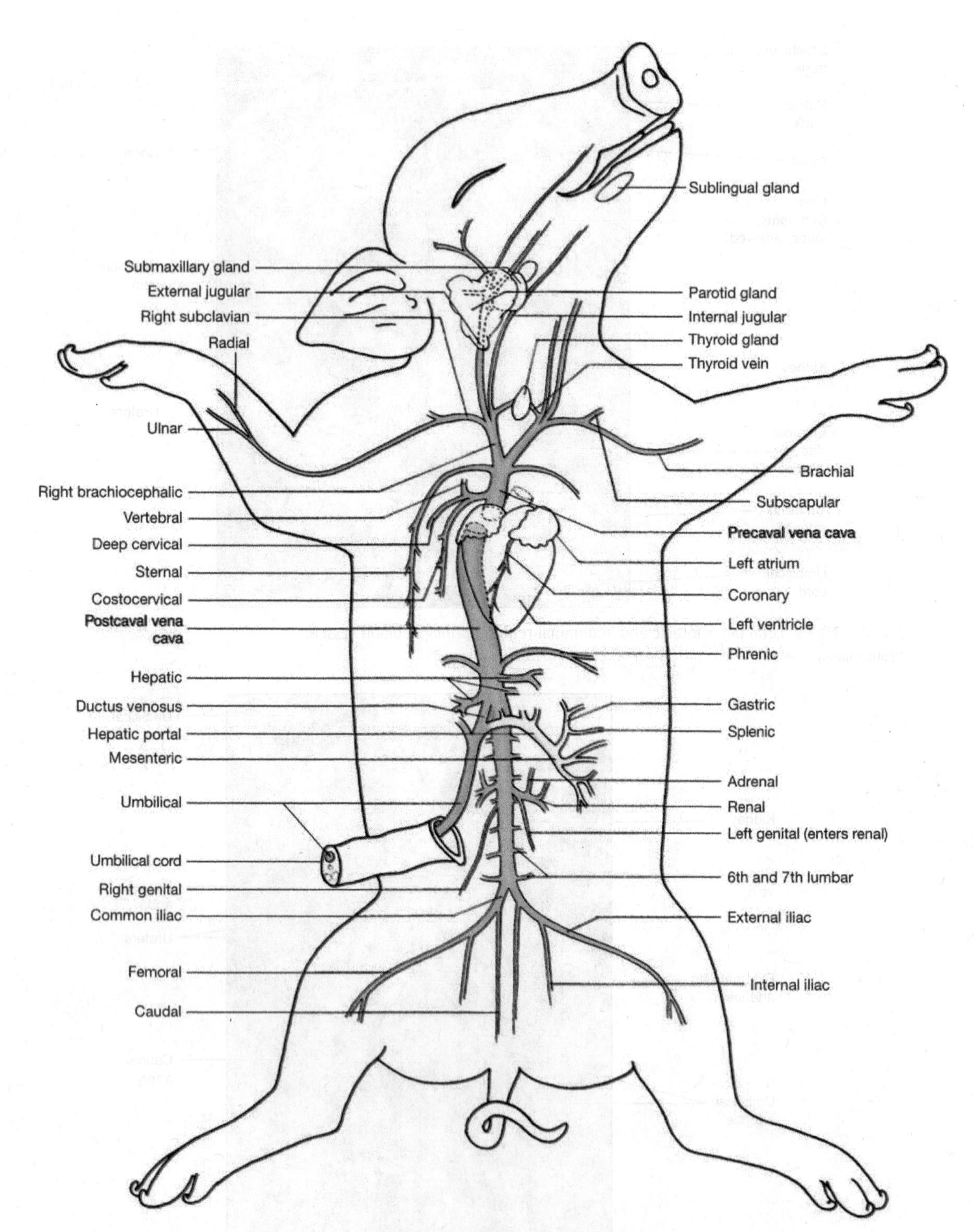

FIGURE 19.20 Fetal pig, venous system. Systemic veins are blue, and the hepatic portal system is yellow.

Trace each of the arteries listed from its origin along the route to the chief organs that it supplies with blood and observe its main branches. When you complete your study of the arterial system, you should be able to identify each of the main arteries and to describe the chief organs supplied by each.

After you have completed your study of the systemic arteries, you should next study the pulmonary circulation. Locate the **pulmonary trunk** leading from the right ventricle (figures 19.16 and 19.17) and trace it to the point where it branches into the **right** and **left pulmonary arteries.** These arteries carry the blood to the lungs and are relatively small in the fetus, but they become much larger at birth when the lungs become functional. Note also the short duct connecting the pulmonary trunk with the aorta. This duct is called the **ductus arteriosus** (figure 19.17); it is a **special fetal structure.** The duct ceases to function after birth and gradually degenerates to form a fibrous cord, the **ligamentum arteriosum.** Remember, in the adult pig, all of the blood in the pulmonary circulation passes to the lungs for oxygenation.

Another special feature of the fetal circulation is also related to the mixing of blood from the pulmonary and systemic divisions. Within the heart, blood passes from the right atrium into the left atrium through a temporary opening, the **foramen ovale,** located in the median wall of the heart. Blood returning to the heart through the postcaval veins is shunted through this opening and directly back into systemic circulation, thus bypassing the lungs. The foramen ovale becomes sealed at birth, and all of the blood from the right atrium thereafter goes to the lungs via the right ventricle, pulmonary trunk, and the remainder of the pathway. The former site of the foramen ovale appears as an oval depression in the medial wall of the heart of the adult pig and is called the **fossa ovalis.**

Venous System

Study figures 19.15 and 19.20 and identify the major veins of the pig, including the **vena cava,** the **right** and **left brachiocephalic veins,** the **subclavian veins,** the **brachial veins,** the **external** and **internal jugular veins,** the posterior vena cava, and the **hepatic veins.** Locate also the **hepatic portal vein** with its three main branches: the **gastric vein** carrying blood from the stomach, the **splenic vein** carrying blood from the spleen, and the **mesenteric vein** carrying blood chiefly from the small intestine. Find the **renal veins** (figure 19.19) leading from the kidneys, and from the hindlimbs find the **common iliac veins** into which blood flows from the **internal** and **external iliac veins.**

Locate the **umbilical vein** in the umbilical cord once again and trace it forward. Note the various branches that carry blood to the **hepatic portal vein** and into the liver. Note also that part of the blood from the **umbilical vein** is shunted directly to the **posterior vena cava** through the short **ductus venosus** located in the liver. Shortly after birth, the umbilical vein and the ductus venosus cease to function and begin to degenerate. Like the umbilical arteries, the ductus arteriosus, and the foramen ovale, the umbilical vein and the ductus venosus are functional only in the fetus.

- Compare the blood of the umbilical vein with that of the umbilical arteries in regard to the content of (1) oxygen, (2) nitrogenous wastes, (3) carbon dioxide, and (4) nutrients.

Observe the several **pulmonary veins** arising from the lungs and uniting to form two main **pulmonary veins,** which enter the left atrium of the heart.

Heart Anatomy

- You should supplement your study of the heart with a demonstration specimen of a dissected adult pig or cow heart. The structure of the four-chambered heart is similar in all mammals (figure 19.21). On the dissected heart, find the four chambers (two atria, two ventricles), and the valves between the atria and ventricles. Review the connection of the heart with the major blood vessels and the pattern of circulation in the pig with the aid of figures 19.15, 19.16, and 19.20.

After you have completed the review of adult heart structure, remove the heart of your fetal pig by severing the large blood vessels (about 1–2 cm from the heart) and by freeing the heart and the blood vessels from the surrounding tissues. Make a clean cut across the heart to expose the interior of the two ventricles. Wash out the clots of blood with cold water and note the interior structure of the heart.

Insert the tips of your scissors into the cavity within the right ventricle and cut through the ventricular wall to expose the cavity within the right atrium. Note the opening, or orifice, between the right ventricle and right atrium guarded by the **tricuspid valve.** Make a similar incision on the left side of the heart and observe the orifice between the two left chambers of the heart guarded by the **bicuspid valve.** ***How does the bicuspid valve differ in structure from the tricuspid valve?*** Next locate the opening from the left ventricle into the aorta guarded by the **semilunar valves.** Note that these semilunar valves consist of three pouchlike structures.

Fetal Circulation and Changes at Birth

The circulatory system of the fetal pig provides a good model for the organization of the circulatory system of an adult mammal, but it also exhibits certain adaptations of circulation in a developing fetus. These differences are due to the internal development of mammalian embryos within the uterus of the adult female. During embryonic development, the mammalian fetus is dependent upon the placenta for its supply of oxygen and nutrients and for the

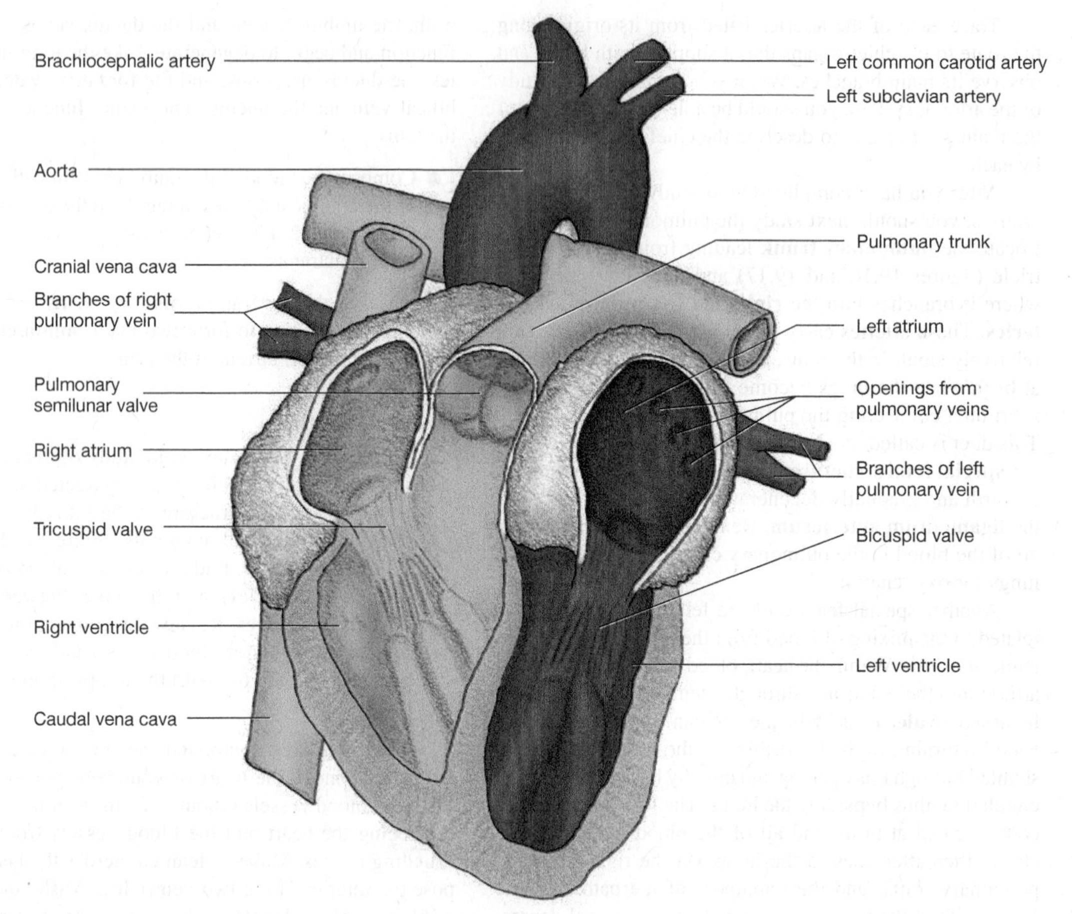

FIGURE 19.21 Mammalian heart, dissected, anterior view. Diagram based on human and rat models.

removal of carbon dioxide and other waste materials. At birth and shortly thereafter, several important changes occur in the circulatory system and in the pattern of circulation of the pig and other placental mammals. These changes are summarized in table 19.3.

Prior to birth, the fetus receives oxygen and discharges wastes through blood circulated via the **umbilical arteries** and **vein** connecting with the placenta. Indirect exchange of gases and nutrients occurs between the fetal blood and the maternal blood in the placenta. This exchange takes place between capillary beds of the fetal circulatory system by diffusion between the fetal and maternal capillary beds. Normally there is no actual mixture of fetal blood and maternal blood in the placenta unless there are ruptures of some of the small vessels in the placenta.

At birth, dramatic changes in fetal circulation take place. The umbilicus connecting the fetus with the mother is cut or broken, and removed from the circulatory pathway. The fluid-filled lungs of the fetus discharge the fluid and are filled with air. Blood flow through the lungs increases as the vascular resistance caused by the fluid surrounding the alveoli in the lungs is discharged. The lungs expand.

The **umbilical vein** no longer functions and the **ductus venosus** in the liver, which had previously carried blood from the aorta and the pulmonary artery closes.

The **foramen ovale,** an opening in the heart between the right and left atria that previously passed most of the blood incoming from the vena cava from the right atrium to the left atrium, closes shortly after birth.

The **ductus arteriosus,** which earlier had carried blood from the pulmonary trunk to the aorta, closes shortly after birth. This causes blood to flow through the pulmonary artery to the lungs and return to the left atrium.

Nervous System

Review figure 19.22, and observe the location and relative position of the principal components of the central nervous system of the fetal pig. Note the large anterior

Table 19.3 Changes in Fetal Circulation at Birth

Structure	Fetal Condition	After Birth
Placenta	Provides oxygen and nutrients and removes wastes via umbilical arteries and vein	Placenta lost, umbilical arteries and vein closed
Lungs	Fluid filled; fluid pressure restricts blood flow through lungs	Air filled after discharge of fluid; blood flow through lungs increases; lungs expand
Umbilical Vein and Ductus Venosus	Delivers oxygen-rich blood to fetus from placenta; blood passes through portal system in the liver through the ductus venosus and then into inferior vena cava	Vein and ductus venosus close at birth
Foramen Ovale	Blood from inferior vena cava passes through foramen ovale into left atrium of heart with some also entering right atrium	Foramen ovale closes after birth redirecting blood to systemic circulation
Ductus Arteriosus	Connects pulmonary trunk to aorta; blood enters right ventricle then to pulmonary through ductus arteriosus to aorta. Most blood bypasses lungs to enter systemic circulation.	Ductus arteriosus closes at birth redirecting blood to lungs via the pulmonary artery
Umbilical Arteries	Carry oxygen-poor blood from aorta back to placenta for oxygenation	Close at birth but remnants vascularize urinary bladder

brain enclosed in the bony **braincase** of the skull and the **spinal cord** within the **neural canal** of the vertebral column.

- Cut away the skin, muscles, and dorsal half of the skull to expose the brain.

Note the three membranes or **meninges** surrounding the brain. The tough outer layer that adheres to the skull is the **dura mater;** underneath is the delicate **arachnoid layer,** and the thin layer that dips into the crevices called **sulci** (singular: sulcus) of the brain is the **pia mater.** The same three meninges also cover the spinal cord but are more difficult to observe than on the surface of the brain.

Cut away the left side of the skull and carefully remove the dura mater. Locate and study the five main regions of the pig brain listed in table 19.4 and illustrated in figures 19.23 and 19.24.

It is often more convenient and satisfactory to study the anatomy of the mammalian brain with a specimen that has been preserved and prepared for this purpose. The brain of the sheep and the cat are most commonly used. The principal parts of the sheep brain are illustrated in figure 19.24. Figure 19.25 shows the cranial nerves on the ventral surface of a dissected cat brain.

You should also review your previous studies of the brains of the shark and the frog, and compare the brain of the pig with those of these two "lower" vertebrates. ***What are the principal differences in brain structure among these animals? Which parts of the brain are more highly developed in the pig? Which structures are less well-developed or absent in the pig? Can you relate these structural differences to differences in behavior of the three species?***

- Dissect away the skin, connective tissue, and muscles to expose the anterior portion of the vertebral column. Next remove the neural arches of several of the cervical (neck) vertebrae to observe the spinal cord within the neural canal.

Note in figure 19.22 the **enlargements** in the spinal cord in the **brachial** and in the **lumbosacral regions** and the **nerve plexuses** (networks) associated with these enlargements. Locate the **cervical** and **lumbosacral nerve plexuses** on your specimen by carefully dissecting away the tissues in these two regions of the vertebral column. After you have removed the muscles and connective tissues from the brachial region of the vertebral column, you will find an interconnected network of tough, whitish nerves that connect with the spinal cord. This is the **brachial plexus;** it is made up of branches from several of the spinal nerves of the pig.

There are 33 pairs of spinal nerves in the pig; other mammals may have more or fewer spinal nerves. The cat, for example, has 38 pairs of spinal nerves. The 33 pairs of spinal nerves in the pig include eight pairs in the cervical region, 14 pairs in the thoracic region, seven pairs in the lumbar region, and four pairs in the sacral region.

- Remove the skin, connective tissue, and muscles in the lumbosacral region to locate the **lumbosacral plexuses.**
- Use your scalpel or scissors to cut out a section of the spinal cord about 3 cm in length from the cervical region. Sever also the nerves attached to the spinal cord in this region, leaving enough nerve fiber to study the attachment of the nerves to the spinal cord.

Observe that each spinal nerve is formed by the union of **two roots,** one **dorsal** and one **ventral.** The former carries principally **sensory fibers,** and the latter carries principally

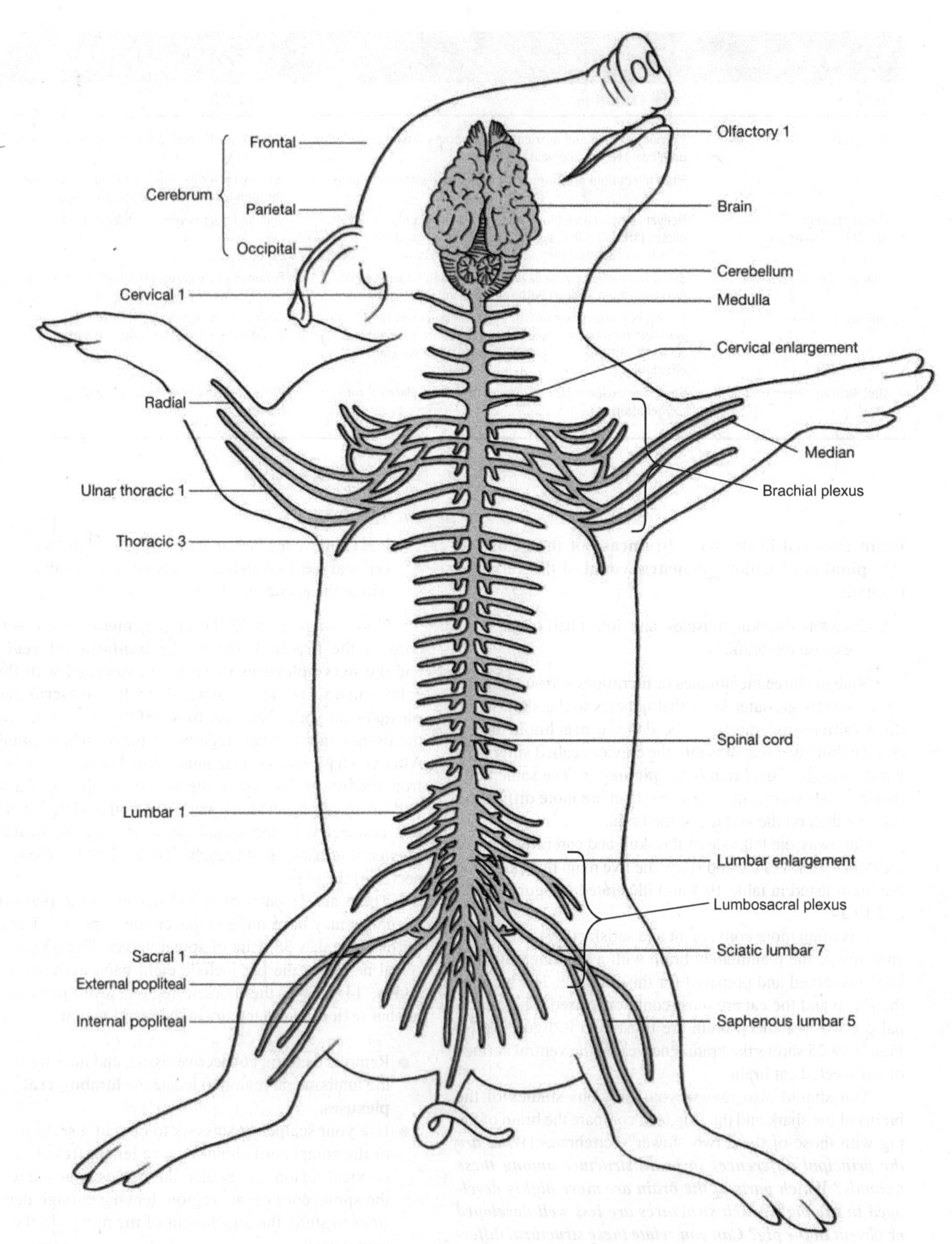

FIGURE 19.22 Fetal pig, central nervous system.

Table 19.4 Principal Regions of the Mammalian Brain

Region	Principal Structure(s)
Telencephalon	Cerebral hemispheres
Diencephalon	Anterior lobe of pituitary gland
Mesencephalon	Corpora quadrigemina
Metencephalon	Cerebellum
Myelencephalon	Medulla oblongata

motor fibers. Also observe the microscopic demonstration illustrating a cross section of a mammalian spinal cord.

The pig has 12 pairs of cranial nerves, the same number found in cats (figure 19.25) and humans. Nerves I–X correspond with those of the frog and the shark studied in previous exercises. Posterior to these are Nerve XI, the **spinal accessory nerve,** and Nerve XII, the **hypoglossal nerve** (figure 19.23). The spinal accessory nerve (XI) arises from several roots on the lateral surface of the spinal cord and medulla, and is made up of several motor and sensory fibers connected with the shoulder muscles. The hypoglossal nerve arises from several roots on the ventral surface of the medulla, and contains sensory and motor fibers connected with the tongue.

A helpful mnemonic device used by many students to help them remember the names of the cranial nerves of the pig and other mammals is:

On **O**ld **O**lympus' **T**owering **T**op, **A** **F**inn **A**nd **G**erman **V**iewed **A** **H**op.

Note that the first letter of each word corresponds to the first letter in the names of the twelve cranial nerves of mammals:

I. **O**lfactory
II. **O**ptic
III. **O**culomotor
IV. **T**rochlear
V. **T**rigeminal
VI. **A**bducens
VII. **F**acial
VIII. **A**uditory
IX. **G**lossopharyngeal
X. **V**agus
XI. **A**ccessory
XII. **H**ypoglossal

- Carefully free the brain, starting from the severed spinal cord in the cervical region and proceeding anteriorly, cutting nerves close to the skull on each side and gently freeing the brain from the skull. As you work, identify as many of the cranial nerves as possible, using figures 19.23, 19.24, and 19.25 as guides.

Find the **pituitary gland** and **optic chiasma** on the ventral surface, and with a *sharp* scalpel carefully bisect the brain from front to rear into equal right and left halves. Locate and study the various brain structures on your specimen with the aid of figures 19.23 and 19.24.

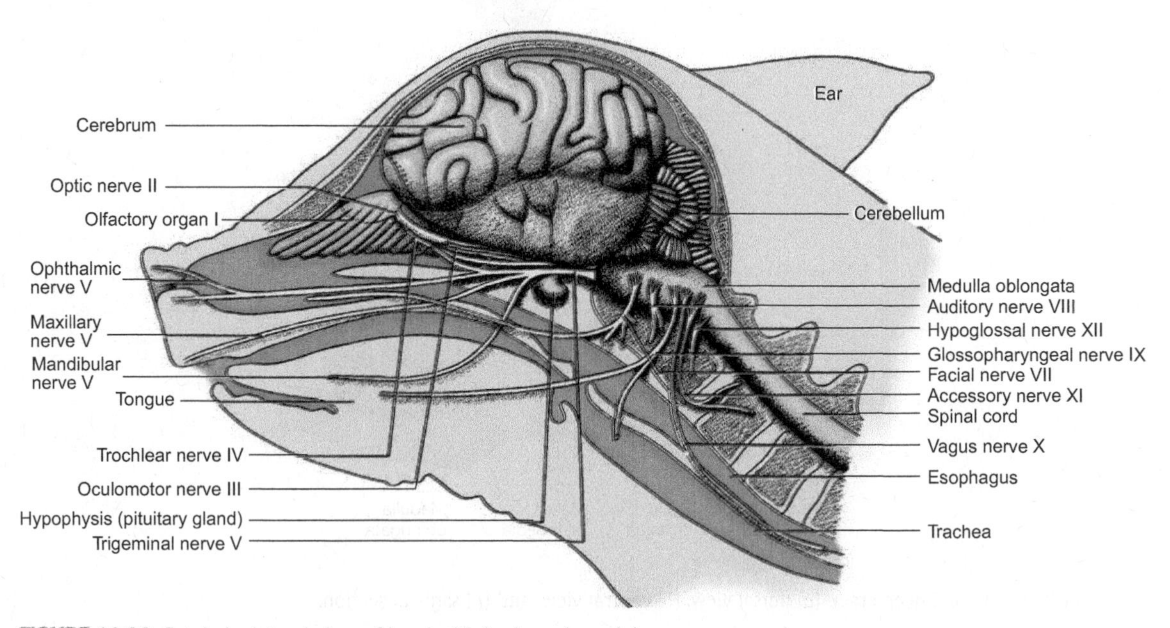

FIGURE 19.23 Fetal pig, lateral view of head with brain and cranial nerves.

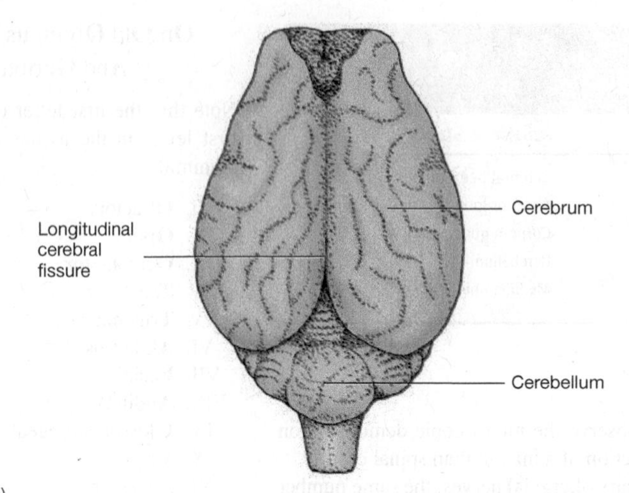

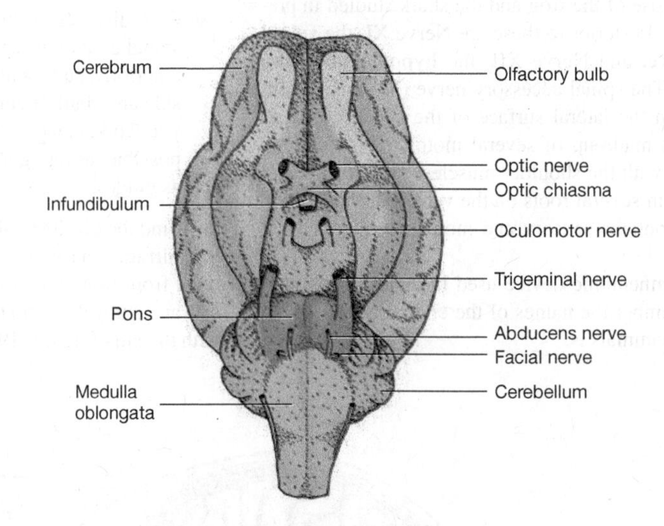

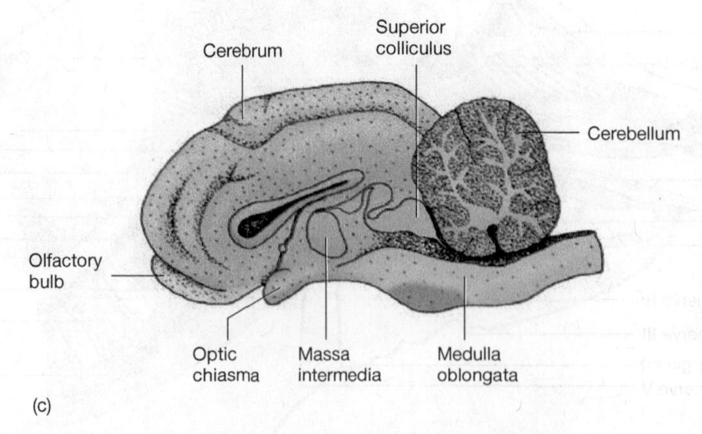

FIGURE 19.24 Sheep brain: (*a*) dorsal view, (*b*) ventral view, and (*c*) sagittal section.

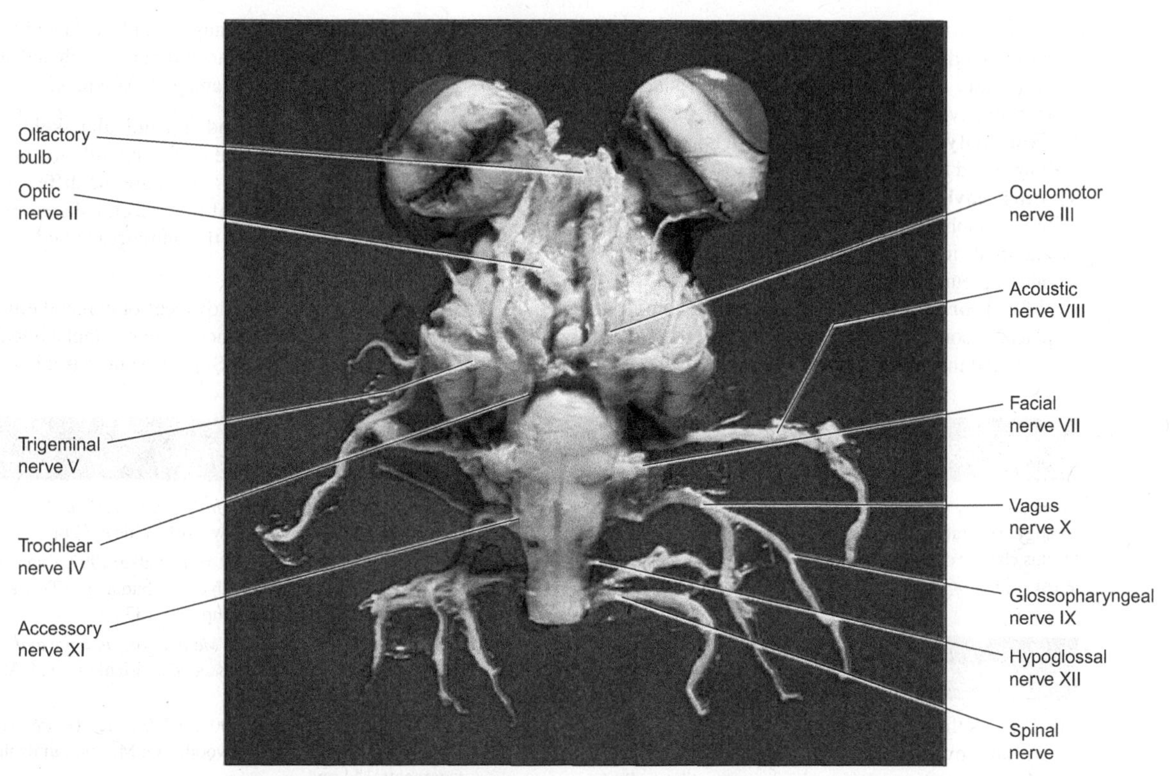

FIGURE 19.25 Cat brain, dissected, ventral view showing cranial nerves.
Courtesy of Carolina Biological Supply Company, Burlington, NC.

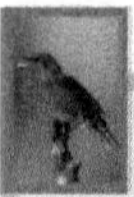

Demonstrations

1. Placentae of pig and other mammals
2. Skeleton of fetal pig
3. Mounted cat skeleton
4. Fresh or dried specimens of mammalian lung
5. Microscope slides of mammalian lung tissue
6. Microscope slide showing germ cells in mammalian ovary
7. Microscope slides showing mammalian testis tissue and developing spermatozoa
8. Microscopic demonstration of stained mammalian sperm
9. Dissection of fetal pigs to illustrate injected arterial and venous systems
10. Dissected pig, sheep, or beef hearts to illustrate interior chambers and valves
11. Preserved sheep brains, whole and sections
12. Dissected fetal pigs to show cranial and spinal nerves
13. Microscope slide to show cross section of mammalian spinal cord

Key Terms

Appendicular skeleton portion of the skeleton of the pig and other vertebrates that includes the bones of the appendages and the girdles, which provide articulation of the limbs with the axial skeleton.

Axial skeleton portion of the skeleton that provides support for the main axis of the body, including the skull, the vertebral column, and the tail.

Mesentery a double sheet of mesodermal epithelium that serves to support various internal organs in the coelom of vertebrates.

Pericardial cavity portion of the coelom that encloses the heart; lined by the pericardium.

Peritoneal cavity (Abdominal cavity) portion of the coelom lying posterior to the diaphragm; contains the abdominal organs.

Placenta structure formed by a combination of maternal and embryonic tissues that attaches the developing embryo to the inside wall of the mammalian uterus. It also enables the passage of nutrients and gases to the embryo and removal of wastes from the embryo without providing direct connection between the two separate circulatory systems. The placenta also produces certain reproductive hormones.

Pleural cavity portion of the coelom that encloses a lung; lined by the pleura.

Thoracic cavity portion of the coelom lying anterior to the diaphragm; subdivided in the pig and other mammals into the medial pericardial cavity and two lateral pleural cavities.

Umbilical cord the cord that attaches the fetus to the placenta contains the umbilical arteries, the umbilical vein, and the allantoic duct.

Internet Resources

There are many valuable Internet sites with information about zoology. Several sites containing pertinent zoological information for this chapter can be found on the McGraw-Hill Zoology web site at http://www.mhhe.com/zoology. Just click on this text's title.

Questions for Critical Thinking

1. Discuss the advantages of mammalian circulation (if any) over those of amphibians and fish. Again, does greater complexity always mean "better"? Is a four-chambered heart better than a three-chambered heart, or is a three-chambered heart better than a two-chambered heart? Explain your reasoning.
2. Discuss the changes that occur in circulation in the pig at birth. Specifically, the fetal pig emerges from an almost marine environment in the uterus to a terrestrial environment (in a matter of minutes). What would happen if any of these changes did not occur?
3. What is the advantage (if any) of a placenta and live birth, rather than a marsupium or pouch and an undeveloped fetus as in marsupial mammals?
4. Compare the vertebral and appendicular skeletons of the fetal pig with those of the amphibian and fish. What are the similarities? What are the differences? Are there any remnants of the visceral skeleton of the fish identifiable in the amphibians and mammals?
5. Discuss the evolutionary advances of homeothermy versus poikilothermy. Which system is metabolically "more expensive." Why? Support your reasoning.

Suggested Readings

Allen, C., and V. Harper. 2003. *Fetal Pig Dissection: A Laboratory Guide,* 2d ed. New York: Wiley. 32 pp

Chaisson, R.B., Odlaug, T.O., and W.J. Radke. 1997. *Laboratory Anatomy of the Fetal Pig,* 11th ed. Dubuque, IA: Times Mirror Higher Education Group, Inc. 133 pp

Donnelley, P.J. 1997. *Laboratory Manual for Anatomy and Physiology with Fetal Pig Dissections.* Menlo Park, CA: Benjamin Cummings. 662 pp

Smith, D.G., and M.P. Schenk. 2003. *A Dissection Guide and Atlas to the Fetal Pig.* Englewood, CO: Morton Publishing Company. 144 pp

Walker, W.F., and D.G. Hamburger. 1998. *Anatomy and Dissection of the Fetal Pig,* 5th ed. New York: W.H. Freeman. 120 pp

Notes and Sketches

Chapter 20

Rat Anatomy

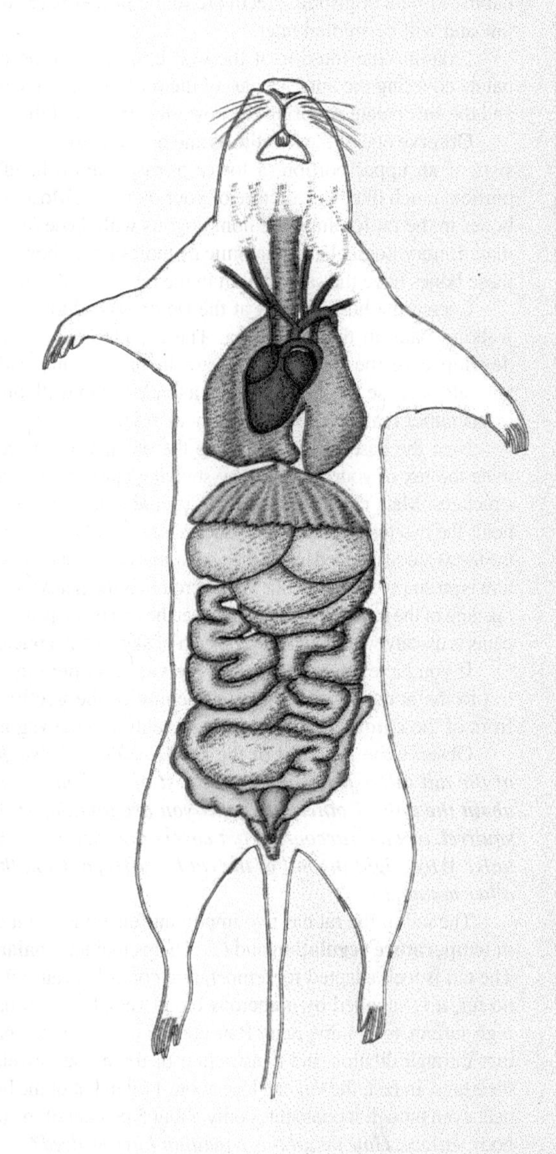

Objectives

After completing the laboratory work in this chapter, you should be able to perform the following tasks:

1. Locate and identify the principal external features of the rat. Explain which external features distinguish it as a mammal.
2. Identify the main elements of the axial and appendicular skeletons of the rat.
3. Locate five major skeletal muscles of the rat and identify their origins, insertions, and actions. Distinguish among extension, flexion, adduction, and abduction.
4. Locate and identify the principal organs in the digestive system of the rat.
5. Explain the general pattern of circulation in the rat and other mammals, and discuss the significance of the four-chambered heart.
6. Describe and identify the principal organs of the urogenital system of the rat. Demonstrate differences between male and female rats.
7. Identify the main parts of the brain of the rat or sheep, and discuss the basic organization of the mammalian nervous system.

Introduction

The white laboratory rat, an albino form of the Norway rat *Rattus norvegicus,* is often studied as a typical representative of the organs and organ systems found in mammals. Among the distinguishing features of mammals (Class Mammalia) are a body surface covered with hair, an integument (skin) with mammary and other types of glands, a skull with two occipital condyles, seven cervical (neck) vertebrae, teeth borne on bony jaws, movable muscular lips and fleshy external ears (pinnae), a four-chambered heart, a persistent left aorta, and a muscular diaphragm separating the thoracic and abdominal cavities. All mammals are homeothermic ("warm-blooded"). Their high metabolism generates enough heat for them to maintain a constant body temperature that is warmer than the surrounding air or water.

The young develop within the uterus of the female, have a placental attachment for nourishment, and are enveloped by special fetal membranes (amnion, chorion, and allantois). Milk to nourish the young after birth is produced by mammary glands. Other common representatives of this group include moles, bats, whales, mice, deer, monkeys, horses, cattle, and humans.

Materials List

Preserved specimens
- White rat, double- or triple-injected
- Sheep brain
- Beef heart

Plastic or liquid mounts
- Rat dissection

Prepared microscope slides
- Rat testis, cross section
- Rat kidney, median section
- Rat heart, longitudinal section

Other
- Mounted rat skeleton
- Mounted cat skeleton

External Anatomy

- Select a preserved specimen and place it in a dissecting pan. Observe the two principal external features that distinguish the rat as a mammal: the **hair** covering most of the body and the paired **mammary glands.** Locate the paired nipples or **teats** on the ventral surface of the trunk between the forelimbs and hindlimbs. ***How many pairs are present? Are they present on both male and female rats?***

The body of the rat is divided into an anterior **head** connected to a cylindrical **trunk** by a short, thick **neck.** A long **tail** extends posteriorly from the trunk. Internally, the trunk is divided by a muscular **diaphragm** into an anterior **thorax** and a posterior **abdomen.** The diaphragm is another distinguishing feature of mammals; it is a muscular sheet that divides and separates the coelom into the paired **anterior pleural cavities,** a single **pericardial cavity,** and the **posterior abdominal cavity.** The diaphragm is lacking in birds, reptiles, and other lower vertebrates.

Study the cone-shaped head, which has an elongated (prognathous) face. On the head, find the two eyes and two ears. The eyes have upper and lower eyelids and a reduced third eyelid or **nictitating membrane,** which can be found on the medial portion of the eye opening beneath the two outer lids.

Note that each ear has an external fold of tissue called the **pinna,** which aids in directing sound waves to the opening of the ear, the **external auditory meatus.** Other important sense organs located on the head are the **vibrissae**—long sensory hairs that provide the rat with a very effective sense of touch. The vibrissae can tell a rat in an instant if a hole is large enough for him to crawl into. Note that most of the vibrissae are attached to the upper lip. ***Where else do you find vibrissae?***

Observe the well-developed upper and lower **lips** surrounding the mouth. In the center of the upper lip, find the **philtrum,** a groove or cleft separating the lip into right and left halves. Locate the two **external nares** above the upper lip. Inside the mouth find the two long, sharp **incisor teeth** that are characteristic of rodents. These two incisors grow continuously, and the rat wears them down by its gnawing habits. Molars or grinding teeth are found farther back in the jaw and will be studied later.

Examine the interior of the oral cavity. Find the hard palate covering the anterior part of the roof of the oral cavity and the soft palate covering the posterior portion of the roof.

Observe one of the **forelimbs** and note that the limb consists of an upper portion, a lower portion, and a handlike portion, much like the structure of your own arm. Most of the bones in the rat forelimb are homologous with those of a human. Embryological and anatomical studies have shown that these bones have the same origin in the rat and in the human.

Locate the horny **claws** at the tip of each digit and the walking pads in the palm area. These pads are less well-developed in the rat than they are in the cat and catlike animals because rats have evolved a tendency to walk on the digits rather than on the entire palm or sole.

Near the base of the tail locate the **anus.** You can determine the sex of your specimen by studying adjacent urogenital structures. Male rats have a **scrotum** ventral to the anus, which holds the two **testes** during reproductive season. (At other times, the testes are suspended inside the abdominal cavity, and the scrotum is an empty sac.) Anterior to the scrotum is the **penis** with the opening of the male urogenital system at the end of the penis. The penis is usually withdrawn into a sheath of skin, the **prepuce.**

If you have a female rat, find the **vaginal opening** ventral to the anus and the separate opening of the **urethra** in front of the **clitoris,** which is located ventral to the vagina.

Observe the long **tail** of the rat. ***How does the surface of the tail differ from that of the rest of the body? Think about the tails of other mammals you are familiar with, a squirrel, a cat, a raccoon. What covers the surface of their tails? Why might the tail of the rat be different from these other mammals?***

The tail of the rat has two important functions: (1) it aids in **temperature regulation** and (2) it helps maintain **balance.** The tail is well adapted for temperature control because it has no fur, it is supplied by numerous blood vessels, and it has a high surface-to-volume ratio. Rats control their body temperature through dilation and constriction of the blood vessels in their tails. In fact, the tail can lose about 17 percent of the body heat even though it constitutes only about 5 percent of the total body surface. ***How might this condition have evolved?***

Rats are very quick and agile. They can walk on wires, fences, and small tree branches; they can climb ropes and the anchor cables of ships. Next time you see an ocean liner or freighter, look, for the **rat guards** on the lines holding the ship to the dock. Rats use their long tails for balance in much the same fashion as an acrobat uses a balancing pole in crossing a high wire. The tail helps the rat keep its center of gravity directly above a wire or rope, thus preserving its balance.

Fact File

Rats

- Laboratory rats and mice are special breeds, raised for use in medical and scientific research. Many strains are now available with carefully selected genetic traits important for specific kinds of research projects. Most rats sold for laboratory dissection, however, are from those raised for pet food for reptiles and carnivorous mammals because they are much cheaper than the specially selected strains raised for research.

Skeletal System

The skeleton of all vertebrates exhibits a common basic design, and an examination of the skeletons of a frog, a rat, a cat, and a human reveals many fundamental similarities. Since the bones of the rat are small, biology and zoology students often use a cat skeleton to learn more about the organization of the mammalian skeleton.

- Compare a mounted cat skeleton to the rat skeleton illustrated in figure 20.1. ***How are they similar? How are they different? Can you relate these differences to the mode of life for each type of mammal?***

Muscular System

A thorough study of the muscular system of the rat would take more time than is available in most introductory zoology courses. In this exercise, we will limit our study to a few of the major muscles of the ventral surface of the thorax and shoulders to provide an example of the organization of the muscular system.

Skinning

- Prior to the study of the muscles, you must first skin the appropriate portion of the body. Most preserved and injected rats have an incision on the midventral surface of the neck where they were injected. Start with this opening and free the skin from the underlying muscle and continue the incision posteriorly, stopping just anterior to the anus as shown in figure 20.2. Extend the incision anteriorly to the mouth and then cut around the side of the mouth.

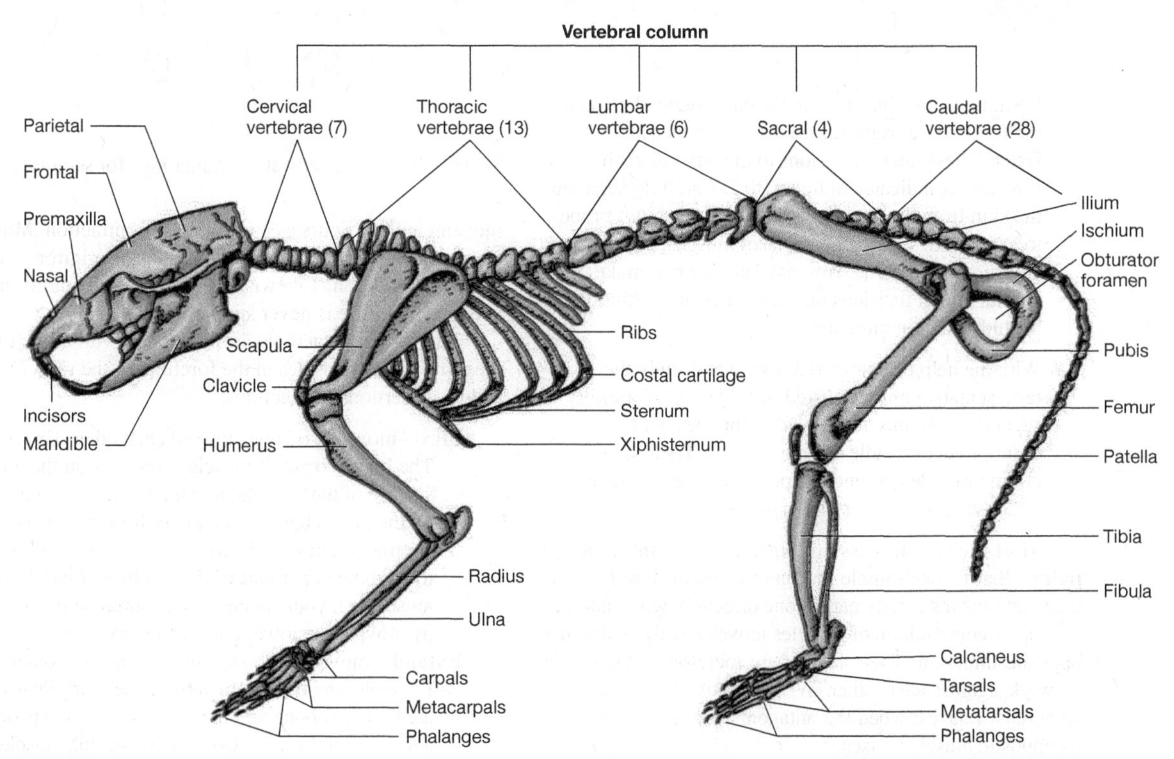

FIGURE 20.1 Rat skeleton.

Fact File

- Rats are social animals and live in colonies with well-defined territories that they mark with urine and glandular secretions. Rats are aggressive and actively defend their territories; social conflicts are most common during mating season, at feeding sites, prime nesting sites, and territorial boundaries. Females fiercely defend their nests and young from other rats.
- With a healthy food supply and adequate shelter, a female Norway rat can produce more than 50 offspring per year. Gestation takes 21–23 days, average litter size is about seven, and there may be six to eight litters per year. The young are born blind, naked, and helpless. Their eyes open in 14–17 days and they are weaned about 3 weeks after birth.
- Despite its name, the Norway rat (*Rattus norvegicus*) is probably native to Japan or the northeastern mainland of China. As a stowaway on sailing ships, it traveled the world, spreading into Europe during the sixteenth century and later arriving in North America around 1775. Today, this species is a common pest associated with human dwellings, factories, wharves, and other structures in all major cities of the world.
- The Order Rodentia is by far the largest order in the Class Mammalia. With over 2,000 species in 30 families, rodents make up more than 40 percent of all mammalian species.

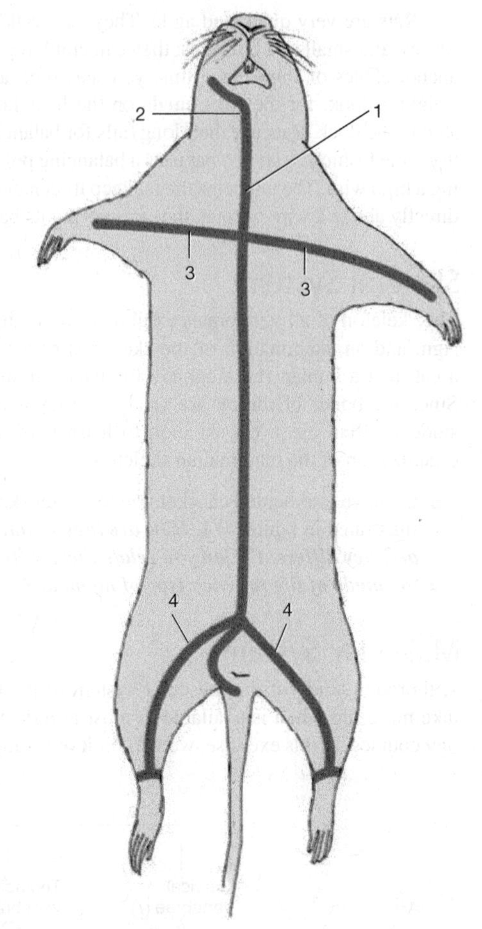

FIGURE 20.2 Rat, ventral view, incisions for skinning.

- Next, make two incisions in the chest region between the forelimbs at right angles to the first incision. Extend these lateral incisions to the wrist of each forearm, as indicated in figure 20.2. Carefully separate the skin from the underlying tissue with a blunt probe to expose the superficial (surface) muscles of the trunk and shoulder regions. You may later wish to make similar lateral incisions on the surface of the hindlimbs to study them in more detail.
- With the help of figure 20.3, locate and study the representative muscles listed in Table 20.1. Carefully separate each muscle from adjacent ones with a blunt probe or handle of your forceps. Try to avoid cutting muscles as much as possible except to reveal underlying muscles or other tissues.

Most skeletal muscles are organized into **antagonistic pairs**—that is, one muscle or combination of muscles contracts and moves a body part in one direction, while another muscle or combination of muscles moves a body part in the opposite direction. Their actions are therefore antagonistic or work against each other. When a muscle contracts, its antagonist relaxes; when the antagonistic muscle contracts, its opposite muscle relaxes.

Skeletal muscles have a fixed end, called the **origin,** and a movable end, called the **insertion.** A muscle's **action** is the movement that occurs as a result of its contraction. Muscles may actually have more than one point of origin and more than one insertion. Likewise, they may have more than a single action. Life is never simple.

The following actions are illustrated with some examples from major muscles in the forelimb of the rat with their origins, insertions, and actions.

Flex—moves two bones toward each other, Example: The **biceps brachii** muscle originates on the medial surface of the scapula near the humerus and inserts on the radius bone of the lower forelimb. This contraction moves the lower part of the forelimb toward the upper part of the forelimb. Find these muscles on your specimen and simulate the action by moving the lower part of the forelimb.

Extend—moves two bones away from each other. Example: the **triceps brachii** muscle originates on the humerus bone and the scapula and inserts on a tendon on the ulna. Contraction of this muscle moves the lower part of the forelimb away from the upper portion of the forelimb.

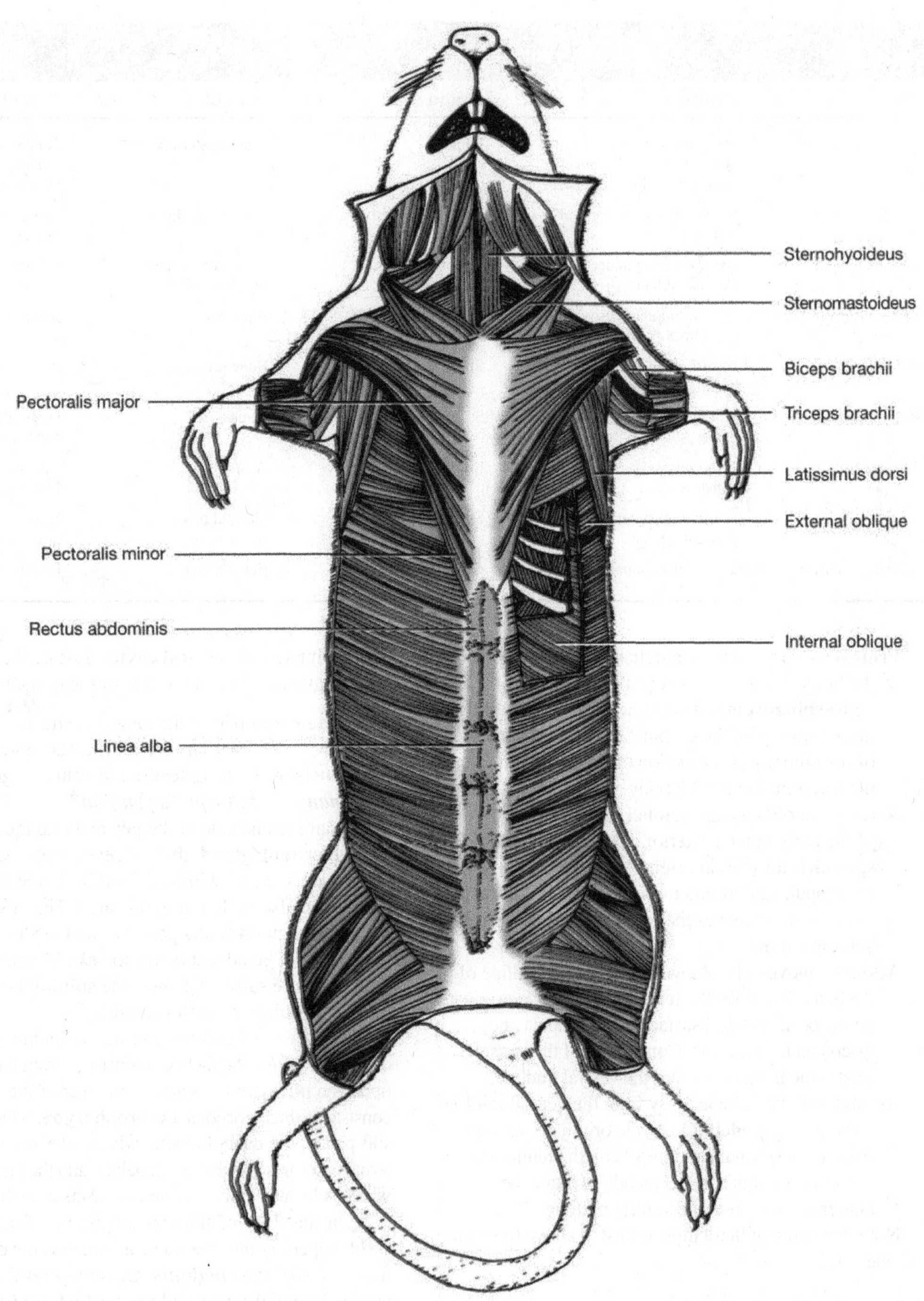

FIGURE 20.3 Rat, muscles, ventral view.

Table 20.1 Major Muscles of the Ventral Surface and Shoulders

Muscle	Location	Origin	Insertion	Action
Rectus abdominis	Paired muscles adjacent to midventral line of connective tissue (linea alba)	Pubis	First rib, sternum	Supports abdominal organs; flexes vertebral column
External oblique	Flat sheet covering ventral surface of trunk	Ribs and spinal fascia	Ilium, pubis, linea alba	Supports abdominal organs
Internal oblique	Flat sheet lying under external oblique	Spinal fascia, ilium	Cartilages of false ribs, linea alba	Supports abdominal organs
Pectoralis major	Large triangular muscle lying midventral to shoulder girdle	Sternum	Humerus	Adducts forelimb
Pectoralis minor	Posterior and beneath pectoralis major	Sternum	Humerus	Adducts forelimb
Biceps brachii	Anterior ventral surface of humerus	Scapula	Radius	Flexes lower part of forelimb
Triceps brachii	Posteriodorsal surface of humerus	Humerus	Ulna	Extends lower part of forelimb
Sternomastoideus	Side of neck anterior to pectoralis major	Sternum	Mastoid process of skull	Turns head sideways or back of head down
Sternohyoideus	Medial to sternomastoid	Occipital bone of skull	Hyoid bone	Elevates hyoid

Protract—moves a bone parallel to the long axis of the body in the anterior direction. Example: the **supraspinatus** muscle originates on the anterior lateral surface of the scapula and inserts on the head of the humerus. Contraction of this muscle moves the forelimb toward the body and the head.

Retract—moves a bone parallel to the long axis of the body in the posterior direction. Example: the **spinodeltoid** muscle originates along the spine of the scapula and inserts on the humerus. Contraction of this muscle moves the forelimb toward the body and toward the tail.

Adduct—moves a bone toward the ventral midline of the body. Example: the **teres major** muscle originates on the posteriodorsal surface of the scapula and inserts on the humerus. Contraction of this muscle rotates the humerus toward the ventral midline.

Abduct—moves a bone away from the ventral midline. Example: **spinodeltoid** muscle originates along the spine of the scapula and inserts on the humerus. Contraction of this muscle tends to rotate the humerus away from the ventral midline.

Note that some of these muscles listed above have more than one action.

Internal Anatomy

Mouth and Pharynx

We will begin our study of internal organs with the mouth and pharynx.

- Cut through the muscles, skin, and bones at the corner of the mouth on each side with a pair of scissors or bone cutters. Push down the lower jaw and examine the interior of the **oral cavity.** Locate the muscular **tongue** attached at the rear of the oral cavity.

The anterior part of the tongue is attached ventrally with a thin sheet of tissue, the **frenulum.** Rats have two types of teeth: **incisors** (cutting teeth) and **molars** (grinding teeth). ***How many of each type do you find?***

Examine one side of the jaw and find the large **extraorbital lacrimal gland** that releases tears. Other lacrimal glands are located within the orbit. Three paired salivary glands can also be found in this area. The largest of these is the **parotid gland** found posterior and medial to the extraorbital lacrimal gland just below the ear. Medial to the parotid, find two more salivary glands, the **sublingual** (anterior) and the **mandibular** (posterior) **gland.**

Find the **hard palate** covering the anterior part of the roof of the oral cavity and the **soft palate** covering the posterior portion. The **pharynx** is located at the rear of the oral cavity and consists of three portions: the **oropharynx,** which is below the soft palate; the **nasopharynx,** which receives air from the external nares lies above the soft palate; and the **laryngopharynx,** which is located at the rear and connects with the esophagus.

On the floor of the laryngopharynx find the **glottis,** a slitlike opening into the trachea, which is covered by a small flap of tissue, the **epiglottis.** Opening into the nasopharynx are the **internal nares** and the **eustachian tubes;** the latter connect with the middle ears. The nasopharynx leads to the more posterior laryngopharynx.

The Coelom

The coelom of the rat is divided into two major regions by a muscular partition, the **diaphragm** (figures 20.4 and 20.5). Anterior to the diaphragm is the **thoracic cavity,** which is subdivided into **two pleural cavities** containing the lungs; and the **pericardial cavity** (mediastinum), which contains the heart and the large blood vessels connected with the heart. The thin epithelial lining of the pleural cavity is the **pleura;** that of the pericardial cavity is the **pericardium.** Posterior to the diaphragm is the large **peritoneal** (abdominal) **cavity** within which the abdominal organs are suspended. Each of the abdominal organs is enclosed in a thin layer of mesodermal epithelium **(visceral peritoneum);** a similar layer also covers the inner surface of the body wall **(parietal peritoneum).** The double sheets of peritoneum that support the abdominal organs are called **mesenteries.**

- To study other features of internal anatomy, you must cut through the ventral muscles to expose the organs within the coelomic cavity. Make an incision through the muscles just to the right of the linea alba (ventral midline). Start just anterior to the anus and continue anteriorly to the neck region. Cut through the pectoral girdle and pull apart the ribs in the chest region to expose the heart and lungs. Make two transverse cuts through the skin and superficial muscles on each side just behind the pectoral girdle and another pair of cuts just in front of the pelvic girdle as shown in figure 20.4. Take care not to damage the organs inside the abdominal cavity beneath the superficial muscle layers. Free the diaphragm from the body wall and pin aside the skin and body wall to expose the abdominal organs as in figure 20.4.

Digestive System

The digestive system consists of the mouth, oral cavity, pharynx, esophagus, stomach, small intestine, large intestine, and rectum, and terminates at the anus. Several parts of

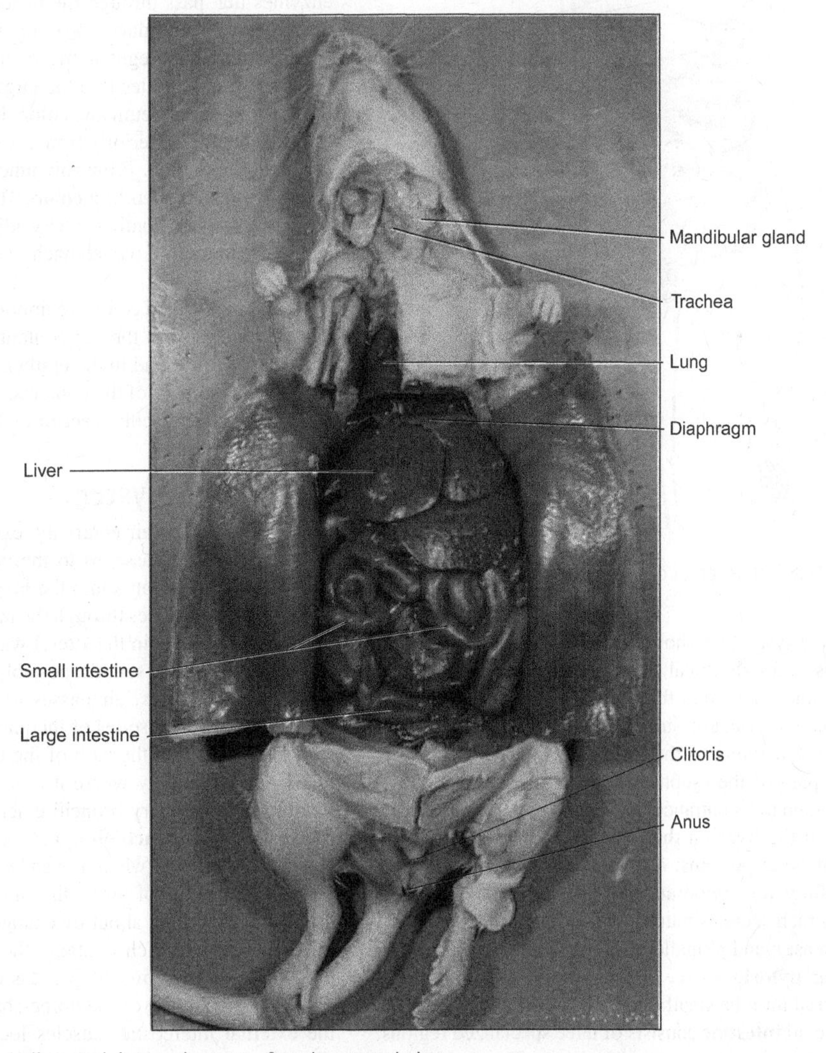

FIGURE 20.4 Rat, dissected, internal organs, female, ventral view.
Photograph by Ken Taylor.

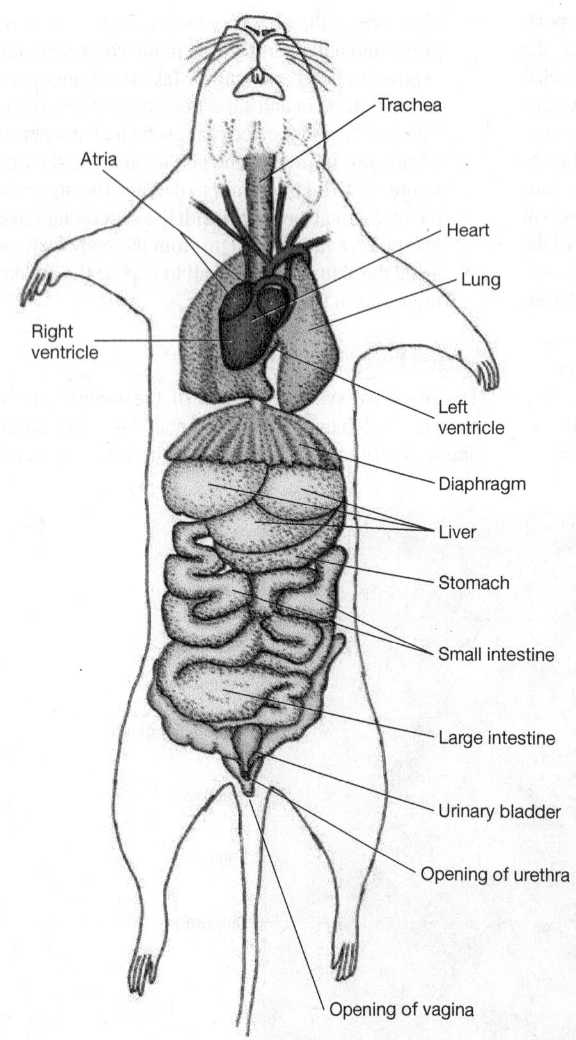

FIGURE 20.5 Rat, internal organs, female.

the digestive system are shown in figures 20.4 and 20.5. The **esophagus** is a cylindrical tube leading from the pharynx to the stomach. It enters the **stomach** on the **lesser curvature** (medial concave surface). The **greater curvature** of the stomach refers to the larger convex lateral surface. Trace the path of the esophagus (dorsal to the trachea) on your specimen to its entrance into the stomach. The stomach lies beneath the liver on the left side of the abdomen and consists of three portions: a **forestomach** (fundus), which serves mainly for temporary storage; a glandular **middle portion,** which secretes mucus, hydrochloric acid, and pepsin (a protease); and a smaller posterior **pyloric region** with a muscular **pyloric valve,** which controls the passage of food material into the small intestine.

The small **intestine** consists of three specialized regions: the **duodenum,** the **jejunum,** and the **ileum.** The anterior portion of the small intestine is the duodenum. It is 25–30 cm in length and receives ducts from the liver and pancreas. Most digestion continues here. Distal to the duodenum is the shorter jejunum. The portion of the small intestine posterior to the jejunum is the ileum. The exterior surface of the ileum often feels lumpy because of the presence of numerous lymph nodes embedded in its walls. These lymph nodes are involved in the absorption of lipids from the intestine.

Locate the large **liver,** which has four lobes. Bile ducts from each of the lobes join to form a single duct carrying bile to the duodenum. Note that rats typically lack, or have a reduced gallbladder. Anterior to the liver, locate the flat, muscular **diaphragm** (figure 20.5), which partitions the coelom into the thoracic and abdominal cavities.

The **pancreas** is not a discrete organ in the rat, but consists of a diffuse pinkish or tan tissue embedded in the **mesentery** (the thin sheet of connective tissue) between the duodenum and the stomach. The pancreas secretes digestive enzymes that pass through the pancreatic duct to the small intestine. It also produces two important hormones, **insulin** and **glucagon,** that regulate the level of blood glucose.

The ileum empties into the large intestine, or colon. At the junction of the ileum and colon, find the **caecum,** a blind sac that extends posteriorly from the **ileocolic valve** between the ileum and colon. From this junction, the colon extends anteriorly as the **ascending colon.** The **transverse colon** extends across the abdominal cavity adjacent to the diaphragm, and, in the region of the stomach, it loops posteriorly as the **descending colon.**

The caecum and colon are important in the resorption of ions and water from the gut contents. Also, numerous mucous glands are found in the epithelium of the colon, which facilitate movement of the contents. The terminal portion of the colon is the muscular **rectum,** which leads to the **anus.**

Respiratory System

During respiration, air enters the external nares and passes through the nasal passages to the nasopharynx and downward through the glottis into the **larynx.** The air is warmed and filtered as it passes through the nasal passages. The presence of vocal cords in the lateral walls of the larynx allows rats to make audible sounds, possibly for communication.

From the larynx, air passes to the **trachea,** a hollow tube supported by a series of incomplete cartilaginous rings (figure 20.6). Trace the path of the trachea from the larynx to the thoracic cavity where it branches into **two primary bronchi.** The primary bronchi enter the lungs, branch further into many **branchioles,** and terminate in many highly vascularized **alveoli** where the gas exchange takes place.

The exchange of air in the lungs of the rat and other mammals is brought about by changes in the volume of the thoracic cavity, which contains the lungs. During normal breathing, **inspiration,** the process of taking in air, results from the contraction of the dome-shaped diaphragm and of the external intercostal muscles located between adjacent ribs. These contractions enlarge the thoracic cavity, lower the pressure in the cavity below the external atmospheric

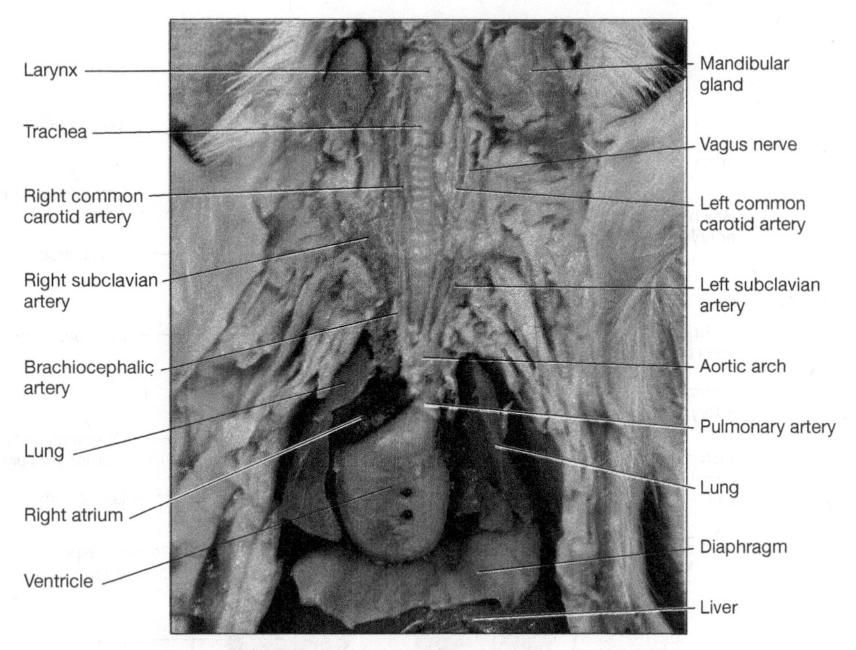

FIGURE 20.6 Rat, dissected, neck and heart regions, ventral view.
Photo courtesy of Betty Black.

pressure, and allow outside air to enter the lungs. **Expiration,** the expulsion of air from the lungs, occurs when the diaphragm and the external intercostal muscles relax, and the volume of the thoracic cavity decreases.

Mammals are therefore **negative pressure breathers** since air enters the lungs because of the negative pressure created inside the thoracic cavity by muscular contractions. In contrast, frogs and other amphibians are **positive pressure breathers** because in these animals air is forced into the lungs by swallowing-like movements produced by the muscles of the mouth cavity and throat, thus increasing the pressure within the lungs above that of external atmospheric pressure.

Circulatory System

The circulatory system of the rat exhibits the same basic pattern characteristic of all mammals, including humans. There are two major features: (1) a **four-chambered heart,** and (2) two separate divisions for circulation of the blood—a **systemic circulation** and a **pulmonary circulation.**

The systemic division carries oxygenated blood from the heart through a branching system of arteries to most organs and tissues of the body, collects deoxygenated blood from these organs via a system of capillaries and veins, and brings the blood back to the heart. The pulmonary division carries deoxygenated blood from the heart to the lungs for oxygenation and returns the oxygenated blood to the heart.

This pattern of circulation shown by mammals represents a major advance over the patterns exhibited by "lower" vertebrates such as the shark or the bony fish, which have a two-chambered heart and a single circulatory circuit, and the frog, with its three-chambered heart and an incomplete double circuit, which allows some mixing of oxygenated and deoxygenated blood in the heart.

In mammals, deoxygenated blood from the body (except the lungs) returns to the right atrium by the large cranial vena cava and caudal vena cava (figure 20.7). From the right atrium, blood goes to the right ventricle and is pumped via the pulmonary arteries to the lungs to be oxygenated.

Oxygenated blood from the lungs passes through the pulmonary veins to the left atrium of the heart. From the left atrium, the blood goes to the left ventricle and is pumped through the aorta and its branches to all parts of the body except the lungs.

The best specimens for study of the anatomy of the circulatory system are preserved rats that have been injected with colored latex. Specimens may be single-, double-, or triple-injected. Single-injected specimens have red latex injected into the arterial system. In double-injected specimens, the arteries are red and the veins are blue; triple-injected specimens are similar to double-injected specimens, but also have yellow latex injected into the hepatic portal system.

Heart and Arterial Circulation

- Carefully remove the pericardial membranes surrounding the heart. Take care not to rupture the blood vessels attached to the heart. Study figure 20.7 and locate the two small, saclike **atria** lying on the surface of the two muscular **ventricles.** Between the right atrium and the right ventricle is the tricuspid valve, and between the left atrium and the left ventricle is the bicuspid valve. Note that the ventricles

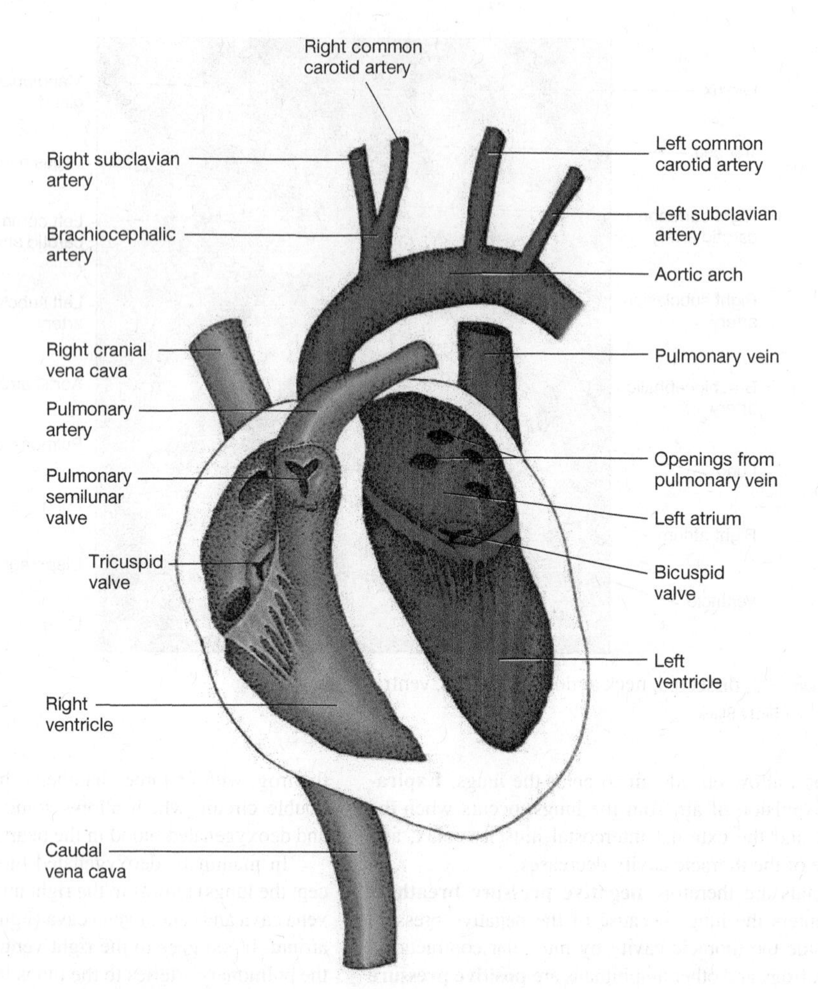

FIGURE 20.7 Rat, heart, longitudinal section.

are unequal in size; ***which ventricle is the larger?*** Examine a microscope slide of heart tissue to observe the difference between the thick, muscular ventricles and the thin-walled atria.

Extending anteriorly from the heart, locate the large **aorta,** which exits from the left ventricle, curves to the left, and passes dorsally to the heart (figure 20.6). This large curvature forms the **aortic arch,** which gives off five branches. The first two branches of the aortic arch are two **coronary arteries,** which supply the muscles of the heart. They arise within the heart and are often difficult to locate. The other three branches of the aortic arch are more prominent and are easier to locate. These branches are the **brachiocephalic artery,** the **left common carotid artery,** and the **left subclavian artery** (figures 20.6, 20.7, and 20.8).

The brachiocephalic artery extends forward and divides to form the **right subclavian artery,** which supplies the right forelimb and shoulder, and the **right common carotid artery,** which supplies the head and brain. The **left common carotid artery** branches from the aortic arch to the left of the brachiocephalic and passes anteriorly along the left side of the trachea. It gives rise to several branches that supply blood to the left side of the neck and head. The third branch from the aortic arch is the **left subclavian artery,** which supplies the left forearm, the chest, and the neck through several branches.

Posterior to the heart, the aortic arch continues as the **dorsal aorta,** which runs along the dorsal wall of the coelom. Along its course from the arch to the posterior end, the dorsal aorta gives off several major branches. ***How many of the following branches can you locate on your specimen?***

Branches of the Aorta Posterior to the Heart

Intercostal arteries—supply intercostal muscles between ribs

Phrenic arteries—carry blood to the diaphragm

Lumbales arteries—supply back muscles and adrenal glands

Coeliac artery—unpaired artery supplying the spleen, liver, stomach, pancreas, and duodenum

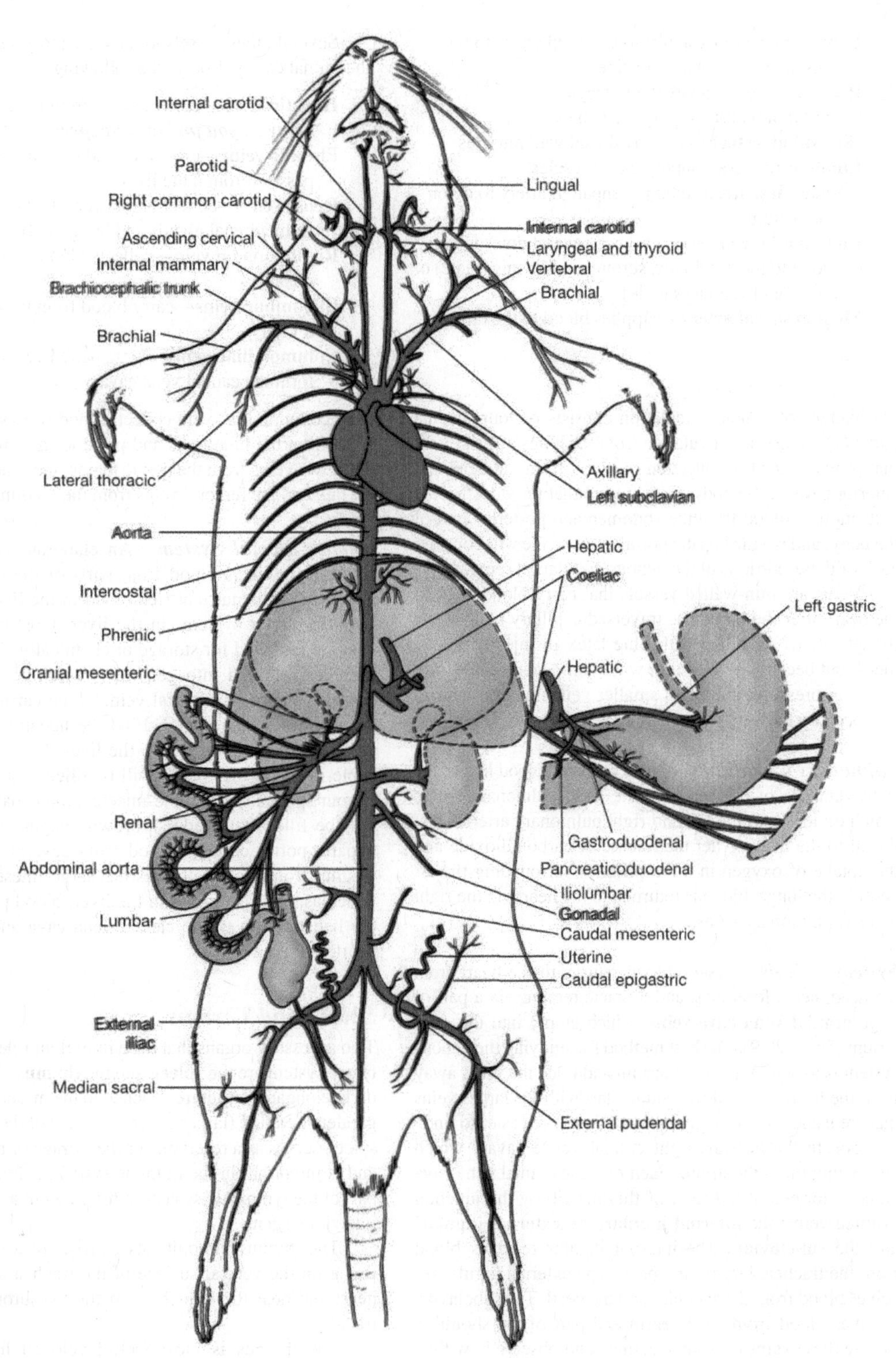

FIGURE 20.8 Rat, arterial system, ventral view.

Cranial mesenteric artery—unpaired artery to the mesentery and small intestine
Renal arteries—supply the kidneys
Gonadal arteries—supply the gonads
Iliolumbar arteries—supply dorsal wall muscles
Lumbar arteries—supply back muscles
Caudal mesenteric artery—unpaired artery to colon and rectum
Common iliac arteries—branch several times to carry blood to the hindlimbs, scrotum, and penis (male) or uterus and vagina (female)
Median sacral artery—supplies blood to the tail

Venous Circulation

The pattern of venous circulation consists of four distinct parts: (1) pulmonary circulation from the lungs, (2) a system that provides for the collection of blood from the head and anterior parts of the body, (3) a system that provides for the collection of blood from the abdomen and posterior part of the body, and (4) the hepatic portal system. We will consider each of these portions of the venous circulation separately.

Veins are thin-walled vessels that carry blood back to the heart after the blood has traversed capillary beds in the tissues. Veins are filled with blue latex in injected specimens, but because of their thin walls some veins burst from injection pressure, and often smaller veins may be hard to find because they are not well-injected.

Pulmonary Circulation Deoxygenated blood leaves the right ventricle of the heart via the large pulmonary trunk, which divides into the left and right pulmonary arteries just dorsal to the heart. After the release of carbon dioxide and the uptake of oxygen in the capillaries surrounding the alveoli in the lungs, blood is returned to the heart via the right and left pulmonary veins.

Systemic Veins System blood returns to the heart from the head, neck, forelimbs, and thoracic regions via a pair of large **cranial vena cava veins,** which empty into the right atrium (figure 20.9). The best method for studying the venous system is to start at the right atrium and trace the veins away from the heart. This allows you to start with the larger veins that are usually the best injected and therefore easier to find.

Locate the left and right cranial venae cavae, which empty into the right atrium. Each of these cranial vena cava veins is formed at the level of the first rib by the junction of three veins: the **internal jugular,** the **external jugular,** and the **subclavian.** The internal jugular receives blood from the trachea, larynx, and neck. The external jugular receives blood from the shoulder and the head. The subclavian receives blood from the forearm and part of the shoulder. Locate these veins on your specimen and observe how they come together to form the cranial vena cava.

The **caudal vena cava** is a single vein extending from the posterior end of the abdominal cavity to the heart. Locate the attachment of this vein to the heart.

Several other vessels join the posterior vena cava in the abdominal cavity. Locate the following:

Hepatic veins—collect blood from the liver. ***How many do you find in your specimen?***
Phrenic veins—enter the caudal vena cava where it passes through the liver
Renal veins—collect blood from the kidneys (also from adrenal glands and left gonad)
Right gonadal vein—collects blood from the right gonad
Iliolumbar veins—carry blood from the dorsal lumbar region
Common iliac veins—large paired veins that join to form the caudal vena cava

Each common iliac vein collects blood from several smaller veins from the hindlimbs and posterior abdomen, including the pelvic veins from the pelvic region, the caudal veins from the tail, and the femoral veins from the hindlimbs.

Hepatic Portal System An elaborate venous system in the rat collects blood from parts of the digestive tract and carries this nutrient-rich blood to the liver. This is the **hepatic portal system.** In the liver, most of these nutrients are removed for storage or chemically changed before they are released into general circulation. The principal vessel is the hepatic portal vein, which carries blood from the intestines to the liver. Find the hepatic portal vein in the mesenteries posterior to the liver. If your specimen is triple-injected, this vessel will be filled with yellow latex. If your specimen is double-injected, the hepatic portal vein will be filled with reddish-brown coagulated blood. The hepatic portal collects blood from several smaller veins leading from the small intestine, large intestine, stomach, pancreas, and spleen. From the liver, blood passes through the hepatic vein to the caudal vena cava and back to the heart, as we have seen.

Spleen and Thymus

Two accessory organs that have important roles in the circulatory system are the **spleen** and the **thymus.** The spleen is a dark, elongated structure attached to the mesentery along the greater curvature (larger convex surface) of the stomach. The spleen serves as a reservoir for the storage of red blood cells and is important in the immune system. It functions also as part of the lymphatic system in filtering out and neutralizing infectious agents.

The thymus in adult rats consists of a small mass of tissue on the ventral surface of the trachea anterior to the heart and near the branching of the two bronchi from the trachea.

The thymus is most well-developed in young rats, where it serves to produce white blood cells. It decreases in size as rats age. The thymus is made up largely of lymphatic tissue and functions in immunity and other forms of defense against infectious agents.

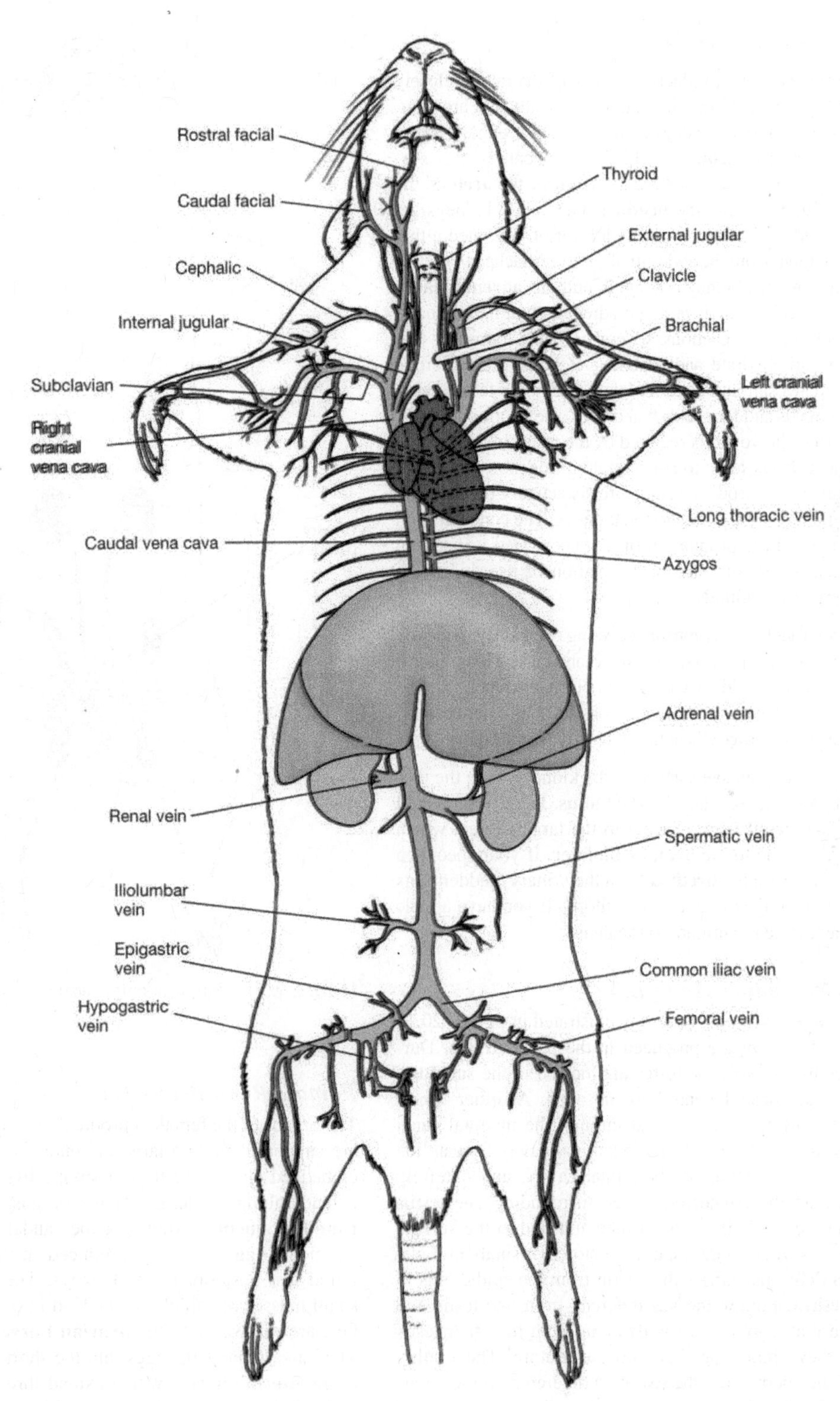

FIGURE 20.9 Rat, venous system, ventral view.

Urogenital System

The excretory and reproductive organs of the rat are closely related both embryologically and anatomically and are commonly considered together as components of a single system, the **urogenital system.** The chief components of this system involved with excretion are the **kidneys,** the **ureters,** the **urinary bladder,** and the **urethra.** The paired kidneys are closely held to the dorsal wall by a layer of peritoneum that separates them from the abdominal cavity. Attached to the anterior end of each kidney is a small, bulbous **adrenal gland.** These endocrine glands produce adrenalin and noradrenalin that regulate stress reactions, several corticoid hormones that regulate carbohydrate and protein metabolism, aldosterone that regulates salt and water balance, and some testosterone (in both males and females!) that acts as a sex hormone. Most testosterone, however, is secreted by the testes in males.

The kidneys serve to concentrate nitrogenous wastes of metabolism and produce urine. Urea is actually produced in the liver and is transported to the kidneys. The concentration of wastes and the production of a concentrated urine is an important water conservation adaptation of rats and many other terrestrial animals.

- Cut through the peritoneum covering one kidney, remove the fat deposits around the kidney, and make a longitudinal section of the kidney with a razor blade or scalpel. Observe that the kidney tissue consists of an outer **cortex** and an inner **medulla** area, which is darker in color.

The inner concave surface of the kidney where the ureter and blood vessels attach is the **hilus.** Identify the **renal artery** and **renal vein** adjacent to the large **ureter,** which carries the urine to the **urinary bladder.** If your specimen is a female, trace the **urethra** from the urinary bladder to its external opening adjacent to the clitoris. If you have a male specimen, trace the urethra to the penis.

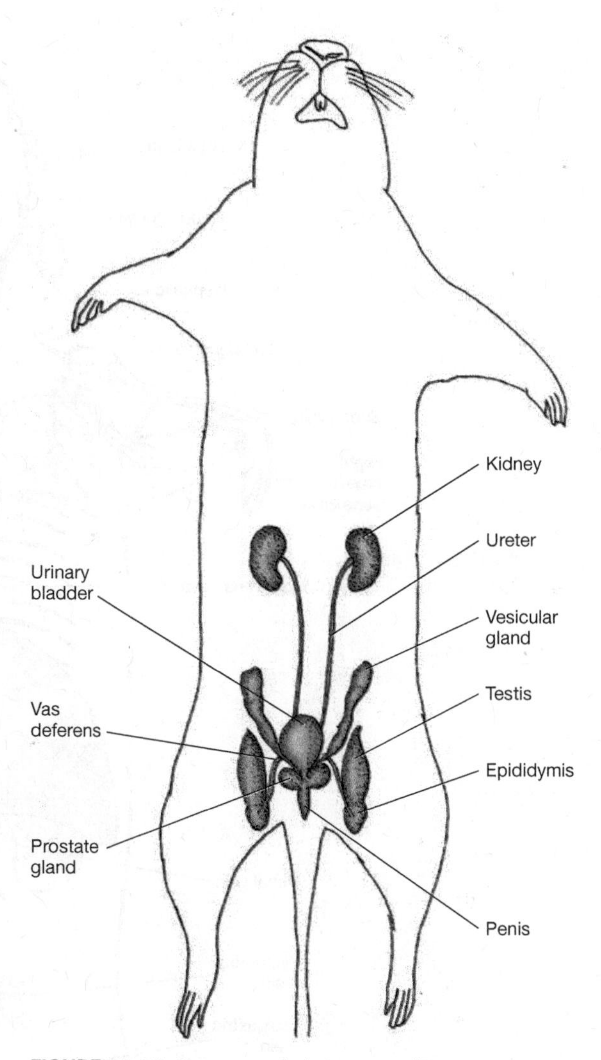

FIGURE 20.10 Rat, urogenital system, male.

Male Reproductive Organs

The male reproductive system is illustrated in figures 20.10 and 20.11. Sperm are produced in the paired **testes.** During breeding season, the testes are located in the **scrotum,** a large sac located ventral to the anus. At other times, the testes are usually retracted through the inguinal canal into the posterior part of the abdominal cavity. Locate the testes in the scrotum or abdominal cavity, and carefully cut through the connective tissue surrounding one testis. Find the **epididymis,** a coiled tube attached to the surface of the testis where mature sperm are stored. A smaller tubule, the **vas deferens,** carries the sperm from the epididymis to the **urethra.** Follow the vas deferens from one testis and observe that it joins the vas deferens from the other testis where they empty together into the urethra. The urethra carries the sperm from the two vasa deferentia to the penis, from which it is deposited in the vagina of the female during copulation. The sperm are suspended in the seminal fluid or **semen** made up of the secretions of several glands associated with the urethra and penis.

Female Reproductive Organs

The organs of the female reproductive system (figure 20.12) are supported by two large mesenteries (containing many embedded clusters of fatty tissue), a **broad ligament** that extends along the length of the organs, and a transverse **round ligament** located near the caudal end of the broad ligament. Eggs **(ova)** are produced in the paired ovaries found just posterior to the kidneys. The rat's ovaries are small masses of follicles embedded in the broad ligament. Ova are released into the **ovarian bursa,** a small pocket. The bursa funnels the eggs into the short **oviducts** leading to the **uterine horns,** which extend through the muscular body of the **uterus** and connect with the **vagina.** The two uterine horns represent a significant morphological adaptation that allows simultaneous development of several embryos and multiple births.

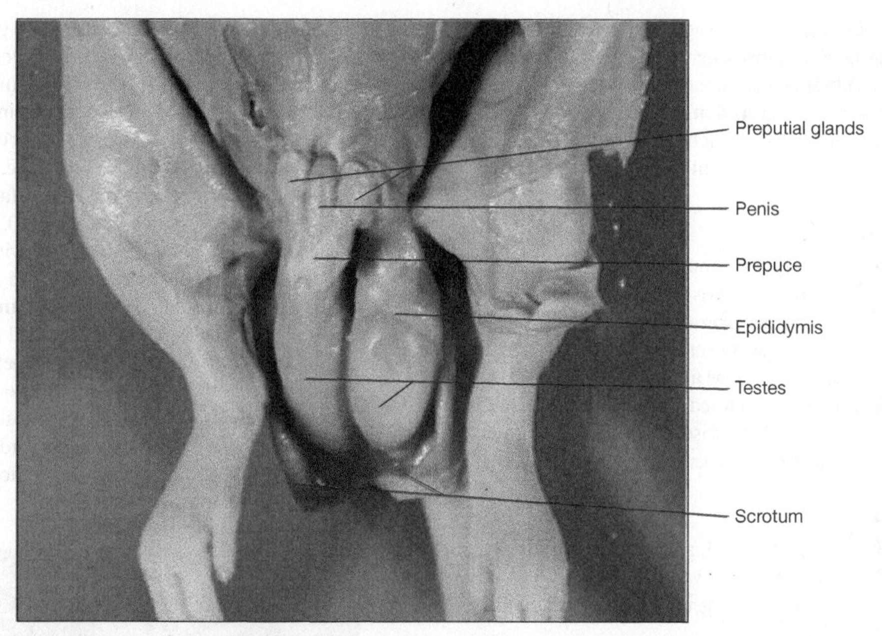

FIGURE 20.11 Rat, dissection, male reproductive system.
Photo courtesy of Betty Black.

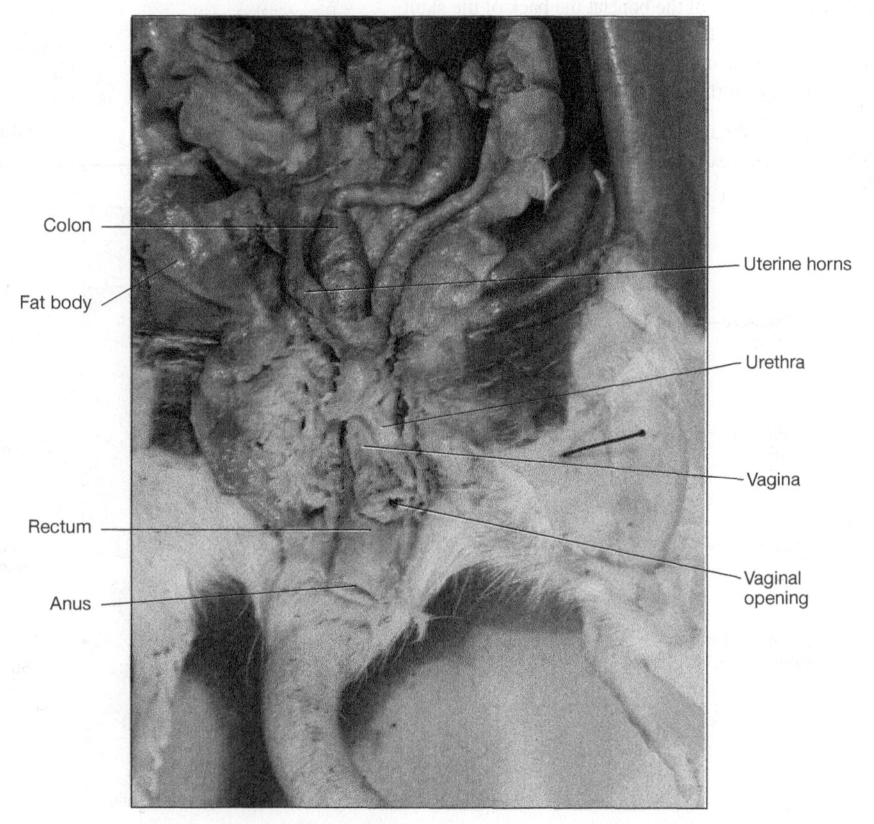

FIGURE 20.12 Rat, dissection, female reproductive system.
Photo courtesy of Betty Black.

During mating, the vagina receives sperm and passes it to the uterine horns where fertilization occurs. Fertilized eggs are subsequently implanted into the richly vascularized uterine walls. The **gestation period** (time of uterine development prior to birth of the laboratory rat) is about 21–23 days with a typical litter of about 6–12 young.

Nervous System

The nervous system of mammals consists of three major divisions: (1) the **central nervous system,** consisting of the brain and spinal cord; (2) the **peripheral nervous system,** made up of the myelinated sensory and motor nerves extending from the central nervous system; and (3) the **autonomic nervous system,** a system of visceral nonmyelinated nerves that regulate the glands and visceral organs. In this exercise, we will confine our study mainly to the brain and the central nervous system (figure 20.13).

Brain

The brain of the rat is small and is more difficult to study than that of larger mammals. We will therefore supplement our study of the rat brain with the larger but similar brain of the sheep.

- Remove the skin and hair from the top of the head between the eyes and ears, and cut away the muscles at the back of the head near the occipital region. Use bone shears to cut away a portion of the bone at the back of the skull to expose the brain. Take care not to cut into the brain. When it is exposed, use forceps to pick away small pieces of bone to further study the brain (figure 20.13).

Observe that the brain is covered by several tough outer layers of connective tissue, the **meninges.** The brain of the rat and other mammals consists of five principal regions that can be traced to its embryological development. These anatomical regions and the principal structures in the adult brain representing these divisions are listed in Table 20.2.

Locate first the two large anterior **cerebral hemispheres** separated by a median groove or **fissure.** Anterior to the cerebral hemispheres find the two smaller **olfactory bulbs** to which the olfactory nerves connect (figure 20.14). At the posterior margin of the cerebral hemispheres within the cerebral fissure is the **pineal body.** Posterior to the cerebral hemispheres on the dorsal surface find the **cerebellum,** consisting of three lobes—a right and left hemisphere and the central vermis—each with a folded surface. The cerebellum is the principal center responsible for coordination of body movements.

The **medulla oblongata** is the most posterior part of the brain and connects with the spinal cord. The medulla regulates such vital processes as respiration, heart rate, blood pressure, and hormonal secretion.

Table 20.2 Principal Regions of the Mammalian Brain

Region	Principal Structures
Telencephalon	Olfactory bulbs, cerebral hemispheres
Diencephalon	Pituitary gland
Mesencephalon	Corpora quadrigemina
Metencephalon	Cerebellum, pons, part of medulla oblongata
Myelencephalon	Part of medulla oblongata

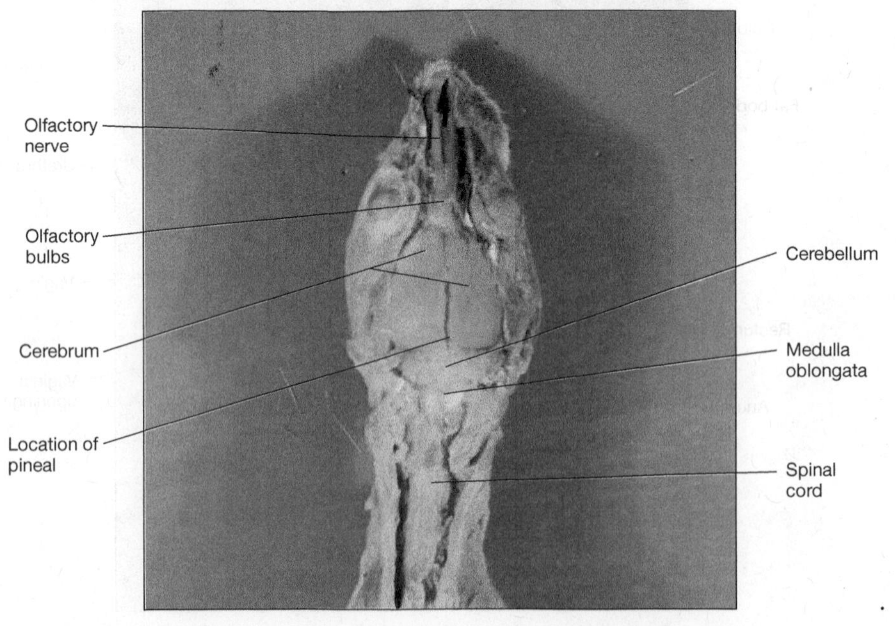

FIGURE 20.13 Rat, dissection, brain, dorsal view.
Photo courtesy of Betty Black.

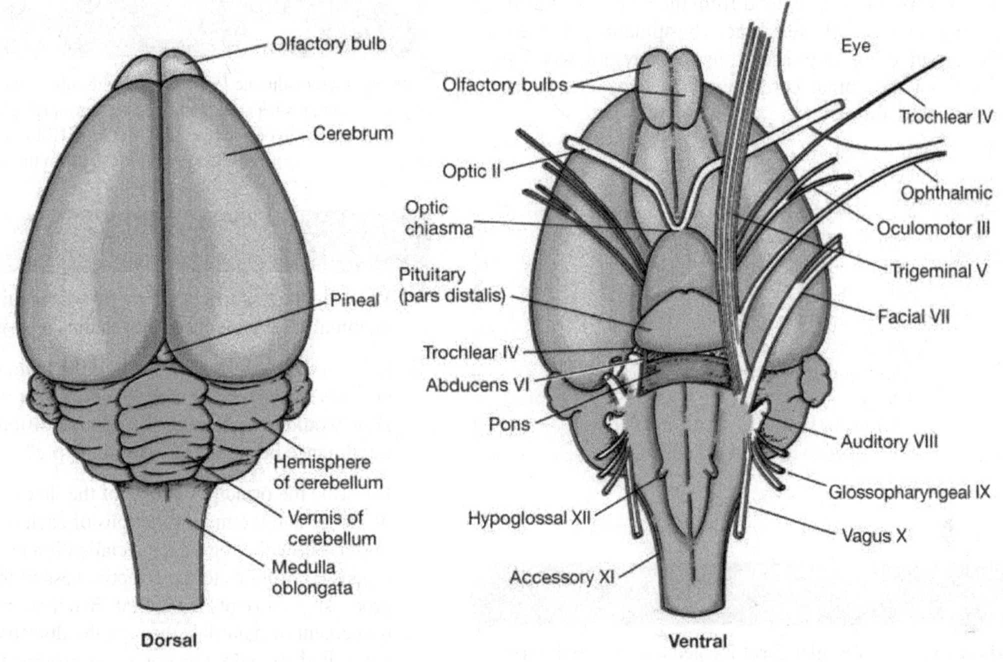

FIGURE 20.14 Rat, brain, dorsal and ventral views.

- After you have studied these structures on the dorsal surface of the brain, carefully detach the brain from the braincase. Try to retain a portion of the several pairs of cranial nerves attached to the ventral surface of the brain so that you can locate where these nerves arise from the brain.

On the ventral surface of the brain, locate the anterior **olfactory bulbs** with two large olfactory tracts leading posteriorly to the cerebrum, and the **optic chiasma** where the two large **optic nerves** cross. Posterior to the optic chiasma is the **pituitary gland,** which may have been left on the floor of the braincase when the brain was removed from the cranium. Behind the pituitary, a transverse tract of nerve fibers, the **pons,** connects the left and right hemispheres of the cerebellum.

Twelve pairs of **cranial nerves** arise from the ventral surface of the brain (figure 20.14). We have already noted olfactory tracts and the optic nerves, which are the first and second pairs of cranial nerves. The 12 pairs of cranial nerves and their locations are listed in table 20.3.

Sheep Brain

For further information on the anatomy of the mammalian brain, study a preserved sheep brain as described in Chapter 19, "Fetal Pig Anatomy" (figures 19.24 and 19.25). Identify the principal structures of the sheep brain and compare them with the rat brain. ***What differences do you find?***

Spinal Cord

The spinal cord extends posteriorly from the medulla oblongata and is covered by meninges. The spinal cord passes through the vertebrae in the vertebral column, and several

Table 20.3 Cranial Nerves of the Rat

No./Name	Location	Function(s)
I. Olfactory	Olfactory bulbs	Smell
II. Optic	Diencephalon	Vision
III. Oculomotor	Ventral surface of mesencephalon	Eye movements
IV. Trochlear	Dorsal surface of mesencephalon	Eye movements
V. Trigeminal	Pons	Innervates skin, vibrissae, jaw muscles, tongue, and teeth
VI. Abducens	Medulla oblongata	Eye movements
VII. Facial	Medulla oblongata	Innervates jaw muscles
VIII. Auditory	Medulla oblongata	Hearing
IX. Glossopharyngeal	Medulla oblongata	Innervates pharynx and tongue
X. Vagus	Medulla oblongata	Innervates larynx, heart, lungs, diaphragm, and stomach
XI. Accessory	Medulla oblongata	Innervates neck muscles and pharyngeal organs
XII. Hypoglossal	Medulla oblongata	Tongue movements

pairs of **spinal nerves** extend from the spinal cord through openings between the vertebrae. The spinal nerves are usually classified in groups according to the region where they arise from the spinal cord: **cervical, thoracic, lumbar, sacral,** and **caudal.**

Demonstrations

1. Liquid or plastic mount of dissected rat
2. Microscope slide with longitudinal section of a rat heart
3. Microscope slide with cross section of rat testis
4. Microscope slide with cross section of rat ovary
5. Microscope slide with median section of rat kidney
6. Microscope slide with rat sperm

Key Terms

Diaphragm muscular sheet located anterior to the liver that separates the thoracic and abdominal cavities in mammals.

Hepatic portal system portion of the venous circulation in many vertebrates that collects blood from capillary beds in the stomach and intestine, and returns it to the liver; here the blood passes through a second capillary bed where nutrients are removed prior to returning the blood to the heart.

Mesentery double sheet of mesodermal tissue that supports various internal organs in the coelom of vertebrates.

Pulmonary circulation portion of mammalian circulatory system that carries blood from the heart to the lungs and returns it to the heart.

Systemic circulation portion of the mammalian circulatory system that carries blood from the heart to and from all organs of the body except the lungs.

Vibrissae long sensory hairs on the face of the rat.

Internet Resources

There are many valuable Internet sites with information about zoology. Several sites containing pertinent zoological information for this chapter can be found on the McGraw-Hill Zoology web site at http://www.mhhe.com/zoology. Just click on this text's title.

Questions for Critical Thinking

1. What are the features of the rat that distinguish it as a mammal? Are these the same features present in the pig?
2. Discuss the basic organization of the mammalian nervous system. What components affect intelligence? How would you measure the relative difference in intelligence between the rat and the pig?
3. Describe the principal organs of the digestive system of the rat and the main functions of each organ. Discuss how the regional specialization of these organs contributes to the effectiveness of the system in processing nutrients for the rat. Briefly explain how the movement of materials through the digestive tract is controlled and why this control is important.
4. Discuss the difference between negative pressure breathers and positive pressure breathers. Does one method have advantages over another in a terrestrial environment?
5. Outline the urogenital systems of both male and female rats. Point out homologous organs wherever you are able (e.g., testis—ovaries).

Suggested Readings

Carleton, M.D. 2005. Order Rodentia. In *Mammal Species of the World: A Taxonomic and Geographic Reference,* 745–752. Baltimore: Johns Hopkins University Press.

Vaughn, T.A., J.M. Ryan, and N.J. Czaplewski. 2011. *Mammalogy,* 5th ed. Sudbury, MA. Jones and Bartlett, 650 pp

Walker, W.F. Jr., and D.G. Homberger. 1998. *Anatomy and Dissection of the Rat,* 3rd ed. San Francisco: W.H. Freeman. 120 pp

Notes and Sketches

Note: Page numbers followed by *f* indicate figures. Page numbers followed by *t* indicate tables.

A

B

D

E

F

G

H

I

J

K

L

N

O

P

Q

R

S

T

U

V

W

X

Y

Z

Credits